Earth

An Introduction to Geologic Change

Earth

...tion to Geologic Cha...

Sheldon Judson
Princeton University

...even M. Richardson
Iowa State University

Prentice Hall Englewood Cliffs, NJ 07632

Library of Congress Cataloging-in-Publication Data

ange/Sheldon Judson, Steve Richardson.

Judson, Sheldon.
    Earth: an introduction to geol s and index.
        p.   cm.   Steven McAfee.   II. Title.
    Includes bibliographical   94-26843
    ISBN 0-13-301193-3   I. F   CIP
        1. Geology.   199
    QE28.J83
    550—dc20

                    Tim Bozik
                   aul Corey
            uction and manufacturing: *David Riccardi*
        Editor: *Kathleen Schiaparelli*
         Editor: *Paula Maylahn*
        r editor: *Deirdre Cavanaugh*
       nt editor: *David Cohen*
       oduction editor: *Jennifer Wenzel*
      g manager: *Leslie Cavaliere*
      ditor: *Jim Tully*
     r and cover design: *Lisa A. Jones*
    r designer: *Barbara Clay*
    er credit: *Photo provided by James Sorenson of Mark Hurd, Aerial Surveys Corporation*
   hoto editor: *Lorinda Morris-Nantz*
Photo reseacher: *Mira Schachne*
Buyer: *Trudy Pisciotti*
Supplements editor: *Mary Hornby*
Editorial assistant: *Veronica Wade*

© 1995 by Prentice-Hall, Inc.
A Simon & Schuster Company
Englewood Cliffs, New Jersey 07632

Printed in the United States of America

10   9   8   7   6   5   4   3   2   1

ISBN 0-13-301193-3

Prentice-Hall International (UK) Limited, *London*
Prentice-Hall of Australia Pty. Limited, *Sydney*
Prentice-Hall Canada Inc., *Toronto*
Prentice-Hall Hispanoamericana, S.A., *Mexico*
Prentice-Hall of India Private Limited, *New Delhi*
Prentice-Hall of Japan, Inc., *Tokyo*
Simon & Schuster Asia Pte. Ltd., *Singapore*
Editora Prentice-Hall do Brasil, Ltda., *Rio de Janeiro*

Observe always that everything is the result of a change, and get used to thinking that there is nothing Nature loves so well as to change existing forms and to make new ones like them.

> Marcus Aurelius,
> Roman Emperor, 161-180.
> Meditations IV-36

Brief Contents

Contents

Preface

The most challenging course in any curriculum is the first one. For a student, it is a leap into the unfamiliar. The teacher and the textbook use strange words to describe stranger ideas. Perhaps more confusingly, they use familiar words in odd ways to reinterpret what the student already knows and to ask questions that the student may not even realize could be asked. Teachers find the first course just as challenging as students, but for different reasons. The language and the ideas are familiar, of course, but must be shaped to reach minds that have never heard them. A teacher in an advanced course may generally assume that students are comfortable with the rhetoric and fundamental tenets of the discipline; a teacher in the first course cannot.

We undertook to write this new textbook for the first college-level course in physical geology only after we had each spent many years in and around the classroom, puzzling about what makes some ideas easier to grasp than others and about how to help students learn how the Earth works. The book we have written comes out of our own experience with geology, but also reflects much of what we have learned with and about our students. One of us (Judson) co-authored his first introductory textbook in 1954, long enough ago to have shared the excitement as the entire profession of geology reinvented itself around the paradigm of plate tectonics. The other (Richardson) has taught the first course in geology since 1977 and now directs a university-wide program aimed at helping faculty members sharpen their teaching skills.

The pervasive theme of this book is *change*. We hope readers will be drawn into discussions of how rocks form, how mountains are raised, how weathering and erosion reshape the Earth's surface, and how geologists can anticipate both the hazards and the benefits that the Earth offers humanity. As we explain at some length in Chapter 1, we have divided the book into four segments that deal with those four broad topics. We have labeled the segments Changes in Earth Materials, Changes Within the Earth, Changes in the Earth's Surface, and Humans and Earth Change. Each segment is introduced by a reflective essay that acts as a bridge to the new block of chapters and, more importantly, invites the reader to explore the theme of change from a fresh perspective. These essays are intended to stimulate discussion and to give students

and their teachers an opportunity to explore a question that is central to the first course: *"Why do geologists care about any of this stuff?"* Because this is such an important question for students who are new to the discipline, we hope that our reflective essays find a receptive home in the classroom.

Other unifying threads are woven through the book. We have highlighted four of these (*Water, Tectonics,* the *Environment,* and *Planetology*) by inserting an appropriate icon wherever each thread appears in the narrative. We have chosen to call attention to the four themes in this active manner because we think it is important to find as many ways as possible for students to carry what they have learned from one end of the course to the other, and out into the world beyond. Instructors may wish to encourage students further by developing classroom exercises or assignments that explore these threads.

Another unifying element is central to the first course in physical geology and appears throughout this book, although we have not assigned an icon to it. The *rock system,* introduced in Chapter 1, is a conceptual framework through which we examine the relationships among rocks and rock-forming processes. In presenting this new framework, we have chosen to de-emphasize the familiar metaphor of the *rock cycle,* which Leet and Judson first presented for first-year students over 40 years ago. Any practicing geologist knows that the Earth is not the same planet that it was 3 or 4 billion years ago, and that rock-forming processes do not actually follow a cycle. Rather than continue to reinforce James Hutton's mistaken view that the Earth has "no vestige of a beginning, no prospect of an end," therefore, we emphasize that Earth's materials evolve along the branching pathways of the rock system.

Our shared fascination with history has introduced another element which, if not a thread, is at least a pedagogical backdrop for this book. We believe that students learn best in the sciences if they see science as a human endeavor. By making frequent reference to the men and women whose ideas have contributed to our understanding of the Earth, we hope to inspire students to think of themselves as potential geologists. Our numerous discussions of key experiments or observations in the history of geology serve an important purpose by "humanizing" the subject, and they address a question that we feel should be at the heart of any science course: *"How do we know?"* In a few places—most conspicuously in Chapters 11 and 12—we have examined failed theories or naive observations of the past in an effort to let students see that scientific knowledge is not a revealed truth, but rather the product of continuous discovery and testing.

We hope that those who use this book will take advantage of other learning tools we have adopted:

- A matched set of focus questions and summary statements open and close each chapter. These identify broad topics that interest geologists and can serve as a guide for systematic study.
- Narrative diagrams in each chapter explore a geological process or identify several related observations that support a theory.
- The list of key words and concepts at the end of each chapter is cross-referenced, by page number, to the place where each of them was introduced as a boldfaced term in the text. Concise definitions of all of these terms are also compiled in the glossary at the end of the book.
- The questions for review and thought at the end of each chapter include some that are identified as Critical Thinking questions. In most cases, these do not have a clear answer. Many, in fact, encourage the student to ask further questions. Instructors may wish to use these as topics for papers or as the starting points for classroom discussion.
- At a time when color dominates the illustrations of our reading matter, we have chosen black and white photographs for the cover of our book and for the opening illustrations of individual chapters. They remind us that often black and white photography can capture the beauty and art of nature with a grandeur and impact that eludes color photography.

Acknowledgments
▲▲▲

Writing this book has been a long and sometimes arduous task that has taught each of us more than we ever wanted to learn about delayed gratification. As in so many projects that are finished one piece at a time, it's hard to appreciate the full effect until all of the pieces have been assembled. We hope that the final product is as satisfying for students and teachers as it seems to us. If so, it is in no small part because we had a lot of help. Our families have been most patient and encouraging and are looking forward to having our lives return to "normal." The editorial team at Prentice Hall, including Ray Henderson, Deirdre Cavanaugh, David Cohen, and Jennifer Wenzel, has been flexible and supportive, and has shaped this into a handsome

book that we may all be proud of. Judson acknowledges with pleasure and appreciation his association with Marvin E. Kauffman through their joint authorship of four editions of a previous text on physical geology, an association that has made the writing of this volume an easier task. We are indebted to many colleagues and friends for providing photographic illustrations; we acknowledge them specifically where their contributions appear. Photographs by the authors are not credited. Finally, we must thank a great number of people who lent a critical ear or eye to the project and have contributed many helpful suggestions. These include:

Carl F. Vondra, Kenneth E. Windom, Carl E. Jacobson, Scott Theiben, and Greg Guyer, all of Iowa State University.
Arthur N. Palmer, State University of New York at Oneonta
E. Kirsten Peters, Washington State University
David King, Auburn University
Victor R. Baker, University of Arizona
Cynthia Lampe, San Diego State University
Dan Prothero, Occidental College
Richard P. Tollo, George Washington University
Carl E. Johnson, University of Hawaii at Hilo
Roger Hoggan, Ricks College
Scott Argast, Indiana University Purdue University at Fort Wayne
J. Douglas Walker, University of Kansas
David R. Dockstader, Jefferson Community College
Anthony Clarke, University of Louisville
Gary Byerly, Louisiana State University
Bradford H. Hager, Massachusetts Institute of Technology
Lauret Savoy, Mount Holyoke College
H. Robert Burger, Smith College
Stephen D. Stahl, Central Michigan University
Estella Atekwana, Western Michigan University
Jack Hall, University of North Carolina
Allen G. Kihm, Minot State University
Lincoln Hollister, Princeton University
Richard Pardi, William Paterson College
Albert M. Kudo, University of New Mexico
Arthur L. Bloom, Cornell University
Bryce M. Hand, Syracuse University
J. Barry Maynard, University of Cincinnati
Vernon Scott, Oklahoma State University
G. J. Retallack, University of Oregon
Mark Johnson, Bryn Mawr College
Charles Scharnberger, Millersville University
Michael Bikerman, University of Pittsburgh
Harry Y. McSween Jr., University of Tennessee
Kevin Burke, University of Houston

William Cornell, University of Texas-El Paso

Jones Baer, Brigham Young University

Paul Tayler, Utah Valley State College

E. Kirsten Peters, Washington State University

Lung Chan, University of Wisconsin-Eau Claire

Douglas Cherkauer, University of Wisconsin-
Milwaukee

Norman P. Lasca, University of Wisconsin-
Milwaukee

David Hickey, Grapolithies Publishing

James Majure, Geographic Information Systems Office,
Durham Computation Center, Iowa State University

Earth

An Introduction to Geologic Change

1

The Teton Range and the Snake River, Grand Teton National Park, Wyoming

The Changing Earth

*O*BJECTIVES

▲▲

As you read through this first chapter it will help your study if you focus on the following questions:

1. What is uniformitarianism? How is it useful in the study of geology?
2. What is the Earth system? How are it and its subsystems important in our study?
3. What are the three major rock families? In what way does each form? How do the rocks change as they progress through the rock system?
4. What is plate tectonics? What are some of its results on the outermost region of the Earth?
5. What are the Earth's surficial systems, what is their source of energy, and why are they geologically important?
6. What are some of the relationships between people and the geologic environment?

*O*VERVIEW

▲▲

The Earth is 4.54 billion years old. Ever since its birth it has undergone many changes and continues to change before our eyes. This book is about these changes.

The transformations going on at present are brought about by the same Earth systems that also operated in the past and produced changes similar to those we see today. Much of what geologists have learned about the Earth and its past has been gained by applying their knowledge of modern-day processes to interpreting the events recorded in the rocks.

To introduce you to these transformations of our globe we have divided this book into four major units. We begin with the description and origin of rocks, how they change, and how they are used to tell geologic time. We then consider how the internal heat of the Earth deforms and moves the outermost regions of our planet. In the third section of the book we concentrate on the processes that mold the surface features of the Earth. Finally, we look at the natural resources produced by geologic processes and at the interaction between humans and the geologic environment.

What Is Geology?

Ge·ol·o·gy, noun, from the Greek *geo,* "Earth," and *logos,* "discourse." The science that deals with the study of the planet Earth—the materials of which it is made, the processes that act to change these materials from one form to another, and the history recorded by these materials; the forces acting to deform the outer layers of the Earth and create ocean basins and continents; the processes that modify the Earth's surface; the application of geologic knowledge to the search for useful materials and the understanding of the relationship of geologic processes to people.

Implicit in this definition is the dynamic nature of our planet. Throughout its existence the Earth has been an ever-changing planet. In fact, it is changing as you read this, and transformations will continue in the future. Sometimes the changes are rapid and violent as earthquakes rumble and volcanoes explode. Just as often they are slow, unnoticed through a single lifetime and perceptible only to the most sensitive instruments.

These changes in the Earth are the focus of this book. As we build an understanding of them we recognize again that the *laws of nature* governing these changes are themselves unchanging. Gravity pulls rocks down steep mountain slopes as surely as it keeps the planets' orbits in order. Rocks are destroyed, but the materials that make them persist in other forms. We can question and test these assertions, but so far, dislodged rocks have always fallen, and when the sun sets in the west it will then rise on a predictable schedule in the east.

Geology has its own expression for our assumption that the laws of nature are constant. It is called the **principle of uniformitarianism.** As it was originally coined in the early nineteenth century, this principle meant that the processes operating to change the Earth in the present also operated in the past at the same rate and intensity and produced changes similar to those we see around us. The meaning of the principle has evolved with time, and today the principle of uniformitarianism acknowledges that past processes, even if the same as today, may have operated at different rates and with different intensities than those of the present. Some geologists now prefer to use the term **actualism** for the second and current meaning of uniformitarianism. We will continue to use the term "uniformitarianism" in our discussions.

Consider the following example of the application of uniformitarianism: Geologists know from modern observations that the movement of glaciers produces some very characteristic results. For instance, present-day glaciers lay down a distinctive deposit of debris made up of rock fragments that range in size from

FIGURE 1.1   Several times in the past glacier ice covered much of Canada, the United States, and Europe. Modern glaciers provide geologists with information on how past glaciers worked and affected the Earth. This is a view of the Antarctic ice cap. *(William Ryan)*

submicroscopic particles to boulders weighing several tons. In the deposits are many different kinds of rocks, some found locally and others brought from great distances. In addition, many of the fragments are broken and scratched. This material is mixed together in apparently random fashion as if it had been churned up in a giant concrete mixer. Geologists know of no process other than the movement of glacier ice that produces a deposit exactly like this. Therefore, when geologists find such deposits spread, for example, across southern Canada and the northern United States, but find no glaciers in the area, they conclude that the debris was left by glaciers now vanished (see Figure 1.1).

Geology as a modern science dates from the introduction of the concept of uniformitarianism. The person most usually associated with its early development was a Scottish medical man, gentleman farmer, and geologist, James Hutton (1726–1797). His ideas first appeared in a series of lectures before the Royal Society of Edinburgh in 1785. In 1795 they were published as *Theory of the Earth with Illustrations and Proofs* (Figure 1.2). It remained for a nineteenth-century geologist, Archibald Geikie, (1835–1924) to state the principle of uniformitarianism most succinctly with **"The present is the key to the past."**

Equipped with the concept of uniformitarianism, geologists were able to explain Earth features, materials, and changes on the basis of modern observations. The logic of the explanation, however, gave rise to a new concept for students of the Earth. It was this: To accommodate all the changes that are found recorded in the rocks, Earth history must span a vast amount of time. It became apparent to geologists that a great deal of time was needed for a river to cut its valley, or for hundreds or thousands of meters of mud and sand to

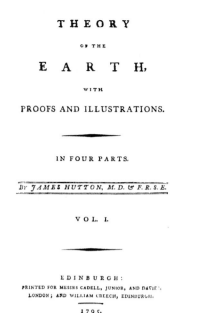

THEORY

OF THE

EARTH,

WITH

PROOFS AND ILLUSTRATIONS.

IN FOUR PARTS.

By JAMES HUTTON, M.D. & F.R.S.E.

VOL. I.

EDINBURGH:
PRINTED FOR MESSRS CADELL, JUNIOR, AND DAVIES,
LONDON; AND WILLIAM CREECH, EDINBURGH.

1795.

FIGURE 1.2   The title page of James Hutton's *Theory of the Earth.* The publication of this work is widely credited with marking the birth of modern geology.

accumulate on an ocean floor, harden into solid rock, and rise far above the level of the sea. Exactly how much time these processes would take, the early geologists could not specify. But they were convinced that it was far greater than the few thousands of years allowed for Earth history by the accepted wisdom of the day. It was not until the 1970s that scientists could demonstrate with confidence that the Earth is about 4.54 billion years old.

Given so much time, even slow processes can achieve what intuition says is impossible. For example, geologists have established that mountains as high as the modern Rocky Mountains once towered where the low hills of northern Wisconsin, Minnesota, and Michigan stand today. Now only the roots of these ancient peaks survive, some of them buried beneath younger rocks. Geologists explain that the mountains were destroyed by rains and flowing water, creeping glaciers, wind, and slowly slipping soil, the same slow-acting processes that go on today. Has there been enough time to accomplish such dramatic change? To answer this, suppose that the mountains were, on average, 5 km high. Various studies suggest that mountains are worn away by erosion at a net rate of about 2.5 cm/1,000 years. At this rate it would have taken approximately 200 million years to reduce the mountains to low hills, a realistic figure given what is known about the geologic history of the region.

If understanding present processes and their results allows geologists to interpret the geologic past, can this understanding let them predict the future? The answer is a qualified yes. We can predict with confidence that earthquakes in California and volcanic eruptions in Washington and Oregon will continue to be commonplace for a few million years to come. However, we cannot yet make firm, short-term predictions of exactly when, or specifically where, the events will occur. Better understanding of the processes, however, increases geologists' ability to make somewhat more accurate short-term predictions. Knowledge of the behavior of rivers makes it possible to predict, at least statistically, how often floods of a given size will occur and how much land will be inundated. Or, if engineers build a dam, geologists can predict how it will affect the behavior of a river and the nature of its channel, both upstream and downstream from the dam.

The Changing Earth

We have divided our study of geology into four separate units, each of which focuses on a major aspect of the science. They are changes in Earth materials; changes within the Earth; changes in the Earth's surface; and Humans and Earth change. This provides a way of grouping topics with a common focus. The following icon will be used throughout the book to represent the four units. The portion being discussed will be highlighted in blue.

Rocks are the foundation of the discipline of geology. Understanding them is basic to our study. Therefore we devote the six chapters in the first section of this book, **Changes in Earth materials,** to a discussion of minerals (the building blocks of rocks); to the processes that form the major rock types; to how one rock is transformed into another; and to how rocks can be used to tell time.

Changes within the Earth are driven by the planet's internal heat and by gravity. We discuss these in a group of five chapters in the second section of the book. The eruption of volcanoes and the shaking by earthquakes are surface expressions of this internal energy. In addition, rocks are bent, broken, and melted and kilometers-thick fragments of the outermost portions of the Earth move laterally over vast distances as part of a system that creates both mountain ranges and ocean basins.

Several surface systems, including running water, glaciers, wind, and underground water, constantly produce **Changes at the Earth's surface.** We address these in the six chapters comprising the third portion of the book. The radiant energy of the Sun, aided by gravity, drives these surface systems. All of them operate to lower the land masses built up by the internal processes of the Earth.

Humans and Earth Change on the Earth have varied during our short span on the globe and are the subjects of the two chapters in the final section of the book. We have long relied on materials and energy from the Earth, and that reliance has been steadily growing. One result of our increasing success in finding and extracting these useful Earth materials and sources of energy has been depletion of a number of resources as well as environmental degradation in many places. Furthermore, humans, as they have bent nature to their uses, have themselves become geologic agents. In doing so they have modified some Earth systems, sometimes with deleterious results.

The fourfold division of this book allows us to compartmentalize our study of geology and cluster chapters focused on closely related aspects of geology. A number of themes, however, keep recurring through these major units. They are *threads* that weave through the chapters and help tie them together. A small icon, or symbol, identifies the thread wherever it appears as a dominant element in the discussion. These are shown in the following list of themes.

- *Water.* This is one of the most common and familiar substances on Earth. It is critical in the creation and destruction of rocks and in modeling the Earth's landscape. It fills the ocean basins, flows in a complex system of river channels, collects in glaciers, and moves slowly through the underground. It supports living things on which weathering and soils depend. Water plays a role in mineral formation, in volcanic activity, and the creation of valuable economic deposits.

- *Plate tectonics.* This process involves the lateral movement, over long distances, of large fragments of the outermost part of the Earth. It accounts for mountain building, volcanic activity, and earthquakes. It is responsible for the growth of continents and the changing shapes and sizes of ocean basins. It sets the stage for the changes in the Earth's surface.

- *Planetary neighbors.* The Earth shares many features and processes with neighboring planets and other solid bodies of the Solar System. This is not surprising when we realize that they all originated about the same time and are ruled by the same fundamental laws of nature. This icon draws attention to shared characteristics between Earth and other planets.

- *Environment.* Humans and the natural environment affect each other in various ways, and we use this icon to mark these interrelations. For example, the icon marks situations in which humans modify a natural process, such as stream flow, or are dependent upon the Earth for such natural resources as fuel or water. We use it also in discussion of events such as earthquakes or volcanic explosions that can be catastrophic to humans but over which they have no control.

In the following pages we outline the topics of our study. Each of these is considered in detail in the appropriate chapter. This preview, however, will provide you with an overall picture of what follows in the major units of the book, as well as the individual chapters in each unit. (See Perspective 1.1.)

Changes in Earth Materials

Rock is the most common material on Earth. We can divide rocks into three large families: **igneous, sedimentary,** and **metamorphic.** When we look at these rocks closely we find that each is made up of smaller units called **minerals.** These are chemical compounds, each with its own composition and physical characteristics, and often visible and identifiable with the naked eye. The specific family to which a rock belongs depends upon the composition, arrangement, and mode of origin of the minerals that form it. Changes in Earth materials ultimately involve changes in minerals.

Minerals   Minerals are composed of combinations of **chemical elements,** or, very rarely, a single chemical element. There are 92 naturally occurring elements each with its own characteristics. The smallest bit of an element that still retains its identifying characteristics is called an **atom.** Beyond this the ordinary techniques of the chemical laboratory cannot further divide an element. Scientists, however, using powerful and sophisticated equipment, have shown that the atom is made up of even smaller particles. These are **protons** and **neutrons,** tightly packed in the center or **nucleus** of the atom, and widely spaced **electrons** moving in a cloud around the nucleus. An atom of each element has a fixed number of protons in its nucleus and an equal number of circling electrons.

The atoms of most elements cannot exist as single particles. Instead, they combine with atoms of other elements to form a chemical compound. They are held together by **chemical bonds,** which result from the interaction of the electrons surrounding the atomic nuclei of the various elements. Geologists describe the chemical compounds that result as minerals (Figure 1.3).

FIGURE 1.3   Minerals are chemical compounds, or, less commonly, single elements. This cluster of crystals includes two different minerals, pyrite and quartz. Pyrite is the brassy yellow, blocky mineral, a compound of iron and sulfur, Fe S$_2$. These quartz crystals are clear to milky, hexagonal prisms. Quartz is a compound of silicon and oxygen, SiO$_2$. Specimen from Spruce Claim, Washington, is 10.6 cm wide. *(Jeffrey A. Scovil)*

Once formed, a mineral can be changed by breaking the bonds between the atoms. This can be accomplished in nature by the same processes that the chemist uses in the laboratory. These processes include heat, pressure, acids, water, and gases, all of which are present in the myriad of natural environments. These same environments account for the different rock types that characterize our Earth.

Igneous Rocks   Igneous rocks take their name from the Latin *ignis,* meaning "fire." They form from a mass of liquid, or molten, material that contains many of Earth's most common elements, including oxygen, sodium, potassium, calcium, iron, manganese, and silicon. As a melt cools, the atoms begin to form compounds, or minerals. Eventually, when the temperature falls far enough, all the liquid has been converted into minerals, which interlock with each other to form a solid rock.

When the molten mass lies beneath the surface it is called **magma.** Magma usually cools slowly over thousands, even millions, of years. The resulting rocks lie hidden from our view until erosion strips away the overlying rock. The cores of many mountain ranges are made of igneous rocks formed from a magma cooling deep below the surface and only later lifted up in the mountain-building process. Volcanic activity occurs when the magma breaks through to the surface. If the magma flows out across the surface it is called **lava** (Figure 1.4) and cools very quickly. If the volcanic activity is explosive it spreads **ash** (Figure 1.5) and **cinders** across the countryside.

Sedimentary Rocks   In contrast with the igneous rocks, sedimentary rocks (from the Latin *sedimentum,*

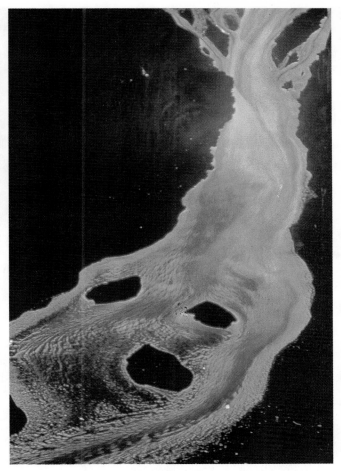

FIGURE 1.4   When molten rock, magma, breaks out at the surface it is called lava. Here, during an eruption in 1959 on the Kilauea volcano, Hawaii, several small lava flows unite to form a larger one that moves toward the viewer. *(U.S. Geological Survey)*

"a settling") form in the comparatively low temperatures at the Earth's surface. They are made up of material derived from preexisting rocks. These are

FIGURE 1.5   An explosive eruption can spread volcanic ash and cinders over a wide area. This landscape is not a winter snow scene, but an ash fall that accompanied the eruption of the Mexican volcano, Paricutin. *(Erling Dorf)*

Perspective *1.1*

Earth Systems and Geologic Changes

The word "system" comes from the Greek *systema*, "a whole made up of several parts." Our planet, itself a member of a system, the *Solar System*, has four parts: the solid Earth, the atmosphere, the oceans, and living things. These components do not exist in isolation, but make up a continuously interacting whole that we call the **Earth system.** The Earth system, in turn, includes an almost countless number of smaller subsystems. These provide an effective framework in which to examine the geologic changes studied in this book. An example is the **hydrologic system,** one of the most important systems in geology (Figure P1.1.1)

Whether in the atmosphere, ocean, rivers, glaciers, or the underground, water participates in a global hydrologic system, also called the *hydrologic cycle.* In it

water evaporates into the atmosphere, primarily from the oceans, and then precipitates as rain or snow. When precipitation falls on land some of it runs off in rivers, some infiltrates into the ground, some turns into glacier ice, some evaporates into the atmosphere, and some transpires (is breathed back by plants) into the atmosphere. We can write these movements as follows:

Precipitation = Runoff + Infiltration + Glaciers + Evaporation + Transpiration

A close look at the diagram of the hydrologic cycle reveals how important a system can be in understanding geologic changes. For example, the glaciers that carve the sharp peaks and steep cliffs in high mountains depend upon the hydrologic system. In fact, we can speak of a smaller subsystem, the glacier

system. Rivers also comprise a system, one of a complex network of channels that not only drains the lands but also changes them by carrying away the products of weathering and erosion.

The more we study geology, the more we realize that geologic changes take place within systems. Furthermore, we should not be surprised to find that a change occurring in one system can bring about change in one or more other systems. For instance, the products of a volcanic eruption may block a nearby river valley. This, in turn can dam the river and form a lake, or it can force the river into a completely new course. At the same time, ash from the eruption can bury the existing soils so that the soil-forming processes must begin anew, changing the freshly fallen ash into soil. Here, then, a change occurring in one system, the vol-

destroyed by the processes of **weathering** (Figure 1.6) that include, among others, climate and the activity of plants and animals. Weathering involves both the mechanical breaking up of rocks and minerals into smaller and smaller bits with no change in their composition, and the chemical change of minerals into new ones. When new minerals are created the chemical bonds of the original minerals are broken, and atoms are rearranged into new compounds, that is, into new minerals.

The products of weathering are moved to new

locations by gravity, water, wind, and ice to form sedimentary rocks (Figure 1.7). Most commonly the materials are solid particles that are deposited in horizontal layers, with stratum after stratum laid down one on top of another like a giant layer cake. For example, beating waves may grind a headland into sand and pebbles, which then settle in layers on the nearby beach and ocean floor. When these deposits harden the result is a sedimentary rock. Very often weathering products are dissolved in water and are transported into new environments where new minerals are

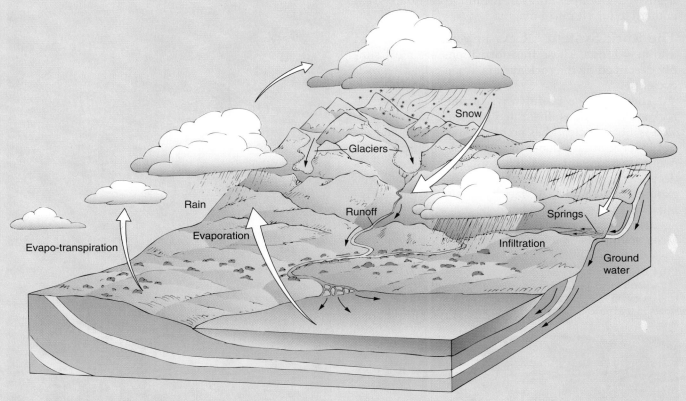

P1.1.1 The hydrologic system depicts the movement of water through the atmosphere from the ocean to the land and back again to the ocean. See text for discussion.

canic system, brings about changes in two other systems, the river system and the soil system.

When we look at the four major parts of this book each part can itself be thought of as a system, albeit a very large and complex one. The individual chapters, into which each part is divided, make it simpler for us to comprehend the whole and deepen our understanding of geology as we identify how particular geologic changes relate to the Earth system and its many subsystems.

formed. For example, calcium, weathered from the minerals of an igneous rock, can be carried by rivers to the ocean where it can be precipitated in a new mineral to form limestone, a type of rock.

Metamorphic Rocks Metamorphic rocks (from the Greek *meta*, "change," and *morphe*, "form") have been changed from their original form into a rock with a distinctly different appearance. Pressure and heat deep beneath the Earth's surface can transform a sedimentary or igneous rock into a metamorphic rock.

These processes can be aided by hot fluids that can transport dissolved material from one place to another. The same processes can also change one metamorphic rock into a different metamorphic rock.

In the process of metamorphism, minerals of the original rock are transformed as chemical bonds are broken and atoms rearranged into the patterns of new minerals. In other cases some minerals will grow larger at the expense of smaller minerals. In many instances metamorphic pressures have been so great that the rock deformed into curving bands, lenses, and stringers like a giant marble cake (Figure 1.8). A metamorphic rock, like an igneous rock or a sedimentary rock, may be destroyed by weathering. It can also be destroyed if temperatures of metamorphism are great enough to turn it into a magma.

The Rock System   Rocks are constantly changed into new forms. These transformations are most simply described by the **rock system,** often called the **rock cycle.**

There are two useful ways of thinking about the rock system. One is illustrated in Figure 1.9. This version is a true cycle. It has no beginning or end, and we can enter it at any point. This discussion begins with magma. An igneous rock that forms as a magma cools deep beneath the surface and may eventually be exposed at the Earth's surface by erosion of the overlying rock. As soon as this happens it is attacked by the agents of weathering. The products of weathering are loose sediments that other surface systems, such as

FIGURE 1.6   Weathering destroys rocks when they are exposed at the Earth's surface. Construction of Finchall Priory, Durham, England, began in 1237. In the approximately 750 years since then the stones in the church wall have been extensively weathered. (*Pamela Hemphill*)

FIGURE 1.7   The erosion of Canyon de Chelly in northeastern Arizona has exposed these sedimentary rocks.

rivers, wind, and gravity, move to new locations, and eventually to the oceans. Here they collect in layers that other processes consolidate into sedimentary rocks. If the path of change follows around the outside of our diagram, the sedimentary rocks are then subjected to deep burial, high pressure, and elevated temperatures, which transform them into metamorphic rocks. Given enough heat, these metamorphic rocks melt to form a magma. Upon cooling, igneous rocks emerge, and we have gone full cycle.

Notice that this system can be interrupted as indicated by the paths through the interior of our diagram. An igneous rock, for example, may form from a magma but never be exposed to weathering. Instead, it can be converted directly to a metamorphic rock by heat and pressure and then turned back into a magma. Similarly, other interruptions can take place if sediments, sedimentary rocks, or metamorphic rocks move into the zone of weathering before they can continue around the outer ring of the diagram.

The rock system as shown in Figure 1.9 provides us with a simple way in which to view the changes

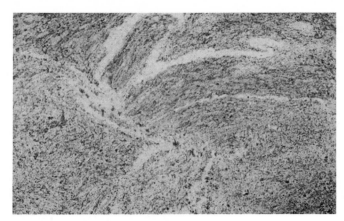

FIGURE 1.8 Metamorphic rocks form from preexisting rocks when they are subjected to high temperature and heat. Swirling patterns and streaks of contrasting types of minerals characterize some metamorphic rocks, as in this example. This polished sample is about 0.5 m wide.

that rocks undergo. In some ways, however, it has its shortcomings, precisely because it is a cycle. In the terminology of systems, it is a closed system. It always comes back to the same place. Nothing is added to the system, and nothing escapes. It is endless. We know, however, that the Earth had a beginning and that sometime it will have an end. In short, it has a history and a future. It has evolved through time and will continue to do so, something the rock system, as we have examined it so far, does not tell us. It lacks the direction of time. We can give it this dimension by breaking the cycle at some place and opening it out in a more or less linear fashion as suggested in Figure 1.10.

As we move from left to right in the diagram of Figure 1.10 we progress forward through time. When we look at the recycling of rocks in this way we can begin to see how complex the system can be. As we examine this linear representation of rock change we can see a number of relationships between and among rocks that are not apparent in the version of the rock system in Figure 1.9. Some of these are listed below.

1. A single rock mass can be transformed into one or more new rocks. Some portion of an igneous rock, for example, may be weathered and produce sediments that eventually form a sedimentary rock. Some of the original igneous rock can be subjected to heat and pressure and transformed to a metamorphic rock. Some of the igneous rock may be heated to form a magma and then cooled to form an igneous rock again.

2. The components of a rock can come from a number of different predecessors. For instance, the sediments forming a single sedimentary rock can be derived from the sediments weathered from several different preexisting rocks. In another example, two or more different rocks may be changed into a single metamorphic rock.

3. A particular rock may persist for a long time before changing. Or it may change very rapidly.

4. A single mineral in a rock can follow a path that takes it through a number of successor rocks. This route can become more complex when a mineral is broken down into its various chemical elements that follow different pathways to contribute to new minerals in different rocks.

The relationships that can develop in this family tree of the rocks become almost endless as the pathways taken by rocks, minerals, and elements in the system create a complicated, fascinating web of relationships through time.

It should be apparent from the rock system that the materials of the Earth take on different forms depending upon their environment. For example, an environment with heat great enough to melt rock produces a magma. Loss of this heat establishes a new environment and creates an igneous rock. Heat and pressure, applied in the proper amounts, turn the igneous rock to a metamorphic rock. In the environment on and near the Earth's surface, weathering destroys all rocks and turns them into accumulations of loose sediments. As these sediments pass through the changing environments of the Earth they reassume the forms of sedimentary, metamorphic, and igneous rocks.

The fact that Earth materials reflect the environments in which they are assembled allows geologists to reconstruct their history. An igneous rock testifies to a melt or magma and to a later cooling stage. A sedimentary rock tells us that weathering has produced the rock's components by the destruction of preexisting rocks, that other processes have accumulated them, and that still others converted them to rock. A rock, therefore, can hold the record of its history. We speak of this as a **rock record.**

As materials pass through the rock system, the hydrologic system plays a most significant role. The material is prepared for movement by weathering, a process usually based on water. For example, water

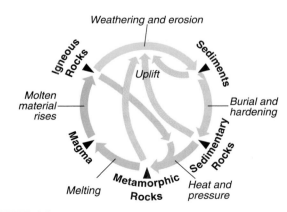

FIGURE 1.9 The rock system (also called the rock cycle) provides a convenient way of tracing the paths that rocks follow as they change from one type to another. See text for discussion.

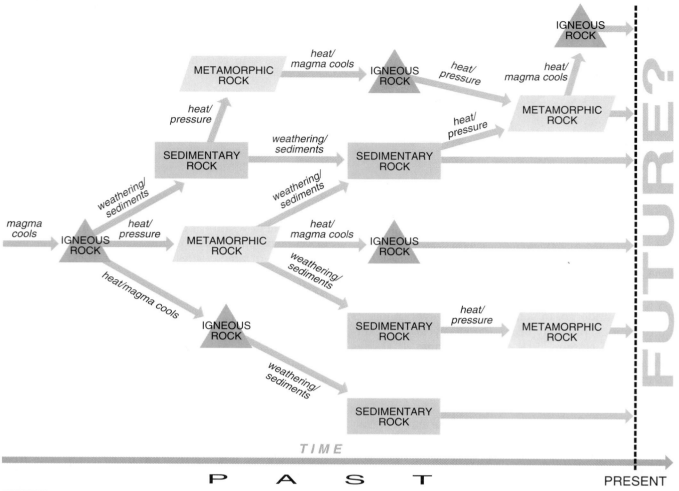

FIGURE 1.10   An alternate way of viewing the rock system, as shown in Figure 1.9, is to spread it out along a time line from the past to the present and into the future. We have begun this version with magma, just as we began the discussion of Figure 1.9 with magma. Figure 1.10 more realistically depicts the different possible pathways Earth materials may follow as rocks change through time. See text for further discussion.

makes possible living things, and water is a major agent in weathering. Likewise it provides a medium in which the chemical reactions of weathering can most easily take place. As new sedimentary minerals form from old, water is included as part of many of them. Rocks themselves contain large volumes of water that seep slowly through the cracks and pores that characterize virtually all rocks. Obviously water plays a major part in moving materials around on the surface, either in streams or in glaciers. These materials are usually deposited and become sedimentary rocks in aqueous environments, such as oceans and lakes. Indeed, it is difficult to envision the rock system without water.

A review of the rock system also makes clear that some mechanism must be available to move Earth materials through the several, very different, Earth environments. This mechanism expresses itself in rock deformation, earthquakes, and volcanic eruptions. Without some process to move the solid Earth the world would be a static, unchanging globe, as is our Moon today.

Geologic Time   Nearly 50 years ago the American geologist, Adolph Knopf wrote:

If I were asked as a geologist what is the single greatest contribution of the science of geology to modern civilized thought, the answer would be the realization of the immense length of time. So vast is the span of time recorded in the history of the earth that it is generally distinguished from the more modest kinds of time by being called "geologic time."

Geologists of the late eighteenth and early nineteenth centuries were convinced that the changes they read in the rock record took a great deal of time to accomplish. Looking today at the various stages in the rock system, we can come to the same conclusion. It must take a great deal of time to cool a magma and form an igneous rock deep beneath the surface, and then to expose it by removing the rock that once buried it. More time is needed to destroy the rock by weathering, and then to move the products of weathering produced from the continents to a collecting basin in the ocean. Still more time is needed to harden the soft sediments into a firm sedimentary rock. The farther we progress through the rock system, the more time is demanded.

Early in the development of their science, geologists began to learn how to arrange in chronological order the events recorded by the rocks. By early in the twentieth century geologists had outlined a worldwide calendar of geologic time. How much time was involved, or what dates might be associated with individual Earth events, they had no way of knowing. The discovery of **radioactivity** at the turn of the present century changed all this and gave us a way of determining the age of Earth materials in terms of years, whether they be billions of years old, less than a hundred, or some age in-between.

Radioactivity, a characteristic of a few elements including uranium, involves a chemical change but one very different from the ordinary chemical change that relies on the interaction of the electrons of atoms. Radioactivity depends on spontaneous decay in the nucleus of an atom. In the process the number of protons in the nucleus of an atom changes and this creates an atom of an entirely different element. What makes this useful to the geologist is that the rate of decay can be measured, and it is constant. Often we can also measure the amount of the new element that is produced. If a radioactive element is included in a mineral at the time of mineral formation, then the measurement of the amount of accumulation of the new element allows geologists to calculate how long ago the radioactive element was incorporated in the mineral. The age of the mineral can then be inferred. We now have literally thousands of age determinations of Earth materials recording the date of a wide variety of Earth events.

Changes within the Earth

For two centuries it has been clear to geologists that the Earth is a dynamic, changing body, as illustrated by the rock system. However, only as the twentieth century progressed have geologists been able to explain the major changes within the Earth: Why volcanoes and earthquakes happen and why they happen where they do; how the Earth is layered from the surface inward to the core; the origin of the forces that bend and break solid rock; how continents and ocean basins formed and how they relate to each other; and how mountains are built.

Heat and Magmas   Heat is a form of energy, and it is heat energy that drives the internal processes of the Earth. It has three major sources within the Earth. Some of it is left over from the birth and early years of the planet, still residing deep in the Earth. Some comes from the continuing solidification of the liquid part of the Earth's **core.** The third major source of heat is the continuing decay of radioactive minerals. As the nucleus of the atom decays it produces not only a new element but also a burst of energy. Most radioactively produced energy is being generated in rocks in the upper few kilometers of the Earth, particularly those immediately beneath the continents.

We see the Earth's heat expressed in several ways. The most obvious is as magma that breaks forth at the surface to form lava flows or as the violence of an explosive volcano. We see it also in hot springs and geysers. We recognize it, as well, in the igneous rocks now exposed in the cores of many mountain ranges.

A Layered Earth   The Earth is not a solid, singular unit, but rather is made up of a series of concentric shells. An onion, though also divided into shells, provides an imperfect analogy. A closer, but still imperfect comparison, is a golf ball, or a baseball, each with its outer cover and inner windings of rubber or string around an inner core. Unlike the shells of an onion, golf ball, or baseball, there is a slow movement of material within the Earth. Our knowledge of the nature of these shells comes in large part from the analysis of earthquake waves. We can describe the shells both on the basis of composition and rigidity.

The outer 50 to 100 km of the Earth is a rigid shell of rock called the **lithosphere** (from the Greek *lithos,* "rock," and "sphere"). Observations from deep mines demonstrate that temperature increases the deeper we go in the Earth. Temperature rises about 15°C (Celsius) with each kilometer of depth. This means that at a depth of about 70 km the temperature will be approximately 1000°C. At this temperature the lithosphere loses its rigidity and deforms slowly, like a sculptor's modeling clay. This "soft" zone lies immediately below the lithosphere and is known as the **asthenosphere** (Greek *asthenes,* "weak"). (See Figure 1.11.)

The lithosphere itself can be subdivided on the basis of compositional changes. The upper portion is called the **crust.** It lies directly beneath the continents and ocean basins. Below the ocean basins it is about 5 km thick. A dark-colored, relatively heavy igneous rock called **basalt** dominates the crust beneath the ocean, which is called **oceanic crust.** Crust beneath the continents is much thicker than the oceanic crust and averages about 35 km in depth. A partial cover of sedimentary rocks overlies a collection of igneous and metamorphic rocks, which together form the **conti-**

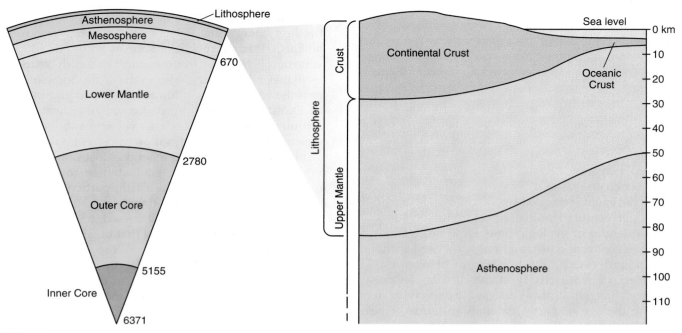

FIGURE 1.11   The Earth is made up of a series of shells. See text for discussion.

nental crust. Most of it is made of the light-colored igneous rock **granite.** Continental crust has a density of about 2.8 g/cm³ (grams per cubic centimeter), somewhat less than oceanic crust, which has a density of about 3 g/cm³. Below the crust lies the zone called the **mantle.** The upper part of the mantle includes the lower lithosphere and the asthenosphere. The section of the mantle just below the asthenosphere and to a depth of 670 km is known as the **mesosphere.** Below the mesosphere the **lower mantle** continues to the core at a depth of 2,780 km. The mantle is made of minerals that increase in density with depth. Geologists believe these are comprised chiefly of the elements iron, magnesium, oxygen, and silicon. The mantle surrounds the **core,** believed to be made chiefly of iron and nickel. The **outer core** is hot enough to be liquid and encloses the **inner core,** which is under enough added pressure at this depth to be solid (Figure 1.11).

Roughly speaking, the zonation of the Earth is a gravitational separation with the lightest-weight material in the crust and the heaviest material in the core. The same sort of differentiation, apparently, has occurred in other solid bodies in the Solar System, although fewer details are available about them than about the Earth. Figure 1.12 compares the Earth's core-mantle zonation with that of the other solid planets and Earth's Moon.

Deformation of Rocks We have found that rocks are not everlasting but can be destroyed by weathering at the Earth's surface. Rocks also change in other ways as they are subjected to Earth pressures. When the pressures are the same from all sides, as

they may be deep beneath the surface, the rock volume changes, becoming smaller as the pressure is applied. When pressure is reduced the rock expands slightly. If the pressure is differential—that is, greater in one direction than another—the shape of the rocks change. This pressure may cause **tension,** stretching a rock; **compression,** squeezing it together from oppo-

FIGURE 1.12   Like the Earth, Mars, Venus and Mercury, as well as the Moon, are layered. Here the comparative thicknesses of their mantles and cores are shown. *(Robert G. Strom "Mercury" in* Geology of the Terrestrial Planets, *edited by Michael H Carr, NASA Special Publication 469, Washington, D.C., 1984, Fig 3.1, p. 15.)*

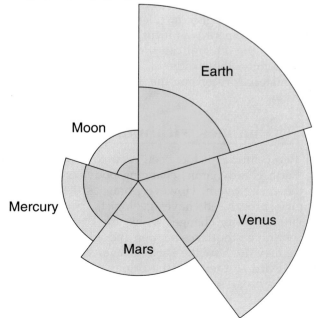

FIGURE 1.13   Faults occur when the rock on either side of a fracture moves. The movement along faults may be thousands of meters down to a few millimeters, or even microscopic. This specimen shows a small fossil fish that is cut by a fault. The movement is between 4 and 5 mm. Specimen from Princeton University Natural History Museum. *(Robert P. Matthews)*

FIGURE 1.14   Under compression rocks may bend as have these originally flat-lying beds of shale along Utah State Road 9.

site directions; or **shear,** deforming it at some angle between the direction of tension and compression.

Changes in rock shapes involve fracturing and can produce **faults** (Figure 1.13), which occur when the rock on either side of the fracture moves. Rocks may also bend or **fold** (Figure 1.14), a process helped along by high temperatures deep beneath the surface.

Plate Tectonics   One of the great advances in geology was the discovery, in the 1960s, that major movement and deformation of the Earth's crust are related to large fragments of the lithosphere moving in relation to each other. The energy that has fragmented the lithosphere and moves the fragments around is the Earth's internal heat. We call the lithospheric fragments **plates.** They are like the pieces of a giant jigsaw puzzle. Unlike a properly assembled puzzle, however, the plates don't always seem to fit, and as they jostle each other, their edges get shoved together or pulled apart. It is this motion of the plates that geologists call **plate tectonics.** The word *tectonics* comes from the Latin *tectare,* "to build," and to geologists signifies the processes associated with mountain building.

The global network of boundaries between plates establishes a set of bands that marks the location of most of the world's earthquakes. Most of the Earth's volcanic activity also takes place along these bands, and along some of these boundaries major mountain chains develop. Between these active bands, or zones, are large areas of the continents and the ocean basins where volcanoes and earthquakes, although not entirely absent, are rare (Figure 1.15).

To illustrate the different motions among plates, imagine moving the fragments of a broken dinner plate in relation to each other on a tabletop. Three types of motion are possible. The pieces may pull apart, they may push together, or they may slip by

each other in opposite directions. Earth's plates behave in this way along their boundaries. Two plates may pull apart along a **divergent boundary,** push together along a **convergent boundary,** or slip by each other in opposite directions on what is called a **transform boundary.**

Down the center of the floor of the Atlantic Ocean, more or less from north to south, runs the Mid-Atlantic Ridge. "Ridge" is somewhat of a misnomer, for it is between 1,500 and 2,000 km wide, with each side rising gently to a central crest. Along the crest is a steep-walled valley. This has formed as the plates on either side have pulled apart, dropping a part of the oceanic crust. Such a pull-apart valley is called a **rift valley.** Here it marks the divergent boundary between the American plate on the west and the African and European plates on the east (Figure 1.16). Along this pull-apart zone earthquakes occur and basaltic magma rises from the mantle to form new oceanic crust, as if to heal the Earth's wound. As the plates continue to pull apart more magma continues to be emplaced along the boundary. This process means that the farther away from the plate boundary a piece of basalt is, the older is the oceanic crust. The seafloor is growing wider as it spreads laterally from the ridge's rift valley. The process is called **sea-floor spreading.**

If plates grow along a divergent boundary, they must be destroyed someplace else to allow space for the new material. The destruction of a plate takes place along a convergent boundary. For example, the part of the American Plate that includes the southwestern Atlantic Ocean and the South American continent (Figure 1.16) is colliding with a large Pacific Ocean plate, the Nazca Plate. The Nazca Plate is made

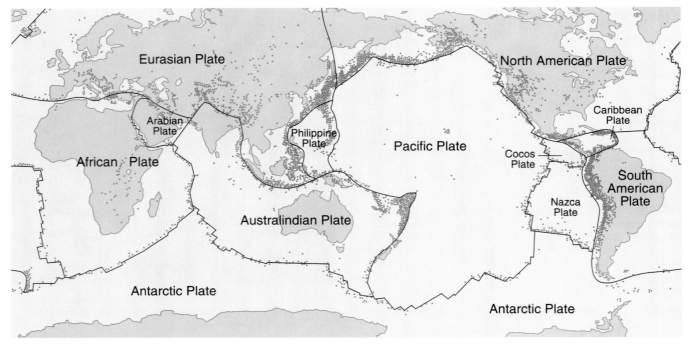

FIGURE 1.15   The lithosphere of the Earth is broken into large fragments, called *plates,* whose margins are marked by extensive concentrations of earthquakes.

of oceanic crust and is therefore denser than the western portion of the American Plate, which is composed of lighter continental crust. The leading edge of the Nazca Plate, being somewhat heavier than continental crust, is forced downward beneath the leading edge of the American Plate. This increases the thickness of the lithosphere along the boundary and marks the site of the still growing Andean Mountains.

As the Nazca Plate moves downward under the Andes, it pulls the seafloor with it, forming a deep oceanic **trench** that borders South America. Along the dipping zone of contact, the leading edges of the two plates overlap. This is called a **subduction zone,** from the Latin *sub + ducere* for "lead under." The mechanical energy generated as the plates scrape by each other produces earthquakes. Volcanism occurs as more and

FIGURE 1.16
Plate tectonics and the relation among the African, American, Nazca, and Pacific plates. See text for discussion.

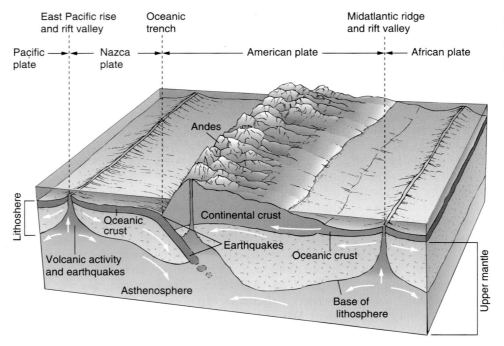

more heat builds up in the thickening crust. As the Nazca Plate moves deeper and deeper its margin becomes hotter and hotter. As it does, it loses its rigidity and eventually melts, becoming a part of the mantle from which it originally came (Figure 1.16).

The third type of plate boundary is the transform boundary, along which plates move parallel to each other. This type of motion produces earthquakes, some of them very large, but volcanism is rare. The San Andreas fault of California is our best-known transform fault. Actually, the San Andreas is one of several, more or less parallel, faults, of which it is the largest. Movement of land on the east side of the fault is south in relation to land on the west side of the fault. Los Angeles is moving north in relation to San Francisco.

Cratons, Orogens, and Plate Boundaries

We do not know when continents first appeared. Their earliest form, however, was in the shape of **cratons,** the cores of our modern continents, portions of which appeared at least 3.8 billion years ago, and probably earlier. They are composed of ancient, metamorphosed lava flows and somewhat younger granite masses. Questions still surround the origin of these cratons, although they seem to have originated as separate, isolated bodies. By about 2.8 billion years ago, the process of plate tectonics became active and began the coalescence of these bodies. Today, accumulations of these bodies, or portions of them, form the cores of our continents. Around the margins of these cratons younger and younger rocks have been assembled by continued plate tectonic action.

The boundaries between plates mark the location of major Earth movement and igneous activity. Along the converging boundaries this action is most intense, and it is here that the great mountain ranges of the past have been built and those of the present day are rising (Figure 1.17). These zones of collision are termed **orogens,** from the Greek *óros* for "mountain" + *genesis* for "origin." Mountain building, in general, is known as **orogeny.**

Changes at the Earth's Surface

Plate tectonics and the internal forces of the Earth create the major irregularities of the globe's surface, including the continents and the ocean basins. The continents, rising above the level of the ocean, are worn away as the products of weathering are eroded and transported to new areas and eventually to the oceans. Erosion modifies the surface of the land and creates many of the features of the landscape. While this is going on, heat-driven internal forces continue to lift large portions of the continents. The form of the land, therefore, is the product of the contest

FIGURE 1.17 Mountain building takes place along compressive plate boundaries. The Himalayan Mountains in Nepal continue to rise along the junction of the Australindian and Eurasian plates. *(Roger Chaffin)*

between the surficial systems of erosion and the Earth's internal systems of continental uplift.

Gravity and the Sun's radiant energy supply the energy that drives the surficial systems. Gravity is the agent that moves water down slopes, causes the flow of glaciers, moves water slowly through the underground, brings on landslides, and drives the world's wind systems. Solar energy passes through the atmosphere as shortwave radiation and warms the Earth. This energy is re-radiated as long-wave radiation and warms the atmosphere. This warming, however, varies from land to sea, from equator to the poles, and from sea level to mountain peak. This differential heating sets up atmospheric circulation that is responsible for major variations in climate, whether it be desert, rain forest, or mountain. The Sun also provides the energy for living things, important in the weathering system. It maintains the Earth's temperature in the range in which water exists as a fluid, a condition unique in the planetary system.

The Role of Water Water, whether in the liquid or solid form, is the single most important agent in shaping the Earth's surface. Flowing in a complex network of stream channels, rivers transport immense amounts of weathered debris from continents to oceans. Water seeps into the underground where it moves slowly through cracks and small pores in the rock before reappearing in springs and rivers. As it moves it dissolves large amounts of solid rock. As glacier ice, water acts as a giant rasp, wearing away solid rock and loose sediments alike. Even in the deserts, streams flow during the occasional and unpredictable rainstorm. Along the shorelines of the world, wind-driven waves erode, deposit, and transport Earth materials, and ocean currents distribute sediments to deeper portions of the basins.

One aspect of the hydrologic cycle, at first difficult to understand, is how a river continues to flow in wet weather and dry. Where does so much water come from, particularly in clear, sunny times? Indeed, even during heavy downpours there hardly seems to be enough water to keep the river flowing. The answer is that most river water is not the result of rain falling directly on the river, or flowing to it off the surface of adjacent slopes. A river's major supply of water has seeped downward from the surface to be stored temporarily in the underground before it re-emerges to feed the stream.

Mass Movement of Surface Material A tremendous amount of Earth material is moved directly by gravity, more or less as a solid mass (Figure 1.18). A catastrophic landslide is an obvious transfer of rock and soil from higher to lower elevation as part of its long journey to the oceans. There are, as well, other less spectacular, but visible movements, caused by gravity. Probably the most important gravity-driven movements, however, are generally invisible to the casual observer. Soils and unconsolidated debris move slowly and steadily down the gentlest of slopes, usually toward the banks of a stream or river where it is picked up by running water and carried onward.

Running Water At any one moment the rivers of the world contain only 0.0001 percent of the world's supply of water. Lakes, glaciers, atmosphere, and water that is underground each contains more. Despite this, rivers are responsible for carrying most of the products of erosion to the great settling basins of the world's oceans and are a major element in the world's landscape (Figure 1.19). Rivers flow in networks of channels that not only carry away the debris of weathering but also drain the continents. Very large stream networks may drain a large part of a continent, and the upstream limits of such networks in many instances mark the location of an ancient plate boundary or are closely associated with present-day boundaries.

Rivers acquire their load of sediments by cutting down into the landmass and, more importantly, by the work of gravity delivering material down valley slopes to the river edge. Rivers not only erode and transport Earth materials but they also deposit them. Rivers in flood escape from their channels and deposit sediments over adjacent low-lying land to build up level stretches of **flood plain.** As streams enter the standing water of a lake or ocean, their flow is stilled, and they no longer have the energy to carry all their sediments. These loads may then be deposited as **deltas.**

FIGURE 1.18 Gravity moves material down slopes. Sometimes the movements are barely noticeable. At other times they are sudden and catastrophic. In the Gros Ventre valley in Wyoming a sudden landslide in 1925 carried the side of Sheep Mountain into the valley below. There it dammed the Gros Ventre River and formed a lake. The scar on the mountain side as well as the debris in the valley bottom are still clear in this photo taken 38 years after the slide. *(Dr. John S. Shelton)*

FIGURE 1.19
Rivers drain the landmasses, move weathered materials toward the ocean, and form major elements of the landscape. Snowdonia National Park, Wales

Ground Water The amount of ground water is only 0.61 percent of the Earth's supply. This is still nearly 7,000 times the amount of water in rivers. It represents that part of the hydrologic system in which water infiltrates from the surface into the ground below. The water seeps through the soil to the **water table,** a level below which all the pores and cracks in the rock are saturated with water. Flowing in such constrained channels, ground water usually flows between 1.5 m/year and 1.5 m/day. By contrast, during average flow, the Mississippi River at Vicksburg, Mississippi, moves at a velocity of about 1.5 m/second.

This vast supply of ground water comes to the surface as springs and swamps. More importantly, it provides the water that keeps most rivers flowing even in dry weather. As water moves from the surface through the rocks and back to the surface it dissolves some minerals and thus contributes to erosion of the continents. When the solution of rock is extensive enough, caves may form (Figure 1.20).

Glaciation Modern glaciers are found on Greenland and Antarctica, and in many high mountains. During the last few thousand years, however, glacier ice has melted from large sections of Europe, North America, and Asia, and has retreated from a formerly much greater extent in most glaciated mountains. This former advance and later retreat has been part of similar alternating advances and retreats of glacier ice during the last 2 million years or so. Throughout much of geologic time, glaciation has been absent or greatly restricted. Between these long stretches of nonglacial climate have been extensive periods of glaciation dating back at least 2.2 billion years.

Glacier ice flows under the influence of gravity. Like streams, glaciers erode, transport, and deposit Earth materials. The glacial system accounts for much of the spectacular mountain scenery (Figure 1.21) of steep-walled valleys, hidden lakes, jagged ridges, and needle-like peaks. On lower lands glaciers have spread a veneer of sediments across much of North America, northern Europe, and Asia, and they have provided

FIGURE 1.20   Water that is underground may dissolve enough rock material to create caves, as it has done here in the Blue Spring Cave, Indiana. Water depth here is about 4.5 m. *(Arthur N. Palmer)*

FIGURE 1.21   Glaciers are responsible for much of the spectacular scenery of mountains. Here in the Beartooth Mountains close to the Montana-Wyoming border, glaciers, now melted, created the basins for "Twin Lakes" and carved the steep canyons beyond.

the base for productive soils over wide areas. In many areas erosion and deposition have left behind the basins for many lakes, including the Great Lakes. In addition, glaciers have disrupted old river courses and channeled them to new locations.

Causes of glaciation are at least twofold. Because glaciers form only on land, landmasses must exist in polar latitudes before extensive glaciation can start. Over the last few million years the mechanism of plate tectonics has been moving continents slowly toward the poles. Once lands are susceptible to glaciation some other mechanism must operate to bring about repeated advances and retreats of glaciers. Geologists now believe that these are due to periodic changes in the amount and distribution of solar radiation received by the Earth.

Shorelines Shorelines are determined, in the first instance, by plate tectonics, and they differ according to the types of plate boundaries that are involved. The transform boundary of California produces cliffed shorelines paralleled by long valleys and a series of hills and low mountains. This is in contrast to the compressional boundary of southern Alaska and the Aleutian Islands where active volcanoes dominate the shoreline. The position of sea level in relation to the land is determined by several different factors, including uplift or subsidence of the land and change in the volume of the ocean as glaciers both advance and melt.

The details of all shorelines are fashioned by the energy of wind-driven waves, indirectly the result of the Sun's energy. In the open ocean, wind piles up water in waves. As waves approach the shoreline the ocean bottom changes their direction, shape, and velocity, and water is thrown forward against the shore. It is the energy developed by this shoreward rush of water that erodes a shoreline, moves sediments, and builds beaches (Figure 1.22).

FIGURE 1.22   Wind-driven waves provide the energy that shapes the shorelines of the world. Here they pound against the distant headland and build a beach in the more protected environment of the Welsh bay, Port Neigwl, near Gwynedd, northern Wales.

FIGURE 1.23   Barren cliffs of black basalt rise above the sandy desert in the northern part of the Faiyum depression west of the Nile River in Egypt. *(Thomas M. Bown and James A. Harrell)*

Wind and Deserts   Deserts are created by the global circulation of air as it is driven by the Sun's energy. They occur in zones of dry, warm, descending air or where mountain masses shut off moisture-bearing winds. In desert areas, without effective vegetative cover, wind becomes a major surface process of the erosion, transportation, and deposition of Earth materials (Figure 1.23). At times the wind is also important in the semiarid regions marginal to true deserts.

Among the most impressive results of wind action are accumulations of sand in dunes that may reach kilometers in length and over 500 m in height. In addition, desert dust is blown for great distances. As it settles from the air it may form sheets of dust several meters thick, as it has in the central United States in the recent geologic past.

Surficial Processes on Other Planets   Modern processes operating on the Earth's neighboring planets, Mercury, Mars, and Venus, and on our own Moon, pale in comparison with those seen on Earth. Among these bodies the surface processes on Mars are most active. Although Mars has no flowing surface water today, dry valleys testify to its former presence. There may be underground water on Mars, and a thin, seasonal cover of ice exists in the regions of the Martian north pole. A still thinner cap of frozen carbon dioxide is seasonally present around the southern pole. A rarefied atmosphere on the same planet moves sand and dust and has produced forms similar to those in the deserts of the Earth. Landslides, some of them of great size, have occurred on Mars, and mass movement may be continuing on a small scale today. Venus, with its heavy atmosphere of carbon dioxide, has modern sand dunes. Mercury, however, displays no observable surface activity (Figure 1.24), nor does the Moon.

FIGURE 1.24   Mercury has no water, no living things, and essentially no atmosphere. As a result its surface has changed little over the last 2 billion or more years, with the exception of craters formed by occasional meteoritic impacts. *(NASA)*

Humans and Earth Change

Human-like creatures began appearing on Earth an estimated 2 million years ago. The current species, *Homo sapiens*, came upon the scene perhaps

FIGURE 1.26   Most of the fuel that humans use has been created by Earth processes. Here an oil well pumps petroleum from the ground in northwestern New Mexico.

FIGURE 1.25   Humans early learned how to put rocks to their own use. This Neolithic ax from Denmark was fashioned about 2500 B.C. from a tough, homogeneous rock called *chert*. *(Robert P. Matthews)*

90,000 years ago and has rapidly made its presence known.

Use of Natural Resources   Early humans sought out suitable rock for stone tools and became increasingly skilled in their fabrication and use (Figure 1.25). The first known use of rock materials, other than stone in implements, took place in South Africa more than 40,000 years ago. There local peoples mined blood-red oxides of iron, presumably for the decoration of their bodies or bedrock walls, or both. Today's humans rely upon the Earth for fuel, both solid and liquid (Figure 1.26), and for metals as familiar as iron and as exotic as germanium, a rare metal used in transistors. The search for these and other useful Earth materials is now guided by specialists (geologists) and has reached immense proportions. The human acquisition of these geologic products, so essential to current civilization, leaves its mark in many ways upon the Earth's surface.

Humans as Geologic Agents   How long people have had a significant impact on the Earth and its systems is unclear. Certainly, by 10,000 years ago agriculture was beginning to increase the rate of soil ero-

FIGURE 1.27   Hoover Dam on the Colorado River of Arizona is an example of the way in which people can serve as geologic agents. The dam creates 185-km-long Lake Mead, and controls the flow of the Colorado River downstream. It produces hydroelectric power and provides water for parts of Arizona, Nevada, California, and Mexico. *(Pamela Hemphill)*

The Changing Earth   **19**

sion and sediment production. Soon afterward the ability to find copper, create bronze, and smelt iron made a more concerted manipulation of the natural environment possible. Now people are able to change the form of the land and to control the flow of rivers (Figure 1.27), changing their behavior and even drying them up. We have been able to pry valuable minerals from the Earth but are unable to avoid the depletion of resources. We can hold back the sea and clear the forest from a continent. As we have used the Earth, however, we have polluted soil, river, lake, ocean, and atmosphere with our waste. We have been able to tap the energy of the atom, but we have not been able to store safely the radioactive debris that has resulted. Furthermore, despite our technological abilities, nature can still overwhelm us with earthquakes, volcanoes, landslides, hurricanes, and meteorites.

▶▶ EPILOGUE ◀◀

The topics previewed in this chapter are examined in detail in the four sections of the text that follow. We begin with Chapters 2–7, which make up the first section, Changes in Earth Materials. There we consider rocks and minerals—how they form, how they change, and how they can be used to tell time. Chapters 8–12, which form the second section of this text, deal with the Changes Within the Earth. These chapters begin with a consideration of the Earth's heat, the energy that forms magmas, causes earthquakes and volcanoes, deforms rock, moves the Earth's plates, and forms mountains. Changes at the Earth's surface are discussed in Chapters 13–18. These chapters deal with mass movements of materials down slopes and with aspects of the hydrologic system including streams, ground water, glaciation, and ocean waves. The section concludes with the work of the wind. The two final chapters, 19 and 20, address the subject of Humans and Earth Change. The first of these chapters discusses the nature and formation of the Earth's fuel and mineral resources. The last chapter focuses on the role of people as geologic agents and on those threatening geologic processes that humans cannot harness.

We start our study in detail with Chapter 2, there examining minerals and how they are made. This will prepare us for succeeding chapters, where we learn how minerals are gathered into rocks, which form the basis for the science of geology.

SUMMARY

1. The concept of uniformitarianism holds that the geologic processes operating today also operated in the past in a similar fashion and produced the same results then as they do now. For example, present day winds lay down characteristic deposits. When similar deposits are found turned to rock, geologists then reason that wind laid down these deposits in the geologic past, before the deposits were turned to stone. This concept allows geologists to interpret the past in the light of their understanding of the present. Without the framework of uniformitarianism there would be little basis for modern geology.

2. A system is a whole made up of several parts. The Earth system is composed of the atmosphere, the solid Earth, the oceans, and living things, which form a continuously acting whole. The Earth system is made up of countless subsystems such as the hydrologic system, plate tectonics system, and the glacial system, to name but three.

3. The three major rock families are igneous, sedimentary, and metamorphic. Igneous rocks form when a magma or lava cools, sedimentary rocks form by the consolidation of the debris of preexisting rocks, and metamorphic rocks form by the application of heat and pressure to preexisting rocks. The rock system traces the members of a rock family as they pass through a new geologic environment and change into members of another family. Igneous rocks are destroyed by weathering, and the products form the basis for sedimentary rocks. These, in turn, can be destroyed by weathering or converted into metamorphic rocks by heat and pressure. Metamorphic rocks can be changed by weathering, further metamorphosed by added heat and pressure, or melted to form a magma for a new igneous rock. The rock system can be described as a circular, closed system. We can turn it into an open system by giving it the direction of time.

4. Plate tectonics is the system in which pieces of the crust (plates) move laterally on the asthenosphere and interact at their edges. They grow along diverging boundaries and are destroyed along convergent boundaries. Motion along convergent boundaries causes earthquakes and explosive volcanic activity, and the boundaries are the focus of mountain building. As plates separate along divergent boundaries quiet volcanism occurs, and here earthquakes are less vio-

lent than they are along compressive boundaries. Blocks of the Earth's crust move parallel to each other and in opposite directions along transform boundaries. Here volcanism is rare, but earthquakes can be very destructive when they occur on land, as they do in California.

5. The Earth's surficial systems are weathering, mass movement, rivers, ground water, glaciation, water waves, and wind. They are driven by gravity and the Sun's radiant energy. Together the surficial systems work to lower the landmasses even as internal systems of the Earth, powered by the globe's internal heat, raise the landmasses. The hydrologic system plays a crucial role in bringing about changes at the Earth's surface. Water in streams moves weathering products from the continents to the oceans, and ground water nourishes stream flow, particularly during times of low precipitation. Water is transferred from ocean to the land as snow forms glaciers, which can transform the landscape of both mountains and plains. Ocean waves, driven by wind, shape the edges of continents.

6. Humans have learned to exploit the fuel and mineral resources of the Earth's crust. We have also developed ways to modify the natural environment and become geologic agents in our own right. For example, we can divert the flow of rivers and change the composition of the atmosphere. We still, however, can be overwhelmed by such natural events as floods, earthquakes, and volcanic eruptions.

KEY WORDS AND CONCEPTS

actualism 2	geology 2	"present is key to the past" 2
asthenosphere 11	granite 12	proton 4
atom 4	hydrologic system or cycle 6	radioactivity 11
basalt 11	igneous rock 5	rift valley 13
chemical bond 4	inner core 12	rock cycle 8
chemical element 4	lava 5	rock record 9
compression 12	lithosphere 11	rock system 8
continental crust 11	magma 5	sea-floor spreading 13
convergent boundary 13	mantle 12	sedimentary rock 5
core 11	mesosphere 12	shear 13
craton 15	metamorphic rock 7	subduction zone 14
crust 11	mineral 4	tension 12
delta 16	neutron 4	transform boundary 13
divergent boundary 13	nucleus (atomic) 4	trench 14
Earth system and Earth subsystems 6	oceanic crust 11	uniformitarianism 2
	orogen 15	volcanic ash 5
electron 4	orogeny 15	volcanic cinders 5
fault 13	outer core 12	water table 17
floodplain 16	plate 13	weathering 6
fold 13	plate tectonics 13	

QUESTIONS FOR REVIEW AND THOUGHT

1.1 Why is the principle of uniformitarianism critical in the study of geology? Illustrate with an example.

1.2 In the rock system, describe how an igneous rock can become a sedimentary rock and a sedimentary rock, a metamorphic rock.

1.3 Figure 1.10 shows some of the pathways that Earth materials can take through time. Extend the diagram from the present into the future.

1.4 What happens when a plate composed of oceanic crust collides with a plate of continental crust?

1.5 Where and how do oceanic plates grow? If the basin of the North Atlantic Ocean has been spreading at the rate of 2 cm/year, and New York and Lisbon are about 5,600 km apart, when did the North Atlantic Ocean come into being?

1.6 What are the major layers of the Earth from the surface to its center? How are the layers arranged according to density? What might be some reasons for this zonation?

1.7 Describe the hydrologic system. What would happen to sea level if glaciers of the world were to melt? What would happen to sea level if glaciers were to expand to cover Canada, the northern United States, and northern Europe?

1.8 What role does the hydrologic system play in the modification of the Earth's surface?

1.9 How does a river keep on flowing in dry weather?

1.10 What are some ways in which present-day civilization is dependent on the geologic environment?

1.11 What are some ways in which people serve as geologic agents?

Critical Thinking Questions

1.12 Would it be possible to form sedimentary rocks if there were no water on Earth? If so, how would they form? If not, why would it be impossible?

1.13 Suppose that the process of plate tectonics were to suddenly cease to operate today. What might the Earth be like tomorrow? What might it be like in a million years? In a billion years?

CHANGES IN EARTH MATERIALS

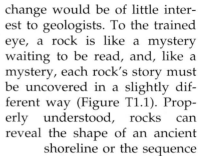

We have an expression in the English language that verbalizes most people's impression of what the Earth is like: "solid as a rock." People use it to say that someone is dependable, unflinching in the face of adversity, or a model of stability in times of change or uncertainty. Being compared to a rock, at least in this sense, is high praise. Most people would probably agree that rocks are, in fact, the least changeable materials in our world. Rocks do not appear to rust or rot, they withstand great stresses and high temperatures with minimal damage, and although they crumble and are eventually eroded by wind and water, they do it so slowly that their destruction is imperceptible most of the time.

Rocks as Records of Change

Geologists have another view of rocks. What they see in a rock is not stability, but rather a record of change. By detecting hints of how the rock formed and what may have happened to it since then, geologists discover clues to geologic settings that may have vanished hundreds of millions of years earlier, in each of which the rock was somehow affected. In fact, a rock that did *not* show such evidence of change would be of little interest to geologists. To the trained eye, a rock is like a mystery waiting to be read, and, like a mystery, each rock's story must be uncovered in a slightly different way (Figure T1.1). Properly understood, rocks can reveal the shape of an ancient shoreline or the sequence of events that followed a disastrous flood while dinosaurs still walked on the planet. With careful study, a geologist may discover that a particular rock was once 20 km below the surface of the Earth and may "read" the temperature that it experienced at that depth. Some rocks may carry such a complex record that a geologist might see in them hints of mountain-building events that took 50 million years or more from start to finish. The primary objective in the next few chapters is to introduce you to this way of looking at "solid" rock so that you too can see it as a dynamic material and a tool for revealing the history of the Earth. To reach this objective, we will explore concepts to which you were introduced in Chapter 1: the building blocks of matter and the ways in which they are assembled to form minerals; the three large families of rocks and the processes by which they are related to one another in the rock system; and the methods by which geologists reckon time and rates of change.

What Is a Rock?

One logical way we might begin this study is to agree on what a **rock** is. Surprisingly, however, geologists have a hard time doing this. One standard source of information, the *Dictionary of Geological Terms*, published by the American Geological Institute, says that a rock is *"strictly, any naturally formed aggregate or mass of mineral matter, whether or not coherent, constituting an essential and appreciable part of the Earth's crust."* Consider, however, the various questions that you might ask a geologist if you were trying to figure out what this definition means. For example, *"What do you mean by an 'aggregate or mass'?"* Well, usually a rock is composed of simpler, or at least smaller, pieces of other materials ("mineral matter"). Of course, a large piece ("a mass") of a single kind of material might also be called a rock. *"How large?"* Large enough to be an "appreciable part of the Earth's crust," whatever that is. A specimen as big as your fist doesn't sound like an "appreciable" fraction of the Earth's crust, but most geologists would agree to call it a rock. *"OK, what is 'mineral matter'?"* In most instances, rocks are made of chemically distinct nonorganic substances known as **minerals,** in each of which atoms are arranged in a regular three-dimensional pattern called a *crystalline structure.* Some nonminerals may also be considered legitimate components in rocks, however. Thus, "mineral matter" may include such things as the fossilized plant tissue in coal (an organic material) or the glass (a noncrystalline material) that forms if lava is cooled rapidly, neither of which fits the more restrictive definition of "mineral" that we just presented. In fact, few naturally occurring substances found in the crust would be out of place in a rock. *"Just the crust? How about the mantle?"* OK, the mantle too, and the core. *"So what* **is** *a rock?"* (Figure T1.2). In the sixteenth cen-

(a)

(b)

(c)

(d)

FIGURE T 1.2
Each of these objects is a rock, although they clearly look quite different. The great variety in rocks is due both to the range of materials from which rocks form and to the many different processes by which those materials change in the Earth. *((a) Breck P. Kent, Earth Scenes; (b)Breck P. Kent, Earth Scenes; (c) Don Fawcett, Visuals Unlimited; (d) Jerome Wyckoff, Earth Scenes)*

FIGURE T 1.3
During nonexplosive eruptions, geologists can take samples of some lava flows while they are still fluid. By comparing information gathered from several locations, or perhaps several different lava flows, geologists can deduce how a volcanic system has evolved through time. *(Katia Krafft, Photo Reseachers, Inc.)*

tury, the German scientist Georg Bauer (commonly known by the Latin version of his name, Agricola) wrote a scholarly treatise entitled *De Natura Fossilium*, meaning loosely, "About Natural Things in the Ground." Rocks. Considering all of the exceptions that we would need to recognize if geologists tried to agree on a formal definition, the title of Agricola's book captures the spirit reasonably well.

We emphasize the ambiguity in this most basic term, "rock," because it calls attention to how much variability there is in the materials that constitute the mass of the Earth, and it suggests indirectly how many different processes can affect them. Rocks are, in fact, composed of many different kinds of "mineral matter," each characteristic of a geologic setting with a distinctive temperature, pressure, level of biological activity, amount of water, and so on. The processes that occur in a particular setting also determine the way these kinds of mineral matter are arranged in

each rock. A rock, therefore, is not simply an object; it is a product of an environment and of changes in that environment. If the environment is altered, so is the rock. If the rock is moved from one environment to another, it is slowly modified to meet conditions in the new setting. It was this strong tie between rocks and change in the Earth that James Hutton and his followers were trying to express through the concept of the rock system which we introduced in Chapter 1.

Studying Changes in Rocks

How does this relationship between rocks and their environments affect the way we have written the opening chapters of this book and, more importantly,

FIGURE T 1.4
An apparatus such as this one simulates the high pressures of the upper mantle by squeezing a small artificial rock sample with a hydraulically driven piston. By heating it simultaneously with an electric furnace (inside the apparatus but not visible in this photo), a petrologist can try to reproduce conditions in a particular environment deep in the Earth. Experiments such as this attempt to produce synthetic rocks that resemble those in the Earth, so that the petrologist can deduce how the "real" rocks formed. *(Connie Bertka)*

FIGURE T 1.5
One simple rule that stratigraphers use to deduce the order in which rocks formed is suggested here. In an undisturbed sequence of layered sand or mud, the layer on the bottom is the one that settled first. Younger layers (late-arriving sands and muds) always settle on top of older layers. *(Photo provided by William E. Ferguson)*

the way we would like you to think about rocks? You have already learned that geologists divide rocks into three major groups: igneous, sedimentary, and metamorphic. It may be tempting to focus on that categorization and to learn the physical attributes that distinguish one group from another. Some of our efforts will be directed toward that goal, of course, but in Chapters 3 through 6 we will pay more attention to how a rock *becomes* igneous, sedimentary, or metamorphic. Magma, created when rocks are heated to their melting temperature, cools and solidifies to form igneous rocks. Rocks exposed on the Earth's surface to weathering are dissolved or reduced to loose particles that are transported and redeposited, and are then bound together again to become sedimentary rocks. Recrystallization, largely due to heat and pressure, produces metamorphic rocks. These processes of change, the framework of the rock system, will be our focus.

As we discussed in Chapter 1, geologists discern the history of the Earth and anticipate the way processes such as those in the rock system will work in the future by assuming that the natural laws they observe in operation today are no different from any other time. For that reason, geologists examine environments where rocks are forming or changing today. A **petrologist** (from the Greek root word *petros*, meaning "rock": a geologist who studies rocks) may travel to observe an erupting volcano and to sample the fresh rock as it solidifies (Figure T1.3), or may sample the mud in a lake bed to see how it differs from the silt entering the lake from streams. If rock-forming environments are inaccessible, as they are anywhere below a few kilometers in the crust, petrologists simulate those environments in the laboratory. They construct elaborate devices to apply pressure and heat, for example, and try by synthesizing rocks to understand how the ones they find in nature may have formed

FIGURE T 1.6
The search for petroleum in the Earth would be much more difficult if geologists had not learned to recognize which rocks usually contain petroleum, how they form, and how they and the petroleum change after they are buried. The successful exploration and evaluation of potential drilling sites both depend on how well geologists can "read" the history of petroleum-forming events in the rock. *(John D. Cunningham, Visuals Unlimited)*

(Figure T1.4). In each of these tasks, the petrologists' goal is not only to understand a rock-forming process but also to recognize physical characteristics that are unique to rocks forming in that way. These might include the sizes and shapes of particles of "mineral matter," the manner in which they are attached to each other, and layering or other patterns of organization among particles, for example. These features then become the clues to recognizing how ancient rocks were formed, the keys of the present that unlock the mysteries of the past.

The final element in our study of change in the rocks is time itself, the topic of Chapter 7. This is not an easy task, because most processes in the rock system are extremely slow and because evidence of specific events, particularly in the distant past, is fragmentary. **Stratigraphers** infer the order of geologic events on a relative time scale primarily by studying layers (in Latin: *strata*), common in sedimentary rocks (Figure T1.5). **Geochronologists** (from the Greek roots *geo,* "earth" + *chronos,* "time") use the extent of radioactive decay, usually in igneous or metamorphic rocks, to determine their ages and construct an absolute time scale of Earth events. The methods of geochronology and stratigraphy thus provide complementary ways of studying the pace and patterns of change in the rocks.

Many of the inferences geologists draw about the rock system depend on understanding when and how rapidly changes have occurred in the Earth.

The Past as the Key to the Future

Why do geologists care about the way rocks have changed through time? Aside from the intrigue of solving a particularly complicated mystery, such as the way the rock system operates, understanding the past makes it possible for geologists to find valuable mineral and energy resources by locating remnants of the environments in which these resources were concentrated (Figure T1.6). Studying other past events such as floods or volcanic eruptions opens the possibility of predicting or perhaps even controlling similar events in the future. These are practical concerns that benefit humanity. The study of rocks, therefore, is a doorway to a fascinating and useful science.

2

A cluster of crystals of the mineral quartz, a common form of SiO$_2$
(Smithsonian Institution)

Minerals and Matter

OBJECTIVES

▲▲

*As you read this chapter, it may help if you focus
on the following questions:*

1. What are the basic building blocks of matter, and how are they combined to make the more complex substances found in the Earth?

2. What are minerals and how do geologists use their appearance and the way they react to physical and chemical tests to distinguish one mineral from another?

3. Why are there so few common minerals, and what are they?

4. What can geologists learn about geologic environments and conditions of change by studying the chemistry and the physical properties of minerals?

OVERVIEW

▲▲

The Earth's mass is nearly 6×10^{21} kg (6 billion billion metric tons). Except for the liquid outer core of the planet, almost all of this mass is in materials that geologists call **minerals.** To investigate the Earth, we must first find out what minerals are and how they behave.

Most geologists who spend their careers investigating minerals have academic training in both geology and chemistry. This overlap of fields will become evident quickly as we enter this chapter. As we have already emphasized earlier in this book, however, a geologist's primary interest is in learning about processes that change the Earth. In this chapter, therefore, we will use the chemist's terminology and some of the fundamental concepts about how matter is organized to study minerals for our own purpose: to begin gathering clues about how rocks form and change.

Atoms and Compounds
▲▲

Atoms

People seem by nature to be curious about how things are put together and how they work. Since prehistoric times, one of the trickiest puzzles to pique our curiosity has been the nature of **matter,** the substance of the physical universe. Even in antiquity, it was clear that the appearance and properties of most materials could be changed. Philosophers suspected, therefore, that some substances were more fundamental than others and were the building blocks of matter. Aristotle, for example, suggested that all matter is composed of combinations of *earth, air, fire,* and *water.* Until the mid-eighteenth century, though, all attempts to identify the basic components of matter and how they form more complicated substances were unsatisfactory.

In the late 1700s, the building blocks of matter became clearer as scientists became more adept at purifying substances in the laboratory. The fundamental materials they identified are now called **chemical elements.** No element can be subdivided further by ordinary chemical processes, such as the application of heat, acids, or electrical energy. For example, table salt, which is not an element, can be broken down (with great difficulty) to release sodium and chlorine, each of which is an element. Both sodium and chlorine are also found in many other substances, but no matter what laboratory methods are used to isolate these two elements, they always have the same properties and are, in fact, the same fundamental materials. Ninety elements are found on Earth. Several others have been synthesized in high-energy experiments.

As the elements were being characterized, scientists debated another question: *Is there a limit to how small a particle can be and still have the characteristics of a specific element, or is matter infinitely divisible?* In 1805 John Dalton, an English chemist trying to figure out why some materials always combine in the same strict proportions, showed convincingly that there is such a limit. The smallest particles of an individual element are called **atoms,** each no more than about 10^{-10} m in diameter (one ten-billionth of a meter). Atoms are so small that their characteristics can only be deduced by indirect experiments. It was not until about 1980 that atoms were first seen by extremely powerful microscopes (Figure 2.1).

Atomic Structure If John Dalton were alive today, he might be surprised to learn that atoms are now known to be made of even smaller particles. An atom *is* the smallest particle that has the properties of any element, but those properties depend on the

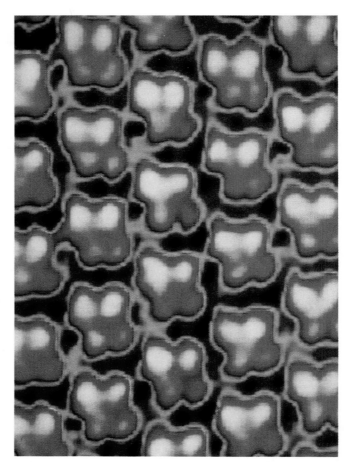

FIGURE 2.1   Tunneling electron image of an array of atoms. *(Peter L. Kresan, Lawrence Berkeley Laboratory)*

arrangement of smaller particles within the atom. At the core of an atom is a tightly packed **nucleus,** which contains particles called **protons** and **neutrons.** The number of protons in an atom, called its **atomic number,** determines what kind of element it is. Thus, scientists can label an atom either by its elemental name or by its atomic number. All carbon atoms, for example, have atomic number 6, and all silicon atoms have atomic number 14.

Neutrons have nearly the same mass as protons, but the number of them can vary from one atom to another so that not all atoms of a single element have the same total mass. Atoms that differ in **atomic mass number,** calculated by adding the number of protons and neutrons, are called different **isotopes** of the element. For example, hydrogen with one proton and no neutrons in its nucleus has an atomic mass number of 1. When a neutron is present, however, the atom is an isotope of hydrogen with a mass number of 2 and is called deuterium (Figure 2.2). Isotopes of an element are indistinguishable from one another by most chemical means. Geologists have found, however, that atoms with even subtle mass differences can be separated during some chemical changes in the Earth. The

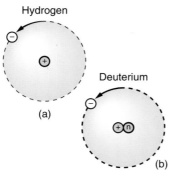

Hydrogen

Deuterium

(a)

(b)

FIGURE 2.2 (a) A hydrogen atom, in its simplest form, consists of one proton and one electron. (b) A hydrogen atom with a neutron in its nucleus has an atomic mass number of 2 and is called deuterium. All hydrogen isotopes, by definition, have one proton.

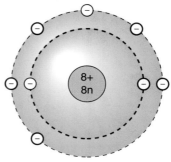

FIGURE 2.3 Electrons orbit in electron shells at precisely controlled distances from the nucleus of an atom. As shown in this schematic drawing of an oxygen atom, the innermost shell, which can hold two electrons, is filled first. The second shell can hold up to eight electrons, but oxygen has only six more. In order to fill the second, outermost shell, it must gain two more electrons.

proportions of isotopes in a rock sample can be valuable clues to how certain changes have occurred in nature. Furthermore, some isotopes are unstable and will disintegrate spontaneously, releasing energy and producing new isotopes. This process, called **radioactive decay,** is responsible for much of the Earth's internal heat production. Geologists can also interpret the age of a rock by measuring the progress of decay in it. We will explore these two topics in Chapters 7 and 8.

Although the nucleus comprises most of an atom's mass, it is an extremely small body at the center of the atom. The rest of the atom consists of a diffuse, roughly spherical cloud of **electrons.** The atom is a miniature electrical system, in which positive charges (one for each proton) are balanced against negative charges (one for each electron). Neutrons carry no electrical charge. Therefore, if an atom is electrically neutral, it has just as many electrons as protons.

The electron cloud has a complex structure. Electrons occupy **shells** in the cloud at characteristic distances from the nucleus, somewhat like the layers in an onion. Chemists have determined that there are rigorous physical "rules" (a result of the energy balance that must be maintained in any atom) that control the distances between shells and the number of electrons that can fit in each of them. Only 2 electrons fit in the innermost shell, no more than 8 fit in the next, and 18 fit in the one beyond (Figure 2.3). Another consequence of these "rules" is that shells must be filled sequentially from the one with the lowest characteristic energy to the one with the greatest energy, so that if there is a partially completed shell it will be one of those farthest from the center of the atom.

Chemical Bonds

When elements combine in strict proportions (for example, when hydrogen and oxygen unite to form water, which has exactly two hydrogen atoms for each oxygen atom), the substances that result are known as **compounds.** Just as the smallest particle of an element is an atom, so the smallest tightly bound assembly of atoms with the chemical characteristics of a particular compound is called a **molecule.** The forces that tie atoms to each other in molecules, called **chemical bonds,** are due to interactions between the electrons in those atoms.

An atom's outermost shell of electrons is most stable when it is completely filled, but few atoms have precisely the right number of electrons to fill their shells. Those that do (helium, neon, argon, krypton, xenon, and radon) are called **noble elements,** because they do not have to interact with other atoms to become more stable. All other atoms must share or exchange electrons to complete each other's outermost shells, thus forming chemical bonds between them. The type of bond depends largely on the number of electrons each atom has and how tightly these electrons are held. Some atoms always form the same type of bond with their neighboring atoms; others may form different bonds in different molecules.

Ionic Bonds If an atom's outer shell is nearly empty, the atom may become more stable by losing the few electrons that are in that shell. If the shell is nearly full, the atom may become more stable by adding extra electrons. Either process upsets the electrical balance in the atom, producing an **ion** (Figure 2.4). If the protons outnumber the electrons, the atom has a positive charge and is called a **cation** (pronounced CAT-ion). If the atom has added electrons to complete a shell, it has a net negative charge and is called an **anion** (pronounced AN-ion). When atoms transfer electrons between themselves by one atom casting off electrons that the other atom can use to complete a shell, their electrical charges cancel each other out. The result is a stable molecule, and the connection between the atoms is called an **ionic bond.**

For example, sodium (Na) readily gives up the single electron in its outer shell to chlorine (Cl), which

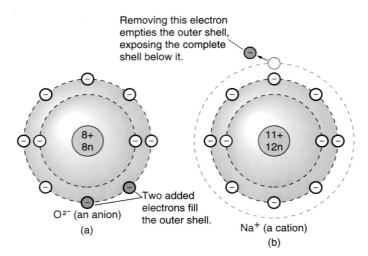

Removing this electron empties the outer shell, exposing the complete shell below it.

8+
8n

O^{2-} (an anion)
(a)

Two added electrons fill the outer shell.

11+
12n

Na^+ (a cation)
(b)

FIGURE 2.4
(a) Because an electron has one negative electrical charge, adding two extra electrons to the oxygen atom in Figure 2.3 would make it an anion with a net electrical charge of −2. (b) Removing the lone electron in sodium's outermost shell makes it a cation with a charge of +1.

needs only one electron to become stable. We indicate that the sodium atom has become a cation with one excess positive charge by writing "Na^+." The negatively charged chloride anion is written "Cl^-." Together, the two form a molecule of NaCl, in which each atom has a complete outer shell and there are no stray electrons left over (Figure 2.5). We know this compound as common table salt.

Covalent Bonds   Instead of transferring electrons between them, some atoms share equally the electrons that each one needs to complete a shell. The attracting force that results is called a **covalent** bond. Carbon commonly forms covalent bonds with atoms of other elements, as in the gases carbon dioxide (CO_2) or methane (CH_4), and with other carbon atoms. Ethyl alcohol, the familiar compound found in wine and other beverages, contains such a chainlike structure of carbon atoms (see Figure 2.6).

To clarify the difference between ionic and covalent bonds, study Figure 2.6. The outer shells of all atoms in the alcohol molecule are complete (carbon and oxygen each have eight electrons in the outer shell, and hydrogen has two). Notice, though, that the *way* an atom's outer shell was completed is not always clear. For example, did each carbon atom lose the four electrons in its partially filled outer shell, or did it gain

four to complete the shell? When electrons are shared equally between atoms in a covalent bond, we cannot tell cations from anions.

As a consequence of the sharing rather than transferring electrons, most covalent bonds are much stronger than ionic bonds. Later in this chapter we pay particular attention to silicates, compounds of silicon and oxygen that are the most common materials in the crust and mantle of the Earth. Because the silicon–oxygen bond is always covalent, silicates are very resistant to change in the Earth.

Metallic Bonds   In a few substances such as iron, copper, or gold, the outer electrons are shared more freely among large groups of atoms than simply between pairs of atoms. As a result, these electrons are free to move over great distances. This arrangement is known as **metallic bonding.** Because the free electrons are highly mobile—they can travel for thousands of miles if "pushed" hard enough by a high voltage through a copper wire, for example—metallic substances usually conduct heat and electricity well.

Van der Waals Bonds   The fourth important style of bonding involves neither the transfer nor the sharing of electrons, but occurs when an electron cloud is not symmetrical. When several electrons are briefly on the same side of an atom or molecule, that

FIGURE 2.5
A sodium atom picks up an electron from a neighboring chlorine atom to form the compound NaCl. By forming this ionic bond, each atom completes its outer electron shell.

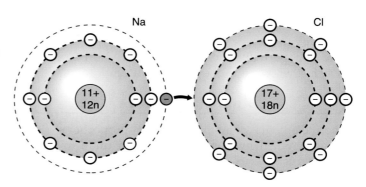

Na

Cl

11+
12n

17+
18n

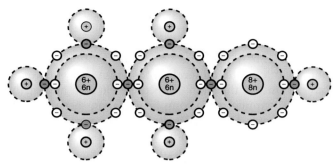

FIGURE 2.6 Schematic representation of a molecule of ethyl alcohol (C_2H_5OH). Electrons shared equally between the atoms form covalent bonds.

side becomes negatively charged. The opposite side takes on a positive charge. The weak electrostatic force between such polarized atoms or molecules is called a **Van der Waals bond.** On a molecular scale, this force is like the static electricity that sometimes makes it difficult to separate sheets in a stack of paper.

Graphite, the black material used to make "lead" pencils, is a common substance containing Van der Waals bonds (Figure 2.7a). Carbon atoms in graphite are bonded in strong, flexible sheets within which all of the electron shells have been completed by covalent bonds. The only connections between sheets, however, are weak Van der Waals bonds. A stack of sheets, therefore, is held together no more tightly than the pages of this book, which you have no trouble separating. This makes graphite one of the softest natural materials and an excellent lubricant. By contrast, *diamond*, in which carbon atoms form a rigid three-dimensional network of covalent bonds, is the hardest natural substance and an excellent abrasive (Figure 2.7b). Yet diamond and graphite both contain only carbon atoms.

States of Matter

Substances can occur in any of three states: solid, liquid, or gaseous. These states differ in the degree of attraction between molecules or atoms. Molecules in a gas are attracted weakly to one another. The attraction between molecules in a liquid is greater, and the attraction in a solid is greater still. Thus, a cloud of gas can assume any shape, and the distances between its molecules may be changed as necessary to fit the cloud in any container. By comparison, the molecules in a liquid are already much closer together and less free to move independently. A liquid adjusts less easily to fit the shape of its container, therefore, and is difficult to compress or expand. A solid is more highly structured than either a gas or a liquid. Its molecules are connected by strong bonds. In many solids, the bonds are arranged so that the atoms form an ordered three-dimensional network—a **crystal structure**—that

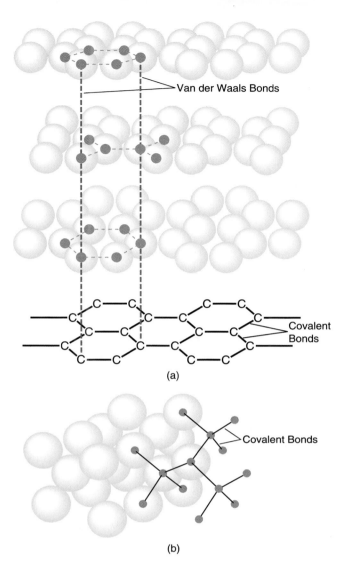

FIGURE 2.7 (a) Graphite has a layered structure. Each carbon atom forms covalent bonds with its neighbor. The resulting sheets lie parallel to each other in a loosely assembled pile. The weakness of Van der Waals bonds between layers allows them to separate easily and accounts for the mineral's softness. (b) In diamond, all of the bonds between carbon atoms are covalent. This makes diamond extremely hard.

is quite rigid. In a crystalline solid, the relationships between atoms within the network are much more important than their relationships in single molecules. Thus, chemists do not use the term "molecule" in this context. A solid without a crystal structure, in which the molecules are arranged randomly, is called an **amorphous** solid. Even in such a material (for example, in a glass), molecules are so firmly linked together that the material resists any effort to change its size or shape.

A substance may exist in any of the states of matter, although we tend to think of the "normal" state as the one that prevails at the pressure and temperature on the Earth's surface. By changing the temperature or pressure, however, we can convert matter from one state to another. Water, for example, evaporates completely as it is heated above 100°C and solidifies as it is cooled below 0°C. In general, substances are gases at high temperatures and under low pressure, and they are solids at low temperatures and under high pressure. In broad terms, the reason we expect substances to behave this way is that heat provides energy to stretch or break bonds, whereas pressure tends to force atoms closer together and, thus, to strengthen bonds.

Minerals

In its earliest usage, centuries ago, the word "mineral" referred to valuable materials dug from the ground. That definition, however, masked a confusion about which substances in the Earth are pure chemical compounds and which are random mixtures of several compounds—hardly surprising, considering the nature of matter was poorly understood—and it ignored the vast number of substances that were not worth much economically. As scientists in the eighteenth and nineteenth centuries began to understand the nature of atoms and bonding, they also developed a clearer and more useful view of minerals. As geologists today use the term, a mineral is understood to be *a naturally occurring inorganic solid that has a well-defined chemical composition and in which atoms are arranged in an ordered fashion.*

The first part of this definition sets a standard that is more a matter of convenience than a necessity. Geologists discovered long ago that they could synthesize in the laboratory substances that are identical to those found in nature. Because some of these synthetic minerals are chemically purer and are more free of mechanical flaws than are natural minerals, geologists often prefer to study geologic processes by experimenting with synthetics. Gemstones produced synthetically may also be identical to "real" ones. By common agreement, though, none of these are minerals. The word "inorganic" is used as a chemist would mean it; that is, in reference to all compounds *except* those made primarily of carbon and hydrogen (which a chemist calls "organic compounds"). Despite its appearance, the word "inorganic" does not necessarily rule out compounds that may be produced by an organism—for example, such things as the crystalline materials found in bones, teeth, and shells.

Composition

The formula for a chemical compound, such as NaCl or CaF_2, tells which elements are in it and how many

atoms of each is in a molecule. If we can provide these two pieces of information for a substance, then it has a well-defined composition. Thus, the formulas NaCl and CaF_2 are vital parts of the descriptions of *halite* and *fluorite*, both minerals. Most minerals, however, have formulas that look both more flexible and at the same time much more complicated than these because ions of many elements are similar enough that they may substitute for each other in crystal structures.

The two factors governing whether an ion may substitute for another are its size and its electrical charge. Ionic size is determined mainly by the number of electrons in an ion. Basically, the more electron shells an ion has, the bigger it is. Also, when electrons are added to form an anion from a neutral atom, the atom grows larger because the attractive force exerted by its nucleus is now applied to more electrons and is thus spread thinner. In contrast, cations become smaller as electrons are removed, because the nucleus can now exert more pull on the remaining electrons. The largest ions are about 10 times the size of the smallest ones (Figure 2.8). Still, many ions are roughly the same size. The **ionic radius** (the characteristic distance from the center of the ion to its outermost shell) of an Mg^{2+} ion, for example, is 0.66 Ångstroms (about 7×10^{-11} m) and the radius of an Fe^{2+} ion is 0.74 Ångstroms, only slightly larger. Because they have roughly the same size, and the same charge (+2), these two ions can fit with equal ease into crystal structures.

FIGURE 2.8 The ionic radii of the elements. Anions are bigger than cations. Ionic radii increase slowly and irregularly as atomic number increases. The more electrons are removed to make a cation, the greater its charge and the smaller its radius.

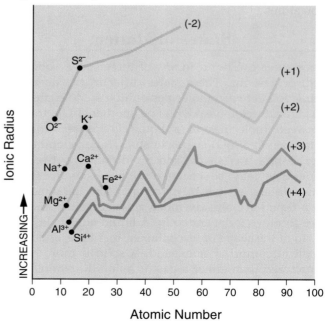

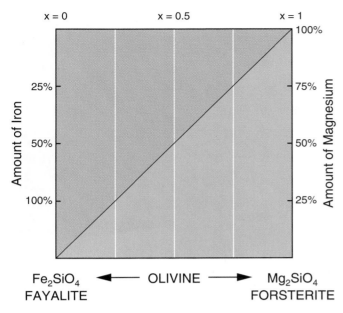

x = 0   x = 0.5   x = 1

Amount of Iron: 25%, 50%, 100%

Amount of Magnesium: 75%, 50%, 25%

Fe_2SiO_4 ◄─── OLIVINE ───► Mg_2SiO_4
FAYALITE   FORSTERITE

FIGURE 2.9 An olivine molecule may contain either Fe^{2+} or Mg^{2+} ions. Because a specimen of olivine consists of an extremely large number of molecules, some of which may contain Fe^{2+} ions and other Mg^{2+} ions, the composition of the specimen as a whole may lie anywhere between the two extremes of Fe_2SiO_4 and Mg_2SiO_4. The only requirement is that the total (Fe+Mg) must be equal to 2. Therefore, an olivine that is one-quarter Mg_2SiO_4 and three-quarters Fe_2SiO_4 would have the formula $Mg_{0.5}Fe_{1.5}SiO_4$.

Some of the molecules in a sample of the mineral *olivine*, for example, may have the composition **Mg_2SiO_4**, whereas others may be **Fe_2SiO_4**. "Olivine" is, in a sense, a family name describing a range of possibilities. If most of the molecules in an olivine are Mg_2SiO_4, the sample is given the name *forsterite;* if most molecules are Fe_2SiO_4, the mineral sample is called *fayalite*. Most olivine samples have a mixture of both kinds of molecules. To indicate that the olivine "family" of minerals as a whole varies between these definite limits, geologists write the formula $Mg_xFe_{(2-x)}SiO_4$ (where *x* is any fractional quantity between 0 and 2) or simply as $(Mg,Fe)_2SiO_4$ (Figure 2.9). These formulas set limits on the kinds of atoms (Fe and Mg) that are acceptable in an olivine molecule and maintain the rules about how many of either there must be (2), while leaving the proportions of Mg_2SiO_4 and Fe_2SiO_4 molecules in a particular sample free to vary from 0 to 100 percent.

Even when ions have different charges, they may substitute for each other if the overall electrical charge balance of the molecule is preserved in some other way. A good example is the common mineral plagioclase feldspar, which has the formula $Na_{(1-x)}Ca_xAl_{(1+x)}Si_{(3-x)}O_8$, where *x* is a fraction between 0 and 1. Because a calcium ion (Ca^{2+}) has a greater positive charge than a sodium ion (Na^+), every substitu-

tion of one for the other must be matched by a substitution of aluminum ions (Al^{3+}) for silicon ions (Si^{4+}) so that the total of all positive charges in the molecule stays the same. This **coupled ionic substitution,** although complicated, maintains the rules for a well-defined chemical composition and is very common among minerals.

Internal Structure

The reason that minerals, by definition, must be solid is that only solids can have chemical bonds that are short and strong enough to hold atoms in a rigid three-dimensional structure. To qualify as a mineral, however, that structure must have a pattern that repeats regularly, so that one region in the structure looks just like another (Figure 2.10a). Figure 2.10b shows how the same pattern can be seen in the internal structure of halite (NaCl). Sodium and chloride ions alternate and are evenly spaced in three mutually perpendicular directions. The precise placement of the ions is an invariable characteristic of halite, as reliable as a fingerprint.

As we will show shortly, the physical properties of a mineral can reveal information about its ordered structure. To describe the arrangement of ions and measure their spacings precisely, however, we use a technique called **X-ray diffraction,** first demonstrated by German chemist Max von Laué, who won a Nobel Prize for his discovery in 1914. Von Laué showed that regularly spaced planes of atoms in a crystalline structure scatter a beam of X-rays as though it were a light beam broken up by a ball of mirrors. By using a photographic film to record the scattered beams as spots, a scientist can determine precisely how molecules are arranged (Figure 2.11).

Because the ordered arrangement of ions is a vital characteristic of each mineral, it is not enough to define a mineral by simply writing a chemical formula for it. Halite is more than just "NaCl," for example. A complete description of halite also includes information about the arrangement and spacing of Na^+ and Cl^- ions, as they are shown in Figure 2.10b. Such information is especially important in cases where a single well-defined collection of atoms can be arranged in more than one way. We found earlier, for example, that carbon atoms may be arranged to form either graphite or diamond, with the very different crystal structures shown in Figure 2.7 a, b. Clearly, it would be insufficient to identify either mineral by its chemical formula alone. This fairly common situation, in which two minerals with different crystal structures have the same formula, is known as **polymorphism** (from the Greek roots *poly-*, "many" + *morph*, "form"), and the minerals are called **polymorphs** (Figure 2.12).

(a)

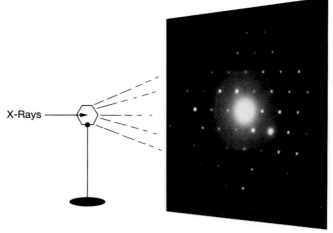

diffraction study of its internal structure. Such investigations reveal crucial details that help a geologist understand how the mineral formed and may have been changed within the Earth. In most instances, however, these tools are not needed merely to identify a mineral. For that purpose, geologists commonly look instead for a few characteristics of the mineral's appearance or its response to easy chemical or physical tests, each of which is a consequence of its composition or crystal structure. It is unnecessary to determine all the properties of a mineral in order to identify it; one or two characteristic properties for each is usually sufficient. The trick is to learn which ones are useful for each mineral you are likely to encounter.

Crystal Form

Most mineral particles have irregular shapes, either because they grew in competition with other mineral particles or because they were damaged since they formed. When a mineral grows in a liquid or into an open space where no other solid interferes with it, it develops symmetrically arranged planar surfaces called **crystal faces.** The word **crystal,** describing the resulting form, comes from the Greek word for cold, because, until recent centuries, people believed that crystals were made of water that had somehow frozen too stiff to remelt.

The appearance or **habit** of the crystal is both a product of the environment where it grew and an external expression of its internal crystalline structure.

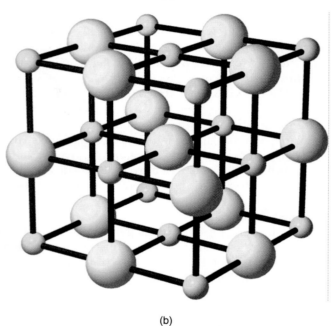

(b)

FIGURE 2.10   (a) The structure in this drawing consists of a pattern of equally spaced boxes and lines that repeats infinitely in three perpendicular directions. *(M. C. Escher)* (b) Ions of sodium (Na⁺) and chlorine (Cl⁻), shown here as small and large spheres, are arranged in the same pattern depicted in Figure 2.10a to form the ionic compound NaCl, also known as halite or common table salt.

Mineral Properties
▲▲▲

A geologist who picks up an unfamiliar mineral sample does not usually have the sophisticated equipment necessary to perform a chemical analysis or an X-ray

(a)

(b)

FIGURE 2.12 (a) Diamond and (b) graphite are polymorphs of carbon. Each mineral has the same simple chemical formula: namely C. Because the atoms are arranged in different ways, as shown in Figure 2.7, the two minerals have very different physical characteristics. *((a) Fred Ward, Black Star (b) M. Claye/Jacana, Photo Reseachers, Inc.)*

FIGURE 2.13 Some of these quartz crystals are short and fat; others are long and thin. The angles between comparable faces on all specimens are identical, however, because all quartz specimens must have the same ordered internal structure. *(Roberto de Gugliemo/ Science Photo Library, Photo Researchers, Inc.)*

Some crystals of the common mineral quartz (SiO_2), for example, are long and slender; others are stubby and short (Figure 2.13). These characteristics are controlled by how fast the crystals formed and by subtle chemical differences in their surroundings. Whether an individual crystal is only a millimeter long or a meter long, however, quartz forms crystals that are always six-sided and in which similar faces always come together at the same angle. As early as the seventeenth century, long before atoms were known to exist, mineralogist Nicolas Steno deduced from this observation (which geologists now call the law of **constancy of interfacial angles**) that each mineral has a characteristic internal structure (see the discussion in Perspective 2.1).

Cleavage

The orderly arrangement of atoms in a mineral can also be seen in its **cleavage**, the tendency of a mineral to break along smooth planes in specific directions. These planes are called **cleavage planes.** One way to view the ordered structure of a mineral is to think of the ions or atoms as being stacked in similar layers. In fact, as suggested in Figure 2.14, there are many ways to imagine stacking the ions or atoms to build the same structure. When a mineral cleaves, it breaks where layers are farthest apart and have the fewest bonds connecting them. Because the layers in any direction must be parallel and evenly spaced, similar cleavage planes can form anywhere in the stack.

Some minerals only cleave in one direction. Others, in which more than one set of layers is weakly bonded, may cleave in two or more directions. Geologists can commonly use the number of cleavage directions and the angles between them to distinguish one mineral from another. For example, graphite and the family of minerals known as the micas cleave in only one direction. Atoms in those minerals are connected by covalent bonds to form sheets that can be separated

Cleavage, Crystal Faces, and Ordered Structure

Nicolas Stensen (usually known by his latinized name, Steno) was a well-educated physician of the seventeenth century, a contemporary of Sir Isaac Newton. Geologists recognize him for making several key observations, one of which was the Law of Constancy of Interfacial Angles. Not content merely to observe the way that crystal faces were related, Steno tried to understand why the angles between them were the same from one specimen to the next. Without any direct evidence, he postulated that crystalline solids are composed of submicroscopic particles and hypothesized that the angles between crystal faces mimicked the angles on the particles.

A century later, René Just Häuy, Abbé of Notre Dame in Paris and an avid mineral collector, was also fascinated by the shapes of crystals. As tradition has it, Abbé Häuy was admiring a friend's prize specimen of calcite when it slid from his grip and smashed to pieces on the floor. As he stooped to pick up the pieces, he was startled to find that they all had the same shape, and that the shape was not the same one as the crystal had been. (The calcite specimen shown in this box is just like the one Häuy broke. Notice that cleavage planes visible within the specimen are not parallel to the external faces on it.) Häuy immediately concluded that a crystal is built of tiny particles bounded by cleavage planes, and

that crystal faces are an incidental result of the way the particles are stacked. We have reproduced here the drawing that Häuy made to explain his idea. To test the hypothesis, he rushed home and quickly shattered dozens of his own calcite specimens, all of which cleaved the same way regardless of their external form.

Steno and Häuy had learned something about the ordered structure of crystalline solids and had tried to explain why a crystal's form and cleavage are reliable tools for identification. Within a few years after Häuy published his findings, chemists recognized the existence of atoms. They soon deduced that the particles Steno and Häuy had tried to envision were not angular

but not easily broken, so that the only possible cleavage direction is parallel to the sheets (compare Figure 2.15a with Figures 2.7 and 2.22). In contrast, the common mineral *amphibole,* which we explore later in this chapter, has two equally distinct sets of cleavage planes close to 120° apart (Figure 2.15b). Calcite cleaves in three non-perpendicular directions, forming rhombohedral fragments (Figure 2.15c). Fluorite has four distinct cleavages (Figure 2.15d).

Some minerals, however, do not exhibit cleavage at all. Such minerals break to produce a rough, non-planar surface called a **fracture** instead. Quartz, for example, develops curved, irregular fractures because

a network of covalent bonds gives it uniform strength in all directions.

Hardness

Another physical property governed by a mineral's structure is **hardness,** defined as a measure of resistance to scratching. Minerals differ widely in hardness. Some are so soft that they can be scratched with a fingernail, whereas others are so hard that a steel knife is required to scratch them. Diamond, the hardest mineral known, cannot be scratched by any other natural substance. To apply a standard for comparing mineral

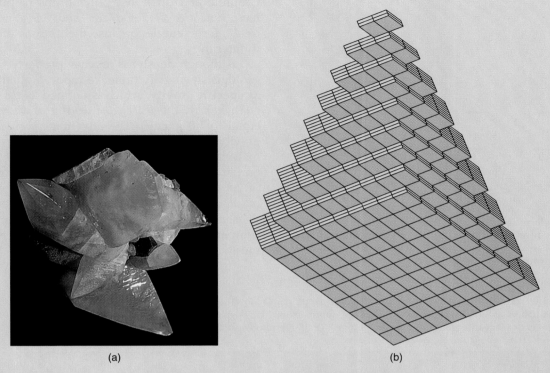

P2.1.1 (a) Photo of a museum-quality specimen of calcite, showing perfect crystal faces and excellent internal cleavages, clearly not parallel to each other. (b) René Just Häuy's drawing of the same crystal (actually one identical to it), showing how he thought the cleavage and crystal faces were related.

solids, but were arrangements of atoms. Crystal forms and cleavage fragments do not necessarily take the shape of a molecule, although each is influenced strongly by the way in which atoms in the crystal structure are arranged. By making some shrewd guesses based on hand specimens, Steno and Häuy recognized that influence and laid an important foundation for our modern theory of matter.

samples, geologists estimate hardness from 1 through 10 on what is known as the **Mohs scale** (named after Friedrich Mohs, a mineralogist who lived from 1773 to 1839). The Mohs scale is merely a relative ranking of 10 common minerals, not a rigorous, quantitative way to measure "scratchability." In practice, geologists rarely compare specimens to each other, or to samples of the minerals on the Mohs scale. Instead, geologists compare samples to the hardness of a fingernail, a piece of window glass, a knife blade, or some other familiar material (see Table 2.1). We can determine the hardness of any mineral by scratching its smooth surface with another mineral or test material, but we must be sure that the mineral tested is actually scratched. Sometimes loose particles crumble off the specimen, fooling us into thinking it has been scratched.

Specific Gravity

The **density** of a mineral is a measure of how much matter there is in a given volume. A cubic centimeter (cm^3) of halite, for example, weighs 2.16 g; a cubic centimeter of pure gold weighs 19.3 g. The two minerals differ mainly because atoms of gold are much heavier than atoms of sodium or chlorine.

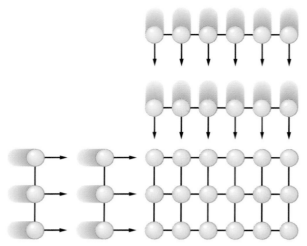

FIGURE 2.14 One way to characterize a mineral structure is to identify parallel planes of ions that repeat to fill space. Two of these are suggested here, but many others could also be defined. Mineral structures are weakest between planes that are farthest apart. Therefore, these are directions of preferred breakage (cleavage).

Density is also affected, however, by how tightly atoms are packed together in a mineral's structure. Graphite and diamond, polymorphs of pure carbon, have densities of 2.3 g/cm^3 and 3.5 g/cm^3, respectively, for example, because their atoms are arranged differently.

Although density is an easy property to measure, geologists generally prefer to report a quantity known as **specific gravity,** which is defined as *the density of a given mineral divided by the density of water* (which is equal to 1.0 g/cm^3). The advantage of using specific gravity is that it compares the density of an unfamiliar material to that of one of the most familiar substances on Earth. Instead of saying that graphite has a density of 2.3 g/cm^3, for example, a geologist reports that it has a specific gravity of 2.3, meaning that it is 2.3 times denser than water. Notice that because specific gravity compares one density to another, it has no units of measurement itself.

FIGURE 2.15 (a) The micas all have one direction of cleavage, parallel to sheets of ions. *(William E. Ferguson)* (b) Amphiboles have two directions of cleavage, approximately 120 degrees apart. (c) Calcite has three distinct cleavage directions (see also Perspective 2.1). (d) Fluorite's four cleavages make it possible to break the mineral into eight-sided fragments.

(a)

(b)

(c)

(d)

TABLE 2.1 The Mohs Hardness Scale for Minerals

HARDNESS	STANDARD MINERAL	COMMON TEST MATERIAL
1	Talc	
2	Gypsum	
		←2.5 Most fingernails
3	Calcite	←Copper penny
4	Fluorite	
5	Apatite	
		←5.5 Penknife
6	Orthoclase	←Plate glass
7	Quartz	←Steel file
8	Topaz	
9	Corundum	
10	Diamond	

FIGURE 2.16 The streak of a mineral can be observed by powdering it. If the mineral is soft enough, powder can be produced by rubbing it against a piece of unglazed porcelain, as shown here. *(William E. Ferguson)*

Color

Color is a striking property of many minerals. These include the intense azure blue of *azurite,* the bright green of *malachite,* and the pale yellow of *sulfur.* The color of a mineral is usually determined by its chemical composition so that, for example, minerals containing iron are commonly dark gray, dark green, or black. Minerals that contain aluminum as a predominant element are usually "light-colored," a term that includes purples, deep red, and some browns as well as "normal" whites, yellows, and similar hues. These general relationships are most reliable for minerals with very simple chemical formulas, however, and are least valid for minerals made of many different elements. This means, unfortunately, that color is not useful for identifying the most common minerals in the Earth, which are not only chemically complex but also allow extensive ionic substitution. More than a small amount of iron in an olivine, for example, turns it from yellow-green to almost black. Even quartz, which has a very simple chemical formula, can vary in color from clear to yellow to violet as the result of almost undetectable impurities.

Streak

The **streak** of a mineral is the color it has when finely powdered—a color that may be quite different from the original specimen. Although the color of a mineral may vary between wide limits in response to subtle changes in chemical composition, its streak is usually the same color consistently. For example, *hematite,* Fe_2O_3, may be reddish brown to black; its streak, however, is always blood-red. *Goethite,* $FeO(OH) \cdot nH_2O$, sometimes known as "brown hematite" or "bog-iron ore," has a color that is dark brown to black but a streak that is yellowish brown. *Cassiterite,* SnO_2 (tin-

stone), is usually brown or black, but it has a white streak.

One of the simplest ways of determining the streak of a specimen is to rub it across a piece of unglazed porcelain known as a streak plate (see Figure 2.16). The color of the powder, or streak, left behind helps to identify some minerals. Because the streak plate has a hardness of 7, it cannot be used to identify harder minerals. These have to be crushed with a hammer or some other tool.

Luster

Luster refers to the intensity and quality of light that a surface reflects. The size and spacing of surface irregularities, the type of bonding in a mineral, and its degree of transparency can all contribute to its luster. Many subjective terms have been used to describe this property. For example, a mineral such as quartz, whose surface looks like glass, has a *vitreous* luster; one that is shiny like polished metal has a *metallic* luster; and one that is dull like clay has an *earthy* luster.

Other Physical Properties

Other properties may also help in identification. If a mineral is magnetic or conducts electricity easily, for example, those properties could be useful guides. Geologists commonly identify transparent minerals under a microscope by studying the way that light is bent or otherwise transformed as it passes through them. Some minerals like halite dissolve readily in water. Others like calcite fizz when a drop of acid is placed on them. Geologists use these and many other specific tests to distinguish one mineral from another. (See Perspective 2.2.)

What Makes a Gemstone Valuable?

*L*ike many people of his time, Pliny the Elder, a Roman naturalist of the first century A.D., was convinced of the magic and medicinal powers of gems. Possession of amethyst, the lavender variety of quartz, was said to be a cure for both snakebite and intoxication. Diamonds were believed to ward off fear and madness. Other stones could calm the sea, hold back high winds, tame wild animals, or (as was said of one magic gem owned by King Solomon) let the owner learn of distant events. Even today, some people attribute mystical powers to gems, although there is no more reason to believe in them now than there was centuries ago.

For most people, though, the greatest appeal of gemstones is their beauty. This implies a standard of clarity and freedom from imperfections not met by most mineral specimens. Civilizations have always prized stones with exotic

Common Minerals

Working in a laboratory, a chemist could easily produce a huge number of synthetic compounds from the array of elements found in nature. You might imagine, therefore, that there are a vast number of different minerals in the Earth. In fact, there are fewer than 3,000, and most of them are quite rare. A handful of minerals or well-defined mineral families constitutes over 90 percent of the Earth's crust (Figure 2.17). Why is this the case?

- First, the elements are not equally abundant. About 90 percent of the total mass of the Earth consists of four elements: iron, oxygen, silicon, and magnesium. Four other elements—nickel, calcium, aluminum, and sulfur—amount to about 1 percent each, and another seven elements—sodium, potassium, chromium, cobalt, phosphorus, manganese, and titanium—make up most of the remainder. Within the crust of the Earth, the selection is even smaller. Nine elements make up 99 percent by weight. These are oxygen, 45.2 percent; silicon, 27.2 percent; aluminum, 8.0 percent; iron, 5.8 percent; calcium, 5.1 percent; magnesium, 2.8 percent; sodium, 2.3 percent; potassium, 1.7 percent; and titanium, 0.9 percent. The other elements may concentrate in rare minerals but more commonly substitute for one of these nine in the structures of a very few common minerals.

- Second, oxygen is the only abundant element in the Earth's crust that can become an anion. Because all compounds must be electrically neutral, molecules of each must contain both anions and cations. Oxygen's status as the most abundant anion guarantees that it will be found in the most abundant minerals. Furthermore, the other eight common crustal elements (with the occasional exception of iron) are classified chemically as **lithophile** elements, meaning that they combine more readily with oxygen, if available, than with other anions. Iron can form compounds easily with either oxygen or sulfur, but there is comparatively little sulfur in the Earth's crust, so iron sulfides are not as common as minerals containing iron and oxygen.

- Third, silicon is the most common lithophile element, and it invariably bonds directly to oxygen. As a result, almost all common minerals contain both elements and are called **silicates.**

- Finally, on the basis of volume rather than weight, oxygen makes up almost 92 percent of the crust. The oxygen anion (O^{2-}) is not only abundant, but it is also by far the largest of the common ions (refer to Figure 2.8). The arrangement of atoms to form minerals, therefore, is limited by the number of ways that large oxygen ions can be packed together. To maintain electrical balance, anions cluster symmetrically around smaller cations to form three-dimensional geometric arrangements known as **coordination polyhedra** (see Figure 2.18). In silicates, silicon cations are always surrounded by four oxygen anions, forming a four-sided polyhedron with the overall chemical formula $(SiO_4)^{4-}$. Each of the oxygen ions helps to fill the outermost electron shell of the silicon ion by sharing one electron in an exceptionally strong covalent bond. An oxygen anion has an electrical charge of –2 and a silicon cation has a charge of +4, so the polyhedron has

colors, those with bright lusters, and those that have striking optical properties. Some varieties of sapphire and ruby (both are forms of the mineral corundum), for example, scatter light in a bright six-pointed star pattern when they are polished. *Tanzanite,* a variety of the uncommon mineral *zoisite,* can change from a deep violet to a rose color depending on the type of light in which it is viewed. Diamond is known for the "fire" within it—a product of its ability to bend light rays and separate them into many colors.

Beauty contributes to value but is not the only factor. We divide gemstones into precious and semi-precious categories, largely on the basis of rarity. Some varieties of common minerals like *garnet,* olivine, and orthoclase can be very attractive yet surprisingly inexpensive. In other times, these same stones were valued more highly because they were rarely mined in gem quality. Gemstones are also chosen for their resistance to scratching and cleavage, although there are valuable stones *(opal* and *lapis lazuli)* that are quite soft and others (diamond, *beryl)* that are quite easily shattered despite their scratch resistance.

Why are gemstones valuable? In the final analysis, value, like beauty, is in the eye of the beholder.

P2.2.1 Clear crystals of any durable mineral can be shaped as gemstones. The only basic requirement that all gems must meet is beauty, a very personal criterion. The faceted stones in this photograph include varieties of quartz, beryl, topaz, sapphire, and garnet.

Few commodities fluctuate as wildly in price as do gems or are as subject to unexpected shifts in popularity. A gem is worth whatever you are willing to pay for it.

a net charge of -4 ($[4 \times \{-2\}] + 4$). Because silicon ions are never found in any other arrangement, with any other anion, this particular coordination polyhedron (called the **silica tetrahedron**) is a basic building block for the most common minerals.

Silicate Minerals

From this brief overview, it should be clear why a few groups of closely related minerals, all of them based on $(SiO_4)^{4-}$, are predominant in the Earth's crust. The silica tetrahedron behaves like a single very large anion. The way it bonds with other ions determines what kind of silicate mineral forms.

Island Silicates In the mineral olivine, the extra electrons on silica tetrahedra are all transferred to magnesium or iron cations, forming ionic bonds. Each silica tetrahedron, then, is like an island separated from the others by a sea of Fe^{2+} or Mg^{2+} ions, thus the name "island silicates." Some geologists refer to these minerals as *nesosilicates,* from the Greek root *nesos,* meaning "island." Tetrahedra are arranged geometrically so that each Fe^{2+} or Mg^{2+} ion in the structure is surrounded by six oxygens, forming an MgO_6 or FeO_6 octahedron (see Figure 2.18). Silicate minerals that, like olivine, are based on Fe^{2+} and Mg^{2+} are known collectively as **ferromagnesian** minerals, regardless of how their silica tetrahedra are connected.

Another important, though less abundant, family of island silicates is the garnet group. Garnets have the general formula $A_3B_2(SiO_4)_3$, in which *A* can be any mixture of Fe^{2+}, Mg^{2+}, Ca^{2+}, or Mn^{2+} ions surrounded by eight oxygens, and *B* can be Al^{3+}, Fe^{3+}, or Cr^{3+} surrounded by six oxygens. Garnets are very hard and have no cleavage; therefore, they are commonly used as abrasives. Clear, flawless garnets can also be beautiful gemstones with colors that range from deep red to green to yellow.

In all silicates except the island silicates, the silica tetrahedra are connected directly to each other by sharing oxygen ions at their corners. The only invariable rule is that two tetrahedra may never share more than one oxygen between them. That is, tetrahedra may share corners but never edges or faces.

Single-Chain Silicates All silicates in the family called **pyroxenes** contain parallel chains of silica tetrahedra (Figure 2.19). Covalent Si–O bonds make each chain like a sturdy fiber. Not all oxygen atoms are shared between tetrahedra, however, so the remaining electrical charge on each $(SiO_4)^{4-}$ unit must still be balanced by bonding it to ions like Fe^{2+}, Mg^{2+}, or Ca^{2+}. Coupled ionic substitution is also possible. The most common pyroxene, *augite,* has the formula $Ca(Mg,Fe,Al)[(Si,Al)O_3]_2$, in which some silicon atoms in the chain are replaced by aluminum atoms as necessary to balance electrical charges elsewhere.

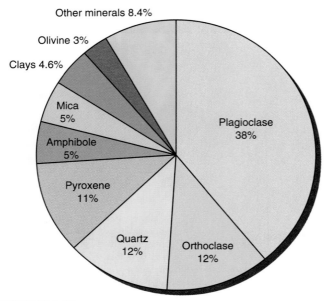

FIGURE 2.17 As this pie diagram shows, more than 90 percent of the Earth's crust by weight is made of eight minerals or mineral families, all silicates. Feldspars alone account for more than half of the crust. The wedge marked "Other Minerals" includes almost 3,000 types of minerals, but much of it consists of the few minerals described in this chapter.

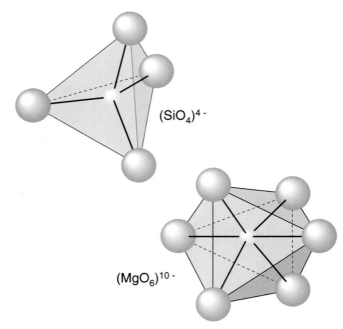

FIGURE 2.18 Anions cluster around smaller cations in regular geometric arrangements called coordination polyhedra. Two of the most common ones are the tetrahedral arrangement, typified by $(SiO_4)^{4-}$, and the octahedral arrangement, shown here for $(MgO_6)^{10-}$. Especially in complex structures, it is easier to notice repeating patterns if you look for symmetrical groups of ions rather than individual ions.

Double-Chain Silicates The **amphibole** family has much in common with the pyroxenes. Its silica tetrahedra are also arranged in parallel chains. In amphiboles, however, chains are paired (Figure 2.20) to make a "fiber" that is twice as wide. The selection of ions that may fit between chains is much larger than for the pyroxenes. The general formula for an amphibole is $A_2B_5(Si_4O_{11})_2(OH)_2$, but many kinds of coupled ionic substitution are possible; thus the actual formulas for most amphiboles are much more complicated.

Because of structural similarities, the single- and double-chain silicates can look very much alike. *Hornblende*, the most common amphibole, is dark green to black and, like augite, has two cleavages. Because the strong covalent bonds make chains difficult to break, the cleavages in both minerals are parallel to chains. Figure 2.21 shows, however, that the pyroxene cleavages are about 90° apart, whereas those in amphiboles are about 120° apart.

Sheet Silicates If single chains can join to form double chains, is it possible to form silicates based on three- or four-chain units? Except in a few very rare minerals, that does not happen. Instead, the next major group of silicates consists of sheets of silica tetrahedra—in essence, an infinite number of parallel chains linked together by covalent bonds, as in Figure 2.22. Only one oxygen ion in each tetrahedron has an extra electron to transfer. Clays, micas, and chlorite, the

three types of sheet silicate minerals, all have one very distinct cleavage parallel to the silicon-rich sheet. They differ, as you might expect, in the kinds of cations they have between sheets and in the way their atoms are arranged.

Many **clays** contain a lot of aluminum, the ultimate example being *kaolinite*, which has the formula $Al_4Si_4O_{10}(OH)_8$. They are commonly formed as other silicate minerals decompose by chemical reactions in soil; thus clays are particularly easy to find at the Earth's surface. Crystals of the clays are usually submicroscopic; hence, characteristics such as cleavage and crystal form are not apparent in a hand specimen. Instead, geologists use X-ray diffraction or chemical analyses to distinguish one kind of clay from another. Clays are used industrially to make ceramics.

Micas are minerals in which aluminum and either sodium or potassium are commonly found between the Si-rich sheets. One abundant mineral of this type is *muscovite*, a clear mica with the formula $KAl_2(AlSi_3)O_{10}(OH)_2$ (Figure 2.23). Large sheets of muscovite, called isinglass, were sometimes used as window panes before factory-made glass became readily available. Isinglass windows are still commonly used in furnaces and other applications where heat resistance is important. The other abundant mica is *biotite*, a dark green to black sheet silicate with the formula $K(Mg,Fe)_3(AlSi_3)O_{10}(OH)_2$.

FIGURE 2.19 The structure of a pyroxene is based on chains of silica tetrahedra. The view in this drawing is down the length of the chains. To simplify the drawing, each silica tetrahedron is shown as an angular, four-sided pyramid. Imagine that each of the four points on a pyramid is an oxygen ion and that the silicon ion is inside the pyramid. In each tetrahedron, two of the four oxygen ions are shared with adjacent tetrahedra. The other two bond to cations like the Mg^{2+} or Fe^{2+} shown here as spheres.

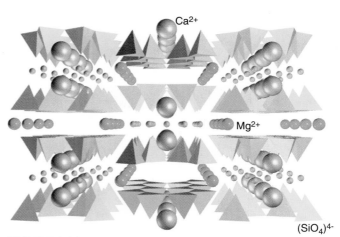

FIGURE 2.20 This drawing views an amphibole from the same perspective that was used in Figure 2.19 for the pyroxene structure. Notice that here the chains of silica tetrahedra are joined in pairs. The pairing is accomplished by sharing oxygen atoms between tetrahedra in adjacent single chains. In an amphibole, double chains are connected to each other through ionic bonds to cations, just as in a pyroxene.

The *chlorites* have the general formula $(Mg,Fe,Al)_6(Al,Si)_4O_{10}(OH)_8$. As you can see, the chlorites contain no potassium and thus are much richer in iron or magnesium than biotite. Like most other ferromagnesian minerals, chlorite is green. In fact, its name is derived from the Greek root *chlor-*, meaning "green."

Network Silicates If silica tetrahedra share oxygen atoms in all directions, the result is a network structure. One of the simplest minerals of this type is *quartz*, pure SiO_2, whose structure is shown in Figure

FIGURE 2.21 Pyroxene and amphibole each have two perfect cleavages. Pyroxene, however, breaks into blocky pieces with sides at right angles. Amphibole breaks into more splintery pieces with an acute angle between their sides. This difference makes it easy to distinguish between the two minerals.

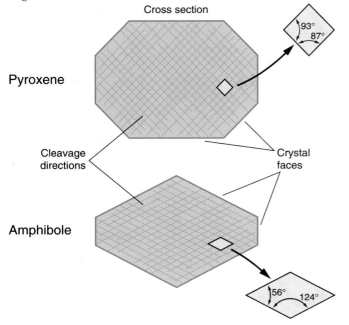

FIGURE 2.22 Each silica tetrahedron in a sheet silicate is connected to others by three of its four oxygen ions. The remaining oxygen forms ionic bonds with cations between sheets. As suggested by the dashed lines, a sheet can be visualized as though it were actually made by connecting a large number of parallel chains.

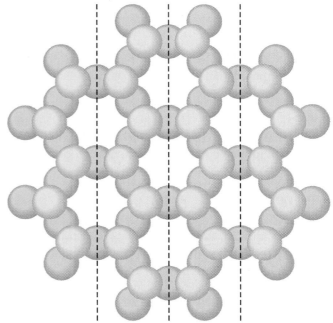

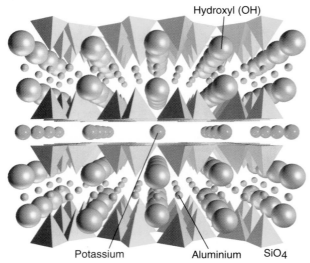

Hydroxyl (OH)

Potassium　　　Aluminium　　SiO₄

FIGURE 2.23　Perspective view through the structure of muscovite mica. As in the drawings of pyroxene and amphibole structures, individual atoms are not shown in the tetrahedral layers so as to minimize confusion. The micas all have a perfect cleavage parallel to this sheet structure.

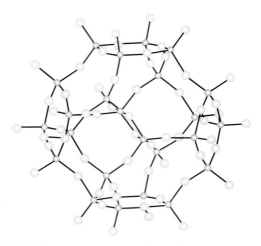

FIGURE 2.24　Bonds between Si^{4+} and O^{2-} ions are greatly exaggerated in this expanded view of the network of (SiO_4) tetrahedra in quartz. Each tetrahedron shares all of its oxygen ions with other tetrahedra. Because all bonds are covalent, there are no evident planes of weakness in this structure. Quartz, therefore, has no cleavage.

2.24. Quartz can be found in many different habits. Transparent crystals of *amethyst* (violet-colored) and *citrine* (yellow) have been used as gems, and varieties of *chalcedony* known as *jasper*, *agate*, and *flint* are also valued for their beauty and durability.

The most abundant network silicates—in fact, the most abundant minerals in the Earth's crust—are the **feldspars.** In these, and in all network silicates other than quartz, Al^{3+} ions take the place of some Si^{4+} ions in the tetrahedra. The resulting difference in charge is balanced by ions elsewhere in the structure. As described earlier, the composition of a *plagioclase* feldspar can range from $NaAlSi_3O_8$, the variety called albite, to $CaAl_2Si_2O_8$, known as anorthite. The most common polymorph of potassium feldspar, $KAlSi_3O_8$, is called *orthoclase.* It is not easy to tell the feldspars apart. As the Greek roots in its name suggest, orthoclase ("perpendicular-breaking") has two cleavages at 90°. The cleavages in plagioclase ("oblique-breaking") are not at right angles. Neither usually has a good cleavage, however; thus this distinction is not always easy to make. Color can be used to identify some common pink, orange, or green varieties of orthoclase, which contain minor impurities of iron.

Carbonate, Halide, Sulfate, and Phosphate Minerals

Carbonate Minerals　All of the carbonate minerals are based on the complex cation $(CO_3)^{2-}$, which consists of a carbon ion surrounded by a triangular array of oxygen ions. The two most common carbonate minerals are *calcite* ($CaCO_3$) and *dolomite*

($CaMg(CO_3)_2$). A third and less abundant carbonate is *aragonite*, a ($CaCO_3$) polymorph.

Calcite and dolomite look very much alike, have the same cleavage, and are soft (they have a hardness of 3). In the field, geologists often distinguish between them by the way they react to a drop of dilute hydrochloric acid: Calcite fizzes vigorously, but dolomite reacts only sluggishly.

Sulfate Minerals　The structural unit common to all sulfates is the complex $(SO_4)^{2-}$ ion. *Gypsum* ($CaSO_4 \cdot 2H_2O$) is formed when seawater evaporates, leaving the previously dissolved salts behind, and thus is found in places that once were shallow marine basins. *Anhydrite* ($CaSO_4$) differs chemically from gypsum only by the lack of water in its formula. It is usually formed when gypsum is buried and subjected to high pressure. Both minerals are mined for use in plaster and wallboard. The pure white variety of gypsum called *alabaster* is a favorite medium of sculptors.

Halide Minerals　Minerals in which the dominant anion is F^-, Cl^-, Br^-, or I^- are called halide minerals. Most are highly soluble in water and are rare. *Halite* ($NaCl$), however, is widespread in the same environments where gypsum and anhydrite are commonly found because it, too, is formed by evaporation of seawater. *Fluorite* (CaF_2), illustrated in Figure 2.25, is much less common, but is mined worldwide for use in steelmaking and foundry operations, where it is added to molten metal to make it more fluid.

Phosphate Minerals　Phosphorus is necessary for all living things. Animals obtain it by eating plants,

FIGURE 2.25   Fluorite (CaF_2) is a halide mineral. Specimens may be a deep purple, like this one, but may commonly be yellow, green, blue, or colorless. These crystals are about 3 cm across.

FIGURE 2.26   These crystals of galena (PbS) are about 1 cm across. The metallic luster, cubic form, and high specific gravity of galena make it easy to identify.

and plants obtain it by drawing it from the Earth, where it is usually bound in the mineral *apatite* ($Ca_5[PO_4]_3[F,OH]$). Ironically, the greatest portion of the phosphorus in the human body is not involved in metabolic processes, where it is indispensable, but is in bones and teeth—in the form of apatite. Apatite is found in trace amounts in most rocks, but can sometimes be concentrated in rocks called phosphorites, which are mined commercially for use in fertilizer.

Sulfide Minerals

Minerals formed by combination of metals with a sulfide ion (S^-) are not abundant, but many are major metallic ores. Several, including *pyrite* (FeS_2) and *chalcopyrite* ($CuFeS_2$), have been known colloquially as "fool's gold," because their color and luster look somewhat like gold to the untrained eye. Unlike gold, however, sulfides decompose easily in contact with water and the atmosphere. Their waste product, sulfuric acid, is a major environmental problem in some mining districts.

Other important sulfides include *galena* (PbS) (Figure 2.26) and *sphalerite* (ZnS), the primary ores for lead and zinc, respectively.

Oxide Minerals

An oxide is a simple chemical compound in which oxygen ions (O^{2-}) bond with ions of a metal such as iron, aluminum, or titanium. Like the sulfide minerals, many oxides are also major sources for metallic elements. Our primary sources for iron, for example, are the minerals *hematite* (Fe_2O_3) and *magnetite* (Fe_3O_4), both of which are widespread in the crust. *Corundum* (Al_2O_3), *cassiterite* (SnO_2), and *rutile* (TiO_2) are much less abundant, but valuable oxides. Easily the most abundant oxide, however, and in fact one of the most abundant minerals on the Earth's surface, is *ice* (H_2O).

The oxides also include minerals based on the complex $(OH)^-$ ion. Of these, the most common is *goethite*, which gives soils a distinctive yellow-brown color. Aluminum hydroxides formed in tropical soils and collectively known as *bauxite* are the primary natural sources for aluminum.

►► EPILOGUE ◄◄

Minerals are commonly found far from the environments where they formed and under very different conditions of temperature and pressure. Halite and gypsum, for example, form as seawater evaporates, leaving a thick layer of salts behind. We mine these from the Earth in places like Kansas or Siberia long after the seas have receded and the salts have been buried under a cover of silt and sand. Because we understand the conditions under which these minerals form, however, we can apply the geologic principle of uniformitarianism to help reconstruct in our minds what the ancient shore must have looked like. Halite and gypsum, therefore, are chemical guides to the workings of the Earth. In the same way, minerals like hematite and the clays, which form by weathering ferromagnesian minerals and feldspars at the Earth's surface, can be used to tell us about those weathering processes. These are topics we will return to in Chapters 5 and 6.

Environments within the Earth's crust are much harder to reconstruct than are many on the surface, because they are beyond our direct observation. By experimenting in the laboratory, however, we can determine the pressure at which diamond, rather than

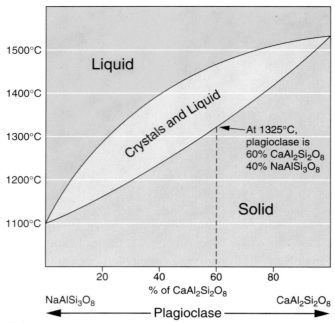

FIGURE 2.27 This diagram shows how the composition of plagioclase changes with temperature, which increases upwards. The composition of a crystal forming at any temperature of interest can be read by looking at the lower line, which separates the region marked "solid" from the one marked "crystals and liquid." For example, at about 1325°C a newly crystallized plagioclase has a composition 60% of the way between $NaAlSi_3O_8$ and $CaAl_2Si_2O_8$, or $Na_{0.4}Ca_{0.6}Al_{1.6}Si_{2.4}O_8$.

graphite, is stable or the high temperature at which magnetite crystallizes from molten rock. Each mineral, then, is a clue to the origin of the rocks in which geologists find it in nature.

Sometimes it is not merely the presence of the mineral that counts, but rather its chemical composition. Ionic substitution is not a random process, but a very controlled, sensitive one. Heat and pressure distort mineral structures, subtly changing the sizes and shapes of the coordination polyhedra. A space inside a polyhedron of oxygen ions that may have been just the right size for a Ca^{2+} ion may shrink slightly but still be big enough to hold a Na^+ ion. For this reason, the proportion of sodium to calcium in a plagioclase is a result of the temperature at which the plagioclase formed. Figure 2.27 is a preview of a topic we will investigate further in Chapter 3: the solidification of magma—molten silicate rock. Even in this quick glance, however, you should see that calcium-rich plagioclase forms at a higher temperature than does sodium-rich plagioclase. In a way, each plagioclase is therefore like a recording thermometer, retaining a memory of the temperature at which it crystallized.

In the chapters ahead, you will also find that combinations of minerals, termed **assemblages** by geologists, can be more informative than individual minerals. Sometimes this is simply because they allow geologists to feel the type of reassurance you get when you check more than one clock to find out what time it is. More generally, however, it is because minerals react chemically with each other as they form. Each mineral in an assemblage, therefore, influences the evolution of the others and carries clues about their common history.

SUMMARY

1. All matter consists of minute particles called *atoms.* Atoms of each element differ in the way they are put together internally and in the ways they combine with each other to form more complex substances. Atoms combine by transferring or sharing electrons, forming various types of chemical bonds. Ionic and covalent bonds are most common in minerals.

2. Minerals are naturally occurring inorganic solids that have well-defined chemical compositions and in which constituent atoms are arranged in an ordered fashion. The appearance of a mineral and the way it responds to a physical or chemical test are determined by its unique composition and crystal structure. Among the physical properties used for mineral identification are crystal form, cleavage, hardness, specific gravity, color, streak, and luster.

3. Ninety-nine percent of the Earth's crust is made of only nine elements, mostly oxygen and silicon. Of these elements, oxygen is the only anion. The abundance of O^{2-}, the ion most available for bonding, also limits the ways in which common mineral structures can be built. For these reasons, there are few abundant minerals or mineral families. Most common minerals are silicates. These all have structures based on a tetrahedral arrangement of O^{2-} ions around a Si^{4+} ion. Tetrahedra can join to form single or double chains, sheets, or networks. Other common minerals are carbonates, halides, sulfates, phosphates, sulfides, and oxides.

4. Minerals can yield information about geologic conditions of temperature and pressure or about processes in the Earth. The existence and composition of minerals, or of mineral assemblages, can be clues to how and where these minerals were formed.

KEY WORDS AND CONCEPTS

assemblage 48
atom 30
atomic mass 30
atomic number 30
bond (ionic, covalent, Van
 der Waals, metallic) 32
cleavage 37
constancy of interfacial
 angles 37
crystal 36
crystal structure 33

electron 31
electron shell 31
element 30
ferromagnesian 43
habit 36
hardness 38
ion (cation, anion) 31
ionic radius 34
ionic substitution 35
lithophile 42
luster 41

mineral 29
Mohs scale 39
molecule 31
nucleus (neutron, proton) 30
polymorph 35
silica tetrahedron 43
specific gravity 40
streak 41
X-ray diffraction 35

IMPORTANT MINERAL NAMES

amphibole 38
apatite 47
calcite 38
clay minerals 44
corundum 47
diamond 33
feldspars (plagioclase,
 orthoclase) 46

fluorite 34
garnet 43
goethite 41
graphite 33
gypsum 46
halite 34
hematite 41
ice 47

micas (muscovite, biotite,
 chlorite) 44
olivine 35
pyrite 47
pyroxene 43
quartz 38
talc 41
topaz 41

QUESTIONS FOR REVIEW AND THOUGHT

2.1 How do atoms of one element differ from atoms of another?

2.2 How do isotopes of one element differ from isotopes of another?

2.3 How do the kinds of chemical bonds differ from each other? Describe the common kinds of chemical bonding and give an example of each.

2.4 What is meant by the term "well-defined chemical composition"? How can a mineral still have a well-defined composition if ionic substitution takes place?

2.5 What is an "ordered internal structure"? What physical evidence indicates that minerals have a crystal structure?

2.6 What chemical rules must be followed if one ion substitutes for another in a mineral's structure? When Al^{3+} substitutes for some Si^{4+} ions in silicate minerals, how is electrical neutrality maintained? Give a specific example.

2.7 What is cleavage? Describe how cleavage can be used to tell one mineral from another.

2.8 What is "hardness" and how is it measured?

2.9 Why are there only a few common minerals or mineral families in the Earth's crust?

2.10 What is the basic building block of the silicates? Describe how building blocks can be arranged to produce each of the major families of silicate minerals.

2.11 How do minerals or mineral assemblages reflect the conditions at the time they were formed?

Critical Thinking

2.12 Make a list of the white minerals we have named in this chapter. Suppose you found a white mineral and wanted to know which one of the minerals on your list it was. What specific tests would you perform to find out?

2.13 Suppose that you found something that looked like a mineral, but might be a piece of glass. How might you tell the difference? What method could a geologist use that you probably could not?

3

Volcanoes and lava flows in the San Francisco Peaks, near Flagstaff, Arizona
(Dr. John S. Shelton)

Igneous Rocks

*O*BJECTIVES

▲▲

As you read through this chapter, it may help if you focus
on the following questions:

1. What is magma and how does it become igneous rock? Specifically, what controls the types of minerals that form, and how can a magma evolve to produce several different types of igneous rock?

2. What are the major igneous rock families, and what characteristics can be used to tell them apart?

3. What do bodies of igneous rock on and below the Earth's surface look like, and how do they form?

*O*VERVIEW

▲▲

All igneous rocks were once molten. To a geologist, that statement is on a par with the biologist's recognition that all mammals are warm-blooded. It's a good place to begin. Igneous rocks, however, are as different from each other as mammals are. In this chapter we explore some of those differences. On one level, this is a guide to how geologists tell one igneous rock from another—a practical and fundamental skill to develop. The more important task, however, is to understand how the igneous rocks form. As we have suggested several times already, the fun and challenge in geology come with understanding how the Earth works. Much of our attention, therefore, will be on events that change both the molten parent material and the growing solid rock as they cool. We will pick up this discussion again in Chapters 8, 11, and 12, where the focus will be on heat, plate tectonics, and mountain building.

What Is Magma?

Molten rock in the Earth is called **magma;** when it is expelled, or extruded, onto the Earth's surface, it is known as **lava;** when solidified pieces are blown into the air, they are called **pyroclastic debris.** All three are dynamic terms, taken from Greek or Latin roots that mean, respectively, "kneaded or squeezed," "sliding," and "broken by fire." In any form, molten rock is a changeable, complex material.

Chemical Composition

Magma is a liquid that contains dissolved gases and suspended solid particles. The composition of the liquid is dominated by oxygen and the eight abundant cations identified in Chapter 2 (Si, Al, Fe, Ca, Mg, Na, K, and Ti). Because these are distributed unevenly in the Earth's crust, however, magma compositions vary widely. As we examine magmas, we find that most of them fall into three broad, but distinct groups (see Figure 3.1). The common magmas with the highest proportion of Fe, Mg, and Ca (together averaging about 20 percent of a typical magma by weight) are called **mafic** or **basaltic** magmas. Ones with the lowest proportion of these three elements (totaling about 3 percent by weight) are **sialic** or **granitic** magmas. The proportion of Fe, Mg, and Ca in the third group, known as **andesitic** magmas, is intermediate (about 12 percent by weight). A fourth group, insignificant in the Earth's crust but important to consider in the mantle, are the **ultramafic** magmas, so named because Fe, Mg, and Ca can make up almost 30 percent of their weight.

Another way to characterize the magma groups chemically is by comparing the amounts of SiO_2 in each. Ultramafic magmas are less than 50 percent SiO_2 by weight, basaltic magmas roughly 50 percent, andesitic magmas 60 percent, and granitic magmas 70 percent.

A third way to describe magmas chemically is on the basis of the gases they contain. Water (as steam) and carbon dioxide are the most abundant gases in magma. Gaseous compounds of sulfur, chlorine, and other elements are also common, although much less abundant. Some of these "minor" gases in magma, however, can have a devastating effect on the local environment by contributing to acid rain or other toxic hazards when they escape into the atmosphere. On average, gases in a magma amount to less than 3 percent of its mass, but the proportion can be as high as 10 percent or more. Granitic magmas typically have the highest proportion, and basaltic and ultramafic magmas the least.

Appearance and Behavior

The chemical composition of a magma is more than a curiosity. Since the 1970s, for example, geologists have discovered that many of the chain and network structures of silica tetrahedra we discussed in Chapter 2 are also present in magmas and can have a large effect on the physical properties of these magmas. For example, granitic magmas, which have a higher number of silica tetrahedra, tend to be much stiffer than basaltic magmas. In essence, the more tetrahedra in a magma, the more likely they are to share oxygen atoms, thus forming large, patchy structures that stiffen the magma and impede its ability to flow smoothly. Because there are few of these structures in a typical basaltic lava, it flows much like water in a stream; its internal resistance to flow (a property called **viscosity**) is very low. Conversely, a granitic magma at the same temperature

FIGURE 3.1   As shown in these pie diagrams, basaltic magmas contain much less silica and more iron, magnesium, and calcium than do granitic magmas. Andesitic magmas have intermediate compositions.

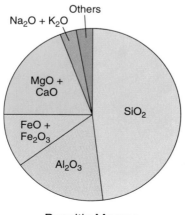

Basaltic Magma

Andesitic Magma

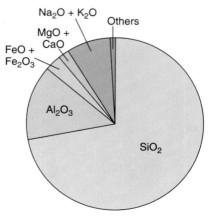

Granitic Magma

(a)

(b)

FIGURE 3.2   (a) A lava flow with a ropy (pahoehoe) surface texture, in Craters of the Moon National Monument, Idaho. This lava had a low viscosity and could therefore flow rapidly. (b) Viscous lava, with a blocky (aa) texture, slowly advancing on a villa in Zafferana Etnea, Italy. *(AP/WideWorld Photos)*

has such a high viscosity that it flows with great difficulty. Andesitic magmas, which have a silica content between those of basaltic and granitic magmas, are neither as fluid as basaltic magmas nor as stiff as granitic ones.

If magma reaches the surface, these differences affect the appearance of lavas. The lava flow in Figure 3.2a has a ropy texture, given the Polynesian name **pahoehoe** (pronounced "PA-HÓ-EE-HÓ-EE"). Basaltic lava, which flows readily, is most likely to have this surface texture. The lava in Figure 3.2b is more andesitic and, thus, is less fluid than the basaltic flow. Consequently, it has developed a sharp, blocky texture called **aa** (pronounced "ÁH-AH"). As we discuss later, viscosity also depends greatly on temperature, so the surface texture of a lava flow is not a reliable guide to its chemical composition.

Together with the viscosity of a magma, the amount of gas it contains also has a profound effect on the style of eruption that occurs. Much like the CO_2 in a bottle of soda, the gases are fully dissolved in a magma while they are under pressure. Bubbles of gas only form as the magma erupts and pressure is released. The rapid growth of the bubbles can greatly increase the violence of an eruption, in the same way that gases can escape from a carbonated beverage with surprising force when the bottle is shaken before being opened. Thus, a granitic or andesitic magma containing a lot of gas is likely to erupt more vigorously than a basaltic magma, which generally has less gas in it. In addition, because the granitic or andesitic magma is also less fluid, it is difficult for the fast-growing gas bubbles to move through it. Trapped and still growing, the bubbles finally escape explosively, throwing

lava and rock fragments high into the air. For this reason, eruptions of granitic or andesitic magma commonly produce clouds of pyroclastic debris, while basaltic eruptions release "gentle" lava flows.

Temperature

It is impossible to measure the temperature of a magma directly while it is still within the Earth, although geologists can estimate it by studying the heat escaping from the planetary interior (Chapter 8). Magmas in the crust probably range from as low as 600°C or 700°C to as high as 1500°C. Those in the upper mantle may be as hot as 2000°C.

It is much easier to measure the temperature of magma once it reaches the Earth's surface. Geologists measured basaltic lava temperatures in Hawaii, for example, by using Seger cones (pieces of ceramic material constructed to melt at various specific temperatures and commonly used by potters to measure kiln temperature). The cones were sealed in pipes, and the pipes were then lowered into the lava. At a depth of 13 m, the lava had a temperature of 1175°C. Fresh lava flows are typically 1000°C to 1200°C. As a lava cools, it becomes more viscous until, at a temperature between 600°C and 800°C, it finally solidifies.

Where Does Magma Come From?

Until the beginning of the twentieth century, the prevailing theory among geologists was that magmas rose from a largely molten layer below the crust of the Earth. As instrumentation for studying the Earth's interior became more sophisticated, geologists discovered that no such molten layer exists. This realization was an important part of a greater revolution in geology that we will return to address at length in Chapters 8–12. At that point, we will also look more closely at the processes that generate magma. For the purpose of this chapter, however, it is important for you to know that magmas form in three types of geologic settings:

- Basaltic and ultramafic magmas come from the upper mantle, although not from a totally molten zone, as was once suspected. Geologists now understand that as much as 2 or 3 percent of the asthenosphere may be molten in places. That quantity of magma does not constitute a separate zone, but merely a small amount of melting along the edges of individual mineral particles within a very large area.
- Granitic magmas form within the continental crust, commonly either where the crust is being reshaped by forces within the Earth that raise mountains, or where the crust is heated by basaltic magma rising from the asthenosphere below. The processes that deform the crust and raise mountains are a major focus of Chapters 8–12.
- Andesitic magmas form primarily along the active boundaries between the continental and the ocean crust, places that we labeled subduction zones in Chapter 1.

As magmas move within the Earth, their chemical compositions change. One simple example of how a magma may change involves **assimilation:** the process of altering a magma by melting other rocks with which it comes in contact. A basaltic magma, for example, rarely retains the composition it had when it left the asthenosphere, but is changed by the addition of rocks that it surrounds or passes through as it rises toward the Earth's surface. In another simple process of compositional change, called **magma mixing,** two magmas from different sources blend as they rise. Much more complicated changes occur when the temperature or pressure around a magma change, as when it begins to solidify.

Crystallization from a Magma

▲▲▲▲▲▲▲▲▲▲▲▲▲▲▲▲▲▲▲▲▲▲▲▲▲▲▲▲▲▲▲▲▲▲▲▲▲

We are accustomed to liquids like water and candle wax that solidify at a specific temperature. Complex liquids like magma, however, rarely solidify all at once. Instead, they solidify over an interval of temperature: a few crystals of one or a few minerals at the highest temperatures, and others as the temperature drops. It is important to understand that the amount of solidification has nothing to do with *how long* it takes the magma to cool or crystals to grow. Rather, it is dependent on the temperature to which a magma

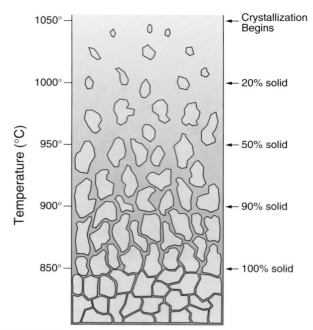

FIGURE 3.3   Magmas crystallize over a temperature interval. In this schematic drawing, crystallization begins at 1050°C, but is not complete until the magma has cooled to 850°C. At 1000°C, crystals make up 20 percent of the magma; the remainder would stay liquid if the magma never cooled further.

cools. If the magma used as an example in Figure 3.3 cooled to 1000°C and no further, it would never be more than 20 percent solid.

It may be more surprising to learn that as a magma solidifies, it changes composition, and so do the crystals that form. To see how this works, follow the example in Figure 3.4, which describes a plagioclase feldspar magma much simpler than any we could find in nature (this is a modified version of Figure 2.27). The composition of a plagioclase feldspar, you will recall, may lie anywhere between a pure sodium variety and a pure calcium one. For purposes of illustration, we ask you to imagine a magma made by melting a plagioclase with a composition halfway between these extremes. That is, the magma contains just as much sodium as calcium. The X in the region of Figure 3.4 marked LIQUID is where we begin. Here the magma's composition is 50 percent $NaAlSi_3O_8$ and 50 percent $CaAl_2Si_2O_8$ and the temperature is 1600°C.

If the magma cools to 1450°C, crystals of plagioclase begin to form. The surprise is that these first crystals are not half $NaAlSi_3O_8$ and half $CaAl_2Si_2O_8$ like the magma. Calcium ions fit into plagioclase better than do sodium ions at this temperature, and so the first crystals are much more calcium-rich. To find out how much more, we look at the line marked "crystal composition" in Figure 3.4. At 1450°C, we find that new plagioclase is 80 percent $CaAl_2Si_2O_8$.

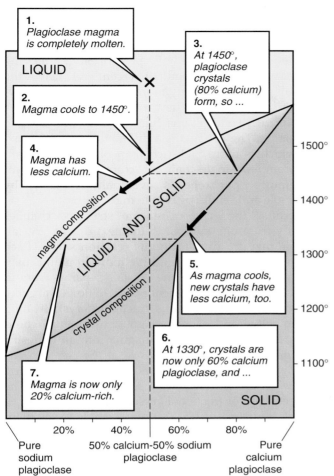

1.
Plagioclase magma is completely molten.

LIQUID

2.
Magma cools to 1450°.

3.
At 1450°, plagioclase crystals (80% calcium) form, so ...

4.
Magma has less calcium.

magma composition

LIQUID AND SOLID

crystal composition

5.
As magma cools, new crystals have less calcium, too.

6.
At 1330°, crystals are now only 60% calcium plagioclase, and ...

7.
Magma is now only 20% calcium-rich.

SOLID

1500°
1400°
1300°
1200°
1100°

20%   40%   60%   80%

Pure sodium plagioclase

50% calcium-50% sodium plagioclase

Pure calcium plagioclase

FIGURE 3.4   This diagram shows how a magma made only of molten plagioclase should crystallize. Temperature increases vertically, and the amount of calcium in the magma (or crystals) increases to the right. The seven-step sequence of events indicated in this diagram is, in fact, a continuous process of gradual crystallization during cooling. The point here is that because the magma and newly formed crystals at any temperature have different compositions, cooling causes a gradual change in both the magma and the accumulating mass of crystals. This is a continuous reaction series.

We get a second surprise when we look back at the plagioclase magma's composition. The act of crystallization has begun to change it. The new crystals consumed some calcium and a much smaller amount of sodium, leaving the magma with proportionally less calcium than it used to have. As the magma continues to cool and crystallize, more and more calcium is removed from it. At any temperature, we can always find out how much is left in the magma by looking at the upper ("magma composition") line (Figure 3.4) and how much is in the crystals by examining the lower one. At 1330°C, for example, the magma contains about 20 percent $CaAl_2Si_2O_8$ and the plagioclase crystals about 60 percent.

Fractional Crystallization

Figure 3.4 shows the results of experiments performed in the early half of the twentieth century by Norman L. Bowen, a Canadian-born geochemist on the staff of the Geophysical Laboratory at the Carnegie Institute of Washington. Because plagioclase retains the same network silicate structure while its calcium content gradually drops, Bowen referred to the process in Figure 3.4 as an example of a **continuous reaction series.** Using this example, he was able to explain observations that had puzzled petrologists for a long time. It was known, for example, that plagioclase crystals are usually zoned; that is, they are calcium-rich on the inside but increasingly sodium-rich toward the outside (Figure 3.5). Bowen showed that this type of zoning should be expected: The calcium-rich core of each crystal forms first and then is gradually covered with calcium-poor plagioclase as the magma cools and evolves.

Plagioclase crystallization is only one example of a process called **fractional crystallization,** by which crystals and magma split the magma's original mixture of ions between them. Bowen's experiment implied that if crystals and liquid were physically separated before the magma solidified completely, the result would be two different igneous rocks. Dense crystals might sink to the bottom of the liquid and form a massive layer, for example, or the liquid might be squeezed into cracks in the surrounding rock and escape from the area completely. If either of these happened, the remaining liquid would solidify to form an igneous rock quite different from that of the early

FIGURE 3.5   Photomicrograph of a zoned plagioclase crystal. The calcium-rich core of this crystal has been covered by successive layers of increasingly sodium-rich plagioclase, which separated it from the surrounding magma. The magma, therefore, must have become more sodium-rich.

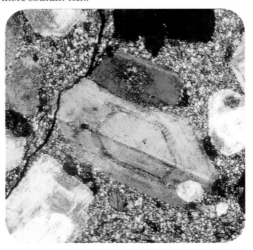

formed crystals. Geologists refer to the physical separation of a magma to form two or more kinds of rock as **differentiation.**

Bowen's study of plagioclase was part of a larger project. Intrigued by the fact that basaltic magma is much more common than either andesitic or granitic magma, Bowen suggested that basaltic magma is the parent from which the other two magmas differentiate. The lower calcium content of granitic magma, for example, might be due in part to early removal of calcium-rich plagioclase crystals from the original basaltic magma.

To test this idea, Bowen performed similar experiments using a basaltic magma instead of a simplified plagioclase magma to determine how minerals other than plagioclase crystallize. He discovered through these experiments that ferromagnesian minerals follow a more complicated crystallization scheme than does plagioclase (Figure 3.6). At a very high temperature, the earliest ferromagnesian mineral to form in

basaltic magma is the island silicate, olivine. Just as in the plagioclase example, fractional crystallization then causes the magma to evolve. Specifically, because basaltic magma is roughly 50 percent SiO_2, removing olivine, which is only about 40 percent SiO_2, leaves the remaining magma slightly more SiO_2-rich. Here, however, the process takes a new twist. As basaltic magma cools further, solid olivine finally reacts with SiO_2 in the magma. The product is a different mineral, pyroxene, which is more SiO_2-rich than olivine and has a more complex single-chain structure. At a still lower temperature, pyroxene is eventually replaced by amphibole and finally by biotite mica.

Because the type of silicate structure changes abruptly from one mineral to the next during crystallization, this is a **discontinuous reaction series.** Each step in the series removes a little more iron and magnesium and, if at least some crystals separate from it, leaves the magma a little richer in SiO_2. Bowen argued, therefore, that simultaneous fractional crystallization of plagioclase and of ferromagnesian minerals could ultimately change a basaltic magma into an andesitic or even a granitic one.

Bowen's two reaction series provide an interesting framework for understanding igneous processes. The broad concept of fractional crystallization, however, is difficult to grasp at first. You may be able to see helpful analogies in nongeological activities such as filtering the sediment out of old motor oil or skimming the fat off of a pot of stew, each of which purifies a liquid by removing parts of it. Going back to review the silicate structures in Chapter 2 will give you another perspective, and we will return to this subject in Chapters 4 and 8.

One final observation before we move on: Geologists are confident that fractional crystallization is the dominant process by which magma evolves within the Earth and can account for many of the differences among igneous rocks, but it cannot do all that Bowen hypothesized. Geologists are now convinced that most granitic magmas do not evolve directly from basaltic ones. For one thing, unless crystals are removed very efficiently from the cooling basaltic liquid, the magma solidifies long before it can become granitic. For another, the extent of differentiation needs to be so extreme that only a tiny amount of granitic magma could form this way. Finally, granitic magmas are found only on the continents, while basalts dominate the ocean basins. If granitic magmas come from basaltic parents, it is hard to explain why they are unheard of where basalts are most common. Most granitic magmas, therefore, are apparently unrelated to basalts. They form instead by melting average rocks of the continental crust or by a process called **partial melting,** which we will examine more closely in Chapter 8.

FIGURE 3.6 Norman Bowen demonstrated that fractional crystallization in a basaltic magma takes place simultaneously in a continuous series, involving plagioclase, and a discontinuous series, involving ferromagnesian minerals. If minerals are systematically removed from the magma as it cools, the magma gradually becomes more granitic.

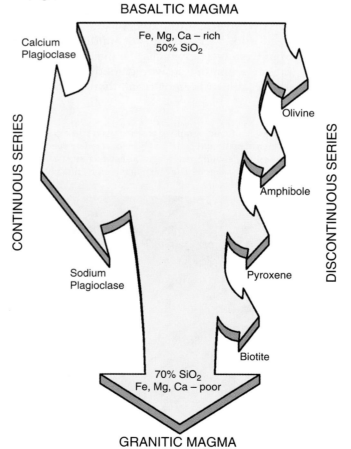

BASALTIC MAGMA

Calcium Plagioclase

Fe, Mg, Ca – rich
50% SiO_2

Olivine

Amphibole

CONTINUOUS SERIES

DISCONTINUOUS SERIES

Sodium Plagioclase

Pyroxene

Biotite

70% SiO_2
Fe, Mg, Ca – poor

GRANITIC MAGMA

Igneous Textures

When geologists use the word **texture** to describe a rock, they are referring to the size and shape of mineral grains and to the way they are connected to each other. Many of these characteristics are easy to see in a hand sample. Others require study under a microscope, as we describe in Perspective 3.1. Different sets of terms describe the textures of igneous, sedimentary, and metamorphic rocks.

In igneous rocks, the primary textural terms refer to grain size, and they are applied in a straightforward way. If the mineral grains in a rock are large enough that they can be distinguished without the aid of a magnifying glass or a microscope, geologists say that the rock has a coarse-grained or **phaneritic** texture

(from the Greek word *phaneros,* meaning "visible") (Figure 3.7a). If the grains are too small to tell apart, the rock texture is fine-grained or **aphanitic** (from the same Greek root, but adding the prefix *a-,* meaning "not") (Figure 3.7b). These are loosely defined textural terms. Allowing for differences in eyesight, geologists generally agree that a mineral grain is "visible" if it is at least 0.5 mm across.

These two terms are enough to describe most igneous rocks, but a few have distinctive characteristics that call for special terms. Rocks with mineral grains more than 2 cm across have **pegmatitic** textures (from the Greek *pegmat-,* "fastened together") (Figure 3.7c). In some of these rocks, crystals of feldspar and other minerals several meters across have been found. A crystal of the mineral spodumene in the Eta pegmatite mine in the Black Hills of South Dakota was

FIGURE 3.7   (a) Phaneritic texture; (b) aphanitic texture; (c) pegmatitic texture; (d) glassy texture. *(The photo inset into part (d) was provided by H.Y. Mc Sween Jr.)*

(a)

(b)

(c)

(d)

Perspective 3.1

Taking a Closer Look at Rock Texture

When geologists step into the field to look at rocks, they identify minerals and describe rock textures with the same degree of precision we are using in this chapter, employing tools no more sophisticated than a hammer, a pocket knife, and a simple magnifying lens. In 1831, however, Scottish geologist William Nicol suggested a way to make more detailed observations with a microscope when we bring rocks back to the laboratory.

Following Nicol's procedure, a geologist glues a chip of rock to a glass slide and grinds it to a thickness of 0.03 mm (thinner than most paper). In a **thin section** like this, most minerals are translucent or transparent, so it is quite easy to observe how they have grown against each other. In a thin section of a granite, for example, you see that feldspars and quartz interlock to form a "jigsaw puzzle" of mineral grains. This feature, typical of all igneous and metamorphic rocks, allows them to support great weight without being crushed.

Nicol's greatest contribution, however, was his suggestion that we should study thin sections with a specially designed microscope, using polarized light (that is, light constrained to vibrate in only one direction). Each mineral bends or rotates polarized light in a distinctive way. If a thin section is sandwiched between two carefully oriented polarizers, its minerals take on distinctive colors that a trained observer can use for identification.

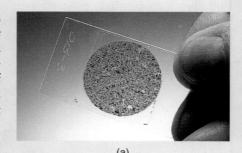

(a)

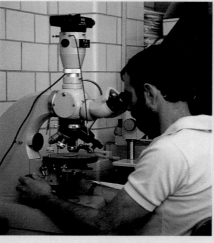

(b)

measured to be over 11 m long! However, some igneous rocks have no crystals at all, but are composed of volcanic glass, and are therefore said to have a **glassy** texture (Figure 3.7d). Some glassy rocks are solid, while others contain many bubbles (called **vesicles**) that form by the rapid escape of gas from the magma. These are said to have a **vesicular** texture. Vesicles in some rocks may later be filled with minerals deposited as groundwater circulates through the cooled lava. These filled vesicles are called **amygdules,** and the rock's texture is **amygdaloidal** (from the Greek word *amygdale*, "almond," a reference to the shape of these bubbles) (Figure 3.8).

Why do grain sizes differ? Mineral grains take time to grow; the slower a magma cools, the longer the grains have to grow before the entire magma solidifies. Phaneritic and pegmatitic textures, therefore, are typical of rocks that cool slowly, often over thousands

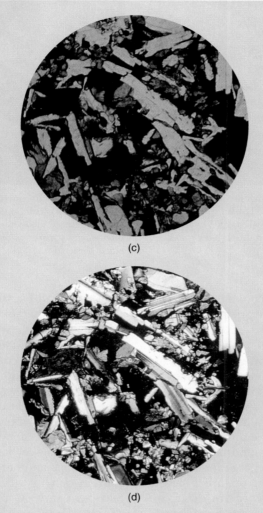

(c)

(d)

(a) A thin section of a basalt. The slice of rock glued to this glass slide has been ground and polished so that it is transparent and can be viewed under a microscope. A thin glass cover slip has been glued on top of this sample to protect it from accidental scratching. (b) A geologist views the thin section with a special microscope in which light can be polarized (much as it is by high-quality sunglasses). (c) If light is polarized, so that it is constrained to vibrate in only one direction, the basalt sample looks like this. (d) When a second polarizing filter is placed above the sample (a configuration that geologists call "cross-polarized"), the same basalt sample looks like this. The colors and textures that become visible in this arrangement are very helpful for identifying the minerals in this rock.

or even hundreds of thousands of years. This long cooling period is likely to be the case when magma cools within the Earth, where heat escapes slowly through the insulating blanket of overlying rock. Igneous rocks in these environments are **intrusive** or **plutonic** rocks (after Pluto, Greek god of the underworld). Aphanitic and glassy textures, conversely, are typical of **extrusive** or **volcanic** rocks (after Vulcan, the Roman god of fire), solidified at the Earth's surface. In

these rocks, the magma cools rapidly (sometimes within a minute or less) when it contacts air or water. You should not conclude, however, that all fine-grained rocks are extrusive, or that all extrusive rocks are fine-grained. The important control on grain size is cooling rate. Magma injected into a fracture in cool rock (an intrusive environment) usually cools as rapidly as if it had flowed out onto the surface and forms a fine-grained rock that looks extrusive. In con-

FIGURE 3.8   An amygdaloidal lava flow along the New England coast. These ancient vesicles, once formed by escaping volcanic gas, are now filled with feldspar and chlorite, deposited over the past 250 million years as water has percolated through the rocks.

FIGURE 3.9   Porphyritic texture in a rock with basaltic composition. This rock is composed of phenocrysts of plagioclase (lighter lath-shaped crystals) set in a matrix of finer-grained plagioclase, olivine, and pyroxene. (*The photo inset into Figure 3.9 was provided by Kenneth E. Windom*)

trast, a very thick lava flow may cool so slowly that rocks in its middle become surprisingly coarse-grained.

When we examine an igneous rock, we usually find that the crystals of a single mineral are roughly the same size, although crystals of one mineral may be slightly larger than those of another. Sometimes, however, a magma may start to cool under conditions that permit large mineral grains to form, and then it may move into a new environment where more rapid cooling freezes the large grains in a groundmass of finer-grained texture (see Figure 3.9). The large minerals are called **phenocrysts,** and the resulting texture is said to be **porphyritic** (from the Greek word for "purple," originally applied to rocks containing phenocrysts in a dark red or purple groundmass). In rare cases magma may be suddenly expelled at the surface after large mineral grains have already formed. Then the final cooling is so rapid that the phenocrysts become embedded in a glassy groundmass.

Classification of Igneous Rocks

▲▲

Scientific schemes for naming things can be very complex depending on how carefully we need to distinguish one object from another. For a first look at igneous rocks, we use a fairly simple classification system based on properties that can be observed in a hand sample.

The chemical composition of a magma determines which minerals will crystallize from it and how abun-

dant each will be. Conveniently, there are only a few dominant minerals in any igneous rock: those in Bowen's reaction series plus quartz and potassium feldspar (orthoclase or a polymorph). We have arranged magma compositions (granitic to ultramafic) along the horizontal axis in Figure 3.10, and have shown the relative proportions of quartz, potassium feldspar, plagioclase, micas (biotite and muscovite), pyroxene, amphibole, and olivine that should crystallize from each along the vertical dimension. Granitic magmas, we see here, should yield about 5 percent amphibole, another 5 percent micas, 25 percent sodium-rich plagioclase, 35 percent potassium feldspar, and 30 percent quartz. Basaltic magmas should produce about 35 percent olivine, 30 percent pyroxene, and 35 percent calcium-rich plagioclase. Using this diagram, it should be an easy matter to figure out which mineral mixture corresponds to which magma composition.

Of course, mineral identification is not always easy, especially in fine-grained rocks. Notice, though, that the progression from left to right in Figure 3.10 is also a trend from light (white to tan or red-brown) to darker rocks (dark green or brown to black). Even when it is hard to tell what minerals are in a hand sample, you can usually make a first guess on the basis of color.

Depending on the cooling rate, mineral grains crystallizing from any magma may vary quite a bit in size, as we have already shown. The upper portion of Figure 3.10 takes this variability into account. Names along the top row correspond to rocks with aphanitic textures; those on the bottom row refer to phaneritic ones. We see, for example, that the coarse-grained rock formed from a granitic magma is called **granite.** A

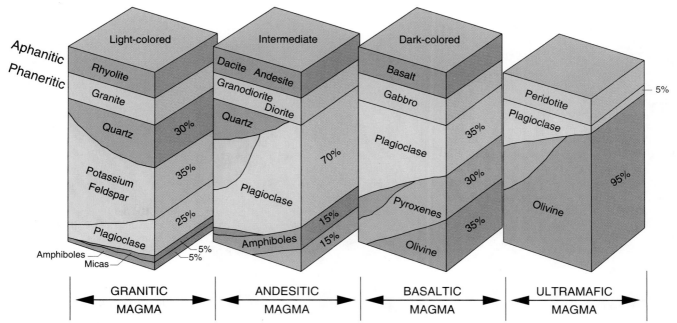

FIGURE 3.10 A simple classification of igneous rocks, based on rough proportions of key minerals. Magma compositions, from granitic to ultramafic, are shown across the bottom of this diagram; textures (aphanitic or phaneritic) are indicated at the top. The broad divisions shown here are meant to be suggestive, not precise. In nature, the boundaries between granitic and andesitic rocks, or between andesitic and basaltic ones, are not sharp.

fine-grained rock with the same mineral composition is **rhyolite.** Three other aphanitic rocks (**dacite, andesite,** and **basalt**) and their phaneritic equivalents (**granodiorite, diorite,** and **gabbro**) are also shown. At the far right, we have added **peridotite,** the coarse-grained ultramafic rock that makes up most of the mantle and, as we will show in Chapter 8, is the parent of many crustal rocks. There is no aphanitic rock similar in composition to peridotite.

Dark-Colored Igneous Rocks

The most abundant igneous rock is basalt. It is common on continents and is the only major igneous rock in ocean basins. Exploration elsewhere in the solar system has shown that basalt is, in fact, the most common rock of any kind on the Moon, Mercury, Venus, or Mars. Gabbro, the phaneritic equivalent of basalt, is much less abundant. In both, the dominant minerals are pyroxene and calcium-rich plagioclase, usually accompanied by olivine. Quartz and potassium feldspar are never present.

A popular name for basalt and some gabbro is **trap rock,** from a Swedish word meaning "step." This name refers to the fact that basaltic magma commonly cools and shrinks so fast that it cracks into gigantic columnar blocks that look like stepping stones (see Figure 3.11).

Intermediate Igneous Rocks

The rocks that form from andesitic magmas are usually aphanitic. If they contain potassium feldspar, geologists call them dacites; if not, they are known as andesites (named for the Andes Mountains of South America, where volcanic rocks of this type are abundant). Granodiorites and diorites, respectively, are the phaneritic equivalents of dacites and andesites. The intermediate igneous rocks rarely contain quartz or olivine. They consist largely of plagioclase and either pyroxene or amphibole. Many andesites have porphyritic textures (Figure 3.12).

Light-Colored Igneous Rocks

The second most common igneous rock is granite. In fact, most of the igneous rocks in continental regions are granitic to granodioritic in composition. Granite, whose aphanitic equivalent is called rhyolite, consists primarily of quartz and potassium feldspar, with some sodium-rich plagioclase and minor amounts of mica and ferromagnesian minerals frequently mixed in. The glassy equivalent of granite is **obsidian.** A light-colored rock with a pegmatitic texture is called a **pegmatite.** These rocks frequently contain gem-quality crystals of rare and beautiful minerals such as topaz or tourmaline.

Classification of Igneous Rocks **61**

FIGURE 3.11
Columnar jointing in basalt. This exposure at Giant's Causeway in Northern Ireland is often visited by geologists and was prominent in local legends of past centuries. The columnar prisms form as magma cools and shrinks. *(Wolfgang Kunz, The Stock Market)*

Pyroclastic Rocks

The igneous rocks we have discussed to this point all form as the result of solidifying a body of magma in the Earth or on its surface. Other important igneous rocks, known as **pyroclastic rocks,** are made of the pyroclastic debris that is blown into the air and subsequently deposited on the ground as some magmas are expelled violently from the Earth. Most of this debris is derived from granitic or andesitic magma, which is both highly viscous and enriched with dissolved gases, as we discussed earlier. Although some crystalline matter is included, most pyroclastic fragments are made of volcanic glass.

FIGURE 3.12 Andesite porphyry. The phenocrysts visible in this sample from Mount Lassen in northern California are crystals of plagioclase.

The term **tephra** (Greek, "ash-colored") covers all the pyroclastic debris that settles from the air (Figure 3.13). The finest of these types of debris constitute volcanic dust, which is made up of particles of the order of 10^{-4} cm in diameter. When volcanic dust is blown into the upper atmosphere, it can remain there for months, traveling great distances. Tephra is categorized in three size ranges:

- **Ash.** Fragments less than 2 mm across, consisting of sharply angular glass particles.
- **Lapilli.** Pieces 2 to 64 mm across.
- **Bombs.** Rounded masses more than 64 mm across that congeal from magma as it travels through the air.

Several other terms are used to describe the texture or composition of tephra. The word **cinder,** for example, is often used to describe any small, irregularly shaped particle of glass. Frothy pieces of lava that trap magmatic gases as they are ejected are called **pumice.** After these solidify, they are honeycombed with gas-bubble holes (vesicles), which give them enough buoyancy to float on water. **Scoria** is like pumice, although usually darker, more crystalline, and denser. **Blocks** are coarse, angular pieces of rock that may have blocked the volcanic vent and been ejected as solid pieces.

The debris may become pyroclastic rock in either of two ways. In the first and most common way, rainwater percolating through ash dissolves silica from glassy particles and redeposits it as a cementing material between other particles. Other materials dissolved in groundwater, particularly calcium carbonate

FIGURE 3.13 Tephra exposed in the Jemez Mountains near Taos, New Mexico. Rhyolitic ash and lapilli in this outcrop have not been cemented to form a solid rock. Volcanism in this area has been active within the past thousand years.

(as the mineral calcite), may serve the same purpose. In the other way, tephra is hot enough that soft, glassy particles actually fuse together as they accumulate. The general term for igneous rocks derived from pyroclastic material is **tuff**. A rock in which glassy particles of ash or lapilli are fused by heat is called a **welded tuff**. A tuff containing bomb-sized particles is an **agglomerate**. Any of these names can be made more specific by adding a description of the magma from which the rock formed—for example, a welded andesitic agglomerate or a rhyolitic tuff.

Extrusive Igneous Rock Bodies

▲▲

Now that we have discussed the various types of igneous rocks, let us turn our attention to the processes that create them. Geologists divide these processes into two groups: those that occur in extrusive igneous environments and those that take place in intrusive environments. The most familiar of these to most people are the extrusive environments, specifically volcanoes.

Volcanoes

Volcanoes are vents through which lava, pyroclastic debris, and gases erupt from the Earth. More generally, a volcano is a mountain produced by the accumulation of igneous rock around such a vent. The igneous matter rises toward the Earth's surface in part because it is less dense than the surrounding crust and in part because it is under pressure from the weight of rock

FIGURE 3.14 Nighttime eruption from the crater at the summit of Stromboli, a volcanic island north of Sicily, often called the "Lighthouse of the Mediterranean" because of its continuous activity. *(Andrea Borgia)*

above or from hot, expanding gases. It usually emerges in a pit, which may be either a **crater** or a **caldera,** at the summit or along the sides of the volcano. A crater (Figure 3.14) is a steep-walled depression out of which volcanic materials are ejected, usually with explosive force. Its floor is seldom over 2 km in diameter; its depth may be as much as 200 m. A crater may be at the top of a volcano or on its flank. A caldera is a much larger basin-shaped depression at the summit of a volcano, generally associated with outpourings of lava rather than pyroclastic debris. It is more or less circular, with a diameter many times greater than that of the volcanic vent or vents that feed it. Most calderas, in fact, are more than 1,500 m in diameter; some are several kilometers across and several hundred meters deep (Figure 3.15).

FIGURE 3.15 Caldera at the summit of Fernandina Volcano, the youngest peak in the western Galápagos Islands. This caldera is more than 6 km across and 750 m deep. *(Bert E. Nordlie)*

A volcano is considered **active** if there is some historical record of its having erupted. If no such historical record exists, and yet the volcano shows only minor wearing from erosion, it is considered **dormant,** or merely "sleeping," and capable of renewed activity. If it has been so long since the last eruption of a volcano that there are no records of its activity and no signs of future revival (such as escaping steam or local earthquakes), and the volcano has been substantially altered by erosion, it is considered **extinct.**

Depending on the chemical composition of the erupting magma and the amount of gas in it, a volcanic eruption may be a relatively quiet outpouring of lava or a violent explosion accompanied by showers of volcanic debris. As a result, not all volcanoes look alike. Geologists classify volcanoes on the basis of differences in their appearance and eruptive behavior.

Classification of Volcanoes

Shield Volcanoes When the extruded material consists almost exclusively of basaltic lava poured out in quiet eruptions from a central vent or from closely related **fissures** (extensive cracks in rock), a mountain builds up that is much broader than it is high, with slopes seldom steeper than 10° at the summit and 2° at the base. Such a mountain is called a **shield volcano.** The largest volcano yet found anywhere is Olympus Mons ("Mount Olympus"), a shield volcano on Mars. Photographed originally from the Mariner 9 spacecraft, it is over 23 km high, with a caldera 65 km across (see Figure 3.16).

The volcanoes of Hawaii are the best examples of shield volcanoes on Earth. The Hawaiian Islands stretch out along a line extending roughly 2,400 km in a northwesterly direction. At the northwestern end of the Hawaiian chain are the low Kure and Midway islands. At the southeastern end is the island of Hawaii, the largest of the group (140 km long and 122 km wide) and the tallest deep-sea island in the world, nearly 10 km above the surrounding ocean bottom and much taller than Mt. Everest, the tallest mountain on the continents. Hawaii is composed of five volcanoes—Kohala, Hualalai, Mauna Kea, Mauna Loa, and Kilauea—of which three are active (see Figure 3.17). Each volcano is made of thousands of lava flows that spread out in thin, fluid sheets over a period of a million years or more.

Geologists monitoring Kilauea have recognized that the level of the ground rises as much as 40 to 135 cm over a period of about 16 months before an eruption, presumably because of pressure from the rising magma. The swelling extends over an area about 10 km in diameter and is centered on, or just south of, the summit. An eruption of the volcano then causes a sudden drop in the ground level, amounting to 20 to 50 cm over a period of days.

This drop is most visible at the summit itself, where great quantities of magma escape, support for the summit is withdrawn, and large blocks of it fall in, forming a large circular depression. The caldera on Kilauea was formed and is maintained in this way. When Kilauea is active, lava floods onto the floor of the caldera and gradually raises its level, forming what is termed a **lava lake** (Figure 3.18). This lava lake may last for years and then disappear completely for equally long periods. The level of the lake drops when lava flows from openings on the flanks of the volcano or on rare occasions when the lava lake rises high enough to breach one wall of the caldera and overflow. From time to time the system is drained, and the caldera floor collapses. Then the magma rises again,

FIGURE 3.16 Olympus Mons, the tallest volcano in our solar system, is one of several shield volcanoes in the northern hemisphere of Mars. The summit has collapsed several times to produce the compound caldera visible in this image. Geologists estimate that Olympus Mons has been extinct for over 1 billion years. *(NASA)*

FIGURE 3.17 This map shows the five shield volcanoes that have been built up from the seafloor to merge and form the island of Hawaii. The longest lava flows since 1750 are also indicated.

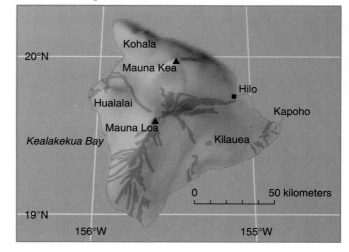

FIGURE 3.18 Because a lava lake loses heat to the atmosphere above, its surface solidifies easily to produce a thin crust. Movement below the surface stretches the crust continually, however, producing a network of cracks through which the lava is visible.

lava floods into the caldera, and the process is repeated (Figure 3.19).

Cinder Cones In striking contrast to the shield volcanoes, other types of volcanoes called **cinder cones,** consisting mostly of rhyolitic or andesitic pyroclastic debris, are both small and violent. They achieve slopes of 30° to 40° and seldom exceed 500 m in height (Figure 3.20). The shape of the cinder cone is produced when a cloud of pyroclastic debris is ejected from the crater by gas pressure, and then settles from the air into a loose pile. Many cinder cones have flows of basalt issuing from their base, commonly produced at the end of their short eruptive lives. Unlike shield volcanoes, cinder cones are rarely active for more than a decade or two. Paricutín, in Mexico, is an example of a cinder cone that has developed in modern times. About 320 km west of Mexico City, Paricutín sprang into being on February 20, 1943. Nine years later it had ceased activity, but during its life it was studied more closely than any other newborn vent in history, until the eruption of Mount St. Helens in 1980.

Stratovolcanoes The most abundant volcanoes, called **composite volcanoes** or **stratovolcanoes,** share characteristics with both shield volcanoes and cinder cones. They are cone-shaped mountains built up of a combination of pyroclastic material and basaltic to andesitic lava flows and characterized by slopes of close to 30° at the summit, tapering off to 5° near the base. Vesuvius in Italy, Stromboli on the island of Sicily, Krakatoa in Indonesia, Pinatubo and Mayon in the Philippines, and the many volcanoes of the Japanese, Kurile, and Aleutian islands and the Cascade Range in Washington and Oregon are all stratovolcanoes (Figure 3.21).

As we have already shown, a volcano's eruptive style depends on the composition of the magma extruded from it. The basaltic lavas typical of shield volcanoes are rich in iron, magnesium, and calcium

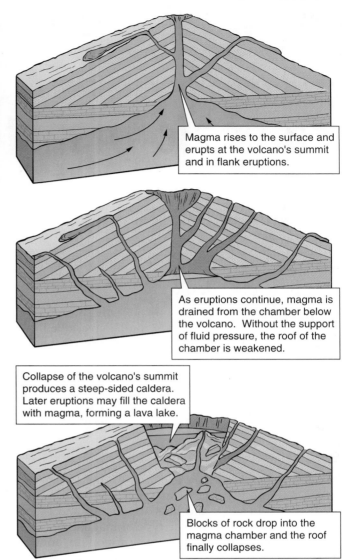

Magma rises to the surface and erupts at the volcano's summit and in flank eruptions.

As eruptions continue, magma is drained from the chamber below the volcano. Without the support of fluid pressure, the roof of the chamber is weakened.

Collapse of the volcano's summit produces a steep-sided caldera. Later eruptions may fill the caldera with magma, forming a lava lake.

Blocks of rock drop into the magma chamber and the roof finally collapses.

FIGURE 3.19 This series of drawings shows one interpretation of how a caldera develops.

FIGURE 3.20 Sunset Crater, an extinct cinder cone in eastern Arizona. This peak, about 300 m high, is one of many in the San Francisco volcanic field.

Volcanic Disasters

At a few minutes before 8:00 A.M. on May 8, 1902, a gigantic explosion occurred through one side of Pelée. A fiery cloud at temperatures around 800°C swept down the mountainside and engulfed the city of St. Pierre, Martinique, wiping out its 28,000 inhabitants and many refugees from other parts of the West Indian island who had gathered there during the preceding days, when the eruption was building up with minor explosions and earthquakes. Some estimates of the death toll ran as high as 40,000.

Eruptions of this kind are unfortunately common, although usually not as disastrous. Stratovolcanoes, in particular, can be highly explosive and are notoriously difficult to predict. Nevertheless, people in many heavily populated corners of the world live in the shadow of these volcanoes, some of which have killed many times. Mt. Vesuvius, on the outskirts of the modern city of Naples, Italy, buried the Roman city of Pompeii with a fiery cloud in the year 79, and it has since erupted more than 50 times. Mt. Asama in central Japan has erupted over 100 times since the year 685, often burying communities in hot ash or with lahars. One of the world's most active volcanoes, Sakura-zima in southern Japan, has sent tephra clouds across the countryside thousands of times since the first recorded eruption in the year 708. Even there, however, people regularly shovel fresh volcanic ash from their streets and turn back to their daily lives as though the next major eruption would never come.

Because so many people worldwide live near stratovolcanoes, geologists try to anticipate volcanic

but low in silica, so they tend to flow relatively quietly. The more rhyolitic lavas typical of cinder cones are enriched in silica, and thus are viscous and explosive. It is not unusual for the type of magma in a stratovolcano to change from one eruption to another because of magmatic differentiation or mixing of different magmas below the surface. For that reason, a stratovolcano may on one occasion produce a gentle lava flow and on another release an explosive burst of pyroclastic debris.

Even during a single eruption, a stratovolcano may change character. Rising gases can accumulate beneath the summit of a volcano if the vent is choked with rock congealed from a past eruption. Pressure builds up until it can no longer be confined. With temperatures of 1000°C or higher, the gases expand several thousand times as they escape, shattering the rock blocking the vent and throwing the rock fragments and magma into the air. After the explosion, there is less gas pressure in the magma still below the ground, although it may be fluid enough to pour out as lava. As an eruption gradually ends, the remaining magma commonly has little gas left in it and has cooled enough to become quite viscous. Instead of flowing easily, it forms a sticky mass called a **lava dome** in the crater, as it has at the summit of Mount St. Helens in

Washington state, perhaps blocking the vent again (Figure 3.22).

During some eruptions a great avalanche of white-hot ash mixed with steam and other gases is extruded. Heavier than air, this highly heated mixture rolls down the mountain slope. Masses of such material are called **fiery clouds** (pyroclastic flows), sometimes referred to by the French equivalent **nuées ardentes.** Fiery clouds have characterized eruptions at Mount Pelée, a stratovolcano on the island of Martinique in the West Indies, to such an extent that they have come to be known as **Pelean** types of eruption.

As the pyroclastic material accumulates on the volcanic slopes, it commonly becomes unstable. Mudflows or debris flows, called **lahars,** result from either the mixing of the pyroclastic material with water derived from melting snow and ice, heavy rains, rivers or lakes, or simply by being shaken loose by the tremors that often accompany volcanic activity. (See Perspective 3.2.)

Submarine Volcanism

Enormous amounts of basalt are extruded on ocean floors. These submarine lavas ooze out from fissures in the sea floor and take on characteristic bulbous or

(a)   (b)

P3.2.1   One of the many victims of the eruption of Mt. Vesuvius, Italy, in 79 A.D., killed by poisonous gases from the volcano. This person was then buried by volcanic ash. During the centuries since then the body decayed, leaving a mold in the volcanic material. Excavators have made a cast from the mold with Plaster of Paris. (b) Residents near Sakura-zima, Japan, live daily with the threat of burial by volcanic ash. Several times a year, clouds of ash descend over the city, slowing traffic and limiting outdoor activities. Long accustomed to their volcano, however, these people sweep away the ash as if it were a snowfall and go back to business as usual.

activity by monitoring ground swelling, earth tremors, and other hints of impending eruptions. Thanks to their warnings at Sakura-zima in 1914 and Pinatubo (Philippines) in 1991, for example, thousands of people were evacuated who would otherwise have died. Such successes are overshadowed each year, however, by dozens of eruptions on unmonitored volcanoes and in places where geologists still do not know enough to make reliable predictions.

rounded **pillow** shapes. A thin, hard crust develops as the molten lava is rapidly cooled by the cold seawater. As pressure from rising material within these pillows builds to the breaking point, the crust fractures, molten material bursts forth, and a new pillow rolls gently to one side of the vent. When it comes to rest on the sea floor, it forms a hardened crust with rounded top and irregularly shaped bottom where the pillow conforms to the round-shaped tops of previously produced pillows.

Basalt Plateaus

Similar eruptions occur on land, although much less often than on the seafloor. On land, they are called **lava floods,** or **fissure eruptions,** because the lava is extruded from a long fissure rather than from a volcano of any kind. The rocks produced by lava floods are known as **flood basalts,** or **plateau basalts,** because of the tendency to form great plateaus built of successive flows (Figure 3.23). The low viscosity required for lava to flood freely over such great areas is characteristic only of lavas that have basaltic composition.

The eruption that began on June 11, 1783, near Mount Skapta on Iceland illustrates how lava floods occur. After a series of violent earthquakes in early June, lava began pouring out along a 16-km line, the Laki Fissure. Lava poured into the Skapta River, evaporating the water and overflowing the stream's canyon, which was 150 to 200 m deep and 60 m wide in places. Soon the Skapta's tributaries were dammed, and many villages in adjoining areas were flooded. The lava flow on June 11 was followed by another a week later and a third on August 3. So great was the total volume of the lava flows that they filled a former lake and an abyss at the foot of a waterfall. They spread out in great tongues 20 to 25 km wide and 30 m deep. As the lava flows diminished and the Laki Fissure began to choke up, 22 small cones formed along its length, relieving the waning pressures and serving as outlets for the final extrusion of debris.

Iceland itself is actually a remnant of extensive lava floods that have been going on for over 50 million years and that have blanketed 0.5 million km². The congealed lava is believed to be at least 2,700 m thick in this area and forms an enormous plateau that extends well into the North Atlantic sea. The Antrim Plateau of northeastern Ireland, the Inner Hebrides, the Faeroes, and southern Greenland are also remnants of this great North Atlantic, or Britoarctic, Plateau.

FIGURE 3.21   Mt. Mayon, a stratovolcano in the Philippines, has erupted over 40 times since 1616. Many of its eruptions have been explosive, sending fiery clouds and lava flows across the countryside. At least eight eruptions have killed people in the densely populated area. *(Carl F. Vondra)*

Of equal magnitude is the Columbia Plateau in Washington, Oregon, Idaho, and northeastern California. In some sections more than 1,500 m of rock have been built up by a series of fissure eruptions. Individual eruptions deposited layers ranging from 3 to 30 m thick, with an occasional greater thickness. In the

canyon of the Snake River, Idaho, granite hills from 600 to 750 m high are covered by 300 to 450 m of basalt from these flows. The Columbia Plateau has been built up during the past 17 million years. The principal activity took place between 16.5 million and 13.5 million years ago, but some flows in the Craters of the

FIGURE 3.22   The lava dome in Mt. St. Helens is a massive plug of volcanic rock that formed after the major eruption in May 1980. Geologists monitor ground tremors and the release of volcanic gases around the dome with the hope of anticipating the next eruption. *(Rick Snyder)*

FIGURE 3.23   Basaltic lava flows extruded from fissures in the Columbia Plateau of Oregon, Washington, and Idaho have buried the earlier land surface to a depth of 3,000 m or more. Here, stream erosion has exposed a cross section of several flows at Pelouse State Park in western Washington.

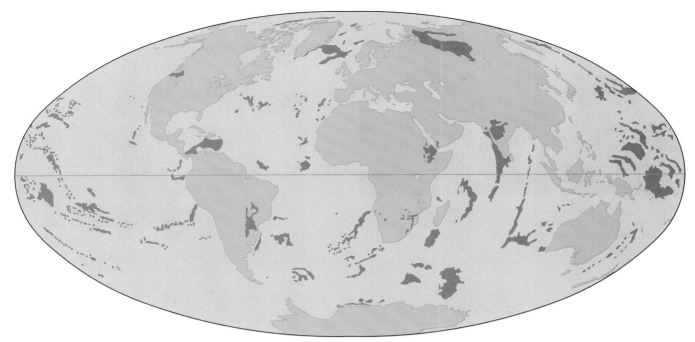

FIGURE 3.24   Basaltic lava flows from fissure eruptions cover very large portions of the Earth's surface, as shown on this map. *(From: Large Igneous Provinces by Coffin and Eldholm. Illustration by Ian Worpole. Copyright © 1993 by W.H. Freeman and Company. Used with permission.)*

Moon National Monument in southern Idaho, probably the most recent of fissure eruptions in the United States, are believed to have occurred within the last 250 to 1,000 years.

 Other extensive areas built up by fissure eruptions include north-central Siberia, the Deccan Plateau of India, Ethiopia, the area around Victoria Falls on the Zambezi River in Africa, and parts of Australia (Figure 3.24). Some of the most extensive flood basalts found anywhere occur on the Moon (Figure 3.25), where there are no volcanoes, and on both Mars and Mercury.

FIGURE 3.25   Extensive flood basalts have filled most of the large craters on the Moon, creating the dark areas we recognize as the eyes of the "Man in the Moon." Galileo, observing these areas through a telescope in the seventeenth century, called them *maria* (the Latin word for "seas"). Fissure eruptions were common on the Moon until about 3.2 billion years ago, but there were never any volcanoes, probably because lunar basalts contain no water or other potential gases.

Intrusive Igneous Rock Bodies

When magma within the Earth's crust solidifies, it forms igneous rock masses of varying shapes and sizes, called **plutons.** These can be seen where previously overlying rocks have been worn away by erosion. As an extinct volcano is worn away, for example, the channels through which magma moved to the surface are gradually exposed (Figure 3.26). These are shallow plutons called **pipes.** In some regions, magma in deeper bodies may never have reached the surface, yet when the crust has been elevated and eroded, these plutons may also be revealed. Plutons are classified according to their size, shape, and relationship to surrounding rocks (Figure 3.27). When magma intrudes into layered rocks, geologists say that the pluton that develops is **concordant** if its boundaries are parallel to the layering, or **discordant** if its boundaries cut across the layering.

FIGURE 3.26 This isolated peak near Kayenta, Arizona, is the eroded remnant of a volcano that once stood here. Thousands of years of erosion have removed the rocks and loose pyroclastic debris of the cone, leaving only the relatively erosion-resistant pipe.

Tabular Plutons

Sills   A pluton with a thickness that is small relative to its other dimensions is called a *tabular* pluton. A tabular concordant pluton is called a **sill**. Sills range in size from sheets less than 1 cm thick to tabular masses hundreds of meters thick. At first glance, a sill might be confused with a lava flow that was later buried by other rocks. Both are tabular and parallel to enclosing rocks. There are fairly reliable ways of distinguishing between the two types, however. A buried lava flow usually has a rolling or wavy top, pocked by the scars of vanished gas bubbles and showing evidence of erosion, whereas a sill has a more even and unweathered surface. Where it is possible to examine the rocks just above a suspected sill, you should find evidence that they were heated by the intruded magma; rocks deposited on an old lava flow should not show evidence of having been heated.

The Whin Sill in northern England (Figure 3.28) is a well-studied example, and one that had a great influence on the history of that region. Intruded between layers of sedimentary rock that are now gently inclined to the south, magma formed a sheet up to 60 m thick in places. Calcite in the rocks above and below the sill was recrystallized and chemically decomposed as it was baked by heat from the magma. For a great distance from the Pennine Mountains toward the coast of the North Sea, the Whin Sill is now exposed in an imposing north-facing cliff. In the first century A.D., Roman occupying forces took advantage of this natural defensive line by using it as the base for a fortified structure, Hadrian's Wall, which still stands today.

FIGURE 3.27   Plutons and landforms associated with igneous activity. (a) Batholiths and stocks form as successive granitic magma bodies are emplaced in a single area. The magma intrudes slowly by entering cracks and by rising to fill holes left as blocks of the surrounding rock fall into the magma. (b) Basaltic and andesitic magma rise along cracks, sometimes reaching the surface to form volcanoes or fissure flows. Discordant and concordant intrusive bodies form as magma solidifies in various openings beneath the surface.

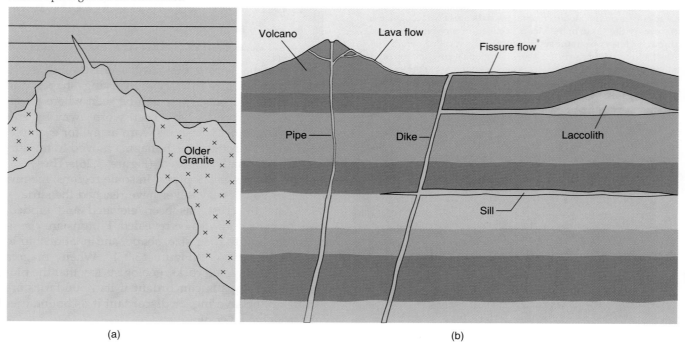

(a)

(b)

FIGURE 3.28   The Whin Sill in Northumberland, northern England, is a prominent feature on the landscape. Hadrian's Wall, built along the crest of the sill, was designed to protect Roman forces from hostile peoples to the north. It was used for less than a century before the Romans retreated southward, but still stands almost 2,000 years later. This view, looking eastward, shows the gentle southward tilt of the sill and surrounding rocks.

Dikes   A tabular discordant pluton is called a **dike** (Figure 3.29). Dikes form when magma forces its way through fractures in rock. The width of individual dikes ranges from a few centimeters to many meters. The Medford dike near Boston, Massachusetts, is 150 m wide in places. Just how far we can trace the course of a dike across the countryside depends in part on how much of it has been exposed by erosion. In Iceland, dikes 15 km long are common, and many can be traced for 50 km; at least one is known to be 100 km long.

Laccoliths   If magma is intruded concordantly in shallow rocks, fluid pressure may be greater than the weight of the overlying rocks. The structure that forms is not strictly tabular; it thickens toward the middle, bulging upward as if it were an object poorly hidden under a rug. Geologists no longer call it a sill, but a **laccolith**.

Massive Plutons

Batholiths and Stocks   A large discordant pluton that increases in size as it extends downward is called a **batholith**. The term "large" in this connection is generally taken to mean that its outline, as seen on a map, includes more than 100 km². A smaller pluton that is otherwise identical to a batholith is called a **stock**. The downward extent, or "thickness," of many batholiths is 10 to 15 km, although there is growing evidence that some batholiths may extend to the base of the Earth's crust, as much as 30 km below the surface. Actually, very little is known about the bottoms of batholiths.

Batholiths are composed primarily of granite or granodiorite, never of gabbro or other mafic rocks, and they contain a great volume of rock. The Sierra Nevada mountain complex of California, known to geologists as the Sierra Nevada Batholith, is 650 km by 60–100 km. A partially exposed complex of plutons in southern California and Baja California, possibly all parts of a single batholith, is probably 1,600 km by 100 km. In these two instances and many others, geologists have found that both the chemical composition

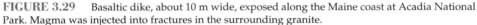

FIGURE 3.29   Basaltic dike, about 10 m wide, exposed along the Maine coast at Acadia National Park. Magma was injected into fractures in the surrounding granite.

FIGURE 3.30
Xenolith of diorite in granite. Its sharp edges suggest that the encompassing magma cooled quickly, before it could melt and assimilate the xenolith.

and the age of the rock vary irregularly across a batholith. This suggests that many batholiths are formed not in a single event, but as the result of several intrusions over a long period of time.

Batholiths have irregular dome-shaped roofs, with fingerlike extensions that reach out into the overlying rock. This characteristic shape is related to **stoping,** one of the mechanisms by which magma moves upward into the crust. As magma intrudes rock, it loosens blocks of overlying rock, which then break off and begin to sink in the magma. Deep in the crust, where the magma is still very hot, the stoped blocks may melt and be digested by the magma body. Higher in the crust, as the magma cools and becomes more viscous, the stoped blocks may become frozen in the intruding magma as **xenoliths** (from Greek roots meaning "foreign rocks") (Figure 3.30). At either depth, stoping allows magma to carve a new "opening" by which it moves gently upward.

Layered Mafic Complexes To a geologist, perhaps the most intriguing plutons are bodies of basaltic composition that have distinct layers of minerals. Most of these **layered complexes** are apparently funnel-shaped, narrowing at their bottoms to a dike or pipe that was once a "feeder" for the body (Figure 3.31). The Bushveld complex in South Africa, the largest layered complex in the world, is exposed across 65,000 km^2 of the land surface, and is as much as 7 km thick in places. Other bodies, such as the Skaergaard complex in Greenland or the Stillwater complex in Wyoming, are more typical in size: Each has about 200 km^2 of exposure and a volume of about 500 km^3.

Unlike batholiths, which have a fairly uniform texture throughout, layered complexes consist of layers of olivine, pyroxene, plagioclase, and a variety of frequently mined minerals. A 1-m thick layer in the west-

FIGURE 3.31   Idealized cross sections through two layered complexes. (a) The Skaergaard intrusion in east Greenland was emplaced as an inclined funnel, and the layered rocks are stacked like saucers within it. (b) The Muskox intrusion in northern Canada is shaped like the keel of a sailing ship. *(After L.R. Wager and G.M. Brown, Layered Igneous Rocks. Copyright © 1967 W.H. Freeman and Company. Used with permission.)*

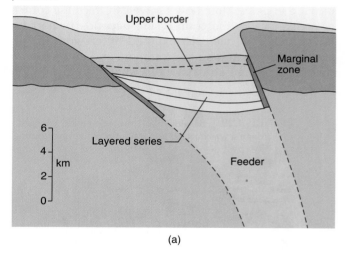

(a)

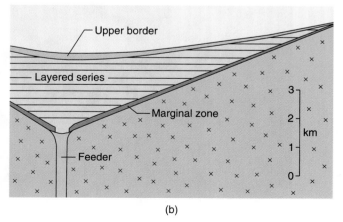

(b)

ern Bushveld complex, exposed in a band over a distance of 220 km, contains an estimated 6,200 tons of platinum, as well as high concentrations of chromium and gold.

Layered complexes are produced by differentiation and **crystal settling** in large magma reservoirs, thousands of meters below the Earth's surface. As crystals form, they are denser than the surrounding magma and collect on the floor of the reservoir as a **cumulate** layer or crystal mush, in much the same way that sand settles on the bottom of a pond. Geologists are uncertain about the details of this process and, in particular, how it generates the hundreds of layers in a typical complex.

▶▶ Epilogue ◀◀

Two hundred years ago, geologists, led by the example of James Hutton, waged a protracted and sometimes vigorous battle of words against Abraham Werner, one of the most respected geologists in Europe, over the question of where basalt comes from.

Hutton's followers, who became known as "Plutonists," insisted that basalts were once molten rock. Werner's adherents, however, were equally convinced that basalt forms from particles that settle out of the ocean, much as salt collects as a pan of seawater evaporates. The "Neptunists," as they were dubbed by some geologists, cited the fine particle size and layered nature of basalts in partial support of their point of view. Geologists settled this debate by about 1830, gradually recognizing most of the characteristics of igneous rocks and of the intrusive and extrusive bodies we have discussed in this chapter.

Geologists today use experiments like those that Norman Bowen performed in order to understand more about the complicated processes of fractional crystallization and magmatic differentiation. They descend carefully into calderas to sample fresh lava and volcanic gases, and they dive to the seafloor to watch pillow lavas extruded from the active mid-ocean ridge. Their goal remains the same as Hutton's and Werner's, however: to learn to recognize mineralogical and textural clues in igneous rocks that can tell us how they formed.

Summary

1. Magma is molten rock, also containing suspended crystals and dissolved gases. Magmas are divided into four groups (ultramafic, basaltic, andesitic, and granitic) on the basis of increasing silica content and decreasing proportions of iron, calcium, and magnesium. Magma crystallizes over a temperature range rather than at a single temperature. If crystals separate from the cooling magma, in a process called fractional crystallization, the magma's composition changes. The magma is said to have differentiated to form two or more rocks with different compositions. Norman Bowen's continuous and discontinuous reaction series demonstrate a trend for basaltic magmas to become more granitic during fractional crystallization.

2. Igneous rocks are categorized on the basis of mineral content and texture. The most abundant rocks are basalt (composed of plagioclase and ultramafic minerals) and granite (composed of feldspars and quartz). Pyroclastic rocks are formed by the accumulation of debris ejected from volcanoes during violent eruptions. Igneous rocks have fine-grained textures if they formed by rapid cooling; coarser textures indicate slow cooling. Fine-grained (aphanitic) igneous rocks are generally found in extrusive environments at the Earth's surface; coarse-grained (phaneritic) rocks are usually intrusive.

3. The most prominent igneous rock bodies on the Earth's surface are volcanoes. These vary in shape and size, depending on the type of magma that formed the volcano, its viscosity, and its dissolved gas content. Magma may also be extruded in lava floods. Plutons, igneous rock bodies formed within the Earth, are categorized on the basis of size and shape as well as their physical relationship to surrounding rocks.

KEY WORDS AND CONCEPTS

aa 53
amygdule (amygdaloidal) 58
Bowen's reaction series 55
caldera 63
concordant 69
crater 63
crystal settling 73
cumulate 73
differentiation 56
discordant 69
extrusive 59
fiery cloud (nuée ardente) 66
fissure eruption 67
fractional crystallization 55

intrusive 59
lahar 66
lava 52
lava dome 66
lava flood (plateau basalt) 67
lava lake 64
mafic 52
magma (andesitic, basaltic, granitic, utramafic) 52
pahoehoe 53
pelean eruption 66
phenocryst 60
pillow 67

pluton (batholith, dike, laccolith, layered complex, pipe, stock, sill) 69
pyroclastic debris 52
sialic 52
stoping 72
tephra 62
texture (aphanitic, glassy, pegmatitic, phaneritic, porphyritic, vesicular) 57
viscosity 52
volcano (cinder cone, shield volcano, stratovolcano) 63
xenolith 72

IMPORTANT ROCK NAMES

agglomerate 63
andesite 61
basalt 61
dacite 61
diorite 61

gabbro 61
granite 60
granodiorite 61
obsidian 61
pegmatite 61

peridotite 61
rhyolite 61
tuff 63
welded tuff 63

QUESTIONS FOR REVIEW AND THOUGHT

3.1 What chemical characteristics do geologists use to distinguish among ultramafic, basaltic, andesitic, and granitic magmas?

3.2 What are the differences in viscosity among the three major kinds of magma? How can you explain those differences?

3.3 Why does the magma in Figure 3.4 become gradually sodium-rich as plagioclase continues to crystallize from it?

3.4 What is the difference between a continuous reaction series and a discontinuous one?

3.5 How can fractional crystallization lead to magma differentiation?

3.6 Why is it unlikely that very much granite is ever produced by fractional crystallization of basaltic magma, as Norman Bowen originally thought?

3.7 Why are phaneritic igneous rocks usually associated with intrusive bodies and aphanitic rocks with extrusive bodies? Why isn't this generalization always true?

3.8 What proportions of common rock-forming silicates are typical for each of the three major groups of

igneous rocks (granite/rhyolite, diorite/andesite, and gabbro/basalt)?

3.9 How is a vesicular texture produced? How is a porphyritic texture produced?

3.10 What is the difference between a crater and a caldera? On what kind of volcano would you expect to find each of them?

3.11 What are the major differences in eruptive style, physical appearance, and rock type associated with shield volcanoes, stratovolcanoes, and cinder cones?

3.12 How is tephra formed, and what variations in its physical appearance are common? How are rocks made of tephra formed?

3.13 How can a geologist distinguish between a sill and a buried lava flow?

3.14 What are the common types of concordant and discordant plutons?

3.15 What features are characteristic of batholiths and stocks? How are these different from the characteristics of layered complexes?

3.16 Suppose the magma in Figure 3.4 cooled to 1300°C. How calcium-rich should it be then? How calcium-rich should newly formed plagioclase crystals be at that temperature?

By locating points on the upper and lower curves at 1300°C, we estimate that the magma is 16 percent $CaAl_2Si_2O_8$ and the new crystals are 54 percent $CaAl_2Si_2O_8$. Given the size and accuracy of this diagram, your answers might differ from ours by as much as 2 or 3 percent.

Critical Thinking

3.17 Suppose that you landed on the surface of the Moon, somewhere within the area shown in Figure 3.26. Describe the kind of igneous rock you would find (giving its name, texture, and mineral composition) and describe the process by which it erupted on the Moon's surface.

3.18 Suppose that a geologist decided to study a single dike very closely. What differences might the geologist expect to see between the center of the dike and its edges, where it is in contact with the surrounding rock?

4

Weathered granite boulders in Joshua Tree National Monument, California

Weathering and Soils

OBJECTIVES

▲▲▲

*As you read through this chapter, it may help if you focus
on the following questions:*

1. What are the basic processes by which rocks are weathered?
2. How does weathering change the most common minerals in the Earth's crust?
3. Why do some materials weather more readily than others?
4. How does a residual soil form, and how do geologists categorize soils?

OVERVIEW

▲▲▲

The combined effects of climate and biological activity at the Earth's surface gradually make rocks crumble and decompose to form the loose accumulation of mineral matter called soil. Even the strongest granite is vulnerable. Weathering goes on around us every day, but it seems like such a slow and subtle process that we easily underestimate its importance. Compared to such geologic processes as mountain building, however, weathering actually proceeds rather quickly. Newly erupted basalt in tropical climates, for example, can decompose to form a fertile soil layer in just a decade or two.

Rock may change physically as cracks form and pieces are removed. It may also change chemically as water, acids, and oxygen react with minerals, dissolving some and altering others to form new earth materials. In these ways, rock is converted into soil, which slowly erodes, is redeposited and buried, and becomes new rock. Weathering, then, is a key process in the rock cycle.

Energy and Weathering

Weathering is the process of change that takes place in surface or near-surface material in response to the action of air, water, and living matter, converting rocks and their mineral components into soil and various ions that dissolve in water. The energy that drives the weathering process comes from both inside and outside the Earth. Some of it is **chemical energy,** released as minerals react with substances in water or air. From time to time, motions originating inside the Earth elevate some parts of the Earth's surface above others. The energy that is used to raise the crust is stored in it as **potential energy.** This may be transformed to **kinetic energy,** the energy of motion, if a hillside is unstable (Chapter 13). The force of gravity overcomes friction on the slope and a rock falls. As it chips and shatters against other rocks, weathering has begun.

Far more energy is received at the Earth's surface as heat from the Sun. Uneven distribution of this energy causes unequal heating of the atmosphere and of the oceans. This, in turn, causes winds and currents, generates our weather, and influences the distribution of life on the planet. All of this process helps to modify the Earth's surface materials—in short, to determine the process of weathering.

Weathering Processes

Weathering involves many kinds of events that lead to the **disintegration** and **decomposition** of rock. These are not separate processes, because one rarely can take place without the other. The physical events that make small rocks out of big ones set the stage for chemical reactions that break them down further. At the same time, chemical agents that attack a rock weaken it, making it more likely to break into smaller pieces and eventually into soil. For our purposes, however, it is convenient to discuss the two types of processes separately.

Disintegration

Disintegration, also called *physical* or *mechanical weathering,* is the process by which rock is broken down into smaller and smaller fragments by physical forces. For example, when water freezes in a fractured rock, the pressure caused by expansion of the frozen water may split off pieces of the rock. Or a boulder moved by gravity down a rocky slope may be shattered into smaller fragments.

Impact and Abrasion If you were asked to make a rock smaller, you would probably choose to hit it with a hammer or some other hard object. That obvious method, and innumerable natural variations of it, is the simplest form of disintegration. Whether a rock is falling down a hillside, being tossed by waves, or getting bounced along the bed of a stream, impact is breaking it into smaller pieces. However, in all of these situations rocks are being disintegrated by another process as well. Rocks dragged over each other, such as those falling down a slope or bouncing along the bed of a stream, carve scratches and loosen chips by abrasion. In later chapters we will show that both processes take place whenever earth materials move.

Certain instances provide geologists with unique examples of disintegration by impact. For instance, on the lunar surface we find many examples of rock debris resulting from meteorite impact (Figure 4.1). The entire surface of the Moon has been stirred to a depth of 100 m or more by the impact of objects of all sizes, down to the microscopic. Lunar scientists refer to this extreme process as **gardening.** The "soil" that results, a loose aggregate of rock fragments that have not been altered chemically, is called **regolith.** We have no comparable material on Earth, although some soils in glacial and desert environments come close because they are either too cold or too dry for chemical reactions to be very effective in them.

Frost Action The force exerted as water freezes provides one of the most effective means of disintegrating rock. When water trickles down into

FIGURE 4.1 The lunar surface is covered with rock fragments (regolith) to a depth of several hundred meters. These have been created by 4.5 billion years of meteorite impacts. *(NASA)*

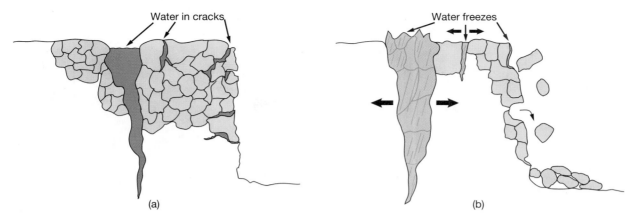

Water in cracks

Water freezes

(a)   (b)

FIGURE 4.2   As water freezes, it expands. In a crack, or in the small open spaces between some mineral grains in a rock, the forces generated as ice forms can be strong enough to shatter rock.

the cracks, crevices, and pores of a rock mass and then freezes, its volume increases by about 9 percent. This expansion of water as it passes from the liquid to the solid state sets up pressures that are directed outward from the inside of the rock, and **frost wedging** results. These pressures can dislodge fragments from the rock's surface (Figure 4.2).

These dislodged fragments of rock are angular, and their size depends largely on the type of rock from which they have been displaced. Usually the fragments are only a maximum of a few centimeters in any given dimension, but in some places they may be much larger.

Certain conditions must exist before frost wedging can take place: There must be an adequate supply of moisture; the moisture must be able to enter the rock in a large enough quantity; and temperatures in and around the rock must move back and forth past the freezing point. As you might expect, frost action is more pronounced in high mountains and in moist regions where temperatures fluctuate across the freezing point either daily or seasonally.

Heating and Cooling   Large, rapid changes in temperature also may cause the mechanical weathering of rock. In areas where bare rock is exposed and is unprotected by a covering of soil, forest or brush fires can generate enough heat to break up the rock. The rapid and violent heating of the exterior zone of the rock causes it to expand. If it expands enough, flakes and larger fragments of the rock can split off. Because it expands and then immediately contracts, a hot rock cooled rapidly may also crack. For centuries before modern explosives were invented, mining engineers used this method to weaken solid rock and make it easier to excavate (see Figure 4.3).

Extreme heating and cooling is known to damage rock, as these examples show. Geologists are still not

sure whether smaller variations in temperature from day to night or from summer to winter are enough to cause mechanical weathering, however. Theoretically, they should be. For instance, we know that the minerals in a granite expand and contract at different rates as they react to rising and falling temperatures. Geologists suspect that even minor expansion and contraction of adjacent minerals over long periods of time would weaken the bonds between mineral grains and that it would be possible for disintegration to occur along these boundaries.

Laboratory evidence to support these speculations is inconclusive. In one experiment, coarse-grained granite was subjected to temperatures ranging from 14.5°C to 135.5°C every 15 minutes. This alternate heating and cooling eventually simulated 244 years of daily heating and cooling, yet the granite showed no signs of disintegration. Perhaps experi-

FIGURE 4.3   Mining engineers in the Middle Ages used fire to weaken rock causing it to crumble and become easier to excavate. *(The Granger Collection)*

FIGURE 4.4   Exfoliation of the Black Hill norite, a Cambro-Ordovician intrusion in southern Australia. *(William C. Bradley)*

FIGURE 4.5   Exfoliation plays a major role in forming most dome-shaped mountains. This exfoliation dome, called Sugar Loaf, is a well-recognized landmark in the city of Rio de Janeiro, Brazil. *(Luis Villota, The Stock Market)*

ments extended over longer periods of time would produce observable effects. If fluctuations such as these bring about the disintegration of rock, they must do it very slowly. For example, some geologists believe that some weathering in the lunar regolith may be due to the effects of millions of heating and cooling cycles.

Exfoliation   **Exfoliation** is a mechanical weathering process in which curved slabs of rock are stripped from a larger rock mass as if they were the layers in an onion being removed (Figure 4.4). The slabs are only a few centimeters thick near the rock's surface, but become thicker—up to several meters—deeper inside the rock. Under certain conditions, one after another of the curved slabs peels off and a large domelike hill, called an **exfoliation dome,** develops. Well-known examples of exfoliation domes are Stone Mountain, Georgia, the domes of Yosemite Park, California, and Sugar Loaf in the harbor of Rio de Janeiro, Brazil (Figure 4.5).

Most geologists agree that the slabs form in response to the gradual removal of weight around the previously solid mass of rock, a process called **unloading** (Figure 4.6). As erosion strips away exposed material, the downward pressure on underlying rock is reduced. Because the rock is no longer under as much pressure as it once was, it expands upward. Even a moderate expansion is too much for brittle rock, however, so it fractures, marking off the slabs that later fall away.

Other Disintegration Processes   Although plants principally weather rocks chemically (as we will show shortly), they also break rocks apart mechanically. The roots of trees and shrubs growing in rock crevices sometimes exert enough pressure to dis-

lodge previously loosened fragments of rock, much as tree roots heave and crack sidewalk pavements (Figure 4.7). More important, though, is the continual abrasion of large and small rock fragments by ants, worms, rodents, and other small animals and, of course, by people as they dig in the soil. Human activity, from tilling fields to excavating for new buildings, is a major cause of rock disintegration in many parts of the world.

Decomposition

Decomposition, sometimes called *chemical weathering,* is a more complex process than disintegration, which merely breaks rock material down into smaller and smaller particles without changing it chemically. Decomposition actually transforms the original rock material into something chemically different. Sometimes the products of decomposition are new minerals, and sometimes they have no mineral form at all, but are substances dissolved in water.

Dissolution   Some weathering reactions involve **dissolution** of solid material in natural fluids. As an example, gypsum undergoes the reaction

$$CaSO_4 \cdot 2H_2O \rightarrow Ca^{2+} + SO_4^{2-} + 2H_2O$$

when it is placed in water. In other words, the water dissolves the gypsum into its component parts. This leaves no residue behind. The calcium and sulfate ions end up in solution, to be washed away in groundwater or, eventually, in a stream.

Some important dissolution reactions involve carbon dioxide as an agent. CO_2 is a minor but significant constituent of the atmosphere. However, it is more abundant in many soils. In forested regions, for example, the decay of plant and animal matter can release

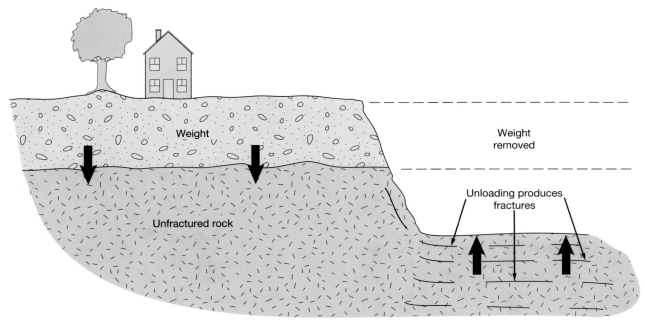

FIGURE 4.6 The weight of overlying material causes pressure on rocks in the Earth. If that pressure is released (for example, if streams carry the overlying rock and soil away, or if people remove it with bulldozers), then the rock expands slightly. This process, called unloading, makes the rock fracture.

100 times as much CO_2 in soil as there is in the air above. When CO_2 is combined with water, it forms H_2CO_3, which is a weak acid. We can write an equation to show this step and a second one that follows as H_2CO_3 separates into hydrogen ions (H^+) and bicarbonate ions (HCO_3^-):

$$H_2O + CO_2 \rightarrow H_2CO_3 \rightarrow H^+ + HCO_3^-.$$

Hydrogen ions are the key product of this reaction because they may go on to decompose minerals.[1] Calcite ($CaCO_3$), for example, reacts with H^+ to leave nothing but soluble calcium ions and bicarbonate:

$$H^+ + CaCO_3 \rightarrow Ca^{2+} + HCO_3^-.$$

Where rocks containing calcite are exposed to acidic rainwater or buried in soil that contains a lot of water and CO_2, they gradually become corroded and eventually dissolve completely.

Hydrolysis Some weathering reactions involve the structural addition of hydrogen ions to a solid, replacing other ions in it and creating a new residual product. Water is also important in these reactions, known collectively as **hydrolysis** reactions. A good example is the action of water and H^+ on potassium feldspar (orthoclase) to form a clay mineral plus dissolved silica (H_4SiO_4) and potassium ions (K^+)

$$2KAlSi_3O_8 + 8H_2O + 2H^+ \rightarrow 2K^+ + Al_2Si_2O_5(OH)_4 + 4H_4SiO_4.$$

FIGURE 4.7 A tree root growing in a crack gradually grows and exerts pressure on the surrounding rock, thus widening the crack. This kind of disintegration is familiar to homeowners who have seen foundations, walls, or driveways damaged by growing root systems.

---

[1]As a result of this reaction, rainwater is always slightly acidic. Reactions with other gases such as SO_2 and HCl also release H^+ and thus make rainwater even more acidic. Some of these gases come from volcanoes, but many are produced in automobile exhaust or in smokestacks. Even when the acid produced by these gases is included, however, rainwater is not as corrosive as the acidic water in many soils.

Oxidation   Some reactions involve oxygen. For example, in the **oxidation** of such iron-bearing minerals as olivine

$$6H_2O + 2Fe_2SiO_4 + O_2 \rightarrow 4FeO(OH) + 2H_4SiO_4$$

oxygen reacts with, or "rusts," the iron to form a new residual mineral. In this particular example, the new mineral is goethite (FeO(OH)), which is one of the minerals that gives soils a yellow or orange color. Goethite may later **dehydrate** (lose water) to form hematite (Fe_2O_3), which makes soils red:

$$2FeO(OH) \rightarrow Fe_2O_3 + H_2O.$$

Agents of Decomposition   These chemical weathering processes involve common agents of weathering: *water, oxygen,* and *acid* (H^+). These may be found anywhere on the Earth, although not always in the same abundance. Climate plays a key role in determining the rate of decomposition and the relative importance of these agents. Moisture is most abundant in coastal regions and in the areas near the equator and the mid-latitudes. Vegetation is most abundant in these same areas. This is important because biological processes affect the balance of oxygen, carbon dioxide, and certain acids that enter into chemical reactions with earthen materials. By contrast, subtropical climates and the interiors of continents at any latitude tend to be dry, and the rate of decomposition is correspondingly low there.

Temperature is the other important climatic factor. Chemical reactions are always more rapid when the climate is warmer.

The interrelation of climatic factors and decomposition is shown in Figure 4.8, which shows how precipitation, temperature, and amount of vegetation vary from the equator to either the North or South Pole. Figure 4.8 shows the relative depth of chemical weathering. Decomposition is most pronounced in the equatorial zone, where the factors of precipitation, temperature, and vegetation reach a maximum. Weathering is least in the desert and semidesert areas of the subtropics and in the far north. A secondary zone of maximum decomposition exists in the zone of temperate climates. Here both the precipitation and the vegetation reach secondary maxima.

Combined Effects of Disintegration and Decomposition

Particle Size   The size of the individual particles of rock being weathered is an extremely important factor in chemical weathering because substances can react chemically only where they come into contact

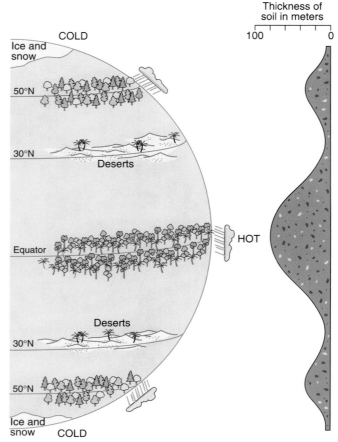

FIGURE 4.8   Decomposition is most intense near the equator, where the average daily temperature and rainfall are both high. Abundant vegetation in the tropics also provides acids that promote decomposition. In subtropical latitudes, where major deserts exist because of persistent dry weather, and in the cold polar regions, where chemical reactions are very slow, decomposition is least intense. There is also less vegetation in these regions. The temperate mid-latitudes are warm and humid enough to support abundant vegetation and promote decomposition. The average thickness of soil reflects these differences.

with one another. The greater the surface area of a particle, relative to its volume, the more vulnerable it is to chemical attack. If we were to take a pebble, for example, and grind it up into a fine powder, the total surface area exposed would be greatly increased. As a result, the materials that make up the pebble would undergo more rapid chemical weathering.

Figure 4.9 shows how the surface area of a 1-cm cube increases as we cut it into smaller and smaller cubes. The initial cube has a surface area of 6 cm² and a volume of 1 cm³. If we divide the cube into smaller cubes, each 0.5 cm on a side, the total surface area increases to 12 cm², although, of course, the total volume remains the same. Further subdivision into 0.25-cm cubes increases the surface area to 24 cm². And if we divide the original cube into units 0.125 cm on a side, the surface area increases to 48 cm². As we have seen, mechanical weathering performs this same

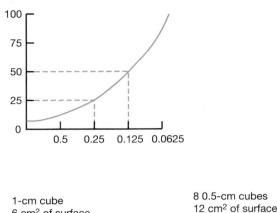

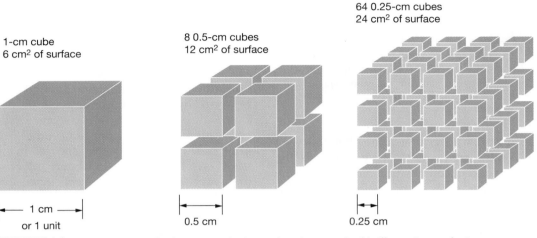

1-cm cube
6 cm² of surface

8 0.5-cm cubes
12 cm² of surface

64 0.25-cm cubes
24 cm² of surface

|← 1 cm →|
or 1 unit

0.5 cm

0.25 cm

FIGURE 4.9 Relationships of volume, particle size, and surface area. In this illustration a cube 1 cm on a side is divided into smaller and smaller units. The volume remains unchanged, but as the particle size decreases, the surface area increases. Because chemical weathering is confined to surfaces, the more finely a given volume of material is divided, the greater is the surface area exposed to chemical activity and the more rapid is the process of chemical weathering.

process by reducing the size of the individual particles of rock, increasing the surface area exposed, and thus promoting more rapid chemical weathering.

Spheroidal Weathering The boulders in Figure 4.10 have been rounded by flaking off a series of concentric shells of rock. They look like products of exfoliation, but here the shells develop from pressures set up within the rock by chemical weathering rather than unloading. When certain minerals are chemically weathered by hydrolysis or oxidation, the resulting products occupy a greater volume than the original material, and this increase in volume creates the pressures responsible for spheroidal weathering. When water and acids penetrate a granite, for example, the feldspar hydrolyzes to form clay, which expands and weakens the rock. Edges and corners of a boulder are most vulnerable because they can be attacked by weathering agents from more than one direction. As a result, angular pieces of rock are gradually smoothed and rounded until they resemble the boulders in Figure 4.10.

Not all spheroidal weathering looks like Figure 4.10. If the cohesive strength of the rock is low, as it might be if the physical processes of disintegration

have produced joints or loosened mineral grains, the rock simply crumbles away. The exposure in Figure 4.11 is in a roadcut (a section of rock cut away to make room for a road) where granite may have been weak-

FIGURE 4.10 Spheroidally weathered granitic boulders are almost completely separated from the bedrock on the eastern side of the Sierra Nevada.

FIGURE 4.11   Granite exposed in a roadcut in northern California has weathered from large fractures to form rounded residual masses. This type of spheroidal weathering has proceeded by loosening one grain at a time (granular or sugary weathering), rather than releasing a rock shell.

ened by blasting or by the unloading that followed excavation. Weathering agents have entered the largest joints first and decomposition has rounded the blocks between joints. Instead of spalling off shells, however, the rock has followed a **granular** or **sugary** style of decomposition, loosening one mineral grain at a time.

Weathering of Common Rocks and Minerals

In Chapter 2 (Figure 2.17) we showed that the most common minerals in the Earth's crust are quartz, the feldspars, a handful of ferromagnesian minerals, and the clays. At the Earth's surface, these can be divided into two informal groups on the basis of their resistance to weathering. The feldspars and ferromagnesian minerals, on the one hand, weather so readily that we usually find them only in freshly exposed igneous or metamorphic rocks. These are referred to as the **vulnerable minerals.** Quartz and the clays, on the other hand, resist weathering or are themselves the products of decomposition, and are referred to as the **resistant** or **residual minerals.** They are the major components of the blanket of soil covering the Earth's surface and of the sedimentary rocks we will discuss in Chapter 5.

Vulnerable Minerals

Feldspars   We have already used the decomposition of potassium feldspar (orthoclase) as an example of hydrolysis, by which water and acid (hydrogen ions) alter a mineral. In fact, the plagioclase feldspars

are also decomposed by hydrolysis. Potassium, sodium, and calcium ions are all highly soluble in water and are easy to coax out of the feldspar crystal structures under mildly acidic conditions. They each also have a moderately good cleavage; thus they disintegrate easily to expose fresh mineral surfaces to chemical weathering.

Because the feldspars comprise about half of the mineral bulk of the Earth's crust, it is hardly surprising that clays, their direct weathering products, are the most common minerals in soil. We will look at clays more closely later in this section. The process by which they are formed, however, is an interesting one to examine now.

Aluminum silicate, derived from the chemical breakdown of original feldspars, combines with water to form hydrous aluminum silicate (Figure 4.12). Doing this requires a total reconstruction of bonds between silica tetrahedra and other ions or ionic groups. The feldspars, as we showed in Chapter 2, are based on a three-dimensional network of silica tetrahedra, whereas the clays are sheet silicates, related to the micas. Disassembling the network leaves some of the silica free to dissolve in water as H_4SiO_4 and sets potassium, sodium, and calcium free as well. Aluminum, which is far less soluble, stays behind with most of the silica.

New clay minerals develop from a suspension of aluminum silicate, sometimes of colloidal size, variously estimated as between 0.2 and 1 micrometer (0.0002 to 0.001 mm) across. Immediately after it is formed, the aluminum silicate may be amorphous; that is, its atoms are not arranged in any orderly pattern. It seems more probable, however, that even at this stage the atoms are arranged according to the definite pattern of a true crystal. In any event, as time passes, the small individual particles join together to form larger crystals, which, when analyzed by such

FIGURE 4.12   The crystal structure of a feldspar is based on a network of silica tetrahedra. As the feldspar decomposes, the network is destroyed and replaced by sheets of silica tetrahedra, which are the foundation of a clay's crystal structure. Ca, Na, and K ions that were once part of the feldspar dissolve in water and wash away, as does a small amount of silica.

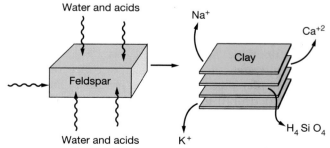

means as X-ray diffraction, exhibit the crystalline pattern of true minerals.

We might expect that the potassium, sodium, or calcium ions released from feldspars would be carried off by water percolating through the ground and eventually find their way to the rivers and finally to the sea. That *is* what happens to the sodium and at least some of the calcium, yet analyses show that not nearly so much potassium is present in river and ocean water as we should expect. Much of it is used by growing plants before it can be carried away in solution, and some of it is absorbed by clay minerals. In some climates, calcium combines with dissolved CO_2 and precipitates as calcite. We will study this more closely when we return to look at how soils form, near the end of this chapter.

Ferromagnesian Minerals   Most of the ferromagnesian minerals have a pronounced cleavage, which gives water, acids, and oxygen easy access to fresh mineral surfaces. Like the feldspars, they are also readily attacked by chemical agents. The decomposition of pyroxenes, amphiboles, olivine, and biotite produces the same products as the weathering of the feldspars: clay, soluble ions, and dissolved silica (H_4SiO_4). The presence of iron and magnesium in the ferromagnesian minerals makes possible certain other products as well.

Iron may be incorporated into one of the clay minerals or into an iron carbonate mineral like *siderite* ($FeCO_3$). Usually, however, it unites with oxygen to form an iron oxide or with water to form a hydroxide. Many oxides and hydroxides are possible; the one that actually forms depends on climate and on other chemical conditions in the weathering environment. Hematite (Fe_2O_3) is one of the most common iron oxides. It commonly has a deep red color, and in powdered form is always red; this characteristic gives it its name, from the Greek *haimatites*, "bloodlike." As we indicated earlier, hematite may combine with water to form goethite ($FeO(OH)$), which is generally yellow-brown to orange in color. (Goethite was named after the German poet Goethe, who had lively scientific interests.) Another possible mineral product is limonite, which is yellow to brown in color and is referred to in everyday language as just plain "rust." Limonite is not a true mineral because its composition is not fixed within narrow limits, but the name is universally applied to the iron oxides of uncertain composition that contain a variable amount of water.

What happens to the magnesium produced by the weathering of the ferromagnesian minerals? Some of it may be removed in solution, but most of it tends to stay behind as a constituent of newly formed minerals, particularly in clays.

Resistant or Residual Minerals

Clays   The clay minerals are sheet silicates like the micas. There are many different minerals in the family, and each has its own chemical behavior, physical structure, and evolution. Most fall into one of three major groups: *kaolinite, montmorillonite* (or *smectite*), and *illite.* Kaolinite is named for the Chinese *Kao-ling* ("High Hill"), the name of the mountain from which kaolinite was first shipped to Europe for ceramic uses. The mineral montmorillonite was first described from samples collected near Montmorillon, a town in west-central France. And the name illite was selected by geologists of the Illinois Geological Survey in honor of their state. Like the mica shown in Figure 2.23, the clay minerals are built up of silica tetrahedra linked together in sheets. These sheets combine in different ways with sheets composed of aluminum atoms and hydroxyl ions (OH^-). (For this reason we refer to the clay minerals as hydrous aluminum silicates.) In addition, montmorillonite may contain magnesium and some sodium and calcium, and illite contains potassium, occasionally with some magnesium and iron. Some clay minerals alter to become other clay minerals. For example, under deep burial, illite is produced from other clays.

Many factors interact to determine which clay minerals will form when a feldspar is weathered. Climate is important, for geologists know that kaolinite tends to form as a result of the intense chemical weathering in warm, humid climates and that illite and montmorillonite seem to develop more commonly in cooler, drier climates. The dominant clay mineral in a region may also depend on the composition of subsurface water. If iron or magnesium ions are abundant because many ferromagnesian minerals have decomposed, an illite or montmorillonite is more likely to form than a kaolinite.

Clays may continue to change once they form, but usually they do not because they are *already* weathered. For example, when a soil forms from a sedimentary rock made of clay, we often find that the soil contains the same type of clay as the parent rock.

Quartz   Weathering affects quartz very slowly. When a rock like granite decomposes, a great deal of unaltered quartz is left behind. The quartz grains found in the weathered debris of granite are the same as those that appeared in the unweathered granite. When these grains are first exposed, they are sharp and angular. Because quartz is a hard mineral and has no cleavage, it resists disintegration and survives as the major component in sand. Abrasion gradually wears off rough edges, and even quartz responds slowly to chemical weathering, so the grains become more rounded as time passes.

Summary of Weathering Products

If geologists know the mineral composition of a rock, they can determine in a general way the products that weathering of that rock will yield. Conversely, when they examine weathering products in a soil, they can sometimes figure out what kind of rock those products have come from, although that is not as easy. The chemical weathering products of common rock-forming minerals are listed in Table 4.1. These products and a handful of physically resistant minerals like quartz make up most of our sedimentary rocks, and we will discuss them again in Chapter 5.

The reactions listed in Table 4.1 are assumed to continue until the original minerals are completely weathered. Because these reactions take much time, that often does not happen. Erosion commonly removes rock debris before minerals in it have fully decomposed. In fact, the debris may even be incorporated into sedimentary rocks before weathering processes have gone to completion.

In summary, we can make the following general observations:

- Chemical weathering commonly yields both dissolved constituents (potassium, sodium, silicon, magnesium, and calcium ions) and residual solids.
- The only common mineral that resists both disintegration and decomposition is quartz.
- Any silicate mineral that contains aluminum will decompose to yield a clay. The most common of these parent minerals are the feldspars.

- Iron in a mineral will oxidize to form iron oxides and may combine with water to form iron hydroxides.

The Progress of Weathering

▲▲

Some rocks weather very rapidly and others only slowly. The rate of weathering is governed by the type of rock and a variety of other factors, from minerals and moisture, temperature and topography, to plant and animal activity.

Depth of Weathering

Most weathering takes place in the upper few meters or tens of meters of the Earth's crust, where rock is in closest contact with air, moisture, and organic matter. We have already shown in Figure 4.8 how the typical extent of chemical weathering varies with latitude. Factors operating well below the surface of the Earth, however, may permit weathering to penetrate to much greater depths. For instance, when erosion strips away great quantities of material from the surface, the underlying rocks are free to expand. As a result, fractures—the joints that we spoke of earlier in the chapter—can develop hundreds of meters below the surface. Large volumes of water may move through these fractures, transforming some of the materials there long before they are ever exposed at the surface. Rock salt, a sedimentary rock composed of halite located far below the surface, often undergoes exactly this transformation. If a large quantity of underground

TABLE 4.1 Decomposition Products of Common Rock-Forming Silicate Minerals

COMMON ROCK-FORMING MINERALS		IMPORTANT DECOMPOSITION PRODUCTS	
MINERAL	COMPOSITION	MINERALS	DISSOLVED IN WATER
Quartz	SiO_2	Quartz grains	Minor H_4SiO_4
Feldspars:			
Orthoclase	$KAlSi_3O_8$	Clay	K^+, Some H_4SiO_4
Albite (Na plagioclase)	$NaAlSi_3O_8$	Clay	Na^+, Some H_4SiO_4
Anorthite (Ca plagioclase)	$CaAl_2Si_2O_8$	Clay	Ca^{+2}, Some H_4SiO_4
Ferromagnesians:			
Biotite	Fe, Mg, Ca, Al	Clay, hematite,	Ca^{+2}, Mg^{+2}, Some
Pyroxene	silicates	limonite	H_4SiO_4
Amphibole			
Olivine	$(Fe,Mg)_2SiO_4$	Hematite, limonite	Mg^{+2}, Some H_4SiO_4
Muscovite	$KAl_3Si_3O_{10}(OH)_2$	Muscovite flakes, clay	K^+, Some H_4SiO_4
Clays	Hydrated Al silicates	Clay, bauxite (little or no weathering)	Some H_4SiO_4
Calcite	$CaCO_3$	None	Ca^{+2}, HCO_3^-

water is circulating through, the salt is dissolved and carried off long before erosion can expose it.

Ease of Mineral Weathering

Field observations and laboratory experiments make it possible for geologists to arrange the minerals commonly found in igneous rocks according to the ease with which they are chemically decomposed at the Earth's surface. Geologists make the following general observations:

- Quartz is highly resistant to chemical weathering.
- The plagioclase feldspars decompose more rapidly than does orthoclase feldspar.
- Calcium plagioclase (anorthite) tends to decompose more rapidly than sodium plagioclase (albite).
- Olivine is less resistant than augite (pyroxene), and in many instances augite seems to decompose more rapidly than hornblende (amphibole).
- Biotite mica decomposes more slowly than do the other ferromagnesian minerals, and muscovite mica is more resistant to decomposition than is biotite.

Notice that these observations suggest a pattern (Figure 4.13a) nearly like Bowen's reaction series for crystallization from magma (Figure 4.13b), which we discussed in Chapter 3. Weathering and crystallization, of course, are very different processes. In decomposition, for example, the successive minerals formed do not react with one another or with a parent magma as they do in a reaction series, so nothing analogous to fractional crystallization can happen. Still, the fact that

minerals appear in the same order in these two diagrams is a tantalizing clue that mineral formation and mineral weathering are related.

The similarity between diagrams a and b in Figure 4.13 suggests that minerals that form at low temperatures resist decomposition best; minerals that form at high temperatures are least stable during weathering. In general, minerals decompose most readily if conditions at the Earth's surface during weathering are very different from the conditions under which they formed. Olivine, for example, forms at a high temperature (and commonly at a high pressure), early in the crystallization of a magma. Consequently, as we might expect, it is extremely unstable under the low temperatures and pressures that prevail at the Earth's surface, so it decomposes quite rapidly. Conversely, quartz forms late in the reaction series, at a considerably lower temperature and pressure. Because these conditions are more similar to those at the surface, quartz is relatively stable and is very resistant to decomposition.

This is a rather simplistic way to look at the relative stabilities of minerals during weathering because there are many exceptions to the tendency we have just described. A mineral that resists disintegration, as we discussed earlier, also decomposes slowly. Garnets, for example, form at fairly high temperatures, and yet they are among the most resistant minerals commonly found in beach sands because they are difficult to scratch or break. Other minerals, such as corundum, are not only extremely hard but nearly insoluble, so they remain unaltered even when

FIGURE 4.13 Minerals toward the top of diagram (a) weather more readily than those near the bottom. Note that these minerals are in the same order as they appear in Bowen's reaction series (b). These diagrams suggest that minerals decompose more readily when the weathering environment is very different from the environment in which they formed.

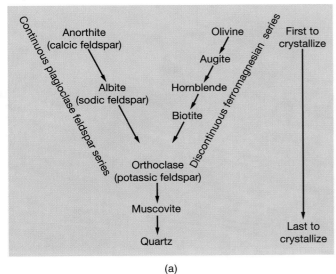

(a)

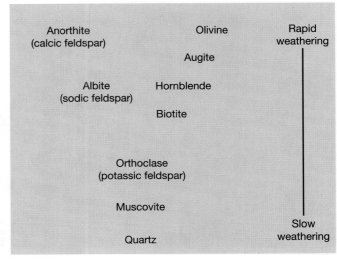

(b)

Wearing Down the Land

Weathering alters and loosens material at the Earth's surface. Agents of erosion carry that material from one place to another, eventually depositing it where it can be reassembled to form sedimentary rock. These subsequent steps will occupy our attention in Chapter 5 and in each of the chapters in the final third of this book. To get an early perspective on those chapters, this is a good place to ask how the continents as a whole gradually wear down over time. This is a topic that geologists call **denudation.**

Sometimes ancient ruins provide an index to the rates of denudation. The photograph in this box shows the remains of a cistern built 60 km north of Rome, Italy, roughly 1,700 years ago. From the 1.6 m of foundations and underlying volcanic ash now exposed at the base of the finished wall, we estimate that the surrounding land has eroded at an average rate of 94 cm/1,000 years since the structure was built.

This, however, is only a spot measurement at a specific place. What method can we use to measure the rates over larger areas? One way to start is by determining the amount of material carried by a stream each year from its drainage basin (all the area drained by the stream and all its tributaries). This is not easy to do precisely, but we can usually make measurements that are close enough to be useful. The major stream near the Roman cistern is the Tiber River. At Rome, geologists find that it is on average removing about 7.5 million metric tons of solid material each year from the area upstream. This sounds like a lot, but it amounts to eroding a layer 0.17 mm thick (about as thick as your thumbnail) from the Roman countryside each year, or 17 cm/1,000 years—much less than the rate we estimated at the cistern.*

How well do either of these numbers compare to global averages? Measurements in the United

P4.1.1   Ruins of a cistern built in the second century A.D. on a Roman farm located 60 km north of Rome. The exposed footings of the building measure 1.3 m, as indicated by the staff. The volcanic ash at the bottom of the building measures 30 cm. Therefore, the amount of denudation since the cistern was built is 1.6 m.

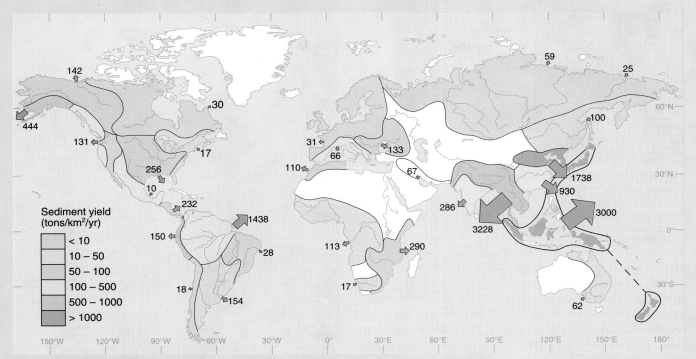

P4.1.2 Annual discharge of suspended sediment from world drainage basins to the ocean (in millions of metric tons per year), as indicated by arrows. The various colors indicate relative rates of erosion (in metric tons per square kilometer per year) from the continents. Pale-yellow areas discharge only negligible amounts of sediment to the oceans. *(After J.D. Milliman and R. H. Meade, "World-wide Delivery of River Sediment to the Oceans,"* Journal of Geology, *vol. 91, p. 16 (1983) Used with permission of The University of Chicago Press.)*

States indicate that denudation rates vary from region to region but average approximately 6 cm/1,000 years. Data from the Amazon River, the world's largest river, indicate that it is removing material from its basin at the rate of 4.7 cm/1,000 years. Another large tropical river, the Congo, is carrying enough material out of its basin each year to reduce it by approximately 2 cm/1,000 years. The adjacent map gathers information from each of the world's large river systems in order to make an estimate for the Earth as a whole. As you can see, the amount of sediment leaving the continents varies dramatically from one place to another, with the greatest being in Southeast Asia, where tropical climates accelerate weathering and where large rivers drain the high Himalayas. Altogether, one estimate places the global sediment load at 24 billion metric tons per year. If correct, that is equivalent to roughly 6.2 cm of denudation per 1,000 years.

Why are the rates that we estimated for the area around the Roman cistern so much greater than these average continental values or even the regional estimate in central Italy? When people occupy an area intensively and turn it to cropland, they increase the erosion rate 10 to 100 times over that in a naturally forested or grassed area. (In fact, before human civilization, the average continental denudation rate was probably no higher than 2.5 cm/1,000 years.) Destroying the natural cover exposes soil to much more rapid erosion and removal by running water, wind, and animals. The fields around the Roman cistern, which have been farmed intensely for thousands of years, are just such an area. We speculate that this is why denudation has been so rapid.

*Here's how we estimated: The average density of river sediment is about 2.6 g/cm^3. The Tiber River drains an area of roughly 17,000 km^2. Express all quantities in grams and centimeters instead of kilograms and kilometers. Divide the mass of eroded sediment by its density and by the drainage area, and the answer is: 7.5×10^{12} g ÷ (170×10^{12} cm^2 × 2.6 g/cm^3) = 0.017 cm = 0.17 mm.

FIGURE 4.14 These adjacent blocks of granite and basalt have been exposed to identical weathering conditions in the mountains above Hell's Canyon, Idaho, yet the granite has decomposed much more than the basalt. Among other factors, the relatively larger mineral grains in the granite may be more vulnerable than grains of the same minerals in the basalt.

exposed to conditions quite different from the ones under which they were formed. Finally, as Figure 4.14 suggests, the average crystal size and physical integrity of a rock may affect its susceptibility to weathering, regardless of how stable its individual minerals may be. Still, the trend suggested by Figure 4.13 is a useful guide in a great many cases.

Now we can offer an alternative version of the definition of weathering given at the beginning of this chapter: *Weathering is a collection of destructive processes by which earth materials change when exposed to conditions at or near the Earth's surface that are different from the ones under which they formed.*

Graveyards provide many fine examples of weathering within historic time and illustrate how much more readily some rocks weather than others. Marble is a metamorphic rock derived from calcite-rich sediments that formed in the ocean (Chapter 6). Calcite in marble headstones weathers easily because the acidic nature of rainwater is very different from the nonacidic nature of the seawater from which the calcite formed originally. The marble headstone pictured in Figure 4.15 has weathered so rapidly that the inscription carved in 1796 was only partially legible some two centuries later.

The slate headstone shown in Figure 4.16 provides a contrast to the strongly weathered marble one in Figure 4.15. Although it too was carved nearly two centuries ago, the inscription and delicate decoration on the tombstone are still plainly visible. Slate (Chapter 6) is a metamorphic rock composed largely of clay minerals, which form in contact with acidic water and thus are not affected by more of the same. Metamorphism converts some of the clay minerals to muscovite, but muscovite also is highly resistant to weathering, as indicated in Figure 4.13.

FIGURE 4.15 Weathering of a marble headstone, Madison Cemetery, Connecticut. The monument illustrates the instability and rapid weathering of calcite (the predominant mineral in marble) in a humid climate (Photographed in 1980.)

Differential Weathering

Rock masses do not weather uniformly, nor should we expect them to. Within even a small area, the spacing of fractures, the average size or composition of mineral grains, or any number of other factors may vary. For that reason, rock develops an irregular surface as it decomposes. We refer to this phenomenon as **differential weathering**. The results vary in scale from the slightly uneven surface of the marble tombstone in Figure 4.15 to the boldly sculptured forms of the Grand Canyon or Monument Valley (Figure 4.17). Unequal rates of weathering are chiefly caused by variations in the composition of the rock. The more resistant rocks stand out as ridges, ribs, or pinnacles above the more rapidly weathered rock on either side (Figures 4.18 and 4.19), or as the steeper slopes on an irregular hillside (Figure 4.20). A resistant caprock may protect much less resistant rocks from being removed. This may produce features called **hoodoos** (Figure 4.21).

FIGURE 4.16   Weathering of a slate headstone in the old bury-ing ground at Braintree, Vermont. The clays and muscovite in slate resist weathering much more effectively than does the calcite in the marble headstone of Figure 4.15. (Photographed in 1976.)

A second cause of differential weathering is sim-ply that the intensity of weathering varies from one section to another in the same rock. This may result

FIGURE 4.17   Differential weathering of sandstone and shale sequence in Monument Valley, Utah. (Catherine K. Richardson)

from differences in the amount of sunlight hitting the rock, causing variation in the effect of moisture and frost action on parts of the rock exposure, or from dif-ferences in the amount of acid available to decompose the rock. Differential weathering is particularly appar-ent if we examine a rock that has been partly buried so that the portion below ground has been exposed to the high CO_2 concentrations brought about by plant decay.

On a larger scale, differential weathering affects the shape and overall look of the land, as shown in the cross section in Figure 4.22 and the accompanying satellite photo in Figure 4.23. The section is drawn across a portion of southeastern Pennsylvania. The highest ridges are underlain by conglomerate or sand-stone, rocks that weather very slowly. Most of the valleys are underlain by shale, a sedimentary rock that, although resistant to chemical weathering because it is made of clay minerals, is so soft that it erodes easily. In these same valleys are still lower spots underlain by a dolomitic limestone, which is sus-ceptible to rapid chemical weathering in this humid

FIGURE 4.18   Differential weathering of sandstones in Bryce Canyon National Park, Utah. (Catherine K. Richardson)

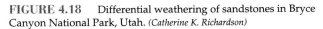

FIGURE 4.19   Differential weathering in volcanic ash at Crater Lake National Park, Oregon. (Catherine K. Richardson)

FIGURE 4.20   The steepest slopes in the Grand Canyon, Arizona, are in sandstone, which is relatively resistant to weathering. Gentler slopes form where shales (made of clay) are exposed, because they are more vulnerable to disintegration.

FIGURE 4.21   "Hoodoos" produced by differential weathering in the San Juan Formation west of Regina, New Mexico. Resistant sandstone forms the cap, and less resistant claystone forms the pedestal beneath.

climate. To the east lie the New Jersey Highlands, held up by resistant units of both igneous and metamorphic rocks.

Soils

Solid rock or **bedrock,** as a geologist would call it, underlies all parts of the land. In some places it is exposed at the surface, but in other areas it is covered by **soil,** the surface accumulation of weathering products and decaying plant material (called **humus**) that sustains plant life. **Residual** soils, which we discuss in the remainder of this chapter, develop in place on the bedrock from which they are derived. **Transported** soils, including those derived from glacial or stream action, for example, lie on a bedrock other than the one from which they formed. We will discuss various transported soils in Chapters 13–18.

We can understand a farmer's interest in soil, but why is it important for a geologist to understand soils and the processes by which they are formed? There are several reasons. First, soils provide clues to the environment in which they were originally formed. By analyzing an ancient soil buried in the rock record, we may be able to determine the climate and physical conditions that prevailed when it was formed. Second, some soils are sources of valuable mineral deposits (see Chapter 19), and the weathering process commonly enriches otherwise low-grade mineral deposits, making them profitable to mine. An understanding of soils and soil-forming processes, therefore, can serve as a guide in the search for ores. Third, because a soil reflects to some degree the nature of the rock material from which it has developed, we can sometimes determine the nature of the underlying rock by performing an analysis of the soil.

Most important of all for geologists, soils are the source of many of the sediments that are eventually converted into sedimentary rocks, which we will begin to look at in Chapter 5. These in turn may be transformed into metamorphic rocks or, following another path in the rock system, may be converted into new soils. If we understand the processes and results of soil formation, we are in a better position to interpret the origin and evolution of many rock types.

FIGURE 4.22   Geologic cross section in southeastern Pennsylvania, between Harrisburg and the Delaware River, showing the relationship of topography to the different rock types below the surface.

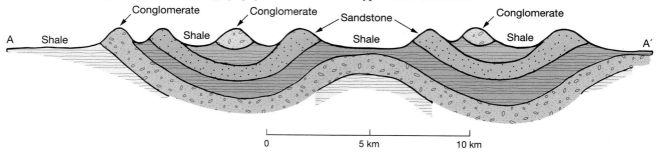

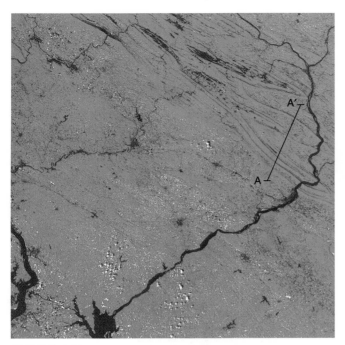

FIGURE 4.23　Satellite photo of the area illustrated in Figure 4.22. This image was recorded with infrared-sensitive film, on which vegetation appears red. *(NASA)*

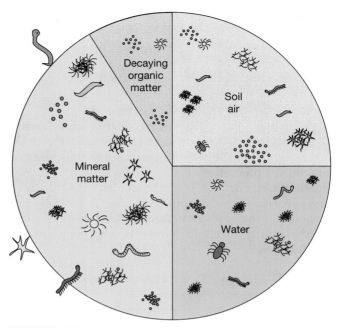

FIGURE 4.24　Plants grow best in a soil that is about half solid material and half pore space. Most of the solid materials are minerals; a small portion consists of decaying organic matter. The pore spaces are half-filled with water and half with soil air, which can contain abundant CO_2. These proportions vary considerably from one soil to another. In addition to these five components, all soils also contain a thriving community of living organisms.

What Is Soil?

Soil is a complex mixture consisting of five major components: mineral matter, living organisms, decaying organic matter, water, and soil air (Figure 4.24). It is a dynamic material from both a geological and biological point of view, one in which the inorganic processes of mineral weathering are balanced against the forces that sustain the richest community of living things on the planet. It is distinct in that regard from regolith, which we defined earlier in this chapter as a layer of loose mineral debris formed by disintegration alone. This means that Earth is alone among the planets in our solar system in having a true soil, since it is the only planet on which life is known to exist. In addition to plant roots, earthworms, insects, and other visible life forms, soil is home to a teeming variety of microorganisms: A single ounce (28 g) may contain over 100,000,000,000. Together, the inhabitants of soil not only churn and aerate it, but they yield waste products that nourish larger plants and promote chemical weathering.

In the early years of soil study in the United States, researchers thought that the parent material almost wholly determined the type of soil that would result from it. Thus, they reasoned, granite would weather to one type of soil and limestone to another.

It is true that a residual soil reflects to some degree the material from which it developed, and in some instances we can even map the distribution of underlying rocks on the basis of the types of residual soil that lie above them. Bedrock, however, is not the only factor determining soil type. In fact, it is not even the most important factor. Russian soil scientists, following the pioneering work of V.V. Dokuchaev, have demonstrated that different soils develop over identical bedrock material in different areas when the climate varies from one area to another.

The idea that climate exerts a major control over soil formation was introduced into North America in the 1920s by C.F. Marbut, for many years chief of the United States Soil Survey in the Department of Agriculture. Since that time soil scientists have discovered that still other factors exercise important influences on soil development. For instance, the height and shape of the land surface plays a significant role. The soil on the crest of a hill is somewhat different from the soil on the slope, which in turn differs from the soil on the level ground at the foot of the hill; yet all three soils rest on identical bedrock. The passage of time is another factor that affects soil: A soil that has only begun to form differs from one that has been developing for thousands of years, although the climate, bedrock, and topography are the same in each instance. Finally, the vegetation in an area influences the type of soil that develops there. One type of soil will form beneath a pine forest, another beneath a forest of oak or maple, and yet another on a grass-covered prairie.

Soil scientists characterize soils in many ways. The **color** of a soil is controlled largely by its chemistry. Iron oxides and hydroxides make a soil red or yellow-brown, respectively. Organic matter turns it black. Where intense weathering has removed all but the clays and quartz, a soil is gray or white. Soil **texture** refers to the mixture of particle sizes, an indirect measure of the type of residual minerals and the extent of mineral disintegration. Soil particles may also clump together. Organic matter or sticky clays bind them in larger masses whose shapes and sizes are described as elements of the soil's **structure.** Other factors, including the **acidity** or **alkalinity** of the soil, may also be worth noting.

Residual Soil Development

The composition of a residual soil varies with depth, usually in **zones** or **horizons,** each recognizably different from the one above. To see how these develop, consider the hypothetical experiment illustrated in Figure 4.25. A cylinder, open at the top and closed with a porous screen at the bottom, is filled with regolith containing the common crustal minerals: feldspars, quartz, and ferromagnesian silicates. On the top is a layer of humus (what a gardener would call "compost"): decaying leaves, grass, and other organic debris, plus the microorganisms to help them decay. To run the experiment, we simply sprinkle a steady supply of water ("rain") on this layer and let it trickle through the regolith and out the bottom of the cylinder.

If we could watch long enough, we would see several things happen, and they would begin at the top of the cylinder, where there is always a fresh supply of water, acids, and oxygen. First, the dissolution, hydrolysis, and oxidation reactions we discussed earlier will start to *decompose* vulnerable minerals. Next, soluble ions (calcium, sodium, magnesium, and potassium) will be *leached* downward, possibly to be washed completely out the base of the cylinder. As fine-grained clay minerals and iron oxides begin to form, they will be *translocated* or flushed down through open spaces in the regolith until they get stuck and start to accumulate in lower, smaller spaces. Finally, especially if the flow of water is too weak to wash all of the dissolved ions out of the cylinder, some of them may recombine with bicarbonate, H_4SiO_4, or other substances and *precipitate* new minerals.

This experiment would take much too long to do in a laboratory, but we can see that it is exactly what happens in nature. As weathering proceeds, the soil gradually develops distinct zones, each of which penetrates farther into the Earth with time (Figure 4.26a,b).

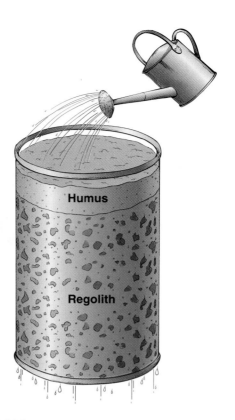

FIGURE 4.25 This hypothetical experiment simulates the development of a residual soil in nature. The cylinder is filled with regolith, covered with a layer of decaying organic matter. As water is continually added at the top, weathering proceeds downward, developing the characteristic A, B, and C horizons.

The *O* Horizon The uppermost zone, perhaps containing little mineral matter at all, is called the *O* horizon. It is the humus layer, which may be a thick blanket of leaf litter or no more than a thin skin of dried sagebrush fragments. This is the most active horizon biologically and one that must be renewed constantly if there is to be a continuous source of the acids required in deeper horizons.

The *A* Horizon The *A* horizon is what most people call the topsoil—the layer into which we sink a spade when we dig a garden, and the one in which seeds germinate. A geologist sometimes calls this the **zone of leaching** because it is the first to be attacked by corrosive waters from above and the first to lose its soluble components. Varying amounts of organic material tend to give the A horizon anywhere from a gray to a black color, but most other materials that might color the soil have been removed. Ferromagnesian minerals are particularly vulnerable, and the iron in them is easily oxidized, to be translocated or precipitated in lower horizons. Feldspars decompose rapidly as well, leaving clays that also translocate downward. The A horizon may ultimately contain little more than quartz in some regions.

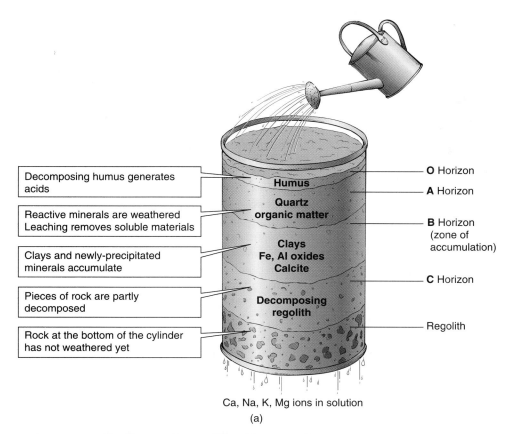

Decomposing humus generates acids

Reactive minerals are weathered
Leaching removes soluble materials

Clays and newly-precipitated minerals accumulate

Pieces of rock are partly decomposed

Rock at the bottom of the cylinder has not weathered yet

Humus

Quartz
organic matter

Clays
Fe, Al oxides
Calcite

Decomposing
regolith

O Horizon
A Horizon

B Horizon
(zone of
accumulation)

C Horizon

Regolith

Ca, Na, K, Mg ions in solution

(a)

(b)

FIGURE 4.26

(a) Decomposition reactions in the A horizon leave only the resistant minerals (mostly quartz) behind. Soluble ions are leached downward and are accompanied by fine particles of clay and other residual minerals, which accumulate in the B horizon. The C horizon contains fragments of the parent rock that have not yet been completely decomposed, as well as new residual minerals. (b) This photograph of an exposed soil profile in Quebec shows the O horizon (black) overlying the A horizon (gray) and the B horizon (red-yellow). The C horizon does not appear in this photo. *(USDA-Soil Conservation Service)*

The *B* Horizon   The *B* horizon, sometimes known as the **zone of accumulation,** lies directly below the A horizon. In moist climates the B horizon contains an accumulation of clays and iron oxides delivered by water percolating downward from the surface. In dry climates we generally find, in addition to the clay and iron oxides, deposits of more soluble minerals, such as calcite. These are precipitated mainly from water carrying dissolved salts downward from the A horizon, but some is brought into the B horizon from below as soil water is drawn upward by high evaporation rates.

In many soils a dense subsurface layer, called a **fragipan,** develops in the B horizon. Its near-impermeability to fluids is due chiefly to its extreme compactness (as in a **claypan**) or to cementation (as in a **hardpan**). When cemented by calcite, it is called **caliche** (a Spanish word derived from the Latin *calix,* "lime") (see Figure 4.27). When cemented by iron oxides, it is called an **ironpan.** In some instances it may be cemented by silica or other materials. The pan may be so hard that it behaves like firm rock.

The *C* Horizon   The *C* horizon is a zone of partially disintegrated and decomposed rock material. Some of the original bedrock minerals are still present, but others have been transformed into new materials. The C horizon grades downward into unweathered material, which may be bedrock or unconsolidated sediments (glacial or stream deposits, for example).

Some Soil Types

▲▲

The pattern of residual soil development just described can be observed in all corners of the Earth. This does not mean, however, that all soils are alike. Because climates, rock types, vegetation, and the availability of weathering agents differ from place to place, soils do as well. In addition, erosion by natural or human processes may disturb or totally remove upper horizons of some soils, or deposition may bury others under less weathered material.

The USDA Soil Classification System

Until the 1960s, soil scientists used a classification system developed by the U.S. Department of Agriculture (USDA) to identify climatic, landform, or vegetative controls on soil formation. Vestiges of the USDA system are still in use today, although more commonly by geologists than by soil scientists. At its most basic level, the USDA system divides soils into two gross categories: the **pedalfers** (iron- and aluminum- accumulating soils) and the **pedocals** (calcium-accumulating soils). Figure 4.28 shows how these two soil types are distributed across the United States. Pedalfers and pedocals are subdivided into great soil groups characteristic of different climatic regions. We will describe only one of these, the **laterites,** which are tropical pedalfers.

Pedalfers   A pedalfer is a soil in which iron oxides, clays, or both have accumulated in the B horizon. The name is derived from *pedon,* Greek for "the ground," and the symbols *Al* and *Fe* for aluminum and iron, respectively. In general, soluble materials such as calcium carbonate or magnesium carbonate do not collect in the pedalfers but remain in groundwater or surface waters, to be completely removed from the soil-forming environment. Pedalfers are commonly found in humid climates, usually beneath a forest vegetation.

There are several varieties of soils in the pedalfer group, including the red and yellow soils of the southeastern states as well as the gray-brown soils of the northeastern quarter of the United States and of southern and eastern Canada (Figure 4.29a,b). Prairie soils

FIGURE 4.27   Caliche (calcite cemented hardpan) in an arid soil. *(William E. Ferguson)*

FIGURE 4.28   In the United States, pedocal soils are found almost exclusively west of the Mississippi River and pedalfers to the east of it. The two major soil groups are divided in their distribution by differences in the amount of rainfall. Laterites do not occur anywhere in the continental U.S., although they are common in Hawaii.

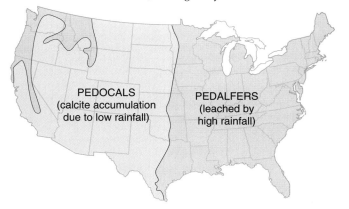

PEDOCALS
(calcite accumulation
due to low rainfall)

PEDALFERS
(leached by
high rainfall)

FIGURE 4.29 The term "pedalfer" describes a wide range of humid climate soils, which vary in color and texture. Both of the above soils are pedalfers. *(John D. Cunningham, Visuals Unlimited)*

are transitional varieties between the pedalfers of the East and the pedocals of the West.

Laterites The name "laterite" is applied to many tropical pedalfers that have a very deep B horizon rich in hydrated aluminum and iron oxides. The name itself, from the Latin for "brick," suggests the characteristic red color produced by the iron in these soils. Laterites are the products of extremely intense weathering, promoted by the year-round high temperatures and continuous rainfall of the tropics. Despite the fact that these are typically jungle soils, they have a very low organic content because the rate of bacterial decay in them is staggering. These soils have inherently low fertility, which is made dramatically obvious when the forest cover is destroyed by logging or ill-advised farming practices. Without a plant community to restore the extremely thin O horizon, organic nutrients are quickly lost, and the soil, unanchored by plant roots, erodes catastrophically (Figure 4.30).

In this extreme weathering environment, even the clays are not stable in many cases. Instead, silica is removed in water as H_4SiO_4, and the aluminum remains in a collection of hydroxide minerals called **bauxite,** an ore of aluminum. In some laterites the concentration of hydrated iron or nickel oxides is so great that it is profitable to mine them as well.

Pedocals Pedocals are soils that contain an accumulation of calcium carbonate. Their name is derived from a combination of *pedon* and the symbol *Ca,* for calcium. The soils of this major group are found where the temperature is relatively high, the rainfall is low, and the vegetation is mostly grass or brush. As

FIGURE 4.30 Deforestation in the tropics can have catastrophic effects on the landscape. Laterites (oxisols) are the least fertile soils in the world. Without a continuous supply of decaying plant matter to replenish their nutrient content, they rapidly lose the ability to support vegetation and are susceptible to rapid erosion in the heavy tropical rains. *(Walt Anderson, Visuals Unlimited)*

What Does a Map Show?

aps are basic tools of geology. Thumb through this book and you will find maps at all scales from the local to the global. There are maps to illustrate stream systems and to show the extent of glaciers during the past 100,000 years; there are maps to show things we can see (like shorelines) and things we cannot (like the distribution of earthquakes). We turn to maps to make sense of puzzling observations in our own backyard or to search for familiar patterns in unfamiliar places. For example, there is more than an even chance that you stopped to find your home region on the soil map that accompanies this short essay before you started reading.

Unfortunately, any map can turn out to be misleading, because a mapmaker makes necessary simplifications by leaving out distracting information or by combining scattered observations to show an average picture. If you were curious enough to wonder what kind of CSCS label fits the soil in your region, you may have found the map below disappointing. It indicates, for example, that the soils in Florida are ultisols, spodosols, and entisols. Look at the other three maps, however, and you will see that there are alfisols, mollisols, and histosols in Florida as well.

As we note in the text, soil types respond not only to regional climate but also to the local landscape, vegetation, and a host of other factors. Within a mile of your house you may find not only the dominant soil type in your area but also an entisol on a nearby floodplain or a histosol around a marshy pond. A map that shows the dominant soil type will not show these. The first map in this box, therefore, is only broadly suggestive of trends across the United States. If you are really curious about the soils in your own area, you should contact the closest office of the USDA Soil Conservation Service or your county extension office. Similar government agencies operate in most countries around the world.

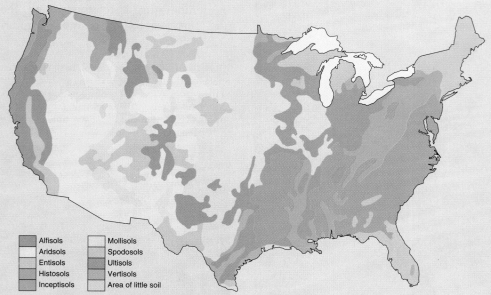

Alfisols
Aridsols
Entisols
Histosols
Inceptisols

Mollisols
Spodosols
Ultisols
Vertisols
Area of little soil

P4.2.1   Generalized soil map of the continental United States, based on the Comprehensive Soil Classification System (CSCS). *(After Tom L. McKnight,* Physical Geography: A Landscape Appreciation, *2nd ed., Englewood Cliffs, NJ: Prentice Hall, 1984. Readapted by permission.)*

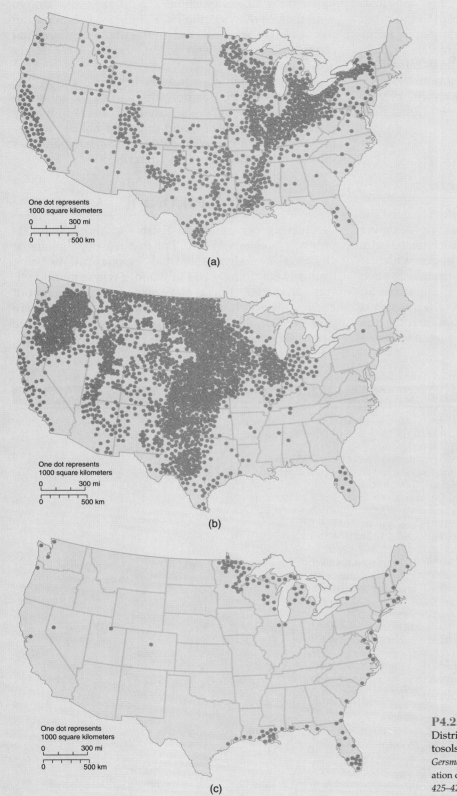

One dot represents
1000 square kilometers

0   300 mi

0   500 km

(a)

One dot represents
1000 square kilometers

0   300 mi

0   500 km

(b)

One dot represents
1000 square kilometers

0   300 mi

0   500 km

(c)

P4.2.2
Distribution of (a) alfisols, (b) mollisols, and (c) histosols in the continental United States. *(After Philip J. Gersmehl, "Soil Taxonomy and Mapping,"* Annals of the Association of American Geographers, *vol. 67 (September 1977), pp. 425–427.)*

described earlier, water evaporates before it can remove carbonates from the soil, leaving them to precipitate in the B horizon, where they often appear as caliche (refer again to Figure 4.27). Except in the driest desert climates, pedocal soils are generally quite fertile because organic nutrients tend to accumulate in them rather than being leached out. The low rainfall in pedocal-forming climates and the common tendency for caliche and other salts like sodium chloride to accumulate in the soil may limit their agricultural potential, however.

The Comprehensive Soil Classification System

During the 1960s, soil scientists gradually replaced the USDA system with another system that was designed to describe the physical and chemical characteristics of a soil rather than the conditions under which it forms. The Comprehensive Soil Classification System (CSCS) divides soils into 10 orders, summarized in Table 4.2. A careful description of each order, which is beyond the scope of this book, would involve the color, mineral and organic content, texture, and struc-

ture of a typical profile through the soil. A map of soil distribution based on the CSCS (see Perspective 4.2) looks more like a patchwork quilt than one drawn with the USDA system in mind, because it depends on the appearance of soil rather than on environmental characteristics that tend to be uniform over larger areas.

Several other classification schemes are in use around the world, but not one is ideal for all situations. Humanity depends heavily on the soil, mainly for agricultural reasons, and it should not be surprising that such a complex natural material challenges our ability to understand and describe it.

▶▶ EPILOGUE ◀◀

Weathering occurs at the Earth's surface, where its progress is clearly visible. For that reason, a geologist studying a modern soil (that is, one that is being formed today) rarely has to wonder where and how it was produced. Rocks have weathered throughout Earth's history, however, and many ancient soils are preserved under the modern ones (Figure 4.31). These **paleosols** (from the Greek root *paleo-*, meaning

TABLE 4.2 Soil Orders of the Comprehensive Soil Classification System

SOIL ORDER	SOIL CHARACTERISTICS	DEGREE OF WEATHERING
Oxisol	The most excessively weathered soils, commonly more than 3 m thick. Low fertility. Red; contains Fe and Al oxides. Tropical jungle soil.	High
Ultisol	Strongly acid, well-weathered soils of tropical and subtropical climates. Reddish to orange, A horizon. Clay-rich, B horizon. Low fertility.	High
Alfisol	Leached gray-brown, A horizon; red-brown, B horizon with Fe oxides and clays. Fertile, mildly acidic soil, typical of deciduous low and mid-latitude forests.	High
Spodosol	Light-colored, heavily leached, sandy A horizon; B horizon is red-brown, rich in Fe and clays. Acidic, fairly infertile soil, typical of cool coniferous forests.	High
Mollisol	Black, organic-rich A horizon; the most fertile soils. Only gently leached. High Ca concentration. Grassland soils.	Moderate
Aridisol	Indistinct horizons, dispersed organic matter, accumulation of Ca, Mg, and K ions. White to gray throughout; salt layers. Formed in dry climates.	Moderate
Inceptisol	Weakly developed horizons showing just the "inception" of soil development. Most common in tundra and mountain environments and on river banks.	Moderate
Vertisol	Abundant shrinking and expanding clays. Wide, deep cracks when dry. Very fertile, but hard to cultivate. Typical of tropical savannas.	Moderate
Entisol	Thin, sandy soils with no recognizable horizons. Formed on many recent river floodplains, volcanic ash deposits, and recent sands.	Low
Histosol	Black, acidic soils composed largely of organic matter, including peats and mucks. Waterlogged soil typical of bogs and swamps.	Little or no mineral matter to be weathered.

After Tom L. McKnight, *Physical Geography: A Landscape Appreciation*, 4th ed. Englewood Cliffs, NJ: Prentice Hall, 1993.

FIGURE 4.31   Buried soil zones in the Hondo Formation, Magdalena Valley, Colombia. Human figures indicate the thickness of these dark soils, which developed on successive units of stream deposits. *(Franklyn B. Van Houten)*

"ancient," and the Latin *solum,* meaning "soil") have been covered by various transported materials (mud or sand deposited by a stream or by wind, for example) and are too far below the present surface of the Earth for weathering agents to penetrate effectively. Such buried soil zones are useful indicators of past periods of erosion and weathering. In fact, some paleosols formed over one billion years ago have provided

invaluable clues to the weathering cycle and to the composition of the atmosphere during that time. For example, a 2.5-billion-year-old paleosol in South Africa contains unweathered pebbles of the mineral uraninite (UO_2), which decomposes very quickly in today's oxygen-rich weathering environment and consequently is never found in a modern soil (Figure 4.32). Geologists infer from observations like this that

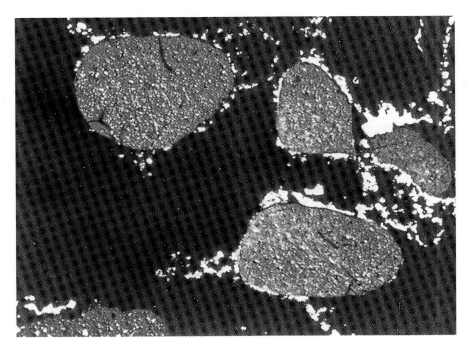

FIGURE 4.32

If these pebbles of uraninite were exposed to weathering today, they would be oxidized quickly. When they were on the Earth's surface 2.5 billion years ago, however, there was almost no oxygen in the atmosphere, so they were not oxidized at all. Geologists studying paleosols have revealed many important clues such as this about the Earth's surface environments in the past.

the atmosphere 2.5 billion years ago contained very little oxygen.

Comparison of paleosols with modern soils is yet another example of the way in which geologists use their understanding of processes occurring today as a guide to interpreting materials that formed in the past.

In Chapter 5 we will show how geologists apply what they have learned about disintegration, decomposition, and the development of soils to understand how sedimentary rocks, composed of residual weathering products, are formed.

SUMMARY

1. Weathering is a collection of destructive processes by which earthen materials change when exposed to conditions at or near the Earth's surface that are different from the ones under which they formed. Disintegration (mechanical or physical weathering) involves a reduction in the size of rock and mineral particles but no change in composition. Impact processes and frost action are the major causes of disintegration. Decomposition (chemical weathering) involves a change in the composition of mineral matter. The agents of decomposition are water, acids, and oxygen.

2. Vulnerable minerals, including the feldspars, calcite, and most ferromagnesian minerals, disintegrate readily because of cleavage, and decompose by dissolution, hydrolysis, or oxidation. Their weathering products are clays, oxide and hydroxide minerals, and various ions dissolved in water. The solid weathering products, plus minerals such as quartz that resist weathering, are called residual minerals.

3. Minerals weather most readily when the weathering environment is very different from the one in which they formed. The rate of chemical weathering increases as the size of particles decreases and as moisture and temperature increase. Differential weathering may be due to variations in the vulnerability of different minerals to weathering, or to fluctuations in the abundance of weathering agents or the intensity of disintegration.

4. Soil is a mixture of mineral matter, living organisms, decaying organic matter, water, and soil air, and it is also the product of both biological and geological activity. Residual soils develop zones or horizons as weathering agents percolate downward, leaching soluble elements and translocating fine particles of clay, iron oxides, or other mineral matter. The USDA Soil Classification System characterizes soils in terms of the environmental conditions under which they form. Pedalfers form in moist climates and pedocals in dry climates. Laterites are pedalfers typical of tropical climates. The Comprehensive Soil Classification System is based on the chemical and physical characteristics of soils, rather than how they form.

KEY WORDS AND CONCEPTS

bauxite 97
bedrock 92
caliche 96
decomposition (chemical weathering) 78
dehydration 82
denudation 88
differential weathering 90
disintegration (mechanical weathering) 78
dissolution 80
exfoliation 80
exfoliation dome 80

fragipan (claypan, hardpan, ironpan) 96
frost wedging 79
gardening 78
hoodoo 90
humus 92
hydrolysis 81
laterite 96
oxidation 81
paleosol 100
pedalfer 96
pedocal 96
regolith 78

residual (resistant) mineral 84
residual soil 92
soil horizon 94
soil structure 94
soil texture 94
transported soil 92
unloading 80
vulnerable mineral 84
weathering 78
zone of accumulation 96
zone of leaching 94

4.1 What processes are usually responsible for the disintegration of rock? Why is water particularly important in the physical weathering process?

4.2 How do geologists theorize that exfoliation happens? How is exfoliation different from spheroidal weathering?

4.3 What are the general types of chemical processes by which rock decomposes? What chemical agents are involved in these processes?

4.4 How does the intensity of weathering vary from pole to equator? What factors are responsible for those variations?

4.5 Why does decomposition usually take place much faster below the ground surface than above it, especially in humid climates?

4.6 How does the rate of chemical weathering of the minerals of an igneous rock compare with the pattern of crystallization from its original magma? What are the reasons for this relationship?

4.7 What happens to sodium, potassium, calcium, magnesium, and iron that are released from silicate minerals as they decompose?

4.8 Why is quartz the only common rock-forming mineral that usually survives the weathering processes?

4.9 Besides the composition of bedrock, what factors influence the kind of soil that develops in a given area?

4.10 What are the five major components of a soil, and how are they related to each other and to the way that soil forms?

4.11 What physical and chemical characteristics are useful in describing a soil?

4.12 How does a residual soil develop on bedrock or on unconsolidated sediment?

4.13 What materials are typical in the O, A, B, and C horizons of a residual soil?

4.14 How do pedalfers (including laterites) and pedocals differ? How do they differ in agricultural utility?

Critical Thinking

4.15 Suppose that a granite and a basalt were each exposed to the same chemical weathering environment. What residual minerals would you expect to accumulate around each of these rocks?

4.16 How could you modify the experiment shown in Figures 4.26 and 4.27 to make sure that the soil produced in it would be a pedocal? How could you modify it to produce a laterite?

4.17 Suppose that you cannot see the bedrock in a particular area because it is covered with soil, but you know that it must be either a basalt or a granite. If you were challenged to find out which kind of rock it was, using only chemical clues in water samples from wells in the area, which chemical clues would you look for? Why?

5

Tilted sedimentary rocks form the Maroon Bells, near Aspen, Colorado. (Myron Wood)

Sedimentary Rocks

*O*BJECTIVES

▲▲

As you read this chapter, it may help in your study of sedimentary rocks if you focus on the following questions:

1. What is a sedimentary rock?
2. What are the more important minerals of the sedimentary rocks? What is "texture" in relation to sedimentary rocks and how is it expressed?
3. What is diagenesis? What part does it play in the formation of sedimentary rocks?
4. What are the most abundant sedimentary rocks, and how is each distinguished?
5. What are the different features that characterize sedimentary rocks?
6. What are depositional environments, and how are they applied to the interpretation of sedimentary rocks?

*O*VERVIEW

▲▲

The weathered debris of preexisting rock provides the material for the creation of sedimentary rocks. Water, wind, ice, and gravity sweep the sediments to new places, both on land and in the world's oceans.

Once sediments are deposited, the process of turning soft accumulations into firm rocks begins. Squeezing the deposits into a smaller space as more sediments accumulate above, combined with the addition of cementing materials, works to produce a new set of rocks with such homely names as sandstone, mudstone, and limestone.

Each sedimentary rock carries the record of the environment in which it formed, a record preserved as the soft sediments turned to stone.

What Is a Sedimentary Rock?

▲▲▲

A **sedimentary rock** is composed of the weathering products of preexisting rock and is formed under the relatively low temperatures and pressures at and close to the surface of the Earth. Sedimentary rocks are in some ways easier to understand than the igneous and metamorphic rocks. This is in part because many of the processes involved in forming sedimentary rocks take place in environments more familiar to us than the environments in which igneous and metamorphic rocks form. Regardless, sedimentary rocks certainly contain a rich record of Earth's history and life for all to read who will take the time to look.

The term *sediment* comes from the Latin *sedimentum* and means "a settling." The dissolved and solid products of weathering that we discussed in the last chapter are transported to areas where they can settle out and accumulate as loose sediments. These sediments are then converted into sedimentary rocks.

Sediments collect in widely different places, and the conditions under which they accumulate vary accordingly. Thus, the conditions of deposition on a barely submerged delta are very much different from those of the deep sea. Geologists refer to the particular conditions of the areas in which sediments collect as **depositional environments.** These environments are reflected in the sediments and in the rocks that form from them. The depositional environment along a zone of breaking waves will be much different from a quiet mountain lake (Figure 5.1a, b); thus the sediments deposited there will record these differences and so will the rocks that form from them.

Many sedimentary rocks are made up of the minerals and rock fragments, or the detritus, produced by the weathering of preexisting rocks. Deposits of this type are called **detrital** (from the Latin *attritus* for "rubbed away"). Sedimentary rocks formed from this debris are called **detrital sedimentary rocks.** For example, the grains of quartz and feldspar released by the weathering of a granite may be moved by a stream to the ocean, where they settle out into layers of sand. Eventually these layers will be converted into the solid rock that we know as sandstone.

Some weathering products are dissolved in water. Given the proper conditions, these may later be removed in amounts large enough to form recognizable layers. These are **chemical sediments,** and the rock they form is a **chemical sedimentary rock.** Beds of salt provide an example. Ions released during weathering reach the ocean in dissolved form. A portion of the ocean basin may become isolated from continuous contact with the main water body. Ongoing

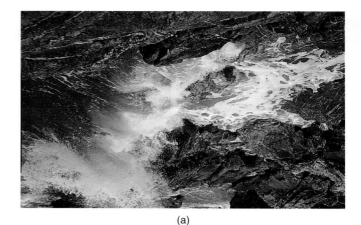

(a)

(b)

FIGURE 5.1   (a) Exposed to the direct attack of the surf, the environment of this rocky shore in Elba favors the deposition of coarse sand and gravel. (b) The sheltered environment of Blue Mountain Lake, in the Adirondack Mountains of New York, favors the deposition of sand along its shores and mud in its deeper waters.

evaporation may then concentrate the dissolved elements so much that they precipitate and form layers of salt. The most familiar of these is rock salt, which begins to precipitate after a little over 90 percent of the water has evaporated. This is halite (NaCl), which we know as common table salt.

Because you know their role in weathering, which we saw in the last chapter, you should not be surprised that living organisms are involved in the formation of some sedimentary rocks. When plants and animals are the major contributors to sediments, they are called **biogenic sediments,** from the Greek for *life* and *origin*. The rocks they form are **biogenic sedimentary rocks**.

Although we distinguish among three general groups of sedimentary rocks—detrital, chemical, and biogenic—most sedimentary rocks are mixtures of these types of sediments. For instance, a largely detrital rock typically contains smaller amounts of chemical or biogenic material.

Distribution of Sedimentary Rocks

If you were to walk across the North American continent you would be on sedimentary rocks, or separated from them by only a thin, weathered zone for three quarters of your trek. The rest of your trip would lead you across metamorphic and igneous rocks. On average, the total thickness of the various sedimentary rocks is about 1.8 km. In a few places the accumulation reaches 10 to 12 km, more or less. Elsewhere it tapers to a feather-thin edge and disappears altogether as the igneous and metamorphic components of the continental crust appear. In effect, then, sedimentary rocks in the Earth's crust form a discontinuous veneer over members of the other two rock families. Although the sedimentary rocks cover most of the continental surface, they make up a bare 5 percent of the crust by volume. The metamorphic and igneous rocks account for the other 95 percent (see Figure 5.2).

Most of the sedimentary rocks of the world have formed in the ocean, chiefly along its shallow margins. Today we find many sedimentary rocks exposed on the continents as the result of internal Earth forces that have raised them above sea level or because of the retreat of the seas in which they were formed. Of the total of about 375,000 km^3 of sedimentary rocks on the Earth, about 70 percent are on the continents and 30 percent still in the oceans. The age of the rocks exposed on the continents varies widely. Some are very recent; others are a billion years in age and older. We would expect to find more younger rocks than older ones, both because younger rocks bury older rocks and because erosion has had more time to remove the older rocks than the younger ones (Figure 5.3).

Mineralogy and Texture of Sedimentary Deposits

Mineral Composition

The most abundant minerals in sedimentary rocks are quartz, clay, and calcite. Recall from the last chapter that the clays are layered silicate minerals that form from the chemical weathering of feldspars and the ferromagnesian silicates found in most igneous and metamorphic rocks. Weathering of quartz-bearing rocks loosens individual quartz grains, thereby allowing them to be moved. Along with clay, quartz is the most common component of detrital sedimentary rocks. One estimate suggests that clay and quartz make up 80 to 85 percent of the sedimentary rocks.

Calcite and the closely related mineral dolomite are both carbonate minerals, and between them they form 10 to 15 percent of all sedimentary rocks. Of the two, calcite is more common and is predominately biogenic. The rest of the sedimentary rocks, 5 percent or less, are made of such other minerals as, for instance, halite and gypsum, or the iron oxides (such as hematite and goethite). In addition to individual minerals, chemically unweathered fragments of preexisting rocks are found in many sedimentary rocks and can be the predominant component. These may be small flakes, a millimeter or less in maximum dimension, or pieces of rock ranging from pebbles a centimeter in diameter up to blocks the size of a small house.

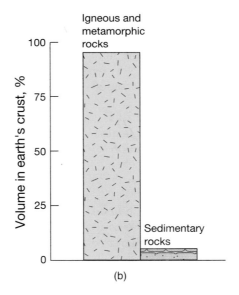

FIGURE 5.2
Relative abundance of sedimentary rocks in relation to the crystalline rocks (a) by surface area on the continents, and (b) by volume in the continents.

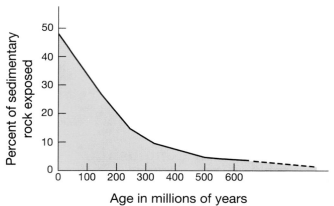

FIGURE 5.3 The older the sedimentary rock the less of it is exposed at the surface of the lands. *(Data from H. Blatt and R.L. Jones, "Proportions of Exposed Igneous, Metamorphic, and Sedimentary Rocks," Geological Society of America, Bulletin., Vol. 86, p. 1087, 1975.)*

Texture

Texture, from the Latin *textura,* meaning "woven," refers to the size, shape, and arrangements of the materials that make up a rock and give it a particular look. The rock's texture will vary with the size of the particles that make it up and how they are arranged in the rock. Geologists recognize two different types of texture in sedimentary rocks, clastic and crystalline.

Clastic Texture
The term **clastic** is derived from the Greek for "broken" or "fragmental." Rocks that have been formed from the detritus of mineral and rock fragments are said to have clastic or detrital texture. The size and the shape of the original particles, or clasts, have a direct influence on the resulting texture. A deposit of fine muds has a much different texture from that of a boulder beach (Figure 5.4a, b). Both deposits have clastic textures, but one is fine and the other is coarse. The texture in which the particles are dominantly of one size will have a different look from one in which the particles have a wide range of sizes, a characteristic called *sorting.* A biogenic deposit may also show a clastic texture. The size, shape, and sorting of different particles in a clastic texture are discussed below.

Size of Particles The size of particles in a rock is important in a description of a rock texture. It is also, as we will see, fundamental in the classification of many sedimentary rocks. As a standard of measurement of particles, geologists have devised the scale of particle sizes given in Figure 5.5.

The size of a particle does not tell us its mineral composition. Thus, a sand-sized particle may be composed of any mineral, but most commonly of quartz or calcite. Clay is a term that can cause confusion. It applies to a particular size, given in Figure 5.5. Clay,

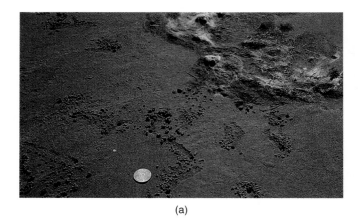

(a)

(b)

FIGURE 5.4 Very different textures are illustrated by (a) the fine mud deposits along the Arabian Gulf shore of Saudi Arabia and (b) a boulder beach in Acadia National Park, Maine.

however, is also a mineral term. To avoid confusion geologists will often use the term "clay-sized" when referring only to the size of these small particles.

Sorting of Particles Clastic rocks commonly have a variety of different-sized particles mixed together. We refer to the range of sizes present as the particles' **sorting.** A rock with a narrow range of particle sizes is called "well-sorted"; one with a very wide range of particle sizes is considered to be "poorly sorted" (Figure 5.6a). For example, the sands of a bathing beach are generally well-sorted, whereas the debris in a landslide are poorly sorted.

Shape of Particles The shape of individual particles is another component of the texture of sedimentary rocks. Geologists use the terms **roundness** (degree to which the particle's edges and corners are rounded) and **sphericity** (how close the shape of a particle is to that of a sphere) in describing a particle's shape. Figure 5.6b shows some of the variations in roundness and sphericity that are used.

Crystalline Texture
Most of the sedimentary rocks with a **crystalline** texture have an interlocking structure resembling that of igneous rocks because,

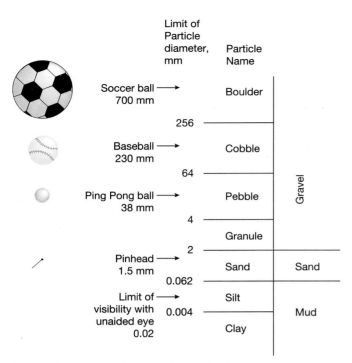

Limit of Particle diameter, mm	Particle Name	
Soccer ball → 700 mm	Boulder	
256		
Baseball → 230 mm	Cobble	
64		Gravel
Ping Pong ball → 38 mm	Pebble	
4		
	Granule	
2		
Pinhead → 1.5 mm	Sand	Sand
0.062		
Limit of → visibility with unaided eye 0.02	Silt	Mud
0.004		
	Clay	

FIGURE 5.5 A scale of particle sizes for clastic sediments.

like igneous rocks, they contain minerals that form at the same time from a liquid. Igneous rocks form from a magma, whereas the crystalline sedimentary rocks form from water. The crystals range from less than 1 mm to 5 mm or more in diameter. Such textures occur both in the biogenic sedimentary rocks and the chemical sedimentary rocks.

Diagenesis

Thus far we have spoken of sedimentary rocks as end products in the weathering stage of the rock system, as indeed they are. A great deal happens to a sediment, however, from the time that it is deposited as unconsolidated material to the time it becomes a true rock and finally reaches the temperatures and pressures that convert it into a metamorphic rock. All the physical, chemical, and biologic changes undergone by sediments from the time of their initial deposition, through their conversion to solid rock, and subsequently to the brink of metamorphism, are referred to as the process of **diagenesis,** from the Greek for "during" and "origin." Diagenesis is important in the study of the sedimentary rocks for, among other reasons, it includes **lithification,** the conversion of unconsolidated sediments to solid rock. The two most important processes in lithification are compaction and cementation.

Compaction

As more and more sediments are piled on top of one another, they all push down on lower sediments, thereby increasing pressures throughout the deposits. This leads to **compaction** of the original deposits as they are pressed closer and closer together. Compaction is most effective in the fine-grained deposits, mud and silt, and least effective in coarse-grained deposits, sand and gravel. The reason sand and gravel compact with difficulty is because the individual particles, particularly if they are quartz, support each other. In contrast, many clay and silt particles deform more easily under the pressure of overlying deposits. Sand deposits will compact most easily if grains are weaker than quartz or if quartz grains are mixed with other particles.

The deeper a sediment is buried, the greater the compaction and the firmer the deposit. This compaction also reduces the amount of space between particles, known as *pore space.* At the same time, this drives out water previously trapped in the sediments and increases the density (the mass of material per unit volume) of the deposit (Figure 5.7a, b).

Compaction promotes lithification in another way, particularly in deposits of fine-grained sediment. As pressure increases, not only are the grains pressed closer together but so are the atoms in the minerals that compose them. This increasing closeness, combined with the heat of increasing depth and water remaining in the original sediments, can bring about creation of new minerals, most commonly feldspars, clays, and iron-bearing minerals. This is called **authigenesis,** the formation of new minerals within a sedimentary accumulation after its deposition. It can give added firmness to the deposits.

Cementation

Although compaction affects coarse-grained deposits to some extent, **cementation** is more effective in lithification. Cementation is the process that fills, or partially fills, the pore spaces in a rock with secondary deposits of mineral matter, which cements individual grains together. Water circulating through the sediment carries dissolved materials. These materials can precipitate as a cement in pore space and turn an originally loose deposit of sand or gravel into rock. The process is similar to making concrete, in which sand and gravel are mixed together with water and cement to form an artificial rock, concrete. The most common cement is calcite. Other cements include silica and the iron oxides (hematite and limonite). The introduction of cement into a sedimentary deposit further reduces its porosity and increases its density. In fact, some of the changes

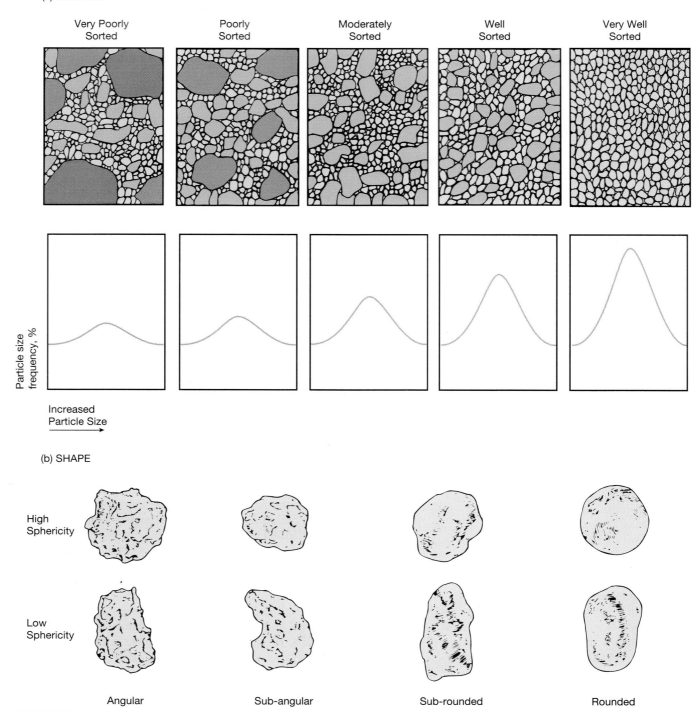

(a) SORTING

| Very Poorly Sorted | Poorly Sorted | Moderately Sorted | Well Sorted | Very Well Sorted |

Particle size frequency, %

Increased Particle Size →

(b) SHAPE

High Sphericity

Low Sphericity

Angular   Sub-angular   Sub-rounded   Rounded

FIGURE 5.6   Sorting and shape of clastic particles. (a) Sorting relates to the range in size of individual clasts. The more tightly they are clustered around a particular size the better the sorting is said to be. The idealized graphs show the percentage distribution of various grain sizes for each of five categories of sorting. (b) Shape is defined on the basis of roundness and sphericity. Roundness increases as corners become more and more smoothed off. Sphericity defines how closely a grain approaches the shape of a sphere.

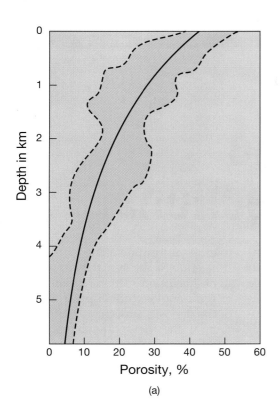

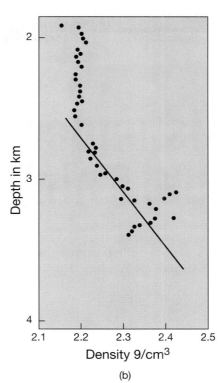

FIGURE 5.7
Changes in porosity and thickness of rock with depth. (a) Decrease in porosity of carbonate deposits with depth measured in 15 wells in south Florida. All data fall within the patterned zone. The solid line is an average curve that fits all data in the patterned zone. (b) Increase in density of rock with depth based on measurements in wells in various localities. (*After Gerald M. Friedman, et al., Principles of Sedimentary Deposits.* New York: Macmillan, 1992, pp. 116, 117.)

reported in Figure 5.7a, b are undoubtedly due to cementation.

Bioturbation

We found in our study of the weathering process that plants and animals are important in working and reworking the soil. Similarly, living organisms may churn recently deposited sediments, a process called **bioturbation,** meaning "stirred by organisms." For instance, worms will burrow through soft sediments, foraging for nourishment, working and reworking the mud. The sediment acquires a mottled appearance, and this is preserved when the sediment eventually turns to rock.

   Microscopic examination of a sedimentary deposit can reveal many of the details of bioturbation and other processes of diagenesis. Perspective 5.1 gives several examples.

The Most Abundant Sedimentary Rocks

Four general rock types make up 95 percent of all the sedimentary rocks. They are **conglomerate, sandstone, mudstone,** and the **carbonate rocks.** Conglomerates, sandstones, and mudstones are detrital rocks and are grouped according to their dominant grain size. The carbonate rocks, **limestone** and **dolostone,** are biogenic or chemical. The remaining 5 percent of the sedimentary family is made up of numerous other less common rock types.

Conglomerate

A conglomerate is the most easily recognized but least abundant of the major sedimentary rocks. It starts out as a deposit of sand and gravel and is turned into a rock by cementation. The largest particles in conglomerates are over 2 mm in diameter and generally much larger. Smaller-grained particles fill the space between the gravel-sized grains in most conglomerates. The particles are most commonly the fragments of preexisting rock, more or less rounded because the process of abrasion during transportation has worn down sharp corners (Figure 5.8). When the gravel-sized particles are sharp and angular the rock is called a **breccia** (Figure 5.9).

Sandstone

Sandstones are composed of particles that range in size from 0.062 mm to 2 mm. At their largest, individual sandstone particles are about the size of a match head and are therefore not difficult to see with the naked eye. For the smaller sandstone particles one

Perspective *5.1*

Diagenesis under the Microscope: Thin Sections

Much of evidence for diagenesis exists at a microscopic level, beyond the ability of the naked eye to observe. The microscope reveals a wealth of information including, for instance, how individual grains are cemented together, how much compaction has taken place, the kind of cement present, the composition of individual grains, how much pore space remains, and the presence of authigenic minerals. Some examples are given in Figure P5.1. These are photographs of 30-micron-thick slices of sedimentary rock viewed under low magnification. The slices are thin enough so that they transmit light. Each thin section has been cut at right angles to the upper surface of the deposit, with the top and bottom of the view oriented accordingly.

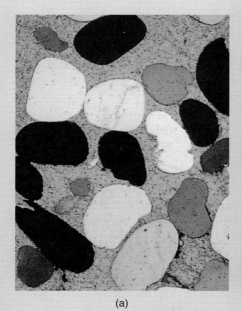

(a)

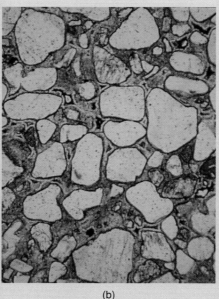

(b)

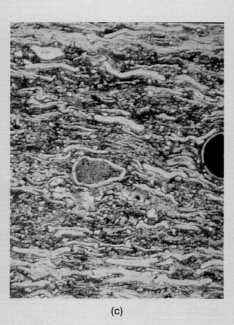

(c)

needs a hand lens or low-power microscope to tell very much about the individual grains.

To the touch, sandstone often has a rough, "sandpapery" feel. In the less tightly cemented sandstones, individual grains may rub off fairly easily on your fingers. Sandstone is relatively resistant to weathering and therefore often forms cliffs (Figure 5.10). It occurs in well-defined layers or beds. The most common kind of sandstone, one in which quartz predominates, is a type that a geologist will call a **quartz arenite** (from the Latin *arena* for "sand"). If the sand-sized grains are made up of calcite—for instance, if they are small frag-

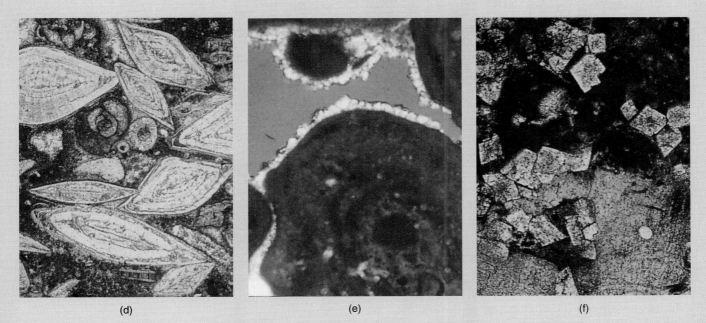

(d)   (e)   (f)

P5.1   (a) Rounded to subrounded and broken quartz grains are cemented together by calcite. Tiny crystals of calcite account for the granular texture of the cement. These are set in larger crystals of calcite whose angular outlines are faintly visible, particularly in the right half of the photo. In places, the calcite cement invades the quartz grains as in the sharp re-entrant in the black quartz grain in the lower left of the photograph. The photograph was taken in polarized light as described in Perspective 3.1. This accounts for the different colors in the small calcite crystals and in the quartz crystals.

(b) Another type of cementation is seen in this photograph. The sand grains are subangular to subrounded and mostly quartz. Lithification has been by an opal cement that is visible between the sand grains.

(c) Effects of compaction appear in this specimen. The irregular lenticular bodies are flattened, spherical algae. These are set in a matrix that includes tiny quartz grains, clay, mica and pyrite flecks. The circular sphere at the right is an algal sphere that was filled with pyrite before compaction affected the deposit and testifies to the amount of compaction that has occurred.

(d) This is a biogenic deposit, and the spaceshiplike forms are the tests of a one-celled animal, the bottom-dwelling foraminifer, *Amphistegina*. The matrix is a calcareous silt in which are contained smaller fragments of other marine organisms. Bioturbation of the deposit is indicated by the random orientation of the foraminifera.

(e) In this photograph the large, dark, calcareous bodies were formed by algal growth. The blue area is pore space. Clear calcite crystals have formed around the edges of the algal masses and project into the pore space.

(f) Dolomite has begun to replace limestone in this photomicrograph. The original deposit was composed of mollusk shells and a dark calcareous mud. The rhombolds of dolomite crystals can be seen replacing the original sediment. *(Albert V. Carozzi, Sedimentary Petrography, Englewood Cliffs, NJ: Prentice Hall, 1993. (a) Plate 3.A; (b) Plate 3.D; (c) Plate 14.E; (d) Plate 21.E; (e) Plate 23.C; (g) Plate 26.E.)*

ments of shell—a geologist will call the rock a **calcarenite** (meaning calcareous sandstone). In many sandstones the cement that holds the sand grains firmly together is calcite. Less often it is hematite, limonite, or quartz. If the original deposit is composed predominantly of quartz and feldspar, the sandstone is called an **arkose.** Another variety of sandstone, called **lithic sandstone** or **graywacke,** is characterized by its angular sand-sized grains of quartz and feldspar and by small fragments of dark rock, all set in a matrix of finer particles. Arkosic and lithic sandstones result from more rapid sedimentation and burial than do

FIGURE 5.8   The sand, rounded pebbles, and boulders in this conglomerate in Baja California were deposited in a torrential flood. *(Franklyn B. Van Houten)*

quartz-rich sandstones. The sediments in these sandstones have not undergone extensive weathering. Their composition also suggests that their component minerals and rock fragments were not carried any great distance from their source.

Mudstone

Mudstones are the most abundant of the sedimentary rocks. In many ways they are more complex than the conglomerates and sandstones, mainly because they are so fine-grained, being made chiefly of particles of clay and silt size. Geologists must often rely on optical microscopes, X-ray diffraction, electron microscopy, and sophisticated chemical techniques to decipher the details of their history.

Mudstone, often called **shale,** may feel and look smooth. Because it is made of compacted mud, it can be dusty or just plain dirty. With a little moisture the less compact mudstones will leave your hands muddy. Unlike with the conglomerates and sandstones, it is difficult for the naked eye to discern any distinct variations in the look of mudstones, particularly in small samples or exposures. It is just possible to make out small quartz grains with a hand lens. Even when this isn't possible, there is a simple way to get a rough estimate of the amount of quartz in a mudstone. Nibble a bit between your teeth. The silt- and clay-sized quartz grains are gritty, and the clay minerals are slippery. Much the same minerals make up both sandstone and mudstone. As you would expect, however, the proportions of the different minerals vary markedly, particularly for the clay minerals and quartz (Figure 5.11).

Carbonate Rocks: Limestone and Dolostone

 Limestones are made up of deposits of calcium carbonate, calcite. The deposits from which most

FIGURE 5.9   Angular fragments set in a fine-grained groundmass characterize this breccia from southern New Mexico.

limestones have formed accumulated in the ocean and are biogenic in origin. That is, plants and animals have played a key role in their creation. Calcium carbonate is brought in solution by rivers to the ocean where it is extracted from seawater by organisms and incorporated into their skeletons. When the organisms die, they leave behind a quantity of calcite, and over time thick deposits of this material build up. Other calcareous marine deposits are inorganically precipitated from seawater that is supersaturated with calcium carbonate. This most commonly occurs in warm, tropical waters of slightly lower than average salinity.

Various marine environments provide the setting for the deposition of calcareous sediments. Reefs, ancient and modern, are well-known examples of major accumulations of calcareous deposits (Figure 5.12). The most important builders of modern reefs are algae, mollusks, corals, and one-celled animals, many of the same animals whose ancestors built the reefs of ancient seas—reefs, now old and deeply buried.

In the warmer oceans, microscopic organisms manufacture tiny calcareous shells. These tiny creatures float and swim in the upper level of the sea. When they die their shells, called **tests,** settle slowly to the seafloor, to form a calcareous ooze. Some shallow tropical seas are so saturated with calcium carbonate that it precipitates inorganically and settles to form a calcareous mud.

All these deposits form limestone. In some places the deposits are rocklike from the beginning as organisms build on one another, and their skeletons and

FIGURE 5.10 Sandstone commonly forms cliffs as do these beds in the Canyon de Chelly, Arizona.

shells accumulate layer upon layer. In other places, however, the deposits are calcareous muds and soft accumulations of the shells of one-celled animals. These deposits eventually lithify by compaction and by the precipitation of additional calcite crystallized from the waters in the mud.

Although all calcareous rock is limestone, specific names are given to various types of limestone. **Chalk** is partly made up of calcite from the skeletal fragments of microscopic oceanic plants and animals. These organic remains mix with very fine-grained calcite deposits of either biogenic or inorganic chemical origin. A much coarser type of limestone composed of organic remains is known as **coquina** (from the Spanish for "shellfish" or "cockle") and has a clastic texture characterized by the accumulation of many large fragments of shells (Figure 5.13). Some limestones are termed **oolitic limestones** when they contain sand-sized, spheroidal, calcareous grains called **oolites.** These grains, whose name comes from the Greek for "egg"—because an accumulation of oolites resembles a cluster of fish eggs—are found on the floors of modern oceans and in rocks formed in oceans long gone. They are thought to form by the inorganic precipitation of calcium carbonate from seawater. Cross sections show that many oolites have grown outward around a mineral grain or around a small fragment of shell that acts as a nucleus about which calcite deposition begins (Figure 5.14). Some limestones are made up largely of oolites. One, widely used as a building stone, is the so-called Indiana or Spergen limestone.

Not all limestones form in marine environments. Minor deposits of freshwater limestones exist, the products of deposition in caves, springs, and streams.

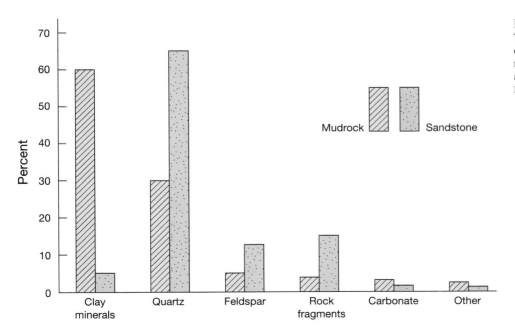

FIGURE 5.11
The mineral composition of the detrital particles in sandstone and mudstone. *(Data from H. Blatt, Sedimentary Petrology, 2d ed. New York: W.H. Freeman and Co., Table 1.1, 1992.)*

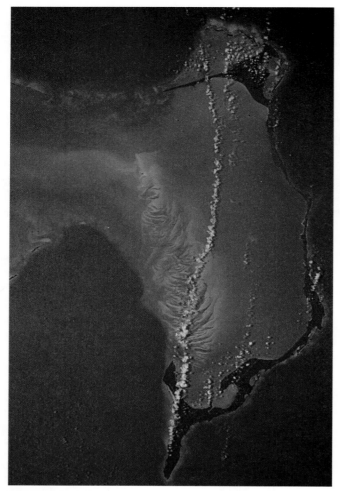

FIGURE 5.12   The island of Eleuthera on the Grand Bahama Banks is the site of modern calcite and future limestone production. Light reflects off the shallow limestone floor of Exuma Sound on the left. Lines of fluffy, white clouds string out over the sound. *(NASA)*

FIGURE 5.13   This coquina from Shark Bay, Western Australia, is made up almost exclusively of shells and shell fragments. Individual shells average about 2.5 cm in size. *(William E. Ferguson)*

One of these is the limestone formed by inorganic chemical processes in caves and known as **dripstone.** Ground water, saturated with calcium carbonate, come to the surface as springs, both hot and cold, around which deposition of limestone takes place (Figure 5.15). Names applied to these deposits are **tufa,** from the Italian for "soft, spongy rock," and **travertine,** from Tivertino, the Roman name for the Italian town of Tivoli, near which extensive deposits of the rock are quarried.

Not as abundant as limestone, **dolostone** nevertheless occurs in significant volume. The two rocks are very similar. Dolostone, however, contains the carbonate mineral dolomite ($CaMg(CO_3)_2$) instead of calcite. The origin of dolostone is not altogether clear and is still the subject of discussion. What is clear, however, is that there is more dolostone in relation to limestone in old rocks than in geologically recent ones. If geologists look back 800 million years in time they find three times as much dolomite as limestone in the continental crust. Five hundred million years ago, about the time fish began to appear, the ratio is reversed: three times as much limestone as dolostone. Carbonate rocks dating from the time the dinosaurs disappeared (about 66 million years ago) show 80 times more limestone than dolostone. Today limestone production overwhelms that of dolostone.

There are two reasonable answers to this variation in the dolostone-to-limestone ratio. Geologists can argue that conditions in the world's oceans have been changing from those favoring dolostone deposition to those favoring limestone deposition. Alternatively, we can suggest that environments of deposition have not changed, but that some process that converts limestone to dolostone has been at work. Close examination of dolostone in both field and laboratory studies suggests the latter explanation. Geologists generally agree that most dolomite is produced by a diagenetic process in which magnesium-rich waters circulate through limestone and convert calcite to dolomite by the replacement of some calcium ions with magnesium ions. An example of this replacement of calcite by dolomite is shown in the thin section in Figure P5.1f. The older the rock, the more time has been available for the conversion of limestone to dolostone.

At first glance dolostone and limestone are difficult to tell apart. The quickest and surest way to do so is to observe the different reaction of the two rocks to a few drops of dilute hydrochloric acid. Limestone will fizz freely as the acid releases carbon dioxide. Unless it is powdered, the dolostone will not react visibly to the acid. Limestone is also a bit softer than dolostone. Dolostone tends to be browner than limestone on weathered surfaces. This is due to the small amount of iron commonly present in dolostone, which weathers to an iron oxide.

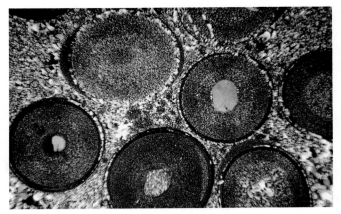

FIGURE 5.14   Oolites, a little over 1 mm in diameter, are here set in a fine-grained matrix of calcite mud. The oolites have grown outward from centers of crystallization by the chemical precipitation of radial calcite fibers.

FIGURE 5.15   Travertine being deposited from hot springs at Thermopolis, Wyoming.

Other Sedimentary Rocks

Evaporite Rocks   An **evaporite** is a sedimentary rock composed of a mineral that precipitates from solution when the liquid in which it was dissolved evaporates. These rocks form by the evaporation of seawater and the brines of salt lakes. They generally have a crystalline texture.

Seawater contains, on average, 3.5 percent dissolved salts, 78 percent of which is halite (NaCl). One might expect, therefore, that halite would be the first salt to form. When, however, we evaporate a bucket of

FIGURE 5.16   The sequence of precipitation of salts from the continuing evaporation of seawater.
(After Bryan L. Skinner, Earth Resources, 3rd ed. Englewood Cliffs, N.J.: Prentice Hall, 1986, Figure 7.2, p. 133.)

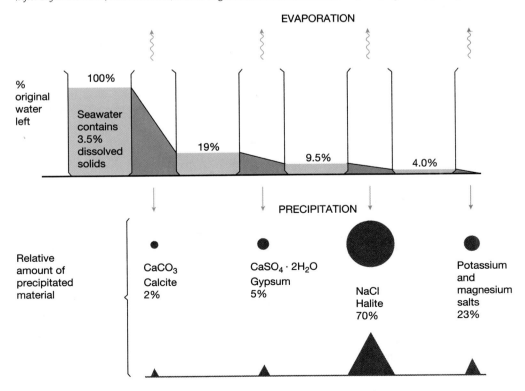

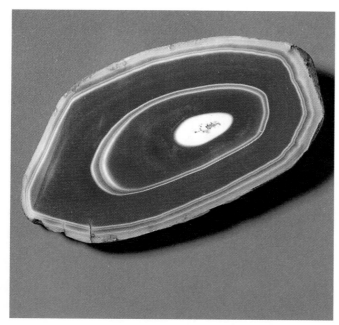

FIGURE 5.17 Agate is a banded deposit of chalcedony and chert. Long dimension, 15 cm. *(John Simpson)*

seawater, halite turns out to be the third salt to precipitate and is preceded by both calcite ($CaCO_3$) and gypsum ($CaSO_4$), in that order. The explanation is simple. Each salt has a different solubility. Less water needs to be evaporated from our bucket to precipitate a salt with low solubility than does a salt of high solubility. This means that the salts will precipitate in a definite sequence as more and more water evaporates. The least soluble, calcite, will precipitate first. As more water is evaporated, gypsum will precipitate, then halite. Finally, as the water is almost all evaporated and only 4 percent of its original volume remains, some complex salts of potassium and magnesium will form (Figure 5.16).

Siliceous Rocks Silica occurs in sedimentary rocks in forms other than the abundant clastic grains of quartz. It occurs in a cryptocrystalline (from the Greek *kryptos*, "hidden," and "crystalline") structure that is so fine that it cannot be seen under most ordinary microscopes. A high-power microscope does reveal two different forms of cryptocrystalline silica. One has a *granular pattern* and the other a *fibrous pattern*. **Chert** is the name usually applied to the granular form of silica and **chalcedony** is the general name for the fibrous form. To the naked eye, chert has a somewhat duller luster than chalcedony, whose luster is higher and more waxy. **Flint** is a form of chert. In many cases it is black (because of included organic matter). **Jasper,** a red variety of chert, is colored by microscopic flakes of hematite. Chert commonly

occurs as lumps or nodules in limestone or mudstone. It also occurs in relatively thin continuous layers. **Agate** is a variegated form of silica, its bands of chalcedony alternating with bands of some variety of granular cryptocrystalline silica, such as jasper (Figure 5.17).

Features of Sedimentary Rocks

Bedding and Bedding Planes

The most common feature of sedimentary rocks is a layering called **bedding** or **stratification.** Individual beds are bounded by **bedding planes.** Layering is sometimes present in igneous rocks, such as basalt flows, but it is a rare sedimentary rock that lacks bedding of some sort.

As gravity pulls sediments downward, they settle and collect in layers or beds on the floor of a lake, ocean, stream channel, or floodplain. The upper surface of each bed will be horizontal. Successive beds will also have horizontal upper surfaces. As a result, any two beds will be separated by a horizontal surface called a **bedding plane.** Each layer or bed records a more or less continuous period of deposition. The bedding plane between two layers tells us that something changed and interrupted the depositional process. Perhaps, for instance, deposition of mud replaced the deposition of sand. The two layers lithify to a bed of sandstone overlain by a bed of mudstone. Because the sandstone and the mudstone have very different appearances, the plane between the two layers, the bedding plane, will be clear and distinct. Furthermore, the two rocks will break most easily along the bedding plane that separates them (Figure 5.18).

Primary Features of Bedding

To the trained eye of a geologist, beds convey a great deal of information about the conditions under which sediments were deposited. A number of characteristics, called **primary features** of bedding, were produced as the beds were being formed. These include such features as cross-bedding, graded bedding, ripple marks, mud cracks, and sole marks. These are important in deciphering the history of the rock. For example, a geologist familiar with how ripple marks are produced by currents in modern sediments can apply

FIGURE 5.18   Well-defined bedding planes mark off the layers in this exposure on the eastern flank of the Big Horn Mountains, Wyoming.

this knowledge to the formation of ripples in ancient sediments and determine the direction of the ancient ripple-forming current. Notice that this is an application of the principle of uniformitarianism discussed in Chapter 1. We use the features that are known results of a present process as an analogy to deduce the origin and meaning of similar features formed in the past. As we proceed in this section you will be able to see how many of these modern features can be applied to similar features in sedimentary rocks. All of them become clues that help the geologist reconstruct the depositional environments that have long since disappeared.

Cross-Bedding

Geologists have found that when first formed, bedding planes are almost always horizontal. Therefore, if we find a sedimentary rock in which the bedding is tilted, bent, or broken we know that some Earth forces have deformed the rocks at some time after their deposition. Recognizing these postdepositional changes in a sedimentary rock pro-

vides clues to the mountain-building history of a region. Some depositional environments, however, produce **cross-bedding,** or *inclined-bedding,* in sediments and in the rocks that develop from them. One such example is found in sand dunes. On the lee face (the slope facing away from the wind) of an advancing sand dune, sand is deposited at an angle of about 30° from the horizontal, and this angle is preserved when the sediments are lithified (Figure 5.19a). Some stream deposits exhibit inclined bedding but on a smaller scale (Figure 5.19b).

Graded Bedding

When the particles in a sedimentary bed vary continuously from coarse at the bottom to fine at the top, the bedding is said to be **graded.** Such bedding usually occurs as a supply of sediments suddenly settles. Then a period of time elapses before another pulse of sediments comes along. The largest particles, because of their weight, settle first. Successively smaller and lighter particles follow. Once the sediments from a pulse of sediments have settled out, some period of time elapses before the next pulse. Therefore, in a sequence of graded beds, the finer, topmost sediments terminate abruptly beneath the coarse material of the next graded bed (Figure 5.20). A succession of graded beds presents a rhythmic, repetitive formation when exposed in cross section.

Graded bedding occurs in many depositional environments. Figure 5.20 shows very coarse graded bedding formed from periodic stream-flood deposits. In the ocean, graded bedding characterizes deposits known as **turbidites,** from the Latin *turbidus* for "disturbed," referring to the stirring up of sediments and water. Turbidites form from sediments carried by muddy currents moved down the slopes of the continental margins into the deep sea. These currents,

FIGURE 5.19   Two types of inclined bedding. (a) Cross-bedding in an ancient sand dune, Utah. Wind blew from right to left. (b) Inclined bedding in stream-laid sand deposits in the 80-million-year-old Raritan formation, New Jersey. Current was from left to right.

(a)

(b)

FIGURE 5.20   Graded bedding in torrential deposits in Baja California, Mexico. Material is very coarse at the bottom of the bed and becomes finer toward the top where it is abruptly terminated by another coarse bed. *(Franklyn B. Van Houten)*

FIGURE 5.21   Turbidites near Zumaya in northern Spain have been tilted on end. The rhythmic nature of the bedding is characteristic of turbidites. *(Franklyn B. Van Houten)*

called **turbidity currents,** can be set in motion by floods on major rivers draining to the ocean, by earthquakes that jar loose marine sediments, or just by instability of the ocean slopes. They travel at high speeds, generally faster than rivers on the land. They descend to the deep seafloor, rapidly lose their momentum, and begin to deposit sediments, dropping the coarsest material first. Turbidite beds are typically a few centimeters to a meter in thickness (Figure 5.21).

Ripple Marks, Mud Cracks, and Sole Marks
Ripple marks are the little waves of sediment—sand or silt—that commonly develop on the surface of a sand dune, or in moving water. **Mud cracks** are common on the dried surface of mud left exposed by the subsiding waters of a river, pond, or tidal flat. **Sole marks** are the casts, or fillings, found on the bottom sides of some beds.

Ripple marks (Figure 5.22a, b) preserved in sedimentary rocks furnish clues to the conditions that pre-

vailed when a sediment was deposited. For instance, if the ripple marks are symmetric, with sharp or slightly rounded ridges separated by more gently rounded troughs, we know from observing the same features today that they were formed by the back-and-forth movement of water in waves such as we find along a seacoast outside the surf zone. These marks are called **oscillation ripple marks.** If, on the other hand, the ripple marks are asymmetric, we can be sure that they were formed by air or water moving more or less continuously in one direction. These marks are called **current ripple marks**.

Mud cracks form when a deposit of mud dries out and shrinks. The cracks outline roughly polygonal areas, making the surface of the deposit look like a section cut through a large honeycomb. Eventually another deposit may bury the first. When the deposits are later lithified, the outlines of the cracks are preserved. Then, when the rock is split along the bedding plane between the two deposits, the cracks will be found much as they appeared when they were first formed, providing evidence that the original deposit underwent alternate flooding and drying (Figure 5.23a, b).

Sole marks develop as irregularities on the bottom of a bed. The upper surface of a bed of soft sediment may be irregularly scoured by subsequent currents. As a later bed is laid down on this surface its sediments will fill the scour marks and give the later bed an irregular lower surface. These irregularities are sole marks. They are casts of the irregular molds on an underlying bed. Very often sole marks have a linear, streamlined aspect that provides information on the direction of current flow. They are very often features of turbidite beds.

Are the Rocks Right Side Up?   As we have
already shown, sedimentary rocks form in horizontal

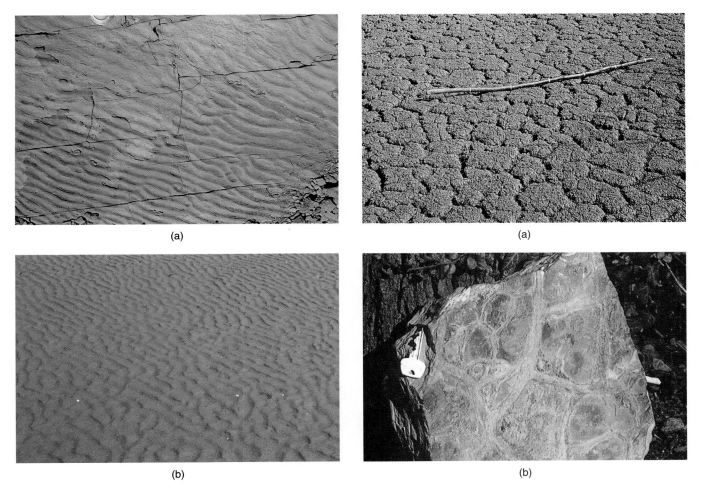

FIGURE 5.22   Ripple marks. (a) Current ripple marks exposed at low tide in Morecombe Bay, England. Current was from right to left. (b) Current ripple marks in sandstone near Moab, Utah. Flow was from upper right toward lower left.

FIGURE 5.23   (a) Modern mud cracks in a dry lake bottom, eastern Sicily. (b) Mud crack preserved in a 450-million-year-old limy mudstone at Rheems, Pennsylvania.

layers. Therefore, in a succession of beds the youngest bed is on the top. Suppose, however, that after formation the rocks are tilted and turned to the vertical and beyond, not an altogether unusual situation. How do you tell which is the top and which is the bottom of a bed? This is not a trivial question when geologists try to work out the relative ages of rock layers in a sequence of strata. Among the criteria a geologist can use to tell which side of a sedimentary bed is up are some of the primary features of bedding that we have just discussed. For instance, if we find a bed in which the particles grade upward from fine to coarse, rather than from coarse to fine, we should consider whether the rock has been overturned and the top of the bed is now on the bottom. Look again at Figure 5.20. The beds are right side up in the picture, but turn it upside down and the beds coarsen upward, the opposite direction in which graded bedding is deposited. If we discover a sole mark on a bedding plane we would be sure that we were looking at the bottom of the bed

(Figure 5.24). Cross-bedding in a sedimentary rock can indicate which side is up. Look carefully at the cross-bedding of the lithified sand dune deposit in Figure 5.19. The rock, as photographed, is right side up. The inclined beds are concave upward. If the beds were tilted 90° or more you could still tell in which direction the top of the deposits lay.

Secondary Features of Bedding

Many sedimentary rocks display distinctive features created after the original beds and their bedding planes formed. Some obviously are parallel to bedding planes while others are not clearly associated with them. Among such features are **nodules, concretions,** and **geodes.** These can hold important clues about the diagenesis of a deposit.

Nodules and Concretions   A nodule is an irregular, knobby-surfaced body of mineral matter that differs in composition from the sedimentary rock in

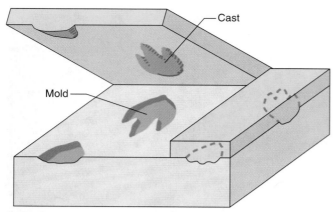

FIGURE 5.24   A cast of a footprint can be made when the impression of the print on the top of one deposit is filled with sediments from an overlying deposit. The original footprint serves as a mold for a cast on the overlying deposit. The cast is referred to as a *sole mark*.

FIGURE 5.25   Giant concretions from Roosevelt National Park, North Dakota.

which it has formed. As you should recall, chert often forms in nodules. Nodules tend to lie parallel to the bedding planes of the enclosing rock and average about 30 cm in maximum dimension. We can have, for example, chert nodules in limestone or pyrite nodules in coal. Adjoining nodules can coalesce to form a nearly continuous bed. Most nodules are thought to be a replacement of the original sediment.

A concretion is a compact mass of mineral matter, usually spherical or disk-like in shape and embedded in a host rock of different composition. They form by precipitation of mineral matter about a nucleus such as a leaf or a piece of shell or bone. Concretions range in size from a few centimeters to spheres up to 3 m in diameter (Figure 5.25). They form during diagenesis of a deposit, usually shortly after deposition of the enclosing sediment.

Geodes

Geodes, more eye-catching than either concretions or nodules, are roughly spherical hollow structures up to 30 cm or more in diameter (Figure 5.26). In some geodes, an outer layer of chalcedony is lined with crystals that project inward toward the hollow center. The crystals, often perfectly formed, are usually quartz, although crystals of calcite and dolomite, and, more rarely, crystals of other minerals are found. Geodes occur most commonly in limestone, but we find them in mudstones as well.

How does a geode form? Here is one way. First, a water-filled pocket develops in a sedimentary deposit, probably as a result of the decay of some plant or animal that was buried in the sediments. As the deposit begins to lithify into a sedimentary rock, a wall of silica with a jellylike consistency forms around the water and isolates it from the surrounding material. As time passes, water with lower concentration of dissolved material may enter the sediments while the water inside the pocket maintains a higher concentration than does the water outside. To equalize the concentrations, there is a slow mixing of the two liquids through the silica wall or membrane that separates them. This process of mixing, called *osmosis*, continues as long as pressure is exerted outward toward the surrounding rock. The original pocket expands bit by bit until the salt concentrations of the liquids inside and outside are equalized. At this point osmosis stops, the outward pressure ceases, and the pocket stops growing. Now the silica wall dries, crystallizes to form chalcedony, contracts, and cracks. If, at some later time, mineral-bearing water finds its way into the deposit, it may seep into the void through

FIGURE 5.26   A half of a geode has been polished and shows the outer rind of chalcedony that usually lines the geode wall. Inside the rind, crystals of quartz have grown into the cavity of the geode. (*A.J. Copley*)

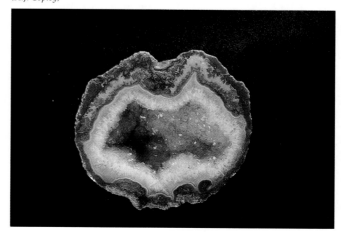

the cracks in the chalcedony. There, where the minerals are precipitated, the crystals are free to grow inward, toward the center, from the interior walls. Finally, we have a crystal-lined geode imbedded in the surrounding rock.

Fossils

The word **fossil** (derived from the Latin *fodere*, "to dig up") originally referred to anything that was dug from the ground, particularly a mineral. Today, however, fossil generally means any direct evidence of past life—for example, the bones of a dinosaur, the shell of an ancient clam, the footprints of a long-extinct animal, or the delicate impression of a leaf.

Fossils are usually found in sedimentary rocks, although they turn up from time to time, severely distorted, in igneous and metamorphic rocks. They are most abundant in mudstone and limestone, but are also found in sandstone, dolostone, and conglomerate. Fossils account for almost the entire volume of certain rocks, such as the coquina and limestones that have been formed from ancient reefs.

The remains of plants and animals are completely destroyed if they are left exposed on the Earth's surface; but if they are somehow protected from destructive forces, they can become part of the sedimentary deposit, where they will be preserved until erosion brings them to the surface again. A fossil can be part of the original organism. Mineral-bearing water circulating through a sedimentary deposit may replace the organic matter with minerals, commonly calcite, silica, or iron oxide. The fossil may disappear before it can be preserved but leave a mold of its original form to record its former presence. This mold may also serve to collect sediments that form a cast of the original organism. It can be the trail or burrow of a once-living thing (Figure 5.27).

(a)

(c)

FIGURE 5.27 Fossils. (a) Silica, dissolved in circulating groundwater, replaced the wood of these ancient trees, which lived about 225 million years ago in Petrified Forest National Park. (b) Trilobites are a group of extinct crustaceans. This one, named *Paradoxides hicksi*, lived 530 million years ago in the seas where Newfoundland is now located. (c) Tracks and trails of animals preserved in the rocks are called trace fossils. Dinosaurs, in what is now Utah, left these footprints in soft mud, which was lithified over the next 160 million years. *((b) Smithsonain Institution. (c) Francois Gohier, Photo Researchers, Inc.)*

(b)

Color of Sedimentary Rocks

Throughout the western and southwestern areas of the United States, bare cliffs and steep-walled canyons provide a brilliant display of the great variety of colors exhibited by sedimentary rocks. The Grand Canyon of the Colorado River in Arizona cuts through rocks that vary in color from gray, through purple and red, to brown, buff, and green. Bryce Canyon, in southern Utah, is fashioned of rocks tinted a delicate pink, and the Painted Desert, farther south in Arizona, exhibits a wide range of colors, including red, gray, purple, and pink. Sedimentary rocks in more humid climates have similar colors, but vegetation and soil usually cover them. Examples of some of the colors in sedimentary rocks are seen in Figure 5.28a–h.

The most important sources of color in sedimentary rocks are the iron oxides. Hematite, for example, gives rocks a red or pink color, and limonite or geothite produces tones of yellow and brown. Some of the green, purple, and black colors are also caused by iron, though in the reduced (low oxygen) form. Only a very small amount of iron oxide is needed to color a rock. In fact, few sedimentary rocks contain more than 6 percent iron, and most contain very much less. Organic matter, when present, also contributes to the coloring of sedimentary rocks, usually making them gray to black. The size of the individual particles in a

FIGURE 5.28 Sedimentary rocks exhibit a wide range of color as shown in this sampling. (a) Mudstones in Petrified Forest National Park. (b) Muddy sandstones in Bryce Canyon, Utah. (c) Limestone in an abandoned quarry near Bandsea, England. (d) Sandstone in the northern Apennine Mountains of Italy. (e) North rim of the Grand Canyon of the Colorado, Arizona. (f) Badlands carved in mudstone, north of Tuba City, Arizona. (g) Coal (black) with sandstone and shale in West Virginia. (h) Sandstone near Jemez Pueblo, New Mexico. (i) Sandstone in northwestern Pennsylvania.
((i) Franklyn B. Van Houten)

(a)　　　　　(b)　　　　　(c)

(d)　　　　　(e)　　　　　(f)

(g)　　　　　(h)　　　　　(i)

rock influences the color, or at least the intensity of the color. For example, fine-grained clastic rocks are usually somewhat darker than coarse-grained rocks of the same mineral composition.

Depositional Environments

From the days of the earliest geologists, it was the allure of the story locked in dusty sandstones, limestones, and other sedimentary rocks that fascinated researchers. It still does. Walking back through time, deciphering from the rocks the letters, words, sentences, paragraphs, and chapters of Earth history, continues to captivate. We can do this by applying the test of uniformitarianism introduced in Chapter 1. By studying the processes of the present we can see the results of similar processes preserved in rocks. With this tool we can reconstruct ancient geographies, replete with shorelines, changing climates, advancing glaciers, belching volcanoes, fluctuating stream ways, and shifting life zones. What we are reading in the rocks is the nature of past depositional environments, and we can only do so by being familiar with those of the present.

Types of Depositional Environments

There is probably no place on Earth were sediments do not settle, if only very briefly. Of importance for the formation of sedimentary rocks are the localities where sediments can accumulate to considerable thickness and eventually turn into rock. Geologists often refer to such areas as **depocenters,** a term shortened from "depositional centers."

The environments represented by these areas of deposition vary widely—from a high alpine lake fed by waters of a melting glacier, to the floor of an oceanic deep (Figure 5.29). Each of these environments can leave its imprint on the sediments deposited there and be reflected in the rock that may form.

The environments on land in which sediments accumulate in significant amounts include the glacial environments associated with not only the ice itself but also the streams and lakes that are fed by the melting ice. Coarse clastic sediments can collect in mountain basins. In arid regions, sheets and dunes of wind-blown sands can reach tens of meters in thickness. In some desert environments salty lake deposits occur. River-formed plains are the sites of fine-grained flood deposits and the sands and gravels of the river channels themselves. The sediments of some rivers, on reaching the sea, accumulate in

FIGURE 5.29   Some major depositional environments. See text for discussion.

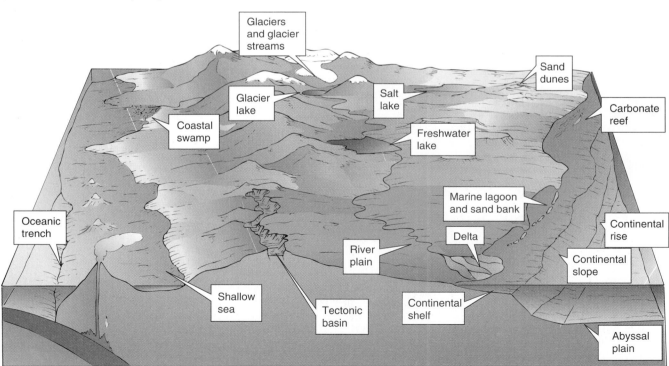

broad, swampy deltaic plains such as those of the Mississippi and Nile rivers.

The margins of the ocean basins provide major depositional environments. They differ depending both on latitude and on the relation of the margin of the basin to plate boundaries. Consider first the Atlantic Ocean. No plate boundaries mark its margin, except for a short distance around the Caribbean Sea (Figure 1.15). Elsewhere the margin of the Atlantic basin is divided into three different depositional zones, the **continental shelf, continental slope,** and the **continental rise.** The continental shelf extends from the shore seaward as a gently sloping feature 20 to 100 km wide. It averages about 100 meters in depth at its outer edge where it slopes away abruptly to form the continental slope, which in turn grades into the gentler gradient of the continental rise. Beyond, at depths of 3 km and more, are the **abyssal plains** of the ocean. It is along the shelf, slope, and rise of the ocean that the greatest amount of clastic sedimentation takes place. On the shallow shelf a large percentage of the land-derived sediments are laid down. Here, too, in the warmer latitudes limestone forms, particularly in

association with continent-fringing carbonate reefs. In geologically recent times continental ice sheets have invaded the more northerly continental shelves. Sediments carried across the shelf move down the continental slope, often in the form of turbidity currents that come to rest along the continental rise or out onto the abyssal plain. The abyssal plain also receives the very fine material that travels for great distances before settling to the deep ocean floor. Added to this material are the tests of microscopic plants and animals that live in the upper waters of the ocean. When they die they settle to the seafloor.

Along coastlines at convergent plate boundaries depositional environments are very different from those of the Atlantic-type margins. A deep oceanic trench occurs just seaward of an arc of volcanoes. Here deep-sea sediments are carried into the trench by the moving oceanic plate to which are added volcanic sediments and landslide debris. Between the volcanoes and the land lies a shallow sea that collects gravel, sand, and mud from the landmass and pyroclastic material from the active volcanoes. In some situations the volcanic islands and shallow sea are missing, and

FIGURE 5.30   Diagram to illustrate sedimentary facies. Fine-grained muds are being deposited in a lagoonal facies that is separated from the open sea by a sandy, beach facies. A sandy, shallow water marine facies lies just offshore from the beach facies and grades into a muddy sand facies, and finally, in deeper water, into a mud facies.

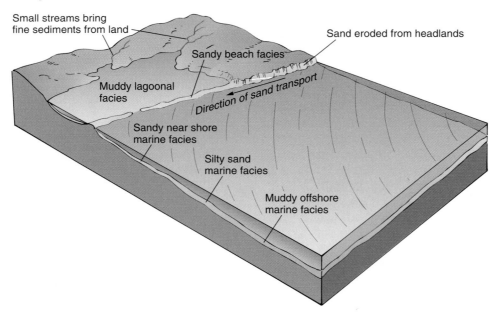

Small streams bring fine sediments from land

Sand eroded from headlands

Sandy beach facies

Muddy lagoonal facies

Direction of sand transport

Sandy near shore marine facies

Silty sand marine facies

Muddy offshore marine facies

the oceanic trench lies immediately next to the continent as it does along the western coast of South America (Figure 1.16).

Sedimentary Facies

We have found that depositional environments differ from place to place, grading laterally from one to the next. Thus, the freshwater environment of a swift-flowing river may change downstream to the quieter, brackish-water realm of a delta that then gives way to the completely saltwater environment of the open ocean. As the environment changes, the sediments that are laid down also change. Deposits in one environment display characteristics, such as mineral composition, grain size and shape, and primary bedding features, different from those of deposits laid down at the same time in another environment. This change in the *look* of the sediments is called a change in **sedimentary facies,** the word "facies" from the Latin meaning "aspect" or "form." We define a sedimentary facies, therefore, as an accumulation of deposits exhibiting specific characteristics, reflecting a particular depositional environment, and grading laterally into other sedimentary accumulations formed at the same time but exhibiting different characteristics.

Figure 5.30 suggests the relationships among several common facies along a shoreline. A lagoonal facies is separated by a beach facies from an offshore sand facies that grades seaward into silt-sand facies and finally into a mud facies. The sediments are all contemporaneous from one facies to the next. If these sediments were lithified and entered into the geologic record we would have a brackish water lagoonal mudstone, giving way fairly abruptly to a beach sandstone and grading gradually through a shallow water marine sandstone into deeper water mudstone. Were we to have this record exposed on the land we would be able to reconstruct the conditions under which the original sediments were deposited.

▶▶ EPILOGUE ◀◀

We can now see how sedimentary rocks fit into the rock system. Our story of the sedimentary rocks began with the transfer of the products of weathering from their places of origin to new places, depositional environments, most commonly in the ocean. These sediments, through varied lithification processes, turn to sedimentary rock, which may exist for hundreds of millions—even billions—of years. Sedimentary rocks vary widely in appearance and characteristics, but can generally be distinguished from igneous and metamorphic rocks by their bedding. Their bedding and bedding features, along with their textures, mineral composition, fossil content, and facies variation allow geologists to determine the geologic history of an area underlain by sedimentary rocks. Thus, while they survive, sedimentary rocks provide a record of past physical environments and the history of life. Destruction, however, is their eventual fate. They may be exposed at the surface, weathered, and converted back into sediments. Alternatively, they may be subjected to temperatures and pressures high enough to create metamorphic rocks. Thus, it is evolution of the metamorphic rocks that we study in the next chapter.

SUMMARY

1. Sedimentary rocks result from the consolidation of loose sediments. They may be detrital, (composed of solid particles weathered from preexisting rocks), biogenic (in which plants and animals provide the material for the rock), or chemical (in which inorganic processes precipitate the material). Most sedimentary rocks are formed in the ocean and then incorporated onto the continents.

2. Roughly 80 to 85 percent of sedimentary rocks are made of quartz and clay. Calcite and dolomite form most of the remaining sedimentary rocks. Texture of a sedimentary rock includes the size, shape, and arrangement of component particles. It may be either clastic or crystalline.

3. Diagenesis includes all those changes a sediment undergoes from the time it is deposited to the time it is destroyed by weathering or metamorphism. It includes lithification, the process by which sediments are turned to rock through compaction, cementation, and crystallization. Diagenesis also includes bioturbation and authigenesis.

4. The four most abundant sedimentary rocks are mudstone, sandstone, conglomerate, and the carbonate rocks—limestone and dolostone. The first three are distinguished from one another on the basis of the size of individual particles in the rocks. The carbonates are distinguished from each other by composition.

5. A diagnostic feature of sedimentary rocks is bedding (stratification). Primary features of bedding include cross-bedding, graded bedding, ripple marks, mud cracks, and sole marks, which allow us to tell the top from the bottom of a sedimentary rock. Secondary features of bedding include nodules, concretions, and geodes. Other characteristics of sedimentary rocks include fossils and color.

6. Depositional environments are the environments in which the sediments that make up a given sedimentary rock were deposited. Sedimentary facies reflect these environments and grade laterally from the environment of one facies to that of another and to other contemporaneous facies. An understanding of the depositional environments associated with different sedimentary rocks allows researchers to determine the geologic history of these rocks.

KEY WORDS AND CONCEPTS

abyssal plain 126	continental rise 126	nodule 121
authigenesis 109	continental shelf 126	oolite 115
bedding 118	continental slope 126	ripple marks 120
bedding plane 118	current ripple marks 120	ripple marks of oscillation 120
biogenic sediment 106	depositional environment 106	roundness 108
biogenic sedimentary rock 106	detrital sediment 106	sedimentary facies 127
bioturbation 111	detrital sedimentary rock 106	sedimentary rock 106
carbonate rocks 114	diagenesis 109	silica 118
cementation 109	fossil 123	sole marks 120
chemical sediment 106	geode 121	sorting 108
chemical sedimentary rock 106	graded bedding 119	sphericity 108
clastic 108	inclined bedding 119	stratification 118
compaction 109	lithification 109	texture 108
concretion 121	mud cracks 120	

IMPORTANT ROCK NAMES

agate 118	coquina 115	mudstone 111
breccia 111	dolostone 111	quartz arenite 112
calcarenite 113	dripstone 116	sandstone 117
chalcedony 118	flint 118	shale 114
chalk 115	graywacke 113	travertine 116
chert 118	jasper 118	tufa 116
coal 127	limestone 111	turbidite 119
conglomerate 111	lithic sandstone 113	

QUESTIONS FOR REVIEW
AND THOUGHT

5.1 Describe how the sedimentary rocks fit into the rock system.

5.2 What are the differences among detrital, chemical, and biogenic sedimentary rocks?

5.3 How do sedimentary rocks compare in terms of volume and area with the igneous and metamorphic rocks of the continental crust?

5.4 What are the dominant minerals in sedimentary rocks?

5.5 What is texture? What determines texture in a sedimentary rock?

5.6 Why are their fewer old rocks exposed on the continents than younger rocks?

5.7 What is diagenesis? Name several kinds of process that might be included in diagenesis.

5.8 What is lithification? How does it take place?

5.9 What are the four main sedimentary rocks? How are they distinguished from one another?

5.10 What is bedding (stratification)? How does it form?

5.11 In addition to bedding what are some characteristics of sedimentary rocks?

5.12 What is a sedimentary facies? Give some examples.

5.13 What is the ultimate fate of sedimentary rocks?

Critical Thinking

5.14 How can limestone be derived from basalt? You should be able to answer this question because you know the composition of limestone, the minerals that make up igneous rocks, and how these minerals weather.

5.15 Suppose you find a rock made up of coarse sand grains of quartz and feldspar that have been cemented together by calcite. What is the general history of the rock?

5.16 Assume you find a mudstone in a sequence of sedimentary rocks. As you look down at the top of the rock you see the several raised imprints of a dinosaur. What has been the history of the rock since the dinosaur left tracks in the soft mud?

6

The Precambrian Vishnu Schist near the mouth of Pipe Creek in the Grand Canyon of the Colorado River. (Dr. John S. Shelton)

Metamorphism and Metamorphic Rocks

OBJECTIVES

As you read this chapter, it may help if you focus on the following questions:

1. What kinds of change take place during metamorphism, and how could you tell that a rock has been metamorphosed?
2. What are the agents that cause metamorphic change, and what is their relative importance in metamorphism?
3. How do geologists map metamorphosed regions, and what criteria do they use to infer the conditions of metamorphism in an area?
4. What conditions characterize each of the five kinds of metamorphic environments (burial, contact, regional, cataclastic, and shock), and in what ways are the rocks that form in each of them distinctive?

OVERVIEW

With this chapter our overview of the rock system will be complete. This is an exciting place to end the tour. Not all geologists will agree, of course, but there is good reason to say that metamorphic rocks are the most enigmatic of the three rock families. They are very common, by some estimates making up a quarter or more of the crust by volume. They also include the oldest rocks yet found on the Earth and some of the most attractive ones used for building or decorative purposes. Unlike volcanism and sedimentation, however, metamorphism takes place out of sight and thus cannot be studied directly. Moreover, metamorphism is almost always a slow process. It is difficult to simulate in the laboratory, although many painstaking experiments have helped to uncover its secrets.

If we truly understand what happens during metamorphism, we can gain a better understanding of the rock system. When the metamorphic process has finished, we sometimes see not only a camouflage of new minerals and new textures but also a trace of what the parent rocks used to be. If we are clever, we can learn to read history in both the new and the old features. Metamorphic rocks can sometimes be a window through which we see all the rest of the rock system.

What Is Metamorphism?

▲▲▲

Metamorphism is a "change of form," if we translate the Greek roots of the word literally. A direct translation doesn't help much, though. Other rock-forming processes also cause changes in form, and the words "igneous" and "sedimentary" at least tell something about the type of change that takes place. What, then, is special about metamorphism?

Mineralogical Changes

During metamorphism, new minerals may grow at the expense of old ones, or existing minerals may change chemically by interacting with surrounding materials. These changes all take place below the Earth's surface but without the presence of magma; minerals grow during metamorphism by reactions that involve only solid materials and heated water (or other fluids). Some of the new minerals are ones that form only in metamorphic environments and are therefore a good way to recognize metamorphic rocks. Among these are:

- *Chlorite,* a sheet silicate of magnesium, aluminum, and iron. The characteristic green color of chlorite was the basis for its name, from the Greek *chloros,* "green." Chlorite has a cleavage similar to that of muscovite or biotite mica, but the small scales produced by the cleavage are much less flexible. Chlorite occurs either as aggregates of minute scales or as individual flakes scattered throughout a rock.

- *Staurolite,* a silicate composed of independent tetrahedra bound together by positive ions of iron and aluminum. It has a unique crystal habit, which is striking and easy to recognize when twinned, with six-sided prisms that intersect either at 90°, forming a cross, or at 60°, forming an X (see Figure 6.1).

- *Andalusite, kyanite,* and *sillimanite,* three polymorphs of Al_2SiO_5. Each has independent silica tetrahedra bound together by positive ions of aluminum, but in its own unique crystalline structure. Sillimanite develops in long, slender crystals that are typically white. Kyanite forms clear to bluish bladelike crystals (Figure 6.2). Andalusite forms white, coarse prisms with nearly square cross sections. Although all three of these minerals are usually associated with metamorphism, sillimanite and andalusite are known to form in some igneous environments as well.

- *Epidote,* a complex silicate of calcium, aluminum, and iron in which the silica tetrahedra are in pairs independent of each other. This mineral is pistachio green or yellowish to greenish black.

Other minerals are less diagnostic, because they may also grow in igneous or sedimentary environments—for example, quartz, calcite, biotite and mus-

FIGURE 6.1 Staurolite ($Fe_2Al_9Si_4O_{22}$ (O, OH)$_2$) grows as porphyroblasts in some muddy sediments that have been metamorphosed to become medium-grade schists (amphibolite facies). It can be recognized easily by its prismatic form and by its tendency to form twins (intergrown crystals) at 60° or 90° to each other. The name staurolite comes from the Greek word *stauros,* "cross," a reference to the twinning. In some parts of the world, these are worn as religious amulets.

covite mica, feldspars, pyroxenes, amphiboles, and garnet. As new minerals form, however, others disappear. Minerals that were produced by chemical weathering, particularly the clays, are usually altered in the creation of new minerals during metamorphism. Their disappearance may also be a sign that metamorphism has occurred.

Chemical changes in existing minerals are not easy to detect without analytical equipment unless they cause a change in color or some other physical property. Sometimes they are the result of renewed growth on extremely small "seed" crystals that were in the parent rock. Changes may also occur when an existing crystal exchanges chemical components with fluids or with other minerals around it. Both kinds of chemical change are slow and are commonly interrupted before they affect the entire parent crystal. As a result, many minerals develop a **reaction rim** (Figure 6.3) in which their own composition is subtly altered or grow a **corona** or halo of some new mineral during metamorphism.

Changes in Grain Size

Very small mineral grains, such as those in a mudstone, are readily changed during metamorphism because a large portion of each grain is exposed to the chemical reactions taking place on its surface. Small grains are gradually reassembled to form a smaller number of large grains, with a much smaller combined surface area. The smaller their total surface area is, the more stable the grains are. One striking result of meta-

FIGURE 6.2   Kyanite is one of three polymorphs of Al_2SiO_5. It is found primarily in sediments that have been metamorphosed at high pressure. Kyanite gets it name from the distinctive blue ("cyan") color of its bladed crystals. *(William E. Ferguson)*

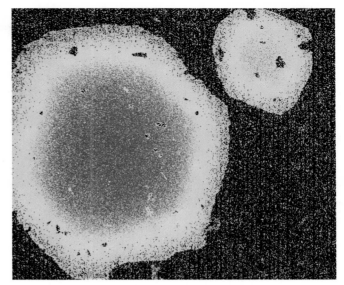

FIGURE 6.3   Although fairly uniform inside the crystal, the manganese content of this garnet decreases near its rim, probably as a result of an incomplete chemical reaction with a neighboring ferromagnesian mineral during metamorphism. This image is not a photograph. It is an x-ray fluorescence map produced by an electron microprobe. For ease of interpretation, high-manganese areas are shown in red and lower-manganese areas in yellow. The larger of these two garnets is about 0.6 mm across. Geologists can use subtle chemical clues like these to figure out how each of the agents of metamorphism functioned as the changes occurred. *(Lincoln Hollister)*

morphism, then, is an overall increase in average grain size.

A few minerals, like garnet and staurolite, tend to grow much larger than other minerals in the same rock, as illustrated in Figure 6.1. These larger crystals, called **porphyroblasts,** are prominent features that may easily attract the eye of a mineral collector. Because they grow slowly during metamorphism, they may also yield valuable information about events during the growth process, as we will see later in this chapter.

Changes in Rock Texture

One universal effect of metamorphism is that recrystallizing minerals grow together to form a tightly interlocked texture that increases the strength of many rocks. Figure 6.4, for example, contrasts the texture of a sandstone, in which rounded quartz grains are held together by the cementing agent between them, with a quartzite (its metamorphic counterpart), in which quartz grains have grown so that there is no longer any space between them and they are firmly interlocked. A nongeologic example of this same process can be seen in pottery, which is fired at a high temperature to increase its strength (Figure 6.5).

In rocks dominated by a single mineral, the only physical changes may be a general increase in grain size and interlocking texture. In some very common metamorphic environments that we will discuss later in this chapter, though, metamorphic minerals are aligned to form new planar structures that geologists call **foliation** (from the Latin *foliatus,* "leaved" or "leafy," hence consisting of thin sheets) or linear patterns called **lineations.** These take many forms, depending on which minerals grow and on how intensely the original rocks were deformed during metamorphism. When geologists describe metamor-

phic rocks, they commonly use foliation or lineations as primary distinguishing features (Figure 6.6). These newly oriented structures sometimes obliterate evidence of bedding or other features in the parent rock. Usually, though, they just overprint earlier features, so that it may be possible to see relics of a rock's history even after repeated metamorphic events.

Agents of Metamorphism

▲▲

Heat

Heat is the most essential agent of metamorphism. It is the energy that breaks chemical bonds in minerals so that their constituents can move independently and recombine to form new crystals. Just as a roast cooks faster in a hot oven than in a cooler one, heat-driven chemical changes are more rapid as the heat available during metamorphism increases. Thermal energy also releases water, another of the important agents of metamorphism, from within clays and other minerals. Metamorphism is most likely to occur, therefore, near natural sources of heat: around igneous bodies, at the boundaries of tectonic plates, and deep in the lithosphere.

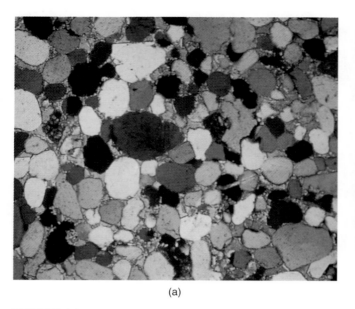 

(a)   (b)

FIGURE 6.4   (a) Quartz sandstone. Individual quartz grains are rounded to subrounded and are held together by the cementing agent between them. There is a small amount of open space (porosity) between grains as well. (b) Quartzite. During metamorphism, individual grains have grown together, forming a much stronger rock. There is no cementing agent or porosity.

Pressure

When rocks are buried under the weight of overlying material, they gradually deform, potentially changing either their size or shape. The effect of pressure on individual mineral grains is a major factor in metamorphism. It may cause some minerals to disappear and favor the growth of others. If pressure is not applied from all directions with equal intensity, it can also influence the shape and orientation of mineral grains within a rock.

If pressure is applied uniformly from all directions, matter is forced into a smaller volume without changing shape. Under these conditions, called **lithostatic** pressure, materials with a high density are more stable than those with a lower density. For example, if we squeeze calcite, the low-density polymorph of $CaCO_3$, its atoms rearrange into the more tightly packed crystal structure of aragonite, the high-density polymorph. If we squeeze a rock that contains aragonite, there is little change because the atoms are

FIGURE 6.5   In this photomicrograph, you can compare the textures of clay pottery before and after firing. These two pieces of pottery are made of the same clay. The lower one was fired in a kiln; the upper one was not. Firing has metamorphosed the clay, increasing its grain size and altering its color. Although it is not apparent in this photo, the fired sample has an interlocking grain texture and is much more durable than the unfired one.

FIGURE 6.6   Foliation is a planar feature that is due to the parallel alignment of platy or elongated minerals, primarily micas, in response to directed pressure during metamorphism. In this photomicrograph, the foliation is largely due to the orientation of the clear, prismatic crystals of sillimanite, and to the brown flakes of biotite mica.

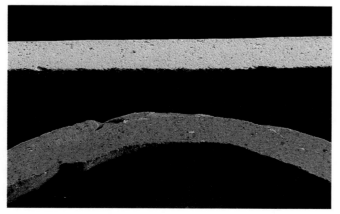

already close together. In the same way, andalusite, the low-density polymorph of Al_2SiO_5, is less stable under pressure than either kyanite or sillimanite, whose crystal structures are more compact.

An assemblage of two or more minerals may also be compressed to form a denser aggregate of new minerals, especially if water, carbon dioxide, or some other fluid escapes during the reaction. For example, metamorphism can transform a rock containing the minerals *hornblende* (an amphibole), biotite, and quartz into a rock composed of pyroxene, plagioclase, and orthoclase, plus water:

$$Ca_2Mg_3Si_6Al_4O_{22}(OH)_2 \ + \ KMg_3AlSi_3O_{10}(OH)_2 \ + \ 4SiO_2 \rightarrow$$
$$\text{(hornblende)} \qquad \text{(biotite)} \qquad \text{(quartz)}$$

$$6MgSiO_3 \ + \ KAlSi_3O_8 \ + \ 2CaAl_2Si_2O_8 \ + \ 2H_2O\uparrow.$$
$$\text{(pyroxene)} \ \text{(orthoclase)} \ \text{(plagioclase)} \quad \text{(water)}$$

The new minerals take up less room than the original ones, especially if the water is squeezed out of the rock through pore spaces or tiny cracks, and thus they constitute the stable mineral assemblage when pressure increases during metamorphism.

When pressure is greater in some directions than in others, as it may be in an active mountain-building region, existing mineral grains can be stretched or flattened. As a crystal is squeezed by this **directed** pressure, layers of atoms within its crystal structure slide past each other like playing cards do when you push gently on one edge of a deck (Figure 6.7). As a result, the crystal changes shape. If the layers of atoms slide slowly enough to keep them from being stretched or broken, then they can reattach to each other seamlessly as the crystal's shape changes. Unlike a deck of cards, then, the deformed crystal is left with no visible cleavages to indicate exactly how its atoms slid. Under directed pressure, it is also possible for parts of a mineral grain to dissolve while other parts are growing. In Chapter 16 we will see that snow gradually turns into glacial ice and then is deformed slowly by directed pressure as the glacier moves, so that a glacier can correctly be called a body of metamorphic rock.

While these and other processes slowly reshape existing grains, crystals of new minerals grow as well. These minerals grow in the shape and direction that is most stable according to the given pressure. The minerals most likely to respond in this way are the micas, whose silicate structures are based on sheets of silica tetrahedra. Flakes of mica grow most readily when their flat surfaces are perpendicular to the greatest

FIGURE 6.7 (a) A mineral may deform in many different ways in response to a directed pressure. One way, illustrated here, takes place at submicroscopic scale as layers in the crystalline structure slide slowly past each other. (b) Without ever fracturing, the mineral gradually changes shape like a disrupted stack of playing cards. *(From: IGNEOUS AND METAMORPHIC PETROLOGY by Best. Copyright © 1982 by W.H. Freeman and Company. Used with permission.)*

FIGURE 6.8 As directed pressure is applied, new mica flakes grow with their flat surfaces perpendicular to the pressure. Because many grains of other minerals survive the recrystallization, it is usually possible to see earlier planar features (such as bedding, igneous flow structures, or even previous foliation) as well as the new foliation. An astute observer, therefore, can "read" the history of events that have modified a metamorphic rock as though they were drawn on acetate overlays. *(From: IGNEOUS AND METAMORPHIC PETROLOGY by Best. Copyright © 1982 by W.H. Freeman and Company. Used with permission.)*

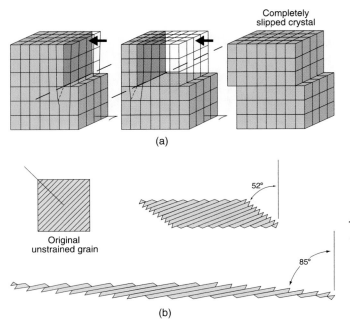

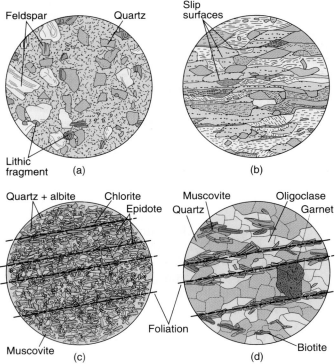

Marble from a Gun

The first laboratory experiments to simulate the Earth's subsurface conditions were conducted by Sir James Hall (1761–1832), a Scottish chemist and geologist who was a younger colleague of James Hutton.

Hutton's theory of the Earth held that heat had acted on all rocks at some past time and turned the original materials into new forms—that is, metamorphosed them. This troubled Hall, who reported that he had "almost daily warfare with Dr.

Hutton on the subject of his theory."

Hall argued that if Hutton's ideas were correct, then there should be quite a lot of metamorphosed limestone in the Earth. By the late 1700s, however, it was com-

---

pressure. Old and new mineral grains, therefore, contribute to the development of foliation and lineations (Figure 6.8; see also Figure 6.6).

Reactive Fluids

Water plays a very important role in metamorphism. Often charged with dissolved ions and gases (commonly CO_2), water may join the metamorphic process from any of four common sources (Figure 6.9). First, **formation waters** are the fluids trapped in the pores of a buried sediment. In most cases, these began as seawater but have been slowly altered as diagenesis changed not only the sediments but also the waters in them. Second, we hinted at the progressive **dehydration** of hydrous sedimentary minerals in discussing the effects of heat and pressure. Certain minerals, like the clay minerals, chlorite, and micas have water in their chemical makeup. During metamorphism the water is driven off in exactly the same way that it was released in the chemical reaction we used as an example earlier, producing anhydrous (waterless) forms of these minerals. Third, **juvenile hydrothermal fluids** (Latin *juvenis* "young" + Greek *hydro* "water" + *therme* "heat") released in the solidification of magma often percolate beyond the boundaries of the magmatic body and react with surrounding rocks. Sometimes these fluids remove ions and substitute others; at other times they add ions to the rock minerals to produce new minerals. Finally, any of the first three sources may be supplemented by deep-circulating **groundwater** or, on the ocean floor, by **seawater.**

These sources vary in significance, depending on the setting of metamorphism. Careful study of meta-

morphic rocks from many environments suggests that fluids do not migrate more than a meter or so from their source area. Geologists speak of the processes in which rocks "stew in their own juices" as **isochemical**

FIGURE 6.9   Formation water (1) is a brine that was trapped between sediment particles during sediment burial. Water of dehydration (2) was originally part of the crystalline structure of clays, micas, or minerals like gypsum, but has been driven out by heating. Hydrothermal fluids (3) are released from some water-rich magmas during cooling and crystallization. Groundwater or seawater (4) may percolate downward through pores or open fractures. Each of these may be a potential reactive fluid as metamorphism progresses. Because fluids may escape and pore spaces gradually close during metamorphism, reactive fluids are usually less effective agents during retrograde metamorphism.

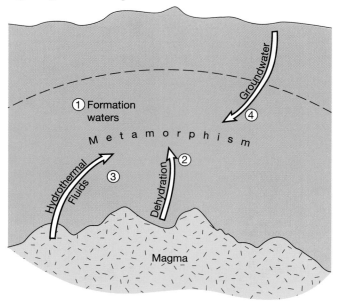

mon knowledge that heat turns calcite ($CaCO_3$) to a powdery product we know as lime (CaO), which bears little resemblance to firm rock. Because geologists never find lime in the Earth, Hall insisted, Hutton's concept of metamorphism must be wrong. Hutton, however, had visualized heat operating deep in the crust not as a lone agent, but together with the confining pressure of overlying rock. Once Hall realized this, he viewed Hutton's ideas "with less and less repugnance."

In 1801, four years after Hutton's death, Hall began a series of experiments that were effective, although certainly crude by today's standards. He loaded powdered limestone into the breech of a gun, and then completely sealed the gun. He then stuck the breech into a furnace and heated it. Thus the calcium carbonate was heated under pressure. In several tries the breech of the gun ruptured, ruining the experiment. In others, however, the powdered limestone turned to a solid mass. In one, a small amount of the powder "was found to be completely crystallized, having acquired the rhomboidal fracture of calcareous spar."

Hall had created marble, now recognized as a metamorphic rock, from powdered limestone by applying two of the agents of metamorphism—heat and pressure.

reactions, meaning that chemical constituents are simply reshuffled within a small volume of rock. In these settings, formation waters and water of dehydration are probably most important, although fluids may make up less than one-tenth of the rock volume. Where larger amounts of fluid are available, or where the rocks are more permeable, fluids may travel great distances. During this kind of metamorphic process, called **metasomatism,** the overall ratio of water to rock may be as high as 10 to 1 or even 100 to 1. Hydrothermal fluids and seawater can be particularly important in metasomatism. Rocks may be changed dramatically, losing some chemical constituents to the migrating fluid and gaining others. In many regions of the crust where metasomatism occurs, the rocks are enriched in valuable metals such as copper, lead, zinc, or silver derived from nearby igneous bodies (Figure 6.10).

Most of what geologists know about metamorphic fluids has been inferred from the chemical compositions of metamorphic rocks. When you collect a sample of rock, of course, the fluid that was once in it has long since been consumed in mineral reactions or has simply escaped along grain boundaries or through minute fractures. Geologists speak of two stages in a metamorphic episode: a **prograde** stage during which rocks heat up and are subjected to increasing pressure, and a **retrograde** stage in which they cool and decompress. There is much more fluid around during the prograde stage, as heat and pressure encourage dehydration and before recrystallization seals open passages in the rock. The fluid has been sampled in a few deep wells in places like the Salton Sea basin in Southern California. The only other "direct" samples we have are tiny droplets trapped inside metamorphic minerals as **fluid inclusions** (Figure 6.11). These few traces, surrounded by the growing crystals, are all that remain once the retrograde stage has ended.

FIGURE 6.10   The veins in this rock are filled with quartz and with sulfide minerals, they were deposited from a hydrothermal fluid that carried dissolved mineral matter through the rocks. *(Peter Kresan, Lawrence Berkeley Laboratory)*

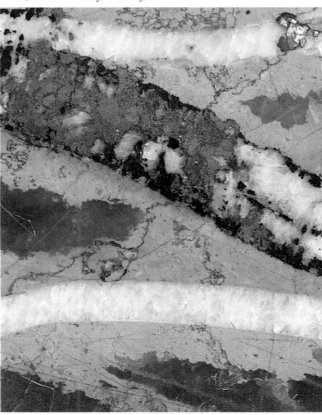

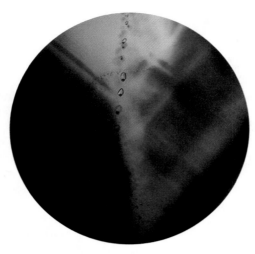

FIGURE 6.11   Fluid inclusions are droplets of the metamorphic fluid that were surrounded as a mineral grain grew. If these are undisturbed by later fracturing or recrystallization, geologists can analyze these minute samples to learn more about the conditions during metamorphism. *(Catherine K. Richardson)*

Analyses of these trace samples confirm what geologists deduce from the rocks themselves. The residual fluids are brines of sodium, potassium, and calcium chlorides and sulfates, from one to ten times as salty as seawater (about 3.5 percent salt by weight). They also bear impressively high concentrations of dissolved silica, carbon dioxide, and, on occasion, metallic elements.

Time and Metamorphism

Metamorphism can occur over a wide range of time scales. At the impact of a meteorite, a metamorphic event is virtually instantaneous, whereas during a mountain-building episode, metamorphism may take more than a million years. In most cases, unfortunately, geologists have little way of telling how long a rock was subjected to high temperatures, pressures, and reactive fluids. When we see large crystals in a metamorphic rock we feel intuitively that they must have undergone metamorphism over a long period of time. This may very well have been true, but it is by no means always the case. Geologists cannot yet confidently equate the size of metamorphic minerals with time, as they can with crystal size in igneous rocks.

Geologists have tried to get better estimates by using dating methods based on radioactive decay, which we will discuss in the next chapter. These applications generally rely on the fact that products of

radioactive decay, which usually escape easily from a mineral when it is hot, are trapped and begin to accumulate in it once the mineral cools and its crystalline structure contracts. Each mineral begins accumulating decay products at a different temperature, thus a careful comparison of the absolute ages of different minerals in the same metamorphic rock can indicate how rapidly temperature changed as the metamorphic event came to a close.

Styles of Metamorphism

The Range of Metamorphism

Figuring out where the upper and lower limits of metamorphism lie is not a simple matter. The boundary between advanced diagenesis and **low-grade metamorphism** is an arbitrary one, customarily defined by the appearance of minerals that geologists have agreed to classify as metamorphic products. This occurs at about 200°C at pressures equivalent to a depth of a few thousand meters in the crust (Figure 6.12).

The limit of **high-grade metamorphism,** at the other end of the scale, is just as hard to pin down. By definition, any process that produces magma is igneous. Therefore, the upper limit of meta-

FIGURE 6.12   The intensity of metamorphism can be described in relative terms as "low-grade," "intermediate-grade," or "high-grade." The diagram illustrates how these general labels are related to the range of metamorphic pressures and temperatures. The boundaries between diagenesis and low-grade metamorphism and between high-grade metamorphism and igneous processes are not sharp, but gradational.

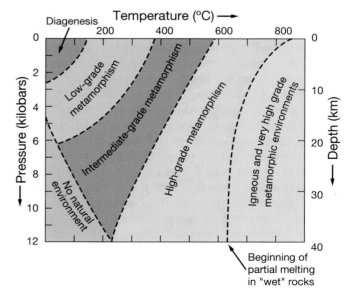

morphism should occur where rock is hot enough to melt. The melting temperature of rock, however, depends heavily on the kind of minerals and the amount of water in it. A "wet" rock or one that contains a high proportion of quartz and feldspars may begin to melt at about 600°C, whereas a relatively "dry" one may remain solid to 750°C or higher. The ferromagnesian minerals, quartz, and feldspars also melt across different temperature ranges as a rock is heated, thus their relative proportions typically vary depending on how much total melting has occurred.

Studies indicate that melting starts first around hydrated minerals or along selected grain boundaries where a small amount of fluid may be present. Most of a rock may be experiencing metamorphism, then, while parts may already be igneous. In 1907, Swedish geologist J.J. Sederholm first called attention to rocks he named **migmatites,** which have zones of quartz, micas, and feldspars that have segregated from the surrounding rock and have clearly been melted. These "mixed rocks," formed across the fuzzy range of temperatures shown in Figure 6.12, are at the upper limit of metamorphism.

Curiosity about high-grade metamorphism leads to a question with much broader implications. James Hutton believed firmly that granite is produced when minerals crystallize from a molten mass. Ever since, most geologists have assumed that granite is an igneous rock, and that batholiths form by the stoping process that we described in Chapter 3. Some geologists, however, have wondered what happens to the great masses of rock that must be displaced when magma intrudes to form a batholith. This so-called space problem has led those geologists to speculate that some batholiths are actually metamorphic bodies and that the granite in them is a product of metasomatism, sometimes called **granitization** in this context. Field observations around some batholiths support this theory. Sedimentary rocks grade into metamorphic rocks and then into migmatites. Closer to a batholith the migmatites, in turn, give way to rocks that contain large, abundant feldspars characteristic of granite but that also seem to show shadowy remnants of foliation. Geologists observe that these rocks ultimately grade into granite without ever exhibiting a sharp boundary.

Some geologists contend that as much as a quarter of the granite exposed at the Earth's surface is metasomatic. Most insist on a much lower proportion. Because batholiths occur at the core of every major mountain chain, an understanding of how they form and of the relative importance of igneous and metamorphic processes is a key element in plate tectonic theory.

Metamorphic Facies

The first geologists to make a systematic attempt to map exposures of metamorphic rocks were faced with an unfamiliar problem. Igneous and sedimentary rock units are reasonably uniform. If you were to examine a single shale bed where it is exposed across the countryside, for example, you would find its grain size and mineral makeup to be nearly the same everywhere. Not so with metamorphic rocks. The mineral assemblage and texture in a *metamorphosed* shale may vary quite a bit from place to place, although its average chemical composition remains the same. Clearly, although a metamorphic rock inherits its bulk chemical composition and various features like bedding planes from its premetamorphosed parent, the minerals at a particular spot are determined by the intensity of the metamorphism there.

Studying metamorphic rocks across Scotland in the 1800s, these geologists devised a way to indicate the intensity of metamorphism by mapping **metamorphic zones** (Figure 6.13). Each zone was marked by a characteristic **index mineral** that was not present in zones of lower-grade metamorphism. In the metamorphosed shales of the Scottish Highlands, the index minerals appeared in the following order of increasing metamorphic intensity: chlorite, biotite, garnet, staurolite, kyanite, and sillimanite. On a map, the lines separating one metamorphic zone from another are called **isograds.** It is now common practice to speak, for example, of the "garnet isograd" as an imaginary line on the ground where garnet first appears as we walk into the "garnet zone."

The only problem with this method is that a sequence of index minerals that works for metamorphosed shale doesn't necessarily apply in any other kind of rock. A mineral like staurolite can only grow in a metamorphic rock with the proper chemical composition. Staurolite may grow in a metamorphosed shale, for example, but will never appear in a metamorphosed basalt or limestone. Basalts and limestones do not contain enough aluminum to make staurolite. We face similar problems with any index mineral or mineral assemblage.

By 1915, it was clear that although index minerals are helpful field guides, it is best to think of levels of metamorphic intensity in terms of the conditions it takes to reach them. Finnish geologist Pennti Eskola used this concept to define **metamorphic facies.** A facies is based on not one but several mineral assemblages that are stable within the same specific range of temperatures and pressures, each in a different kind of rock. Table 6.1 shows, for example, that shaly rocks in the *greenschist* facies may typically contain chlorite, muscovite, plagioclase, and quartz, whereas basaltic

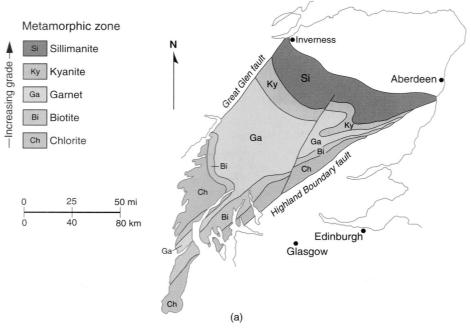

Metamorphic zone

↑ Increasing grade

Si — Sillimanite
Ky — Kyanite
Ga — Garnet
Bi — Biotite
Ch — Chlorite

(a)

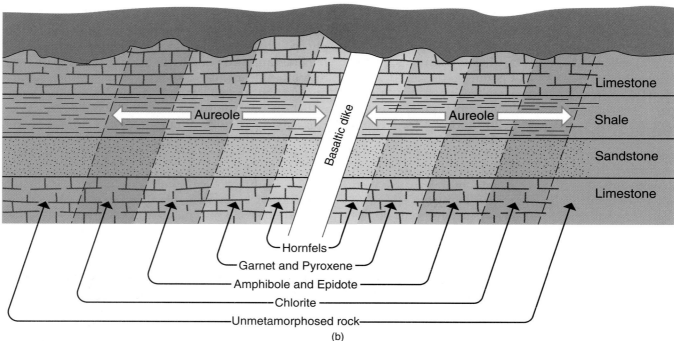

Limestone

Shale

Sandstone

Limestone

Aureole ← → Basaltic dike ← Aureole →

Hornfels
Garnet and Pyroxene
Amphibole and Epidote
Chlorite
Unmetamorphosed rock

(b)

FIGURE 6.13 (a) Metamorphic zones in the Scottish Highlands. Each zone is marked by the first appearance of an index mineral, and is separated from the next by an imaginary line called an *isograd*. The staurolite zone is so narrow that it does not appear on a map at this scale. (b) The same method identifies narrower metamorphic zones in an aureole adjacent to an intrusion of magma.

rocks contain chlorite, amphibole, plagioclase, and epidote.

Because Eskola was most familiar with basaltic rocks, many of the facies names that we now use— *greenschist, eclogite,* and *amphibolite,* for example—are ones that originally described metamorphosed basalts.

These, and the range of pressures and temperatures they represent, are shown in Figure 6.14.

Metamorphic Environments

Geologists identify five kinds of metamorphic environments, each characterized by a unique set of geo-

TABLE 6.1 Characteristic Mineral Assemblages in Selected Metamorphic Rocks

| FACIES NAME | MINERAL ASSEMBLAGES DEVELOPED FROM | |
	SHALE	BASALT
Zeolite	Zeolites, pyrophyllite, clays	Calcite, chlorite, zeolites
Greenschist	Chlorite, muscovite, plagioclase, quartz	Chlorite, amphibole, plagioclase, epidote
Epidote-amphibolite	Garnet, chlorite, muscovite, biotite, quartz	Amphibole, epidote, plagioclase, garnet
Amphibolite	Garnet, biotite, muscovite, sillimanite, quartz	Amphibole, garnet, plagioclase
Granulite	Biotite, K-feldspar, quartz, sillimanite	Pyroxene, plagioclase
Blueschist (Glaucophane schist)	Muscovite, chlorite, quartz, garnet, lawsonite	Glaucophane, lawsonite, chlorite
Eclogite	Not observed	Jadeite, Mn-garnet, kyanite
Hornfels	Andalusite, biotite, K-feldspar, quartz	Pyroxene, plagioclase

logic conditions and indicated in Figure 6.15. These are the environments of **burial, contact, regional, cataclastic,** and **shock** metamorphism.

Burial Metamorphism In places where great thicknesses of sediment accumulate, the temperature 7

FIGURE 6.14 Geologists have redrawn this diagram many times, so there is no standard set of facies names, nor are the exact boundaries of facies (in terms of pressure and temperature) universally accepted. This simplified version, however, is close to the one proposed by Eskola in 1915 and is in reasonably common use. The two labeled lines indicate "pathways" of pressure-temperature change during metamorphism in (a) subduction zones and (b) continental mountain-building regions.

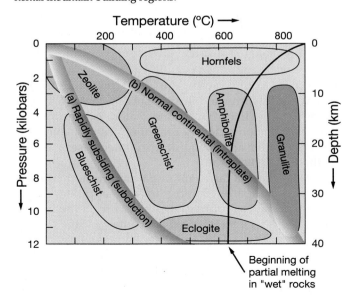

or 8 kilometers below the surface may approach 300°C or more. Some recrystallization takes place at shallower depths, during diagenesis. The appearance of new, unique minerals here, however, tells us that metamorphism has begun. The most distinctive of these are the *zeolites,* silicates that have many chemical traits in common with the feldspars but that contain water.

Pressures applied during *burial metamorphism* are fairly low and uniform. Formation waters trapped in the spaces between mineral grains speed up the pace of recrystallization and may produce larger crystals. High fluid pressure between grains, however, also holds individual particles slightly apart and absorbs some of the stress on them. As a result, these rocks show few signs of deformation. In fact, unless the new minerals grow large enough to appear unusual we may have a hard time recognizing that metamorphism has occurred.

Contact Metamorphism Baking of rocks immediately adjacent to a body of magma is called *contact metamorphism.* Heat is the primary agent of metamorphism in these rocks and is effective only in a zone called an **aureole** ("halo"), which may be anywhere from a few centimeters to a few hundred meters wide (see Figure 6.13b). Aureoles are found mainly bordering plutons, although surface rocks may be metamorphosed immediately below a lava flow.

During contact metamorphism, temperatures may range from 300°C to 800°C. The narrowness of contact aureoles, however, suggests that the rocks are not hot for very long. Lithostatic pressure, as deduced from index minerals, is very low. Geologists conclude from these conditions that contact metamorphism typically

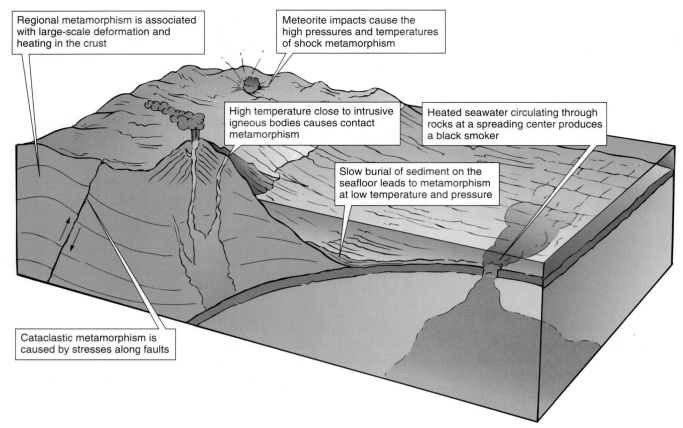

Regional metamorphism is associated with large-scale deformation and heating in the crust

Meteorite impacts cause the high pressures and temperatures of shock metamorphism

High temperature close to intrusive igneous bodies causes contact metamorphism

Heated seawater circulating through rocks at a spreading center produces a black smoker

Slow burial of sediment on the seafloor leads to metamorphism at low temperature and pressure

Cataclastic metamorphism is caused by stresses along faults

FIGURE 6.15 Geologists recognize five kinds of metamorphic environments, suggested in this summary illustration.

occurs within one kilometer of the Earth's surface. There is little directed pressure, thus no reorientation of minerals normally occurs. A textural term commonly applied to the nonfoliated, fine-grained products of contact metamorphism is **hornfels,** a word that geologists also modify for use as a rock name and as a facies label.

Chemical reactions during contact metamorphism depend on the temperature, the composition of the intruding magma, and the properties of the intruded rock. Without a high lithostatic pressure to confine them, metamorphic fluids escape easily from this environment; hence, the rocks of a contact aureole may boil dry like a pot left on the stove. This is particularly likely if the magma providing the heat is basaltic, therefore probably containing little water itself. The result may be an assemblage of anhydrous minerals. For example, when an impure limestone is subjected to thermal contact metamorphism, its carbonate minerals, clay, and quartz may be changed to new minerals. Calcite and quartz may combine to form *wollastonite* (a pyroxene-like mineral) and release carbon dioxide:

$$CaCO_3 + SiO_2 \rightarrow CaSiO_3 + CO_2.$$
(calcite) (quartz) (wollastonite)

Dolomite may react with the quartz to form *diopside* (a pyroxene) plus carbon dioxide:

$$CaMg(CO_3)_2 + 2SiO_2 \rightarrow CaMg(Si_2O_6) + 2CO_2.$$
(dolomite)   (quartz)   (diopside)

In more complex reactions, Al_2O_3 and silica in clay will react to form andalusite, spinel, or garnet, and release water. If carbonaceous materials are present, they may be converted to graphite, possibly allowing not only water and CO_2 but also other gases such as methane (CH_4) to escape.

By contrast, a large volume of hydrothermal fluid may be released as granitic magma approaches the surface and the surrounding rocks are bathed in it. Under these circumstances, a contact aureole may develop an assemblage of unusual minerals containing water, chlorine, fluorine, carbon dioxide, or sulfur. These include serpentine, amphiboles and micas, gemstones like tourmaline and topaz, and frequently mined metal sulfides.

Regional Metamorphism Regional metamorphism results from intense compression and deep burial and is commonly associated with batholiths or other large intrusions—all of which are common during mountain-building and other events that occur near convergent plate boundaries (see Chapter 1). Regionally metamorphosed rocks, therefore, are found in the root regions of old mountains, commonly in belts that extend over thousands of square kilometers. Usually, thousands of meters of overlying rock must be eroded away in order to expose these rocks.

In many ways, regional metamorphism is similar to contact metamorphism but on a larger scale and at a greater depth in the crust. In fact, it would be difficult to decide exactly where one of these environments grades into the other. In most cases, however, the conditions of regional metamorphism produce rocks that look very different from most contact-metamorphosed rocks.

Regional metamorphism may take place within a wide range of temperature and pressure conditions. These conditions are met in very different tectonic environments, each giving rise to a particular series of metamorphic facies (compare Figures 6.14 and 6.15). We have suggested two common pathways of progressive regional metamorphism by the dashed lines in Figure 6.14. Blueschist and eclogite facies conditions (Figure 6.14a) occur as rocks are drawn deeper into subduction zones. Because subduction is a relatively rapid process but rocks are poor conductors of heat, this is a pathway of high pressure but fairly low temperature. Greenschist and amphibolite facies conditions (Figure 6.14b), however, occur during the slow compression that forms continental mountain belts. Along that pathway, temperature may rise steadily for several million years, keeping pace with the gradual increase in pressure even to the limit of partial melting.

In any regional metamorphic environment, rocks are subjected to directed as well as lithostatic pressure. In general, the characteristic foliation that results is not parallel to bedding in the parent sedimentary rocks or to flow banding in igneous rocks. The direction and intensity of foliation give geologists a way to interpret the direction and intensity of the forces that prevailed during metamorphism.

Geologists define three kinds of foliation, loosely correlated with different average-grain sizes in regionally metamorphosed rocks:

- **Rock cleavage** is sometimes called **slaty cleavage** because it is characteristic of the rock known as **slate** (from the Old French *esclat,* "fragment" or "splinter"). Rock cleavage is not the same as mineral cleavage, which we discussed in Chapter 2. Mineral cleavage takes place parallel to planes of atoms *within* a single crystal. Rock cleavage occurs parallel to and *between* crystals of flat minerals such as micas and clays. In very fine-grained metamorphic rocks, typically those that were originally shales, mudstones, or tuffs, the foliation occurs along flat surfaces separated by distances of microscopic dimensions (Figure 6.16). This makes it possible to cleave slate into thin, strong sheets for use as a building stone or, in decades past, for blackboards or billiard tables. Slate is composed predominantly of small mica flakes formed during low-grade metamorphism but may also contain carbonaceous material, hematite, or other minerals that produce a wide variety of colors.

FIGURE 6.16   (a) The upper surface of this cylinder of slate is a rock cleavage surface. It is at an angle to the bedding planes in the original sediment, still visible on the sides of the cylinder. Notice the many cleavage fragments of slate nearby. (b) Slate's excellent cleavage makes it useful as a building stone. A skilled workman can cleave uniform sheets of slate 1 cm thick or less.

(a)

(b)

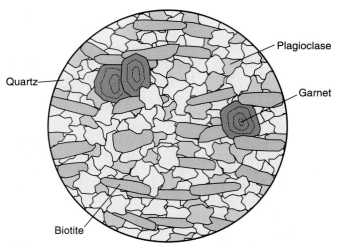

FIGURE 6.17   As flakes of mica grow during regional metamorphism, they tend to align perpendicular to the directed pressure, thus producing foliation. Because the micas and other minerals in the rock are moderately large, however, they drape around each other and do not form perfectly flat layers. This "crinkly" foliation is called *schistosity*.

- **Schistosity** is also a planar feature, but is more crinkly. Instead of lying perfectly flat, flaky or fibrous minerals grow at a slight angle to each other or are disrupted by directed pressure that continues during metamorphism (Figure 6.17). Rocks with a schistosity, therefore, rarely cleave perfectly. When slate is subjected to continued metamorphism, the micas in it develop large flakes, giving the resulting rock a silky sheen on freshly broken surfaces. If the micas are barely visible to the unaided eye, we call the rock a **phyllite** (from the Greek *phyllon*, "leaf") (Figure 6.18). This rock usually contains many of the same minerals as slate, but sometimes a new metamorphic mineral, such as tourmaline or manganese garnet, makes its appearance. A **schist** (from the Greek *schistos*, "divided" or "divisible") contains mica flakes that are clearly visible (Figure 6.19). This is by far the most common kind of regional metamorphic rock. Both muscovite and biotite are abundant in metamorphosed mudstones, as are large quantities of quartz and feldspar and lesser amounts of pyroxene, amphibole, garnet, epidote, and magnetite. Chlorite and epidote, both green minerals, are more common in metamorphosed basalts; hence the name "greenschist." Blueschists are named for the abundance of *glaucophane*, a blue, prismatic amphibole formed in metamorphosed oceanic basalts.
- **Gneissosity** (from the Greek *gneis*, "spark," for the luster of some of the components), is a much coarser foliation than schistosity, in which the new minerals occur in bands or layers. **Gneiss,** a granular metamorphic rock (pronounced "nice"), has bands of quartz, feldspar, and micas that alternate with bands of ferromagnesian minerals (Figure 6.20a). These are the product of high-grade metamorphism, possibly associated with granitization, as we discussed earlier in this chapter. Under the extreme directed pressures of mountain building, the banding of gneisses may be highly contorted, as though they had once been the consistency of putty or toothpaste (Figure 6.20b).

FIGURE 6.18   The silky appearance of a phyllite is due to light reflected from nearly parallel flakes of chlorite or other micas that are a fraction of a millimeter across. *(William E. Ferguson)*

Many foliated rocks display other features that indicate their history of intense deformation. The garnet porphyroblasts in Figure 6.21, for example, were evidently rolled as they grew between mica-rich layers that were sliding past each other. Some regional metamorphic rocks may also display a characteristic lineation. One of these is **amphibolite,** in which elongated amphibole and plagioclase crystals are aligned in a more or less parallel direction. Amphibolites may be green, gray, or black and sometimes contain such minerals as epidote, pyroxene, biotite, and garnet. They are products of the medium-grade to high-grade metamorphism of ferromagnesian igneous rocks and of some calcareous sediments that have other types of sediment mixed in.

Finally, some regional metamorphic rocks show few effects of directed pressure at all. This is likely to be so in rocks that contain few platy or elongated minerals. These include:

- 🌳 **Marble,** a familiar metamorphic rock, differs from its parent limestone in having larger, more interlocked mineral grains. The calcite or dolomite grains in a

FIGURE 6.19   A schist.

FIGURE 6.20   (a) Gneiss has a granular texture, marked by bands of light and dark-colored silicates. (b) The banding in gneiss may be highly twisted and folded as a result of directed pressure in heated rocks at the roots of evolving mountain belts.

marble recrystallize during metamorphism, but because they are not platy like the micas, no foliation develops. The purest variety of marble used as a decorative building stone is snow white (Figure 6.22). Small percentages of other minerals account for the wide variety of color in marble. Black marbles are colored by bituminous matter; green marbles by diopside, hornblende, serpentine, or talc; red marbles by hematite; and brown marbles by limonite. Marble occurs most commonly in areas of regional metamorphism, where it is often found in layers between schists or phyllites, but it may also form during contact metamorphism.

- **Quartzite** is formed by the metamorphism of quartz-rich sandstone (compare Figures 6.4a and 6.4b). During diagenesis, quartz grains in the original sand are firmly bonded by the entry of silica into pore spaces. Under the higher temperatures of regional metamorphism, the original sand grains and the silica cement are recrystallized. The result is a much stronger, coarser-grained rock with greatly decreased pore space. Like calcite, quartz is neither platy nor elongated; thus, quartzite is generally nonfoliated. Pure quartzite is white, but iron or other impurities sometimes give the rock a reddish or dark color.

- **Granulites** are very high-grade metamorphic rocks, usually containing pyroxenes, garnet, feldspar, and quartz, most of which are nonplaty. Micas, amphiboles, or other minerals that have water in them are absent.

Cataclastic Metamorphism   This fourth kind of metamorphism results from the grinding that takes place in fault zones, where minerals and rocks are broken and even pulverized. Coarse-grained breccias or finer-grained varieties called **mylonites** may be produced. These rocks all display a foliation, but one that is produced purely by mechanical forces rather than by chemical recrystallization. Temperature is a less important agent in cataclastic metamorphism than in the other four kinds of metamorphism.

Shock Metamorphism   A shock metamorphic environment is created by the impact of a meteorite with the planetary surface and lasts for only a fraction of a second, during which rocks reach an extremely high pressure and temperature. Shock

metamorphism was long overlooked by geologists, who until the 1950s underestimated the significance of meteorite impact on planetary surfaces. They have now learned to recognize shock features in many samples collected from impact craters on the Earth and those returned from the Moon (Figure 6.23).

A shock wave generated by impact passes through a rock in less than a second, creating pressures that can be hundreds of times greater than any normally produced in the crust. Rocks that are far from the impact site may be only slightly fractured. Those closer, however, show some features similar to cataclastic metamorphic rocks, as well as bands of distortion called **shock lamellae** within individual mineral grains (Figure 6.24). Sometimes the shock wave can refocus behind a hard pebble, so that locally intense pressures produce a **shatter cone** that records the direction the wave was traveling (Figure 6.25). Closest to the impact, the passing shock wave causes so much friction between and within grains that melting can occur. The high pressures can also make some minerals such as quartz recrystallize as rare high-density polymorphs. Geologists use their knowledge of these

FIGURE 6.22 This nearly pure white marble is in a regionally metamorphosed sequence of marine limestones and shales widely quarried in Vermont as building stone. Marble from this quarry in the heart of Dorset Mountain in Danby, Vermont, has been used for monuments and public buildings around the world. *(Vermont Marble Company)*

FIGURE 6.23   Meteor Crater in Arizona is the youngest and best preserved of nearly 200 craters recognized on Earth. The 1-km-diameter crater was formed 50,000 years ago by the impact of an iron meteorite estimated to have been 150 meters across. Shock metamorphic features can be observed in samples of the surrounding sandstones. *(John Sanford, Photo Researchers)*

FIGURE 6.24   Photomicrograph of shocked quartz grain from the Manson Impact Structure, Manson Iowa. Note planar deformation features. The sample is from an impact breccia at 675 ft. depth on the flank of the central uplift. Plane light, field of view 1.3 mm. *(Photo courtesy of L. Crossey, Department of Earth & Planetary Sciences, University of New Mexico.)*

FIGURE 6.25   Shatter cones near Sudbury, Ontario, formed as the shock wave from a major impact struck pebbles in this sediment. The energy was focused immediately behind each pebble, causing shock metamorphism.

various features to recognize impact debris from craters that completely eroded away long ago.

Other Metamorphic Environments   During the 1970s and 1980s, undersea exploration revealed yet another style of metamorphism. Teams of geologists in submersible vehicles making detailed surveys along the Mid-Atlantic Ridge and the East Pacific Rise (the midocean ridge in the Pacific Ocean) discovered dramatic hot-spring activity. At each hot-spring location, which corresponds with a major divergent plate boundary, seawater penetrates fissures in young, warm basalts, dissolves minerals from them, and is reemitted from vents known as **black smokers** (Figure 6.26). The black plume visible in Figure 6.27 is a cloud of very small crystals of metal sulfide miner-als, precipitated from the hot water as it mixes with much cooler seawater. These settle on the seafloor, where they build up a small but rich ore deposit. Accumulations of metal sulfides have now also been found in sediments on the floor of the Red Sea, another divergent plate boundary.

As seawater dissolves metals from the midocean ridge basalts, it leaves behind other elements. Magnesium, sodium, and some calcium that had been dissolved in the seawater are added to the rocks, forming new minerals and greatly altering the appearance of older rocks. Because an immense volume of seawater must circulate through rocks along the midocean ridge system, geologists are convinced that this style of metamorphism has a significant effect on the chemical composition of the world's oceans.

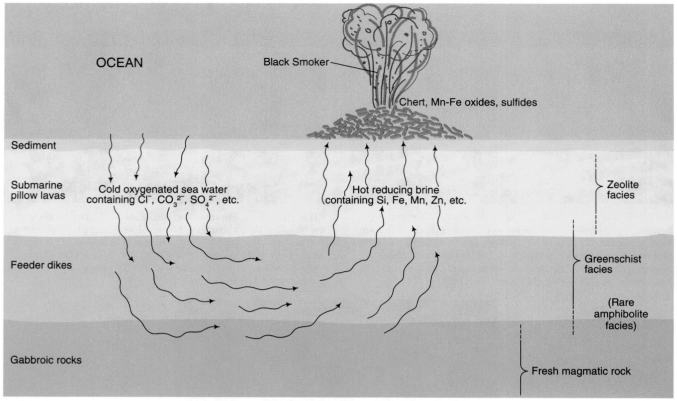

OCEAN

Black Smoker

Chert, Mn-Fe oxides, sulfides

Sediment

Submarine pillow lavas

Cold oxygenated sea water containing Cl^-, CO_3^{2-}, SO_4^{2-}, etc.

Hot reducing brine containing Si, Fe, Mn, Zn, etc.

Feeder dikes

Gabbroic rocks

Zeolite facies

Greenschist facies

(Rare amphibolite facies)

Fresh magmatic rock

FIGURE 6.26 According to this interpretation of metamorphism at midocean ridges, cool seawater penetrates fresh pillow lavas on the seafloor and even into the system of dikes below, from which the lavas once emerged. As the water is gradually heated, it reacts with the igneous rocks, metasomatizing them and altering its own chemical composition. When the hot water finally reemerges in a hot spring, it carries metals that it has extracted from the rock. Upon cooling and reacting with fresh seawater, these metals form a black cloud of fine particles (metal sulfides and oxides) that eventually settle on the seafloor. *(Readapted from:* IGNEOUS AND METAMORPHIC PETROLOGY *by Best. Copyright © 1982 by W.H. Freeman and Company. Used with permission.)*

FIGURE 6.27
A "black smoker" on the East Pacific Rise is a vent from which hot fluids escape on the seafloor. Seawater circulates through fractures surrounding magmatic intrusions at the midocean ridge crest. This metasomatizes the surrounding rock, removing some elements and adding others that were previously dissolved in the seawater. As the hot fluids finally leave the vent and combine with cold seawater, fine particles of metal sulfides precipitate, producing a black cloud. Many sediment-hosted ore deposits may have been formed in this way. *(Peter Ryan/Scripps/Science Photo Library, Photo Researchers)*

▶▶ EPILOGUE ◀◀

As we have shown throughout this chapter, the kinds of minerals and the specific textures that develop differ greatly from one metamorphic environment to another. A geologist familiar with the temperature, pressure, and fluid conditions under which particular minerals or assemblages of minerals grow can recognize those minerals and specific textures as "fingerprints" for inferring environments in which rocks formed.

In previous chapters we have shown how geologists can use similar kinds of information to infer environments of igneous, weathering, and sedimentary activity. Metamorphic rocks, however, may yield more information than is usually available from materials in other parts of the rock system. Because regional metamorphism in particular is a very slow process, geologists commonly find evidence that metamorphism ended before all the mineral and textural changes were complete. Thus, it is not unusual for a geologist to discover minerals or textural features (such as bedding) that are clues to a premetamorphic environment, not fully replaced by products of metamorphism. It is even fairly common to find partially concealed evidence that a rock has been metamorphosed two or three times (a pattern called **polymetamorphism**). Such evidence might include foliation at an angle to a previous foliation, or a mineral assemblage typical of one facies that appears to have partially overgrown a different assemblage. Rocks such as these may be very difficult to interpret, but they offer geologists a window through which to see a succession of earlier changes in the Earth. This is one of many methods that can be used to piece together a record of events in the Earth's history, a subject we will pursue in the next chapter.

SUMMARY

1. Changes during metamorphism may include the growth or disappearance of minerals, a general coarsening of grain size, and the development of new textures such as foliation. Some of the minerals that form are also common in igneous or sedimentary rocks, but others such as chlorite or sillimanite (for example) are only found in metamorphic rocks and provide evidence of metamorphic activity.

2. The agents of metamorphism are heat, pressure, and chemically reactive fluids. Heat may be associated with nearby igneous events, with various changes at the margins of lithospheric plates, or simply with increasing depth in the crust. Pressure, due to the weight of overlying rock or to compressive forces in the crust, may cause mineralogical changes. Directed pressure will cause the development of foliation or lineations. Chemically reactive fluids consist largely of water containing gases and dissolved mineral matter. They may be derived from formation waters, dehydrating minerals, cooling magmas, or groundwater. In many settings, fluids are more readily available during the prograde than during retrograde stage of metamorphism.

3. Geologists map metamorphosed regions by identifying index minerals that grow within specific ranges of pressure and temperature. Assemblages of minerals that are stable within a given range of conditions define a metamorphic facies. If rocks are foliated, the orientation and degree of foliation can reveal the direction and intensity of directed pressures that prevailed during metamorphism.

4. Burial metamorphism occurs where thick accumulations of sediment begin to recrystallize at low temperatures. Contact metamorphism is due to heat from a nearby intrusive body of magma, usually at shallow depths (and thus low pressures) in the crust. Contact and burial metamorphic rocks typically have no foliation. Regional metamorphism is associated with large-scale movement in the lithosphere, commonly at convergent boundaries between plates. The range of metamorphic temperatures and pressures is very great. Most regional metamorphic rocks are foliated; rocks with few platy or elongated minerals have no foliation. Cataclastic metamorphism occurs in fault zones, where rock is pulverized and recrystallized by directed pressure at low temperatures. Shock metamorphism occurs quickly at extremely high pressures and temperatures, such as when the compressional shock wave from a meteorite impact passes through rock.

KEY WORDS AND CONCEPTS

aureole 141
black smoker 147
burial metamorphism 141
cataclastic metamorphism 141
contact metamorphism 141
directed pressure 135
fluid inclusion 137
foliation 133
formation water 136
gneissosity 144
granitization 139

index mineral 139
isochemical reaction 136
isograd 139
juvenile hydrothermal fluid 136
lineation 133
lithostatic pressure 134
metamorphic facies 139
metamorphic zone 139
metasomatism 137
polymetamorphism 149
porphyroblast 133

prograde 137
reaction rim 132
regional metamorphism 141
retrograde 137
rock cleavage (slaty cleavage) 143
schistosity 144
shatter cone 146
shock lamellae 146
shock metamorphism 141

IMPORTANT ROCK AND MINERAL NAMES

amphibolite 144
andalusite 132
blueschist 144
chlorite 132
diopside 142
eclogite 140
epidote 132
garnet 139
glaucophane 144

gneiss 144
granite 139
greenschist 140
hornblende 135
hornfels 142
kyanite 132
marble 144
migmatite 139
mylonite 145

phyllite 144
quartzite 145
schist 144
sillimanite 132
slate 143
staurolite 132
wollastonite 142
zeolites 141

QUESTIONS FOR REVIEW AND THOUGHT

6.1 Name some minerals that a geologist might observe in a metamorphic rock, but would not expect to find in an igneous or sedimentary rock.

6.2 What conditions of pressure and temperature mark the approximate limits of metamorphism in the crust? What other geologic processes take place above or below these limits?

6.3 What effect do lithostatic and directed pressure have during metamorphism?

6.4 What is a metamorphic fluid made of and where does it come from? How do geologists know?

6.5 When geologists map metamorphic zones, why are some minerals better choices as index minerals than others? Why isn't it possible to use the same set of index minerals for mapping metamorphosed shales that are used for mapping metamorphosed basalts?

6.6 Where would a geologist be most likely to find rocks undergoing burial metamorphism today?

6.7 In what kind of physical setting does contact metamorphism typically occur? What general characteristics are typical of contact metamorphic rocks?

6.8 How could you tell the difference between a shale and a slate? How about a quartz sandstone and a quartzite? List as many characteristics of each as you can.

6.9 How are the three different kinds of foliation different from each other? What features do they have in common?

6.10 What is a mylonite, and under what metamorphic conditions does it form?

6.11 What rock characteristics could you use to determine whether shock metamorphism had occurred?

6.12 What is a black smoker? What changes occur as water circulates through fractures in midocean ridge basalt?

Critical Thinking

6.13 Suppose that the high-grade metamorphosed shales in a particular area had been gently heated and subjected to directed pressure a second time. Describe a possible set of mineral and textural clues to each metamorphic event that a geologist might see in the rocks. [Hint: Look at Table 6.1.]

6.14 Many scientists have been looking for a good way to store high-level radioactive waste. One suggestion is that it should be encased in large blocks of concrete. Critics worry, however, that heat generated by radioactive decay could metamorphose the concrete. If you were expected to evaluate this proposal, what kinds of questions would you want to ask? What kind of information would you need in order to address the critics' concerns about metamorphism?

7

The Grand Canyon from Tavapai Point, Grand Canyon National Park, Arizona

Geologic Time

OBJECTIVES

▲▲▲

*As you read through this chapter, it may help if you focus
on the following questions:*

1. What are the rules of stratigraphy on which a geologic chronology is based, and how are the rocks of one area correlated with those of another? How are gaps in the geologic record shown in the rock record?

2. What is the geologic column? The geologic time scale?

3. How do radioactive elements decay, and how is this process used to determine the age of Earth materials?

4. How are absolute ages of Earth events fitted into the geologic time scale based on relative time?

5. How do some minerals record the changing magnetic polarity of the Earth, and how is this useful in magnetic stratigraphy?

OVERVIEW

▲▲▲

Geologic time encompasses all Earth history since the birth of the planet 4.54 billion years ago. Rocks are the archives of this history. Arranging these rocks, and the events they describe, in the correct chronological order has been one of the great achievements of geology. When the task began over 200 years ago, geologists had no way of telling how old a rock was in terms of real years. Yet they managed to develop methods that allowed them to determine comparatively the order in which the rocks of the Earth's crust formed. Using these methods, geologists have developed a geologic calendar, or sequence of geologic events and ages, accepted and used throughout the world. Since the establishment of this calendar, radioactivity has emerged to provide the means for determining a rock's age in actual years. Today, thousands of radioactively based rock ages have confirmed the calendar developed by geologic methods.

Developing a Geologic Chronology

▲▲▲

The birth of the Earth is only one event, albeit the most important one thus far, in the history of the planet. Between its beginning and the present, our globe has undergone countless changes. An impressive accomplishment by geologists has been the identification of these events and the arrangement of the rocks that recorded them into a chronological sequence. When geologists began this task in the late eighteenth century they could determine only **relative time,** that is whether one rock was older or younger than another. They could tell whether, for instance, a volcanic eruption had occurred before or after a rise in the level of the sea. They could not, however, attach dates to these events in terms of years before the present. Relative time has been the basis for the development of the geologic calendar, which geologists refer to as the **geologic time scale.** That scale, put together largely in the nineteenth century, is still valid and in universal use. Today, however, we can attach absolute dates to the chronology based on relative time. As an example of **absolute time** we can say that the dinosaurs became extinct a little over 66 million years ago and that 11,000 years ago the glacier ice of the Great Ice Age began to recede from New England and the lands bordering the Great Lakes. Before we can discuss absolute time and how it relates to the geologic time scale, we need to discuss how relative time, the basis for the geologic time scale, has been established.

FIGURE 7.1   Clearly defined layers of limestone alternating with darker shale beds in this cliff face near Allumiere, Italy, illustrate the rule of superposition. The oldest bed is on the bottom and the youngest is on the top, which is always true unless the original sequence has been overturned.

Stratigraphy

Stratigraphy, from the Latin for "strata" or "layers" plus the Greek for "writing," refers to the succession and age relations of layered rocks, most of which are sedimentary rocks. The raw materials of stratigraphy are the countless rock exposures scattered around the world, as well as the samples of rock brought up from holes drilled for oil, collected from mines, and dredged from the ocean floor. The task of geologists is to place all these in correct relative order and to relate sequences of rocks in one area with those in another. At the simplest level, let us begin at the outcrop where several layers of rock are exposed. There are some very simple, common-sense rules to get us started.

Superposition   A basic rule of stratigraphy is the rule of **superposition.** Here is an example: A deposit of mud laid down this year in, say, the Gulf of Mexico will rest on top of a layer that was deposited last year. Last year's deposit, in turn, rests on successively older deposits that extend backward into time for as long as

deposition has been going on in the gulf. If we could slice through these deposits, we would expose a chronological record, with the oldest deposit on the bottom and the youngest on top. This sequence would illustrate the law of superposition, which states that if a series of sedimentary rocks has not been overturned, the topmost layer is always the youngest and the bottommost layer is always the oldest (Figure 7.1).

The rule of superposition was first derived and used by Nicolaus Steno (1638–1687), the physician and naturalist we first mentioned in Chapter 2. His geologic studies were carried on in Tuscany, in northern Italy, and published in 1669. During these studies he was able to distinguish between rocks containing marine fossils and those containing terrestrial fossils. In studying these sedimentary rocks Steno concluded that, because fossiliferous beds are laid down in water, they must be horizontal when first formed and that as each new bed is laid down it is younger than the bed immediately beneath it.

Original Horizontality   In Chapter 5 we discussed the fact that layers of sedimentary deposits accumulate horizontally. We cited some exceptions, but overwhelmingly, sedimentary layers possess an **original horizontality.** Steno first recognized this and appreciated that any departure of a sedimentary rock from the horizontal said something about its history subsequent to deposition. Therefore, when a sedimentary rock does not lie horizontally, the first thing geologists suspect is that some event has intervened to move it out of its original position. If the rocks are upside down we will obviously get our local history backwards unless we recognize what has happened. In Chapter 5 we described some clues that can help geologists determine whether sedimentary beds have been overturned.

Lateral Continuity In a cliff face we see only a part of the lateral extent of a sedimentary rock. When first laid down it stretches over a much wider area, defined by the extent of the depositional environment or basin in which it formed. Therefore, when we encounter an exposure of sedimentary rock we should suspect, as did Steno, that we are seeing only a part of the formation. We are then justified in asking whether we can find the same rock elsewhere. Does it, for example, extend beneath the hill and reappear on the other side or across the valley on the opposite hill slope? The odds are greatly in favor that our small sample is part of a larger unit that has a **lateral continuity** of considerable, although definite, extent.

Cross-Cutting Relationships Another simple rule is that of **cross-cutting relationships.** Figure 7.2a presents one example. A dike of igneous rock has been injected into beds of sedimentary rocks. The sedimentary rock had to be present before the dike was injected and cut across the sedimentary rock. The igneous rock, therefore, must be younger. A fault,

FIGURE 7.2(a) In Iceland a lava flow overlies finely bedded, horizontal pyroclastic sediments. A dike has been intruded across the lower three fourths of the sediments and therefore must be younger than the beds it cuts. The dike is about 40 cm across at its widest point. *(Alan Rubin)*

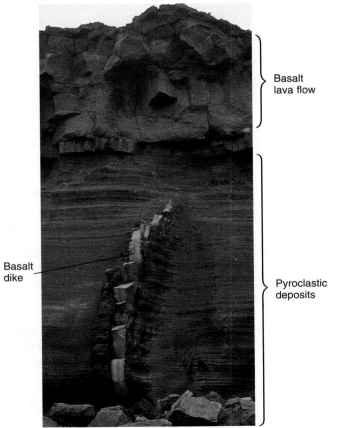

Basalt lava flow

Basalt dike

Pyroclastic deposits

(a)

(b)

FIGURE 7.2(b) Faults cut this bed of white volcanic ash in New Mexico. The ash is therefore older than the fault. *(Steven Elston)*

breaking a rock, offers another example. The rock had to be present before the fault could break it. Therefore the fault is younger (Figure 7.2b). The general rule of cross-cutting, then, is that a geologic feature, whatever it may be, is younger than the feature that it affects.

Inclusions **Inclusions** of one rock in another also provides evidence about the relative age of the two rocks. For instance, a fragment of rock caught up in a lava flow must be older than the volcanic event. A granitic melt intruding a sequence of sedimentary rocks may include fragments of the sequence in its magma. If the fragments are not melted and digested by the melt they will survive as inclusions and tell us that the sedimentary rocks are older than the granite. The rule of inclusions is really a variation of the rule of cross-cutting relations and can be stated as follows: A rock enclosed in another is older than the rock in which it is included.

Correlation of Rock Units

It may be fairly straightforward to tell the relative age of rocks in a single exposure. However, it becomes more difficult to determine what the relative age of a rock in one area is to a rock in another area. Is it the same age? Or younger? Or older? How do you arrange the rocks of one area and those of another area in proper relative chronologic position? The process of tying a rock sequence in one place to another in some other place is known as **correlation,** from the Latin for "together" plus "relate." Correlation has applications ranging from the search for mineral deposits, to the exploration for oil and gas, to the location of beds that may provide a source of underground water, and to the identification of geologic environments that can safely serve to contain toxic wastes.

Correlation by Physical Features When sedimentary rocks show constant and distinctive features

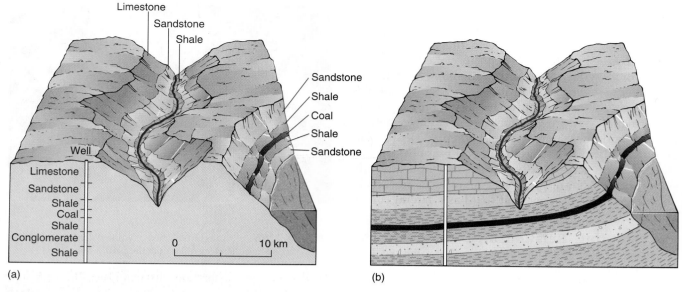

FIGURE 7.3 (a) Diagram to illustrate the data that might be used to correlate sedimentary rocks in a sea cliff (right) with those in a stream valley (center), and those encountered in a well-drilling operation (left). (b) Similar lithologies and sequences of beds in the three different localities of (a) suggest the correlation of rock layers as shown in this diagram. See text for discussion.

over a wide geographic area, we can sometimes connect sequences of rock layers in one locality with those in another. Figure 7.3a illustrates how this is done. It shows a series of sedimentary rocks exposed in a sea cliff. The topmost, and hence the youngest, is a sandstone. Beneath the sandstone we first find a shale, then a seam of coal, and then more shale extending down to the level of the modern beach. We can trace these rock layers for some distance along the cliff face, but how are they related to other rocks farther inland?

Along the rim of a canyon that lies inland from the cliff, we find that limestone rocks are exposed. Are they older or younger than the sandstone in the cliff face? Scrambling down the canyon walls, we come to a ledge of sandstone that looks very much like the sandstone in the cliff. If it is the lateral continuation of the cliff sandstone, then the overlying limestone must therefore be younger. The trouble is that we cannot be certain that the two exposures of sandstone are part of the same layer. Therefore, we need to continue down to the bottom of the canyon where we find some shale beds very similar to the shale beds exposed in the sea cliff beneath the sandstone. We can feel fairly confident that the sandstone and the shale in the canyon are the same beds as the sandstone and upper shale in the sea cliff, but we must admit the possibility that we are dealing with different sandstone and shale layers.

In searching for further data, we find a well being drilled farther inland, the drill bit cutting through the same limestone that we saw in the canyon walls. As the bit cuts deeper and deeper, it encounters sandstone, shale, a coal seam, more shale, and then a bed of conglomerate before the drilling finally stops in

another shale bed. This sequence is more continuous than anything we have seen so far. Part of it duplicates the one we first observed in the sea cliff, and another part duplicates the sequence we saw in the canyon. In addition, it reveals an underlying conglomerate and shale that we have not seen before. The repetition of the rock sequence should give us confidence that the sandstone in the canyon and the sea cliff are of the same age. The limestone turns out to be the youngest rock in the area; the conglomerate and lower shale are the oldest rocks. This correlation is shown in Figure 7.3b.

Many sedimentary formations are correlated in just this way, especially when physical features are our only keys to rock correlation. As we extend the range of our correlation over a wider and wider area, however, physical features become less and less useful. For example, individual rock units thicken, thin, and disappear, as suggested in Figure 7.4. Changing facies introduce another complication. As you'll recall from the discussion of sedimentary facies in Chapter 5, lagoonal muds can grade into beach sand, which then grades into a marine sand, silty mud, and finally a mud (see Figure 5.30). All of these deposits are of the same age but exhibit different lithologies. Thickening, thinning, and facies changes, therefore, all place limits on the lateral continuity of sedimentary rocks. Fortunately, we have another method of correlation, a method that involves the use of fossils.

Correlation by Fossils Around the turn of the nineteenth century, an English surveyor and civil engineer named William Smith (1769–1839) became

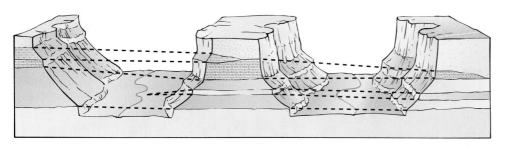

FIGURE 7.4
Sedimentary rock units thicken, thin, and pinch out. This complicates their correlation, as suggested in this figure.

impressed with the relationship of rock strata to the success of various engineering projects, particularly the building of canals. As he investigated rock strata from place to place, he found that many of them contained fossils of marine invertebrate animals. Furthermore, he observed that no matter where these same rock layers were found they contained identical fossils, whereas the fossils in rock layers above or below were different. Eventually Smith became so skillful an observer that when confronted with a fossil, he could identify the rock from which it had come.

At about the same time, two French geologists, Georges Cuvier (1769–1832) and Alexandre Brongniart (1770–1847), were studying and mapping the fossil-bearing strata that surround Paris. They had also used the rule of superposition to arrange the rocks of the Paris area in chronological order, just as William Smith had done for the rocks of southern England. Cuvier and Brongniart also arranged their collection of fossils in the same order as the rocks from which the fossils had been dug. They discovered that the fossils in each layer differed from the fossils in every other layer. Then, when they compared the fossil forms with modern forms of life, they found that the fossils from the higher rock layers bore a closer resemblance to modern forms than did the fossils from the older rocks lower down.

As geologists reported more and more observations of the sort made by Smith, Cuvier, and Brongniart, it became increasingly evident that the relative age of a layer of sedimentary rock could be determined by the nature of the fossils that it contained (Figure 7.5). This fact has been verified time and again by other workers throughout the world. It has become an axiom in geology that fossils are a key to correlating rocks and that rocks containing the same fossil assemblages are similar in age.

Through countless observations of the occurrence of fossils in different rock units throughout the world, geologists find that groups of fossil animals and plants have succeeded one another in a definite and discernible order, and each geologic time period can be recognized by the fossils found in rocks of that age. This is known as the **principle of faunal and floral succession.**

Superposition and faunal succession go hand in hand in the determination of the geologic history of a region. In fact, superposition is basic to the demonstration of faunal succession and proves its validity. When we examine the assemblages of fossils, we find that older rocks commonly have one association of fossils and that younger rocks have another. Faunal and floral succession shows that life has changed with the passage of time. By examining fossil assemblages from rocks in one area and comparing them with fossil assemblages from rocks in another area, geologists are able to demonstrate the correlation of rock formations, with their included fossil assemblages, from region to region and even from continent to continent.

Missing Records

In a basin of deposition, sediments are laid down one after another, each bed in conformity with the horizontal beds above and below. There may be slight breaks in time between the deposition of one bed and the next, but these breaks are not significant in the context of geologic time. There are many localities, however, where large gaps do appear in the continuity of the geologic record. A stratigraphic sequence of rocks that would normally be present is missing, either because it was never laid down or because it

FIGURE 7.5 Sequences of sedimentary rocks may be correlated by similarities in the fossils found in them. In this diagram a sequence of sedimentary rocks has been put together at three widely separated localities. The similarities of the contained fossils allow the correlation shown by the dashed lines. Bed 2 is present at localities A and C but missing at B. It was either eroded away before the deposition of bed 3, or was never laid down.

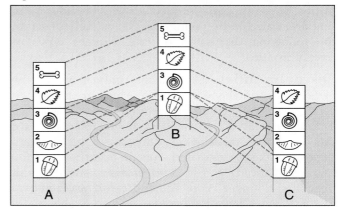

Siccar Point—Looking into the Abyss of Time

Two centuries ago James Hutton, John Playfair, and Sir James Hall discovered the gently dipping strata of a formation, known today as the Old Red Sandstone, resting upon older, nearly vertical graywackes at Siccar Point, Scotland (Figure P7.1.1). From an historical point of view the unconformity at Siccar Point is of great significance because it provided solid evidence of the vast amount of time that many geologists were just beginning to recognize as characteristic of Earth history. The exposure showed that a dark sandstone, a graywacke (Playfair called it a schistus), had been deposited, then lithified, then tilted on end, lifted from the water, and attacked by erosion. The vertical beds were later submerged in a depositional basin in which sand was deposited, then lithified, and finally uplifted. Erosion again followed. No one then knew how many years were involved, but the eighteenth-century geologists grasped the fact that an enormous amount of time, perhaps millions of years, was most certainly involved. Here are portions of Playfair's account of the discovery in the prose of the day.

. . . We made for a high rocky point or head-land, the Siccar, near which, from our observations on shore, we knew that the object we were in search of was likely to be discovered. On landing at this point, we found that we actually trode on the primeval rock, which forms alternately the base and the summit of the present land. It is here a micaceous schistus, in beds nearly vertical, highly indurated, and stretching from southeast to northwest. The surface of this rock runs with a moderate ascent from the level of low water, where the schistus has a thin covering of red horizontal sandstone over it. . . . The rugged tops of the schistus are seen penetrating into the horizontal beds of sandstone, and the lowest of these last form a breccia containing fragments of schistus, some round and others angular, united by an arenaceous cement. Dr. Hutton was highly pleased with appearances that set in so clear a light the different formations of the parts which compose the exterior crust of the earth. . . . On us who saw these phenomena for the first time, the impression made will not easily be forgotten . . . . We often said to ourselves, "What clearer evidence could we have had of the different formation of these rocks, and of the long interval which separated their formation, had we actually

---

was eroded before the deposition of overlying beds. This gap in the record is a **hiatus.**

Hiatuses show up in the rock section as **unconformities.** These appear as eroded surfaces on older rocks that are sealed in by younger rocks, as we will see in the following discussion of the different types of unconformities.

Disconformities An unconformity in which the beds below and above the unconformity are parallel to each other is called a **disconformity.** Figure 7.6a shows sedimentary layers separated by an erosion surface. In this example marine sedimentary beds were deposited, the sea retreated, erosion occurred, and the sea readvanced across the erosion surface and deposited younger beds. The end result was removal of some pages from the geologic record at this place.

The fact that the beds above the unconformity are parallel to those below tell us that whatever caused the retreat and advance of the sea did not also deform the lower and older beds. Because beds above and below the unconformity are parallel with each other, a disconformity may be difficult to recognize at first. An eroded surface on the beds directly beneath the disconformity, however, should be expected. This surface may also be weathered (see Figure 7.7). Fossil assemblages above and below a disconformity may change abruptly, indicating a hiatus between the two periods of deposition.

Angular Unconformities An **angular unconformity** differs from a disconformity in that the beds above and below it are at an angle to each other. Figure 7.6b illustrates this. The sequence of

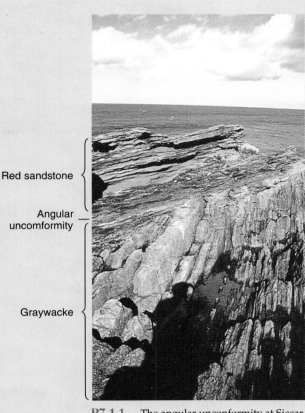

Red sandstone

Angular unconformity

Graywacke

P7.1.1 The angular unconformity at Siccar Point, Scotland, shows gently dipping beds of sandstone overlying the eroded stubs of much older layers of graywacke. The time represented by the unconformity is measured in tens of millions of years. *(Edward A. Hay)*

seen them emerging from the bosom of the deep?" We felt ourselves necessarily carried back to the time when the schistus on which we stood was yet at the bottom of the sea, and when the sandstone before us was only beginning to be deposited, in the shape of sand or mud, from the waters of the superincumbent ocean. An epoch still more remote presented itself, when even the most ancient of these rocks instead of standing upright in vertical beds, lay in horizontal planes at the bottom of the sea, and was not yet disturbed by the immeasurable force which has burst asunder the solid pavement of the globe. Revolutions still more remote appeared in the distance of this extraordinary perspective. The mind seemed to grow giddy by looking so far into the abyss of time. . . .

["Biographical Account of the Late Dr. James Hutton," Transactions, Royal Society of Edinburgh, Vol. 5, pp. 39–99, 1805].

events surrounding an angular unconformity is more complicated than the sequence of events surrounding a disconformity. In Figure 7.6 we can read the sequence as follows: Sedimentary rocks were first laid down, then tilted, uplifted, and eroded. A new sea flooded across the erosional surface and in it were laid down a younger sequence of beds. A famous angular unconformity, at Siccar Point in Scotland, is discussed in Perspective 7.1 and pictured in Figure P7.1.1.

Nonconformities The sketch in Figure 7.6c shows a third type of unconformity, a **nonconformity.** Nonconformities separate two profoundly different rock types. In the example shown, the two types are granite and sedimentary rocks. As depicted, the record reads that a granite, formed at depth beneath the surface, was later exposed by ero-

sion and then covered by sedimentary rocks. An example from Wyoming is pictured in Figure 7.8.

Unconformities in the Grand Canyon Walk down into the Grand Canyon of the Colorado River and you walk back in time. In absolute time, we can see back 1.7 billion years to the metamorphic event recorded in the tough, black Vishnu Schist at the bottom of the canyon. We realize that we are seeing further back than 1.7 billion years, however, when we recall that the Vishnu Schist was metamorphosed from an older sequence of sandstone, mudstone, and lavas. The canyon is indeed an impressive slice through Earth time. Large chunks of time, however, are unrecorded. They are hiatuses represented by unconformities. Four of them are major gaps in the geologic record in the canyon walls and

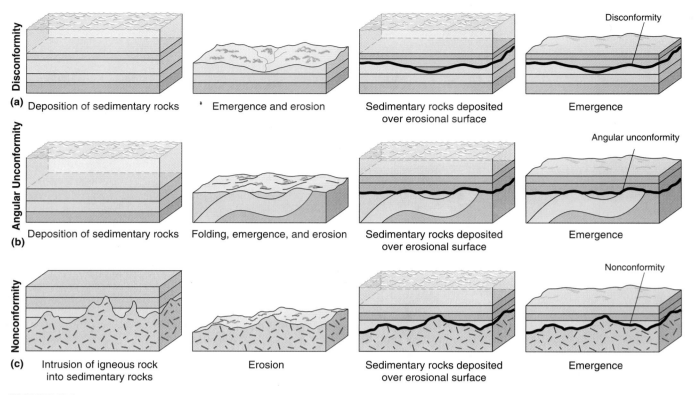

Disconformity

(a) Deposition of sedimentary rocks • Emergence and erosion | Sedimentary rocks deposited over erosional surface | Emergence — Disconformity

Angular Unconformity

(b) Deposition of sedimentary rocks | Folding, emergence, and erosion | Sedimentary rocks deposited over erosional surface | Emergence — Angular unconformity

Nonconformity

(c) Intrusion of igneous rock into sedimentary rocks | Erosion | Sedimentary rocks deposited over erosional surface | Emergence — Nonconformity

FIGURE 7.6 Sequences of events that turn surfaces of erosion into unconformities. (a) A disconformity. (b) An angular unconformity. (c) A nonconformity.

together account for over a billion years, probably more than half of the Grand Canyon's history. As Figure 7.9 shows, the oldest unconformity is a nonconformity, the next younger is an angular unconformity, and the two youngest are disconformities.

The Geologic Column and Geologic Time Scale

▲▲

Using the rules of stratigraphy and the techniques of correlation, geologists have arranged the world's sedimentary rocks into chronologic order. They have pictured the rocks as forming a great **geologic column** with the oldest at the bottom and the youngest at the top. The column has been spliced together from thousands of local rock sequences. As the column grew, geologists began subdividing it where they found what looked like sudden changes in the record. For example, if a particular group of fossils was present in one sequence of rocks but did not appear in successive sequences, the level at which the fossils disappeared might be chosen as a boundary.

By the time the twentieth century began, geologists had subdivided geologic time into units of vary-

ing lengths as shown in the geologic time scale in Figure 7.10. The Earth began about 4.54 billion years ago (see Perspective 7.2). All Earth time is divided into four units called **eons.** The youngest is the **Phanerozoic,** from the Greek for "visible life," for fossils are prevalent in these rocks. Older rocks have their own informal name of **Precambrian,** a general term applied to all the rocks that lie beneath the Cambrian rocks, which begin the Phanerozoic. More formally we

FIGURE 7.7 Horizontal beds of eroded and weathered marine clay are overlain by flat-lying beds of volcanic ash to create this disconformity near Tarquinia, Italy. It represents an hiatus of about 1 million years.

Volcanic ash

—Disconformity
}Pliocene clay

FIGURE 7.8 This nonconformity is in the Shoshone Canyon, near Cody, Wyoming. An erosional surface developed on granite is overlain by bedded sandstone. Hundreds of millions of years are represented by the nonconformity.

divide Precambrian time into three older eons, **Proterozoic,** meaning "earlier life," **Archean,** meaning "ancient," and **Hadean,** after Hades, the mythical underworld of the dead. Divisions smaller than the eon are, in decreasing size, the **era,** the **period,** and the **epoch.**

Telling Time in Years

In our daily lives we use two basic units of time: (1) the day, the interval required for our planet to complete one rotation on its axis; and (2) the year, the interval required for the Earth to complete one revolution around the Sun. If we want to date geologic events in terms of years our problem, of course, lies in the fact that no one was around to count and record the revolutions of the Earth around the Sun. As the American geologist George D. Louderback wrote 60 years ago:

Geology needs an independent time clock that runs at a uniform rate, just as we need it in our daily life, and the physicist needs it in his laboratory.

That clock has turned out to be radioactivity.

Radioactivity

Atoms of a single element that vary in mass (because of the difference in the number of neutrons) are called *isotopes* of that element. The abundant isotopes of most elements are stable. Some combinations of protons and neutrons, however, are unstable: They transform or decay into other elements, giving off energy at the same time (Figure 7.11). Those isotopes that spontaneously decay in this manner are said to possess **radioactivity.** There are 70 naturally occurring radioactive isotopes. A number of these have turned out to be useful in dating geologic materials.

Types of Radioactive Decay

To see how the decay process works, we need to look

FIGURE 7.9 Four major unconformities account for over half of the history of the Grand Canyon of the Colorado River. See text for discussion.

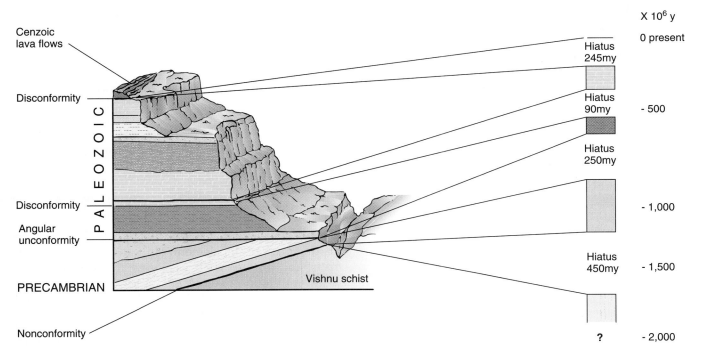

How Old Is the Earth?

Early on geologists were convinced that the Earth was very old. During the nineteenth century and on into the twentieth a number of different estimates emerged (Figure P7.2.1). Not until well after the discovery of radioactivity, however, and how it could be used to date Earth rocks and extraterrestrial materials, were geologists able to say just how old. By the last quarter of the twentieth century several lines of evidence combined to indicate an age of 4.54 billion years for the Earth and our planetary system. The information has come from Earth rocks, from rocks brought back from the Moon, and from objects that have fallen to the Earth from space.

Several radioactive elements have given dates of over 3.5 billion years for rocks from each of the continents. Geologists have dated rocks from Antarctica at 3.93 billion years and from northwest Canada at 3.96 billion years. From southwestern Australia crystals of the mineral zircon, found in ancient sedimentary rocks, have radiometric (radioactively determined) ages of 4.0 to 4.3 billion years. Evidence from the Earth, therefore, says that the planet must be older than 4 billion years.

Geologists believe that the Moon formed by the accumulation of the debris produced from a collision between an early Earth and an asteroid, a small celestial body orbiting the Sun. If so, the minimum age of the Moon is also a minimum age for the Earth. The oldest rocks returned from the Moon have ages between 4.4 and 4.5 billion years, making the Earth–Moon system about 4.5 billion years old.

Meteorites come primarily from the belt of asteroids between Mars and Jupiter. They are fragments of asteroids bumped into Earth-crossing orbits by inter-asteroid collisions. All seriously considered hypotheses of the Earth's origin hold that the planets, including Earth and asteroids, condensed in a very short time from material orbiting the early Sun. Geologists feel justified, therefore, in giving 4.5 to 4.6 billion years, the oldest radioactive dates from the meteorites, as the birth date of the Earth.

The evidence for the now generally accepted age of the Earth is compelling, and most geologists have confidence in the estimate of 4.54 billion years. Remember, though, that earlier scientists, as well, had faith in their data and results. The fact that they were proved wrong reminds us that the conclusions of science are tentative. Future evidence may modify today's estimate of the antiquity of the Earth. For the moment, however, our answer of 4.54 billion years for age of the Earth is the most convincing available.

We divide the Phanerozoic into three eras. From oldest to youngest they are the **Paleozoic** (old life); the **Mesozoic** (middle life), and the **Cenozoic** (recent life). There are no generally agreed upon divisions of the three older eons. For the oldest, the Hadean Eon, this is quite understandable given the fact that there are almost no Hadean rocks on Earth. If we draw the boundary between the Hadean and next younger eon, the Archean, at 3.9 billion years, then only a very few specimens can be called Hadean. There are a few rocks from Antarctica and northwestern Canada and some examples of the mineral zircon from southwestern Australia that date from the very top of the Hadean. Geologists have difficulties

inside the nucleus as one element turns into a different one. Recall in the discussion of atomic structure in Chapter 2 that the nucleus of an atom is made up of protons and neutrons. A proton is positively charged. A neutron has no charge, being made up of a positive proton and a negative electron. The number of protons determines the atomic number, or element. The number of neutrons along with the protons make up the atomic mass of the element. How the number of neutrons and protons changes during the three different types of radioactive decay (which is important in radioactive dating) is discussed briefly below and illustrated in Figure 7.12.

Beta Decay Beta decay is the most common type

subdividing the Archean and Proterozoic Eons in part because the most common rock types of those eons, igneous and metamorphic, are not easily subdivided using the stratigraphic principles applied to the sedimentary rocks so prevalent during Phanerozoic. Furthermore, the further back in time we go the more extensive the destruction of the rock record; thus the number of samples from these eons is limited.

Rocks found in Britain, Germany, Russia, Switzerland, and the United States form the basis for subdividing the Paleozoic, Mesozoic, and Cenozoic eras into smaller units, called periods. The period names have been assigned without any particular plan. Some have geographic contexts, such as Devonian, after Devon in southern England; Mississippian, after the region of the upper Mississippi River valley in midwestern United States; and

Jurassic after the Jura Mountains of Switzerland and nearby France and Germany. Other names have lithologic connotations, such as Cretaceous, from the Latin *creta*, for chalk, in reference to the chalk cliffs of southern England and northern France. The epochs of the Cenozoic all use *ceno* from the Greek for "recent," combined with prefixes that are meant to describe how recent.

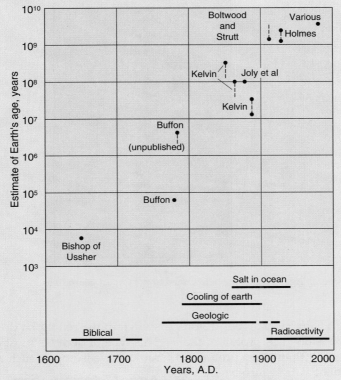

P7.2.1 Estimates of the age of the Earth have varied with different methods of age determination. In general, the estimates have increased during the last 350 years.

of radioactive decay. In an unstable nucleus a negatively charged particle, a beta particle (physically identical with an electron), may be ejected from the nucleus. This negative charge is lost from the neutron, and the loss changes the neutron to a positively charged proton. The nucleus now has one more proton than it originally had, and the atomic number has thus

increased by "1" to give us a new element, as shown in the blue beta decay quadrant of Figure 7.12. The atomic mass does not change, however, because there is still the same *total* number of protons and neutrons.

Electron Capture During **electron capture,** the nucleus acquires an electron from the inner zone of the atom's cloud of electrons. The negative charge of

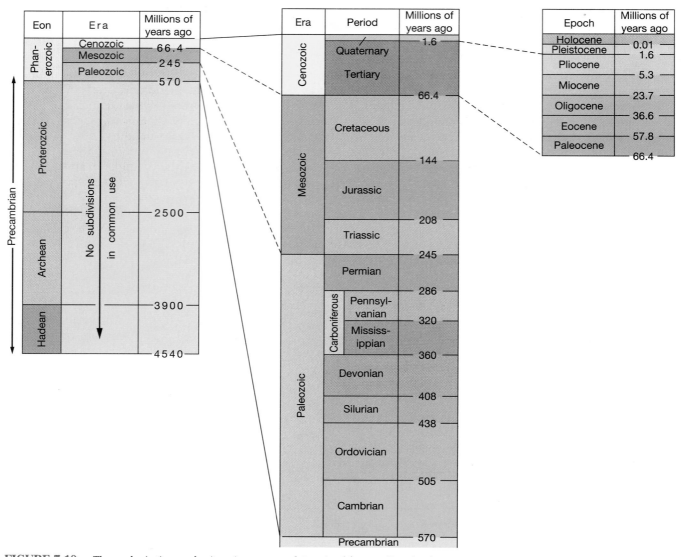

FIGURE 7.10 The geologic time scale. Ages in years are determined from radioactive isotopes.

the electron combines with a positively charged proton to create a neutron. The result is one less proton in the nucleus and one more neutron. With one less proton in the nucleus the original element is transformed into a new element whose atomic number is lower by "1" than the original element. Again, as with beta decay, the total number of protons and neutrons in the nucleus remains the same, thus the mass remains constant. (See green electron capture quadrant, Figure 7.12.)

Alpha Decay An **alpha particle** is made of two protons and two neutrons, which is the nucleus of a helium atom. **Alpha decay** involves the ejection of an alpha particle from the nucleus. As a result the element loses two protons and therefore two places in the periodic table. At the same time the nucleus also loses two neutrons. This loss, combined with the loss of two protons, reduces the mass of the isotope by four, as

shown in the red alpha decay quadrant of Figure 7.12. This type of decay is most common among the heavier radioactive isotopes, such as those of uranium.

Age Determination and Radioactivity In the process of decay, the original isotopes, called the **parent** isotopes, are transformed into stable isotopes of a new element, the **daughter** isotopes. The rate at which radioactive isotopes decay, the **decay rate,** can be measured in the laboratory. Scientists know of no physical or chemical conditions that change this rate of transformation, which means that decay happens at a uniform rate.

Scientists frequently express the decay rate in terms of a **half-life.** This is the time needed for one half of the original amount of a radioactive isotope to decay. One half of the original amount of the isotope remains after one half-life, one fourth after two half-

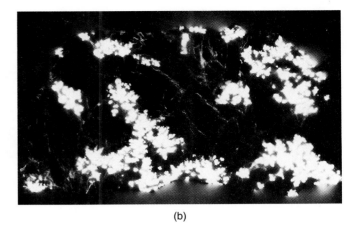

(a)

(b)

FIGURE 7.11 Radioactivity was discovered a century ago when the French scientist A. Henri Becquerel found that uranium compounds emitted a form of energy that exposed a photographic plate. Here, in (a) a rock specimen from Grafton, New Hampshire, the dark areas are made up of the uranium-bearing mineral gummite. In (b) the same specimen was placed on a photographic plate in the darkroom. Radiation from the uranium has exposed the plate, of which this is a print. The white areas mark the location of the uranium-bearing minerals.

lives, one eighth after three half-lives and so on. At the same time, the isotope of the new element is accumulating (Figure 7.13). The history of a gram of ^{238}U (uranium), recorded in Table 7.1, will help to illustrate this concept. The decay of ^{238}U involves a series of intermediate steps that lead to a stable end product, ^{206}Pb (lead). At the end of 4,500 million years half of the original isotope of ^{238}U will have disappeared; thus we say that the half-life of ^{238}U is 4,500 million years. During the same time 0.433 grams of the stable daughter product, ^{206}Pb, have accumulated. As this process continues there is less and less of the original ^{238}U isotope remaining and more and more of the ^{206}Pb. A specific ratio exists between parents and daughters at any stage in the decay process. This tells us how long the decay process has been going on. Thus, if we find a mineral with ^{238}U in it, we can determine the age of the mineral by measuring the amount of ^{238}U remaining and the amount of ^{206}Pb accumulated and knowing the rate of decay. Knowing these three pieces of information allows geologists to calculate back to a time when

^{238}U began to decay and before the daughter products began to accumulate.

Some Widely Used Isotopes for Age Determination

The heavy radioactive isotopes of ^{238}U, ^{235}U, and ^{232}Th (thorium) are widely used in dating metamorphic and igneous rocks. So also are the radioactive isotopes of rubidium, ^{87}Rb (which decays to strontium-87), and potassium, ^{40}K (which decays to argon-40 and calcium-40).

Although the radioactive isotope of potassium decays to argon-40 and calcium-40, only the ratio between potassium-40 and argon-40 is used in radioactive dating. The ^{40}K/^{40}Ar method is a widely used radioactive age determination method, and there are good reasons for this. In the first place, potassium is a very abundant element and forms a major portion of a number of minerals. Furthermore, it is well-suited to determine the ages of rocks from the very old to the very young because the half-life of ^{40}K (1.25 billion years) allows accumulation of measurable amounts of decay products in rocks of most all ages. It has even been used to date some minerals as young as 50,000 years in age, although with less reliability than for older materials. As with many radioactive isotopes, the method is widely applicable to igneous and metamorphic rocks.

In addition, one of the great virtues of the ^{40}K method for geologists is its use in determining the age of some sedimentary rocks directly. Some sandstones and, more rarely, shales contain glauconite, a silicate mineral similar to biotite. Glauconite, which forms in certain marine environments when the sedimentary layers are deposited, contains ^{40}K. The age of the glauconite thus gives us the date the sedimentary rocks

TABLE 7.1 History of 1 g of Uranium-238

AGE (MILLIONS OF YEARS)	^{206}Pb FORMED (g)	^{238}U REMAINING (g)
100	0.013	0.985
1,000	0.116	0.825
2,000	0.219	0.747
3,000	0.306	0.646
4,500 (1 half-life)	0.433[a]	0.500
9,000 (2 half-lives)	0.650	0.250
13,500 (3 half-lives)	0.758	0.125

[a]Amount of daughter product formed does not equal the amount of parent element remaining because ^{238}U decays through a number of intermediate daughter stages before it reaches the stable, final daughter product, ^{206}Pb.

Changes in Nucleus

Decay System	Protons ◯	Neutrons ●	Atomic number	Mass change
Electron Capture	-1	+1	-1	0
Alpha	-2	-2	-2	-4
Beta	+1	-1	+1	0

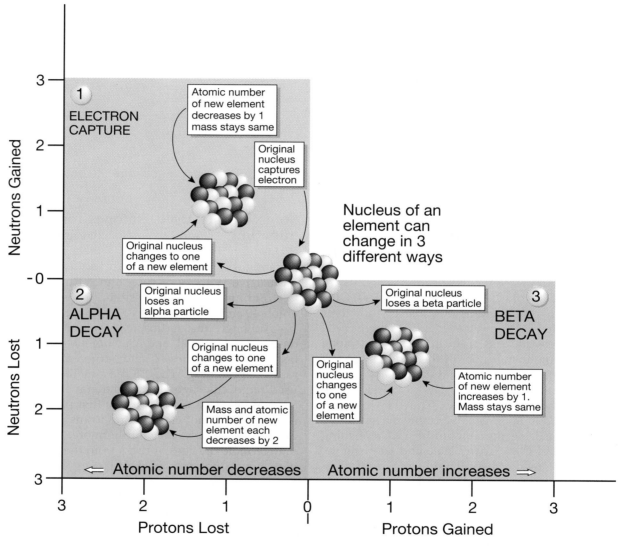

FIGURE 7.12 The three types of decay associated with radioactive age determination. Electron capture (green quadrant) decreases the atomic number of an element by "1", but the mass remains the same. Alpha decay (red quadrant) decreases the atomic number of the original element by 2 and the mass by 4. Beta decay (blue quadrant) increases the atomic number of the new element, but, as with electron capture decay, the mass is constant. Changes in the nucleus are summarized in the table at the the upper right.

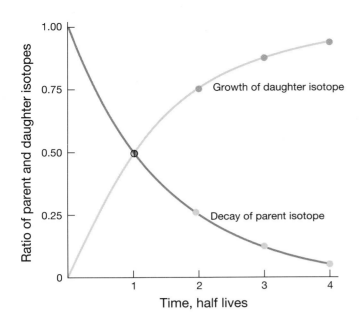

FIGURE 7.13 Graph showing the change in ratio of a radioactive parent isotope and its daughter isotope plotted against time as half-lives.

were laid down. Volcanic ash included in sedimentary deposits can also be useful. Fine ash, carried by the winds far from its point of origin, may settle in thin, but widespread, beds in both the marine and terrestrial environment. When the ash contains minerals with ^{40}K, their age is close to the age of the enclosing sediments. Data on some often-used radioactive isotopes are given in Table 7.2.

You may ask whether two or more radioactive isotopes give the same ages when used to date the same material. In many instances we are able to make just such a check. Because decay rates are unique to each unstable isotope, the probability that two or more methods will agree randomly is very low. The fact that several isotope methods regularly agree indicates that

TABLE 7.2 Some Radioactive Elements Used in Age Dating

PARENT ISOTOPE (RADIOACTIVE)	DAUGHTER ISOTOPE (STABLE)	PARENT'S HALF-LIFE (YEARS)	DECAY SYSTEM
Carbon-14	Nitrogen-14	5,730 ± 30	Beta
Potassium-40	Argon-40	1.25 billion	Electron capture
Rubidium-87	Strontium-87	48.8 billion	Beta
Thorium-232[a]	Lead-208	14 billion	Alpha & Beta
Uranium-235[a]	Lead-207	704 million	Alpha & Beta
Uranium-238	Lead-206	4.47 billion	Alpha & Beta

[a]Like uranium-238, these isotopes decay through a series of unstable isotopes before reaching a stable daughter isotope.

radioactive dating is self-consistent and reassures geologists that they are measuring real ages. It also indicates that geologists are able to keep consistent time with several different radioactive clocks.

Radiocarbon (^{14}C)

The isotope carbon-14 (^{14}C), known also as **radiocarbon,** is unique among the radioactive elements used in age determination. Its relatively short half-life of 5,730 ±30 years makes it invaluable for dating geologically very young material and even historical events. Its presence in all living things makes it possible to date organic material, something that cannot be done with the other radioactive isotopes. The method is practical for dating material back to 40,000 years. With special techniques it can be used for dating material 70,000 years.

Carbon-14 is continuously produced by cosmic ray bombardment of nitrogen, the major component of the atmosphere. Most nitrogen is the stable isotope ^{14}N, with 7 protons and 7 neutrons in its nucleus. However, a nucleus of ^{14}N may capture a neutron from cosmic radiation entering the atmosphere and eject a proton. This produces the radioactive isotope ^{14}C, with 6 protons and 8 neutrons. Radiocarbon decays by a beta emission and transforms back to ^{14}N (Figure 7.14). If this were all that happened we would probably have no interest in radiocarbon. We need to trace the path it follows.

Radiocarbon combines with oxygen, the other major component besides nitrogen in the atmosphere, and forms radioactive carbon dioxide, $^{14}CO_2$, which diffuses rapidly through the atmosphere. Because the

FIGURE 7.14 Radiocarbon is created by cosmic ray bombardment of nitrogen in the atmosphere; ^{14}N acquires a neutron and ejects a proton. Radioactive carbon reverts to nitrogen as it loses a beta particle and thus gains a proton and loses a neutron.

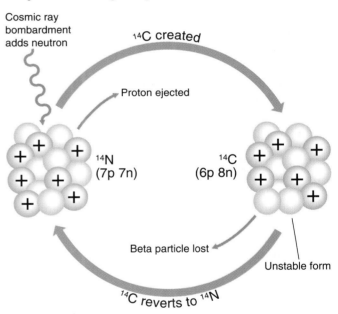

Telling Time in Years **167**

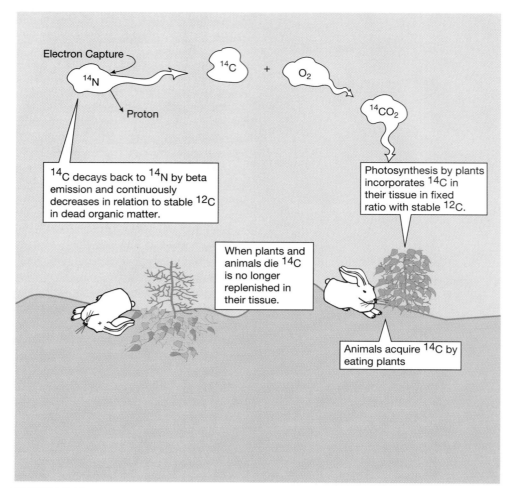

Electron Capture

^{14}N

Proton

^{14}C + O_2

$^{14}CO_2$

^{14}C decays back to ^{14}N by beta
emission and continuously
decreases in relation to stable ^{12}C
in dead organic matter.

Photosynthesis by plants
incorporates ^{14}C in
their tissue in fixed
ratio with stable ^{12}C.

When plants and
animals die ^{14}C
is no longer
replenished in
their tissue.

Animals acquire ^{14}C by
eating plants

FIGURE 7.15   The path of radiocarbon in the food chain begins when cosmic radiation converts
^{14}N to ^{14}C in the atmosphere. The radiocarbon then combines with atmospheric oxygen to form $^{14}CO_2$
(carbon dioxide) in a constant ratio with the carbon dioxide of stable carbon, ^{12}C and ^{13}C. Photosynthesis by plants uses this $^{14}CO_2$, water, and solar energy and thus brings radiocarbon into the plant
tissue. As long as the plant lives there is a constant ratio between the radiocarbon and the stable
forms of carbon, ^{12}C and ^{13}C, in their tissue. Animals acquire radiocarbon by eating plants, or eating
animals who have fed on plants. When the plants and animals die their supply of radiocarbon is no
longer replenished. As it decays back to ^{14}N, less and less ^{14}C remains in the plant and animal, but the
amount of stable carbon does not change. As a result, the ratio between ^{14}C and stable carbon
changes. This tells us how long it has been since the death of the plant or animal.

rate of formation of radiocarbon is constant, as is its
decay, the amount of $^{14}CO_2$ is constant in the atmosphere. There it maintains a constant ratio with the carbon dioxide made of the stable isotopes of carbon, ^{12}C
and ^{13}C. Carbon dioxide is used by plants in their
process of photosynthesis. When the plants take in carbon dioxide it has the same ratio of carbon-12 to car-
bon-14 as in the atmosphere. Thus some radiocarbon is
found in all living plants. Some animals eat plants, and
thereby directly acquire radiocarbon in their tissue.
Those animals that are not plant eaters get their radiocarbon by eating those that are. While the plant or animal lives, it acquires ^{14}C at the same rate that the ^{14}C
decays. With radiocarbon firmly and universally

established in the food chain, the stage is set for its use
in age determination of organic material (Figure 7.15).

As long as an organism lives it maintains a constant ratio of ^{14}C with the more abundant stable carbon, chiefly ^{12}C. However, as soon as the animal or
plant dies, the decaying ^{14}C is no longer renewed by
the life process. The longer the time since the death of
an organism the less radiocarbon remains in its tis-
sues. This changing ratio between radiocarbon and
stable carbon forms the basis for radiocarbon dating.

Radiocarbon dating is a marvelously useful tool in
a wide variety of fields, and requires only very small
samples (a gram to a hundredth of a gram and less).
Radiocarbon dating has given precision to the calen-

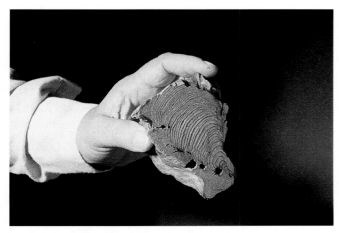

FIGURE 7.16 This fragment of wood has been cut from a spruce log found in a buried forest bed at Two Creeks, Wisconsin. It clearly shows annual rings decreasing in width out to the bark layer when the tree died as the result of an advancing ice sheet. A count of the tree's growth rings showed that it lived for 79 years. Radiocarbon dating give the time of death as about 11,530 radiocarbon years ago, which in turn dates the advance of the ice. *(John Simpson)*

eral can be determined in the laboratory, and this, combined with the number of fission tracks, is an index to age.

Sedimentation and Absolute Time Another way of establishing absolute dates for sedimentary strata is to determine their rate of deposition. Certain sedimentary rocks show a succession of thinly laminated beds. Various lines of evidence suggest that, in some instances at least, each one of these beds represents a single year of deposition. Therefore, by counting the beds, we can determine the total time it took for the rock to be deposited.

Deposition of fine sediments in lakes fed by glacial meltwater may produce distinctive beds called **varves** (Figure 7.17). A varve is a pair of well-defined units in which the sediments grade upward from a light-colored sand or silt into a darker clay-sized material, an example of graded bedding. It usually represents the deposits of a single year and is thus reminiscent

dar for geologically recent advances and retreats of glaciers (Figure 7.16). It has provided dates for human history following the disappearance of glaciers from most of northern Europe and Canada and the northern United States. It can be used to date glacier ice by using the ^{14}C trapped in bubbles of air buried within the ice. Underground water, too, can be dated. Radioactive carbon dioxide, dissolved in water, loses contact with the atmosphere after it has soaked into the ground and begins to move deeper into the Earth. Radiocarbon dating, in this instance, tells us how long the water has been underground. Not all the applications are geological, by any means. Art historians, for instance, have appealed to the method to aid in determining the authenticity of paintings, wooden sculpture, furniture, and tapestries. For example, suppose a wooden statue is said to have been carved in the year A.D. 1,000. A radiocarbon date of A.D. 1850, obtained for a small sample taken from the base of the statue, would disprove the date of carving claimed for it.

Fission Track Dating Fission track dating is another method of radioactive dating particularly used when the relatively abundant heavy isotope uranium-238 is present in a mineral. The method receives its name because *fission* is another term for radioactive decay, and the decay process damages the mineral containing the uranium, leaving tiny *tracks* behind. These tracks result from the emission of alpha particles ejected from the nucleus of the uranium during decay, and they can be observed under the microscope after first enhancing them by treating the mineral with hydrofluoric acid. The amount of uranium in the min-

FIGURE 7.17 Each of these varves in a Massachusetts clay pit records one year of deposition in a now-extinct glacial lake. *(Joseph H. Hartshorn)*

of the annual growth ring of a tree. The varve forms in the following way: During the period of summer thaw, meltwaters carry large amounts of sand, silt, and clay out into the glacier-fed lakes. The coarse material sinks rapidly and blankets the lake bottom with a thin layer of sand and silt. As long as the lake remains unfrozen, the wind creates currents strong enough to keep the finer clay particles in suspension. When the lake freezes over in the winter, these wind-generated currents cease, and the fine particles sink through the quiet water to the bottom, covering the coarser summer layer. A varve is usually a few millimeters to 1 cm thick, although thicknesses of 5 to 8 cm are not uncommon. There are rare instances of varves 30 cm or more thick.

Geologists have been able to link this kind of information to our modern calendar in only a very few places, such as the Scandinavian countries. Here the Swedish geologist Baron de Geer counted the varves that formed in extinct glacial lakes. These varves enabled geologists to piece together some of the geologic events of the last 20,000 years or so in the countries ringing the Baltic Sea.

Much longer varve sequences have been found in other places, but they tell us only the total length of time during which sedimentation took place, not how long ago it happened in absolute time. One excellent example of such a sequence is recorded in the Green River shales of Wyoming (Figure 7.18). Here each bed, interpreted as an annual deposit, is less than 0.02 cm thick, and the total thickness of the layers is about 980 m. These shales, then, represent approximately 5 million years of time. The sequence, however, is not tied

FIGURE 7.18   The Green River shales of Wyoming are very finely layered. This specimen has been cut into a series of seven small steps. Each step contains 100 layers, or varves, each representing one year. The tred on the large step on the right exposes the partial remains of a fossil fish. Height of specimen 6.25 cm. Specimen from the Princeton University Museum of Natural History. *(Robert P. Matthews)*

to the modern calendar, so we don't know which 5 million years of time they record.

Fitting Absolute Dates into the Geologic Time Scale

As we have found, the geologic time scale is made up of units of relative time, and these units can be arranged in proper chronological order without assigning dates in absolute time. This relative time scale has been constructed on the basis of sedimentary rocks. When a radioactive isotope is intimately associated with the sedimentary rock, as, for instance, is the case with glauconite, it is an easy matter to insert the radioactive date into the geologic column. Many of the radioactive dates available, however, have been determined for igneous and metamorphic rocks. How can the dates obtained from these rocks be matched with the relative-time units determined from sedimentary rocks?

Geologists use the absolute age of igneous rocks to bracket the probable age of sedimentary units. To do this, we must know the relative-time relationships between the sedimentary and igneous rocks. Recall the rule of cross-cutting relationships, which states that a rock is younger than any rock that it cuts across, and the rule of superposition, which states that in a sedimentary sequence the oldest rock is at the bottom. Now consider the example given in Figure 7.19, which shows a portion of the Earth's crust with both igneous and sedimentary rocks. The sedimentary rocks are arranged in three assemblages, numbered 1, 3, and 5, from oldest to youngest. The igneous rocks are 2, 4, and 6, also from oldest to youngest.

The sedimentary rocks labeled 1 are the oldest rocks in the diagram. They record that, after deposition, they were folded by Earth forces, and then a dike of igneous rock was injected into them. Because the sedimentary rocks had to be present before the dike could cut across them, they must be older than the dike. After the first igneous intrusion, erosion beveled both the sedimentary rocks and the dike, and across this surface were deposited the sedimentary rocks labeled 3. At some later time the batholith, labeled 4, cut across all the older rocks. In time, this batholith and the sedimentary rocks labeled 3 were also beveled by erosion, and the sedimentary rocks labeled 5 were laid across this surface. Finally, a basalt flow partially covered the youngest sedimentary rock. We have now established the relative ages of the rocks, from oldest to youngest, as 1, 2, 3, 4, 5, and 6.

Now, if we can date the igneous rocks by means of radioactive minerals, we can fit these dates into the

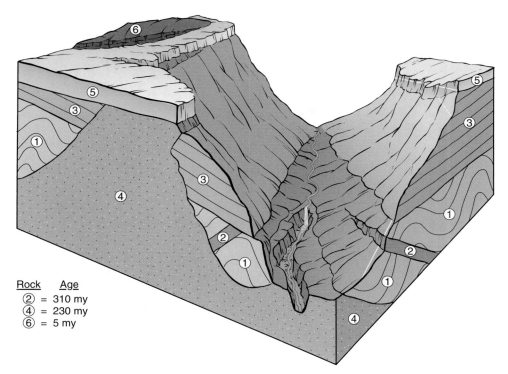

Rock   Age
② = 310 my
④ = 230 my
⑥ = 5 my

FIGURE 7.19
Radiometric dates obtained on igneous rocks can be fitted into the geologic time scale based on sedimentary rocks. This is done by bracketing the sedimentary rocks with the igneous rocks. In some instances radiometric dates can be assigned directly to the sedimentary rocks. See text for discussion.

relative-time sequence. We establish that the batholith is 230 million years old, that the dike is 310 million years old, and that the basalt is 5 million years old. The ages of the sedimentary rocks now may be expressed in relation to the known dates, as in the table below.

In some situations sedimentary rocks can be dated directly by radiometric means. For example, let us

Telling Time with Magnetism

▲▲

If you have ever used a compass you know that the needle points generally north. We accept this as a constant in our world, even as we accept the fact that the

ROCK UNIT	RADIOMETRIC AGE (MILLIONS OF YEARS)	BRACKETED AGE (MILLIONS OF YEARS)
6	5	
5		older than 5, younger than 230
4	230	
3		older than 230, younger than 310
2	310	
1		older than 310

assume that a close examination of the youngest sedimentary rock in Figure 7.19, unit 5, reveals two bits of information that we had overlooked before. At the very base of 5 we find deposits containing the mineral glauconite. At the very top of the unit is a thin bed of volcanic ash. The glauconite is dated by the $^{40}K/^{40}Ar$ method as 40 million years. Some mica from the volcanic ash is dated, using the same method, as 30 million years old. Unit 5 now turns out to have been forming between at least 30 million and 40 million years ago. Between the top of unit 5 and the volcanic rock above it there is a hiatus of at least 20 million years.

Sun sets in the west and we presume it always has and will. If suddenly our compass needle were to point in some direction other than north we would be—quite literally—lost. Nevertheless, contrary to what we have come to expect, geologists have found that the compass has not always pointed north. Sometimes it has swung around 180° and pointed south, before swinging back to point north again. The direction in which the compass needle, or any other magnetic needle, points is called the **magnetic polarity.** When it points north, as it does today, we call the polarity **normal.** When it points south the polarity is termed **reversed**

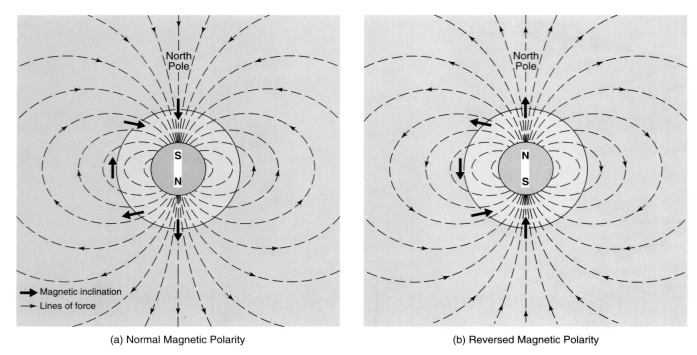

(a) Normal Magnetic Polarity   (b) Reversed Magnetic Polarity

FIGURE 7.20   We can picture the Earth's magnetic field as a series of lines of force, shown here in dashed lines around an imaginary bar magnet in the core. The dark arrows indicate the inclinations of a magnetic needle free to swing vertically. In (a) the Earth's magnetic polarity is the same as today's and is called *normal polarity*. The compass needle points along a line of magnetic force toward the North Pole. In (b) the Earth's magnetic field is reversed and the compass needle points in directions opposite to those of a normal field.

(Figure 7.20). This has opened the way to development of a **magnetic polarity time scale** and what is known as **magnetic stratigraphy.**

Earth as a Magnet

We can picture the Earth's magnetic field as a series of lines of force, arching around the Earth from pole to pole. The field is best described as if its cause were a giant bar magnet located in the core of the Earth (Figures 7.20 and 7.21). The magnetized needle of the familiar compass will point to the north following one of these lines of force. If the needle is allowed to swing freely in a vertical plane, as well as in a horizontal plane, it will take on the dip of the magnetic line of force as well. At a point north of Prince of Wales Island in the Canadian Arctic, at about 75° north latitude and 100° west longitude, the north-seeking end of the magnetic needle will dip vertically downward. This is the **north magnetic** or **dip pole.** Near the coast of Antarctica, at about 67° south and 143° east, at the **south magnetic** or **dip pole,** the same end of the needle points directly skyward. Halfway between the two poles at the **magnetic equator** the magnetic needle is horizontal. Between the dip poles and the magnetic equator a magnetic needle assumes positions of intermediate tilt. The angle of dip of the compass needle, as

FIGURE 7.21   The geomagnetic poles are defined by an imaginary magnetic axis passing through the Earth close to its center and inclined 11.5° from the axis of rotation. They are also called the *dip poles* because a compass needle allowed to swing freely will point vertically, upward at the South Magnetic Pole and downward at the North Magnetic Pole. A plane halfway between these two poles defines the magnetic equator and here the compass needle is horizontal.

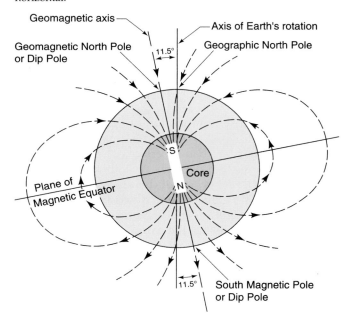

it varies from vertical to horizontal, is the **magnetic inclination.** If the inclination is zero the needle is at the magnetic equator. If it points downward it is in the Northern Hemisphere and the size of the angle at which it dips is a magnetic latitude. Likewise, if the needle points upward it is in the Southern Hemisphere and the angle from the horizontal determines a magnetic latitude. We will find these relationships particularly important in our later study of plate tectonics.

We must specify that the compass points to the magnetic poles because these poles do not correspond with the true geographic poles, which are located at the ends of the Earth's axis of rotation. The discrepancy occurs because the axis of the Earth's magnetic field is tilted about 11.5° from the axis of the Earth's rotation. As a result, the direction of the magnetic needle in most instances diverges from the true geographic poles. The angle of this divergence between true north and magnetic north is called the **magnetic declination.** It is measured in degrees east and west of the direction toward geographic north (Figure 7.22).

Cause of the Earth's Magnetism

The cause of Earth's magnetism has remained one of the most vexing problems in Earth study. Scientists are still trying to find a completely satisfactory answer to the question.

William Gilbert (1540–1603), physician to Queen Elizabeth I, first showed that the Earth behaved as a magnet. He suggested that the Earth's magnetic field resulted from a large mass of permanently magnetized

FIGURE 7.22 Because the dip poles and the magnetic poles do not coincide, the compass needle does not point to true north. The angle of divergence of the needle from the geographic pole is the magnetic declination and is measured east and west of true north.

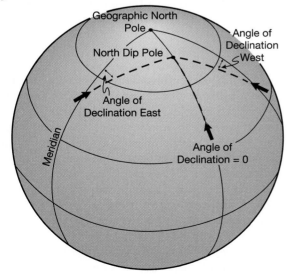

material beneath the surface. The idea was attractive then because large quantities of magnetic minerals had been found in the Earth's crust, and later because it was shown that the core of the Earth is largely iron, a highly magnetic element.

Although the idea of a bar magnet existing at the Earth's core describes the Earth's magnetic field very well, it cannot possibly be an explanation. Experiments show that magnetism is lost when the temperature of a material rises above a certain temperature. This is the material's **Curie point,** named for Pierre Curie. It varies for different materials. Iron, for instance, loses its magnetism above 760°C, and the mineral magnetite (Fe_3O_4) loses its magnetism at temperatures above 573°C. Scientists know that the **thermal gradient** of the Earth, the rate at which heat increases downward, is great enough so that at a depth of 20 to 30 km all Earth materials lose their magnetization. We therefore conclude that there can be no permanently magnetized mass at the center of the Earth.

If there is no permanently magnetized mass in the center of the Earth, is it possible that the source of magnetism lies in the upper regions of the lithosphere, where temperatures are generally above the Curie points of most minerals? Here we run into another problem, however. If the magnetism were to lie in the upper lithosphere, it would have to be 80 times stronger than it actually is, thus ruling out another explanation of Earth's magnetism.

Geologists have recently turned again to the interior of the Earth to find the cause of magnetism. Although the details are not agreed upon, and are vague at best, they hypothesize that Earth's magnetic field is generated by a **self-exciting dynamo** within the fluid metallic outer core. According to this theory, the metallic fluid in the Earth's outer core is in constant motion because of thermal convection and the rotation of the Earth. As the fluid moves, it passes through weak magnetic fields generated by the Sun or by sources elsewhere in the Earth and produces an electric current. This current, in turn, creates a more intense magnetic field. The electric and magnetic fields of the outer core reinforce each other in much the same way that they do in the generator at your local power plant. Scientists generally accept the broad outlines of this theory today, although they disagree about important details and still have no good way of explaining why the magnetic field varies.

Paleomagnetism

In 1906 Bernard Brunhes, a French geologist working in the volcanic region of central France, discovered that some of the Pleistocene basalt flows there had reversed magnetism. Later, in 1929, Motonori

Matuyama found that some of the volcanic rocks of Japan also had reversed magnetism. As magnetic studies of Earth materials expanded it became clear that the magnetism of many rocks was different from that of the Earth's present magnetic field and seemed to be left over from some previous time when the rocks were formed. This ancient magnetism is called **remanent magnetism** and its study is called **paleomagnetism.**

Geologists do not yet know what caused the shift in Earth polarity back and forth between normal and reversed. We know more, however, when it comes to understanding how a mineral acquires its magnetization. Magnetite provides an example. It is the iron component of the magnetite that makes it magnetizable. After the mineral crystallizes but before it drops below its Curie point, the thermal agitation of the atoms is so great that they are randomly oriented and the mineral is without magnetism. As the mineral cools, however, the thermal agitation of the iron atoms decreases. As they cool through the Curie point the atoms are easily influenced by an external magnetic field, like the Earth's. If the Earth's field is normal, the iron atoms in the mineral are magnetized in the direction of the present field. If the Earth's field is the reverse, then the magnetization is the reverse of today's field. The mineral thus becomes a tiny compass that records the Earth's magnetic field at the time and place it cooled below its Curie point. In an igneous or metamorphic rock, this condition is known as **thermoremanent magnetism.** Sedimentary rocks can develop a weak **depositional remanent magnetism** if magnetic particles settle gently through water and orient themselves with the Earth's magnetic field. Other magnetic minerals may grow during diagenesis, aligning themselves with the global field to produce a **chemical remanent magnetism.** No matter how the remanent magnetic field is created, it remains a permanent feature of the rock, even if the rock is moved or the Earth's magnetic field changes. It is, in a sense, a geophysical fossil, telling geologists what the globe's magnetic field was like when the rock formed. The only way the magnetism of a rock is lost is when it is destroyed by weathering, or heated above the Curie points of its magnetic minerals. In the latter case, if the rock subsequently cools down, it will acquire a new magnetization, that of the Earth's field at the new time of cooling.

When Brunhes and Matuyama made their discoveries that the Earth's magnetic field had at one time been reversed from that of the present, they ascribed the cause to a local phenomenon. There was, however, the possibility that the reversals had been worldwide. By the 1960s, three Californians—Allan Cox, G. Brent Dalrymple, and Richard R. Doell—were addressing this possibility. They reasoned that the best way of approaching the question was to analyze rocks from all over the world to see whether rocks of the same age had the same polarity. They determined the magnetic polarity of many individual lava flows, and it soon became clear that there were about as many flows with reversed polarity as there were with normal polarity. They also dated the flows using the $^{40}K/^{40}Ar$ method, and a pattern began to emerge in which there were extended periods of time when the rocks recorded a normal polarity for the Earth's magnetic field, and other similarly long periods in which the Earth's field was reversed.

The Magnetic Polarity Time Scale

Following the original work of Cox, Dalrymple, and Doell, geologists have put together the magnetic time scale by determining the age and polarity of extrusive igneous rocks. As a simple example, imagine an accumulation of lava flows. Superposition tells us the youngest is at the top. Radioactive dating of the flows gives us an age in terms of years for each flow. If we now measure the magnetic polarity of each flow we can construct a radioactively dated magnetic time scale (Figure 7.23).

During the last four decades, thousands of age and polarity measurements representing millions of years have been spliced together to construct a magnetic polarity time scale. A portion of this is given in Figure 7.24. Times of predominantly normal polarity, or of predominantly reversed polarity, are called **magnetic chrons.** The four most recent chrons have been named after scientists who have made major contributions to our knowledge of Earth magnetism. We have already met three of them: Brunhes, Matuyama, and Gilbert. The third, Karl Friedrich Gauss (1777–1855), a German mathematician, made important contributions to the theory of electromagnetism. These same chrons are also designated by numbers 1 through 4, a numerical system that is extended backward in time for the older chrons, as shown in Figure 7.24.

One of the difficulties with using the magnetic time scale is that any one reversal looks like any other. You need more information about the chron than just its magnetic signal to know where you are in time. One help is that each chron usually has within it one or more shorter events in which the magnetic field is reversed from that of the dominant polarity of the chron. These are called **magnetic subchrons.** Look at the Matuyama reversed chron in Figure 7.24. Magnetism is reversed during most of the chron. Within it, however, three short events, or subchrons, occur in which the magnetism is normal. Each has been named for the place of its discovery: Jaramillo Creek in New

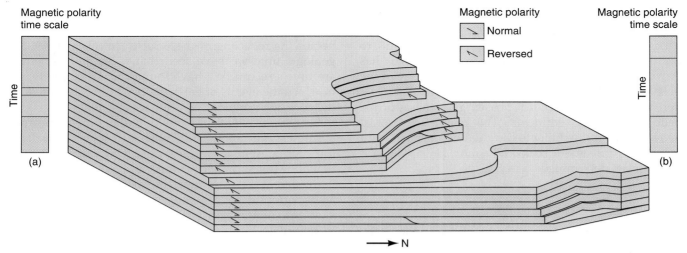

FIGURE 7.23 An imaginary sequence of lava flows records changing magnetic polarities. The inclined arrows indicate normal and reversed polarities of the Earth's field. The angle of inclination indicates a latitude of about 30° north for the magnetic polarities. The magnetic polarity time scale on the right describes only one sequence of reversed polarity because a single bed with normal polarity is missing here that is present in area (a).

FIGURE 7.24 The magnetic polarity time scale back to the beginning of the Miocene. During a magnetic chron, shorter subchrons of the opposite polarity usually occur.

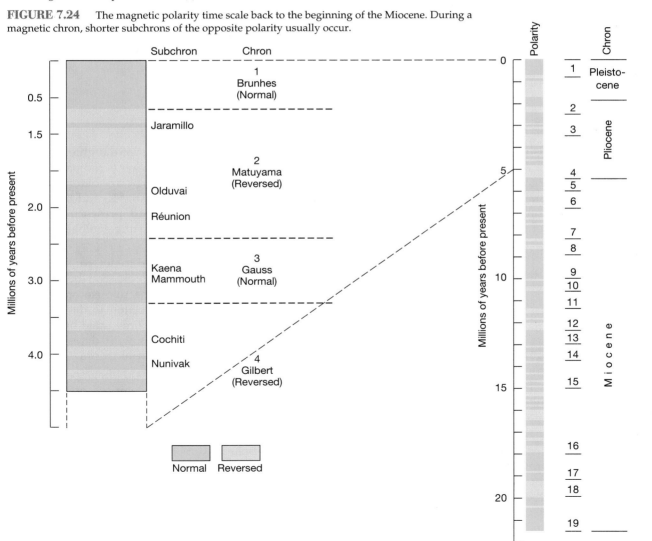

Mexico; Olduvai Gorge in Tanzania; and Réunion Island in the Indian Ocean. These short periods of magnetic reversal give a characteristic "fingerprint" to the longer chron. Another approach, which has proved very successful in studies of the geology of the deep sea, has been to use the magnetic scale in association with the fossil record. Fossils give the approximate age of the sediments. The magnetic reversal pattern from the sediments can then be matched with the standard magnetic scale and a more precise age for the deposits determined.

▶▶ EPILOGUE ◀◀

In Chapters 2–6 we studied the materials of the Earth. We considered the nature of minerals, how they combine to form the three rock families (igneous, metamorphic, and sedimentary), and how weathering begins the recycling of these rocks. In this chapter on geologic time we have seen that rocks and minerals are time's record. We have found that there are ways in which the events of Earth history can be arranged in the correct chronological sequence and that we can tell the age of the events in terms of both relative time and absolute time. Rocks and minerals progress through the rock system at various rates and carry "clocks" that allow us to time those rates and to tell us what has happened at different times in Earth history. With this as background we can begin, in the next chapter, our study of some major Earth processes, from volcanoes to mountain building, each of which has its place in the global recycling of the planet's rocks.

SUMMARY

1. Events in Earth history can be arranged chronologically using the simple rules of stratigraphy, which include the superposition of sedimentary rocks, their original horizontality, their lateral continuity, inclusions, and cross-cutting relationships. Sequences of sedimentary rocks from different areas are correlated by both their physical characteristics and their contained fossils. Superposition of strata (and their fossils) shows that life forms change with time and leads to the concept of faunal and floral succession. Unconformities are gaps in the rock record caused by erosion.

2. The geologic column is a succession of rocks arranged in the correct chronologic order. The geologic time scale is the division of the geologic column into time units.

3. Radioactivity is the spontaneous decay of the nucleus of an unstable isotope. It occurs in the nucleus by beta decay (the nucleus loses a neutron and gains a proton), by electron capture (the nucleus loses a proton and gains a neutron), and by alpha decay (the nucleus loses two protons and two neutrons). Geologists can date a mineral containing an unstable isotope when they know its rate of decay, the number of parent atoms remaining, and the number of daughter or resultant atoms accumulated. Some radioactive isotopes widely used in geologic age determinations include ^{238}U, ^{235}U, ^{232}Th, ^{87}Rb, ^{40}K, and ^{14}C. Radiocarbon's short half-life and presence in living things makes it useful in dating geologically very recent materials.

4. Sedimentary rocks are used to determine relative time. Some sedimentary rocks can be dated in real time directly by radioactive isotopes. Other radioactive dates can be related to the sedimentary rocks in the geologic time scale by bracketing the sedimentary rocks with dates obtained from igneous and metamorphic rocks.

5. The Earth's magnetic field has periodically shifted polarity. Today a compass points to the north magnetic pole (normal polarity). At various times in the past it would have pointed to the south magnetic pole (reversed polarity). The Earth's polarity direction is recorded by magnetizable minerals as they cool below their Curie points and acquire the magnetic polarity of the Earth's field at the time. When these minerals can be dated radiometrically they form the basis for the magnetic polarity scale.

KEY WORDS AND CONCEPTS

^{14}C method 167
^{40}K/^{40}Ar method 165
absolute time 154
alpha decay 164
alpha particle 164
angular unconformity 158
Archean 161
beta decay 162
Cenozoic 162
chemical remanent magnetism 174
correlation 155
cross-cutting relationships 155
Curie point 173
daughter 164
decay rate 164
depositional remanent magnetism 174
dip pole 172
disconformity 158
electron capture 163
eon 160
epoch 161

era 161
fission track dating 169
geologic column 160
geologic time scale 154
Hadean 161
half-life 164
hiatus 158
inclusions 155
lateral continuity 155
magnetic chron 174
magnetic declination 173
magnetic equator 172
magnetic inclination 173
magnetic polarity 171
magnetic polarity time scale 172
magnetic stratigraphy 172
magnetic subchron 174
Mesozoic 162
nonconformity 159
normal polarity 171
north magnetic pole 172
original horizontality 164
paleomagnetism 174

Paleozoic 162
parent 164
period 161
Phanerozoic 160
Precambrian 160
principle of faunal and floral succession 157
Proterozoic 161
radioactivity 161
radiocarbon 167
relative time 154
remanent magnetism 174
reversed polarity 171
self-exciting dynamo 173
south magnetic pole 172
stratigraphy 154
superposition 154
thermal gradient 173
thermoremanent magnetism 174
unconformity 158
varve 169

QUESTIONS FOR REVIEW AND THOUGHT

7.1 What are the rules of stratigraphy?

7.2 How does the rule of superposition work? The rule of cross-cutting relations?

7.3 What is the principle of faunal and floral succession? How was it established?

7.4 How are fossils used in correlation of sedimentary rocks?

7.5 What is a hiatus? What is an unconformity? What are the three types of unconformities? How are they different?

7.6 Suppose you discovered a rock cliff in which a granite was overlain by a series of limestone beds tilted at 25 degrees from the horizontal. What history could be deduced from the exposure?

7.7 What is the geologic column?

7.8 What is radioactivity? Why is it useful in the study of geology?

7.9 What are three types of radioactive decay? How do they work?

7.10 In the laboratory you find that a radioactive element is emitting alpha particles and creating a new element. How much more (or less) is the mass of the new element in relation to the parent element? Does the parent material have a higher or lower atomic number than the daughter product? By how much?

7.11 How are radioactive ages placed in the geologic time scale?

7.12 How is radiocarbon formed? How does it decay? Why is it useful as an age-dating technique?

7.13 How does the ^{40}K/^{40}Ar method of age determination work?

7.14 What is today the best estimate for the age of the Earth? How has it been determined?

7.15 Under what circumstances can the rate of sedimentation be used to date geologic events?

7.16 In what way does the Earth resemble a magnet?

7.17 What is the Curie point of a magnetizable mineral? Why is it important in geologic studies?

7.18 What is the magnetic polarity time scale? How has it been constructed?

Critical Thinking

7.19 How may sediments deposited in water be used to tell geologic time?

7.20 The Moon is about the same age as the Earth. What ways do you think could be used to establish a lunar chronology?

CHANGES WITHIN THE EARTH

According to the popular account of the European "discovery" of the Americas, people at the end of the fifteenth century believed the world to be flat. That part of the story is far from true—scholars as early as the sixth century B.C. had correctly deduced that the Earth is spherical—but explorers during the European Renaissance clearly knew very little else about their planet. Map makers were not only ignorant of prominent land masses (including the Americas, Australia, and Antarctica), but they also rarely gave even the coastlines that were familiar trade routes their proper shapes (Figure T2.1). It is remarkable how much we have learned about the Earth in barely five centuries since then and how many of the questions that concern geologists today would have been totally foreign to our ancestors.

As we approach the end of the twentieth century, geologists have mapped much of the Earth's surface in detail and have begun to explore a new frontier: its interior. In some ways, geologists study the Earth's interior as the Renaissance explorers did, locating new territories and describing new features (although the exploration today is done from a distance, using sophisticated technology). In other ways, geologists' interests are very different. The emphasis today is on how the planet evolves and on how changes within it help to reshape the planetary surface. Beneath the Earth's surface is a dynamic system that we have only begun to investigate.

The chapters immediately ahead, from Chapter 8 through Chapter 12, deal with large-scale changes within the Earth that scholars of the fifteenth century had few explanations for and, in most cases, were not even aware of. Indeed, long after world maps had become reasonably accurate, scientists had given little thought to questions like: "Why are the continents high and the ocean floors low?" or "Why do mountains generally occur in belts, commonly on one side of a continent?" or the question that seems like an obvious one in hindsight: "Why do several of the world's coastlines look as if they could fit together like puzzle pieces?" (Figure T2.2) The modern explanation for each of these lies in **the theory of plate tectonics,** accepted by geologists only a little over a quarter-century ago. What we are about to share with you, therefore, is not only a perspective on changes within the Earth, but also a story of changes in geology.

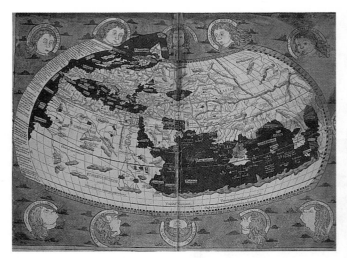

T2.1   This world map, attributed to the geographer Ptolemy around the year 150, was republished and updated in 1482 by an unknown cartographer at Ulm, in southern Germany. Among its many errors is the notion that the Indian Ocean is completely surrounded by land, an idea that was probably already known to be incorrect many centuries before Ptolemy lived. Notice, however, that despite its many peculiarities, this map is drawn with a global grid of latitude and longitude that clearly indicates that its author believed the Earth to be spherical.

T2.2   This image of the South Atlantic Ocean, taken from a weather satellite, shows how nearly congruent the coastlines of South America and Africa are. Maps available in the early nineteenth century were already accurate enough to show this same parallelism and to suggest that the continents had once been together. The idea was not proposed seriously until 1912, however, and has been accepted by most geologists only since about 1967.

A Changing Science

Historically, the greatest obstacle that geologists had to overcome between the fifteenth century and today was their own limited ability to recognize changes in the Earth. Not too surprisingly perhaps, when they began to see evidence of change, they saw it first in small things. For example, in Chapter 7 we explored the laws of stratigraphy that were proposed by Nicolas Steno in the seventeenth century. He based these laws on what he saw in rock samples or in local exposures in hillsides or stream beds. A century later, William Smith used those laws to correlate observations that he made in localities all across southern England. By the end of the nineteenth century, geologists were regularly stretching their minds to understand change on a global scale. Geologists working in widely separated locales compared their observations and began to extract principles common to all of them. We saw this in Chapter 6, for example, in the way that metamorphic petrologists developed the facies concept from field studies of rocks that at first seemed to have nothing in common. To a great degree, this broadening of perspectives was an inevitable outcome of the information explosion that began to accelerate in the 1700s. The more that geologists learned from small-scale observations, the easier it became to see how those observations all fit into a larger framework. This global perspective was also due, however, to

geologists' acceptance of James Hutton's notion that the Earth is a very old planet. Large-scale changes such as the uplift of mountains happen slowly, and geologists could not even imagine how they occurred until they realized how much time had passed since the Earth was formed.

Another kind of change had taken place in the science of geology as well. Although Steno and his contemporaries had been content to document changes in rock for use as tools to study Earth's history, nineteenth- and twentieth-century geologists became more interested in finding out what *causes* change. As we have emphasized in the preceding chapters, the transformation of one kind of rock into another commonly involves moving it to a new geologic environment. The rock system, therefore, also implies a pattern of physical movement, at least in the uppermost part of the Earth. A plutonic igneous rock, for example, cannot be weathered and reconstituted as a sedimentary rock unless it is first raised to the surface and exposed. Regional metamorphism of sedimentary rocks is made possible when compression of the Earth's crust generates new mountains. As we have tried to emphasize in discussing igneous, sedimentary, and metamorphic rocks, the processes of change in the rock system are often more interesting to geologists today than are the rocks themselves. The same is true for studies of the

Earth as a whole. As a result, many of the exciting questions in geology now focus on *how* forces in the Earth do such things as deform rock, cause earthquakes, and rearrange the crust. Less than 50 years after William Smith published his maps, geologists' attention had almost exclusively shifted from simply describing the Earth to asking how it worked.

Heat in the Earth

To help you discover how the Earth works, or at least what kinds of changes happen within it, we will begin with a study of heat in Chapter 8. You should already recognize heat as an important agent of change in rocks, through such processes as volcanism or metamorphism. It is now time to ask what connection these processes have to the physical evolution of continents and ocean basins, and why the rock-forming environments that you studied in the first portion of this book are found where they are. The short answer to these questions is that heat is also the driving force for movement in the crust. The transfer of heat from deep in the Earth

The most dramatic evidence of physical movement in the Earth, and one that certainly commands attention from both geologists and the general public, is earthquakes.

toward the surface involves a wholesale convection, or stirring, of the mantle, and the lithosphere is carried along in pieces on top. Many rock-forming environments occur at the edges of large lithospheric blocks or **plates,** as we introduced them in Chapter 1 (see Figure T2.3). Historically, heat has been a heavily debated topic among geologists, and there continue to be unresolved questions about how heat is produced and transferred inside the Earth, and exactly how the transfer of heat is responsible for changes within the Earth.

Earthquakes

Of course, to engage you fully in thinking about changes within the Earth, we need to provide proof that change actually occurs. The most dramatic evidence of physical movement in the Earth, and one that certainly commands attention from both geologists and the general public, is earthquakes. In Chapter 9 we will discuss what earthquakes are, how they are detected, and the effects they have on both the Earth's surface and human civilization. The more intriguing questions in that chapter, though, will relate to the causes of earthquakes and to

T2.3   Rock-forming environments related to large-scale movement of lithospheric plates. The rock cycle and the pattern of global movement the geologists call plate tectonics are both driven by heat in the Earth.

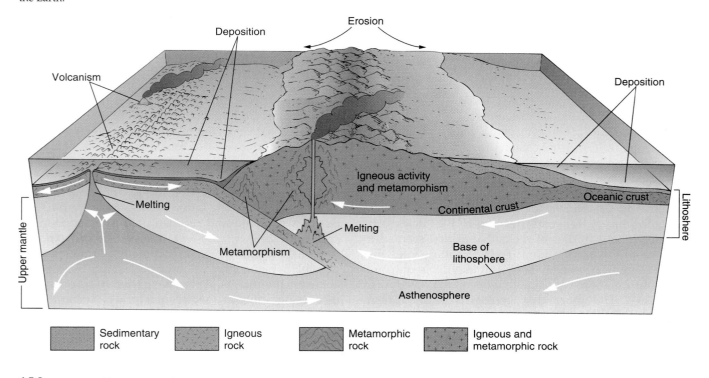

the various ways that earthquakes may be related to other events such as volcanism and mountain-building. As with the discussion of heat, these are questions that we will then follow through each of the remaining chapters in this section of the book. We will also show how geologists have learned to use the energy released by earthquakes to probe the Earth's interior for clues that may reveal not only its structure but the processes that alter it through time.

Deformation
▲▲

At the Earth's surface, geologists see evidence of past movement in rocks that have been bent or broken. Geologists in the subfield of **structural geology** use the information that they gather from analyzing faults and folds to deduce the pattern of forces that has acted on the crust (Figure T2.4). In this way, they relate the small changes that they note in single rock exposures across a region to large-scale events that stretched or squeezed the region in the past. In Chapter 10 we will explore some of the basic principles that geologists apply to interpret deformed rocks and understand how the crust has moved.

Plate Tectonics
▲▲

The ultimate goal in this portion of the book is to share what geologists today feel is the best explanation for large-scale crustal movement: the theory of plate tectonics. The brief discussion of plate tectonics in Chapter 1 presented the fundamental idea behind the theory. Lithospheric plates move horizontally toward, away from, or past each other, carried like boxes on a conveyor belt. The convection of heat in the Earth's interior, to which we referred above, produces slow lateral movements in the asthenosphere that keep the

T2.4   Sediments are deposited in nearly horizontal layers, but forces within the crust break and bend those layers. As mountains form, rock is compressed into complex folds such as the ones in this photo or are fractured and slid over each other. Geologists learn about forces in the crust by studying the extent and style of deformation.

plates in motion. Earthquakes, deformation, volcanism, and many other dynamic processes take place where adjacent plates interact. In Chapter 11 we will trace the evolution of this idea and explore the evidence that supports it, particularly in the seafloor. In Chapter 12 we will consider through case studies some of the ways that geologists have clarified their understanding of the Earth by applying it to specific geologic environments. These two chapters will also serve as a way to address the theme with which we opened this short essay: change within the science of geology itself. The story of how the plate tectonic theory was proposed, tested, modified, and finally adopted by geologists is quite recent and, in fact, continues today. It is one for which the authors of this book have had a front row seat, and which we have found fascinating. We hope to share some of the excitement of discovery and the uncertainty of abandoning old ideas for new ones.

8

Mt. Parícutin in Michoacán, Mexico, in eruption on February 20, 1944
(Tad Nichols)

Heat and Magmas

OBJECTIVES

▲▲

*As you read through this chapter, it may help if you focus
on the following questions:*

1. What are the major heat sources within the Earth?
2. How is heat transported from one part of the Earth's interior to another?
3. What have geologists learned about temperatures inside the Earth by studying surface heat flow, and how do they explain the fact that the temperature beneath the continents is different from that under the ocean basins?
4. How does the process of partial melting produce magma, and how and where are basaltic and granitic magmas formed?

OVERVIEW

▲▲

Heat is a form of energy. That realization, made by scientists in the mid-nineteenth century, was essential to framing one of the basic laws of physics, the *First Law of Thermodynamics*, which says that heat and mechanical work are equivalent ways of bringing about change. For example, we may hold a piece of iron in a flame (thus adding energy as heat) or may pound on it with a hammer (thus adding energy by performing work). Either action changes the temperature of the metal. Recognizing this equivalence was an important breakthrough for scientists because it implied that energy added in one form to a material could be extracted in another form. The First Law of Thermodynamics explains how an engine, for example, can convert the heat from burning fuel into energy of motion. For geologists, it established a way of connecting processes that generate heat in the Earth with mechanical events like the growth of mountains.

In this chapter we will explore the sources of heat in the Earth, discuss how it is transferred from one place to another, and examine how it alters the state of matter and causes movement in the Earth.

Heat in the Earth

Sources of Heat

During Earth's roughly 4.6-billion-year history, the planetary interior has been heated in three major ways: by **gravitational processes**, by **solidification of the core**, and by **radioactivity**. A small amount of heat has come from other sources as well. Quite a lot of heat comes from the Sun, for example, but it barely penetrates the planet's surface, and thus it has no effect on the interior. Tidal energy also is inconsequential in the Earth, although it is responsible for a small amount of heating in the ocean.

Gravitational Processes When an object falls from a high place, it carries mechanical energy that can be converted to heat. For example, a meteor passes through the atmosphere and is warmed as some energy is transformed into heat by friction. It may be heated much more (even to the point of melting or vaporizing) when it strikes the ground. Today, this is a trivial source of heat for the planet because impacts are small and rare. Early in Earth's history, however, it was much more significant. The planet grew by **accretion**, the gravitational capture of particles of all sizes from dust to asteroids (see Figure 8.1). Each new impact added not only mass but heat. Geologists estimate that before it even reached the size of the Moon, Earth's surface was so hot that a **magma ocean** several hundred kilometers deep had formed. Although this is largely a theoretical conclusion, based on what we can deduce about what the early Solar System looked like, it is also supported by field observations on the Moon and by subtle chemical patterns in Precambrian rocks.

Much of the impact-generated heat was lost to space, but some was trapped under rock debris that continued to land on the growing planet, thus heating its interior. A side effect of accretionary heating, however, has turned out to be even more important. Geologists theorize that the igneous process of differentiation led to a planet-wide separation of metallic iron and some nickel from the magma ocean. The dense metallic liquid then sank gradually to the center of the Earth, where it displaced less-dense silicates and became the core. This event may have taken as much as 500 million years to complete. It is sometimes called the **iron catastrophe** and was the second gravitational heat-producing process. To see why, think of the descending liquid metal as if it were a meteor passing through an extremely thick atmosphere, in this case an "atmosphere" of solid silicate minerals in the mantle of the Earth (see Figure 8.2). Friction between the mantle and iron that was headed for the core became a

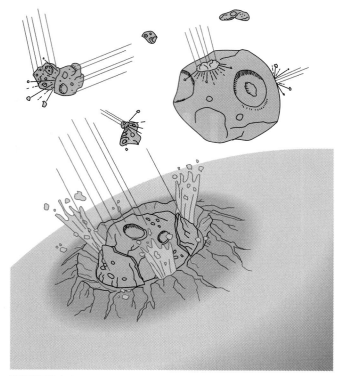

FIGURE 8.1 Four and a half billion years ago a lot of debris, ranging in size from dust to asteroids, still orbited the young Sun. Collisions between these materials, called *planetesimals,* were frequent. As planetesimals gradually accumulated to form larger bodies, these developed stronger gravitational fields and began to attract more particles from nearby space. Each impact added heat to the planet and buried earlier hot rocks under a new blanket of debris. According to some estimates, the Earth could have grown to full size by this process of accretion in a few tens of millions of years. Long before it did, the planet had become so hot that its outer region melted completely to form a global-scale ocean of magma.

major source of heat during the waning stages of planetary formation. Even now, thanks to the insulation provided by almost 3,000 km of rock, some of that heat is still deep in the Earth.

Solidification of the Core A second source of the Earth's interior heat—one that continues to generate heat today—is solidification of the liquid metallic outer core. It may be easiest to see why this is a source of heat if we consider the opposite process first; that is, the process of melting a solid. Atoms in any solid are held tightly to their neighbors by chemical bonds, which can be broken if the solid absorbs energy (heat) from its surroundings. As groups of atoms in the solid gradually break free and become a liquid, therefore, the surroundings must *lose* heat. Solidification has the reverse effect. As groups of atoms in a liquid connect with each other to form a solid they release heat to their surroundings, which become warmer (Figure 8.3). Geologists hypothesize that the core loses heat in this way as its molten outer

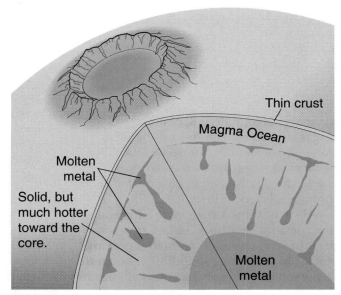

FIGURE 8.2 Liquid iron/nickel, separated from the outer portion of the young Earth, gradually settled toward the center of the planet because of its high density. As it "fell" through the mantle and had to overcome its viscosity, some of the metal's energy of motion was converted into heat. This "iron catastrophe" was a second source of gravitational heating in the young Earth.

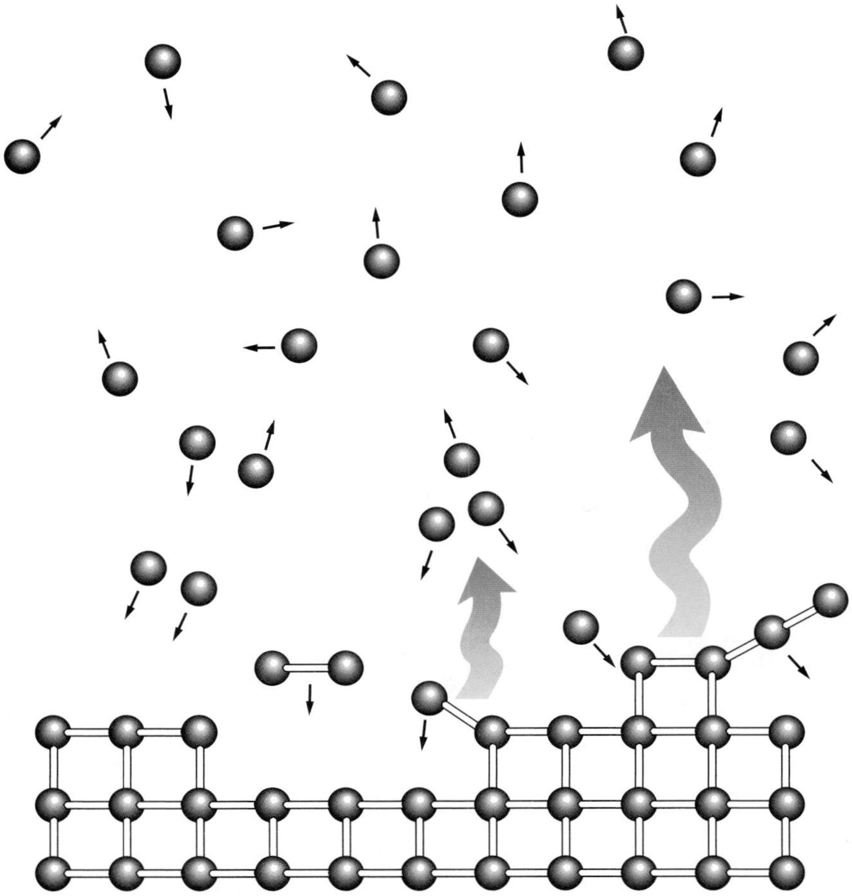

portion gradually solidifies. Given enough time the core will eventually become entirely solid, as has happened on the Moon and Mars, both of which have cores that are much smaller than the Earth's. This will take several billion years. Until then, the solidification process will provide a steady supply of heat from the deepest parts of the Earth.

Radioactivity As we discussed in Chapter 7, radioactive decay has given geologists a valuable tool for dating rocks and geologic events. The discovery that radioactivity is a source of heat, however, may have been even more important to geology. As an unstable atom decays, a burst of high-energy radiation is released, often in the form of X-rays or gamma rays. As this energy passes through surrounding materials it is converted to heat. Alpha and beta particles also generate heat as they crash through other substances, disrupting their crystalline structures. Scientists take advantage of these processes to generate power in nuclear reactors. In the Earth, they provide energy to move tectonic plates and raise mountains.

Three chemical elements are particularly important sources of radioactive heat. The two most abundant isotopes of *uranium*, ^{235}U and ^{238}U,

FIGURE 8.3
Atoms in a liquid move randomly and are not bonded permanently to their neighbors. As the liquid cools to its freezing point, the atoms bond firmly together to form a solid, losing their freedom to move about. The energy of motion that they once had in the liquid is now converted to heat, which may then leave the area. Geologists hypothesize that the inner core of the Earth is gradually growing as the liquid outer core solidifies, and that this process is an ongoing source of heat in the deep planetary interior.

account for 37 percent of the radioactive heat in the Earth. The heat produced by ^{232}Th, the most common isotope of *thorium,* amounts to another 36 percent of the total. The remaining 27 percent comes largely from the radioactive isotope of *potassium,* ^{40}K, which we discussed in Chapter 7. Actually, only one potassium atom in 10,000 is radioactive. An atom of potassium also releases much less energy when it decays than do atoms of uranium or thorium. However, there is about 5,000 times as much ^{40}K in the Earth as there are the isotopes ^{235}U and ^{238}U, enough to make it a substantial source of heat.

These three heat-producing elements are most abundant in the continental crust, less abundant in the oceanic crust, considerably less in the mantle, and probably inconsequential in the core. There are at least three reasons for this uneven distribution. First, uranium, thorium, and potassium are lithophile elements, which means that we are most likely to find them in silicates or in oxides, the primary constituents of the mantle and crust. Second, all three—but particularly uranium and thorium—have large atoms that do not fit easily into the compact mineral structures that are stable under high pressure. For this reason, they tend to concentrate in the lithosphere instead of the mantle. Third, potassium is found primarily in potassic feldspars (orthoclase), in micas, and in clays; most uranium and thorium are found in a handful of simple oxides. As Table 8.1 suggests, these minerals are usually associated with granites, which dominate the continental lithosphere. The basalts that comprise the oceanic lithosphere and the peridotites that make up the mantle are much less radioactive.

Heat Transport

Heat produced within the Earth can pass through it in one of three ways: by **radiation**, by **conduction**, or by **convection**. Of these, radiation, the method by which Earth receives heat from the Sun, is the least effective way to transfer heat through rock. When a hot object such as the Sun glows, it gives off not only light, but also infrared waves, another form of electromagnetic radiation. This energy can travel immense distances at the speed of light (300,000 km/sec in a vacuum). To illustrate why this process is relatively unimportant inside the Earth, consider what happens to infrared radiation from the Sun. When infrared waves encounter matter, they are absorbed and converted to heat. Having crossed 150 million kilometers of open space, about half of the solar radiation reaching the Earth never even makes it through the atmosphere, but is scattered, reflected, or absorbed by dust, clouds, or gas molecules. The rest can be absorbed by even a thin layer of sand. If infrared waves from the Sun can be blocked so easily, then clearly heat produced deep inside the Earth cannot be radiated very efficiently through solid rock either.

Heat is transported more efficiently by conduction, passing by direct contact from one atom to the next. In any substance, no matter how rigid or dense, atoms always vibrate slightly. The hotter the substance is, the more vigorously its atoms vibrate and the more they tend to jostle their neighbors. Getting heat to move from one part of the Earth to another by conduction, then, is like knocking over a row of dominoes by touching the first one. One atom collides with another until even ones far away from the hottest area have more energy than they started with (Figure 8.4).

Conduction is certainly a better way to move heat in the Earth than is radiation, but it is still not very effective. Rock is a good insulator, which is another way of saying that one vibrating atom in it does not easily set the next one moving. The **thermal conductivity** of rock is so poor that if heat could rise through the mantle only by conduction, very little of it would have escaped from the deep interior since the Earth formed. Iron, by comparison, is about 25 times as conductive.

Conduction transports heat through the outermost 5 to 30 km of the Earth (the lithosphere), which is fairly cool and rigid. If conduction were the most efficient heat transport process in the deep interior of the Earth, however, it would cause a serious problem. Not only would much of the Earth's original accretionary heat still be trapped inside, but a lot of extra heat generated by radioactivity over the last 4.6 billion years would also be there. Thus the mantle would actually be much hotter now than when it formed. Everything below the lithosphere would probably be molten.

TABLE 8.1 Radioactive Heat Sources in Dominant Rocks of the Crust and Mantle.

ROCK	AMOUNT OF RADIOACTIVE ELEMENT (IN PARTS PER MILLION)			HEAT PRODUCED (IN MW/KG)[a]
	Uranium	Thorium	Potassium	
Granite	4	13	4	0.0096
Basalt	0.5	2	1.5	0.0016
Peridotite	0.002	0.06	0.02	0.00003

[a]A milliwatt (mW) is 1/1000 of a *watt,* a unit of power equivalent to one Joule per second or about 1/746 of one horsepower. A watt is most familiar as a means of rating the electrical power consumption of light bulbs or appliances. [From various sources]

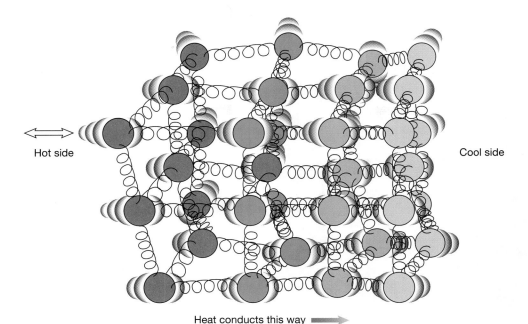

Hot side

Cool side

Heat conducts this way ⟹

FIGURE 8.4   Even in a solid, atoms continually vibrate, pushing or pulling nearby atoms. As matter is heated, atoms vibrate more vigorously. When heat is transferred by conduction, it is because energetic atoms in one location have disturbed their neighbors, passing on some of their own energy. In this way, heat slowly transfers from one place to another even though each individual atom stays in roughly the same spot.

Geologists have good evidence that this is not the case; therefore, we conclude that heat has not accumulated in the deep Earth but must be escaping by a more efficient process than conduction.

That process is convection, by which heated matter becomes buoyant and rises through its surroundings. Most people have heard that "heat rises" and can offer as proof the familiar sight of a hot air balloon climbing through the summer sky (Figure 8.5). In fact, a hot air balloon rises because the air in it is less dense than the cooler air outside it. It is not merely the heat energy that ascends during convection, passing from one atom or molecule to another; the entire mass of low-density air floats upward, dragging the balloon along.

This convection process is what drives winds across the globe. Air rises over heated regions. Cooler, denser air sinks elsewhere. In between, the winds blow horizontally as the "extra" air in one place moves to replace air that has moved from somewhere else. The result is a circulation pattern called a **convection cell**, shown in Figure 8.6. Similar patterns move heat in any body of water, whether it is as large as the ocean or as small as a pot on the stove. For a material to convect, though, it must be fluid enough to overcome internal resistance to motion. In Chapter 3 we called this resistance **viscosity**.

In 1931, British geologist Arthur Holmes was the first to argue convincingly that the mantle is fluid enough to convect, even though it clearly is not liquid.

The behavior of earthquake waves tells geologists that the mantle is rigid when it is subjected to short-term stress. Many observations, however, suggest that it can deform slowly like an extraordinarily viscous liquid if the same stress is applied for long periods. For example, ice sheets that covered large portions of Canada and Scandinavia during the Pleistocene epoch were heavy enough to bow the crust downward, a process that would not be possible if the underlying mantle were not flexible. In the more than 12,000 years since ice sheets melted back, the crust has slowly rebounded, again suggesting that the mantle is not totally rigid but is like an extremely stiff putty.

Computer simulations like the one in Figure 8.7 suggest that hot rock normally rises through the mantle in chimney-like zones that are a few hundred kilometers across. As we first noted in Chapter 3, geologists have identified a large number of these major zones, called **mantle plumes**, around the world. The area at the Earth's surface under which a plume rises is called a **hot spot**. Cooler rock descends over much broader, diffuse regions as well as in narrow, elongated zones associated with oceanic trenches. Recent studies suggest that there may be major downward convective areas under parts of central North America, west-central Africa, northern Asia, and western Australia. As with circulation in the atmosphere, these areas of rising and descending rock must be connected by slow horizontal movements in the asthenosphere and in the deeper mantle. We will return to

FIGURE 8.5   A hot air balloon rises because heat causes the air in it to expand, thus making its density (mass/volume) decrease. This makes the balloon buoyant in the relatively denser air around it. The balloon rises, and so does the heat in it. This is an example of heat transfer by convection, in which both energy and matter are transported upward. *(Tom Martin, The Stock Market)*

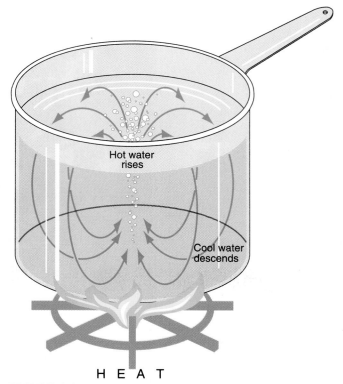

H E A T

FIGURE 8.6   Most substances become less dense as they are heated. If a fluid is heated from the bottom, therefore, it is unstable. Low-density fluid will tend to rise through the higher-density material above it, provided that it meets little enough resistance. As the rising material leaves the heated area, it is replaced by material moving horizontally from nearby areas. This, in turn, is replaced by cool material descending from the top of the fluid. The result is a convection cell.

refine this general picture many times in chapters ahead. Geologists still have many serious questions about how convection works in the Earth, however, so we will not be able to provide as clear an explanation as we would like.

Surface Heat Flow

The only direct method we have for estimating how hot the Earth is inside is to measure how rapidly heat escapes through its surface, a rate geologists call the **heat flow**. In principle, this is simply a matter of determining the local thermal conductivity of the crust and then lowering a thermometer down a hole in the ground to see how temperature changes with depth. The rate at which temperature increases from the top to the bottom of the hole (typically between 10°C/km and 30°C/km, but much greater in some places) is called the **thermal gradient**. Heat flow is the product of conductivity and the thermal gradient (see Figure 8.8):

$$\text{Heat flow} = \text{Conductivity} \times \frac{\left(\begin{array}{cc}\text{"Bottom"} & \text{"Top"} \\ \text{temperature} - \text{temperature}\end{array}\right)}{\text{Depth of hole}}$$

In practice, it is hard to measure heat flow anywhere on the continents. As climates change gradually over the course of decades they disturb the meager outward flow of heat through the crust so much that they can invalidate measurements made even 300 meters below the surface. It is much easier to make reliable measurements in sediments of the ocean floor, where the temperature of seawater is monotonously constant. Several thousand measurements of the Earth's heat flow have been made, mostly in the oceans (Figure 8.9).

The global average heat flow is about 60 milliwatts per square meter (mW/m²). The heat flow from an area the size of a football field, therefore, is roughly equivalent to the energy output of three 100-watt light bulbs. Put another way, if we could set a coffee mug on the ground and magically capture every bit of heat

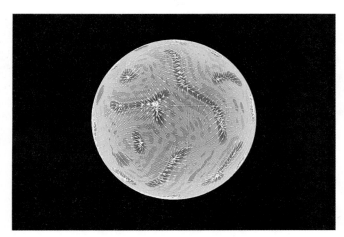

FIGURE 8.7   According to the results of this computer simulation, heat rises through the mantle primarily in narrow plumes at hot spots, like the one shown in dark red. Cooler rock descends into the mantle in sheets, marked here in blue, corresponding to subduction zones. Any computer simulation of a body as complex as the Earth must be greatly simplified. Because that is so, and because we deal with very limited information, images like this must be considered speculative, although thought-provoking. *(Gary A. Glatzmaier, Los Alamos National Laboratory)*

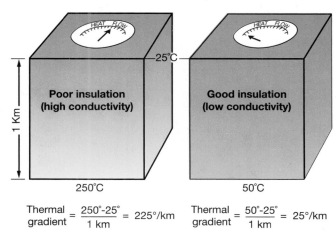

Thermal gradient $= \dfrac{250° - 25°}{1 \text{ km}} = 225°/\text{km}$ Thermal gradient $= \dfrac{50° - 25°}{1 \text{ km}} = 25°/\text{km}$

FIGURE 8.8   If the temperature increases rapidly with depth in the lithosphere and if the rock is a good heat conductor, then the surface heat flow rate will be large. If temperature changes only slightly with depth, or if the rock is a good insulator, then not much heat escapes through the Earth's surface.

escaping from the Earth under it, allowing it to accumulate in the cup, we would have to wait 4-1/2 years for the coffee to warm from room temperature to a boil. That's a long time. The mug could boil the same amount of coffee in about 1 hour if it could instead accumulate all the heat from the Sun that struck it in that period (see Figure 8.10). Taken over the entire surface of the Earth, however, the puny flow of internal heat adds up to 3.1×10^{13} W (31 trillion watts). Over the course of a year, this is roughly the same as the energy yield of 5 million one-megaton hydrogen bombs—hardly insignificant.

Not surprisingly, we find that heat flow varies from one location to another on the continents. Old, geologically quiet regions have low heat flow values; young mountain-building areas have much higher values. In the Archean rocks of central Canada, for example, heat flow is about 40 mW/m², in Arizona, Utah, and Nevada, the region of Cenozoic mountain-building known to geologists as the Basin and Range Province, values are about 80 mW/m². Where does the heat come from? Roughly 25 percent of it is generated by radioactivity in the granitic crust of the continents. The other 75 percent rises from deep in the mantle and is conducted through the crust: 30 mW/m² in Canada, 60 mW/m² in the Basin and Range Province.

The average heat flow in the ocean basins is nearly the same as that on the continents, and so also is the spread of heat flow values. Young portions of the seafloor near the spreading centers have higher heat flow values than do older regions. At the centers, heat flow averages 80 mW/m². Away from the centers, val-

ues are closer to 55 mW/m², and in oceanic trenches they drop to less than 50 mW/m² (Figure 8.11). Overall, these are quite like the continental values. The difference is that oceanic basalts have very little radioactivity; thus, most of the heat coming through the seafloor must be rising from deep in the mantle.

Temperatures Within the Earth

By interpreting the surface heat flow observations, we can make three broad generalizations about the Earth's internal heat that will help us deduce a **geotherm**, or average depth-temperature profile for the crust and upper mantle:

- *Although the lithosphere is the most radioactive part of the planet, most of the heat reaching the Earth's surface comes from much deeper sources.* Geologists cannot yet agree about how much of this "deep heat" was originally released by gravitational processes and how much is due to dispersed radioactivity in the mantle or solidification of the core.

- *Because heat rises through the mantle by convection, some places receive more "deep" heat than others.* Heat flow is greatest above mantle plumes or in places that are volcanically active and lowest where cool rock is descending into the mantle.

- *The balance between "deep" heat and "shallow" heat at any spot depends on what the crust is made of and how thick it is.* Because the crust is too cool and rigid to convect, heat must pass through it by conduction. The continental crust is thick enough to insulate us from the hot mantle below, but generates a modest amount of heat itself through radioactive decay. The oceanic crust is less radioactive, but is also much thinner. As a result, "deep" heat can be conducted through it more easily.

With these three generalizations in hand, we should not be surprised that the relationship between depth and temperature differs from one place to

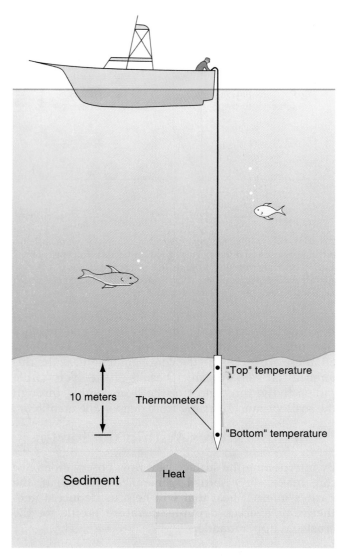

68,000 mW/m² Average solar heat at earth's surface

At this rate, water in the cup will boil in about one hour

60 mW/m² Average heat flow

At this rate, the water will take 4½ years to boil

FIGURE 8.10 This hypothetical coffee mug would be an engineer's dream. It lets heat in, but does not let it escape. Therefore, it can only get hotter, not cooler. If it is filled with water at room temperature and left to stand in the sunshine, the water will start to boil in about 1 hour. Because the amount of heat escaping from the Earth is less than 0.02% of what we receive from the Sun, it would take almost 39,000 hours (4.5 years) to boil water in the cup using only heat that it collects as it flows from the Earth below.

FIGURE 8.9 Heat flow through the seafloor is determined by dropping a weighted probe about 10 meters long into the sediments. Thermometers on the outside of the probe measure temperature at different depths, making it possible to calculate the thermal gradient. When the probe is withdrawn, it contains a sample of the sediment, in which shipboard scientists can measure heat conductivity.

another. If we lowered a thermometer into the Earth in Iceland or Hawaii, which are both volcanically active regions, we would find high temperatures at much shallower depths than we might find in the middle of Australia, a particularly inactive region. This general concept is something we first noted in Chapter 6: The wide range of metamorphic facies implies that rocks at any particular depth vary in temperature from one location to another (refer back to Figure 6.14). Now, however, we go one step further by pointing out a systematic difference between average thermal gradients beneath the two largest subdivisions of the Earth: the

continents and the ocean basins (Figure 8.12). At most places on the continents, we would have to drill a hole 150 km deep to find rock at 1300°C. Because more "deep" heat is rising in the ocean basins, we could find rock at the same temperature a mere 75 km down.

Of course, we cannot really lower a thermometer 75 or 150 km, or even 25 km, into the Earth. The deepest hole yet drilled is about 18 km deep; thus, geologists can barely confirm the uppermost part of the profiles in Figure 8.12. What, then, do they know about the geotherm in the mantle? Not nearly enough. Geologists are fairly sure that the mantle is made of peridotite, the olivine-rich igneous rock we introduced in Chapter 3. The geotherms in Figure 8.13 are estimated from heat flow values and from the way that the asthenosphere transmits earthquake waves (Chapter 9). Lab experiments tell us how the melting tem-

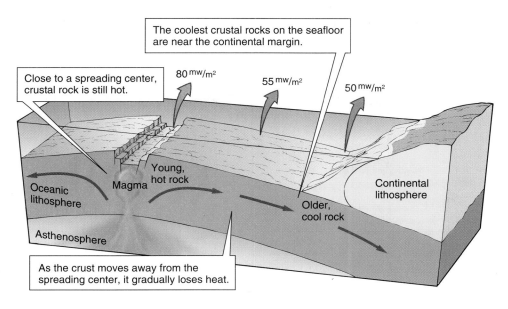

FIGURE 8.11   Heat flow from the seafloor is greatest at the mid-ocean ridges but drops off to minimal values toward deep ocean trenches. This happens because fresh lava erupts at the ridges, underlain by about 5 km of other fresh intrusive igneous rocks, and then must cool largely by conduction as the newly formed seafloor spreads. This is a slow process, so some of this heat is still escaping tens of millions of years later and thousands of kilometers away from the ridges.

FIGURE 8.12   Estimated average geotherms in continental and oceanic lithosphere. The convecting mantle is nearer to the surface in the ocean basins than under the continents, so high temperatures are reached at relatively shallow depths.

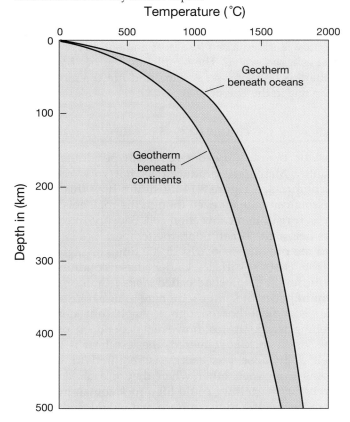

perature of peridotite changes as pressure increases, and evidence from seismic waves assures us that there is very little magma in the mantle. From this, we conclude that the mantle can be no hotter at any depth than the melting temperature of peridotite. Conversely, if it were much cooler, then the mantle would have a hard time convecting. This tells us that the temperature increase with depth, which was 20°C to 30°C/km in the shallow Earth, must be much less in the asthenosphere and below, probably between 1.5°C and 2°C/km. Seismic peculiarities at various depths help refine our guesses, but geologists are still uncertain about how hot the mantle is, especially at its deepest. One reasonable estimate puts the temperature at the edge of the outer core at just slightly above 3000°C. Others range as high as 4000°C.

The geotherm in the core itself is even less well known. Experiments with metal alloys, compared with seismic information, make rough estimates possible. From 1980 onward the methods used to come up with these estimates have become increasingly sophisticated, but they still depend on assumptions that are hard to verify. Among other things, geologists cannot even decide how much of the heat leaving the core is produced by radioactivity or solidification and how much is left over from gravitational processes that occurred more than 4 billion years ago. The temperature at the center of the Earth is probably at least 4300°C, but some geologists have estimated that it may be nearly 6000°C.

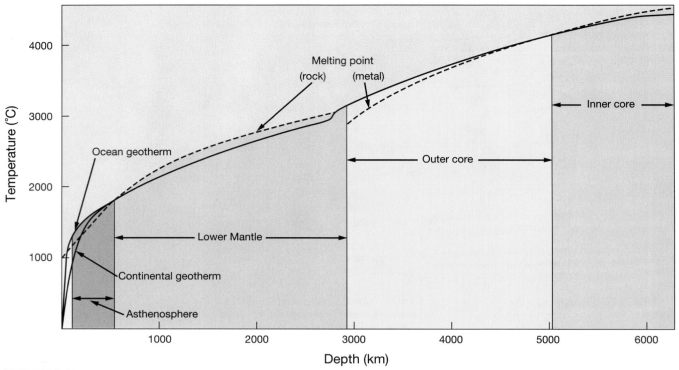

FIGURE 8.13 Estimated geotherms from Figure 8.12, extended all the way to the center of the Earth. The temperatures indicated are more and more uncertain with depth. For comparison, the minimum melting temperatures of mantle and core materials are also shown.

Melting: The Origin of Magma

In Chapter 3 we considered what happens as a magma cools and crystallizes to form igneous rock. At that point, we provided only a brief answer to the question of where magma comes from. Now that we have examined the Earth's heat engine, let us return to igneous processes again. Our reason for studying the origin of magma now is that it will provide additional insights into the distribution and transport of heat within the Earth.

We have already shown that silicate magmas solidify over a broad temperature range rather than at a single temperature (refer back to Figure 2.26 and Figure 3.3 and to the discussion of Bowen's reaction series in Chapter 3). The reverse is true as well. That is, if we heat a rock in a furnace, we find that some of its components melt at much lower temperatures than do others. As a result, the magma that forms has a very different chemical composition from the unmelted minerals around it. In some fundamental ways, therefore, this process, called **partial melting**, is the inverse of fractional crystallization. Like fractional crystallization, it leads to **igneous differentiation**, the evolution of one parent material into two or more daughters through the separation of magma from rock. There is one significant difference, however. It is easy to imagine how a magma can lose heat and solidify, but what makes a rock melt? The simple answer of "Making it hotter" is only one possibility, and not the most common one in the Earth.

Pressure Release

Basaltic magmas are derived from peridotite in the mantle. Many are produced as hot rock rises and decompresses. Figure 8.14 illustrates how this kind of melting can happen even though the rock gets no hotter as it rises in a convection cell. Seventy-five kilometers below the ocean floor, the temperature is 1300°C, and the pressure is nearly 26,000 times as high as it is at the Earth's surface. Under these conditions, the sodium-rich plagioclase called *albite* (a minor component in peridotite, but useful here to illustrate a principle) is almost hot enough to melt, but not quite. Melting temperatures, however, depend on pressure. The less a rock is squeezed, the less heat it takes to melt it, because less energy is required to separate atoms from each other when they are already relatively far apart. If we could lift a rock containing albite from the mantle to the surface, we would find that its

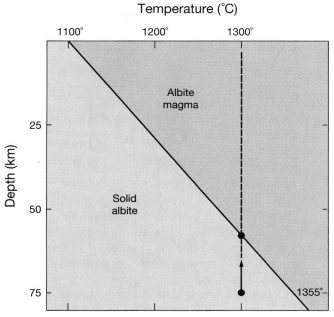

FIGURE 8.14   This line on the graph shows how the melting temperature of albite (NaAlSi$_3$O$_8$) changes as pressure increases. At the Earth's surface, albite remains solid up to 1104°C. At a depth of 75 km, albite will not melt until it is heated to 1355°C. Albite in a rock at this depth would thus be solid if its temperature were only 1300°C. If that solid albite were raised without cooling, it would melt at a depth of about 58 km.

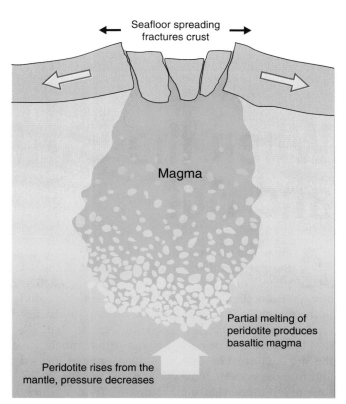

FIGURE 8.15   Partial melting in mantle rocks beneath a mid-oceanic ridge.

FIGURE 8.16   At a depth of about 100 km, the average geotherm exceeds the minimum melting temperature for peridotite. Because the geotherm increases only slowly at greater depths, it eventually crosses back over the minimum melting curve at about 470 km. In the interval between 100 and 470 km, therefore, a small amount of partial melt (2% to 3%) is present. This region, the asthenosphere, is the source for some basaltic magmas. *(This illustration was derived from Figure 4.15 in B.J. Skinner and S.C. Porter's PHYSICAL GEOLOGY, © 1987 by John Wiley & Sons, Inc.)*

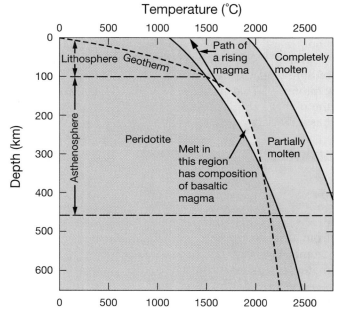

melting temperature had dropped from 1355°C to only 1104°C. Although 1300°C is too cool to melt albite 75 km below the surface, this should be more than hot enough to melt it at a shallower depth. If the rising rock does not cool very much, partial melting is inevitable. When the melting point finally drops below the rock's actual temperature, an albitic magma will start to form.

This is what happens at the spreading centers, where the lithosphere is weakened and split by tension, and warm peridotite rises by convection in the mantle (see Figure 8.15). The peridotite remains solid until it is within 60 km of the surface, where pressures decrease and partial melting, not only of plagioclase but also of pyroxene and olivine, begins. The magma that emerges is basaltic. It does not form because the parent peridotite gets any hotter, but because it is under less pressure as it rises. Basaltic magma forms in the same way at hot spots, where convection in a mantle plume carries peridotite upward.

This quick explanation is much too simple to account for the subtle differences that geologists recognize between one basaltic magma and another. In Chapter 3 we discussed two common processes, assimilation and magma mixing, that can alter the composition of a magma as it moves through the crust. As a result of these and other more complicated

Perspective 8.1

When the Earth Was Young and Hot

Large-scale movements within the Earth and the motion of plates on its surface have always been driven by convection in the mantle, just as they are today. Geologists hypothesize, however, that the Earth was much hotter during the Archean Eon than it is now and that the pace of convection was thus much more vigorous. Estimating the efficiency of the planetary "heat engine" in the past is important for geologists who want to understand how the planet evolved during those early times. Consequently, they have considered several types of information that might make it possible to estimate just how hot the Earth was.

Radioactivity, for example, made the Earth much warmer during the Archean Eon than it does today. The half-life of ^{238}U is 4.5 billion years, and the half-lives of ^{232}Th and ^{40}K are 14 billion and 1.3 billion years, respectively (see Chapter 7). That means that half of the planet's original amount of ^{238}U has now decayed, along with about

20 percent of the ^{232}Th and 91 percent of the ^{40}K. As the total amount of radioactive material in the Earth has decreased, so has its potential for producing heat. Geologists estimate that ^{238}U generated 5.5 times as much heat shortly after the planet formed as it does now, ^{232}Th produced 25 percent more heat than it does today, and ^{40}K produced heat at almost 12 times its current rate.

Much of the Earth's original accretionary heat was also still around to keep the crust and mantle warm, even 2 billion years after the planet formed. The meteorite impacts that had once heated the planet's surface enough to melt it were much more "recent" events during the Archean Eon than they are now.

If we could travel back in time to walk on the Earth then, of course, we would find it as solid as it is now and would not be aware that heat flow from the planetary interior was any greater until we made careful measurements. Those heat

flow measurements, however, would reveal a geotherm that increased very rapidly with depth, making temperatures in the deep crust and upper mantle perhaps hundreds of degrees higher than they are today. We cannot travel back in time, however, and it is quite difficult to calculate how much warmer radioactivity or accretionary heat might have made the crust and mantle, partly because the answer depends on how rapidly heat was transferred through the mantle by convection. Unfortunately, as we indicated above, geologists do not *know* how fast the mantle was convecting; that is one of the reasons why they are estimating ancient temperatures in the first place.

As difficult as this problem of estimating ancient temperatures seems, geologists have made a few rough estimates by other means. They point to the existence of one unusual ultramafic igneous rock, called a **komatiite**, for example. Komatiite lavas, known only from

processes, basalts on continents are chemically distinct from those in the ocean basins, and there are minor chemical variants of each, although all come ultimately from the mantle.

It is also the case that not all basaltic magmas are produced by pressure release. Some partial melting

takes place in the asthenosphere simply because the actual rock temperature is higher than the minimum melting temperature of peridotite at that depth. Figure 8.16 is similar to Figure 8.14, but it describes the melting of peridotite rather than albite and extends to a much greater depth. In it, we can see that the geo-

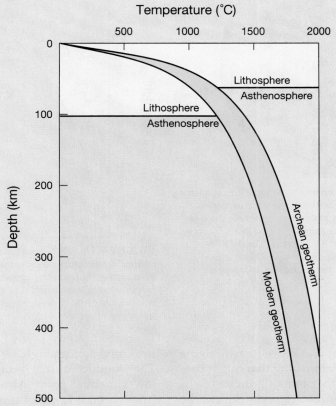

P8.1.1   The present average geotherm is compared in this graph with a hypothetical geotherm that might have existed during the late Archean Eon. Notice that the Archean geotherm exceeds the minimum melting temperature for peridotite at a much shallower depth than the present geotherm does. As indicated in Figure 8.16, this point marks the top of the asthenosphere.

regions of Archean age, were formed by a high degree of partial melting in the mantle and erupted at temperatures of at least 1500°C. That is 400°C hotter than most lavas erupting today. One interpretation for the absence of komatiites in younger regions is that the upper mantle has cooled by 400°C, so that extreme partial melting at the depth where komatiite magmas once originated is no longer possible. Using rough methods such as this to estimate the Archean temperature of the mantle, some geologists have hypothesized that the rate of convection could have been twice as great during the Archean Eon as it is today.

therm crosses the curve of minimum melting temperature about 100 km below the surface. Because temperature rises much more slowly with depth in the mantle than it does in the lithosphere, the geotherm crosses the melting curve again at about 470 km. In between, the actual temperature is never very much above the melting point, so not a lot of partial melting occurs. Geologists estimate that there is probably less than 3 percent liquid in the asthenosphere, and that it is dispersed as a thin film between solid mineral grains in the peridotite (Figure 8.17). Geologists disagree about how much of this sparsely distributed

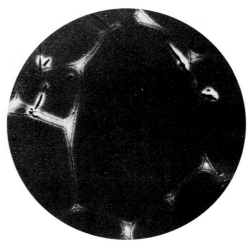

FIGURE 8.17   In this image, taken by a scanning electron microscope, it is clear that partial melting begins along the boundaries between mineral grains, particularly at those places where three or more grains meet. This is a peridotite sample, heated in a laboratory to 1300°C under 10 kilobars of confining pressure. The dark areas are olivine crystals; the light regions now consist of basaltic glass, but were molten during the experiment. *(Harve Waff, University of Oregon)*

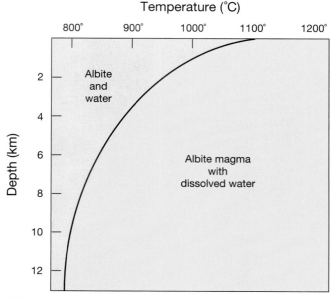

FIGURE 8.18   When water is added to albite (NaAlSi$_3$O$_8$) its melting temperature decreases. The melting temperature decreases even further if pressure is applied because water dissolves best in magma that is pressurized. The more water that dissolves in magma, the cooler it needs to get before it solidifies. Compare this figure with Figure 8.14.

magma may be squeezed out and make its way to the surface. More extensive partial melting that might produce a peridotite magma rather than a basaltic one is very rare, although it was more common during the Earth's early history (see Perspective 8.1).

The Effect of Water

The partial melting process we have just described is most correctly called **dry partial melting** because it is what takes place in the absence of either water or minerals containing water. "Wet" minerals like the amphiboles and micas generally melt at lower temperatures than do minerals containing no water, such as the pyroxenes. Even "dry" minerals—for example, the pyroxenes or the feldspars—melt at lower temperatures when water is present (see Figure 8.18). Furthermore, high pressure forces more water to dissolve in magma and *lowers* the melting temperatures of wet minerals instead of raising them. For this reason, it is easier to melt wet rocks the deeper they are in the Earth.

This behavior, called **wet partial melting**, suggests another way to produce magma without heating a parent rock. Strange as it may seem, it is possible to melt one rock by pushing another cold, wet rock into it. In a subduction zone, wet marine sediments and water-saturated basalts from the ocean floor are forced into the upper mantle, where rocks are

not quite hot enough to melt under dry conditions. The subducted rocks, being wet, naturally have a lower melting point. Also, as excess water is released from them, it lowers the melting point of dry rocks in the vicinity. Some of these rocks are in the mantle, and some are in the overlying lithosphere (Figure 8.19). The magma that results is andesitic, a partial melting product of the subducted basalt, also influenced by the more silica-rich continental lithosphere.

Andesitic magma is unlike basaltic magma because its melting point *increases* as the magma rises through the crust and is depressurized. Also, andesitic magmas contain more silica than do basaltic magmas. Andesitic magmas, therefore, are usually less fluid than basaltic magmas and are more likely to solidify before erupting. If they do erupt, they are likely to do it with great force, because water escaping from the magma under reduced pressure generates explosive quantities of steam. Eruptions from Mt. Saint Helens (northwestern U.S.) to Mt. Pinatubo (Philippines) and Mt. Unzen (Japan) during the 1980s and 1990s attest to the potential violence associated with wet partial melting.

Because the wet partial melting that leads to andesitic magma is associated with subduction, many andesitic magmas erupt around the margins of the Pacific basin, where most subduction zones are located. The **andesite line**, indicated in Figure 8.20, coincides with a nearly unbroken series of subduction

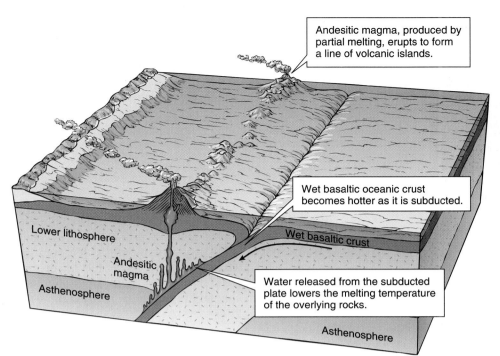

Andesitic magma, produced by partial melting, erupts to form a line of volcanic islands.

Wet basaltic oceanic crust becomes hotter as it is subducted.

Lower lithosphere

Andesitic magma

Asthenosphere

Wet basaltic crust

Water released from the subducted plate lowers the melting temperature of the overlying rocks.

Asthenosphere

FIGURE 8.19
By the time the basaltic rock on the seafloor reaches a subduction zone, it has been saturated with seawater and chemically altered. As it is subducted, partial melting begins in the wet basalt. The result is andesitic magma, which then rises toward the surface because it is less dense than the rocks above it. Excess water released during melting may also lower the melting point of rocks above the subducting slab. These may be incorporated into the andesitic magma. Stratovolcanoes on the surface above may erupt either a fluid lava or an explosive cloud of pyroclastic debris, depending on the degree of partial melting that produced the magma and on how much silica-rich sediment may have melted and mixed with the magma during its ascent.

zones extending clockwise around the basin from Tonga in the southwest Pacific to the southern tip of Chile. It also marks the seaward limit of a "Ring of Fire" marked by stratovolcanoes on the continents and in islands such as Japan and the Aleutian archipelago in Alaska. Andesitic magma is never found on the deep ocean side of the line. Ancient andesites elsewhere in the world were apparently formed under similar conditions, along subduction zones that are no longer active.

Wet partial melting can also produce granitic magma. It is not unusual to reach temperatures above 700°C in the continental crust where plate collisions have crumpled and thickened the lithosphere. As we illustrate in Figure 8.21, these temperatures are more than hot enough to melt wet sediments partially. Sandstones and mudstones, in particular, contain high concentrations of silica and Al_2O_3, which are the two dominant components in granitic magma. For the most part, the magmas produced in this way remain

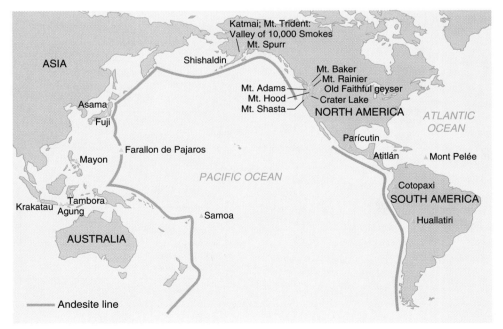

Katmai; Mt. Trident: Valley of 10,000 Smokes
Mt. Spurr
Shishaldin
ASIA
Asama
Fuji
Mayon
Farallon de Pajaros
PACIFIC OCEAN
Krakatau Tambora
Agung
Samoa
AUSTRALIA
Mt. Adams
Mt. Hood
Mt. Shasta
Mt. Baker
Mt. Rainier
Old Faithful geyser
Crater Lake
NORTH AMERICA
Parícutin
Atitlán
Mont Pelée
ATLANTIC OCEAN
Cotopaxi
SOUTH AMERICA
Huallatiri
—— Andesite line

FIGURE 8.20
The Andesite Line is nearly coincident with the system of volcanic mountains and islands called the "Ring of Fire" in the Pacific basin. Andesitic magma rising to form these volcanoes is generated by partial melting of wet basalt in subduction zones. Volcanoes seaward of the Andesite Line, like those in Hawaii or along the mid-ocean ridge, erupt only basaltic lava.

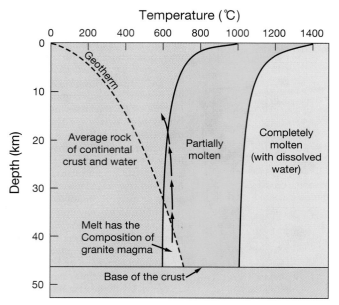

Temperature (°C)

FIGURE 8.21   Unlike the "dry" rocks discussed in Figure 8.16, "wet" rocks have low melting temperatures, and those decrease further as pressure is applied. A granitic magma, which may include as much as 10% water by weight, can form at around 600°C at pressures like those in the deep continental crust. As shown here, the geotherm exceeds the melting temperature for granite at a depth of roughly 30 km. If this magma were to rise toward the surface, however, it would cross the melting point curve on the way and would solidify. For this reason, most granitic magmas remain in plutons instead of forming extrusive features. (*This illustration was derived from Figure 4.16 in B. J. Skinner and S. C. Porter's PHYSICAL GEOLOGY, © 1987 by John Wiley & Sons, Inc.*)

deep in the crust, where they form batholiths or other intrusive bodies. Granitic magma tends to remain deep in the crust because it is too stiff to move easily and because its minimum melting temperature would increase by more than 200°C if it managed to rise as far as the surface. Unless there were some way to add more heat to this magma as it rose, it is unlikely that granitic magma would ever erupt.

That, however, is exactly what does happen in some locations. In Yellowstone Park, for example, several violent eruptions of granitic magma have occurred over the past few million years, producing craters dozens of kilometers across and sending clouds of pyroclastic debris across the northern United States. The heat source for melting is a plume, in which basaltic magma rises from the mantle at temperatures above 1100°C. This magma pools in the lower crust and solidifies, losing heat to the surrounding wet sediments and partially melting them (Figure 8.22).

▶▶ EPILOGUE ◀◀

Why do geologists care how temperature varies inside the Earth, besides satisfying their curiosity? The best answer is that the convecting interior of the Earth is like a massive engine whose power depends on how much heat is available and where it is.

FIGURE 8.22   Beneath Yellowstone National Park, basaltic magma rising above a hot spot has pooled in the lower crust. Because the crust is thick and the basaltic magma easily drops below its melting temperature, it may never rise to the surface. As the basalt loses heat, however, the overlying wet sediments can easily reach their own lower melting temperature and form a granitic magma. Several times in the past few million years, this has led to explosive eruptions that formed collapse calderas tens of kilometers across.

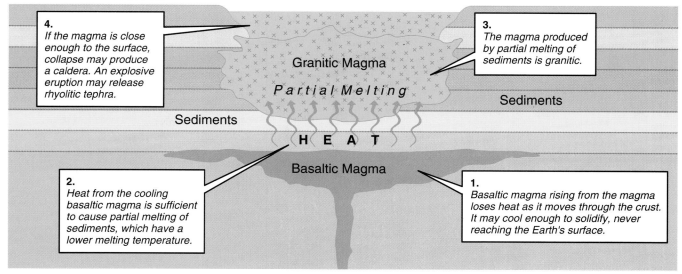

4. If the magma is close enough to the surface, collapse may produce a caldera. An explosive eruption may release rhyolitic tephra.

3. The magma produced by partial melting of sediments is granitic.

Granitic Magma

Partial Melting

Sediments

Sediments

HEAT

Basaltic Magma

2. Heat from the cooling basaltic magma is sufficient to cause partial melting of sediments, which have a lower melting temperature.

1. Basaltic magma rising from the magma loses heat as it moves through the crust. It may cool enough to solidify, never reaching the Earth's surface.

This "convection engine" in the mantle is now understood to be the driving force for plate tectonics. Convection in the core is the most likely source of the Earth's magnetic field, as we discussed in Chapter 7.

Figuring out how heat is distributed in the planet is one vital step in understanding how the engine works. An engineer trying to predict how much power an auto engine will deliver usually knows some fundamental things about the engine, such as how large its cylinders are and how far its pistons travel. The engineer also knows what fuel the engine uses and how efficiently its chemical energy can be converted to heat and mechanical movement. Unfortunately, we cannot lift the hood of the Earth to see clearly how many convection cells it has, how big they are, how they are related to each other, or where they get their energy (see Figure 8.23).

By measuring surface heat flow and deducing geotherms, we can at least get a glimpse at what the engine looks like. The extent of partial melting that takes place in the mantle and crust, along with the kinds of magma that form, provide other clues. By mapping the geographic distribution of basaltic eruptions, geologists can deduce where hot rock is convecting upward at mid-ocean ridges and in mantle plumes. The distribution of andesitic volcanoes at subduction zones similarly marks at least some of the places where cooler rock is descending into the Earth. We will develop these ideas further as we investigate the basis for the theory of plate tectonics in the chapters ahead. In the next chapter, however, we will search for other ways to understand how large-scale movements take place in the Earth.

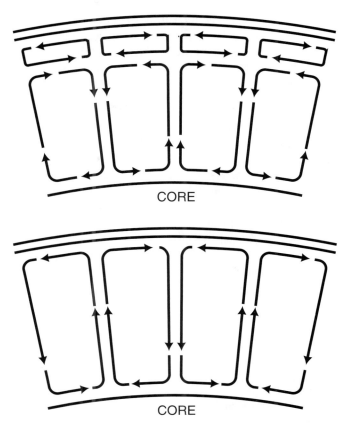

FIGURE 8.23 Geophysicists and geochemists disagree about how deeply convection cells extend into the mantle. One possibility is that they go all the way to the core-mantle boundary. Another is that one set of cells, operating only to a depth of about 670 km, sits on top of a second set that circulates in the deep mantle. Evidence supports each possibility, but the true picture is likely to be more complex than either one of them.

SUMMARY

1. Heat in the Earth's interior is derived from three primary sources: gravitational processes, solidification of the core, and radioactivity. Gravitational heating of the Earth took place as a result of meteorite impacts during planetary accretion. Further gravitational heating occurred as metallic iron sank to form the Earth's core. Heat continues to be released from the core as its liquid outer region gradually solidifies. Radioactivity heats the Earth as rock absorbs high-energy radiation and fast-moving atomic particles. The primary sources of radioactive heat in the Earth are isotopes of uranium, thorium, and potassium. These are concentrated in crustal rocks, largely those of the continents.

2. Heat may be transported by radiation, conduction, or convection. Only conduction and convection are significant processes within the Earth. Conduction involves the direct transfer of energy from one moving atom or particle to another and is the dominant heat transfer process in the lithosphere, which is too rigid to allow internal convection. Convection involves the movement of hot material from one place to another, usually as a consequence of density differences from one place to another. This is the primary heat transfer process in the mantle and core and is the driving force for all large-scale movements in the Earth.

3. The only direct information geologists have about heat in the Earth comes from measure-

ments of surface heat flow. Heat flow values are roughly the same around the world, but they differ from one tectonic setting to another. Most of the heat reaching the Earth's surface comes from the mantle or core. Although the continental crust generates more heat by radioactivity than does the oceanic crust, it is thicker and transmits less heat from below. In the ocean basins, virtually all heat comes from the mantle or core, because the oceanic crust is thin and only slightly radioactive. As a result, high temperatures are reached at relatively shallow depths in the ocean basins but at much greater depths on the continents. Temperatures are less well known with increasing depth in the mantle and core.

4. Dry rock melts at a higher temperature as pressure is applied. As a result, hot rock can melt partially as pressure on it is released. This is the basic method by which partial melting of peridotite in the mantle produces basaltic magma at mid-ocean ridges and in mantle plumes. Some basaltic magma may also be produced by limited partial melting in the asthenosphere. Wet rock melts at lower temperatures than does dry rock. As increasing pressure is applied to wet rock, however, its melting temperature decreases. This is the basic method by which andesitic magma is produced in subduction zones. Granitic magma is sometimes produced by wet partial melting in thick portions of the continental crust. It may also be produced where much hotter basaltic magma has risen from the mantle and pooled within the continental lithosphere, heating it and causing melting.

KEY WORDS AND CONCEPTS

accretion 184	heat flow 188	radiation 186
andesite line 196	igneous differentiation 192	solidification of the core 184
conduction 186	iron catastrophe 184	thermal conductivity 186
convection 186	komatiite 194	thermal gradient 188
convection cell 187	magma ocean 184	
geotherm 189	partial melting (dry and wet) 192	

QUESTIONS FOR REVIEW AND THOUGHT

8.1 How did the process of planetary accretion heat the early Earth? How did the process of core formation heat the Earth further?

8.2 Why are the radioactive isotopes of uranium, thorium, and potassium largely concentrated in the continental crust rather than in the oceanic crust or rocks of the mantle?

8.3 How has the amount of heat in the Earth changed over the course of its history? How do geologists know?

8.4 What makes a convection cell work? That is, what force causes hot material to rise, and what force must be overcome?

8.5 Why is heat conducted through the crust instead of being transported by convection as it is in the mantle?

8.6 What factors control the rate of heat flow through the Earth's surface?

8.7 How is average heat flow on the continents similar to that in the ocean basins, and in what ways is it quite different? How do geologists account for the differences?

8.8 What kinds of information do geologists use to estimate the geotherm in the deep mantle and core?

8.9 How may partial melting lead to igneous differentiation?

8.10 Why is basaltic magma most likely to form at mid-ocean ridges and over hot spots? Why is it less likely in regions of subduction?

8.11 Why is basaltic magma more likely to rise all the way to the Earth's surface than either andesitic or granitic magma?

8.12 Why are there no andesitic eruptions beyond the Andesite Line in the Pacific Ocean?

8.13 Why is basaltic magma not likely to form by partial melting of rocks in either the oceanic or continental crust?

8.14 What effect does water have on the melting temperature of magma and on the way the magma behaves as it rises toward the Earth's surface?

8.15 In what kind of geologic settings are granitic magmas commonly formed?

8.16   The planet Roligt, in another solar system, is identical to the Earth in every way but is almost 10 billion years old. Suppose that you visited Roligt to make heat flow measurements in its continents and seafloor. Would you expect those measurements to be larger or smaller than ones on Earth? Why?

8.17   Suppose you had to draw a geotherm for the planet Roligt (Question 8.16). How do you think it would compare with Figure 8.12? How might the Earth and Roligt differ in the amount of volcanism they experience? Why?

9

Trace of the San Andreas Fault, truncating and offsetting stream gullies on the Carrizo Plain, Southern California. (Dr. John S. Shelton)

Earthquakes and Earth Structure

OBJECTIVES

▲▲

*As you read through this chapter, it will help if you focus
on the following questions:*

1. What causes earthquakes?
2. How does the energy from an earthquake travel through the earth or on its surface?
3. How do geologists locate the source and measure the intensity of a distant earthquake?
4. What methods do geologists use to predict earthquakes?
5. What are the greatest hazards during an earthquake?
6. How do geologists use earthquake waves to investigate the Earth's internal structure?

OVERVIEW

▲▲▲

Few geologic events can be as frightening as earthquakes. Even a mild quake may have the explosive force of a high-yield nuclear bomb, and the destruction it causes can be just as devastating to property. On occasion, as in Anchorage, Alaska, in 1964, a single quake may release as much energy as 50,000 1-megaton hydrogen bombs. Not only are earthquakes destructive, but they also strike with little perceptible warning. In this chapter we will consider what causes earthquakes and will introduce the tools and the vocabulary used to describe them. By examining the events that precede and follow earthquakes, we will explore ways that are now being used to predict them and to minimize the human losses they cause.

Geologists have also learned that earthquakes can reveal secrets about the structure and composition of the Earth's interior. Much of what geologists know about the interior, in fact, could not have been learned in any other way. As they learn more about plate tectonics, they realize how much the processes on the surface of the Earth depend on what happens inside it. We will close this chapter, therefore, with a quick voyage to the center of the Earth.

What Causes Earthquakes?

An earthquake is a release of energy that occurs as portions of the lithosphere slide over, past, or away from each other. This scientific explanation has been refined by geologists during the past century and is at least vaguely familiar to many people outside of the scientific professions. Not long ago, however, many civilizations believed earthquakes were caused by divine or supernatural forces as signs of displeasure, retribution, or simple perversity. Even today, both the suddenness and the intensity of earthquakes are humbling reminders that nature wields forces far beyond human control. To our ancestors, earthquakes were also beyond understanding. Some Native American cultures supposed that the Earth was balanced on the back of a giant tortoise that would move occasionally under its weight, causing earthquakes. In ancient Japan, giant catfish thrashing their tails beneath the ground were thought to be responsible for earthquakes. In Western Europe as late as the eighteenth century, scientists accepted without question Aristotle's idea that earthquakes were caused by hot winds rushing through the Earth. These mystic beliefs helped to make sense of a frightening side of the natural world.

FIGURE 9.1   The San Francisco earthquake of 1906 and the fire that followed destroyed over $400,000,000 worth of property. It remains the greatest natural disaster to strike a U.S. city. For geologists, however, it presented a wealth of new information. "Modern" sensitive instrumentation had been developed little more than a decade earlier, and seismologists were just beginning to understand how earthquake waves travel through the Earth. H.F. Reid, studying the San Francisco event, proposed the elastic rebound theory for the cause of earthquakes. *(AP/Wide World Photos)*

Elastic Rebound

The scientific study of earthquakes, called **seismology** (from the Greek verb *seiein*, "to shake"), evolved rapidly from a descriptive field to a highly quantitative one in the closing decades of the nineteenth century. Many of the basic interpretations we will discuss in this chapter were made within a 25-year period that spanned the turn of the century. One of these was made by Princeton professor Harry Reid in 1906 as he searched for the cause of a disastrous earthquake in San Francisco (see Figure 9.1). Reid hypothesized that the Earth's crust resists continual small forces that try to move it, storing this energy as if in a coiled spring until finally so much stress has accumulated that the crust ruptures. He called this hypothesis **elastic rebound**.

The principle of elastic rebound is illustrated in Figure 9.2. When small opposing forces are applied slowly to a block of the lithosphere, it bends within a narrow zone. If the forces were removed, the lithosphere would return to its unbent shape, just as a diving board springs back once the weight of a diver is removed from it. This particular style of deformation is called **elastic** behavior. When the forces continue to act, as is always the case when earthquakes are involved, the block stores so much energy that it finally can bend no further. Suddenly, where the deformed rocks are weakest (a spot called the **focus**), they break. A fracture spreads from there through the block at about 3.5 km per second (7,200 miles per hour) until all of the stored-up stress is released. On opposite sides of the fracture, rocks snap back to their unbent state and are offset from each other. A fracture of this kind, along which blocks of rock have moved in opposing directions, is called a **fault**. The energy released during the event that produces a fault causes a seismic (that is, "shaking") event we usually call an earthquake.

Energy is released in one seismic episode only to begin building toward the next. The forces that trigger earthquakes generally continue to push in the same direction for long periods of time. Thus, if the rocks on both sides of a fault are "locked" together by friction so that they cannot move, they begin again to accumulate stress and potential energy. When the accumulated energy is great enough to overcome friction, rocks on opposite sides of the fault suddenly "unlock," and elastic rebound produces a new earthquake. In this way, a single fault may be the site for repeated seismic events.

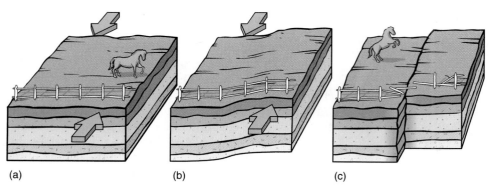

FIGURE 9.2 (a) Small opposing forces act on some parts of the crust. (b) As these accumulate, the crust is deformed elastically. If the forces were removed at this point, the crust would return to its unbent shape. (c) If forces continue to be applied, however, they finally exceed the strength of the rock, which breaks. When it does, stress is relieved suddenly as the two sides of the fracture leap past each other, causing an earthquake and leaving a permanent offset in the rocks.

This intermittent style of movement is called **stick-slip** behavior. In cases where friction is great, episodes of movement are rare but large. If there is less friction, earthquakes are more common and much less intense. You can observe stick-slip behavior at home if you try to move furniture across an unwaxed floor. A light chair skids along. Periods of sticking and slipping are so closely spaced that you hear them as a low groan or squeak. If you shove a heavy refrigerator, instead, it makes less frequent jerks or hops, which may shake the room. The amount of friction controls the frequency and intensity of stick-slip events, including those that cause earthquakes.

Water and Triggering Events

The concept of elastic rebound is a useful one, but is not a complete explanation for why earthquakes occur. In theory, the pressure at even shallow depths within the Earth should be high enough to lock faults and to prevent movement. Friction between blocks increases rapidly with depth, so that it should be no easier to move the Earth along most old faults than to create entirely new ones. Still, most earthquakes do occur on existing faults. *Why?* The elastic rebound hypothesis also fails to account for the mechanism that triggers the actual release of energy. Stick-slip behavior seems to suggest that a mechanical change takes place in the rock or in the fault surface as stress accumulates. *What sort of change finally triggers an earthquake?*

Geologists can find a possible answer to both of these questions by simulating the conditions shown in Figure 9.2 with wet rocks in the laboratory. Under some circumstances, they have found that as stress increases, the volume of a wet rock also increases. This phenomenon is known as **dilatancy** (pronounced *die-LAY-ten-see*). Experimenters hypothesize that the forces on the rock create a large number of microfractures—each less than 0.001 mm wide—which then fill with water under high pressure. Water pressure keeps the microfractures from closing again and gradually weakens the rock as the microfractures extend and interconnect. Geologists have noticed that the ground surface expands subtly before a major earthquake, suggesting that dilatancy is important in nature as well as in the laboratory.

Other observations confirm that water is important in facilitating earthquakes. A typical fault zone contains not only crushed rock but also a lot of clay and other materials that are products of chemical reactions with water. These are known collectively as **fault gouge**. On active faults, gouge commonly forms smoothed and striated surfaces called **slickensides** (Figure 9.3), which are direct evidence of movement by slipping. It is also common to find springs along faults. From these observations researchers infer that water (or water-bearing materials) may lubricate ruptures in the Earth and make them more likely to slip.

Perhaps the best evidence linking water to earthquakes came from an inadvertent experiment performed in Colorado by the United States Army. In 1962, the army began pumping liquid chemical waste products from the manufacture of weapons into a 3,670-meter-deep well at the Rocky Mountain Arsenal. Fluids were injected under pressure from March 1962 to September 1963, and again from September 1964 to

FIGURE 9.3 When crushed rock (fault gouge) is dragged between the walls of a fault, it polishes them and leaves fine grooves that parallel the direction of movement. Slickensides, as these surfaces are called, are also commonly coated with water-bearing minerals like chlorite or epidote, evidence of the role that water plays in activating faults.

September 1965. Within the first period alone, over 700 minor earthquakes were recorded in the Denver area, where tremors are usually rare. The statistical correlation between the volume of fluid injected and the occurrence of earthquakes, shown in Figure 9.4, is striking. Geophysicists studying the Denver earthquakes hypothesized that pressurized liquid reduced the strength of rock by forcing open microfractures, and that it decreased friction along existing faults.

Water, therefore, may serve as both a lubricant and a triggering mechanism for earthquakes. Some earthquakes, however, apparently occur not because of a change in the rocks, but because a rapid change in the level of stress provides enough energy to overcome friction even if the rocks are not lubricated. Water may thus play only an incidental role in these earthquakes. Some earthquakes accompany volcanic eruptions. Collapse events like landslides or cave-ins may also cause earthquakes (although it is more commonly the other way around). So can large explosions, such as underground nuclear weapons tests. It has even been suggested that tidal attractions by the Sun and the Moon can provide enough force to trigger earthquakes.

Seismic Waves

When we throw a pebble into a pond, pluck a guitar string, or strike a tuning fork, we generate waves that carry energy from one place to another. The same thing happens within the Earth as energy is released at the focus of an earthquake. As soon as elastic rebound occurs, **seismic waves** (earthquake waves) begin to spread rapidly in all directions, eventually dissipating the energy throughout the planet. Just as the sizes and shapes of ripples on a pond may tell us about both the pebble that produced the ripples and the obstacles they may have encountered, seismic waves also carry information about how they were produced and what they may have encountered as they moved. Seismologists can interpret the structure and properties of the Earth, or can locate and characterize the source of an earthquake by studying the collection of waves that it generates.

Body Waves

When energy is released at the focus of an earthquake, two kinds of waves immediately begin to speed away (see Figure 9.5). The first of these are **compressional** waves (also commonly called *Primary* or **P-waves**). They are, in fact, low-frequency sound waves. As they move through the Earth, P-waves displace particles forward and backward along their line of travel, alternately compressing and expanding the volume of rock they pass through. The other kind of waves are **shear** waves (also known as *Secondary* or **S-waves**), which advance by displacing particles of the rock at right angles to their line of travel. Viewed from the side, as in Figure 9.5, S-waves appear as a series of traveling crests and troughs. Both S-waves and P-waves are known as **body** waves because they travel inside the earth, rather than on its surface.

Wave Velocity Scientists use the term **elasticity** to describe a material's tendency to spring back to its original shape and volume once it has been deformed. It is a measure of how rigid the material is. The stronger the chemical bonds in a material are, the more readily an atom is pulled back into position once a disturbance—such as an earthquake wave—has passed. Also, the stronger the bonds, the more efficiently each atom transfers an impulse to others near it. That is why you can hear distant footsteps sooner and with greater clarity if you put your ear to the floor than you can if you listen through the air.

Some of the most fundamental information that geologists get from seismic waves has to do with the waves' velocities. Waves slow down or speed up as they meet rocks with different densities and elasticities; thus measurements of wave velocity yield valuable clues about what the Earth is made of. Furthermore, S- and P-waves respond in different ways to the same materials; therefore, geologists can learn a lot by comparing velocities of the two. Shear (S) waves, for example, travel through rock at only about 45 percent of the speed of compressional (P) waves. In a liquid, the difference is even more dra-

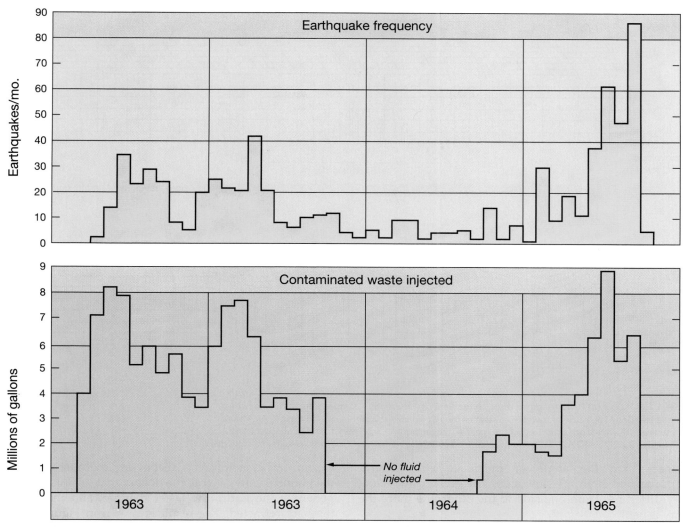

FIGURE 9.4 As the U.S. Army Corps of Engineers injected waste water into deep wells outside of Denver in the early 1960s, the number of small earthquakes in the area increased. This graph shows that the number of quakes varied directly with the volume of water injected, suggesting that the water lubricated faults and triggered the earthquakes. *(The Rocky Mountain Association of Geologists)*

matic. P-waves traveling from rock to water or magma lose over 70 percent of their velocity, and S-waves die out completely. An S-wave cannot travel through liquids because liquids have no resistance to side-to-side movement, and therefore cannot spring back once they are deformed.

Direction of Travel As body waves meet obstacles or abrupt changes in the properties of Earth materials, a complicated set of events takes place. The waves may be **reflected** (bounced) or **refracted** (bent).

Reflection is a familiar process. You face it in the bathroom mirror every morning. In fact, most of your visual impressions are based on the way light reflects from surfaces. Sound, ripples on a pond, and even basketballs bounce off surfaces in an easily predictable way; so does seismic energy. The rule is this: *If an*

*object or a ray[1] of energy reaches a sharp boundary, the angle at which it leaves the boundary (the **angle of reflection**) is the same as the angle from which it approached (the **angle of incidence**).* This rule is illustrated in Figure 9.6.

Refraction, the process by which a ray of energy is bent as it passes from one medium to another, may be less familiar. Suppose that there is a boundary between one substance in which energy can travel at low velocity and another, more rigid one, in which energy can travel at a much higher velocity. Figure 9.7 shows that the **angle of refraction** at which the ray leaves from the fast side of the boundary is always

---

[1]A ray is a line along which energy travels. If you could mark a spot on an earthquake wave and watch it travel through the Earth, you would be following a seismic ray.

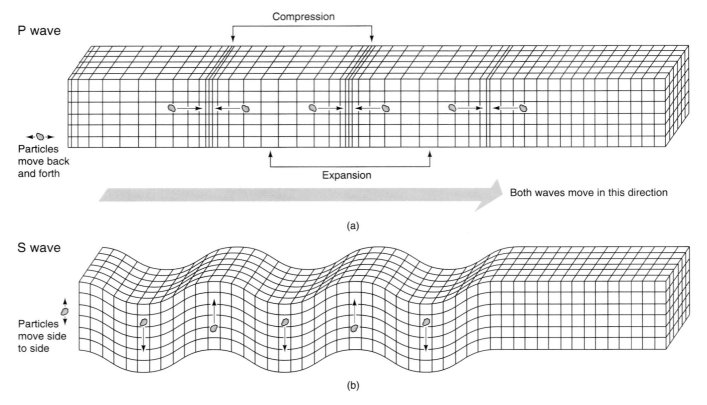

P wave

Compression

Particles move back and forth

Expansion

Both waves move in this direction

(a)

S wave

Particles move side to side

(b)

FIGURE 9.5   (a) P-waves move through the Earth by alternately compressing and expanding materials along their line of travel. (b) S-waves move particles at a right angle to their line of travel. Both types of waves are produced at the focus of an earthquake, and they travel in all directions. *(From: EARTHQUAKES by Bolt. Copyright © 1993 by W.H. Freeman and Company. Used with permission.)*

larger than the angle of incidence at which it approached from the slow side. The greater the velocity difference from one side to the other, the more the ray is bent.

A ray of energy approaching a surface, therefore, can do one of two things: It may either bounce off it, or pass through it and bend. In many cases, it does both (see Figure 9.8). The process that dominates at a particular boundary depends on the ray's angle of incidence and on how different the materials across the boundary are. What happens if the two substances blend smoothly into each other instead of meeting at a sharp boundary? Without a boundary, there can be no reflection. The easy answer, then, is that most of the energy passes smoothly from one material into the other. As Figure 9.9a shows, however, there is a more complicated answer. A ray can behave as though it were traveling through a stack of very thin layers, each slightly more rigid than the one before. The result is that the ray

is bent further as it travels deeper into the "fast" side of the boundary. In general, the deeper that seismic waves penetrate the Earth, the faster they go, and the more they are refracted. The result is shown in Figure 9.9b. All rays eventually curve upward to the surface.

FIGURE 9.7   If a ray of energy speeds up as it crosses a boundary into another material, then its path is bent (refracted). If the refracted ray is faster, then its angle of refraction (r) is always greater than the angle of incidence (i).

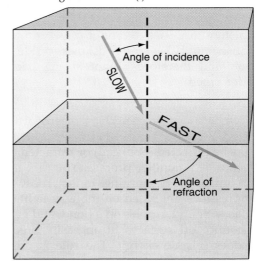

Angle of incidence

SLOW

FAST

Angle of refraction

FIGURE 9.6   An object or a ray of energy reflects off a surface at the same angle from which it approached.

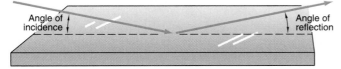

Angle of incidence

Angle of reflection

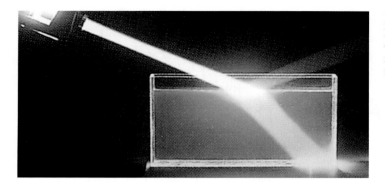

FIGURE 9.8
Most of the energy in this ray of light is refracted when it encounters the surface of the water in the tank, but some is reflected. In the same way, seismic rays are both reflected and refracted as they encounter boundaries within the Earth.

Surface Waves

Seismic waves that travel along the Earth's surface, called **surface waves**, are produced whenever a P- or S-wave reaches the Earth's surface. Surface waves come in two forms. The simplest of these is the **Love wave**, which is very much like an S-wave with its side-to-side vibrations confined to the surface of the Earth (Figure 9.10a). There is no vertical motion associated with Love waves. **Rayleigh waves**, however, involve both horizontal and vertical movement, much like water waves in the open ocean. A particle on the Earth's surface at the crest of a Rayleigh wave moves forward and then downward. As the wave crest passes, the particle retreats and then moves upward, completing an elliptical trajectory (Figure 9.10b).

Because they travel through relatively unconfined materials, surface waves are much slower than body waves, and they are also much bigger. Seismologists of the nineteenth century referred to them as "large waves," in contrast to the "preliminary tremors" that we now recognize as body waves. The complicated mixture of shaking, pitching, and yawing motions that result from Love and Rayleigh waves is responsible for most of the damage that earthquakes cause.

FIGURE 9.9   (a) A ray passing through a stack of layers, each more rigid than the one above it, speeds up at each boundary and is bent further away from the vertical. (b) Rigidity increases gradually downward through much of the Earth, so a ray of seismic energy acts as though it is traveling through a large stack of very thin layers like those in (a). As a result, it curves gently toward the surface.

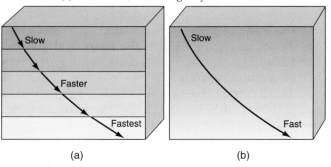

(a)   (b)

Free Oscillations

As waves travel around and through the Earth, they gradually die out. Delicate instruments, however, can detect the strongest waves for weeks after a major earthquake. During that time, the waves have circled the globe many times and have interacted with each other repeatedly. Some of the waves reinforce each other, but most cancel each other out, so that only a few waves remain. These surviving waves, called **free oscillations**, have much in common with the simple resonances of a gong. In fact, seismologists characterize the free oscillations of the Earth in the same terms that physicists use to describe the ringing of a large bell. The tones of a bell depend on its size, its density, and its elasticity. In exactly the same way, the "tones" of the Earth are clues to its density and rigidity. Like body and surface waves, therefore, free oscillations can reveal encoded information about the planet's interior. To interpret seismic information, however, we first need a means of collecting it.

Detecting Earthquakes
▲▲

Seismic Instruments

Early in the second century A.D., a Chinese scholar named Chang Heng developed the first seismic measurement instrument. The sketchy descriptions that have survived suggest that this primitive instrument had no mechanism for making a permanent record of seismic activity. It was instead a **seismoscope**, a simple device for detecting an earthquake and indicating roughly which direction seismic waves had come from. Science historians assume that it was based on a pendulum, which would start to swing toward or away from an earthquake focus when it was disturbed by seismic waves. Around the pendulum were several delicately placed balls. The swinging pendulum would dislodge one of the balls, thus indicating the direction to the earthquake focus (see Figure 9.11). Despite some reputed successes, the instrument fell

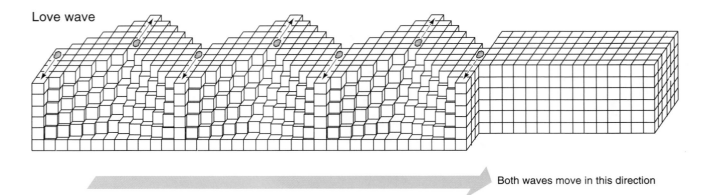

Love wave

Both waves move in this direction

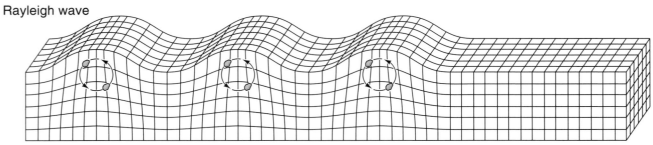

Rayleigh wave

FIGURE 9.10   Surface waves are produced when P-waves and S-waves reach the surface of the Earth. They are much larger and slower than body waves. (a) Love waves are similar to S-waves, but they can only vibrate parallel to the Earth's surface, not up and down. (b) Rayleigh waves are like water waves in the ocean. They push particles in elliptical paths. *(From: EARTHQUAKES by Bolt. Copyright © 1993 by W.H. Freeman and Company. Used with permission.)*

into disuse by the end of the Han dynasty a century later.

A clockmaker named Domenico Salsano, from Naples, is credited with building the first **seismograph**, an instrument capable of producing a permanent record of an earthquake. It too was designed around a long pendulum. A brush attached to it made a tracing in ink as Earth motions set the pendulum

FIGURE 9.11   Chang Heng's seismoscope, as visualized by Wang Chen-to.

swinging. An additional mechanism rang a bell whenever motions were particularly strong. Salsano's instrument was used to record a series of moderate to large earthquakes near Calabria, Italy, in 1783. A similar seismograph was in use in Cincinnati, Ohio, during the earthquakes that shook New Madrid, Missouri, during the winter of 1811–12. While this instrument was a definite improvement over Chang Heng's, it was still not capable of detecting the weak vibrations from a small or distant earthquake, and it made no provision for recording the time when an earthquake was measured.

The first instruments designed to overcome these limitations were built during the last quarter of the nineteenth century. On April 17, 1889, Ernst von Rebeur-Paschwitz used the new instrumentation in Potsdam, Germany, to make the earliest recognized tracing of a distant earthquake, which had occurred in Japan several hours earlier.

The basic components of a modern seismograph are shown in Figure 9.12. At the heart of the instrument is a large mass suspended on a spring or a wire from a supporting frame so that it is free to move in only one direction (E ↔ W, N ↔ S, or Up ↔ Down). In fact, however, it is the frame and the Earth to which that frame is firmly anchored that move during an earthquake while the mass remains very nearly immobile. A heavy mass resists acceleration (a characteristic

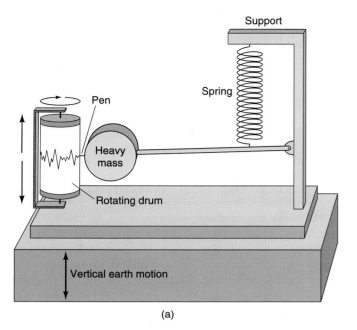

Pen

Spring

Support

Heavy mass

Rotating drum

Vertical earth motion

(a)

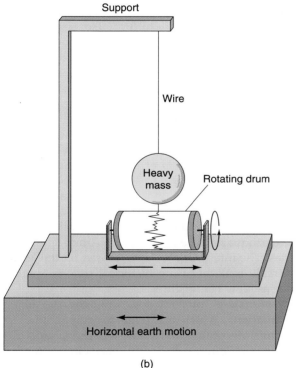

Support

Wire

Heavy mass

Rotating drum

Horizontal earth motion

(b)

FIGURE 9.12 A pendulum seismograph may be designed to detect either horizontal or vertical motion. In either case, it must have the basic elements shown here. In most modern instruments, movements are also recorded electronically so that they can be analyzed by computer. To compare records from several seismographs, each must also have an accurate clock. (*This illustration was derived from Figure 15.3 in B.J. Skinner and S.C. Porter's* THE DYNAMIC EARTH 2/E, © *1992 John Wiley & Sons, Inc.*)

that scientists refer to as **inertia**); thus, it is difficult to set in motion. The connection between the mass and the Earth is also a tenuous one, so movements of the

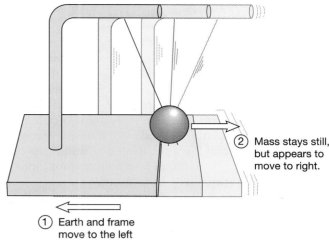

② Mass stays still, but appears to move to right.

① Earth and frame move to the left

FIGURE 9.13 The base and framework of a seismograph are attached to the Earth. Thus, when an earthquake occurs the framework and the Earth move together. The heavy mass, however, resists movement and is only connected loosely to the Earth. Thus, it does not move. To an observer, however, the mass *appears* to have moved while the framework stood still.

Earth are not easily felt by the mass. Hence, as the Earth and the frame move, a recording device attached to the mass traces their relative movement (see Figure 9.13). The recording device has traditionally been a pen, which makes marks on paper on a rotating drum, but today it is often an electronic sensor, whose signal is preserved on a magnetic storage device.

Three other fundamental components are needed to maximize a seismograph's reliability and sensitivity. The first is a *damping device*, used to make the suspended mass stop moving once a seismic wave has passed. Without such a device (commonly a pair of electromagnets that pull weakly on the mass from opposite directions), a seismograph set in motion by a large vibration would keep moving and possibly not detect smaller seismic waves that arrive shortly afterwards.

Second, a useful seismograph needs to have a means of amplifying the very small vibrations that it detects. It is possible to build a seismograph with a modern solid-state *electronic amplifier* that can detect movements even as small as 10^{-8} cm (roughly the width of an atom). Such an instrument would be much too sensitive to use on a planet as noisy as the Earth, so a practical seismograph rarely needs such a strong amplifier. The Apollo astronauts, however, placed several of these sensitive instruments on the Moon in the 1970s to record the impacts of small meteorites.

The final necessary component of a seismograph is an accurate *clock*. Modern observatories check their clocks continuously against widely broadcast radio time signals to be sure that they are precise to within 1 ms. All seismologists report their observations in Greenwich Mean Time (*GMT*, also often called Universal Time), rather than local time.

Locating Earthquakes

When seismologists speak of locating an earthquake, most of the time they really mean that they are marking the place where it occurred on a map of the Earth. This spot is not the earthquake's focus, where energy was released within the Earth, but a point directly above it on the Earth's surface. This point is called the earthquake's **epicenter**. Thanks to the simultaneous generation of S- and P-waves, the job of finding an epicenter with today's global network of seismographs is fairly simple (Figure 9.14a, b).

Using Figure 9.14a as a guide, suppose that an earthquake takes place at point A, and that S- and P-waves begin to travel toward seismographs at points B, C, and D. As the first faster-moving P-wave reaches seismograph B, the S-waves are already behind by 1.5 min. The two curved lines in Figure 9.14b, a **travel-time** diagram, indicate how long it takes P- and S-waves to travel any given distance. By identifying the distance at which the P- and S-curves are exactly 1.5 min apart on that diagram, we can see that seismograph B must be 700 km from the epicenter. When the first P-wave reaches seismograph C, the first S-wave has lagged farther behind (2.1 min), and at station D it is delayed even more (2.5 min). These indicate that points C and D are 1,300 and 2,000 km from the epicenter, respectively. The gap between the arrival times of S- and P-waves at any seismic detection station is a unique measure of its distance from the epicenter.

Using this method, scientists at each seismograph can calculate how *far* it is to the epicenter, but none of them can tell which *direction* the epicenter lies in. By drawing a circle with a radius of 700 km around seismograph B on a map and drawing 1,300 km and 2,000 km circles around the other two seismographs, however, they can locate the epicenter exactly (see Figure 9.15). It is at the point where all three circles intersect. This technique for locating an epicenter is known as **triangulation**. If estimates from other stations fail to meet at the same map location, they may be based on faulty travel-time diagrams, which must then be adjusted. Each time we use a travel-time diagram to locate an earthquake, therefore, we have an opportunity to refine our understanding of the seismic velocities on which it is based.

Earthquakes tend to occur in belts, or zones, many of which are associated with belts of active volcanoes. These also coincide with the places where lithospheric plates meet (see Figure 9.16). The earthquakes that occur in the "Circum-Pacific Belt," around the borders of the Pacific Ocean, account for a little over 80 percent of the total seismic energy released throughout the world. The greatest seismic activity is near Japan, western Mexico, Melanesia, and the Philippines. Another 15 percent of the total energy released by all earthquakes is in a zone that extends from Burma through the Himalaya Range, into Baluchistan, across Iran, and westerly through the Alpine structures of Mediterranean Europe. This is sometimes called the "Mediterranean and Trans-Asiatic" zone. Earthquakes in this zone have foci aligned along mountain chains. The remaining 5 percent of the earthquake energy is released throughout the rest of the world. Narrow belts of activity are found to follow the oceanic ridge systems.

How Big Is an Earthquake?

Intensity The most obvious way to estimate the strength of an earthquake is to find out how much damage it caused. That is what Robert Mallet, an Irish civil engineer and geologist, decided to do when the first reports of a major earthquake in Naples reached London in December 1857. He spent two months making detailed maps of the damage to buildings and tabulating the results of his interviews with local citizens. By connecting points with equal damage or perceived equal intensity, he was able to draw **isoseismal** (from Greek roots meaning "same shaking") lines and to locate the epicenter of the quake. From the degree of maximum destruction and the spacing of isoseismal lines, he estimated its relative strength.

Modern intensity scales follow the same approach. One that is used widely today was devised in 1902 by Italian seismologist G. Mercalli and later adapted for use in North America. The **Modified Mercalli Intensity Scale** details 12 levels of shaking, characterizing each by the damage to human structures, the response of animals, or the disturbance to ground level (Table 9.1). Figure 9.17 shows an isoseismal map of the 1811 earthquake in New Madrid, Missouri—still the most intense earthquake to be felt in North America in modern times. The intensities were estimated on the Modified Mercalli scale by using historical reports from as far away as New Hampshire.

Unfortunately, intensity scales depend on subjective evaluations, which are open to exaggeration or simple misinterpretation. Even where reports are believable, they often reflect differences in population density or construction standards as well as real differences in intensity.

Acceleration An alternative measure is the **acceleration** of ground shaking: the rate at which the Earth speeds up, slows down, or changes direction as it shakes. A description of an earthquake in terms of acceleration conveys how sharply and suddenly it disturbs the ground in a way that intensity measurements do not. This is also a very important measure for engineers who design buildings and bridges. A building is not damaged by movement any more than a car is damaged by traveling at 100 miles per hour. Damage

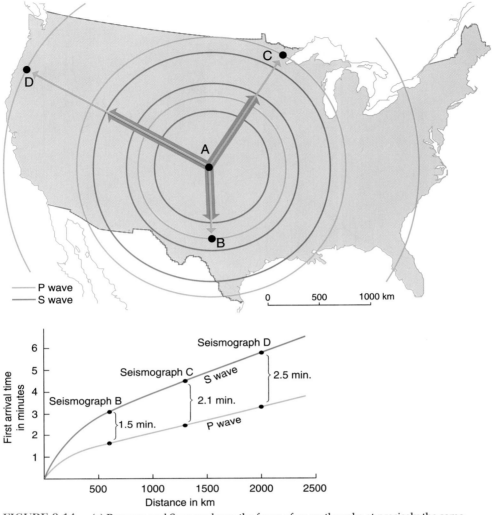

FIGURE 9.14   (a) P-waves and S-waves leave the focus of an earthquake at precisely the same time. Because P-waves are faster, however, they will always reach distant seismic stations first. The farther they travel, the greater the difference in their arrival times becomes. This can be seen in (b), a travel-time diagram. If geologists measure the delay between P- and S-waves at any station, they can use this graph to determine how far away the focus was. In this example, scientists at seismographs B, C, and D in Figure 9.14a each calculate distances from the travel-time curves in Figure 9.14b using the following measurements:

SEISMOGRAPH	ARRIVAL TIME DIFFERENCE (P MINUS S)	CALCULATED DISTANCE
B	3.1 min - 1.6 min = 1.5 min	700 km
C	4.5 min - 2.4 min = 2.1 min	1300 km
D	5.8 min - 3.3 min = 2.5 min	2000 km

is caused by *changes* in motion as a building (or a car) accelerates or decelerates too quickly. Earthquake-resistant buildings are commonly designed to resist gentle to moderate changes in movement, but they cannot withstand the sharp, whiplike accelerations of a violent earthquake.

Measurements of acceleration are compared against the rate at which a falling object gathers speed if it is dropped from a high place, a rate of increase referred to as an acceleration of 1.0 *g*. If we come to a quick stop in a vehicle or have a particularly bumpy airplane trip, we may experience accelerations as high as 0.4 *g* (that is, we change velocity 40 percent as quickly as a free-falling object does). More rapid changes can cause serious injury and, fortunately, are uncommon.

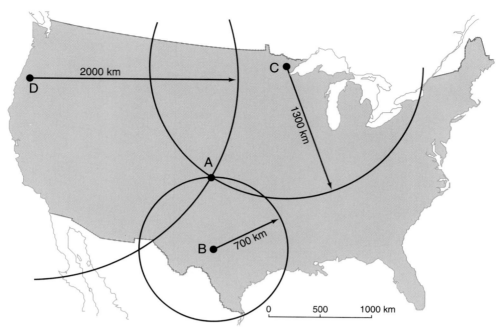

FIGURE 9.15 Locating an epicenter. The seismologist at seismograph B, C, or D can calculate the distance to the epicenter by using the method in Figure 9.14, but has no way to figure out which direction it lies in. To make that determination requires the help of scientists at all three locations. On a map, the seismologists draw a circle of the appropriate size around each seismograph. The epicenter is where the three circles intersect, at point A.

During most earthquakes, the ground within 10 km of the epicenter is shaken with an acceleration of 0.05 to 0.35 *g*. As shown in Figure 9.18, however, there are regions where the probability of peak accelerations higher than 0.4 *g* is greater than 1 in 10 during a 50-year period. When they occur, such intense shocks may continue for 15 sec to a minute or more. Few structures can survive this much shaking, most of which is directed horizontally, where buildings are naturally weakest.

FIGURE 9.16 Locations of earthquake epicenters (between 0 and 100 km deep) during a 9-year-period. Over 80% of all earthquake energy is released in either the Circum-Pacific or Mediterranean and Trans-Asiatic Belt. Most remaining earthquake energy is released along the mid-ocean ridge system. *(Data from the National Earthquake Information Center)*

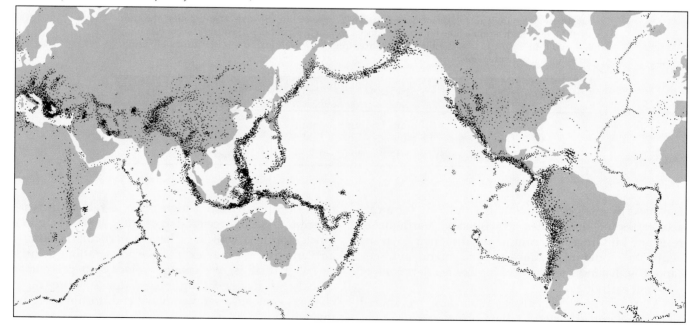

TABLE 9.1 Modified Mercalli Scale of Earthquake Intensity

INTENSITY	DESCRIPTION OF EARTHQUAKE EFFECTS
I	Not felt except by a very few people under especially favorable circumstances.
II	Felt only by a few persons at rest, especially on upper floors of buildings. Delicately suspended objects may swing.
III	Felt quite noticeably indoors, especially on upper floors of buildings, but many people do not recognize it as an earthquake. Vibration like a passing truck.
IV	During the day, felt indoors by many, outdoors by few. At night, some people awakened. Dishes, windows, doors disturbed; walls make creaking sound. Sensation like a heavy truck striking the building.
V	Felt by nearly everyone, many people awakened. Some dishes and windows broken; cracked plaster in a few places; unstable objects overturned. Disturbances of tall objects sometimes noticed.
VI	Felt by all; many people frightened and run outdoors. Some heavy furniture moved; a few instances of fallen plaster and damaged chimneys. Damage slight.
VII	Everyone runs outdoors. Damage negligible in well-constructed buildings, considerable in poorly built or badly designed structures. Some chimneys broken.
VIII	Damage slight in specially designed structures, but considerable in ordinary buildings. Panel walls thrown out of frame buildings. Chimneys, smokestacks, columns, monuments fall down. Heavy furniture overturned. Sand and mud ejected in small "mud volcanoes."
IX	Damage considerable even in specially designed structures, great in large buildings, with partial collapse. Buildings shifted off of foundations. Ground cracked conspicuously. Underground pipes broken.
X	Some well-built wooden structures destroyed; most masonry buildings destroyed. Ground badly cracked. Rails bent. Landslides considerable. Water splashed over river banks.
XI	Few, if any buildings remain standing. Bridges destroyed. Broad fissures in ground. Underground pipes entirely out of service. Earth slumps in soft ground. Rails bent greatly.
XII	Damage total. Wave seen on ground surface. Lines of sight and level distorted. Objects thrown into the air.

FIGURE 9.17 Isoseismal map of the 1811 New Madrid earthquake, reconstructed from historical records. Roman numerals indicate the estimated Modified Mercalli intensities (see Table 9.1). *(Seismological Society of America)*

Magnitude   In contrast to scales that measure the destructive capacity of earthquakes, some scales gauge the total energy release of an earthquake or the absolute amount of ground movement that they cause. In 1935, Charles Richter of the California Institute of Technology developed a scale based on the size, or **amplitude**, of seismic waves recorded on a seismograph. Richter referred to the values on his scale as **magnitudes**.

The Richter magnitude of an earthquake depends on the recorded amplitude (X) of its largest seismic wave and the duration (T) of that wave (see Figure 9.19). To calculate a magnitude, a geologist measures X in units of 10^{-4} cm and divides it by T, measured in seconds. Thus, X/T is an indication of the amount of energy released by the earthquake. To adjust for the fact that some energy is lost as seismic waves travel, the geologist adds a correction factor Y that is derived from the S–P interval. The formula used to complete the calculation is

$$\text{Magnitude} = \log X/T + Y.$$

The scale is logarithmic, which means that a change of one unit of magnitude (say from Richter magnitude 4 to magnitude 5) corresponds to a tenfold increase in X.

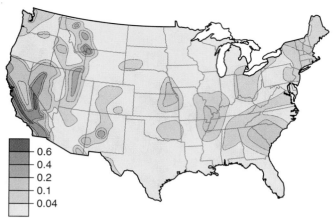

FIGURE 9.18   Map of expectable levels of earthquake shaking hazards. Levels of ground shaking for different regions are shown by contour lines, which express the maximum amount of shaking likely to occur at least once in a 50-year period as a percentage of 1 *g*, the acceleration of gravity. (*Modified from S.T. Algermissen and D.M. Perkins, "A Probabilistic Estimate of Maximum Acceleration in the Contiguous United States," U.S.G.S. Open File Rep. 76-416, July 1976*)

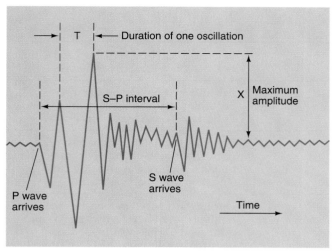

FIGURE 9.19   The Richter magnitude of an earthquake is calculated by measuring the amplitude (*X*) and the duration (*T*) of the largest seismic wave recorded on a seismograph, and applies a distance correction (*Y*) based on the interval between P- and S-wave arrival times. (*This illustration was derived from Figure 15.10 in B.J. Skinner and S.C. Porter's THE DYNAMIC EARTH 2/E, © 1992 John Wiley & Sons, Inc.*)

A magnitude-7 earthquake thus has an amplitude 1,000 times greater than a magnitude-4 earthquake.

Unlike intensity scales, a magnitude scale has no theoretical upper limit. However, the largest magnitude ever recorded was 8.6, which suggests that there is a limit to the strength of crustal rocks and to the amount of energy that can be stored in stressed rocks before they fracture. To appreciate just how much energy that can be, make the following comparison: A magnitude-5.0 earthquake releases roughly the same amount of energy as the atomic bomb dropped on Hiroshima in 1945. As a general rule, the energy of an earthquake increases 30 times for each unit of increased magnitude. A magnitude-7 earthquake, therefore, releases as much energy as 30 magnitude-6 earthquakes or 900 magnitude-5 earthquakes. Thus, the maximum possible yield of a single earthquake is apparently 100,000 to 1 million times as great as the Hiroshima atomic bomb.

Earthquake Prediction

During the twentieth century, population growth has been explosive throughout the world. Oddly enough, some of the Earth's most densely inhabited regions are in areas where earthquake activity is most frequent. The damage that accompanies earthquakes can take many forms. Personal injury and property damage may be caused by the ground shaking itself or by secondary effects that are even more lethal on occasion. In several large countries—the United States, China, Japan, and the former Soviet Union—earth-

quake prediction and public education programs about the need for earthquake-resistant structures and effective disaster relief systems are serious priorities.

Statistical Methods

The most common methods for predicting when and where earthquakes will occur are based on statistical interpretations of past activity. Geologists predict that the next earthquake on a fault will be somewhere between the two previous earthquake epicenters, a region known as a **seismic gap**. A variation on this technique relies on the record of historical activity on a fault to suggest a "normal" time interval between earthquakes. An earthquake is more likely, and will probably be more destructive, if it is overdue. Both kinds of statistical argument illustrate the seismic gap method of prediction.

This method was first used by H.F. Reid following the 1906 San Francisco earthquake. Geologists mapping the San Andreas fault matched and dated offset features, and they estimated that opposite sides of the fault had moved 3.5 m in the previous 50 years. Much of the movement had occurred in a single earthquake at Fort Tejon, south of San Francisco, in 1857. During the 1906 event, the same fault moved 6.5 m in Marin County, north of the city. Reid concluded that the 1906 earthquake had caused roughly the equivalent of 100 years worth of displacement. Assuming that stress accumulates at a constant rate, he predicted that the San Andreas system would continue to move at 6 to 7 m per century. The most likely place for the next earthquake to occur would be in the seismic gap between the epicenters in Fort Tejon and Marin

County. The total movement could take place in frequent small events or in infrequent large ones at least 50 years apart. In the decades since 1906, almost no movement has occurred. With each passing year, therefore, the probability of a seismic disturbance increases as does its likely size.

This method is obviously imprecise. Rates of stress accumulation are probably not truly constant, so that predictions based on past activity may be wrong by several decades. This may be particularly true in regions with many faults; stresses that build up on one fault may be relieved by movements on others. Still, the idea that future earthquakes are likely in the quiet seismic gap between previous epicenters has merit, even if figuring out when to expect them is a bit tricky.

Precursor Events

Following a great earthquake, various **aftershocks** occur as nearby portions of the fault surface slip and readjust. There is then a period of several years during which minimal seismic activity takes place. We can imagine that stress builds slowly during this period and that triggering events are infrequent. As stresses accumulate, however, low-level seismic activity becomes more common on some faults. Finally, in the years (or perhaps only hours) before the next great earthquake, a series of moderate-sized **foreshocks** may occur. This cycle has been recognized on several large fault systems. Unfortunately, however, seismologists have usually noticed the pattern *after* a major earthquake, not before it. Also, foreshocks do not always occur; thus the cycle is not a reliable way to predict earthquakes.

Seismologists now recognize several other types of unusual behavior that take place before a great earthquake. Like foreshocks, each shows that stress is accumulating but does not necessarily predict when it will be released. During the 30 months before the 1971 San Fernando earthquake in Southern California, for example, instruments at the California Institute of Technology recorded a 10 to 15 percent decrease in P-wave velocities. Then, within days before the San Fernando quake, the seismic velocities suddenly returned to normal. This behavior has been noted elsewhere in the United States and Asia and may be a reliable indicator that an earthquake is on the way.

Major changes in ground level, possibly associated with dilatancy below the surface, have also been recorded before some major quakes. Surveying records for the coastline of Honshu, Japan, near the city of Niigata, indicate that the land rose at a steady rate of 2 mm per year from 1898 to 1958. The rate then increased rapidly until 1962. After a large earthquake shook the area in 1964, it was found that the land had dropped by 15 to 20 cm. Seismologists have detected similar shifts in ground level near Parkfield, California. Tilting and other slow changes in ground position suggest that this is a major risk zone for future earthquakes. In fact, the U.S. Geological Survey calculates the probability of a major earthquake on this segment of the San Andreas Fault before the year 2020 to be over 90 percent.

In recent years, field studies have also indicated a gradual decrease in the electrical conductivity of rocks and a measurable increase in the release of radon gas from wells before a major earthquake. Each of these precursor events may be related to the formation of small fractures and the movement of water. All of these tools bring geologists closer to the goal of reliable prediction, but even with the full repertoire of techniques it will probably be a long time before geologists are able to predict the day and hour of a specific earthquake.

The Destructive Potential of Earthquakes

Every year there are about 100 earthquakes with magnitudes greater than 6.0—statistically, about one every three days. Each one of these has the potential for causing moderate to heavy damage if it occurs in a populated region. About once every three weeks, a severe earthquake with a magnitude greater than 7.0 strikes. On average, two of these a year have magnitudes of 8.0 or greater. The possible annual loss of property and lives from seismic activity is great (see Table 9.2). Nearly 1.2 million people are estimated to have died in earthquakes during the first eight decades of the twentieth century—on average, 15,000 deaths per year.

The degree of risk varies considerably from place to place. Compared to your chance of dying in a highway accident or a violent crime, your chance of dying in an earthquake in North America—even in California, where most of this century's earthquake fatalities have been—is insignificant. Property damage from earthquakes is similarly skewed. During the 1970s, when a $16 billion loss to property was recorded worldwide, only $566 million worth of earthquake damage—about 3.5 percent of the total—occurred in North America. The uneven distribution of risk is mainly due to the fact that earthquakes do not occur randomly. Unfortunately, however, much of the variation in risk is attributable to differences in construction standards.

Earthquakes in San Francisco: Is It Time for the Big One Yet?

On October 17, 1989, millions of people in and around San Francisco, California, experienced the chaos and destruction of a magnitude-7.1 earthquake. The epicenter was near Loma Prieta, a minor peak in the Santa Cruz Mountains, nearly 60 miles to the south. Despite the epicenter's location, some of the most catastrophic damage took place in San Francisco and nearby Oakland. Surface waves were magnified in artificially filled land, several buildings toppled into the streets, fires broke out where gas mains had ruptured, and a one-mile section of the elevated Nimitz Freeway collapsed. The acceleration of ground shaking was not as severe as it had been during the 1906 earthquake, but it cost Californians several billion dollars in property damage and over 60 lives (Figure P9.1.1).

Bay area residents, aware of geologists' predictions of another earthquake like the one in 1906, wondered whether this was "the big one," and if not—then when would the big one actually strike?

P9.1.1 During the 1989 Loma Prieta earthquake, a one-mile segment of the double-level Nimitz Freeway in Oakland, California, collapsed, killing several motorists. The support columns were designed to hold the weight of the upper deck but could not withstand the horizontal ground shaking.
(Paul Sakuma, AP/Wide World Photos)

Most of the seismic activity was along the western branch of the San Andreas Fault Zone. The aftershocks, however, were not scattered evenly around the site of the major quake. Instead, they were concentrated to the south. From this, seismologists conclude that the fault was more tightly locked to the north, in the direction of San Francisco. Seismologists believe that stress in the seismic gap between San Jose and Point Reyes is building gradually toward "the big one" that is yet to come.

The United States Geological Survey keeps a constant watch on seismic activity in the San Francisco Bay area and has accumulated a huge database describing the frequency and intensity of earthquakes. The seismic-risk map shown in Figure P9.1.2 is one result of this work. The probability of an earthquake with a magnitude greater than 7 occurring somewhere in the Bay area before 2020, according to this map, is about 60 percent. This answer to when the next big quake will hit is less specific than most San Francisco residents might like, but it means that there is a better than even chance that another earthquake at least as destructive as the 1989 one will strike within your lifetime.

Ground Shaking

The greatest hazard during an earthquake is **ground shaking**. Most buildings are designed to handle vertical stress and do not stand up well to horizontal vibrations. In Japan and the United States, public buildings in earthquake-prone areas are now designed using research on ground motion. These buildings are generally resistant to damage from moderate-sized earthquakes. The Imperial Hotel in Tokyo, designed by Frank Lloyd Wright in 1913, was among the first modern earthquake-resistant structures. It suffered only minor structural damage in the 1923 earthquake that devastated Tokyo, although it was destroyed in the

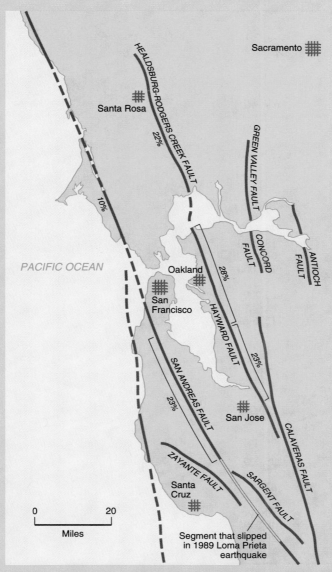

P9.1.2 Seismic risk in the San Francisco Bay area, showing the calculated probability of an earthquake along various segments of the San Andreas and Hayward faults before the year 2020. *(U.S. Geological Survey, "Probabilities of Large Earthquakes Occurring in California on the San Andreas Fault," Open File Rep. 88-398, 1988.)*

subsequent fire. As engineers learn more about how buildings behave during earthquakes, construction methods improve. Particularly in the developed nations, life in seismically active areas is becoming safer. In poorer nations, however, people rarely have access to expensive building materials or the technical skills for building earthquake-resistant structures.

One factor that can intensify ground shaking is that surface waves are magnified as they pass through loose, uncompacted soil or landfill. Seismologists and engineers who studied collapsed portions of the elevated Nimitz Freeway in Oakland after the Loma Prieta earthquake of 1989 have identified this as the cause of the disaster. The dashed segment of highway where

TABLE 9.2 Notable World Earthquakes

YEAR	REGION	DEATHS	MAGNITUDE	COMMENTS
1556	China, Shensi	830,000		
1663	Canada, St. Lawrence River			Maximum intensity X; chimneys broken in Massachusetts
1737	India, Calcutta	300,000		
1755	Portugal, Lisbon	70,000		Great tsunami
1783	Italy, Calabria	50,000		
1811	Missouri, New Madrid	Several		Intensity XI; 2 major aftershocks in early 1812
1857	California, Ft. Tejon			San Andreas Fault rupture; Intensity X–XI
1906	California, San Francisco	700	8.25	San Francisco fire
1920	China, Kansu	180,000	8.5	
1923	Japan, Kwanto	143,000	8.2	Great Tokyo fire
1939	Turkey, Erzincan	23,000	8.0	
1950	India, Assam	1526	8.6	Surface faulting
1960	Southern Chile	5700	8.5	
1964	Alaska, Anchorage	131	8.6	Damaging tsunami
1976	China, Tangshan	About 250,000	7.6	Great economic damage; also perhaps 500,000 injured
1985	Mexico, Michoacán	9500	7.9	More than $3 billion damage; 30,000 injured
1988	Armenia, Spitak	25,000	7.0	13,000 injured; 500,000 homeless; severe damage
1989	California, Loma Prieta	63	7.0	$5.6 billion damage
1990	Iran	Above 40,000	7.7	400,000 homeless; extensive landslides
1993	India, Killari	28,000	6.4	Severe damage to loose masonry buildings

Source: Bruce Bolt, *Earthquakes* (San Francisco: Freeman, 1993.)

the failure took place, shown in Figure 9.20, is on artificially filled land on top of San Francisco Bay mud. Undamaged sections to the north and south were built on older, more stable stream sediments. To test the hypothesis that wave amplification was to blame, geologists set up seismic instruments at points 1, 2, and 3 to record aftershocks. The inset to Figure 9.20 compares records of a single shock measured at all three locations. You can see that on the filled land the shock caused horizontal movements that were eight times more intense than those on solid rock. Similar amplification during the main shock was apparently more than the freeway structure could withstand.

Liquefaction

Adding to the hazard of ground shaking is that intense shaking can turn water-saturated surface material into a fluidlike quicksand. This process of **liquefaction** weakens the material so much that even an earthquake-resistant structure built on top of it can topple. Liquefaction was a major cause of property damage during the Good Friday earthquake in Anchorage, Alaska, in 1964. The magnitude 8.6 disturbance under Prince William Sound was felt over a 1.3 million square mile area, causing $86 million in damages and killing 15 people. The Bootlegger Cove Clay,

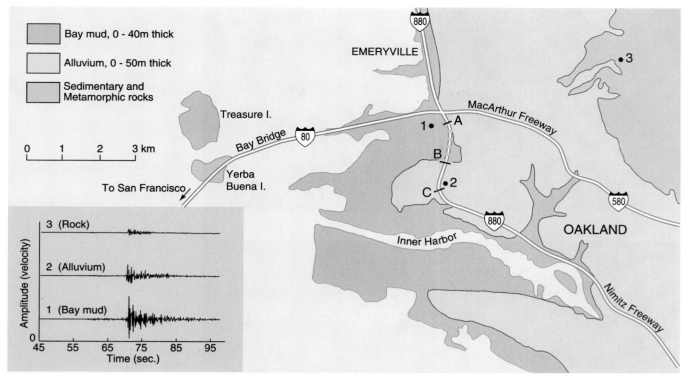

FIGURE 9.20 Map of Oakland, California, showing the surficial geology and the location of the Nimitz Freeway, which was damaged during the 1989 Loma Prieta earthquake. Segment A–B, which was built on uncompacted bay mud, collapsed. Segment B–C, built on firmer sediment, was damaged but did not collapse. The intensity of shaking varied greatly from one kind of material to another, as shown in the record of an aftershock measured by seismographs at points 1, 2, and 3 (see inset). *(Illustration derived from Figure 1 and Figure 2 in "Did Mud Contribute to Freeway Collapse?" EOS, Trans. American Geophysical Union, v. 70, n.)*

which underlies much of the waterfront district in Anchorage, lost virtually all of its strength. Buildings settled unevenly in the liquefied clay, and several major landslides caused further destruction (see Figure 9.21).

Landslides

Landslides can be a significant hazard even without liquefaction. Slopes that have been oversteepened by erosion or by human intervention can collapse with devastating effect. On December 16, 1920, a magnitude-8.5 earthquake in the remote province of Kansu in southwestern China was recorded on instruments around the world. As news filtered out of the region three months later, it contained stories of incredible landslides that had killed 180,000 people. Communities at the base of steep cliffs had been destroyed as the earthquake loosened great masses of rock and soil above them.

Fire

Fire is another major hazard during earthquakes. As ground shaking liquefies loose sediment, it is not uncommon for gas mains and water pipes to rise to the surface and rupture. Fires ignite easily and then spread as firefighters find that water pressure is too low to put out the blaze. Confusion and rubble in the streets also make it difficult for firefighting equipment to get to where it is needed. The San Francisco earthquake of 1906 is remembered by many as the San Francisco fire because most of the 700 deaths and much of the $400 million property damage were caused by fires during the days *after* the quake. Fires consumed 508 city blocks in three days while water poured from ruptured pipes beneath the streets. Modern construction and fire-prevention methods have diminished the risk, but fire remains a major worry in earthquake-prone regions.

Seismic Sea Waves

Finally, coastal communities face the risk of **seismic sea waves**, also known by the Japanese name **tsunami** (spelled the same way as either a singular or a plural noun). The most damaging waves have usually plagued the rim of the Pacific Ocean, but tsunami can occur in any large body of water. They can be caused by submarine landslides or volcanic eruptions, but most often they occur when an earth-

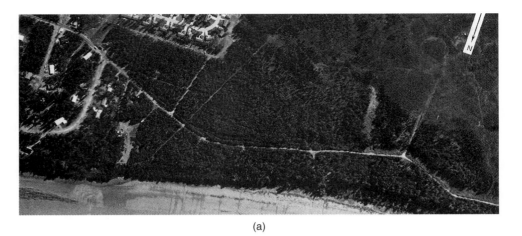

(a)

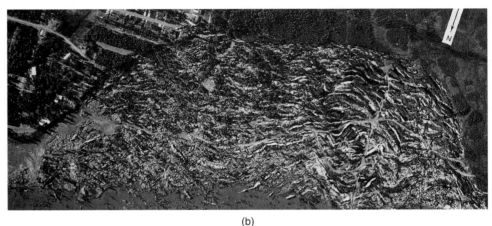

(b)

FIGURE 9.21
Turnagain Heights, Anchorage, Alaska. (a) Photograph taken on August 12, 1961. (b) Photograph taken on July 26, 1964. Bootlegger Cove clay, which has little strength when wet, underlies much of Anchorage. Earthquake vibrations liquefied the clay and triggered multiple slides, covering an area about 2,400 m long and up to 400 m wide. *(ESSA, U.S. Coast and Geodetic Survey)*

quake drops or lifts a segment of the ocean floor abruptly. As the wave that is produced travels across the open ocean, its crests may be as much as 100 km apart and no more than a meter high. As it enters shallow water, though, the wave slows and its height increases dramatically. Coming onshore, the tsunami appears as a wall of water that crushes everything in its path (Figure 9.22).

Crescent City, on the north coast of California, suffered $7.5 million worth of damage when a series of tsunami struck after Alaska's Good Friday earthquake in 1964. People were evacuated an hour before the first

FIGURE 9.22   A seismic sea wave (tsunami) is generated when the ocean floor is abruptly lifted up or dropped along a fault as the result of an earthquake. In the open ocean, the wave is very broad and not unusually high; it would not be noticed by a ship at sea. As it approaches land, however, the wave is slowed by friction with the seafloor and becomes extremely high. *(From: UNDERSTANDING EARTH by Press and Siever. Copyright © 1994 by W.H. Freeman and Company. Used with permission.)*

Waves become taller as they approach the shoreline

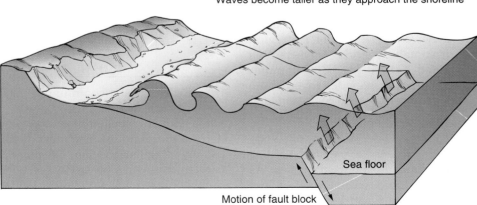

Sea floor

Motion of fault block

tsunami wave hit, but, thinking it safe, several residents returned to the area only to be overcome by later waves. Elsewhere on the California coast, damage was slight. The shape of the harbor and the contour of the seafloor near Crescent City, though, had magnified the waves dramatically.

The best defense against seismic sea waves or any of the other earthquake hazards we have mentioned is public awareness of the risks. Earthquakes cannot be controlled or accurately predicted, but people can avoid injury and property damage by recognizing the danger of building on landfill or in areas where landslides are possible. Buildings can be designed to withstand the acceleration of ground shaking and to limit the spread of fire. By anticipating the destructive potential of earthquakes, we learn to survive them.

Earth Structure

We are trapped on the outside of the Earth, free to travel thousands of kilometers in any compass direction but unable to go more than a few thousand meters below the surface. Most people do not view this as a serious problem. Geologists, however, want to know what the Earth is made of and what goes on in its interior. For them, being confined to the surface is a frustrating limitation. Out of necessity, they have learned to interpret the composition and internal structure of the Earth by studying indirect clues, largely gleaned from the study of earthquakes.

Seismic Clues

One seismologist has suggested lightheartedly that using earthquake waves to describe the Earth is like trying to describe a piano by listening to the sound it makes when you throw it down a flight of stairs. In principle, seismologists use the travel times of seismic waves and their interpretation of reflected and refracted waves to deduce what the planetary interior is like. Because a wave may have been reflected or refracted many times, this is not an easy task (see Figure 9.23). Fortunately, high-speed computers are able to compare the continuous records from a large number of seismographs and overcome most of these difficulties. The goal of these computations is to calculate the probable velocities of P- and S-waves at all depths in the Earth and, from these velocities, to estimate the thickness and elasticity of each layer. The resulting diagram or set of tabulated values is known as a **velocity profile**. Seismologists believe that they now have calculated velocity profiles that are 99 percent correct.

Figure 9.24 is a cross section of the Earth, drawn from a seismologist's perspective. The two colored lines are graphical versions of P- and S-seismic veloc-

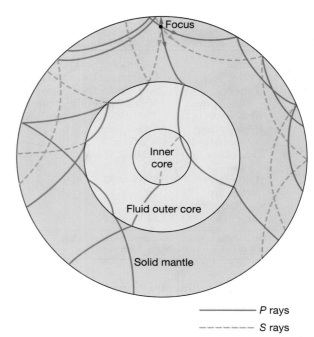

FIGURE 9.23   Body waves traveling from the focus of an earthquake may be reflected or refracted several times by the time they are detected at a distant seismograph. It is possible to determine which path a wave has followed only if we compare the records of a quake from a large number of stations. This is a job for large computer programs. *(From: EARTHQUAKES by Bolt. Copyright © 1993 by W.H. Freeman and Company. Used with permission.)*

FIGURE 9.24   Variations in seismic wave velocity with depth in the Earth. On the basis of changes in wave velocity, seismologists have defined eight distinct layers or "shells." These are labeled and described in Table 9.3.

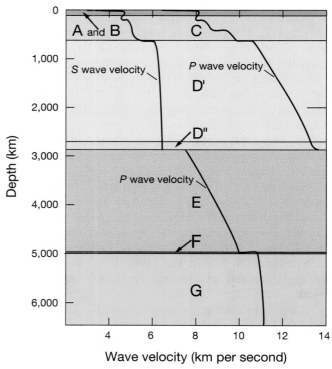

TABLE 9.3 Seismic Shells Within the Earth

SHELL	NAME	DEPTH (KM)
A	Crust	0–5 under the oceans 0–40 under the continents
B	Lower lithosphere	5–60 under the oceans 40–100 under the continents
C	{ Asthenosphere { Mesosphere	100–400 400–670
D′	Lower mantle	670–2780
D″	Transition shell	2780–2885
E	Outer core	2885–4590
F	Transition shell	4590–5155
G	Inner core	5155–6371

Note: The A and B shells, taken together, are called the *lithosphere*. The B and C shells, taken together, are called the *upper mantle*.

ity profiles. Working downward from the surface, seismologists can recognize eight distinct layers or "shells," some of which have less distinct zones within them. These are summarized in Table 9.3.

What Is in the Crust?

In 1909, Croatian seismologist Andreiji Mohorovičić made a dramatic discovery. On a recording of tremors from a shallow earthquake in the Kulpa Valley in Croatia, 800 km away, he found that there were two distinct sets of P- and S-waves. Mohorovičić was among the first generation of scientists to use modern, sensitive seismographs, and he made his discovery within a decade after the existence of body waves was first explained. He hypothesized that one set of waves had traveled directly from the earthquake focus, while the other had followed a longer path through the upper mantle (Figure 9.25). Because the seismic velocity in the upper mantle was evidently much greater than in the crust, waves traveling by the longer path had actually arrived first. Using the techniques described earlier in this chapter, Mohorovičić was able to locate a sharp seismic boundary several dozen kilometers below the surface. Above the boundary, rocks have an average density between 2.6 and 3.1 g/cm³; below it, density ranges from 3.3 to 3.6 g/cm³. Today, the boundary is known as the **Mohorovičić discontinuity** or the **Moho**. It is the base of the A layer: the crust. The depth to the Moho varies. Under the ocean basins, it averages 5 km; under the continents it ranges from 20 km in places like the Great Valley of California to almost 80 km in the Himalayas of Asia. The average thickness of the crust in North America is 33 km.

Wave velocities in the crust also vary. Early in the twentieth century, before the technology for sampling rocks in the floor of the deep oceans was developed,

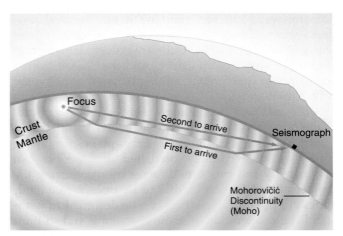

FIGURE 9.25   A. Mohorovičić discovered that two P-waves arrived at his seismograph following the Kulpa Valley earthquake of 1909. He deduced that one of these had traveled a direct path from the focus, while the other had gone into the upper mantle, been refracted twice, and returned to the surface. Because the P-wave velocity in the mantle is much higher than in the crust, the wave that followed the longer path actually arrived before the one that traveled directly. By comparing their arrival times, Mohorovičić was able to locate the base of the crust, now called the Mohorovičić discontinuity in his honor.

seismologists used this information to verify what some geologists had suspected but not been able to prove: that the oceanic crust is made of a different rock than the continental crust. Seismic velocities in continental areas are similar to what might be expected of waves traveling in a granite, a diorite, or a gneiss. Because these rocks are relatively rich in *si*licon and *al*uminum, geologists sometimes refer to continental crust by the name **sial**. Body waves move faster in oceanic crust, where basalt is the dominant rock type. Geologists use the term **sima**, derived from the fact that basalts are *si*licon- and *ma*gnesium-rich, to describe oceanic crust.

We will look much more closely at the crust and upper mantle in Chapters 10, 11, and 12, where the focus will be on changes associated with plate tectonics.

What Is in the Mantle?

Whether you consider its mass or its volume, the crust is a negligible fraction of the Earth. If a geologist could reach into the Earth and pick samples at random, most of them would come from the mantle, which has about 83 percent of the planet's volume and two thirds of its mass. What geologists have learned about the mantle tells them that it is much more uniform than the crust. On the other hand, because they cannot visit the mantle to sample it directly, geologists find it much more difficult to describe.

Seismic waves help geologists to come up with a description. The P-waves leap suddenly from 6–7 km/sec to over 8 km/sec as they cross the Moho and increase speed to 8.3 km/sec at a depth of 100 km. They slow again to slightly under 8 km/sec within a **low-velocity zone** that we have already identified as the asthenosphere and then increase to over 13.5 km/sec by the time they reach the core-mantle boundary. The S-waves follow a similar trend, varying from about 4.5 to nearly 7.0 km/sec. Seismologists agree that the probable errors in these velocity profiles are small. There are great differences of opinion, however, about how to interpret them, especially as they pertain to the deepest parts of the mantle.

Laboratory experiments to determine how rapidly sound waves travel through various kinds of rock have provided some insights. In earlier chapters we indicated that basaltic magma, which is produced in the upper mantle by partial melting, is derived from peridotite, a dark-colored rock consisting mainly of olivine and lesser amounts of pyroxene and garnet (Figure 9.26). Seismic experiments verify that peridotite samples in the lab and rocks in the uppermost mantle share the same "fingerprint" of seismic velocities. Geologists are fairly sure what the uppermost mantle is made of, therefore. The fact that S-waves slow down but do not vanish in the asthenosphere confirms that it can be no more than 2 or 3 percent liquid. The only important difference between the lower lithosphere and the asthenosphere is mechanical, not chemical. Both are made of peridotite, but the lower lithosphere is rigid whereas the asthenosphere is not.

Geologists actually have several samples from the upper mantle. Some of these have come from rare igneous bodies known as **kimberlite pipes** (Figure 9.27), mined as a source of diamonds in South Africa and many other parts of the world. Diamonds form only under extremely high pressures. This suggests that kimberlites erupted violently from the upper mantle. Other rare samples of the upper mantle may be found in **ophiolite complexes,** thought to be slabs of ancient oceanic crust and underlying mantle that have been shoved onto the continental crust. The rocks in both of these unusual settings are varieties of peridotite.

Below the low-velocity zone, geologists are much less certain what the mantle is made of. Seismic information from these deeper layers can be interpreted in more than one way. At the base of the asthenosphere, for example, seismic waves accelerate quickly. Geologists have suggested two explanations for this phenomenon. In laboratory studies, olivine collapses into a denser form known as β-*spinel* when it is placed under pressures and temperatures simulating the mesosphere. These experiments suggest that the velocity increase is due to a polymorphic change in one of the major minerals in peridotite, not a change in the chemical composition of the mantle. Other experiments, however, suggest that rocks below 400 km are not peridotite at all, but a dense form of eclogite, a partial melting product that might accumulate at that depth if conditions were right. Peridotite and eclogite each allow for the wave velocities measured below 400 km. Consequently, a great difference of opinion

FIGURE 9.26
The upper mantle is made of peridotite, an ultramafic igneous rock that contains olivine and pyroxene (green) and garnet (red). This sample is from a kimberlite nodule. *(Martin G. Miller)*

FIGURE 9.27
Mining in this kimberlite pipe at Kimberly, South Africa, produced over 14.5 million carats of diamonds between 1870 and 1914. *(Tad Nichols, Peter Kresan Photography)*

exists about which of these two explanations is correct.

Although geologists disagree about the composition of the mesosphere, there are fewer disputes about the lower mantle below it. This is largely because geologists have no direct samples and few relevant lab results that might allow them to speculate in detail. The transition at the base of the mesosphere, at about 670 km, is marked by another rapid increase in seismic velocities. During the 1980s, research teams demonstrated that olivine goes through another polymorphic change under pressures about 300,000 times greater than atmospheric pressure (or about 4.4 million lb/sq in). *Silicate perovskite,* the highly compressed mineral that replaces the β-spinel form of olivine in this transition, is now thought to be the most common mineral in the lower mantle. If so, because the lower mantle contains 54 percent of the Earth's volume, then silicate perovskite is also probably the most common mineral in the Earth.

What Is in the Core?

The Earth's magnetic field, as we discussed in Chapter 7, is generated within the core. As early as the seventeenth century, scientists deduced from this that the core probably contained iron metal. Further evidence was gathered in the nineteenth century as scientists began to study iron meteorites (actually about 92 percent iron and 8 percent nickel), which they hypothesized were fragments from the core of a shattered planet. Metallurgists in this century have confirmed that unusual textures in these meteorites might be expected to form in the cores of some asteroids (Figure 9.28). It is reasonable to suppose that the Earth's core is similar, although geophysicists have determined since 1950 that seismic velocities in the core do not quite match any simple iron-nickel alloy. They now agree that lighter elements (probably silicon, but perhaps oxygen, or sulfur as well) make up as much as 20 percent of the mixture.

FIGURE 9.28   Meteorites provide clues about the structure and history of our planet. The irons, in particular, give geologists a glimpse of what the core of the Earth might look like. This characteristic appearance is the result of intergrowths of two different minerals, each composed of iron and nickel but with different crystalline structures. *(University of New Mexico)*

In 1906, R.D. Oldham of the Geological Survey of India "discovered" the core by plotting travel-time curves for P- and S-waves from distant earthquakes. Within a decade, German seismologist Beno Gutenberg verified and improved upon Oldham's work by calculating that the depth to the E shell was 2,900 km—within 15 km of the value accepted today. In his honor, the core-mantle boundary is sometimes called the **Gutenberg discontinuity**.

Oldham and Gutenberg found that P- and S-waves from an earthquake could be recorded at all stations up to 103° away from the epicenter. In the example shown in Figure 9.29a, the travel times recorded in Nairobi, Kenya (longitude 36°E), and in the Maldive Islands (73°E) clearly fall on smooth, continuous curves. Beyond 103°E, both P- and S-waves disappear abruptly. Waves that penetrate deeply enough to reach the core-mantle boundary are slowed and therefore refracted into the core (see Figure 9.29b). They appear on seismographs again beyond 143°E (in the western region of Papua New Guinea). The gap between 103° and 143°, known as the **shadow zone**,

FIGURE 9.29   (a) If an earthquake occurred on the equator at 0° longitude, it would be recorded on seismographs as far away as Singapore, roughly 103°E. No station from there to 143°E, within the 40°-wide shadow zone, would record the event. Beyond 143°E, only P-waves would appear, and those would arrive later than expected. (b) The reason is apparent in this cutaway view of the Earth. Rays traveling less than 103° remain in the mantle. Those headed toward points beyond 103° must pass through the core. As they do, they are refracted downward because waves travel slower in the liquid outer core. When they finally reach the surface, they are more than 143° from where they started.

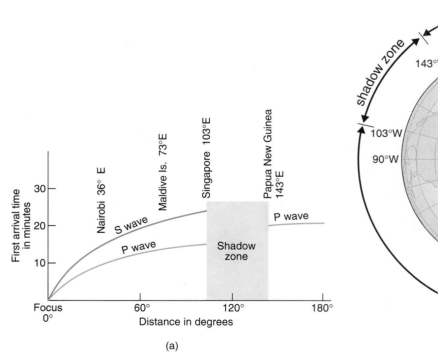

(a)

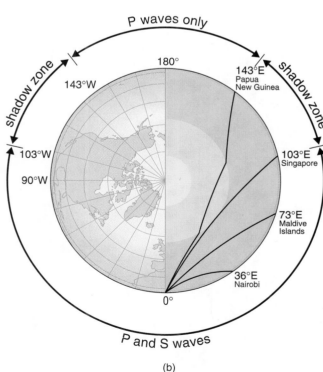

(b)

tells seismologists where the core-mantle boundary is. At 143° from the epicenter, only P-waves reappear on seismic records, and they have travel times that do not fall on an extension of the curve for less distant stations. These P-waves must have been delayed, and their companion S-waves eliminated, by passing through a liquid core.

The final step in determining the gross structure of the core came in 1936. The clue, again, was in the shadow zone. Danish seismologist Inge Lehman became curious when she recorded slow, faint P-waves from several earthquakes in the shadow zone. She hypothesized that these had traveled through the outer core and been refracted by shallow passage through the solid inner core (Figure 9.30). Later calculations by Gutenberg and by English seismologist Harold Jeffreys verified her work, but showed that she had overestimated the radius of the inner core by about 20 percent. The inner core is probably chemically similar to the outer core. It is solid only because it is under extreme pressure.

In the 1980s, seismologists began to map the core-mantle boundary in detail with a technique called **seismic tomography** (from the Greek word *tomos*, meaning "section"), by analogy with the medical tool of computerized axial tomography (CAT scan). Just as the medical technique builds up three-dimensional images of the human body from X-ray information,

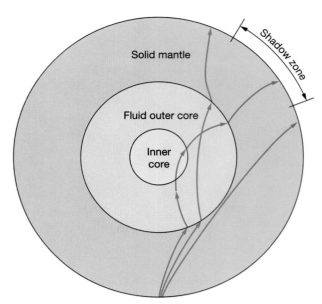

FIGURE 9.30   Because a P-wave speeds up as it enters the inner core, it is refracted upward. A ray crossing into the inner core at a low angle, therefore, shortly reenters the outer core and eventually reaches the surface within the shadow zone. Seismologist Inge Lehman observed these unusual P-waves for the first time in 1936 and correctly inferred the existence of the inner core.

seismic tomography creates three-dimensional views of the Earth's interior from seismic data (see Figure 9.31).

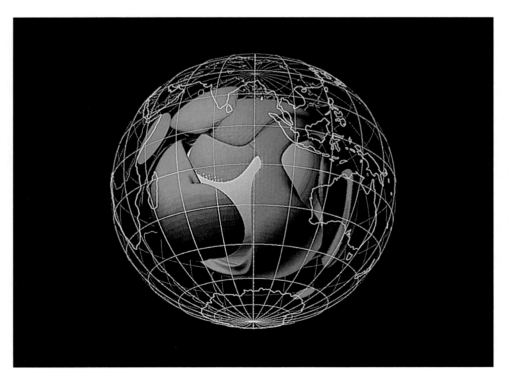

FIGURE 9.31
Seismic velocities vary laterally as well as vertically within the Earth, although it was not possible to evaluate the three-dimensional nature of these variations until computer tomography methods were developed. In this view of the Earth, centered on the Indian Ocean, regions of relatively low seismic velocity are indicated in red and those with high seismic velocity are shown in blue. Low velocity is associated with less than average rigidity and hence with warmer portions of the mantle. Computer analyses such as this one, therefore, offer a way to view the Earth's internal heat engine. *(Adam Dziewonski, Harvard University)*

As you have already seen many times in this book, geologists ask some questions about the Earth simply because they are curious about the planet beneath their feet and about how it changes over time. They ask other questions of a highly practical nature in order to make better use of the Earth's resources and to anticipate or avoid natural catastrophes. Nowhere are these parallel interests more apparent than they are in the study of earthquakes. Geologists have learned most of what they know about the Earth's interior—its structure, its physical properties, and its chemical makeup—by studying the way seismic waves travel through it. Those same studies have helped geologists to understand how and under what conditions movements occur in the Earth, bringing us closer to predicting or perhaps even controlling earthquakes.

In the next three chapters, we will show how this practical and theoretical knowledge has helped geologists to shape the theory of plate tectonics. Much of what geologists understand about the Earth's "heat engine" and the forces that raise mountains and rearrange the lithosphere has been gained from the methods and observations of seismology to which you have just been introduced.

SUMMARY

1. Earthquakes are usually caused by elastic rebound, as stress that has accumulated in the Earth is suddenly released. Movement usually takes place on faults that have been active in the past. Earthquakes typically strike the same region repeatedly as stresses build and are relieved in a stick-slip fashion. Water may enhance or even trigger earthquakes by lubricating or weakening the rocks.

2. Seismic energy may travel through the earth as compressional (P) waves or as shear (S) waves. Both types of body waves are always produced by an earthquake. The S-waves are slower than the P-waves and cannot travel through liquids. Surface waves (Love waves and Rayleigh waves) are produced when body waves reach the Earth's surface. Waves change velocity as they encounter materials with different elasticities, and can be reflected or refracted.

3. A seismograph records the movement of the Earth relative to an inertial mass that is suspended above it. Seismologists can use the delay between the first arrival of a P-wave and the later arrival of the first S-wave to find the distance to the epicenter of an earthquake. To locate the epicenter, they need distance measurements from at least three seismic stations. The strength of an earthquake may be described by the intensity of damage it causes, the magnitude of ground shaking as recorded on a seismograph, the acceleration of ground motion, or the amount of energy released.

4. Most attempts at earthquake prediction are based on a statistical approach called the *seismic gap method*, which assumes that earthquakes in a given location occur between earlier earthquakes and at roughly repeatable intervals. Other methods depend on observations of precursor events such as foreshocks and changes in land elevation, seismic wave velocities, electrical conductivity, or release of gases from deep wells.

5. The major hazards associated with earthquakes are ground shaking, liquefaction, landslides, fires, and seismic sea waves. Damage from these can be reduced by tightening standards for earthquake-resistant construction and by developing warning and disaster preparedness systems.

6. Geologists use measured changes in wave velocities to suggest areas where major changes in the composition or mechanical properties of the Earth occur. This method has allowed geologists to map the Earth's interior and to divide it into eight distinct layers, starting with the crust. The base of the crust is the Mohorovicic discontinuity, a sharp seismic boundary 5 km below the ocean floor or 30 km below the continents on average. The asthenosphere is a seismic low-velocity zone between 100 and 400 km in depth that owes its plastic nature to limited partial melting (probably 3 percent or less). The upper mantle, to a depth of at least 400 km, is made of peridotite. The lower mantle is probably composed of peridotite in which the primary minerals have been compressed to form high-pressure polymorphs. The outer core and the inner core are composed of iron-nickel metal, alloyed with 10 percent of lighter elements. Seismologists

located the core-mantle boundary by measuring the shadow zone on the Earth's surface, from which most P-waves are missing owing to refraction. The inner core boundary can be located by looking for certain P-waves that it refracts into the shadow zone.

KEY WORDS AND CONCEPTS

QUESTIONS FOR REVIEW AND THOUGHT

9.1 Why is movement along faults generally episodic instead of continuous?

9.2 What evidence do geologists have that suggests that water is important in causing earthquakes?

9.3 Why do aftershocks follow a major earthquake?

9.4 How do the two types of body waves and the two types of surface waves all differ in the way they travel?

9.5 Why can't S-waves travel through nonsolid materials?

9.6 What are the basic components of a modern seismograph?

9.7 How do seismologists usually locate the epicenter of an earthquake?

9.8 What is the difference between an earthquake's intensity and its magnitude?

9.9 In what way are measurements of the acceleration of ground shaking useful?

9.10 What is the logical basis for the seismic gap method of earthquake prediction?

9.11 How may various precursor events (changes in electrical conductivity, release of radon as in wells, etc.) be related to dilatancy?

9.12 What are the greatest hazards during an earthquake, and how might their effects be minimized?

9.13 What kinds of geophysical evidence suggest that the Earth is layered, rather than being homogeneous or, possibly, changing only gradually with depth?

9.14 What changes take place in Earth materials at each of the major seismic boundaries in the interior?

9.15 How did Andreiji Mohorovičić locate the base of the crust?

9.16 What evidence do geologists have to support their hypothesis that most of the mantle is made of peridotite?

9.17 What is the shadow zone and how can seismologists use it to locate the core-mantle boundary?

Critical Thinking

9.18 After a major earthquake, geologists commonly receive many phone calls from concerned citizens. There are always a few people who are worried about the possibility that the major earthquakes around the Pacific Ocean are triggered by underground testing of nuclear weapons. What sort of information would you look for in order to find out whether this is a legitimate worry? How might you respond to a person who raised this concern?

9.19 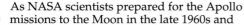 As NASA scientists prepared for the Apollo missions to the Moon in the late 1960s and early 1970s, they designed seismographs to be placed at each of the landing sites. By studying signals transmitted back to the Earth, the scientists hoped to learn more about the structure of the lunar interior. Suppose that you had been in charge of planning those missions. What potential problems might you have had to anticipate in designing seismographs, choosing where and how to place them, and interpreting survey results? In what ways are these problems the same or different from ones that seismologists face on the Earth?

10

Folded rocks in the Lost River Range, north of Borah Peak, Idaho.
(Dr. John S. Shelton)

Deformation

OBJECTIVES

▲▲

*As you read through this chapter, it may help if you focus
on the following questions:*

1. What is the difference between lithostatic and nonlithostatic stress, and how may each be applied to a rock?
2. How are stress and strain related in rocks subjected to elastic, ductile, or brittle deformation?
3. How do temperature, confining pressure, and strain rate affect deformation, and how do these factors influence the type of deformation that occurs in the crust and lithosphere?
4. What kinds of geologic events may cause jointing, faulting, or folding?
5. How do geologists describe faults and folds, and what types of field evidence do they use to interpret faults or folds that may be partly obscured?

OVERVIEW

▲▲

Whether they are on the global scale associated with plate tectonics or on a much smaller scale, movements in the crust deform rock. The earthquakes we discussed in Chapter 9 are dramatic evidence of the forces at work in the Earth and the changes they can produce. Not all deformation occurs so quickly or so visibly, however. Deep in the crust, forces stretch and squeeze rock continually as the Earth's heat engine stirs the mantle below and as erosion, sediment deposition, and other events redistribute the overlying weight on the planet's surface. Millions or even billions of years later, when those rocks are exposed to view by erosion and uplift, geologists can study the deformed rocks and read in them the record of events that shaped the crust.

This line of study is of more than a passing academic interest. Geologists have an additional economic incentive. As they search for petroleum or for mineral wealth, or work with engineers to design dams or tunnels, they use their knowledge of rock deformation to develop three-dimensional "maps" of the subsurface. Rock deformation, consequently, presents both a window on the history of the Earth and a guide to how we may use it for the benefit of humanity.

The Strength of Rock

▲▲▲

Rocks are broken or bent by the application of force. The relationship between a force and the change it produces, however, is not a simple one. Geologists recognize several different ways in which a rock's shape or size can change, depending on the amount and direction of force acting on the rock, the rate at which force is applied, and the temperature and pressure around the rock.

Stress and Strain

Carrying 20 pounds in a backpack is not a hardship for most people. Balancing a 20-pound weight on one of your toes, however, could be quite painful. The major difference between the two situations is how great an area the force is spread across. The total force is less relevant than the amount of *force per unit of area*, a quantity known as **stress.** A geologist uses a metric unit called a **bar** or another called a **pascal** (1/100,000 of a bar) to measure the stress on a rock. One bar is approximately equal to the pressure of the atmosphere at sea level (14.5 lb/sq in). About 1 km below the ground surface, the weight of overlying rock produces a stress of about 300 bars. The stresses that cause permanent change in rock are so large that we use the prefixes *kilo-* (a thousand), *mega-* (a million), or *giga-* (a billion) to describe them. Stresses in the crust are commonly measured in kilobars or megapascals.

Geologists describe two different types of stress, or "pressure." Stress is called **lithostatic** when forces of equal magnitude are applied from all directions. A good analogy would be the pressure that you feel at the bottom of a deep swimming pool. In contrast, a stress that is applied so that it has its greatest effect in a specific direction, may be thought of as a *directed* or **nonlithostatic stress.** In general, rock in the Earth's crust is subjected to both lithostatic and directed stress. To visualize this situation, imagine standing at the bottom of the swimming pool and simultaneously balancing a heavy rock on your head. The vertical force you would feel is the sum of the water pressure (a stress applied with equal intensity from all directions) and the weight of the rock (a stress that is only applied straight down).

Lithostatic stress makes a rock change size; directed stress makes it change its shape. Either change, expressed as a percentage of the rock's original dimensions, is called **strain.** If the length or the volume of a stressed rock decreases (or increases) by 10 percent, for example, a geologist would report that it has experienced a 10 percent strain.

Geologists estimate the amount of strain a rock has experienced and observe the directions in which

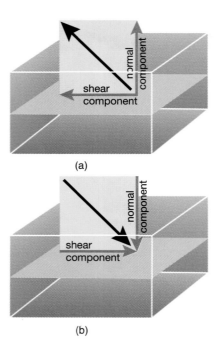

FIGURE 10.1   In general, a force applied to a rock is neither perpendicular to nor parallel to most planar surfaces in the rock, but is at some other angle. A geologist analyzing the force and its relationship to a particular plane treats it as the composite of two force components. One component, parallel to the plane, is the shear stress component. The other, perpendicular to the plane, is the normal stress component. (a) If the normal stress component pulls the rock apart, it is called a tensional stress. (b) If the normal stress component pushes the rock together, it is a compressional stress.

change has occurred. From these two pieces of information, they try to determine the amount of stress that affected the rock and, if the stress was non-lithostatic, the direction from which force was applied. The orientation of the force is always described relative to some planar surface in which geologists are interested, commonly because it represents a direction of layering or of potential weakness in the rock. As indicated in Figure 10.1, this makes it possible to consider two different components of stress: one that is perpendicular to the plane (the **normal stress** component), and the other parallel to the plane (the **shear stress** component). If a normal stress component tends to pull a rock apart across the plane, it is called a **tensional** stress. If the normal stress component tends to push the rock together, it is called a **compressional** stress.

Depending on the angle at which the force is applied to the plane in any particular situation, the normal and shear stress components vary in relative importance. In the case where normal stress dominates, the rock is either extended or compressed in the direction of the normal stress. The result is a set of fractures or other planar features perpendicular to the stress (Figure 10.2). In the case where the shear stress component is large, planar **shear zones** of bent or frac-

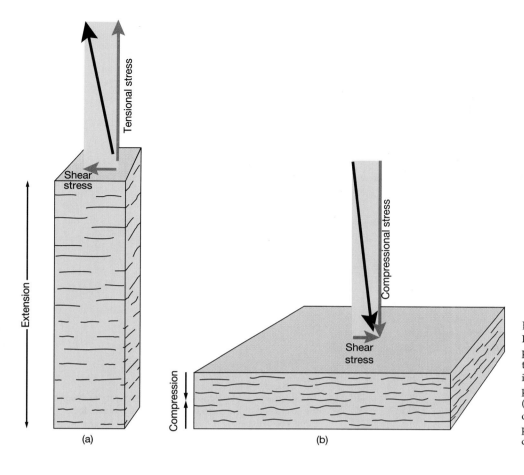

FIGURE 10.2
If most of the force applied to a plane is nearly perpendicular to it, then the normal stress component is larger than the shear stress component. The rock is (a) extended or (b) compressed, and fractures or other planar features develop perpendicular to the normal stress component.

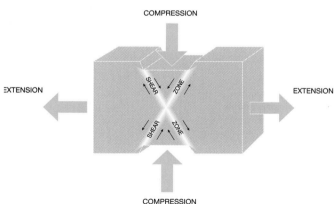

FIGURE 10.3 When the shear stress component is large, shear zones develop at an angle to the direction of extension or compression.

tured rock develop at an angle to the direction of extension or compression (see Figure 10.3).

Modes of Deformation

An **elastic** material is one in which the amount of strain increases in direct proportion to stress that is applied to the material, and that returns to its original dimensions once the stress is removed. You should already be familiar with the second of these characteristics from Chapter 9, where we showed how elasticity allows rock to transmit seismic waves. As a P-wave passes through the Earth, for example, it makes atoms move forward or backward by exerting alternating compressional and tensional stresses (see Figure 10.4). Provided that the stress caused by seismic waves is not too large, the atoms end up just where they were before the waves passed through. The deformation of the rock is temporary and therefore meets one of the two criteria for elastic **deformation.**

Geologists test for the other criterion by seeing whether the strain observed in the material is related to the applied stress in a way consistent with **Hooke's law,** which, written simply, says *stress/strain = constant.* Robert Hooke, a British physicist in the seventeenth century, was the first to describe elastic behavior quantitatively, using experiments that are easy to reproduce with a spring and a set of measured weights. One such experiment is shown in Figure 10.5.

The spring in Figure 10.5 is L_0 centimeters long. It supports a small platform. As weights are added to the platform, they produce a compressional stress and the spring becomes shorter. When one weight is placed on the platform, the spring's length is shortened to L_1; that is, the strain in the spring is equal to

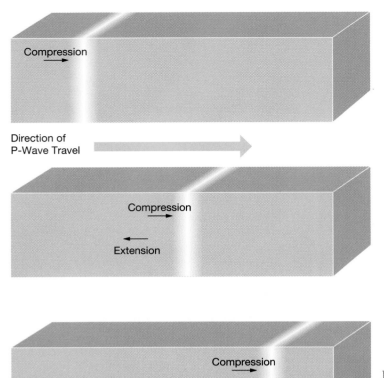

Compression

Direction of
P-Wave Travel

Compression

Extension

Compression

Extension

FIGURE 10.4
The seismic waves discussed in Chapter 9 can be described by the stresses that they produce in rock. A P-wave, for example, travels by subjecting the rock to alternating extension and compression.

FIGURE 10.5
Hooke's law can be demonstrated by measuring how much a spring is shortened as it compressed by a succession of weights. Before weight is placed on the platform, the spring in this experiment is L_0 cm long. When one weight is added, its length shrinks to L_1 cm. The strain in the spring is equal to the percentage by which it has been shortened ($[L_0 - L_1/L_0] \times 100\%$). Adding more weight shortens the spring to L_2 cm, resulting in strain equal to ($[L_0 - L_2/L_0] \times 100\%$).

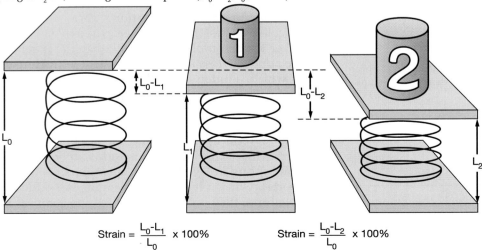

$$\text{Strain} = \frac{L_0 - L_1}{L_0} \times 100\%$$

$$\text{Strain} = \frac{L_0 - L_2}{L_0} \times 100\%$$

$100 \times (L_0 - L_1/L_0)$ percent. By adding extra weight you can verify that the amount of shortening is directly proportional to the total weight on the platform. That is, a graph of stress (weight) against strain (degree of shortening) is a straight line that is the graphical equivalent of Hooke's law (Figure 10.6). The line for a very "stiff" (rigid) spring has a steep slope; a very "soft" spring such as you might find in a small postal scale produces a flatter slope. A statement that a rock or a portion of the Earth's interior is "highly rigid" means that a large stress on it produces only a small amount of deformation (a glass marble, therefore, is "more rigid" than a tennis ball).

There is, however, a limit to the elasticity of rocks. If stress is applied beyond this limit, strain becomes permanent and is no longer proportional to stress. Either one of two things may happen. The rock may rupture, producing a **brittle fracture.** If so, geologists say that the rock has passed its **brittle limit.** The other possibility is that the rock may pass its **yield point** and flow like toothpaste squeezed from the tube, even when only small additional stresses are applied. In either case, the deformation is nonreversible; the rock is permanently broken or folded (Figure 10.7). Rocks that progress almost directly from elastic behavior to brittle fracture are said to be **brittle.** Those that change shape without breaking are **ductile.**

The **strength** of a rock, which is particularly important to know for some engineering or geophysical applications, is equal to the amount of stress that is needed to reach the rock's brittle limit or to produce substantial strain beyond the yield point. A rock has a compressional strength of 2,000 bars, for example, if it breaks or bends when a stress of 2,000 bars is applied (see Perspective 10.1).

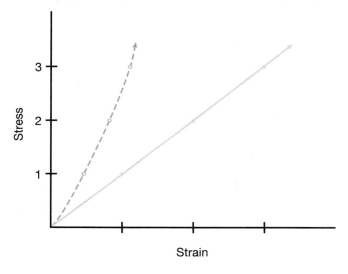

FIGURE 10.6
A graph of stress (weight) vs. strain for the experiment in Figure 10.5 yields a straight line, whose slope is a measure of the elasticity of the spring. The dashed line illustrates an experiment using a stiffer spring.

FIGURE 10.7   (a) A brittle substance fractures when stress is applied beyond its brittle limit. (b) A ductile substance reaches a yield point, beyond which strain is no longer proportional to stress. Both types of deformation are nonreversible.

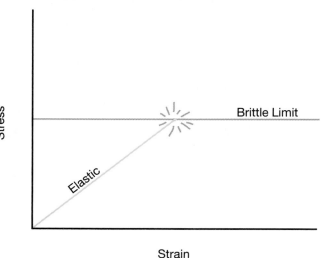

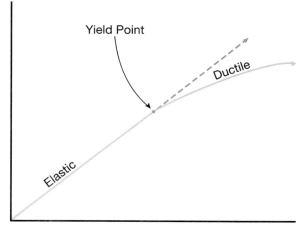

Testing the Strength of Rock

The rough cartoon in this box shows an apparatus for testing the strength of rocks that is common in many laboratories. A rock sample to be tested is cut in the shape of a cylinder and is placed in a flexible but impermeable metal jacket inside the apparatus. Pressure can be applied by opening either valve 1 or valve 2. When hydraulic fluid is pumped through valve 1, it squeezes the jacket and the rock sample toward the central axis of the cylinder; when fluid is pumped through valve 2, a piston applies a force along the cylinder's length. If the pressure is the same in both hydraulic systems, the rock is subjected to a uniform, or lithostatic, confining stress. If the pressure is greater in one system than the other, then a directed pressure results. Applying a greater pressure through valve 1 compresses the rock cylinder from a direction perpendicular to its axis; applying a greater pressure through valve 2 results in extension along the cylinder's axis (you can verify this by looking again at Figure 10.2). In either case, the difference in pressure between the two systems is increased until the rock sample

breaks or flows. The pressure at that point is recorded as the

strength of the rock (see Figure P10.1.2).

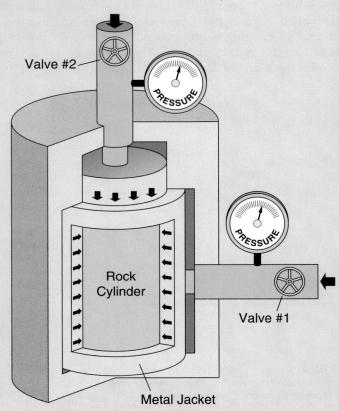

P10.1.1 Cutaway illustration of a hydraulic press used for axial stress testing. Hydraulic pressure can be applied separately through valve 1 or valve 2. If the pressure through valve 2 is greater, then the rock sample is compressed along the cylinder axis. If the pressure through valve 1 is greater, the rock sample is compressed from the sides and therefore experiences a tensional stress along the cylinder axis.

The application of excessive directed stress supplemented by low lithostatic stress in this apparatus usually causes brittle deformation. Diagonal shear zones form within the rock cylinder. The orientation of the shear zones depends on how directed stress was applied. Shear zones develop at about 30° to the cylinder axis (60° from each other) if the sample was compressed from that direction, or at 60° from the cylinder axis (120° from each other) if the sample was compressed at a right angle to the cylinder axis. When geologists see these same angles on fractured rock exposures, they can easily figure out from which direction the rocks were squeezed (see Figures P10.1.2 and P10.1.3).

Strength-testing laboratory experiments are also commonly performed on materials other than rock. Engineers, for example, estimate the strength of concrete being used for buildings, highways, or dams by testing cylindrical samples that were poured from the same concrete mixture used for construction. By analyzing the results of lab tests like these, geologists and engineers learn how materials behave under stress and how to interpret the conditions that cause these materials to fail.

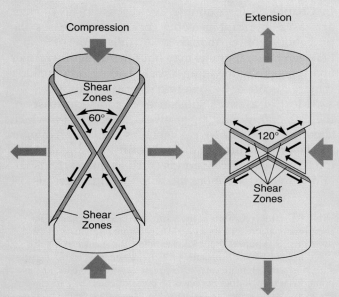

P10.1.2   When the stress on a test sample of rock finally exceeds its brittle limit, a pair of diagonal fractures forms. If the stress is compressional, the angle between the fractures is roughly 60°; if the stress is tensional, the angle is about 120°.

P10.1.3   By applying the results of experiments, geologists examining exposures of fractured rock can determine the primary direction from which they were stressed. Compare the angle between the two major sets of joints in the sandstone in the White-Inyo Mountains of southeastern California with the angles illustrated in Figure P10.1.2 (Robert J. Twiss)

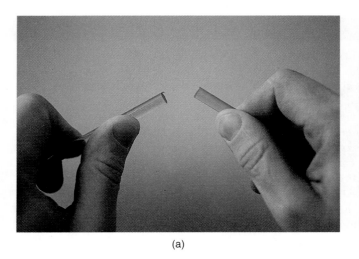
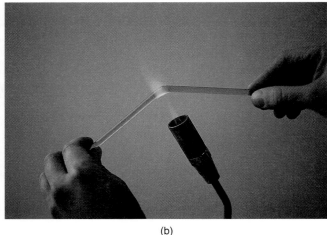

(a)	(b)

FIGURE 10.8   (a) Glass is a brittle solid at room temperature, but (b) becomes ductile when it is heated.

Moderating Influences

Several factors influence whether a rock is brittle or ductile. These include temperature, lithostatic stress, composition, and time.

Temperature   Most rocks are brittle at room temperature. Granite, for example, will crumble under a compressional stress of about 1,800 bars; limestone will fracture at about 800 bars. If either rock is heated, however, it becomes ductile when even a small stress is applied and may not become brittle under any reasonable amount of stress. You may see why heat has this effect on rock deformation if you recall the way we described crystal deformation during metamorphism in Chapter 6 (see Figure 6.7). Heat lengthens and weakens bonds between atoms, making it easier to realign a few of them at a time throughout the material instead of waiting until enough stress has accumulated to move a large number of them at once. Glass behaves this way too, as is easy to demonstrate with a glass rod and a Bunsen burner in the laboratory (Figure 10.8).

Lithostatic Stress   Lithostatic stress has a similar effect because it holds the rock together and inhibits fracturing. A granite confined by a lithostatic stress of 1,000 bars remains elastic until it receives about 4,500 bars of compressional stress, when it becomes ductile. It fractures only when compressional stress is greater than about 6,000 bars (see Figure 10.9). If lithostatic stress were higher, the granite would be ductile under an even wider range of compressional stress conditions.

Composition   A rock's chemical composition also affects its strength and style of deformation.

In Chapter 9, for example, we discussed how water acts as a lubricant in some rocks and increases the probability of earthquakes. A thin film of water between mineral grains may hold them slightly farther apart, generally lowering a rock's strength by making it more ductile or by lowering its brittle limit.

Most of a rock's substance, of course, consists of the mineral grains themselves. Minerals in which chemical bonding is weak (such as graphite, halite, or the layer-structure silicates) deform in a ductile fashion; those with strong bonds (such as quartz, feldspar, and diamond) are more likely to be brittle. Because of

FIGURE 10.9   Granite at the Earth's surface can withstand a compressional stress of about 1,800 bars before it fractures. If the same granite is buried about 3,300 m below the surface, it is subjected to 1,000 bars of lithostatic pressure. Under that confining pressure, the granite remains elastic to stresses up to 4,500 bars, its yield point. Even then, it does not fracture, but is ductile. Only when compressional stress exceeds 6,000 bars does the granite finally become brittle. As confining pressure or temperature increases, the yield point and the brittle limit both generally increase as well.

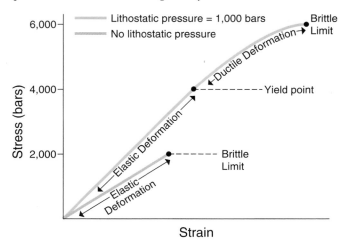

this compositional influence, limestones, marbles, shales, and schists tend to deform in a ductile fashion. Under the same conditions, most igneous rocks, sandstones, and quartzites are brittle.

Sometimes rocks with very different ductility—for example, quartzite and slate—are interlayered. When they deform, the brittle layers may fracture into segments called **boudins** (after a French word for "sausage"). The more ductile layers then flow into the space between boudins, producing a structure known as **boudinage** (Figure 10.10).

Time Because the atoms in a rock do not move easily, the rate at which deformation happens (the **strain rate**) can determine which mode of behavior dominates. If stresses are applied slowly and maintained for a long time, they are distributed through a large volume of the rock and the rock flows or folds. The same stresses applied rapidly may cause fractures because they are concentrated in a smaller volume (see Figure 10.11). The folding visible in Figure 13.26a, for example, is due to slow downhill movement under the influence of gravity. Although rocks near the Earth's surface are usually brittle, the strain rate in this instance is low enough that the rocks instead become ductile.

The effect of strain rate on rock is impossible to demonstrate in a classroom because rocks at room temperature become ductile only at very low strain rates (0.0001 percent deformation per second or slower). Mayonnaise or apple jelly, however, changes from elastic to ductile to brittle behavior within a range of strain rates that you can experiment with at home. A less messy substance that is ductile if deformed slowly but can become brittle if stretched

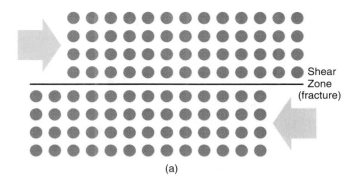

(a)

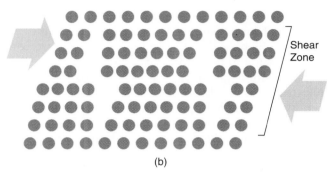

(b)

FIGURE 10.11   As a shear stress is applied to this block of material, its behavior will depend in part on how fast it changes shape. (a) If stress is applied rapidly, then the entire top of the block will move at once. A fracture will develop where the strain is greatest. (b) At a lower strain rate, a few atoms at a time detach from one portion of the crystalline structure, move slightly, and reattach to a neighboring portion. No fracture forms, but the block gradually changes shape.

quickly is the silicone polymer sold as "Silly Putty." A mass of it can be bounced like a rubber ball, and deforms into a shapeless puddle under the gentle pull of gravity, but will break if you pull it apart quickly by hand or strike it sharply with a hammer.

Stress Environments

Rocks at shallow crustal depths, where temperature and lithostatic stress are low, tend to fracture when they are squeezed or stretched too much. At greater depths in the crust or mantle both temperature and lithostatic stress are higher, and thus rock is more likely to be ductile. This difference in behavior explains in large part why most earthquake foci are within 100 km of the Earth's surface. Elastic rebound, the mechanism responsible for producing most shallow earthquakes, happens when accumulated stress finally exceeds the brittle limit of rock (recall Figure 9.2). The increased ductility found at greater depths within the Earth, however, would allow rock to deform smoothly, and no rebound would occur.

FIGURE 10.10   The lens-shaped boudins in this rock formed during metamorphism. The light-colored quartz- and feldspar-rich layers, which now are pinched into boudins, deformed in a ductile fashion. Surrounding shales were more brittle. *(Martin G. Miller)*

Evidence of Deformation

People are rarely aware of deformation in the Earth. They notice fractures that form during earthquakes, of course, and may gradually notice places where the land surface has subsided because of mining or the excessive withdrawal of fluids. Strain rates as low as 10^{-15} per second (about 1 percent strain per million years) are common in nature, however. Fracturing or folding at this rate is much too slow to detect with even the most delicate instruments. Geologists look instead for physical evidence of past deformation.

The most obvious evidence is in rocks that are tilted, bent, or broken. In Chapter 7, for example, we discussed Steno's observation that sediments are usually deposited in horizontal layers. If we find an exposure of sedimentary rock that is not horizontal, some deformation must have occurred (Figure 10.12). One step toward describing the deformation in such a case would be to measure how much the rock's bedding planes have been tilted, and in which direction. Geologists make three measurements for this purpose, illustrated in Figure 10.13. The first of these determines the **angle of dip** of the bedding planes (that is, the angle at which the bedding planes are inclined downward from the horizontal). The second measurement is a rough estimate of the **direction of dip,** the compass direction that a drop of water or a rolling marble would follow down the inclined bedding plane. If the direction of dip could be determined precisely, there would be no need for a third measurement because the angle and direction of dip would leave no uncertainty about the bedding plane's orientation. In practice, however, it is difficult to measure the direction of dip with precision, so geologists measure instead the compass direction of a horizontal line drawn on the bedding plane. That direction, called the **strike** of

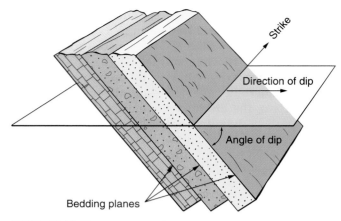

FIGURE 10.13 The *angle of dip* of any surface is the angle between the surface and the horizontal. The *direction of dip* is the compass direction toward which the surface is inclined. The *strike* of the surface is the compass direction of a horizontal line drawn on that surface. All horizontal lines on the surface are parallel, and thus point in the same direction. It is not sufficient to report only the strike and the angle of dip, because those same measurements will also describe a surface that dips in the opposite direction.

the plane, is much easier to determine with the standard field instrument that geologists use. The three measurements (angle of dip, direction of dip, and strike) can be made to determine the orientation not only of a deformed bedding plane but of any inclined surface.

Measurements are ultimately reported on geologic maps, as shown in Figure 10.14. The long side of the T-shaped map symbol points in the direction of strike; the short leg points in the direction of dip. The angle of dip is indicated by the number next to the symbol.

FIGURE 10.14 The T-shaped symbol in the upper left corner of this mapped area is a compact way of indicating the tilt on beds of rock where the symbol is placed. Its long edge indicates the direction of strike of the beds (approximately 30° west of north); the short leg points in the direction of dip (southwesterly); and the number next to the symbol indicates the angle of dip (14°). To record this same information in written form (as, for example, in a notebook), a geologist would write "N30W 14W."

FIGURE 10.12 These sedimentary rocks east of Alamogordo, New Mexico, were once deposited in horizontal layers. They are now tilted as a result of deformation.

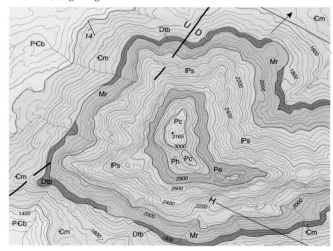

Features of Brittle Deformation

Rocks at the Earth's surface are highly fractured. Some fractures are prominent enough that they can be recognized from space (Figure 10.15). Conversely, some fractures are so small that they can be seen only with a microscope (Figure 10.16). A single cubic centimeter of rock may have millions of microfractures no more than a micron (10^{-6} cm) wide. Fractures that form as the result of shear stress are called faults; if fractures result from tensional stress, they are called **joints.**

Jointing   Joints develop perpendicular to the direction of tensional stress, as shown in Figure 10.17. They are like the fractures that form on a cracked mirror; the opposite sides of the crack move apart, but no movement takes place parallel to the fracture.

Volume change, one common source of tension, is responsible for the columnar jointing that develops in many basaltic rocks during cooling, as we first mentioned in Chapter 3 (see Figure 3.11). Hydration or other chemical reactions during rock weathering may also cause a volume change that produces joints, as is the case in spheroidal weathering, discussed in Chapter 4 (see Figure 4.10). Unloading, illustrated in Figure 4.6, is yet another example of a joint-forming process that involves a volume change.

Rocks within a broad region commonly exhibit several intersecting sets of joints that have developed as tectonic events at different times have stretched the crust in one direction or another. By

FIGURE 10.16   This sample of plagioclase feldspar, a single crystal, was used in a laboratory experiment in which it was heated to 900 degrees Celsius and subjected to shear stress. The plagioclase has deformed by producing a network of microscopic fractures, most of them no more than a thousandth of a millimeter wide. At this temperature some ductile deformation also occurs. Notice, for example, the gentle bending that is evident in the upper left corner of this image. This photograph was taken in a microscope, using cross-polarized light. *(Matthew Kramer, U.S. Department of Energy, Ames Laboratory)*

FIGURE 10.15   The eastern coast of the Mediterranean Sea, from Cairo, Egypt to Beirut, Lebanon, viewed from space. The prominent line running north-south through Israel (left and right in this photo, parallel to the coastline) is the Dead Sea fault. *(Computer-enhanced image, courtesy of NASA)*

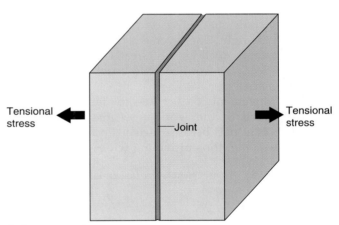

FIGURE 10.17   Joints are produced by tension. The walls of the fracture move directly away from each other.

measuring the orientations of joints across the region, geologists can deduce the pattern of stresses that generated them.

Faulting   Rock fractures along which movement parallel to the fracture occurs are called **faults.** The narrow shear zone in which the movement occurs is called the **fault plane.** Various kinds of faults are characterized by the direction that rocks move along them (Table 10.1).

Strike-slip and Transform Faults   Movement that is purely horizontal (and therefore, by definition, in the direction of the strike of the fracture) produces a

TABLE 10.1 Types of Faults

NAME	DIRECTION OF MOVEMENT	DESCRIPTION
Strike-slip (Transcurrent)	Horizontal	A fault on which the net movement is in the direction of the strike of the fault plane, with no vertical separation of flat-lying strata. A strike-slip fault may have either a *right lateral* or *left lateral* sense of displacement, depending on the relative direction in which rocks on opposite sides of the fault plane have moved.
Transform	Horizontal	A strike-slip fault that serves as a boundary between tectonic plates, linking spreading centers or subduction zones that form adjacent plate boundaries.
Normal	Dip-slip	A fault on which the net movement is in the direction of dip of the fault plane, with no movement in the direction of its strike, such that the hanging wall block drops relative to the foot wall block.
Reverse	Dip-slip	A fault on which the net movement is in the direction of dip of the fault plane, with no movement in the direction of its strike, such that the foot wall block drops relative to the hanging wall block.
Thrust	Dip-slip	A reverse fault whose fault plane has an angle of dip less than about 15 degrees.
Oblique-slip	Combined horizontal and vertical	A fault on which there is net movement in both the direction of strike and the direction of dip of the fault plane
Hinge	Rotational	A fault on which there is both strike-slip and dip-slip movement, but on which the amount of vertical displacement decreases along the direction of strike, eventually dying out at a point.

strike-slip fault, also commonly called a **transcurrent** fault (Figure 10.18). The San Andreas fault, shown in Figure 10.19, is known as a right-lateral strike-slip fault because an observer looking across the fault would conclude that rocks on the opposite side had moved to the right. The reverse is true for a left-lateral strike-slip fault, such as the Great Glen fault, which intersects Scotland from coast to coast and on which Loch Ness lies.

Movement on strike-slip faults generates much of the world's seismic activity each year. This is in part because strike-slip faults are commonly very long, thus providing widespread sites for earthquakes. Strike-slip faults may extend for thousands of kilometers, and several hundred kilometers of movement may have occurred along them. Measurements suggest that the San Andreas fault has moved 580 km since pre-Cretaceous time. Another reason for the high earthquake activity is that some strike-slip faults form the boundaries between tectonic plates, along which there is continual movement. Canadian geologist J. Tuzo Wilson, in 1965, was the first to recognize that strike-slip faults in this special class, which he called **transform** faults, act as links to reconcile competing movements on other plate boundaries. For example, Figure 10.20 illustrates a common

FIGURE 10.18   Transcurrent faults. (a) Left-lateral. (b) Right-lateral.

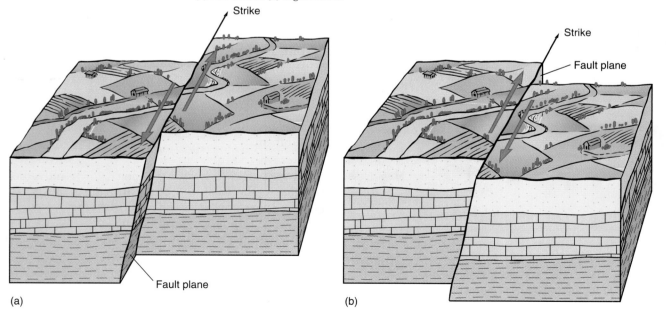

(a)   (b)

FIGURE 10.19
The scarp of the San Andreas fault in the Carrizo Plains, 200 km northwest of Los Angeles, California. Stream valleys are offset to the right where they cross the fault, most clearly near the center of this photo. This is a right-lateral transcurrent fault. *(Robert E. Wallace, U.S. Geological Survey)*

FIGURE 10.20   At first glance, movement along this large fault appears to be sliding one segment of the rift (a plate boundary) to the left relative to the adjacent segment. Assuming that the crust continues to spread apart at each rift segment, however, the only place where relative movement is *guaranteed* to occur on opposite sides of the fault plane is in the short section colored red, where blocks A and D move in opposite directions. Thus, if the crustal blocks labeled A and C were to move at the same speed toward the left, and if B and D moved at the same speed toward the right, the rift segments might not move away from each other at all. Most earthquakes will occur along this part of the fault, where it is known as a *transform fault,* and only a few elsewhere on the fault plane as one rift segment spreads at a slightly different rate than the other.

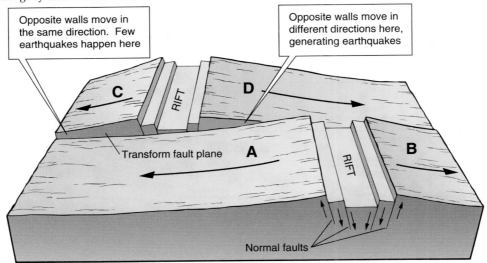

Opposite walls move in the same direction. Few earthquakes happen here

Opposite walls move in different directions here, generating earthquakes

Transform fault plane

Normal faults

setting in which transform faults link offset segments of the ocean floor that are moving in opposing directions at the mid-ocean ridges.

Dip-slip Faults   If all of the movement on a fault takes place in the direction of dip of the fault plane, it is called a **dip-slip** fault. Just as two types of movement (right-lateral and left-lateral) are possible on strike-slip faults, the same is true for dip-slip faults. Geologists describe these possibilities by using the mining terms shown in Figure 10.21. The fault plane divides rocks locally into two blocks, one above the plane (the **hanging wall block**) and the other below (the **foot wall block**). If the hanging wall block has moved down, relative to the foot wall block, the fault is called a **normal** fault (Figure 10.22). If the hanging

FIGURE 10.21   Dip-slip faults. A miner standing in a tunnel that crosses any dip-slip fault would see the rocks on one side of the fault (in the *hanging wall* block) overhead and those on the other side (in the *foot wall* block) underfoot. (a) If offset beds or other features indicated that the hanging wall block had slid down, the miner would recognize this as a normal fault. (b) If the hanging wall block had slid up, it would be a reverse fault. Referring to movement in this way avoids the confusion that might arise if the miner simply described the "left-hand" side or the "right-hand" side of the fault.

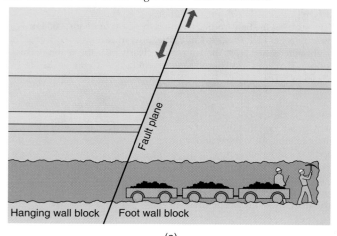

(a)

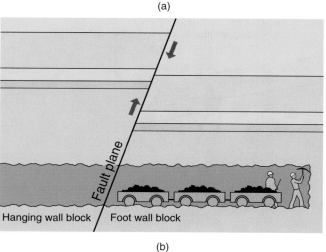

(b)

FIGURE 10.22   (a), (b), (c) Normal faults. In each case the hanging wall block has slid down, relative to the foot wall block, as indicated by offset beds.

(a)

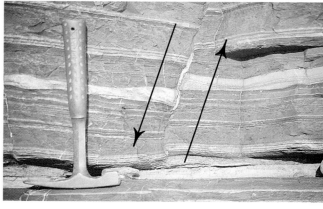

(b)

(c)

wall block has moved up, relative to the foot wall block, it is called a **reverse** fault (Figure 10.23). A reverse fault that dips less than 15° is called a **thrust** fault (Figure 10.24).

Horizontal extension in the Earth's crust commonly produces a series of related normal faults (see Figure 10.25). If a block between paral-

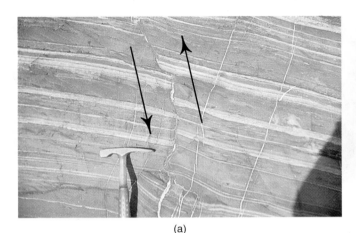

(a)

(b)

(c)

FIGURE 10.24 The Lewis thrust fault, exposed here in Glacier National Park, is a low-angle reverse fault. When movement is this great, it is impossible to recognize offset beds. The direction of movement is inferred from the relative ages of rocks above and below the fault plane. Precambrian granitic rocks in the hanging wall have slid several thousand meters up the fault plane toward the left (east) so that they now lie on top of much younger Paleozoic and Mesozoic sediments.

lel normal faults drops down, it creates an elongated, trenchlike valley called a **rift** or a **graben.** A block that is lifted up is called a **horst.** These features are most common at boundaries where lithospheric plates are moving apart. The rift along the crest of parts of the mid-ocean ridge system is one such example. The largest rift on the continents is in East Africa, where a thin strip of the crust extending from Ethiopia to Zimbabwe is moving slowly eastward relative to the rest of the continent.

By contrast, reverse faults form where the crust has been squeezed horizontally, often in association with places where lithospheric plates move toward each other. The Andes of South America and the Alps of Southern Europe are good examples of this type of region.

Other Faults Faults rarely show pure strike-slip or pure dip-slip movement. Instead, they tend to show both types of movement. The faults that result are called **oblique-slip** faults (Figure 10.26a). Geologists describe these faults by referring to their strike-slip and dip-slip characteristics separately. Because both the amount and the direction of stress, as well as the strength of rocks, vary from one place to another along a fault, it is common for both the orientation of the fault plane and the amount of displacement along it to vary. Eventually, any fault must end, either because it is abruptly terminated by another structural feature (a second fault, for example) or because it gradually dies out. A dip-slip fault on which displacement decreases perceptibly along the strike, finally disappearing at a point (Figure 10.26b), is called a **hinge fault.**

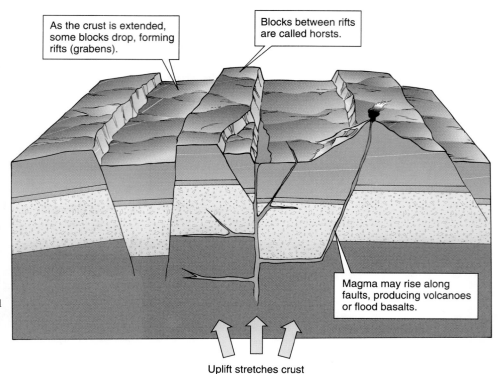

As the crust is extended, some blocks drop, forming rifts (grabens).

Blocks between rifts are called horsts.

Magma may rise along faults, producing volcanoes or flood basalts.

Uplift stretches crust

FIGURE 10.25
When the crust is stretched, as it is in this drawing by gently arching, a series of parallel normal faults can form. A valley bounded by normal faults is a *rift* or *graben*. The high land between faults is called a *horst*. Volcanic activity is common in settings like this, partly because crustal arching typically occurs in areas of high heat flow (see Chapter 8) and partly because the faults offer a pathway along which magma can reach the surface.

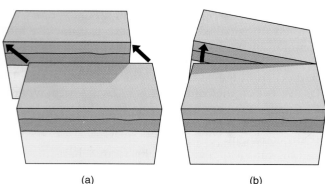

(a)   (b)

FIGURE 10.26   (a) Oblique-slip fault. Part of the movement is in a horizontal direction, and part in the direction of dip. (b) Hinge fault. The block on one side of the fault plane rotates downward, as if around a hinge. As a result, displacement on the fault is small near the hinge but increases away from it.

Evidence of Movement on Faults   The amount of movement along a fault can be determined only by measuring how far it has offset beds or other features. If the displacement is fairly small (a few hundred meters or less), measurement can be simple. If a large offset has occurred, however, or if the rocks are uniform enough that a geologist cannot identify a distinct feature that was disrupted by the fault, it may be impossible to measure displacement. In fact, the geologist may even find it difficult to determine which direction the rocks have moved, or whether movement has occurred at all. In such cases, it is helpful to look for other ways that rock has been deformed in the vicinity of the fracture.

In Chapter 9 we discussed the development of gouge, fault breccia, and slickensides along an active fault. Small **tensile fractures** may also open up as rocks are dragged and torn along a fault, as shown in Figure 10.27. The acute angle between a tensile fracture and the fault plane points in the direction of movement. Another common feature near faults is **drag folding,** which develops in much the same way that wrinkles in a carpet do if you push it across the floor. The orientation of folds on opposite sides of the fault plane can yield a clue to the direction of fault movement.

Features of Ductile Deformation

When rock is ductile, stress makes it fold or bend. **Folds** in rock, like fractures, may be microscopic or may be hundreds of kilometers wide. They may also be very open and gentle or bent as tightly as a hairpin; curved or angular; upright or inclined; symmetrical or highly asymmetrical. Folds rarely occur in isolation. Instead, they are usually found in groups or in association with fractures, as in drag folding. The size and style of folding depends on the direction and amount of stress, and on the strength of rock.

The basic features of a fold are shown in Figure 10.28. Geologists locate the center of a fold by finding either its **axial plane,** an imaginary surface that divides the fold as symmetrically as possible, or its **axis,** a line lying in the axial plane and parallel to the

FIGURE 10.27
Tensile fractures open as rocks shatter and are dragged along a fault. Arrows on this photo indicate the direction of movement.

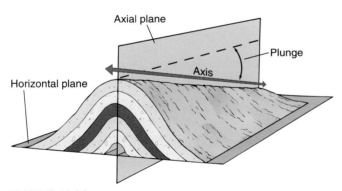

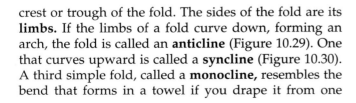

FIGURE 10.28 The parts of a fold, shown on a plunging anticline but applicable to any kind of fold. The axial plane divides the fold as symmetrically as possible. The axis is a line parallel to the crest of the fold and lying in the axial plane. If the axis is not horizontal, the fold is plunging, and the angle between the axis and the horizontal is the angle of plunge. The sides of the fold are its limbs.

Labels on figure: Axial plane, Plunge, Axis, Horizontal plane

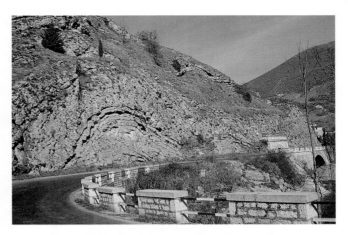

FIGURE 10.29 Anticlines may vary in appearance, as indicated by these three examples, but all arch upward at the center of the fold. *(Bottom photo provided by Peter Kresan Photography)*

crest or trough of the fold. The sides of the fold are its **limbs.** If the limbs of a fold curve down, forming an arch, the fold is called an **anticline** (Figure 10.29). One that curves upward is called a **syncline** (Figure 10.30). A third simple fold, called a **monocline,** resembles the bend that forms in a towel if you drape it from one step to the next on a staircase (Figure 10.31). A folded structure in which the rocks uniformly dip away from a central point, rather than an axial plane, is referred to as a **dome** (Figure 10.32). A similar structure in which rocks dip toward, rather than away from, a central point is called a **basin** (Figure 10.33).

(a)

(b)

(c)

FIGURE 10.30   As with the anticlinal folds in Figure 10.31, synclines may also vary in appearance. Synclines, however, bow downward in the center of the fold. *((a) Maryland Geological Survey)*

FIGURE 10.31   This ridge in southeastern Utah is an exposed monocline. Erosion has carved a toothlike pattern in resistant sandstones along the axis of the fold.

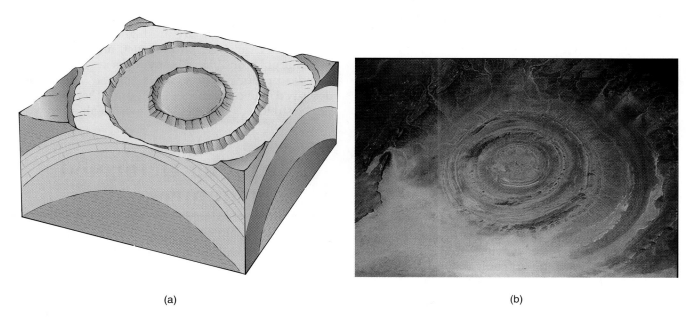

(a)

(b)

FIGURE 10.32   (a) Where horizontal rocks are arched upward so that they dip away in all directions from a central point, the result is a dome. If the beds do not erode with equal ease, they form an outcrop pattern of raised rings. (b) This pattern is particularly obvious in this satellite photo of a dome exposed in the western Sahara Desert, in Mauritania. *(b) NASA)*

FIGURE 10.33

(a) Where originally horizontal rocks sag downward, so that they dip from all directions toward a central point, the result is a basin. As with a dome, the exposed rocks form a pattern of rings as they are eroded. (b) The Michigan Basin is not obvious from a satellite, because of its great size and because it is masked by soils and vegetation. When geologists record the distribution of exposed rock types in this region on a geologic map, however, the ring pattern is clear. From strike and dip measurements and from the relative ages of rocks (younger toward the middle of the pattern), geologists infer that this is a basin.

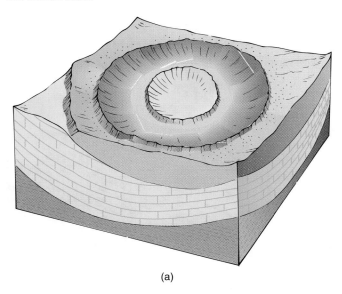

(a)

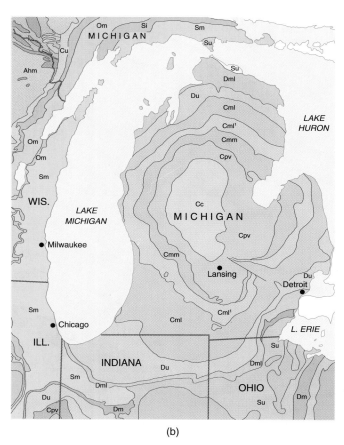

(b)

Although it is easiest to visualize folds in which the axial plane is vertical and the axis is horizontal (see Figure 10.34a), these are rare. In general, either the axis is nonhorizontal, so the fold becomes a **plunging** fold (Figure 10.28), or the axial plane is nonvertical, so the fold becomes an **asymmetrical** fold (Figure 10.34b), or both. An **overturned** fold is one in which the axial plane has been inclined so far that both limbs dip in the same direction, although not necessarily at the same angle of dip (Figure 10.34c). A **recumbent** fold has a nearly horizontal axial plane (Figure 10.34d).

Rocks are most intensely folded where lithospheric plates are forced against each other. In this setting, the stresses that cause folding in ductile rocks are roughly horizontal and are compressional. The axes of anticlines and synclines thus lie parallel to the plate boundary and perpendicular to the direction of greatest compression (Figure 10.35). Most folds in this setting are thus known as **parallel** folds. In them, originally horizontal rocks bend around the fold axis but do not change their thickness or volume. The folds in Figures 10.29 and 10.30 are parallel folds.

Parallel folding occurs where the rocks are strong enough to resist being stretched and thinned, typically in environments where the rocks are only lightly metamorphosed, if they are metamorphosed at all. Where temperature and pressure are highest, the stress pattern is less clear. Deep in the crust, under high-grade metamorphic conditions, many rocks act like very stiff liquids, producing **flow folds** instead of the ones described above. As shown in Figure 10.36, rock layers can be stretched, squeezed, and twisted almost beyond recognition.

Mapping Deformed Structures

A geologist describing the folds and faults in many areas faces a complex problem. Except where folds are small, or are exposed in a vertical excavation or valley wall, they are difficult to recognize. The limbs of some large, open folds may dip only a few degrees. Some folds are so tight that their limbs are parallel (a style known as **isoclinal** folding). In either situation, it may be hard to locate and orient their axes and axial planes. Other folds may have been truncated by faulting or erosion, or may have been refolded in multiple stages of deformation. In addition, vegetation and soil cover

FIGURE 10.34   (a) A symmetrical fold is one whose axial plane is vertical. As the axial plane of a fold tilts progressively farther to one side, any of these other forms is possible; (b) an asymmetrical fold; (c) an overturned fold; (d) a recumbent fold. *(This illustration was derived from Figure 14.22 in B.J. Skinner and S.C. Porter's* THE DYNAMIC EARTH 2/E, © *1992 John Wiley & Sons, Inc.)*

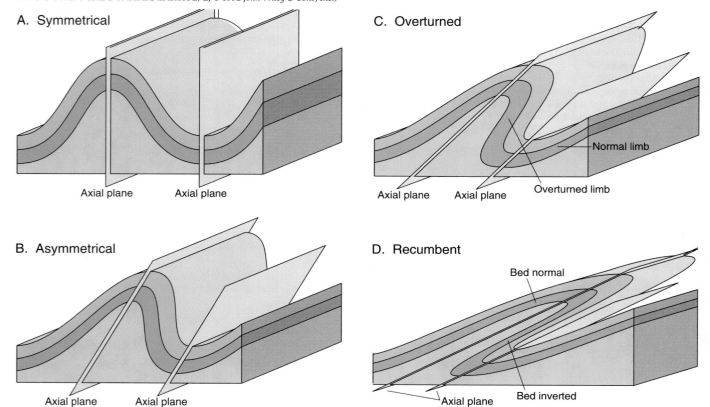

may partly obscure the rocks. Even when exposures are ideal, we can see rocks only at the Earth's surface and must imagine what is below. A geologist, therefore, must learn to interpret faults and folds from limited information.

FIGURE 10.35 Aerial view of McDonnell Range, Australia. Folded beds are late Precambrian to early Paleozoic, with east-west trending axes. Compression in a north-south direction was produced by plate collision. Width of view is about 75 km.

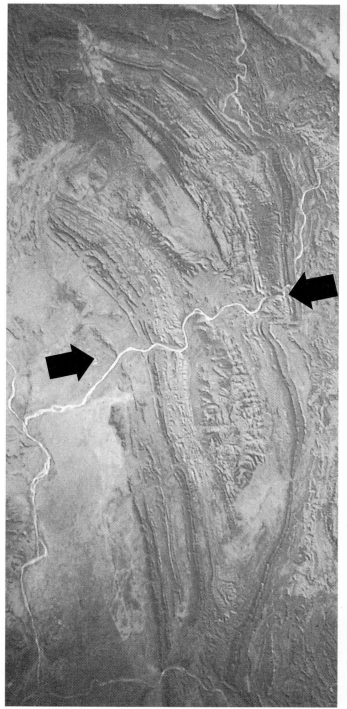

FIGURE 10.36 Flow folds in gneiss near Keene, New Hampshire. These rocks were metamorphosed at a high temperature at least 10 km below the surface. Under those conditions, the beds deformed like putty so that their original thickness and orientation are distorted into a chaotic mess.

How is it done? Largely by making many painstaking measurements of strike and dip and by recording the pattern of rock exposures over a wide geographic area. The job is easiest where sedimentary rocks predominate. Where erosion exposes the limbs of a nonplunging fold, as shown in Figure 10.37, beds of sedimentary rock form stripes parallel to the fold axis. The beds exposed near to the axial plane in an anticline are older than those farther away; those near the center of an eroded syncline are younger than rocks on the limbs. If the fold's axial plane is vertical, the sedimentary stripes are symmetrical around the

FIGURE 10.37 Nonplunging anticlines and synclines, exposed by erosion, form parallel stripes of rock. The corresponding geologic map indicates rock units by color and their orientations with strike and dip symbols (see Figure 10.15).

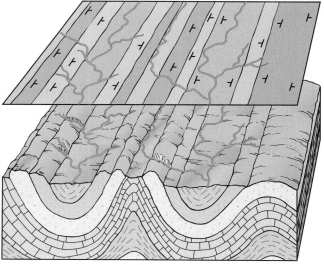

center of the fold; if the fold is inclined, the stripes are wider on one side than the other. A plunging fold, when eroded, yields an outcrop pattern of curved stripes (Figure 10.38).

Like fractures, folds may either terminate in other folds or in faults, or they may gradually die out as their limbs flatten out and become indistinguishable. A monocline, for example, may rupture to become a hinge fault (see Figure 10.39) or may subtly grade into horizontal strata. If you walk the length of many other large folds, you will find that their axes plunge toward both ends and that the fold disappears, perhaps engulfed in another deformed structure or simply grading into undeformed rocks. Such doubly plunging folds in sedimentary rocks have outcrop patterns that are elliptical (Figure 10.40).

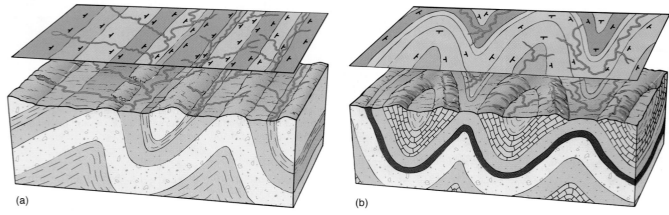

(a)

(b)

FIGURE 10.38 (a) Inclined folds, exposed by erosion, form stripes that are wider on one side of the axial plane than on the other. (b) Plunging folds form exposure patterns that are bent. Curved outcrop stripes formed by a plunging anticline point in the direction that the axis is plunging. Those for a plunging syncline point in the opposite direction.

FIGURE 10.39
A monocline may flatten out and disappear along its strike or may rupture to form a hinge fault.

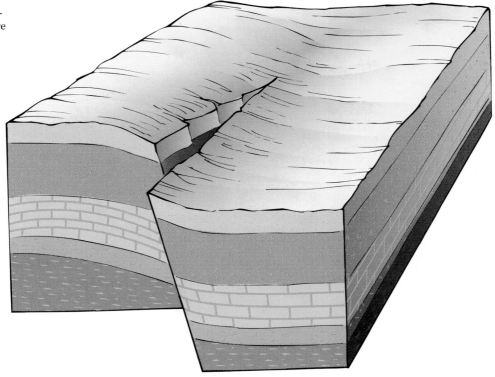

FIGURE 10.40 Sheep Mountain in Wyoming is a doubly plunging anticline. The sedimentary beds exposed here dip away from the center of the exposed fold in all directions. *(Dr. John S. Shelton)*

▶▶ EPILOGUE ◀◀

Because the crust has been deformed continuously throughout its history, it is rarely as easy to interpret subsurface faulting and folding as Figures 10.39 to 10.42 make it look. It is certainly much harder to recognize faults and folds concealed within the Earth than it is to characterize them in exposures such as the ones that have appeared in photographs throughout this chapter. Although they can learn more by studying seismic information or by drilling costly holes to sample the rocks below, geologists mapping a complex area still commonly disagree about what is under their feet. Their work is important, though, because faults and folds conceal and sometimes trap valuable petroleum, coal, copper, or other earth materials. Mining and energy companies depend on the skills and experience of geologists to find these materials.

Tunnels, building foundations, dams, and other structures need to be built where the danger of movement is minimal; thus, engineers also depend on geologists' talents. By locating faults or joints along which fluids may migrate, geologists can sometimes predict the movement of pollutants in the subsurface or anticipate places where landslides or other slope failures may occur. By visualizing complex faults and folds beneath the ground surface, geologists can thus limit some kinds of environmental damage or hazards to property and life.

Even if there were no economic or environmental incentive, though, geologists would still need to understand what the crust looks like to figure out how it changes through time. Plate tectonics, to which we will devote the next two chapters, is the story of deformation on a global scale.

SUMMARY

1. Stress applied to a rock causes strain, which is expressed as a percentage change in the rock's dimensions. Lithostatic (uniform) stress changes a rock's volume; nonlithostatic (directed) stress changes its shape. A nonlithostatic stress consists of a normal component (applied perpendicular to a plane of interest within the rock) and a shear component (applied parallel to the plane). The normal component may be either compressional or tensional.

2. When rock receives a small stress, it develops a strain proportional to the stress. When stress is removed, the rock returns to its original dimensions. This behavior is known as *elastic deformation.* Even a small amount of stress applied beyond a rock's yield point may cause a large amount of strain, as the rock changes dimensions by ductile deformation. Stress applied beyond the rock's brittle limit will cause it to fracture—a style known as *brittle deformation.* Ductile and brittle deformation are both irreversible.

3. When a rock is heated or if stress is applied to it slowly, then strain is distributed through much of the rock's volume. Under these conditions, the rock is likely to deform in a ductile deformation. High lithostatic stress also favors ductile deformation. When a rock is deformed rapidly, especially if the rock is at low temperature or under a small lithostatic stress, the stresses are likely to concentrate in a small volume of the rock, thus favoring brittle deformation. For these reasons,

brittle deformation is more likely in the shallow crust of the Earth. Ductile deformation is more common deeper in the lithosphere.

4. Joints develop perpendicular to a direction of tensional stress. They may be produced by volume change, unloading, or local stretching of the crust. Strike-slip and transform faults are produced where the direction of movement in the fault plane is horizontal. Normal faults are associated with horizontal stretching of the crust; reverse faults with horizontal shortening. In both normal and reverse faulting, blocks of rock move on the fault plane in the direction of dip. Folds are generated by compressional stress. The source of stress, and thus the style of folding, may vary from one tectonic setting to another.

5. Geologists describe a fault by reporting the relative movement of the hanging wall and foot wall blocks on opposite sides of the fault plane. Movement on faults can be determined by direct measurement of offsets, or by observing slickensides, tensile fractures, drag folds, or other local effects of faulting. Geologists describe a fold by reporting the relative movement of its limbs, and by determining the orientation of its axis and axial plane. Where folds or faults are partially obscured, their characteristics can sometimes be deduced by recording strike and dip measurements over a large area, and by studying the geographical pattern of rock exposures.

KEY WORDS AND CONCEPTS

anticline 249	dip-slip fault 244	limb 249
asymmetrical fold 252	dome 249	monocline 249
axial plane 248	drag folding 248	normal fault 246
axis 248	ductile 237	normal stress 234
bar 234	flow folding 252	oblique-slip fault 247
basin 249	hanging wall block 246	overturned fold 252
boudin (boudinage) 241	hinge fault 247	parallel folding 252
brittle 237	Hooke's law 235	pascal 234
brittle limit 237	horst 247	plunging fold 252
compression 234	isoclinal fold 252	recumbent fold 252
dip 242	joint 243	

QUESTIONS FOR REVIEW AND THOUGHT

10.1 What does a geologist mean by the terms "stress" and "strain"? What kinds of information would you need to report in a complete description of stress?

10.2 What is Hooke's law? How could you demonstrate that deformation of a substance is elastic?

10.3 How is the elastic behavior of rock related to the way seismic waves travel?

10.4 How do temperature, lithostatic stress, rock composition, and strain rate affect the strength of a rock and the manner in which it deforms?

10.5 Why do most earthquakes caused by elastic rebound occur within 100 km of the Earth's surface?

10.6 In what ways are joints and faults the same? In what ways do they differ?

10.7 What kinds of geologic events may cause jointing?

10.8 How is the orientation of a planar surface, like a fault plane, measured and reported by a geologist?

10.9 How are transcurrent and transform faults different from one another?

10.10 How do the two types of dip-slip faults differ from one another? What kinds of crustal movement are associated with each?

10.11 What kinds of evidence could you use to determine the direction and amount of movement on a fault?

10.12 How would a geologist describe the difference between an anticline and a syncline?

10.13 What types of information might a geologist need to report in describing a plunging fold or an inclined fold?

10.14 What kinds of stress are likely to produce folds?

10.15 How can a geologist describe a fold if it is only partly exposed to view?

10.16 How can a fault or a fold "disappear"?

Critical Thinking

10.17 The tilted sedimentary rocks shown in Figure 10.13 might be on the limb of a symmetrical syncline. If so, where should the fold's axial plane be? Suppose, instead, that these rocks were on the upper limb of a recumbent syncline. Where would the axial plane be then? (*Hint:* See Figure 10.34.) If you could examine the rocks closely, can you think of a way to determine which kind of fold this is?

10.18 Underground mining is particularly hazardous because of the constant danger that walls may collapse or that rocks may fall from the roof of a tunnel. For that reason miners are very careful to remove loose rock and to check for signs of weakness. Even after they've taken these precautions, however, they find that new fractures form and that rocks occasionally "pop" loose from the walls or the roof of an excavation. How would you account for this environmental hazard?

11

Trace of the Garlock Fault, looking WSW toward Garlock, California. Koehn Dry Lake is in the distance to the left; the El Paso Mountains are to the right.

(Dr. John S. Shelton)

Plate Tectonics: The Growth of an Idea

*O*BJECTIVES

▲▲

*As you read through this chapter it may help if you focus
on the following questions:*

1. What major theories did geologists use to interpret change in the lithosphere before the theory of continental drift was introduced? What questions did these theories answer? Are these theories still held today?

2. What was the basis for the theory of continental drift, and what evidence was offered in support of the theory? What were geologists' major objections to the theory?

3. How does the oceanic crust evolve according to the seafloor-spreading hypothesis? What observations eventually led geologists to formulate and accept this hypothesis?

4. In what basic way is the plate tectonic theory an improvement over previous theories?

*O*VERVIEW

▲▲▲

In the nineteenth century, chemists recognized the existence of atoms, physicists discovered how electricity and magnetism work, and biologists adopted the idea of evolution. Scientists focused their attention in new directions, making sense of old, puzzling observations and opening up unexpected lines of inquiry. Geologists made major discoveries, too, but the greatest change in their discipline took place in the twentieth century. The speculations of German meteorologist and explorer Alfred Wegener in 1912 began a 50-year debate over a theory that he called *continental drift*. The expanded and updated version of that theory, which we outlined in Chapter 1 and in the essay that introduced Chapters 8–12, is called *plate tectonics*. It has become as fundamental to geology as the atomic theory is to chemistry. Our goals in this chapter and the next are to explore the foundations of plate tectonic theory and to show why geologists find it such a useful framework for understanding changes in the lithosphere.

Early Thoughts

Over two thousand years ago, the Greek philosopher Plato wrote about the mythical continent of Atlantis that disappeared into the sea. His story may not have been the first speculation about how Earth's crust evolves, but it is certainly one that has thrived in human imagination for a long time. In the seventeenth century, the French natural philosopher René Descartes also suggested that the elevation of the Earth's crust has changed over time, although not as catastrophically as in the loss of Atlantis. Descartes hypothesized that the Earth was once much like the Sun, but had begun to solidify. The crust, which had crystallized first, was still floating on a fluid interior and becoming wrinkled as the planetary interior slowly cooled and contracted (Figure 11.1). This is an idea that geologists still use today to explain wrinkled ridges on the surface of the planet Mercury.

Before 1800, many writers and philosophers came up with ideas that in some way echoed what Plato or Descartes had proposed. For reasons that seemed more obvious to some theorists than others, the continents and the seafloor were believed to move up and down, creating and destroying mountains through time. When James Hutton set the stage for modern geology in 1795 by writing his revolutionary ideas about uniformitarianism and the age of the Earth, he accepted this premise of vertical movement without question, and it became part of the conventional wisdom of geology. Evidence of vertical movement in the past, in the form of such things as marine fossils found on mountaintops, was plentiful. Until the 1850s, however, most ideas about what *causes* the crust to rise and fall were based more on faith and abstract reasoning than on solid observations.

Geosynclines

In 1859 James Hall, geologist for the State of New York, found ripple marks and marine fossils at all stratigraphic levels in the northern Appalachian Mountains. These suggested to Hall that the entire 12-km-thick pile of sediments had been deposited in shallow water. He noticed further that the same rock layers gradually thinned and became less deformed to the west of the mountains. Hall hypothesized that sediments that later became the Appalachian Mountains had been deposited in a long, narrow basin, which gradually deepened under the weight of sand and mud as it filled. One of Hall's contemporaries, Professor James Dana of Yale University, named this trough-like basin a **geosyncline.** The sediment was derived from a landmass to the east, where the Atlantic Ocean sits today. Once it was deeply buried, the sediment was heated and deformed, then finally raised as a mountain chain. The mountain chain took its shape from the earlier basin, and was confined to a continental margin because that is where shallow marine sediments are deposited. Hall's idea explained why mountains generally lie at the edges of continents, and why the mountains on one continent might be older than those on another, depending on their progress in the cycle of events (see Figure 11.2).

The idea answered some puzzling questions about mountains, and it fit with the accepted notion that continents and ocean basins rise and fall slowly in place. Dana, however, was quick to point out that water-saturated sediments are not dense enough to warp the crust downward. Furthermore, he noted, Hall's hypothesis accounted for the sedimentary pattern but didn't explain what made the geosynclinal rocks buckle upward again. Dana proposed an explanation that recalled Descartes' 200-year-old theory. As the Earth cooled and contracted, he suggested, the ocean crust buckled against the continents (Figure 11.3), raising a narrow offshore landmass (which he

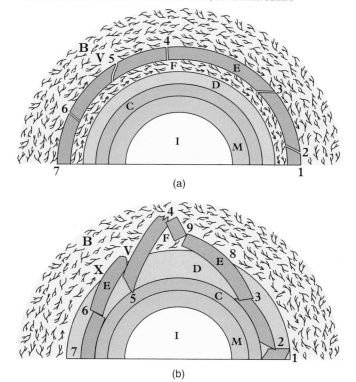

FIGURE 11.1 René Descartes used these two illustrations in 1644 to explain his theory that mountains formed as a result of contraction of the Earth. (a) In his view, the crust solidified early in the Earth's history, trapping a layer of fluid beneath it. (b) As the fluids solidified or escaped through fractures in the Earth's surface, the crust contracted and wrinkled, raising mountains. *(The Granger Collection, from René Descartes [1644] (1984)* Principles of Philosophy *translated by Valentine R. Miller and Reese P. Miller. Dordrecht, Netherlands: Reidel.)*

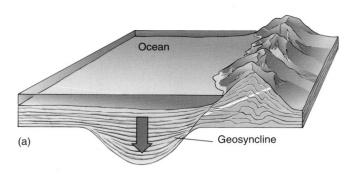

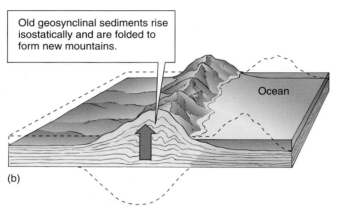

Old geosynclinal sediments rise isostatically and are folded to form new mountains.

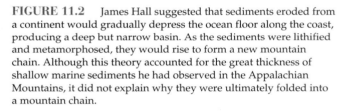

FIGURE 11.2 James Hall suggested that sediments eroded from a continent would gradually depress the ocean floor along the coast, producing a deep but narrow basin. As the sediments were lithified and metamorphosed, they would rise to form a new mountain chain. Although this theory accounted for the great thickness of shallow marine sediments he had observed in the Appalachian Mountains, it did not explain why they were ultimately folded into a mountain chain.

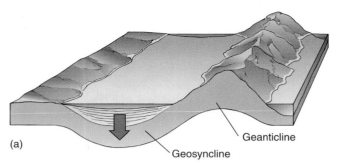

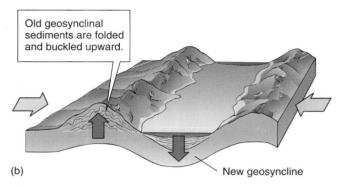

Old geosynclinal sediments are folded and buckled upward.

FIGURE 11.3 James Dana theorized that Hall's sedimentary troughs, which he called *geosynclines*, were produced because the Earth is shrinking. As it shrinks, the thin ocean crust wrinkles to form a geosyncline and a parallel uplifted area, known as a *geanticline*. Contraction continues, trapping sediments in the geosyncline between the continent and the geanticline. They squeeze into folds and rise to form mountains.

called a **geanticline**) and causing the geosyncline to warp downward. Continued contraction would squeeze the geosynclinal sediments between the geanticline and the stable platform of the continent, forcing them to rise as a folded range of mountains. Dana's theory, with further modifications, became doctrine for nearly a century. By the time it was finally displaced by the plate tectonic theory in the 1960s, the geosyncline theory had become very elaborate. It formed the foundation for many other ideas, some of which vanished along with the geosyncline theory, but some of which are still central to geologists' interpretations of change in the crust. We will return to consider why the geosynclinal theory finally collapsed, but first we will examine a second major idea that was more successful in the long run.

Isostasy

The growth of mountains from geosynclines suggested to nineteenth-century geologists that the crust as a whole must be very flexible. It also piqued their curiosity about what lay below the crust. Recall from Chapter 9 that the Mohorovičić Discontinuity (Moho) was not discovered until 1909, and thus earlier geologists could only speculate about how thick the crust was and what the mantle was like. A particularly useful idea was proposed in 1889 by American geologist Clarence Dutton. Using a present-day analogy, if we could stand back and watch the crust slowly rising and falling over tens of millions of years, it would look sort of like a "waterbed" with some regions rising at any given time to compensate for others that are depressed. Dutton referred to this idea that the crust is in dynamic balance with the mantle as **isostasy** (from the Greek *isostasios*, "in balance with").

Dutton's theory was based on earlier explanations for a puzzling set of small errors that kept appearing when surveyors made maps in mountainous areas. Here is the problem that they were trying to solve: A mountain standing high above its surroundings exerts a weak gravitational attraction on nearby objects. You cannot feel this subtle pull, but it can be detected and measured. One device for doing this makes use of a plumb bob—a weight hanging on a string called a *plumb line*. If the Earth were a perfect sphere with uniform density, the plumb bob would be pulled straight

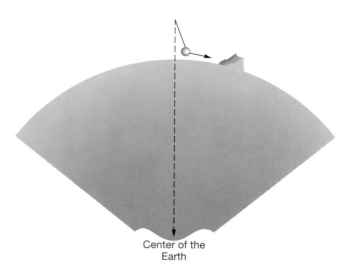

FIGURE 11.4 If the Earth were perfectly spherical, gravity would pull a plumb bob straight toward the center of the planet. The mass of a nearby mountain, however, should attract the plumb bob slightly so that it does not hang vertical. From the amount of deflection (exaggerated here for clarity) a surveyor could calculate the mass of the mountain. Alternatively, if the surveyor thought that he or she knew the mass of the mountain already but was not sure how much the plumb bob had been deflected, it would be possible to calculate the deflection by using fundamental gravity formulas.

down and the line would point directly toward the center of the Earth. If there is any deviation from these perfect conditions, however—that is, if there are large high or low areas on the planet or if the density of the crust is not the same everywhere—the plumb line will not be quite vertical (see Figure 11.4). Knowing this, a surveyor working in an area like the Andes or the Himalayas should be able to estimate the mass of nearby mountains and calculate a correction factor to compensate for its effect on the plumb bob. The accuracy of this calculation is extremely important for performing some types of mapping tasks.

In 1847 a British surveying party was attempting to determine the latitudes of two cities approximately 600 km apart in northern India. They performed the measurements once by standard methods (using a plumb bob) and a second time by a method that did not depend on a plumb bob. The two latitude determinations disagreed by about 150 m. This error may seem small, but it was actually quite large even by nineteenth-century standards. It was soon clear that the difference had occurred because the Himalayan mountain range affected the plumb bob. When the survey team estimated the mountains' mass and calculated their gravitational pull, however, the correction factor they got was much too large. Clearly, something was fundamentally wrong with the way people were thinking about mountains. Scientists recognized that this was an opportunity to learn more about the structure of the Earth's crust.

English astronomer George Airy suggested in 1855 that the Himalayas are like icebergs floating in water. All icebergs have the same density, but a large iceberg extends much farther below water level than does a small one, so its top naturally also rises higher. Similarly, Airy hypothesized, a mountain is high because it has deep roots. The Himalayan surveyors, unaware of those roots, assumed that the dense mantle was closer to the surface than it really is and had therefore overestimated the mass of the mountains.

Mathematician John Pratt offered another possibility: The crust has the same thickness everywhere, but its density varies from place to place. A mountain is high because it is less dense than the low plains around it. In this interpretation, the surveying error was still due to overestimating the mass of the Himalayas. According to Pratt, however, the poorly known density of the mountains was to blame, not their thickness.

Since the 1860s, Pratt and Airy's ideas, illustrated in Figure 11.5, have each been confirmed to a degree

FIGURE 11.5 Why do mountains appear to be lighter than we expect them to be? (a) George Airy theorized that the crust under a mountain is thicker than it is under the surrounding lowlands. The crust, however, is not as dense as the mantle. This means that there should be less mass *immediately beneath* the mountain than if the depth to the mantle were the same everywhere. (b) John Pratt hypothesized that the rocks in a mountain are less dense than the rocks elsewhere in the crust, so they "float" higher on the mantle, just as a cork floats higher in water than a block of denser wood.

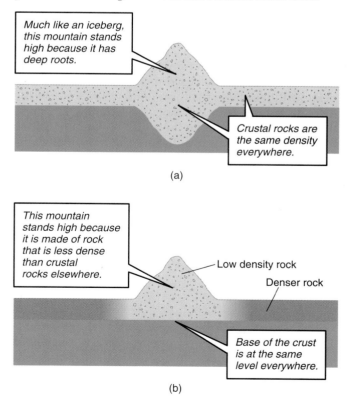

by seismic methods, but Airy's idea explains the structure of the crust and the way it "floats" on the mantle much better. As you learned in Chapter 9, the base of the crust (the Moho) is much further below sea level under mountain ranges than elsewhere. The deep, massive roots of mountains displace the mantle the way your body depresses the springs in a mattress (Figure 11.6). If mass is removed by erosion, the crust displaces less of the mantle and is buoyed up, just as the mattress regains its shape as you shift your weight. Also, like a mattress the crust is flexible and may rise or fall in different places as mass is moved around on it.

Unlike a mattress, of course, the crust does not respond instantaneously to the release of weight upon it. The mantle is much too stiff and the crust is too massive to readjust quickly. Still, vertical movements are fast enough that the crust and mantle are never far from a perfect balance. Around Hudson Bay in Canada, for example, the weight of Pleistocene ice sheets depressed the land until 10,000 to 15,000 years ago, just as glaciers do in Greenland and Antarctica today (see Chapter 16). Ever since the ice melted back, the land has been rising from 5 to 10 m per 1,000 years (Figure 11.7).

From examples such as this, geologists conclude that vertical adjustments in the crust are likely whenever weight is redistributed on the surface. Ice, lava flows, sediments from deltas, water, or other materials deposited on the crust will bow it downward. Conversely, erosion or glacial retreat causes uplift. Because deep-seated rock is continually brought to the surface

FIGURE 11.7   This false-color satellite image shows a region of North America from the Great Lakes (in the lower left quadrant of the image) to Hudson Bay (the darkest blue area in the top center; ice floating near the southern shore of the bay appears lavender in this image.) Shortly after the Pleistocene ice sheet melted, Hudson Bay covered much of the blue area in the upper left quadrant of this image. Its shoreline has receded to its present position over the past few thousand years as the crust, once bowed downward under the weight of ice, has rebounded. *(NASA EROS image)*

by such isostatic adjustments, mountains wear down much more slowly than you might expect (Figure 11.8).

As Dutton and others developed the theory of isostasy, it became apparent that this was a powerful idea that could be used to refine James Dana's geosyncline hypothesis and to explain many large-scale

FIGURE 11.6   According to the principle of isostasy, the crust is in a state of dynamic balance with the mantle below, just as you are in balance with the springs in a mattress when sleeping. The weight of the crust at any point is compensated by a buoyant force from the mantle below.

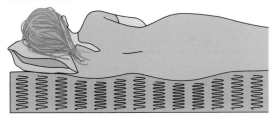

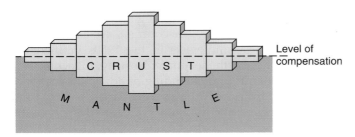

FIGURE 11.8   As weight is moved from one part of the crust to another by erosion, the mantle compensates by adjusting the base of the crust. For example, if erosion removes 1,000 m of rock from a mountain, buoyancy from the mantle lifts its roots by 875 m, so that the mountain is only shortened by 1000 − 875 = 125 m.

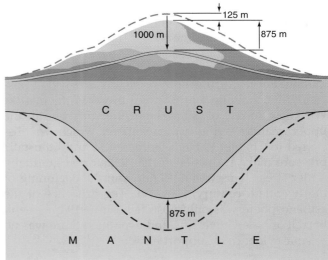

Alfred Wegener, Explorer

Ask any geologist what the German scientist Alfred Wegener is best known for, and the answer will be "continental drift." If you had asked Wegener about his own greatest accomplishments, however, the chances are that he would have mentioned other topics first.

As a young man, Wegener spent much of his free time skiing and climbing mountains, building his endurance for an outdoor career. He dreamed of joining an expedition across the remote interior of Greenland, which had been explored only around its coastline by the start of the twentieth century. The opportunity came in 1906 when the Danish government organized a two-year survey in northeastern Greenland. (See Figure P11.1.1.) Wegener, then 26 years old

and having just received a doctoral degree, left his job and served for the next two years as a meteorologist in one of the world's most inhospitable regions.

Wegener accepted a junior faculty position in Astronomy and Meteorology at the University of Marburg in 1908 and became known as a popular and lucid instructor. Although he taught meteorology, his thoughts were still often about exploring. Perhaps that is why, in the autumn of 1910, he found himself puzzling over an atlas and wondering about the shorelines of the Atlantic Ocean. By his own accounts, it was simply curiosity about the atlas that led him over the next two years to formulate his theory of continental drift. In 1912 he offered his ideas tentatively in a set of papers that

P11.1.1 Arctic expeditions, difficult even today, were perilous in Alfred Wegener's day. Scientific teams crossed the Greenland ice cap on foot and with ponies or dogsleds several times during the opening decades of the twentieth century. Without modern air support or radio communications, a turn in the weather or an accident could quickly become disastrous. *(The Granger Collection)*

were read at seminars in Marburg. That could have been the end of Wegener's hypothesis, for the following year he nearly died in a disastrous expedition across Green-

movements in the Earth like those we have just discussed. One aspect of the theory, however, was hazy: Geologists were not sure just what the crust was floating on. Many believed that the upper mantle was molten, but by 1910 geophysicists discarded that possibility by showing that S-waves travel in the mantle. We know today, of course, that the soft zone in the upper mantle is the asthenosphere, which we discussed in Chapter 9. High temperature and lithostatic pressure make it ductile so that it deforms gradually under stress. This was not known in the beginning of the twentieth century, however. The existence of the asthenosphere was only postulated in 1919 by Joseph Barrell of Yale University, and even then there was no seismic evidence to support his idea.

A Synopsis

The geosyncline theory and the theory of isostasy were standard fare in geology textbooks in the early years of the twentieth century, despite the fact that there were unresolved problems in each of them. As we will discuss in the remainder of this chapter, geologists eventually found that geosynclines and Descartes' old idea of a shrinking Earth were incorrect. The theory of isostasy, in contrast, was finally confirmed when seismologists in the 1950s verified the existence of the asthenosphere. Isostasy, in fact, is very important to our understanding of how plate tectonics works. Regardless of their eventual fates, isostasy and the geosyncline theory both correctly described the

land, and in 1914 he was called to war.

Wegener was called to duty as an officer in the German army during World War I, yet he still managed to write several scientific papers based on his trips to Greenland. He wrote his famous book on continental drift while he was home briefly in 1915, recovering from a battle wound. (See Figure P11.1.2.) When the war finally ended, he moved from Marburg to Hamburg, Germany, where he served for ten years as director of the meteorology department at the Marine Institute. In addition to directing atmospheric research, he made substantial improvements on his original concept of continental drift and revised his book three times. Greenland, however, was still his first love. In 1930, he returned for one last expedition.

The goal was to establish a permanent scientific station, called Eismitte, to monitor weather conditions and perform seismic studies at the heart of the ice cap. Several teams spent the late sum-

mer and early autumn carrying supplies across the ice to prepare for the long winter ahead. Wegener brought the last supplies to the station in late October and celebrated his fiftieth birthday with the men who were preparing to stay. He and a companion left for the coast, 400 km away, on November 1. They never arrived. Wegener's body was recovered months later, showing no signs of starvation or trauma from the cold. Skiing across the rough ice and drifting snow at 40 degrees below zero had apparently been too strenuous. He had died of heart failure, brought on by exhaustion. His companion and their dogsled were never found.

Die Entstehung der Kontinente und Ozeane

Von

Dr. Alfred Wegener

Privatdozent der Meteorologie, prakt. Astronomie und Kosmischen Physik
an der Universität Marburg i. H.

———

Mit 20 Abbildungen

Braunschweig
Druck und Verlag von Friedr. Vieweg & Sohn
1915

P11.1.2 The first edition of Wegener's book, *The Origin of the Continents and Oceans*, was published in Germany during World War I. Only a few copies of the slim 94-page volume were printed, and most of these were never circulated abroad. This copy resides in the Parks Library at Iowa State University. The second edition, published in 1920, and the third edition, translated into English in 1924, attracted much more attention and controversy.

Earth's crust as a mobile and easily deformed portion of the planet. Geologists entered the twentieth century with a growing curiosity about how the crust moves and with confidence that they were on the right track toward understanding this movement.

Continental Drift

▲▲

Because Dana, Dutton, and their colleagues were most interested in how mountains form, they understandably focused on theories that explored vertical movements in the crust. With the concepts of isostasy and geosynclines, they had discovered an explanation for what makes the crust rise and fall, admitting only the

small amount of horizontal compression that was a consequence of planetary shrinkage. No reputable scientist of the nineteenth or early twentieth century seriously discussed the possibility that the crust could move horizontally over great distances. Scientists saw no clear evidence that continents or ocean basins had moved around on the face of the Earth, and because they had theories that seemed to work well without calling for horizontal movement on that scale, they did not look for it.

The evidence was there, however. The most obvious clue was in the shape of the Atlantic Ocean basin, whose east and west shorelines are so similar in places that they look as if the continents on both sides could fit together like pieces from a jigsaw puzzle. Geologists speculated about this observation, but decided that it was nothing more than a coincidence. It received little attention, even in the popular press. Antonio Snider, an American living in Paris in 1858, wrote an imaginative volume entitled *Creation and Its Mysteries Revealed*, in which he declared that the biblical flood of Noah was a record-breaking tidal wave caused by a volcanic explosion that ripped the Eastern and Western Hemispheres apart. His "before" and "after" maps of the world (Figure 11.9) are the earliest published illustrations of movable continents. They were not taken seriously.

A New Theory

Considering how little interest geologists had shown in the idea that continents can move great horizontal distances, it is not surprising that the first serious study of the subject based solely on observations and scientific principles was made by a nongeologist. In 1910, Alfred Wegener, a young meteorology professor at the University of Marburg, Germany, began wondering about the shape of the Atlantic basin and ended up proposing a theory, which he called **continental drift,** that revolutionized the science of geology (Figure 11.10).

FIGURE 11.10   Alfred Wegener, meteorologist and Greenland explorer, proposed the theory of continental drift. *(Stock Montage, Inc.)*

Wegener published several short articles in 1912 and then, in 1915, a book: *The Origin of Continents and Oceans.* In it, he made the radical proposition of continental drift. He hypothesized that the granitic crust of which the continents are made has split into fragments

FIGURE 11.9   Antonio Snider's engravings of the Earth before and after the separation of the Americas from Europe and Africa, an event that he correlated with the Great Flood of the Old Testament. His book received only minor public recognition and was ignored by the scientific community. *(Museum of Comparative Zoology, Harvard University)*

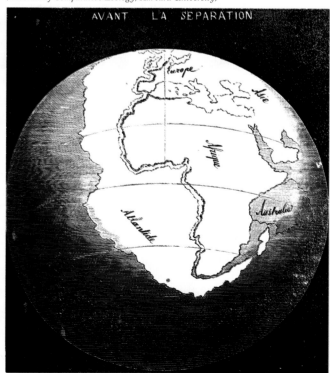

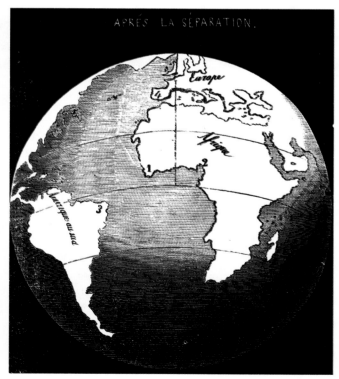

through time. The fragments have moved horizontally, collided with each other and with the crust of the ocean basins, and thickened as a result of folding. Furthermore, in later editions of his book he claimed that the present continental outlines were produced from the breakup of a large supercontinent called **Pangea** (Greek for "all the earth") during the Mesozoic era (Figure 11.11). By the end of the Mesozoic era, Pangea had begun to fragment into two smaller bodies: **Laurasia** in the north and **Gondwana** in the south (Figure 11.12).

The evidence Wegener offered for the Mesozoic supercontinent was extensive. He recognized quite early that it was not sufficient to show that continental coastlines fit together. As you know from assembling jigsaw puzzles, pieces fit properly only when both their shapes *and* their patterns match. Wegener noted that fossils of several Mesozoic plants and animals are found in lands now separated by thousands of kilometers of deep ocean. The fernlike plant *Glossopteris*, for example, had been recognized in fossils of identical age from South America, Australia, Africa, and India (see Figure 11.13). (In 1970, scientists also reported finding them in rocks within 500 km of the South

FIGURE 11.12   This modern reconstruction of Pangea shows the continents of the Southern Hemisphere beginning to separate from those in the north. In 1924, these two landmasses were named Gondwana (after an indigenous people in India) and Laurasia (a name composed from the *Lau*rentian mountains of Canada + Eu*rasia*).

Pole.) Seeds of *Glossopteris* were too large to be wind-borne and too dense to float, so it seemed evident to Wegener that the continents where their fossils are now found must once have been connected. The Triassic reptile *Lystrosaurus* was another good example (Figure 11.14). Cold-blooded land-dwelling, and only 1.5 m long, *Lystrosaurus* could not have traveled to the widely separated sites where it was found in South Africa, Madagascar, and Asia unless those continents had once been joined.

Paleontologists (geologists who study evidence of ancient life) explained the distributions of these fossils

FIGURE 11.11   This drawing from Wegener's 1915 edition of his book shows his reconstruction of the Atlantic prior to the breakup of Pangea. The solid lines are the boundaries of continents. Dashed lines show Paleozoic mountain ranges that Wegener matched from North America to Europe and from South America to the Cape of Good Hope in South Africa.

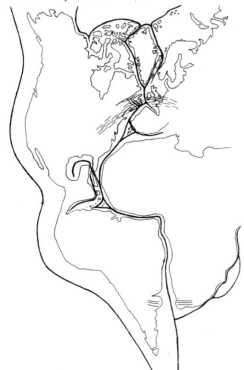

FIGURE 11.13   These leaves of the fossil plant *Glossopteris* came from strata of Permian age in Australia. The *Glossopteris* flora is also found in South America, Africa, India, and Antarctica. The widespread occurrence of this very uniform flora was one piece of evidence that Alfred Wegener offered in support of his theory. The diameter of the detail is 2.5 cm.

FIGURE 11.14 Fossils of the small reptile *Lystrosaurus* have been found on all of the southern continents. It was not an aquatic animal and therefore could not have traveled from one continent to another unless the land masses were connected. Alfred Wegener saw this, like *Glossopteris*, as support for the theory that Pangea had existed. *(Courtesy Department Library Services, American Museum of Natural History)*

by assuming that land bridges like Central America had once spanned the southern continents (see, for example, Figure 11.15). A land bridge is a narrow body of granitic continental crust connecting two larger bodies. If the paleontologists were right, the southern continents had not moved; the land bridges had sunk beneath the waves like Plato's mythical continent of Atlantis, stranding fossil populations on opposite sides of the ocean. Wegener, however, sided with those geologists—still a minority—who hypothesized that there was no granitic rock in the ocean basins. The fundamental difference between the basaltic ocean crust and the granitic continental crust was a key element in the continental-drift hypothesis because it argued against the existence of sunken continents and intercontinental land bridges, which, Wegener claimed, were too buoyant to sink into basalt and disappear. He showed that the distribution of fossils made sense if it was drawn on a map of the reassembled southern continents (Figure 11.16).

As further proof for his theory, Wegener pointed to rocks and structural features that are interrupted on one shore of an ocean and continue on the opposite shore. He listed eight examples, ranging from a 9-km-thick sequence of Permian sediments and plateau basalts found in Brazil and South Africa to a series of red Devonian sandstones identified in the eastern United States, Greenland, Great Britain, Norway, and the Baltic states. In each case, the units are not only similar in terms of structure and rock makeup but are also identical in age and have the same fossil assemblages. Wegener, perhaps a bit brazenly, estimated a probability of one in a million that these similarities were coincidental. He claimed it was like finding two torn pieces of newspaper and discovering that the lines of type on each matched when they were put side by side (Figure 11.17).

Wegener also pointed to evidence of past climate similarities across the continents. In geologic reports from the southern continents of Africa, South America, Australia, Antarctica, and the subcontinents of India and Madagascar, he found descriptions of glacial sediments and of bedrock gouged by the passage of what must have been a massive Permian ice sheet (Figure 11.18). For a similar glacier to cover all of these places today, extending from the South Pole to well beyond the equator, the Earth would need to become a cold planet indeed. In research reports on Permian rocks elsewhere, however, Wegener found that subtropical deserts and coal swamps had been common across the eastern United States, Europe, and central Asia while glaciers covered the southern lands. Evaporites and reef-forming corals, typical of warm climates, are found in late Paleozoic to mid-Mesozoic-

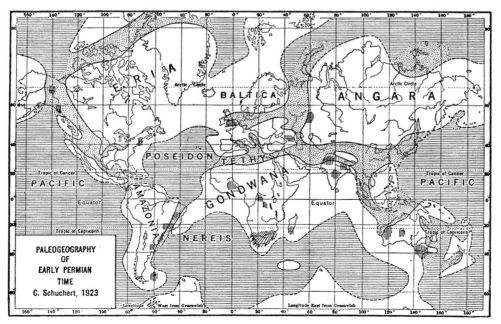

FIGURE 11.15
This map of continents and land bridges of the early Permian period was drawn by paleontologist Charles Schuchert in 1923 to refute the continental-drift theory. The stippled areas are geosynclines and the cross-hatched ones are glacial sediments. Schuchert represented the majority of geologists, who believed that the distribution of fossils such as *Glossopteris* could be explained by land bridges that had since sunk into the oceans. *(American Association of Petroleum Geologists)*

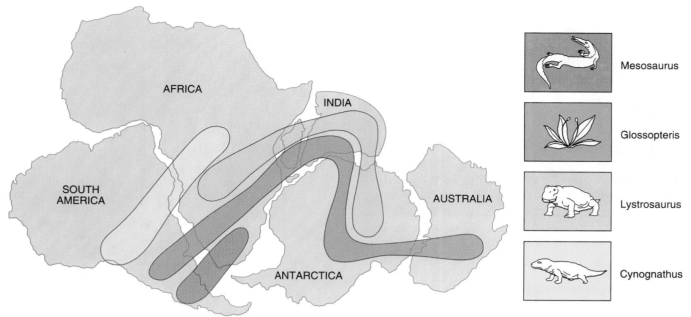

AFRICA

INDIA

SOUTH
AMERICA

AUSTRALIA

ANTARCTICA

Mesosaurus

Glossopteris

Lystrosaurus

Cynognathus

FIGURE 11.16   Instead of assuming the existence of land bridges, Alfred Wegener theorized that the southern continents had once been components of a single landmass. When he marked the locations of fossils on his map of the reconstructed supercontinent, their distribution seemed easy to explain. The plants and animals had once occupied a single region that was later fragmented as the continents began to drift apart. *(Readapted with the permission of Macmillan College Publishing Company from EARTH'S DYNAMIC SYSTEMS 6/E by Kenneth W. Hamblin. Copyright © 1992 by Macmillan Publishing Company, Inc.)*

age rocks as far north as Greenland and Siberia. For these reasons, Wegener argued, the glaciers had covered only the relatively small southern end of a supercontinent while its northern end remained ice free (Figure 11.19).

Why Continental Drift Didn't Catch On

Despite the evidence that Wegener and his supporters amassed, continental drift met with cool indifference from most geologists and outright hostility from others. Some geologists opposed Wegener himself because he lacked credentials as one of them, although he was a well-respected meteorologist. Several prominent geologists declared the apparent match of coastlines to be coincidental and the correlation of fossils and rock units to be poorly substantiated. (In fact, however, most of Wegener's critics had never studied the geology of the Southern Hemisphere, where Wegener found the strongest evidence for his theory.) Others discovered errors that Wegener had copied, uncritically, from other scientific papers. As Wegener was joined by a few distinguished geologists willing to test his hypothesis, most of these objections were resolved. What remained, however, were three serious problems.

First, Wegener's theory assumed that continents drift through the ocean basins like icebreakers moving

through sea ice. That is, continents drift but the oceanic crust stands still. According to Wegener, the great Tertiary mountain ranges were produced by

FIGURE 11.17   If the lines of printing on one scrap of newspaper match those on another scrap, you would conclude that they were once joined. Wegener argued by analogy that geologic features on opposite sides of the Atlantic were evidence that the continents were once together.

FIGURE 11.18   Grooves and scratches like these ones along the coast at Kennebunk, ME, are made by glaciers dragging rock fragments under the ice.

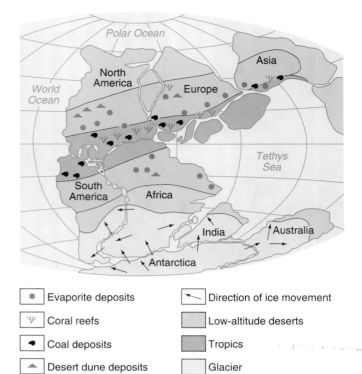

- ● Evaporite deposits
- Ψ Coral reefs
- ● Coal deposits
- ▲ Desert dune deposits
- ↖ Direction of ice movement
- Low-altitude deserts
- Tropics
- Glacier

FIGURE 11.19   Alfred Wegener found references in geologic studies to coal measures, evaporite deposits, and other indications of a tropical or subtropical climate in northern lands during the same period when Permian glaciers were present in the south. Not only could these be matched across continental boundaries, but their mere existence suggested that the Permian glaciation could not have affected a very large part of the Earth. *(Readapted with the permission of Macmillan College Publishing Company from EARTH'S DYNAMIC SYSTEMS 6/E by Kenneth W. Hamblin. Copyright © 1992 by Macmillan Publishing Company, Inc.)*

crumpling that occurred along the edges of continents where they collided with each other (as in the Himalayas) or with resistant oceanic crust (as in the Andes). It was hard for geologists to believe in an oceanic crust that was weak enough for a moving continent to shove aside, yet substantial enough to crush the edge of a continent to raise mountains like the Andes.

Second, Wegener was at a loss to explain where the force that moved continents came from. He suggested tentatively that the gravitational attraction of the Moon could pull continents. Harold Jeffreys, one of England's foremost geophysicists and an outspoken critic of continental drift, calculated that a force to drive continents would need to be a million times stronger than the lunar tides. If such a force existed, Jeffreys said, it could not possibly escape our notice. Clearly, the force did not exist and continents could not move.

Finally, Wegener's story of a supercontinent that broke up in the Mesozoic era contradicted the principle of uniformitarianism. More than a century had passed since James Hutton's declaration that geologic events occur in endless cycles throughout time, and geologists were unwilling to consider what seemed like a return to a science based on one-time catastrophes (recall Chapters 1 and 7). Charles Schuchert, professor emeritus of paleontology at Yale University, asked why Pangea should survive through massive mountain-building periods in the Precambrian and the Paleozoic, only to disintegrate during the relatively quiet Mesozoic. If the continents could rift and scatter across the globe today, why hadn't they done it all through the Earth's history?

Eventually these objections would be answered by showing that the *entire* crust is in motion, not merely the continents. Geologists would find that the crust is driven by convection within the mantle rather than by

tidal attraction. Finally, it would become clear that the crust has been in motion for at least 2.5 billion years and that Pangea was simply the most recent of many supercontinents. These revelations came much later, however, and were not included in Wegener's original continental-drift hypothesis. In the meantime, geologists saw no reason to abandon the concept of geosynclines and vertical crustal movement in favor of a theory of horizontal movement that to them seemed clearly flawed.

New Observations and Interpretations

For several decades, seemingly unrelated observations gradually reinforced Wegener's hypothesis, but most geologists were still not convinced. When geologists finally accepted the idea that the crust was moving horizontally, they did it after a ten-year span that was filled with one dramatic discovery after another. Many of these were made in the ocean basins, where neither Wegener nor his critics had been able to gather evidence to test their ideas. Exploration of the seafloor during the 1950s revealed a complex landscape with immense faults and volcanic features and raised some puzzling new questions. The Deep Sea Drilling Project (DSDP), which followed in the 1960s, gave geologists their first extensive information about the upper portion of the basaltic oceanic crust, finally confirming that there were no submerged land bridges between continents. Ultimately, seismic and magnetic surveys of the oceans yielded the unexpected and exciting observations that opened the door to plate tectonic theory.

To emphasize how crucial the evidence from the seafloor is, we have chosen in the last sections of this chapter to mimic the development of modern tectonic theory by making only light mention of the continents, which will be our focus in Chapter 12.

A Mountain Range on the Ocean Floor

In the mid-1930s a British oceanographic expedition surveyed a canyon over 300 m deep, looking very much like an open fracture, along the crest of a newly discovered ridge in the western Indian Ocean. Very soon after, they found a similar ridge in the Arabian Sea. By comparing their charts with crude maps of earthquake epicenters, they determined that the canyons were "lines along which the Earth's crust is in a condition of instability. . . . "—fault zones. These were the first observations of the mid-ocean ridge system, until then completely unknown to geologists. Even in the North Atlantic Ocean, where scientists had made laborious measurements of water depth for a hundred years, nothing like it had been suspected.

During World War II and the Cold War that followed, sonar technology advanced rapidly, and world governments could see strategic advantages in using it to map the seafloor in detail. The Lamont Geological Observatory at Columbia University was established under the direction of Maurice Ewing, a well-respected oceanographer, as a center for the effort in the United States. In the mid-1950s Marie Tharp, a young assistant at Lamont compiling a map of the North Atlantic seafloor from newly acquired sonar measurements, discovered a mid-ocean ridge like those found by the British survey team before the war. It, too, had a steep-sided central valley. Like those discovered before the war, the ridge coincided with an active earthquake region. Encouraged by this discovery, Tharp and a research team headed by Bruce Heezen searched the rest of the seafloor. By 1957 they had revealed that the North Atlantic ridge and those in the Indian Ocean and the Arabian Sea are parts of a

seismically active ridge system 65,000 km long that extends into each of the ocean basins—the longest geographic feature on the face of the Earth (Figure 11.20). Much of the ridge has a central valley, and in some places it is as much as 50 km wide and twice as deep as the Grand Canyon. Although geologists did not immediately see any connection between this discovery and the continental-drift hypothesis, there was a burst of activity as they searched for other clues to the origin of the ridge system.

Global Seismicity

As important as sonar surveys were for mapping the seafloor, it was actually the study of earthquakes that was most useful in guiding Heezen and Tharp toward their discovery. The World Wide Seismic Network (WWSN), established to monitor atomic weapons testing, gave seismologists a vastly improved means of measuring and locating earthquakes. Geologists during the 1950s rapidly improved their understanding of how and where the crust moves, particularly in the ocean basins, where there had been only spotty seismic information.

On average, about 3,000 earthquakes of magnitude 5 or greater occur each year. Although no place is free of earthquakes, geologists learned that most of them take place in the few linear regions (the Circum-Pacific belt, the Mediterranean and Trans-Asiatic belt, and the mid-ocean ridge system) that we identified in Chapter 9 (see Figure 11.21). Geologists today realize that earthquakes are concentrated in these few narrow belts because they mark the edges of **plates,** adjacent portions of the lithosphere that are moving horizontally in opposing directions. In the 1950s, however, the geologists who discovered the seismic belts did not deduce the existence of plates or their pattern of movement. Instead, geologists initially tried to explain the

FIGURE 11.20   The topography of the seafloor was very poorly known until the 1950s, when oceanographic surveys began in earnest. The most startling discovery was a world-girdling mountain chain, mapped by Marie Tharp and Bruce Heezen at the Lamont Geological Observatory at Columbia University, New York. This submarine chain, known as the mid-ocean ridge system, is seismically and volcanically active over most of its 65,000-km length. *(Cartographic Division © National Geographic Society)*

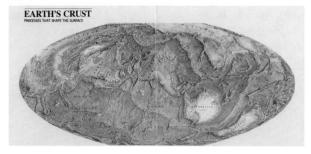

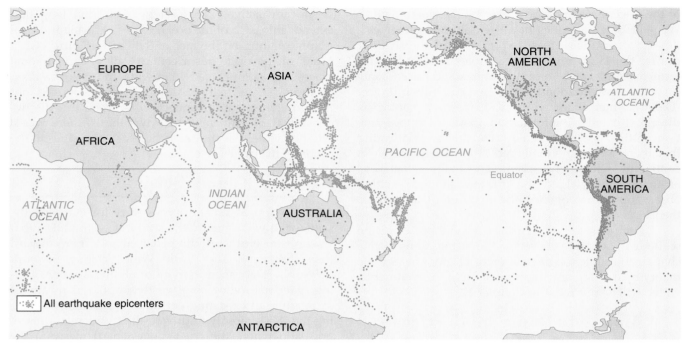

FIGURE 11.21 Earthquake epicenters are concentrated in very narrow belts, as we described in Chapter 9. These belts are now understood to coincide with the boundaries of lithospheric plates. When geologists first recognized this uneven distribution in the 1950s, however, they did not have a good explanation for it. *(The above figure was readapted with permission of the Seismological Society of America.)*

seismic belts through prevailing theories of crustal tectonics that still emphasized vertical, rather than horizontal, displacement.

As seismic information continued to accumulate, it became more apparent that the old theories were inadequate. This became particularly obvious when geologists determined not only the positions of epicenters but also the depths to earthquake foci. Earthquakes are arbitrarily classified as "shallow-focus" if they occur within 70 km of the Earth's surface; "intermediate-focus" earthquakes have foci between 70 and 300 km deep; and a very few "deep-focus" earthquakes occur between 300 and 700 km down. What geologists in the 1950s found was that more than 90 percent of the intermediate-focus activity and virtually all of the deep-focus earthquakes take place in the Circum-Pacific belt (Figure 11.22). Earthquakes elsewhere, particularly those west of Malaysia in the Mediterranean-Asiatic belt, those in the mid-ocean system, and the few isolated ones that take place in the middle of continents, nearly all occur at shallow depths.

Intermediate- and Deep-Focus Earthquakes

Intermediate- and deep-focus earthquakes around the Pacific are associated with major submarine **trenches,** arc-shaped troughs near the margin of the ocean basin. Several of these, including the Java Trench, the Japan-Kurile Trench, and the Aleutian Trench, lie parallel to major island systems (Java, Japan, and the Aleutian Islands) known as **island arcs.** The Middle America Trench and the Peru-Chile Trench follow the coast of Central and South America and one of the world's major continental mountain ranges, the Andes. As you may recall from Chapter 8, this active region around the Pacific basin is also known as the Ring of Fire—a region of high andesitic volcanic activity. By studying earthquakes in the Circum-Pacific belt, geologists soon deduced that most of them were caused by **convergent** horizontal movement (that is, movement produced by compression parallel to the Earth's surface).

To see how they arrived at this conclusion, let us consider an example. Figure 11.23 follows a line from west to east across the Tonga Trench, where over 70 percent of the world's deep-focus earthquakes and a very large number of shallow-focus earthquakes occur. The circles, representing foci of earthquakes, lie on a surface that angles downward and to the west from the Tonga Trench at about 45 degrees. All foci along the Kermadec-Tonga arc system lie within 20 km of this surface, which has been called a **Wadati-**

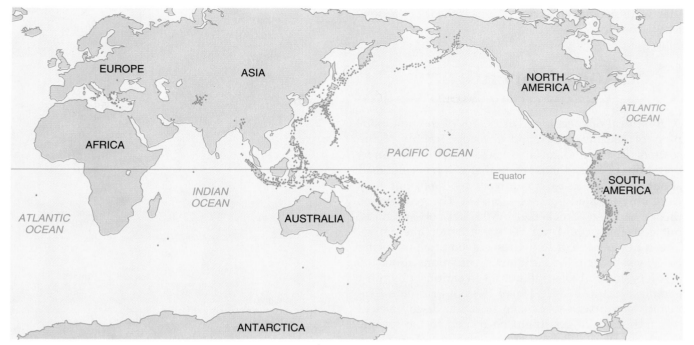

FIGURE 11.22   Over 90% of the world's intermediate and deep earthquakes occur in the Circum-Pacific belt, where they are closely associated with submarine trenches and with island arc systems like Japan and the Aleutian Islands. Geologists who noticed this in the 1950s speculated that these earthquakes occurred for a different reason from the shallower earthquakes in the Mediterranean-Asiatic belt or in the middle of the ocean basins. Compare this map with Figure 8.20, which shows how stratovolcanoes are distributed in the Ring of Fire around the Pacific basin. *(The above figure was readapted with permission of the Seismological Society of America.)*

Benioff Zone in honor of seismologists K. Wadati and Hugo Benioff, who first recognized it in 1954. Other Wadati-Benioff zones follow each of the trenches, dipping outward toward the edge of the Pacific Ocean basin.

Wadati and Benioff speculated that earthquakes along the sloping zone might be the result of friction between blocks of the crust, one of which was sliding under the other along a very large-scale reverse fault. Geophysicists have since confirmed this basic idea and, as you have learned in previous chapters, understand that it is part of the process of **subduction,** by which crustal rock descends into the mantle. Wadati and Benioff did not make that final important deduction, but their observations stimulated a lot of geologists to reexamine their interpretations of stresses and geologic structures around the Pacific basin. In particular, they found the existence of Wadati-Benioff zones very difficult to explain in the context of the geosyncline theory.

Shallow-Focus Earthquakes

In contrast with areas of intermediate- and deep-focus activity, regions with shallow earthquake foci are dominated by extension. Geologists now recognize that these are areas of

divergent horizontal movement or of transform faulting. Most commonly, divergence occurs near the crest of a major mid-ocean ridge, in many cases producing the fault-bounded valleys (rifts) that Marie Tharp, Bruce Heezen, and others mapped. Zones of divergence have also been described on the continents, so it was not difficult for geologists in the 1950s to recognize them once they had been mapped in the oceans. As we illustrated in Chapter 10 (Figure 10.25), however, crustal extension is commonly associated with gentle uplift. Many geologists, therefore, assumed that the mid-ocean ridge system was caused by local vertical movement, not related to a large-scale horizontal movement.

Geologists did not recognize transform faults on the seafloor as readily. In fact, as we discussed in Chapter 10, the concept of transform faulting was only finally proposed in 1965 by J. Tuzo Wilson. Until then, geologists were aware of the characteristic jagged offsets on the mid-ocean ridge system, but identified them as local strike-slip faults. By either interpretation, naturally, geologists found it hard to explain these faults as products of vertical movement in the crust. Together with the existence of Wadati-Benioff zones and the world-girdling system of mid-ocean ridges, they were strong evidence that a new theory of tectonics was needed.

Seafloor Spreading

A great deal of speculation arose out of the first mapping of seismic zones and the topography of the seafloor in the mid to late 1950s. Some of the interpretations of crustal movement that you have just read about were suggested tentatively by 1960, but they were only isolated parts of a puzzle. The geosyncline theory had been weakened by the new observations, but geologists could not yet see a better alternative. The continental-drift hypothesis, even after maturing for 30 years, included very few speculations about the ocean basins and was still not taken seriously by most geologists. Then in 1962 Harry Hess of Princeton University saw a way to explain how the discoveries at the mid-ocean ridges might be related to the seismic activity at Wadati-Benioff zones. He and Robert Dietz, of the Navy Electronics Laboratory in San Diego, introduced a concept they called **seafloor spreading,** describing a cycle in which volcanic activity forms new rock at the mid-ocean ridges, which then diverges. The newly formed rock then moves slowly toward subduction zones, where it plunges back into the mantle (Figure 11.24). Hess and Dietz built their new theory on two notions: (1) that large-scale movements on the seafloor were primarily horizontal, not vertical, and (2) that the *entire* seafloor was moving, not merely small parts near Wadati-Benioff zones or mid-ocean ridges (which were rechristened **spreading centers** or **spreading axes).**

Among the discoveries that impressed Hess were three that we have not yet mentioned, but were important in establishing the new theory:

- First, nearly 200 flat-topped mountains called **seamounts** had been discovered on the floor of the Pacific. The closer a seamount was to a trench, the more its top was submerged. This suggested that seamounts had once been in shallower water but had slowly been moved as if on a conveyor belt. One seamount near the Tonga Trench was not only 370 meters below the water surface but was also leaning conspicuously toward the trench as though it were sliding into the abyss (Figure 11.25).

- Second, the layer of sediment on the basaltic seafloor was very thin—almost nonexistent on the mid-ocean ridges. If the ocean floor were as old as the continents, a layer 10 to 20 times thicker should have accumulated.

- Third, new seismic studies confirmed that the upper mantle was ductile. By 1960, geophysicists had convincing evidence of the low-velocity layer that we discussed in Chapter 9. The asthenosphere was real.

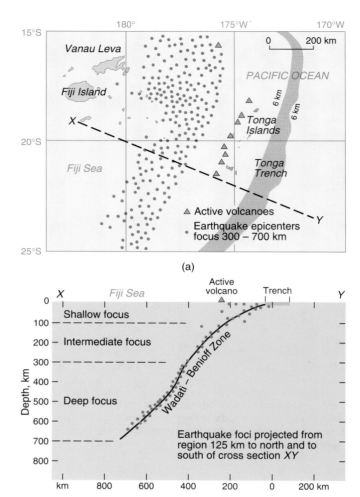

FIGURE 11.23   The most active seismic zone in the world is between the Tonga Trench and the Fiji Islands in the South Pacific. Earthquake foci define a plane (a Wadati-Benioff Zone) that is shallowest near the trench and dips steeply to the west. Wadati and Benioff suggested that the earthquakes were the result of friction along a fault. Geologists today have almost the same interpretation. A Wadati-Benioff Zone is a place where a portion of the oceanic lithosphere is plunging deep into the asthenosphere below. This process is called *subduction. (This illustration was derived from Figure 4.9 in P.J. Wyllie's The Way the Earth Works, © 1976 by John Wiley & Sons, Inc.)*

This last observation was perhaps the most significant, for two reasons. First, it provided the missing information that geologists needed to confirm how isostasy works. The lithosphere *is* "floating" on the asthenosphere. Second, it finally overcame one of the great objections to Alfred Wegener's theory. As we noted earlier, geologists in the 1920s had suggested thermal convection as a force to move the continents. Without seismic confirmation that the asthenosphere existed, however, it seemed unlikely that the mantle could be soft enough to permit convection. With that

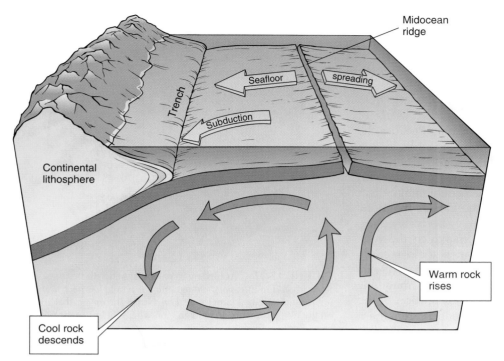

FIGURE 11.24
The theory of seafloor spreading, proposed by Harry Hess and Robert Dietz in 1962, relates the seismic and volcanic activity at the mid-ocean ridges to events at the trenches. Convection in the asthenosphere causes divergence at the ridges, carries the oceanic lithosphere sideways to the trenches, and then causes it to subduct. New volcanic rock is continually formed at the ridges by this process and old rock is recycled down the trenches at the same rate.

problem out of the way, Hess and Dietz theorized that the mantle turned over like a set of gears driving a conveyor belt and that the entire lithosphere, continental and oceanic, was carried along passively on top. The continents no longer needed to crush their way through a stationary oceanic lithosphere, as Wegener had hypothesized. Hess theorized that the convection cell was entirely within the asthenosphere. Geologists today believe that much more of the mantle is convecting, but there is little agreement about how much of it, as we discussed in Chapter 8.

Paleomagnetism

The seafloor-spreading hypothesis thus offered geologists a view of horizontal tectonics in the ocean basins that had several features in common with Alfred Wegener's continental-drift hypothesis, but improved upon it by suggesting answers to problems that continental drift had never been able to resolve. Many geologists accepted the new hypothesis and began looking for ways to apply it. Others remained skeptical because they could see no proof that the seafloor was

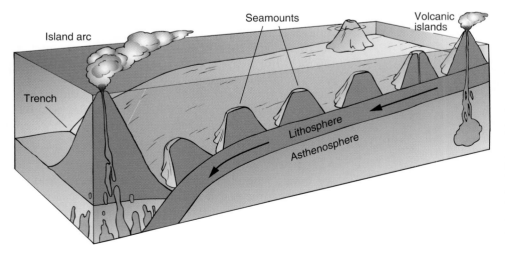

FIGURE 11.25
A seamount is a volcanic mountain on the seafloor. Some seamounts form near a mid-ocean ridge, where the water is commonly less than 3 km deep. Wave erosion gradually carves a flat top on them. As these seamounts are later carried to one side on the back of the moving oceanic lithosphere, they eventually find themselves in much deeper water and are gradually submerged. If such a seamount finally reaches a trench, it tilts and "rides down into the jaw-crusher," as Harry Hess described the subduction process. *(This illustration was derived from Figure 19.20 in B.J. Skinner and S.C. Porter's Physical Geology, © 1987 by John Wiley & Sons, Inc.)*

moving *except* at the ridges or Wadati-Benioff zones. The observations that finally convinced them came from a corner of geophysics that was considered esoteric until the 1950s: **paleomagnetism,** the study of changes in Earth's magnetic field throughout time.

In Chapter 7 we discussed Earth's magnetic field and the influence that it has on minute crystals of magnetic minerals while a rock is forming. Because the crystals align with the global field and are then locked in place as the rock becomes solid, the rock contains a nearly permanent magnetic record of its orientation known as *remanent magnetism*. Remanent magnetism can be used to detect changes in the Earth's magnetic field and in the position of the rock body itself or, rather, of the portion of the crust in which the rock body sits. The inclination of a remanent field records the latitude where the rock body was formed, while its declination indicates the direction to the magnetic north pole. Within limits, therefore, if a portion of the crust has moved, geophysicists can read its remanent field to tell where it used to be. During the 1950s and early 1960s, geologists studying remanent magnetism made two important discoveries, providing overwhelming evidence that Earth's crust is in continuous horizontal motion. One discovery was made on land and the other at sea.

Apparent Polar Wandering
Scientists have long been aware that the magnetic north pole and the planetary rotation axis ("true" north) do not coincide. They have also known that the magnetic poles migrate slowly. In the 1950s, geophysicists studying remanent

FIGURE 11.26   For the last 30 million years, the magnetic north pole has migrated slowly around the rotational ("true") North Pole. These minor movements are random and probably due to irregularities in the behavior of the outer core, not to plate movement. This "true" polar wandering is called *secular drift*.

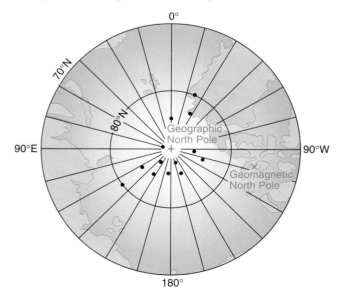

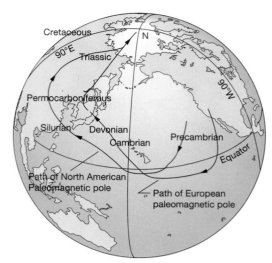

FIGURE 11.27   According to paleomagnetic studies in North America and in Europe, the magnetic north pole has been wandering throughout the Phanerozoic Eon. Notice, however, that studies on the two continents do not agree on where the magnetic pole has been. Because there is no way for the magnetic pole to be in two places at once, this suggests instead that the continents have moved independently.

magnetism in rocks on the continents showed that as far back as the mid-Oligocene epoch (30 million years ago) the magnetic north pole has drifted slowly around true north (Figure 11.26). That small amount of movement was not a subject for much discussion although geologists could not explain it completely (and, in fact, still do not understand it fully today).

When geophysicists began to examine older continental rocks, however, they discovered evidence that the magnetic north pole had apparently wandered as far away as the equator during the past 600 million years (Figure 11.27). Geophysicists then discovered that the record of polar wandering indicated by rocks in Europe was different from the record determined in North America. Remanent fields on the two continents appeared to point in different directions at the same time, and the discrepancy became greater as the age of the rocks increased. Research teams briefly considered the possibility that Earth could have two magnetic north poles but discarded it because the idea was completely at odds with all that was known about magnets. It slowly became clear instead that the pole had remained fixed, but the *continents* had moved horizontally and independently.

Magnetic Reversals: The Final Clue
At the same time, another related mystery was unfolding. Oceanographic research ships towed magnetic instruments across many thousands of kilometers of the seafloor during the 1950s. Transferring the information from those surveys to maps, geologists soon recognized areas of high and low magnetic intensity that formed invisible stripes on the seafloor (Figure 11.28).

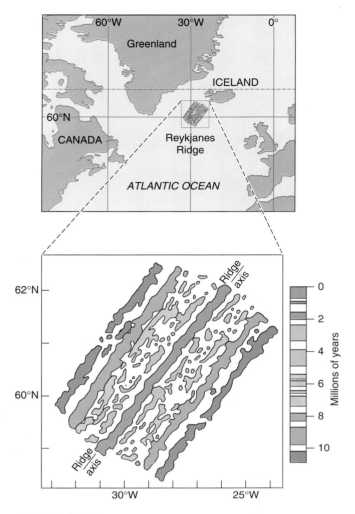

FIGURE 11.28 Stripes of high (colored) and low (noncolored) magnetic intensity on the ocean floor are parallel to the mid-ocean ridge system. The pattern of intensities is the same on both sides of a ridge, as indicated in this portion of the Reykjanes Ridge southwest of Iceland in the North Atlantic. Ages of the various anomalies are based on the paleomagnetic time scale discussed in Chapter 7. *(After F.J. Vine, "Magnetic Anomalies Associated with Ocean Ridges," in R.A. Phinney (ed.), The History of the Earth's Crust, Princeton University Press, Princeton, N.J., 1968.)*

Stripes with a higher intensity than the average global magnetic field were said to display positive **magnetic anomalies;** those with intensities below the global average had negative anomalies. In 1963, Fred Vine, a British graduate student from Cambridge University, and D.H. Matthews, his research supervisor, theorized that the stripes were bands of basaltic crust with alternating magnetic polarity.[1] Where the remanent magnetic field was normally polarized, its intensity added to that of the Earth's global field and led to a positive anomaly. Where the polarity was reversed, the intensity of the rock's own magnetic field subtracted from

---

[1.] Recall the discussion of magnetic polarity stratigraphy in Chapter 7.

the global field, producing a negative anomaly (Figure 11.29). The idea was proposed independently by a Canadian geologist, L.W. Morley.

It was not merely the stripes themselves that made this an important discovery. It was the *pattern* of stripes, which began at the crest of the mid-ocean ridge system and extended in opposite directions with rough symmetry. In it, Vine, Matthews, and Morley saw a way to verify Harry Hess's hypothesis about seafloor spreading. If the lithosphere were continuously pulled apart at the ridge, fresh lavas—new crustal rocks—should form there as magma rose along the fracture from the mantle. The newly erupted lavas, they hypothesized, cooled and adopted whichever polarity of the global magnetic field was current at the time. Continued spreading and occasional reversals in the global field would produce stripes of alternating polarity that would slowly travel in both directions from the ridge. The rocks, in other words, carry a magnetic recording of their own journey across the seafloor (Figure 11.30). By dating magnetic stripes, geologists could determine both the direction and the rate of crustal movement.

▶▶ EPILOGUE ◀◀

Geologists now generally give Vine, Matthews, and Morley credit for finding the final, convincing evidence for seafloor spreading. Presented a half century after Alfred Wegener's continental-drift hypothesis was published, their explanation for the pattern of magnetic stripes on the seafloor made it clear that Wegener had been on the right track, but had only understood part of the picture. The seafloor was revealed as the most dynamic portion of the Earth's surface. Within four years, geologists were adopting the idea of large-scale horizontal movement in the crust as enthusiastically as they had once rejected it. In doing so, they quickly went beyond the hypothesis of seafloor spreading that Hess and Dietz had envisioned.

By the late 1960s it was apparent that parts of the lithosphere move relative to each other at rates between 1 and 10 cm/yr. Paleomagnetic information and modern satellite laser-ranging measurements shown in Figure 11.31 indicate that the North Atlantic Ocean is widening by 2 cm (about twice the width of your thumbnail) per year. As slow as that seems in human terms, it is remarkably fast by geologic standards—fast enough, in fact, that geologists now have no trouble accepting Wegener's idea that much of the current landmass was part of a supercontinent until 200 million years ago. Paleomagnetic studies around the Atlantic basin, for example, confirm that South America and Africa were joined until the early Creta-

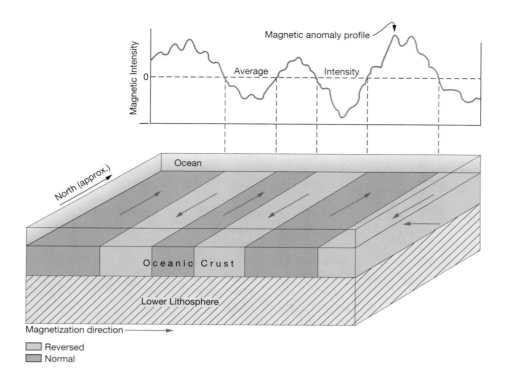

FIGURE 11.29
Vine, Matthews, and Morley interpreted the intensity patterns shown in Figure 11.29 as records of paleomagnetic polarity reversals. A positive magnetic anomaly occurs where the remanent field has the same polarity as the present global field, and the two add together to produce a higher than average field intensity. Where the remanent field and the global field have opposite polarity, they interfere, producing a negative-intensity anomaly. (*This illustration was derived from Figure 10.4 in P.J. Wyllie's* The Way the Earth Works, © 1976 by John Wiley & Sons, Inc.)

ceous period. The Atlantic, thus, is a very young ocean, and the continents on opposite sides of it *have* moved apart. What geologists of the 1960s learned as they studied this and other examples, however, was that neither "continents" nor "ocean basins" moved as distinct, separate bodies.

FIGURE 11.30   The development of magnetic anomalies are related to an oceanic spreading center. New oceanic crust forms from basaltic magma at the axis of the ridge. As it cools below the Curie temperature, the new crust acquires the magnetic characteristics of the Earth's field at that time. As the crust moves laterally away from the ridge crest, changes in the Earth's magnetic field are recorded in stripes of the oceanic crust. Each change in the polarity of the field produces a pair of stripes on either side of the ridge.

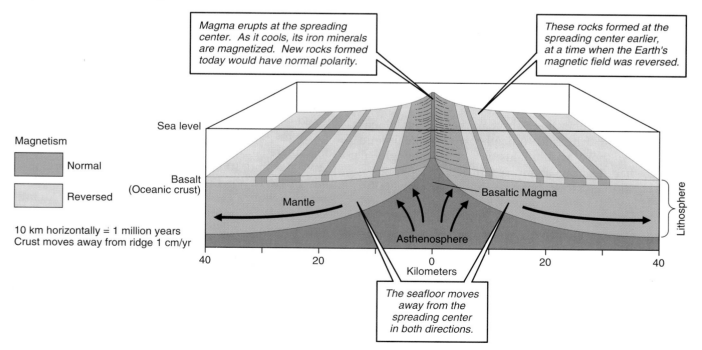

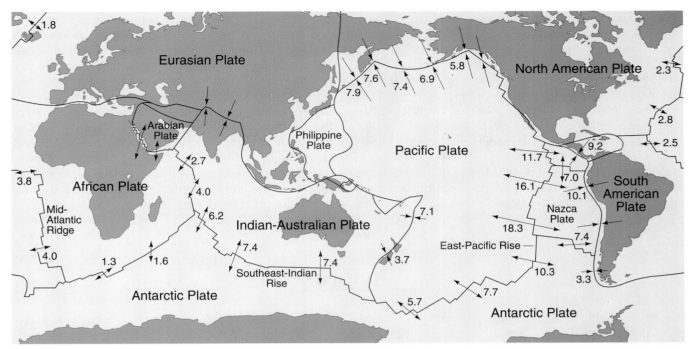

FIGURE 11.31 The average rates of movement shown on this map in cm/yr have been determined by using laser surveying instruments. A beam from a station on one part of the lithosphere is bounced off a satellite and returned to a station on another part. By repeating measurements over a period of time, geologists can determine with high precision how far the lithosphere has moved.

FIGURE 11.32 Unlike old maps of the world, which focused on the differences between continents and oceans, the geologist's map of the world today emphasizes plates and plate boundaries. Coastlines and plate boundaries rarely coincide. *(Readapted by permission from Fig. 1-13 in J.S. Monroe and R.W. Canders, PHYSICAL GEOLOGY: EXPLORING THE EARTH. Copyright © 1992 by West Publishing Company. All rights reserved.)*

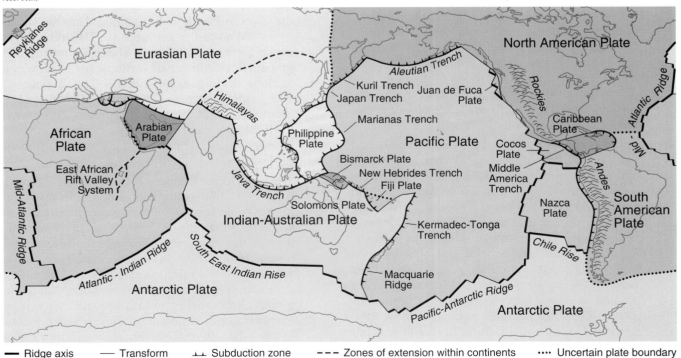

— Ridge axis   — Transform   ⊥⊤ Subduction zone   - - - Zones of extension within continents   ···· Uncertain plate boundary

The most dramatic advances in understanding came as geologists used their observations of seismicity, volcanism, and paleomagnetism to redraw the world map (Figure 11.32). Where all earlier maps divided the planet into continents and ocean basins, the new one recognized only **plates,** rigid portions of the lithosphere whose boundaries rarely coincide with the world's coastlines. Some lie entirely within ocean basins, but none are made completely of continental material.

This revised view of the world began to take shape soon after Vine, Matthews, and Morley published their work in 1963 and was developed through the efforts of a continuously growing community of geologists who saw the seafloor-spreading hypothesis as part of a much larger theory. One of the most influential figures in expanding and popularizing the new tectonic perspective was J. Tuzo Wilson, whose discovery of transform faults has already been mentioned as a key to understanding offsets on the mid-ocean ridge system. In 1971 Jason Morgan of Princeton University and Dan McKenzie of Cambridge University gave the broadened theory a name. They called it **plate tectonics.** In the chapter ahead, we will examine the plate tectonic theory closely and show how geologists use it to study the evolution of Earth's lithosphere.

SUMMARY

1. Before the continental-drift theory was introduced, geologists explained most large-scale movement in the Earth's crust in the context of isostasy and the geosyncline theory. Isostasy, the theory that the crust is "floating" on the mantle, allowed geologists to deduce correctly that the crust varies in thickness from place to place. This theory also helped explain how the elevation of the crust may change as weight is redistributed on it by erosion or other processes. Isostasy is still a viable theory today. The geosyncline theory hypothesized that mountain ranges are formed when sediments in large troughs are compressed as a side effect of gradual cooling and shrinking of the Earth. It did explain why mountain ranges are long and narrow and why they occur parallel to ancient shorelines. Eventually, however, the geosyncline theory was discarded when geologists found that it was incompatible with what was observed in the modern oceans and with evidence of large-scale horizontal movement in the crust.

2. Alfred Wegener claimed that the continents had been joined until the Mesozoic era as a single landmass, which he called Pangea. As fragments of Pangea moved apart, collisions between them and compression against the oceanic crust buckled the continental crust, producing mountains. In support of this continental-drift hypothesis, Wegener offered evidence of glacial sediments, fossils, truncated mountain ranges, and other features. Geologists objected to Wegener's theory because it could not explain how collisions between the thick continental crust and the thinner oceanic crust could deform the continents but not the ocean floor, it could not identify a realistic force capable of moving continents, and it appeared to contradict the principle of uniformitarianism.

3. According to the seafloor-spreading hypothesis, new oceanic crust is produced by the outpouring of lava along spreading axes that coincide with the mid-ocean ridge system. New crust gradually moves away from the ridge system and eventually descends into the mantle at subduction zones, which coincide with marine trenches. Two sets of observations led Harry Hess and Robert Dietz to formulate the hypothesis. One was the discovery that most seismic activity occurs in narrow belts, and that many earthquake foci in the Circum-Pacific belt lie along dipping planar zones (Wadati-Benioff zones). The other was Bruce Heezen and Marie Tharp's confirmation that shallow mid-ocean earthquakes occur along a globe-circling ridge system. The hypothesis was widely accepted when Fred Vine, D.H. Matthews, and L.W. Morley correctly interpreted the pattern of magnetic polarity-reversal "stripes" oriented symmetrically around the mid-ocean ridges.

4. The plate tectonic theory incorporates Wegener's suggestion that continents move sideways, but improves on it by recognizing the importance of seafloor spreading. According to plate tectonic theory, the entire lithosphere is made of plates whose boundaries do not necessarily coincide with continental margins. It is these plates, rather than simply continents or the seafloor, that are in motion. Earthquakes, volcanism, and many other events occur at plate boundaries, where portions of the lithosphere are moving in different directions or at different rates.

KEY WORDS AND CONCEPTS

continental drift 266
convergent movement 272
divergent movement 273
geanticline 261
geosyncline 260
Gondwana 267
island arc 272

isostasy 261
Laurasia 267
magnetic anomaly 277
paleomagnetism 276
Pangea 267
plate 271
plate tectonics 280

seafloor spreading 274
seamount 274
spreading axis (center) 274
subduction 273
trench 272
Wadati-Benioff Zone 272

QUESTIONS FOR REVIEW AND THOUGHT

11.1 What kinds of arguments or observations could you use to support Descartes' hypothesis that the Earth is cooling and shrinking? What counterarguments can you suggest?

11.2 What kinds of arguments or observations could you use to support the theory of isostasy?

11.3 What observations led James Hall to propose the concept of geosynclines? How did James Dana modify the idea?

11.4 What evidence did Alfred Wegener offer in support of his hypothesis of continental drift?

11.5 How does the record of glaciation in Permian-age rocks of the Southern Hemisphere support the hypothesis of continental drift?

11.6 Why did Wegener consider it unlikely that land bridges had ever connected the continents?

11.7 Which of Wegener's original claims for continental drift have since been discarded? Why?

11.8 Where do most of the world's earthquakes happen? Why?

11.9 How is seismic activity at mid-ocean ridges different from activity associated with subduction zones?

11.10 How does the observation of apparent polar wandering support the theory that large portions of the crust have moved horizontally?

11.11 What is a magnetic anomaly? According to the hypothesis suggested by Vine, Matthews, and Morley, how are magnetic anomalies produced?

11.12 How do geologists use magnetic anomalies on the ocean floor to interpret past movements of the lithosphere?

11.13 Where are the youngest rocks in the oceans? Where are the oldest rocks? Why are the oldest rocks in the oceans only about 150 million years old?

11.14 How rapidly are lithospheric plates moving relative to each other? How do geologists know?

Critical Thinking

11.15 Geologists on Albondiga do not yet realize that seafloor spreading is taking place on their planet. Unlike the Earth, Albondiga has no global magnetic field and its seafloor basalts therefore do not have any magnetic reversal "stripes." The evidence that rocks are progressively older from spreading centers toward the margins of ocean basins is just as persuasive as the magnetic evidence, however, so we may assume that Albondigan geologists will eventually figure out that seafloor spreading is taking place. It is just likely to take them much longer than it would if Albondiga had a magnetic field like the Earth's. Why?

11.16 By using seismic reflection methods to determine the thickness of the ice cap on Greenland, geologists have discovered that the land surface under the ice is well below sea level. Suppose that scientists who are predicting a period of gradual global warming are correct, and that the Greenland ice cap melts during the next several hundred years. If you wanted to draw a speculative map to indicate what Greenland might look like to our distant descendants, what kinds of information would you want to gather?

12

Mt. Everest (also known as Chomolungma or Sagarmatha), to the extreme left, is the highest peak in the Himalayas of Nepal. To the right of it is Lhotse, and to the far right is Ama Dablam. (Keith Gunnar, Photo Reseachers, Inc.)

Cratons, Orogens, and Plate Boundaries

OBJECTIVES

▲▲

*As you read through this chapter it may help if you focus
on the following questions:*

1. How do scientists believe the earliest portions of continental crust formed, and how do scientists relate the formation of continental crust to plate tectonics and to the rock system?

2. What are the major subdivisions of the continental crust, and how are they related to each other?

3. What are the five types of continental margins, what geologic features characterize each of them, and how are they related to plate tectonic activity?

4. What is the Wilson Cycle, and what have geologists learned about the role of plate tectonics in the growth and breakup of supercontinents?

OVERVIEW

▲▲

Our view of the world is influenced most greatly by the parts of it that we can see, and so we should not be surprised that geologists once believed that the entire lithosphere was like the parts they could observe on the continents. As the story of plate tectonics has unfolded for geologists, they have learned that the lithosphere of the ocean floors is in fact quite different from the familiar lithosphere of the continents. That realization has made it possible for them to look back to the continents with new understanding. The excitement that followed the discoveries in the ocean basins in the 1950s and 1960s has continued, and is now stimulating geologists to reexamine the basic processes by which continents change, particularly processes of mountain building.

In this chapter, then, we will return to questions raised by Descartes, Hall, Dana, and other scientists of past centuries: *Where did the continents come from? How are mountains formed, and why do they appear where they do?* With the knowledge we have gained in previous chapters, we will add others as well: *How long have plates moved as they do today? Are the continents still growing? How does subduction really work?* Because they are still learning, geologists have more than one answer for each of these questions, which makes these some of the most tantalizing questions of all.

Cratons
▲▲

How has the continental crust changed since the Earth formed? Geologists, of course, cannot find rocks of Hadean age, which would have been part of the crust during the first few hundred million years of Earth's history. Those rocks have long ago been destroyed by weathering or recycled by subduction and melting. Several subtle chemical clues have convinced geologists that the total amount of continental crust was much smaller during the Hadean eon than it is today. Beyond that point of agreement, geologists disagree about crustal formation and have developed two prevailing theories (Figure 12.1). Some geologists argue that continental crust formed quickly during the Archean eon. According to this view, tectonic activity since then has recycled crustal rocks continually, but the total amount of continental material has stayed nearly the same. Those who accept this theory generally think that tectonic processes where lithospheric plates meet destroy continental rocks as efficiently as they form new ones. In contrast to this "early growth" theory, other geologists hypothesize that the amount of continental crust has been increasing gradually for 4.5 billion years. In their view, the process of plate tectonics does not merely recycle the crust. Instead, it is biased slightly so that over time continental rocks are slowly created at the expense of oceanic material.

Why do geologists care about how the crust formed? The reason is, the issue sheds light on the relationship between tectonic processes that use heat from the Earth's interior to drive plate movement, and the rock system, in which Earth materials change

chemical and physical form. The connection between plate tectonics and the evolution of rocks in the crust and mantle is still poorly understood. Geologists would like to know whether one process can exist without the other, and how the continents and ocean basins we see today have been affected by each of them. One way to find out is to see how the oldest and most stable portions of the continental crust, called **cratons,** are put together and to consider how the intensely deformed mountain belts around them, called **orogens,** have formed.

Cratons are regions of the crust in the geographic interior of each continent that are composed largely of granites and metamorphic rocks. The cratons may be deformed, but are not mountainous today. They formed more than 2 billion years ago, during the Archean or very early Proterozoic eon, and have long since been eroded to form low, featureless portions of the crust in perfect isostatic balance with the mantle below. That is, the elevation of their land surface has stayed nearly the same for an extremely long time—at least hundreds of millions of years. Unlike the cratons, orogens are younger regions that contain great thicknesses of deformed marine sedimentary rock, often metamorphosed or even melted during mountain-building episodes. The orogens typically surround the older cratons, as is the case in North America, where the Innuitian, Appalachian, Caledonian, and Cordilleran orogens ring the Canadian Shield (Figure 12.2). As geologists investigate cratons and orogens, they find patterns of dynamic change that have dominated the margins of continents throughout the history of the Earth.

The Archean Crust

Geologists have recognized two subdivisions of crustal material within cratons, each characterized by a different assemblage of igneous and metamorphic rocks. As geologists study these subdivisions, they try to figure out not only where the primitive continental crust came from but also what role the movement of tectonic plates may have played in those early times.

Large areas in the cratons are dominated by ancient basalts that tend to have pillow structures characteristic of submarine lava flows. They are chemically similar to modern basalts on the seafloor, although most have been intensely weathered and at least lightly metamorphosed to become chlorite schists, commonly known among geologists as **greenstones.** Regions in which they are common are known as **greenstone belts.**

Other volcanic rocks in these greenstone belts have no modern counterpart. The komatiites we discussed in Box 8.1 for example, are olivine-rich, much like peridotites. The olivine crystals, however, have

FIGURE 12.1 The total amount of continental lithosphere is much greater today than it was during the first billion years of Earth history, but geologists are uncertain about whether it grew rapidly to its present size or whether it has been growing slowly through the history of the planet. These two alternatives are suggested by the lines on this graph showing the amount of continental lithosphere as a function of time.

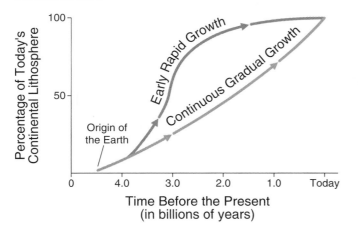

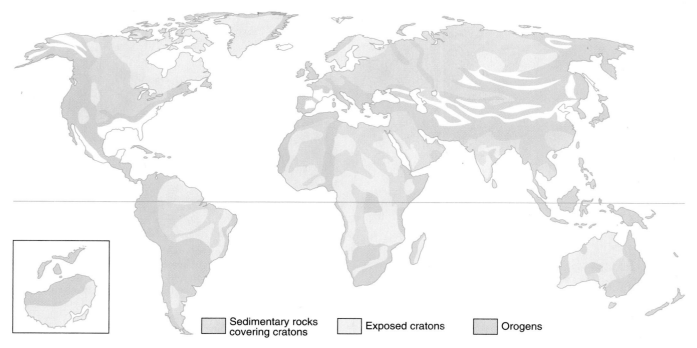

Sedimentary rocks covering cratons	Exposed cratons	Orogens

FIGURE 12.2   Each of the continents has a craton, as shown on this map of the world. These are usually regions of stability, where no mountain building has occurred for extremely long periods in Earth's history. Cratons may be partly covered by a veneer of sediment, as in much of central Asia and the mid-continent in North America. The orogens surrounding the cratons are younger, highly deformed belts of rock that are, or have been, mountainous regions.

unusual textures characteristic of rapid growth that convince geologists that the komatiites were lava flows, not intrusive rocks that solidified in the mantle or deep in the crust. Magmas with such ultramafic compositions solidify as they cool below about 1500°C. Magmas reaching the Earth's surface today are rarely hotter than 1200°C.

A second group of igneous rocks, largely granitic, characterize **granitic belts** in the cratons. These intrude the greenstones here and there as rounded plutons of various sizes. Modern granites, conversely, are commonly found in elongated batholiths associated with mountain belts. From the principle of cross-cutting relationships, we deduce that the granitic belts are younger than the greenstone belts.

The jumble of granites and greenstones within the cratons produces a distinctive exposure pattern. The 3.6-billion-year-old Pilbara Complex in northwestern Australia, shown in Figure 12.3, is typical of cratonic rocks found on each of the continents.

The Growth of Cratons

How did the cratons begin to form? One theory, illustrated in Figure 12.4, holds that cratons were formed as a by-product of extensive submarine volcanism of the kind now seen in places such as Hawaii. As lava flows accumulated one after another on the seafloor to build a shield volcano, the

thickening crust was depressed into the upper mantle under its own weight. In the crust's deepest, hottest

FIGURE 12.3   The Pilbara Complex in northwestern Australia is one of the oldest cratons in the world. Radiometric ages of about 3.7 billion years have been measured. The rounded masses in this photograph, taken from orbit, are granitic belts. These granitic bodies intrude older greenstone belts. *(NASA)*

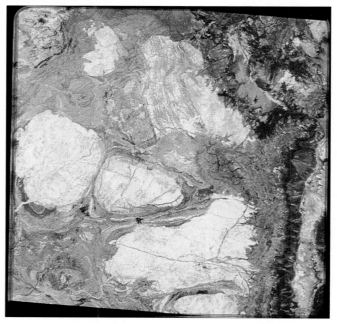

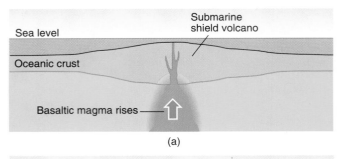

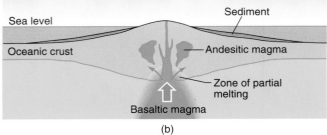

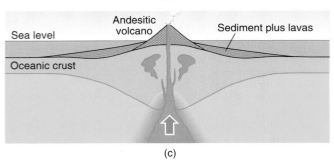

FIGURE 12.4 One hypothesis for the origin of cratons suggests that they were formed as shield volcanoes at hot spots. The first magmas to rise from the mantle would have been basalts. As the crust thickened with successive lava flows, however, partial melting within it would have generated more andesitic magmas. Thus, these "island continents" would gradually become more granitic with time. This hypothesis does not involve any plate movement in the early growth of cratons but, of course, does not rule it out either. *(This illustration was derived from Figure 18.18 in B.J. Skinner and S.C. Porter's Physical Geology, © 1987 by John Wiley & Sons, Inc.)*

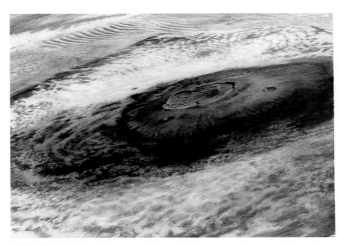

FIGURE 12.5 Olympus Mons, the largest volcano on Mars, has a diameter of about 600 km. Its summit is 27 km above the surrounding plains and contains a caldera almost 80 km across. Space scientists attribute its great size to the fact that it grew over a stationary hot spot. Also, because there is no ductile asthenosphere on Mars, the full weight of the mountain is supported on the lithosphere and the rigid mantle below. It does not have the immense roots that a volcano of its size on Earth would have. *(NASA)*

regions, partial melting of the basaltic rocks produced andesitic magma, which erupted to build stratovolcanoes. These, and the sediments derived from them, became "island continents" from which the early cratons grew. Greenstone belts are remnants of the earlier, basaltic crust; granitic belts are intrusions of later, differentiated magma. This theory allows for only vertical movement in the lithosphere. Although it does not rule out horizontal movement, it is consistent with its proponents' view that the cratons formed *before* the plate tectonic cycle began to work. Some geologists suggest that this theory explains how immense volcanoes were formed on Mars, where it does not appear that lithospheric plates ever moved across the face of that planet (Figure 12.5).

An alternative hypothesis suggests that the early cratons grew from **island arcs** (Figure 12.6). Andesitic

magma around the Pacific rim today is produced in large part by partial melting above basaltic ocean crust in subduction zones. Stratovolcanoes characterizing this modern transition zone between purely oceanic and continental igneous environments are clearly a result of convergent tectonics. Arguing by analogy, then, proponents of this theory claim that the plate tectonic cycle began to operate before cratons existed and actually *caused* the growth of cratons. In this theoretical picture as well, greenstone belts are derived from earlier basaltic crust; granitic belts are products of partial melting during subduction.

It may never be possible to decide which theory is correct because truly ancient portions of the crust are rare. Weathering, metamorphism, and tectonic activity have taken their toll on the original rocks of the crust. The oldest known rocks, near Isua in western Greenland, are not simple igneous rocks, but are metamorphosed conglomerates and volcanic ash, materials commonly found on continents or at continental margins. Was there a continent in the sense that we use the term "continent" today? A youthful craton? The metamorphosed boulders at Isua tell us that older volcanic rocks were exposed to weathering, but that they were not the granites we associate with continents today.

Geologists have found a few other tantalizing clues to the history of cratons and the origin of continents. Detrital grains of zircon (Zr_2SiO_4) 4.2 billion years old have been collected in the Pilbara Complex in Western Australia. Although these are isolated mineral grains deposited in much younger sediments, the fact that zircons commonly crystallize in granitic rocks

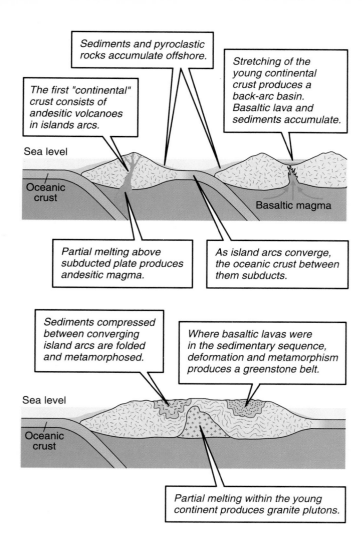

Sediments and pyroclastic rocks accumulate offshore.

The first "continental" crust consists of andesitic volcanoes in islands arcs.

Stretching of the young continental crust produces a back-arc basin. Basaltic lava and sediments accumulate.

Sea level

Oceanic crust

Basaltic magma

Partial melting above subducted plate produces andesitic magma.

As island arcs converge, the oceanic crust between them subducts.

Sediments compressed between converging island arcs are folded and metamorphosed.

Where basaltic lavas were in the sedimentary sequence, deformation and metamorphism produces a greenstone belt.

Sea level

Oceanic crust

Partial melting within the young continent produces granite plutons.

FIGURE 12.6 The alternative theory to the one illustrated in Figure 12.4 holds that early cratons were island arcs, formed as andesitic magma rose to the surface above a subduction zone such as the one near Tonga in the South Pacific Ocean today (see Chapter 11). Unlike the first hypothesis, this one requires that plates be in motion very early in Earth's history. *(This illustration was derived from Figure 18.17 in B.J. Skinner and S.C. Porter's* Physical Geology, *© 1987 by John Wiley & Sons, Inc.)*

suggests that cratons may have begun to form at least 400 million years before the Isua metaconglomerates.

The discoveries at Isua and Pilbara reveal secrets about the evolution of the Archean crust and, perhaps, even give us a glimpse of the rock system in the Hadean eon before. Unfortunately, however, it is risky to hypothesize about the entire crust when all we have are a very few surviving fragments of it. So far, the rocks have given geologists no clear proof that there were stable cratons or that the plate tectonic cycle of subduction and crustal regeneration was operating until much later than the rocks at Isua were formed. We have only ambiguous hints to suggest how either plate tectonics or the rock system began and, as yet, few clues about how they were related in the earliest eons of Earth's history.

In contrast, quite a lot is known about plate tectonics and the development of cratons by about 1 billion years after the sedimentation at Isua, or roughly 2.8 billion years ago. Geologists have concluded that the North American continent, for example, formed as a cluster of smaller continental objects that gradually coalesced during the Proterozoic eon (Figure 12.7). Each of these is comprised of granitic and greenstone belts but has distinctive chemical characteristics or field relationships indicating it was formed as a separate body from the others. The last major event, attaching the Churchill craton in the west to the Superior craton in the east, occurred early in the Proterozoic eon, about 2.2 billion years ago. This evidence that cratons were coming together and fusing into larger masses is widely accepted as proof that plate tectonic processes were already assembling and reassembling the continents more than 2.5 billion years ago.

Orogens and Continental Margins

▲▲

The margins of a continent are modified both by the processes of erosion and deposition and by tectonic activity. As a result, each of the cratons is ringed by younger bodies of rock that are (or were at one time) at the edge, or **margin,** of the continent, where continental and oceanic crust meet. So far there is no uniform terminology used to describe these regions. One popular approach, which we will use here, identifies five different types of continental margin by the kind of plate interaction that influences each (Figure 12.8). These are **passive** margins, **subduction** margins, **collisional** margins, **accretionary** margins, and **transform** margins. The last four of these are continental margins where mobile lithospheric plates meet (called **plate boundaries**), and are therefore known to geologists as **active** margins. At three of these (the subduction, collisional, and accretionary margins) are the elongated regions of intense deformation that we labeled earlier as orogens.

Passive Margins

Because the continental crust is about 30 km thick, on average, it is not easily pulled apart by divergent plate motion. Clearly, however, that is what happened as the Paleozoic supercontinent of Pangea broke up. This process continues today in several parts of the world.

Continental lithosphere begins to split apart when heat rising at a hot spot over a plume deep in the mantle causes the crust to expand. A broad plateau rises, sometimes as much as 2.5 km above sea level. As uplift and horizontal stress increase, incipient rifts

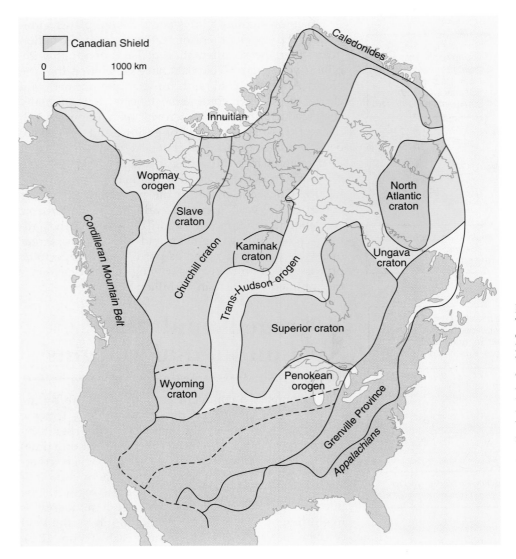

FIGURE 12.7
The North American craton is exposed in Canada, where geologists have concluded that it was assembled from several smaller bodies of continental crust during the Archean and Proterozoic eons. To the south, these Precambrian rocks are covered with a stable platform of sediments blanketing much of the mid-continent. Beyond them, younger orogens form the margins of the continent. The four youngest comprise Phanerozoic mountain belts. The Cordilleran belt is an active orogen today. *(This illustration was derived from Figure 16.19 in B.J. Skinner and S.C. Porter's* THE DYNAMIC EARTH 2/E, © *1992 John Wiley & Sons, Inc.)*

form, as we illustrated in Figure 10.26. Typically, three major rifts radiate from the center of the uplifted region at 120° angles. For this reason, the place where the rifts meet is known as a **plate triple junction.** The junction between the 2,900-km-long East African Rift and others in the Red Sea and the Gulf of Aden, shown in Figure 12.9, is typical. Widespread normal faulting in the region becomes common in and parallel to rifts. Structural valleys (grabens) may develop where crustal blocks settle between the faults.

There is no guarantee that a rift will continue to open once it has begun. In fact, one of the three rifts extending from any plate triple junction is always less successful than the others. Geologists call this one a **failed rift.** The East African Rift, for example, is apparently only opening very slowly, although both the Gulf of Aden and the Red Sea are spreading rapidly. Because failed rift valleys may extend far into a continent, the rivers that flow in them are some of the

world's longest. Many major rivers (the Niger and the Mississippi, for example) lie in failed rift valleys that are older and less obvious than the East African Rift, and are in places where there is no longer any active uplift or rifting (see Figure 12.10).

The floor of the major graben that develops along a rift is commonly several hundred meters lower than the surrounding plateau and is typically a region of intense volcanic activity. Basaltic lavas and volcanic ash accumulate in it, along with sands and coarser sediment eroded from the valley walls. As a rift widens and deepens, one end of it finally opens to the sea. The shallow, salty water that invades the rift may evaporate easily, leaving behind thick layers of evaporites and, finally, normal marine sediments (shales and limestones) (Figure 12.11). The Red Sea and the Gulf of Aden have such a sequence on their floor today, as does much of the Afar triangle at the plate triple junction in Somalia and Ethiopia.

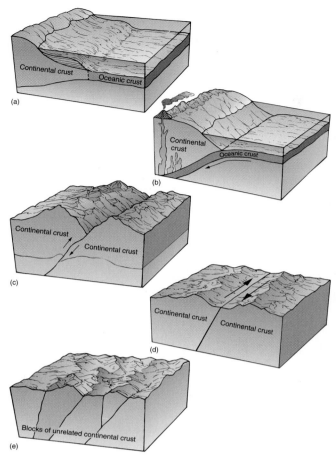

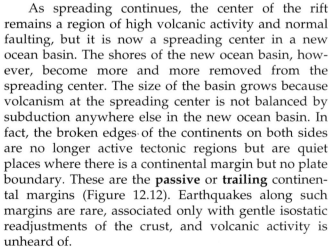

FIGURE 12.8 Continental margins fall into five categories, depending on the tectonic relationships between lithospheric blocks that lie along them. (a) Passive, (b) subduction, (c) collisional, (d) transform, and (e) accretionary margins.

FIGURE 12.9 The plate triple junction in the Afar triangle at the intersection of the Red Sea, the Gulf of Aden, and the East African Rift is clearly visible in the lower left corner of this false-color satellite image. The African plate, to the left, is nearly stationary. The Arabian peninsula, to the upper right, is slowly being compressed against the larger mass of the Asian plate, of which it is a part. The Persian Gulf, in the far upper right, is not at a continental margin but is a large syncline with continental crust at its base. The East African Rift, in the extreme lower left, is a failed rift. *(NASA EROS image)*

As spreading continues, the center of the rift remains a region of high volcanic activity and normal faulting, but it is now a spreading center in a new ocean basin. The shores of the new ocean basin, however, become more and more removed from the spreading center. The size of the basin grows because volcanism at the spreading center is not balanced by subduction anywhere else in the new ocean basin. In fact, the broken edges of the continents on both sides are no longer active tectonic regions but are quiet places where there is a continental margin but no plate boundary. These are the **passive** or **trailing** continental margins (Figure 12.12). Earthquakes along such margins are rare, associated only with gentle isostatic readjustments of the crust, and volcanic activity is unheard of.

Passive margins are characteristic of young oceans, in which spreading is typically less than 4.0 cm/yr. The amount of heat rising at a ridge controls not only its spreading rate but also its shape, because the elevation of the seafloor decreases as it cools. The mid-ocean ridge in a young ocean is narrow because rocks only a short distance from the ridge are already quite cool. Such a ridge also has a characteristic graben or central rift valley reminiscent of the rift valley that formed as the continents initially split apart (Figure 12.13). You will recall from Chapter 11 that this was one of the peculiar features that oceanographers noticed when they first surveyed the Mid-Atlantic and Carlsberg ridges. The East Pacific Rise and the Southeast Indian Rise, both of which are in older ocean basins, are spreading much faster. Even at a great distance from the ridge, therefore, rocks are still warm enough that the seafloor elevation has not decreased much. These ridges are wide, have no central rift valley (Figure 12.14), and are not in basins with passive continental margins.

The eastern coast of North America is an excellent example of a modern passive margin. The land between the Appalachian Mountains and the shore is a broad, low-relief region known as the Atlantic Coastal Plain (Figure 12.15). It is a thick ramp of clastic

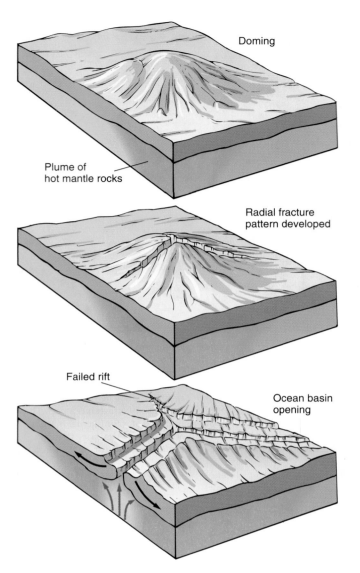

Doming

Plume of
hot mantle rocks

Radial fracture
pattern developed

Failed rift

Ocean basin
opening

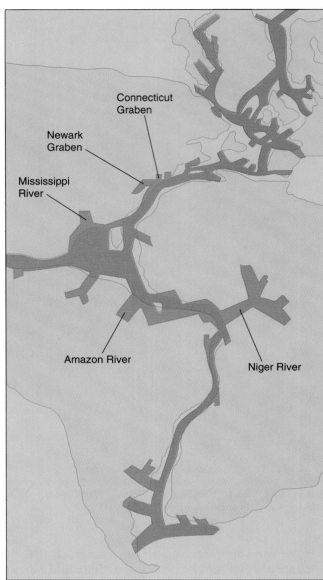

Connecticut
Graben

Newark
Graben

Mississippi
River

Amazon River

Niger River

FIGURE 12.10   (a) Failed rifts are those along which divergence may have begun but it never managed to open a new ocean basin between fragments of the continental lithosphere, as shown in this sequence of drawings. *(This figure was derived from Figure 17.21 in E.J. Tarbuck and F.K. Lutgen's* THE EARTH: AN INTRODUCTION TO PHYSICAL GEOLOGY, 4/E. *Reprinted by permission of Macmillan Publishing Company. Copyright © 1993 by Macmillan Publishing Company, Inc.)* (b) These rifts, however, extend far into a continent, often becoming occupied by major river systems. As this figure illustrates, the Mississippi, the Amazon, and the Niger rivers all have their channels in failed rifts that formed as the Atlantic Ocean opened during the breakup of Pangea during the Mesozoic era. *(This illustration was derived from Figure 16.22 in B.J. Skinner and S.C. Porter's* THE DYNAMIC EARTH 2/E, © 1992 by John Wiley & Sons, Inc.)

sediments, gradually accumulated from the erosion of continental rocks to the west and carried seaward by rivers from Florida to Labrador.

Beyond the shoreline, this wedge of sediment forms a wide **continental shelf** no more than 100 m below sea level that extends seaward for hundreds of kilometers in places like the Grand Banks off of Nova Scotia. These sediments are deposited largely on submerged portions of the craton, which is still part of the North American continent. Further offshore, on the

continental rise, a wedge of fine-grained marine sediments (muds and biochemically produced SiO_2-rich material known as **pelagic ooze**) has collected on the oceanic crust. The true edge of the continent lies between the shelf and the rise, in a narrow region known as the **continental slope.**

Like many of the continental margins we will highlight in this chapter, the margins of the North Atlantic began to take their modern form with the breakup of Pangea. The record of that event is easy to

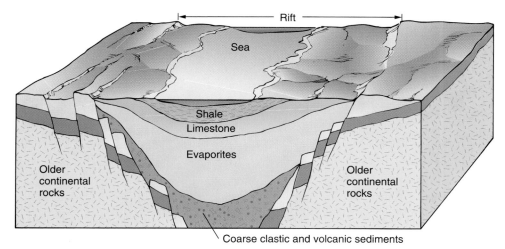

Coarse clastic and volcanic sediments

FIGURE 12.11

This generalized cross section of a typical rift valley illustrates the sequence of rocks that form as the rift gradually widens and is invaded by seawater. Basaltic lava erupting from fissures along the rift forms the base of the valley, as well as ash from volcanic eruptions along its floor. Sand and coarse sediment from the surrounding craton also wash into the rift. As seawater finally enters the expanding rift, it evaporates easily because the new sea is so shallow. Gypsum and salt beds form. As the water gradually deepens, muds and carbonate sediments accumulate. From these, shales and limestones eventually form. Basaltic fissure eruptions, of course, continue at the spreading center that remains at the center of the rift.

FIGURE 12.12

(a) As a new ocean gradually widens, its passive margins are farther and farther from its spreading center. These margins accumulate great thicknesses of sediment that erode from the continents and are deposited at the mouths of rivers and redistributed along the coast by currents and wave action. Because there is no nearby plate boundary, earthquakes are rare on these coastlines, and volcanoes never occur. (b) The spreading center in the North Atlantic Ocean is clearly visible in this aerial photo taken near Thingvellir, Iceland. The island of Iceland formed as the result of continuous volcanism on the Midatlantic Ridge. *(Wolfgang Kaehler This illustration was derived from Figure 16.20 in B.J. Skinner and S.C. Porter's* THE DYNAMIC EARTH 2/E, © 1992 *John Wiley & Sons, Inc.)*

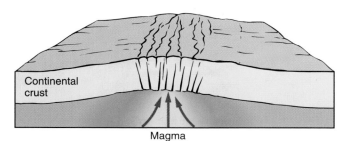

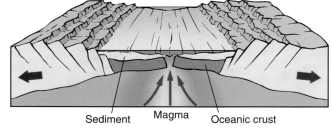

(b)

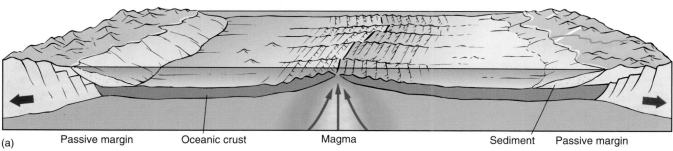

(a)

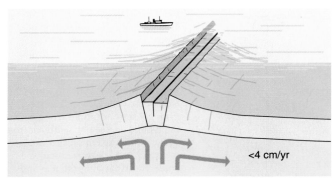

FIGURE 12.13   A mid-ocean ridge like the one in the Atlantic basin spreads slowly (less than 4.0 cm/yr). As a result, the seafloor cools before it has moved very far from the ridge crest, and its elevation decreases over a short distance. The ridge, therefore, is narrow and pronounced. As shown in this profile of a typical ridge of this type, it also has a well-defined central rift valley.

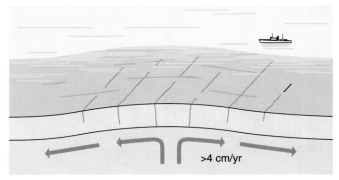

FIGURE 12.14   In contrast to a slowly spreading ridge, a ridge that spreads rapidly is broad and not very pronounced. Even at a great distance from the spreading center the seafloor is warm and thermally inflated. There is less vertical movement on normal faults as well; thus, a ridge of this type rarely has a distinct central valley.

FIGURE 12.15   The eastern margin of North America is a passive margin. A ramp of clastic sediments (sands and muds) extends from the Appalachian Mountains to the outer edge of the continental shelf. These sediments eroded from the continent and deposited near the shore by rivers, are more than 12 km thick in places. At its widest point, on the Grand Banks offshore from Maine and the Canadian Maritime provinces, the upper layers include a thick blanket of Pleistocene glacial sediment. The true edge of the continent, as marked by the transition from a granitic crust to a basaltic one, is approximately at the outer edge of the continental shelf.

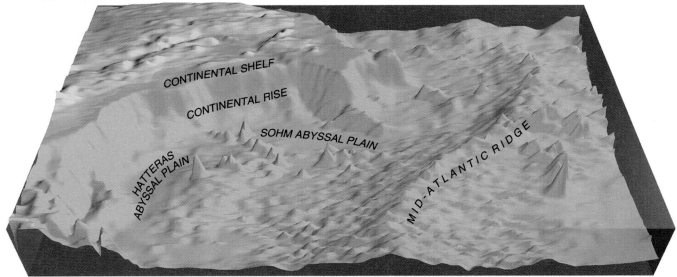

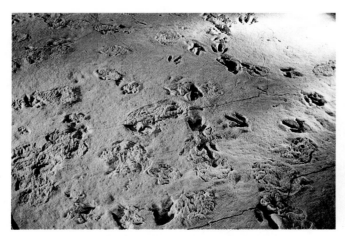

FIGURE 12.16   Dinosaur footprints are preserved in the red sandstones of the Connecticut River Valley near Hartford. During the Triassic period, this graben was produced by the tension that opened the North Atlantic Ocean. *(Dinosaur State Park)*

FIGURE 12.17   The Palisades Sill, exposed here along the west bank of the Hudson River near New York City, was intruded during the rifting that created several large grabens along the east coast of North America in the Triassic period. *(John Serrao, Visuals Unlimited)*

find on both shores of the Atlantic. Geologists working in the northeastern United States and the Maritime provinces of Canada, for example, have mapped an extensive system of Triassic normal faults that strike almost due north. In many places the faults mark the edges of grabens (rift valleys) that once hosted major river systems during the Mesozoic era. One of these is the valley of the modern Connecticut River, which is renowned locally for its steep cliffs of red sandstone, derived from the Triassic stream sediments. Near the city of Hartford, Connecticut, these sandstones are replete with fossils and footprints of dinosaurs and other creatures that once populated the valley (Figure 12.16). In other places, there are no grabens but instead swarms of basaltic dikes that formed as magma rose through the fracture system. From these, lava once flowed on the surface and magma below the ground moved laterally to create thick sills. Today, these beds of basalt have been carved by rivers, producing dramatic cliffs in places such as the Palisades along the west bank of the Hudson River near New York City (Figure 12.17). Similar features can be found through western Europe and the British Isles.

Each of these features is the result of tension that eventually opened the rift valley that became the North Atlantic Ocean during the mid-Mesozoic era. By the Jurassic period, the center of activity was no longer on either North America or Europe. Volcanism and faulting have been unknown along the quiet margins of the Atlantic for over 150 million years.

Subduction Margins

In contrast to passive margins, **subduction margins** are seismically and volcanically active, convergent plate boundaries at which oceanic lithosphere plunges under the continents. Most of these margins are

located around the rim of the Pacific Ocean basin. Subduction also occurs at plate boundaries in the western Pacific basin where there is no continental margin. We discussed one of these boundaries, coinciding with the Tonga Trench, in connection with the concept of Wadati-Benioff zones in Chapter 11.

As we explained in Chapter 11, subduction and seafloor spreading are companion processes in the cycle of crustal formation and destruction. Clearly, the total amount of new crustal rock produced at spreading centers around the world must be exactly the same as the total amount of old crust that is subducted. After all, the Earth does not get any bigger or smaller as it ages. This means that the spreading of the Atlantic Ocean, which has passive margins, must be balanced by subduction somewhere else. Most of that subduction occurs in the Pacific Ocean basin, in which more rock is subducted than is created. In terms of its tectonics, therefore, the Pacific basin is nearly the opposite of the Atlantic: It is an "old" shrinking basin instead of a young growing one. Thus even if we knew nothing else about the Pacific, we should expect it to have very different features around its margins.

How Does Subduction Work?   The ultimate cause of subduction and seafloor spreading is thermal convection in the mantle. Ever since the seafloor spreading hypothesis was proposed in the early 1960s, however, geologists have debated exactly *how* convection moves the oceanic lithosphere without coming to a clear-cut solution. Three hypotheses have been advanced:

1. One possibility is that lithospheric material descends into the subduction zone because it has finally cooled so much as it moves across the ocean basin that it is denser than the asthenosphere below it. As it sinks, the older, denser, and colder basaltic rock pulls the rest of the

plate behind it, eventually stretching the lithosphere so much that it cracks to form a mid-ocean ridge. If this idea proves to be correct, then subduction is the cause of seafloor spreading, and the mid-ocean ridges are a by-product of it (Figure 12.18a).

2. The opposite possibility has also been suggested: Magma rising at the mid-ocean ridges pushes the plates apart, and that force eventually drives the lithosphere further into the subduction zones far away. If correct, then mid-ocean volcanism is the cause of seafloor spreading, and subduction is its consequence (Figure 12.18b).

The best way to find out which of these hypotheses is correct is to look for evidence of major tensional or compressional deformation in the oceanic crust. Unfortunately, neither has been observed. This has led geologists to consider a third possibility:

FIGURE 12.18 Geologists have suggested three hypotheses to explain what makes the seafloor move from a spreading center to a subduction zone. (a) As the lithosphere descends into a subduction zone, it pulls the seafloor behind it. (b) As magma is extruded at a spreading center, it pushes the seafloor ahead of it. (c) Because the seafloor tilts slightly from the ridge toward the subduction zone, it is sliding downhill. Each of these ideas appears unsatisfactory for a different reason, but it is possible that all three processes contribute to plate movement.

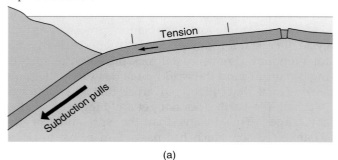

(a)

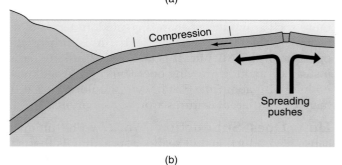

(b)

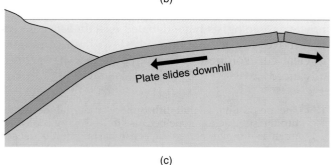

(c)

3. The oceanic lithosphere is simply sliding downhill across the seafloor and falling into the subduction zones. The ocean floor is tilted gently away from the ridges because the lithosphere contracts as it cools on its journey toward the subduction zones (Figure 12.18c). Calculations suggest that if the tilt averages any more than 0.02 degrees, the lithosphere's own weight would be enough to make it slide a few centimeters each year. This hypothesis, however, has yet to be confirmed either, so no one is sure what makes the oceanic lithosphere move toward subduction zones, nor is anyone certain how subduction begins in the first place.

Long before basalts of the oceanic plate reach a subduction zone they have been altered chemically by reactions with seawater much as continental rocks are weathered chemically to produce soils. However, very little mechanical weathering (disintegration) occurs on the seafloor; thus, the rocks remain solid. Other chemical changes in the oceanic crust are due to metamorphism by hydrothermal fluids, largely consisting of seawater that has penetrated cracks and been heated by the still-warm basalts. The olivines and pyroxenes that make up much of a fresh basalt are hydrated to form a new mineral called *serpentine* (Figure 12.19). Some geophysicists have suggested that the shallow and intermediate-depth earthquakes common in subduction zones are caused in part as this serpentine is heated under pressure until it dehydrates explosively. The more conventional view, which we have discussed in Chapter 11, is that deep earthquakes are caused by brittle fracture as the plate is stretched and bent. Because it descends rapidly, relative to the rate at which it is heated by the surrounding asthenosphere, a subducting oceanic plate may remain brittle even to a depth of 600 or 700 km.

As dense oceanic lithosphere plunges beneath the continental margin, the ocean floor bows downward

FIGURE 12.19 Basalt from the seafloor is altered chemically by both cool and heated seawater. The green mineral that dominates in this microscopic view is serpentine. *(Kenneth Windom)*

and a trench forms. Marine sediments on top of the descending plate are drawn into the trench but cannot be subducted easily. Sediments have much lower densities than the basaltic ocean crust and are therefore buoyant. Instead of being drawn into the mantle, therefore, much of the sediment is caught between plates and is sheared and mangled to form a chaotic mass of fractured rock known as a **mélange** or **clastic wedge** (Figure 12.20). A distinctive type of metamorphism caused by high pressure on the relatively cool mélange and adjacent basalt produces blueschists and eclogites that are typical of subduction margins. Geologists disagree about how much of this material eventually descends into the asthenosphere. It is possible that most of it becomes attached to the continental lithosphere. This is one of several possibilities suggested by geologists who believe that the total mass of the continents is still gradually increasing.

Above the descending plate, the overriding continental rocks begin to melt. Partial melting is made easier by the fact that water-saturated basalts and mélange release water as they are heated, and the water then infiltrates the overlying continental rocks. The melting product is andesitic magma, which rises toward the surface.

Two Kinds of Subduction Margins There are two ways that a subduction margin may evolve from this point; both are common. The prevailing theory suggests that the type of margin is determined by whether the continental mass is moving rapidly toward a quiet ocean basin or the ocean floor is moving rapidly toward a stationary continent.

Fast Continent, Slow Ocean Around much of the Pacific Ocean basin today, shallow- and deep-water sediments are compressed against the leading edge of a continent faster than they can be subducted (Figure 12.21). Geologists see evidence in highly deformed regions elsewhere that this has happened at other times in Earth's history as well. When it does, shallow-water sediments may be draped back over the

FIGURE 12.21 According to the prevailing theory, a subduction margin like the one along the west coast of South America develops because the continental lithosphere is moving rapidly toward oceanic lithosphere that is subducting slowly. In this circumstance, offshore sediments and the rocks of the leading edge of the continent are deformed in thrust sheets and nappes. Melting above the subducting plate generates andesitic magma, which rises to form stratovolcanoes. Granitic plutons and regional metamorphism are also common within the mountain range created by this process. *(Readapted with permission of Macmillan College Publishing Company from* THE EARTH: AN INTRODUCTION TO PHYSICAL GEOLOGY 4/E *by Edward J. Tarbuck and Frederick K. Lutgens. Copyright © 1993 by Macmillan College Publishing Company, Inc.)*

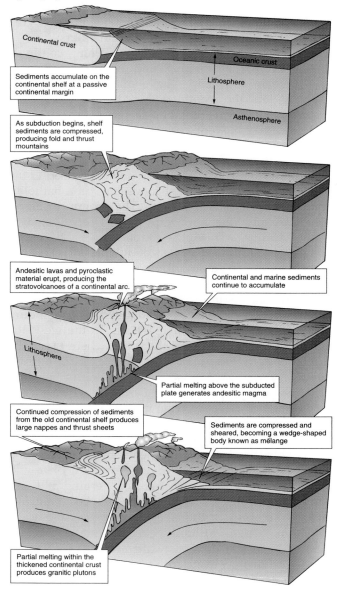

FIGURE 12.20 A mélange is a chaotic mixture of deep- and shallow-water sediments. Too buoyant to be subducted with the basaltic seafloor, sediments of the mélange are scraped off of the descending plate and form a triangular wedge that is thickest at the base of the submarine trench.

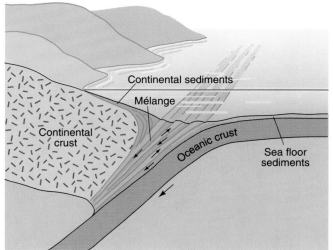

continent in large-scale recumbent folds called **nappes** (a French word for "tablecloth"), or may slide over it in **thrust sheets** (that is, in one or more thin crustal slices that move on low-angle reverse faults). These folds and thrust sheets increase the thickness of the continental crust substantially. The deeper-water sediments, including mélange, are intensely deformed as they, too, are forced onto the advancing continental crust. The result is a belt of **fold-and-thrust** mountains that may be several hundred kilometers wide.

At this first type of subduction margin, magma may rise with difficulty through the thickened and deformed crust to form a **continental arc** of stratovolcanoes. Geologists believe that the volcanoes in the Pacific Northwestern states of the United States (including Mt. Lassen, Mt. Hood, Mt. Shasta, and Mt. St. Helens) are being formed in this way, as similar volcanoes were all along the west coast of South America. Because the leading edge of the continent has been thickened by compression, however, magma rises more easily in some places than in others.

As compression and crustal thickening continue, large-scale partial melting and regional metamorphism begin to change the continental lithosphere itself. When the continental sediments melt, the magma that forms contains more SiO_2 than the andesitic magma that formed stratovolcanoes. It is granitic, more viscous, and therefore much less likely to rise to the surface (recall our discussion of granitic magma in Chapter 8). Instead, it solidifies deep in the roots of the fold-and-thrust mountains as a batholith or other plutonic body of rock. The prominent Sierra Nevada batholith, now exposed by erosion, was formed in this way while subduction was still active along the California coast tens of millions of years ago.

The Peruvian Andes provide one of the best examples of how such a subduction margin evolves. The city of Cuzco lies in the central Andes at an elevation of 6,300 m above sea level, surrounded by high mountain peaks. A little less than 400 km to the west, the land drops rapidly toward the coast, where Lima, the capital of Peru, lies at sea level (Figure 12.22). Beyond lies a very narrow continental shelf and a further drop of 7 km to the bottom of the Peru-Chile Trench. The total change in elevation from mountain top to seafloor is more than 14,100 m—greater than the height of the tallest peak on the continents (Figure 12.23).

Highly precise measurements by Earth-orbiting satellites indicate that the South American plate and the Nazca plate to the west are converging at about 11 cm/yr—one of the highest rates in the world. All along the margin, earthquakes are common. As shown in Figure 12.24, earthquake foci define a Wadati-Benioff Zone that dips eastward at about 25 degrees and is still distinct at a depth of over 250 km, beneath the highest peaks in the Andes.

FIGURE 12.22   Satellite view of the Andes in southern Peru. The shore of the Pacific Ocean, visible at the bottom of this photo, is less than 400 km from the highest peaks in the mountains. *(NASA)*

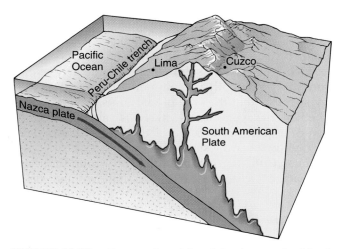

FIGURE 12.23   Cross section of the subduction margin of South America.

FIGURE 12.24   The Wadati-Benioff Zone beneath the Peruvian Andes dips gently to the east, reaching a depth of 250 km under the highest mountains of the region. *(Lubrecht and Cramer, Ltd.)*

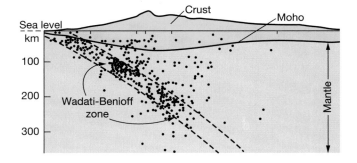

Volcanoes are also common in this area. Between central Peru and northern Argentina are over 600 stratovolcanoes—the greatest number in any single geographic region of the world. Many of these are inactive today, but others such as Mt. Mistí erupt ash continuously. According to some estimates, nearly 2,000 km³ of lava and pyroclastics have been deposited in the central Andes during the past 10 million years.

Geologists believe that this subduction zone has been active ever since the early Mesozoic era, although they disagree about the details of its history in the 200 million years since then. The diagrams in Figure 12.25 show one interpretation. They suggest that the pace of subduction was slow at first and was marked by stretching and uplift of the continental margin as its leading edge was drawn into the subduction zone. Throughout the Mesozoic, sporadic volcanism shaped the continental margin, much as it does from northern California to Washington State today. Marine sedimentation, however, continued to dominate the coast until the close of the Cretaceous period, about 70 million years ago.

Finally, about 58 million years ago, vigorous compression began to thicken the South American lithosphere. The marine sediments were intensely folded and thrust over the continental basement to the east. The rapid growth of the Andes coincided with a 35-million-year phase of regional metamorphism and

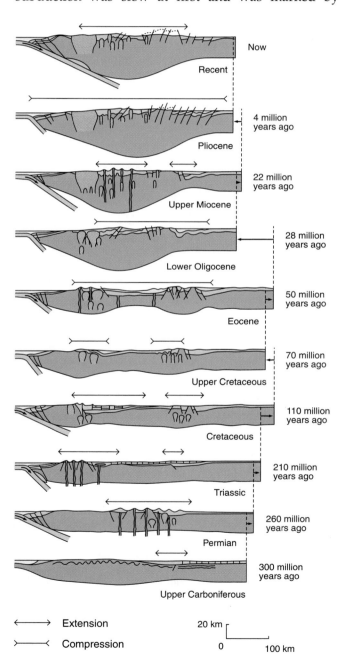

FIGURE 12.25

A geologic interpretation of events that have raised the Andes since the early Mesozoic era. Notice that the rate of plate movement has not been steady, and that for brief periods the continental lithosphere was stretched and faulted while subduction was rapid. This kind of behavior is characteristic of the second type of subduction margin, which can be seen in the Japanese islands. *(Modified from W. Zeil, The Andes: A Geological Review (Gebruder Borntraeger, Berlin, 1979 Lubrecht and Cramer, Ltd.)*

Orogens and Continental Margins   **297**

granitic plutonism. By the close of the Oligocene epoch, the crust was over 65 km thick. Isostatic uplift has led to several minor phases of crustal extension and volcanic activity since then, particularly during the Miocene epoch. Subduction continues today, but even with its occasional earthquakes and its many volcanoes, the Peruvian Andes are quieter than they have been for almost 60 million years.

Slow Continent, Fast Ocean A second type of subduction margin can develop when the ocean crust converges rapidly on a relatively slow-moving continental mass. In this situation, oceanic material is subducted faster than it can pile up against the adjacent plate. The overriding plate does not thicken, but can actually be stretched and thinned, and may even fracture as a result of extension (Figure 12.26). In this case, stratovolcanoes form an island arc separated from the continental mainland by a **back-arc basin.** The back-arc basin is a shallow sea that accumulates both clastic sediments from the continent and volcanic debris from the island arc. Sometimes, basaltic magma rising along

deep fractures may form a small region of oceanic crust in the basin as the island arc is pulled farther and farther away from the mainland. Because the subduction zone is far from shore, under the island arc, there is no volcanism on the continental mainland at all. Separated from the subduction event by the back-arc basin, the mainland experiences no compressional tectonics; hence, there are no nappes or thrust sheets. Its coastline, in fact, is very similar to the coast of a continent with a passive margin.

An excellent, although complex, example of this type of subduction margin can be seen in the Japanese islands. The people of Japan have always lived with geologic uncertainty. Theirs is a fragile country, an island system that, on average, feels one-tenth of the seismic energy and one-fifth of the volcanic energy released in the entire world each year. As Figure 12.27 shows, it is a complex region composed of not one but portions of five island arcs, each associated with an ocean trench that lies between 50 and 100 km east of the coast.

FIGURE 12.26
If subduction is rapid compared to the rate at which the continental lithosphere moves seaward, a subduction margin like the one in the Japanese islands develops. The subducting ocean floor drags the leading edge of the continent toward the subduction zone, causing tension in the continental lithosphere. As a result, rifting forms a back-arc basin between the stratovolcanoes of the island arc and the continental mainland.

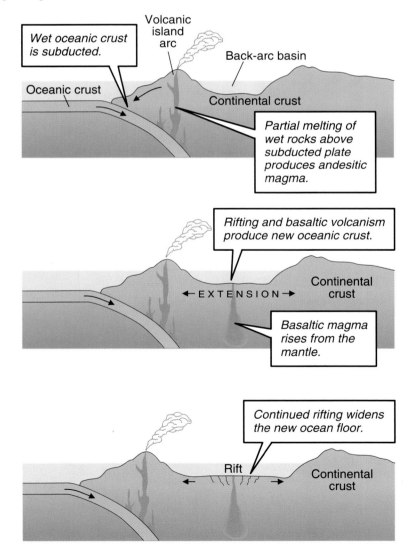

FIGURE 12.27 Tectonic map of Japan and the surrounding region. There are three back-arc basins here: in the East China Sea, the Sea of Japan, and the Sea of Okhotsk. Each has been created by rapid subduction along one or more of the trenches in this region.

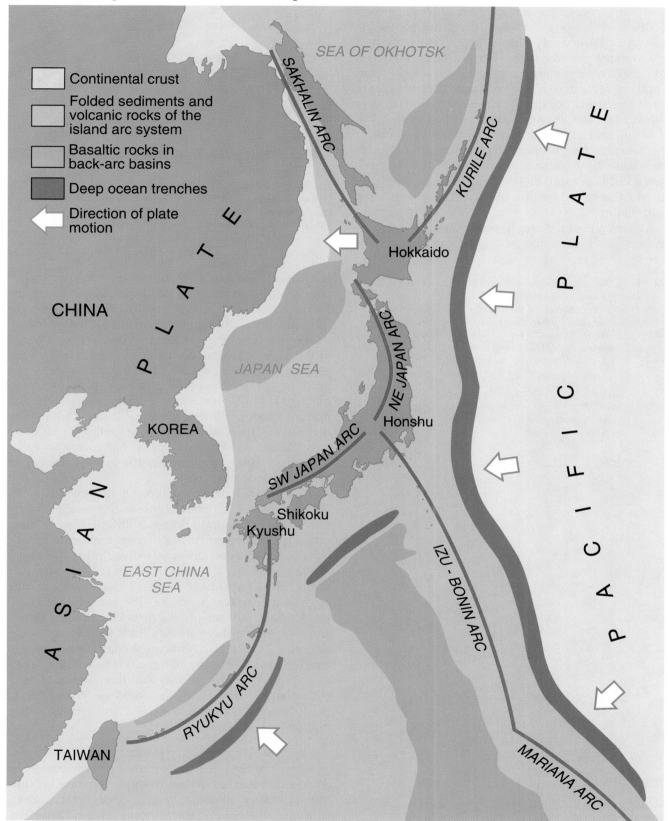

The oldest rocks in Japan are early Paleozoic sedimentary deposits in the southwestern third of Honshu, the largest of the four main Japanese islands. These have been metamorphosed during several lengthy episodes of subduction along the Southwest Japan Arc. Particularly during the Cretaceous and early Tertiary periods, this was a region of high volcanic activity. Beginning in the Miocene epoch about 20 million years ago, however, the centers of activity shifted eastward to the Northeast Japan Arc and southward to the Ryukyu and Izu-Bonin arcs. All of the 40 volcanoes active in historical times lie in one of these two arcs, as do almost all of the roughly 200 volcanoes that have erupted since the Miocene epoch (Figure 12.28). Among these are some of the most violent and the most picturesque stratovolcanoes in the world (Figure 12.29).

At least one of the arcs has been an active subduction zone ever since the late Paleozoic era. Japan's

FIGURE 12.29   Mount Fuji, the Japanese volcano that is most familiar to Westerners, has a picturesque shape that has made it a favorite subject for artists. Mount Fuji is 3,776 m tall and has a circumference of 30 km at its base. This woodblock print by Hokusai (1760–1849) is one of a well-known series of 36 views. *(Giraudon, Art Resource)*

FIGURE 12.28   Quaternary volcanoes in Japan are concentrated in two belts, one associated with the active Northeast Japan Arc and the other with subduction along the Ryukyu Arc, which extends southwestward toward Taiwan. There are no active volcanoes in southern Honshu or Shikoku. *(A map derived from Figure 15.2 in A. Miyashiro,* Metamorphism and Metamorphic Belts, *p. 351, George Allen and Unwin © 1973 by Routledge, Chapman & Hall, Inc.)*

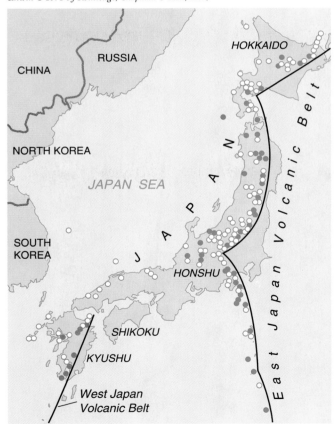

● Active volcano
○ Other Quaternary volcano

active volcanoes sit about 100 to 150 km above westward-dipping Wadati-Benioff zones, along which the Pacific plate and several smaller fragments of the lithosphere are diving into the mantle. Shallow earthquake epicenters are concentrated on the Pacific coasts of the islands. Deeper earthquakes are felt as far west as the coastal provinces of China. Vigorous subduction gradually pulled the island arcs away from the Asian mainland until roughly 5 million years ago and opened a series of back-arc basins: the East China Sea, the Sea of Japan, and the Sea of Okhotsk. All three have floors of continental rocks, but each also has a spreading center at which basaltic magma has begun to form new oceanic crust.

The overall pattern of crustal movement in the Japanese islands is not as simple as this brief overview may make it sound. Because different subduction zones have been active at different times, the islands have changed direction repeatedly. During the past 5 million years, in fact, movements in the northern part of the Sea of Japan have reversed. Geologists attribute large earthquakes west of the major island of Hokkaido to movement on a thrust fault that is slowly carrying the northern islands back toward Asia. The arrows in Figure 12.27 indicate the intricate pattern of relative movements in the region. Given the area's complicated geologic history and the number of competing subduction zones, it would be impossible to predict how the Japanese islands will change in the millions of years ahead.

Collisional Margins

An ocean basin ultimately disappears when subduction at one of the surrounding continental margins consumes the oceanic lithosphere between them (Fig-

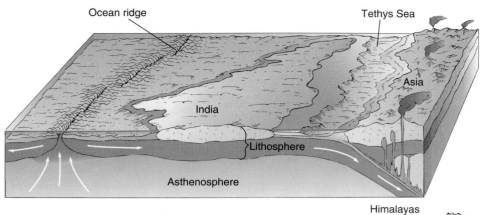

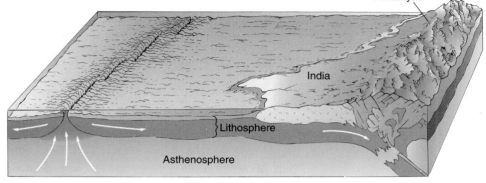

FIGURE 12.30
This series of illustrations shows
how a subduction margin may eventually become a collisional margin if
the ocean floor between two continental masses is completely consumed. *(Readapted with permission of
Macmillan College Publishing Company from
THE EARTH: AN INTRODUCTION TO
PHYSICAL GEOLOGY 4/E by Edward J.
Tarbuck and Frederick K. Lutgens. Copyright
© 1993 by Macmillan College Publishing
Company, Inc.)*

ure 12.30). When opposing segments of continental
lithosphere finally meet, the result is a **collisional margin,** sometimes also called a **suture** because it joins
previously separate continents into a new, larger unit.
The margin is marked by a wide belt of fold-and-
thrust mountains and substantial crustal thickening.

Despite many similarities to the subduction margins from which they evolved, collisional margins differ in two important ways. Because *both* continental
masses at a margin such as this have low densities,
both are buoyant and neither can be subducted below
the other. This means that the andesitic volcanism
associated with subduction zones ceases when the last
bit of ocean crust is consumed. This also means that
deep earthquakes are rarely felt once continental
masses converge, although widely scattered shallow-
to intermediate-focus earthquakes may be common.

The best modern example of a collisional margin is
the suture that extends from the Caucasus Mountains
of Georgia and Azerbaijan to the Himalayas in India
and southern China. The Himalayas alone, which take
their name from the Sanskrit words *hima* ("snow") and
alaya ("abode"), stretch over 2,700 km along the borders of countries from Afghanistan to Burma and
include the world's highest continental mountain
peaks (Figure 12.31). Mt. Everest, known to Tibetans
as Chomolungma ("goddess mother of the world")
and to the Nepalese as Sagarmatha ("sky head"), is the
tallest peak, nearly 9 km above sea level. The base of

the crust is as much as 70 km below these peaks.

Before the supercontinent of Pangea began to fragment in the early Mesozoic era, its northern portion
(Laurasia) and its southern portion (Gondwana) were
partially separated by a narrow arm of the ocean that
geologists have named the Tethys Sea (Figure 12.32).
The modern political boundary between Nepal and

FIGURE 12.31   These peaks in Nepal are among the highest in
the Himalayas. This mountain range at the northern margin of the
Indian subcontinent, the most elevated region of the world, began to
rise 20 million years ago and is still growing at a rate of 2 to 5
mm/yr. The colorful cloths in the foreground are prayer flags. *(Roger
Chapin)*

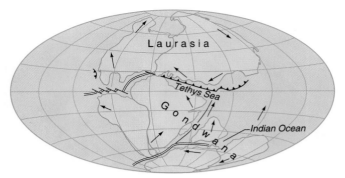

FIGURE 12.32   Pangea as it may have looked as it began to break up 225 million years ago. The northern shore of the Tethys Sea, where it lapped against Laurasia, was a subduction margin. Its southern shore, the coast of Gondwana, was a passive margin. (*This illustration was derived from Figure 16.1 in B.J. Skinner and S.C. Porter's* THE DYNAMIC EARTH 2/E, © 1992 John Wiley & Sons, Inc.)

FIGURE 12.33   By about 200 million years ago, the breakup of Gondwana had already begun. Antarctica, Australia, and the Indian subcontinent moved eastward during the Jurassic period. They separated from each other during the Cretaceous period, and India started moving slowly northward. (*Readapted with the permission of Macmillan College Publishing Company from* THE EARTH: AN INTRODUCTION TO PHYSICAL GEOLOGY 4/E by Edward J. Tarbuck and Frederick K. Lutgens. Copyright © 1993 by Macmillan College Publishing Company, Inc.)

China lies roughly along what was then the northern shore of this sea, a subduction margin along which the Tethys seafloor was sliding beneath Laurasia.

Gondwana, to the south of the Tethys Sea, began fragmenting about 200 million years ago (Figure 12.33). During the Jurassic period, a rift began to divide the southern lands, opening what are now the Somali and Mozambique basins in the western Indian Ocean. The continental fragment to the east of the rift subdivided further during the late Mesozoic era. The two largest pieces eventually became Australia and

Antarctica. A third fragment, the Indian subcontinent, separated from Australia and Antarctica early in the Cretaceous period and began moving northward. The northern shore of the Indian subcontinent at that point was a passive margin.

Then, about 80 million years ago, India picked up speed as the rate of subduction at the northern margin of the Tethys Sea increased. Geologists read the record of India's northward movement in the ages of rock on the Ninetyeast Ridge, a submarine volcanic chain on the floor of the Indian Ocean (Figure 12.34). Each vol-

FIGURE 12.34   Ninetyeast Ridge takes its name from the 90° east longitude line, along which it lies. Geologists believe that it grew as the Indian Plate moved northward over a mantle plume, in much the same way that the Hawaiian Islands are growing today. By determining the ages of volcanic rocks along the ridge, geologists can interpret the progress of the Indian Plate on its northward journey. The inset shows relative positions of India and Asia from 80 million years ago to the present.

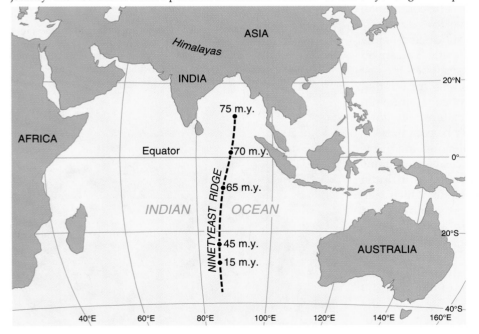

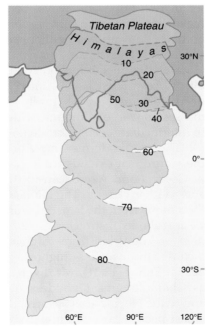

canic peak on the Ninetyeast Ridge formed when it was situated at 35° south latitude, directly above a plume in the mantle below. As India moved northward, the oceanic crust following it was punctured repeatedly by rising basaltic magma, leaving a train of seamounts much like the Hawaiian Islands and the Emperor seamount chain in the north Pacific basin. By examining the isotopic ages of the basalts, geologists have determined that India traveled the last 3,300 km to its collision with Asia at between 10 and 12 cm/yr, comparable to the fastest plate velocities in the world today.

Geologists have determined that the last segment of ocean crust in the eastern end of the Tethys Sea was subducted about 50 million years ago, as India finally collided with the main Asian landmass. All that remains of that seafloor are bands of metamorphosed sediment, called **ophiolites,** found in the high Himalayas. As the Tethys Sea vanished, the subduction margin became a collisional margin. India continued northward, crushing its leading edge and the southern margin of Asia.

The Himalayas began to rise about 20 million years ago as the Indian continental mass tried to follow the Tethys seafloor into the subduction zone. The relatively buoyant granitic crust of India would not descend into the asthenosphere, but was added to the base of the Asian plate, thickening it and elevating both the Himalayas and the broad region in western China known as the Tibetan (Qinghai-Xizhang) Plateau. The Himalayas have risen most rapidly since the beginning of the Quaternary period less than 2 million years ago. Geologists estimate that the Himalayas and the Tibetan Plateau are still rising at the astonishing rate of 2 to 5 mm/yr. Even considering the rate of erosion in the region, Mt. Everest may now be almost a centimeter taller than when it was first scaled in 1953. The crust continues to thicken as India is forced farther northward.

Transform Margins

Transform faults on the seafloor develop because portions of a mid-ocean ridge spread at different rates. The reason that spreading rates vary is that plates cannot move directly apart on the spherical surface of the Earth. Because the Earth's surface is curved, any plate moves along an arc rather than in a straight line. We can picture that movement by imagining that the plate is actually rotating around a **spreading pole** (Figure 12.35a). If the plate rifts, then each of the pieces rotates around the spreading pole and, as Figure 12.35b indicates, the end of the rift that is farthest from the pole inevitably opens faster than the end closest to it. It does not open evenly, however, because the lithosphere deforms more easily along some parts of the rift than others. Where adjacent parts of the lithospheric boundary move at very different rates, transform faults form perpendicular to the rift.

Transform faults are an unavoidable consequence of divergence, yet it is still the rift that marks the boundary between plates. The outstanding feature in Figure 12.35b, for example, despite the presence of several transform faults, is the spreading center. How can a transform fault, then, ever be the dominant tectonic feature at a continental margin? How can a **transform margin** exist?

To find out, let us study the San Andreas fault zone, which extends from Cape Mendocino, north of San Francisco, to the southern end of the Gulf of Cali-

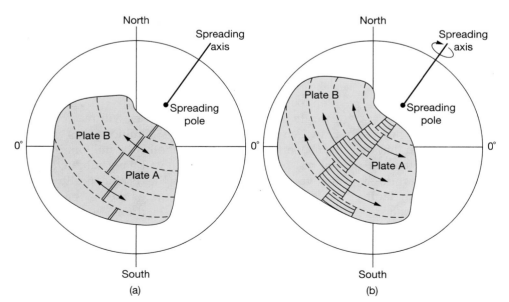

FIGURE 12.35
(a) Plates A and B, moving on the spherical surface of the Earth, are beginning to separate into independent plates along a spreading center. Their movement follows an arc, and can be described as a rotation around a spreading pole. (b) The rift between plates A and B opens faster on the end that is farthest from the spreading pole. Because of variations in the strength of the lithosphere, the rift does not open along one smooth, straight line but breaks into jagged pieces, each bounded by a transform fault. (*This illustration was derived from Figure 5.7 in P.J. Wyllie's THE WAY THE EARTH WORKS, © 1976 by John Wiley & Sons, Inc.*)

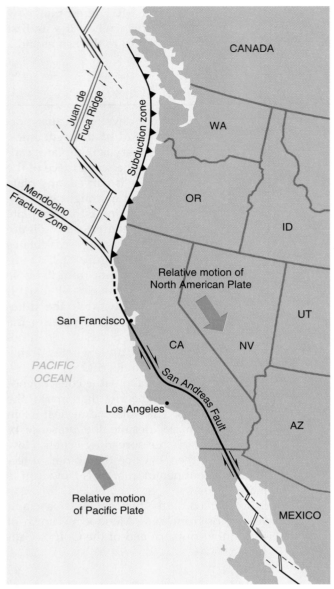

FIGURE 12.36 The San Andreas fault separates the North American and Pacific plates. It is a transform boundary, originally a transform fault on a portion of the East Pacific Rise that has been subducted beneath the continental margin. *(Readapted with permission of Macmillan College Publishing Company from* THE EARTH: AN INTRODUCTION TO PHYSICAL GEOLOGY 4/E *by Edward J. Tarbuck and Frederick K. Lutgens. Copyright © 1993 by Macmillan College Publishing Company, Inc.)*

fornia (Figure 12.36). Along the way the San Andreas fault has several names, but it is so well known that to most Californians it is simply "the fault." It is the best-known example of a modern transform margin—a continental margin dominated by a transform fault. On its eastern side is the North American Plate, a single lithospheric block with its eastern edge at the Mid-Atlantic Ridge, over 6,000 km away. To the west, the narrow sliver of continental material from San Francisco to the southern tip of Baja California is part of the Pacific Plate.

The transform fault itself is actually much longer than the continental margin. From Cape Mendocino, it turns abruptly westward into the Pacific basin, where it is known as the Mendocino Fracture Zone (Figure 12.37). There, hidden under 4 km of water, it is even more spectacular than on the land. Scientists from the Scripps Oceanographic Institute in La Jolla, California, were astonished to discover in 1956 that opposite sides of the fault were offset by more than 1,100 km.

Unlike the Atlantic basin, in which the spreading axis is nearly centered, the Pacific basin is asymmetrical. Its ridge, in fact, is within 1,000 km of the shore along much of the South American coast and finally runs into it in central Mexico near Puerto Vallarta. Geologists have determined, however, that the ridge system (known as the East Pacific Rise) was once much farther from the coast of North and South America than it is today. A single plate—the Farallon Plate—comprised the ocean floor from Central America to Alaska (Figure 12.38a). The trench that still parallels the South American coast continued as far north as the Gulf of Alaska. The California coast at that time, 60 million years ago, was a subduction margin much like the present-day coast of Peru. Proof of this is that the mineral changes of blueschist metamorphism, commonly associated with subduction, can be found throughout the coastal mountains of Southern California. Along this segment of the plate boundary, however, the oceanic lithosphere was subducted much faster than the East Pacific Rise was spreading. As a result, the North American Plate gradually overran the Farallon Plate, which disappeared into the mantle (Figure 12.38b).

About 29 million years ago, the East Pacific Rise itself was drawn into the subduction zone. As shown in Figure 12.38c, the present San Andreas system is thought to have evolved from a transform fault that once connected segments of the rise. Its lateral movement has sheared off a portion of North America, which now travels as part of the Pacific Plate instead. The Farallon Plate has largely been consumed. Only two small remnants, renamed the Cocos Plate in Central America and the Juan de Fuca Plate from Cape Mendocino to Vancouver Island, survive. At its present rate of movement, Los Angeles will reach the San Francisco suburbs in roughly 30 million years and will be 600 km farther north in another 30 million. By then, the rest of the Farallon Plate will have subducted and the volcanic mountains of the Pacific northwestern states will have ceased to erupt as the last part of the subduction margin vanishes (Figure 12.39).

Shallow earthquakes, sometimes quite intense, are common along this transform margin. Because there is no subduction zone, however, volcanism is not common. Mountain building of any kind, in fact, is rare because there are no convergent forces to thicken and

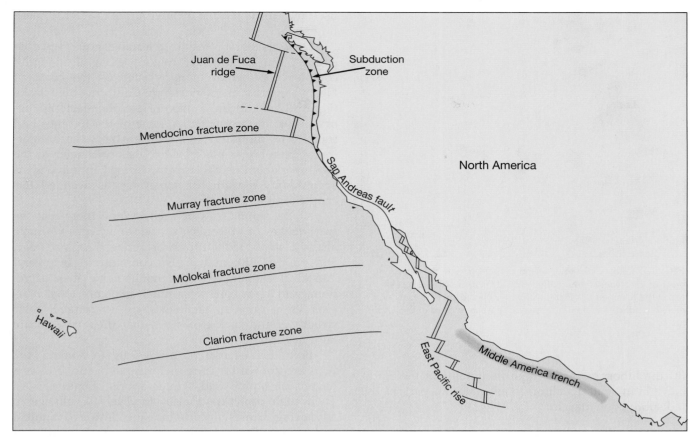

FIGURE 12.37 There are several major parallel fracture zones in the eastern Pacific basin. Each is a transform fault that once offset a portion of the East Pacific Rise. The Mendocino Fracture Zone is the seaward extension of the San Andreas fault, which ceases to be a plate boundary as it leaves the North American coastline.

FIGURE 12.38 (a) Until the early Tertiary period, the western continental margin of North America was a subduction margin. (b) The Farallon Plate, to the east of the mid-ocean ridge, was gradually consumed as North America moved westward. (c) About 29 million years ago, the mid-ocean ridge was finally subducted and the San Andreas fault developed from a transform fault. *(This illustration was derived from Figure 16.25 in B.J. Skinner and S.C. Porter's THE DYNAMIC EARTH 2/E, © 1992 John Wiley & Sons, Inc.)*

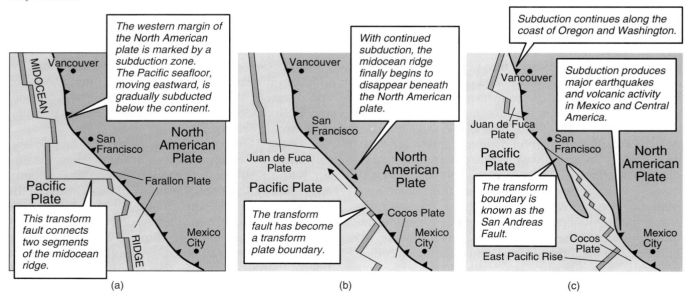

Orogens and Continental Margins **305**

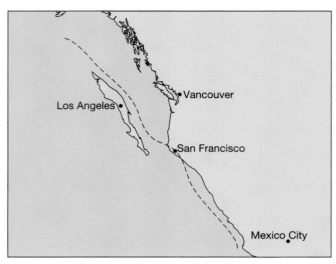

FIGURE 12.39 Several million years from now the remaining piece of the ancient Farallon Plate in the Pacific northwest (now called the Juan de Fuca Plate) will be subducted and the transform plate boundary will dominate the entire western coast of the continent.

lift the lithosphere. The high mountains of the Sierra Nevada in central California are relics, in part, of the old subduction margin.

Accretionary Margins

As geologists have learned more about continental margins, they have found that there are few places in the world that have had simple tectonic histories. The geographic regions we have highlighted throughout this chapter, in fact, demonstrate how complex even some of the best examples of passive, subduction, collisional, and transform margins are. Geologists now recognize that some continental margins have evolved from such a complicated series of tectonic events that they should be considered as yet a fifth kind of margin. These are called **accretionary margins.**

Along some plate boundaries, the lithosphere is weak enough that it breaks into several pieces. Some of these may be fragments pulled off the larger plates by rifting or sliced off by transform faulting, just like the California coast as it slides away along the San Andreas fault. Elsewhere, an island arc may be detached from a continental mass as its back-arc basin grows wider. In still other places, an island arc system like the Marianas Islands forms where one piece of oceanic crust subducts beneath another, creating a miniature "continent" where each volcano erupts above the subducting plate. Each of these many situations produces a small crustal fragment that is too buoyant to be subducted and will eventually collide with other fragments or a larger plate. Geologists have

used the words **terrane** or **microplate** to refer to these fragments.

Geologists first recognized terranes not as separate entities, but as unusual regions along structurally complex continental margins such as the coast of North America from northern California to Alaska (Figure 12.40). In the context of the plate tectonic theory, it became evident that some of the terranes had traveled for thousands of kilometers before accumulating against larger plates. Geologists now refer to the patchwork continental margins formed in this way as **accretionary margins** or sometimes as **accreted terrane margins.**

At the boundaries of each terrane there may be earthquakes or volcanism of the same types we have already associated with other active margins. Fold-and-trust mountains are also common, as is general crustal thickening. To the nongeologic eye, accretionary margins look very much like collisional margins. It is only as geologists try to unravel their structures and rock exposures that their true nature becomes apparent.

From France and northern Italy to Austria, the Alps are known for their picturesque beauty and as a haven for alpine and winter sports (Figure 12.41). Their high peaks have also stood as a challenge to advancing armies since before the days of Hannibal and serve today as a natural defense for the nation of Switzerland. Like the Himalayas, these are mountains raised by massive continental collisions, but their history is far more complicated.

Like the Himalayas, the southern margin of Europe also evolved from Pangea's disintegration, but in a different way. The Atlantic Ocean basin opened from the south to the north, so that Africa drifted eastward while Europe remained attached to North America. Along the boundary between Africa and Europe, several small terranes became detached from the larger continental bodies and were dragged and rotated between them (Figure 12.42). The Iberian Peninsula, on which Spain and Portugal sit, was one of these. Traces of several smaller fragments can be found in the rocks of the western Mediterranean Sea and around Greece. The largest terrane was a lithospheric block originally torn from Africa that geologists have called Apulia.

As the Mesozoic era progressed, the African and European plates changed relative directions several times as a result of the gradual opening of the southern Atlantic Ocean and the simultaneous closing of the Tethys Sea. Apulia consequently changed directions repeatedly as well. West of Apulia smaller terranes caught between southern Europe and Africa also moved first one way and then another, and a set of interconnected marine basins opened slowly around

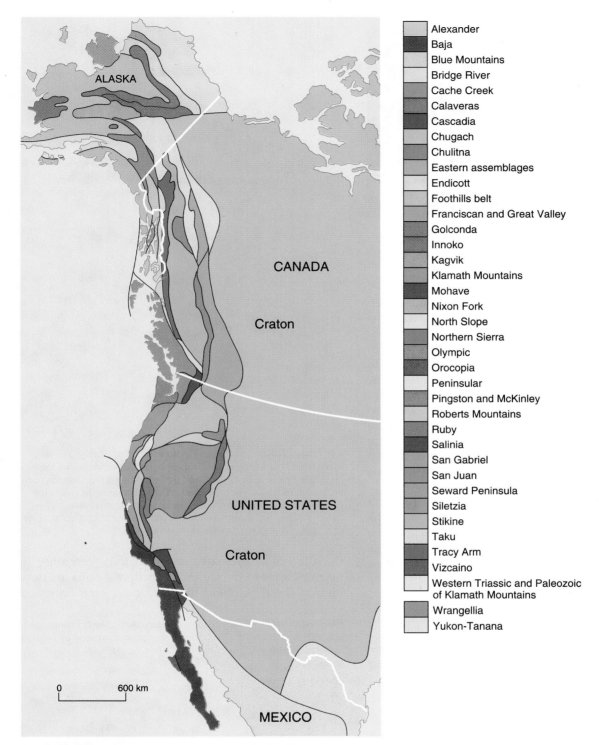

Legend:
- Alexander
- Baja
- Blue Mountains
- Bridge River
- Cache Creek
- Calaveras
- Cascadia
- Chugach
- Chulitna
- Eastern assemblages
- Endicott
- Foothills belt
- Franciscan and Great Valley
- Golconda
- Innoko
- Kagvik
- Klamath Mountains
- Mohave
- Nixon Fork
- North Slope
- Northern Sierra
- Olympic
- Orocopia
- Peninsular
- Pingston and McKinley
- Roberts Mountains
- Ruby
- Salinia
- San Gabriel
- San Juan
- Seward Peninsula
- Siletzia
- Stikine
- Taku
- Tracy Arm
- Vizcaino
- Western Triassic and Paleozoic of Klamath Mountains
- Wrangellia
- Yukon-Tanana

FIGURE 12.40 The coast of North America from northern California to Alaska is composed of a large number of structurally and stratigraphically unrelated terranes, some of which have moved at least 6,000 km from where they were probably formed. The various terranes include island arcs, pieces of the Asian continent and other landmasses, and pieces of North America that have slid along small-scale transform margins. Until its complex nature was gradually revealed in the 1970s, this coastline remained a source of serious conceptual problems for geologists who had accepted the plate tectonic theory. *(This illustration was derived from Figure 16.26 in B.J. Skinner and S.C. Porter's THE DYNAMIC EARTH 2/E, © 1992 John Wiley & Sons, Inc.)*

FIGURE 12.41 Swiss Alps near the town of Chateau-d'Oex, Switzerland. *(R.L. Ciachon, Visuals Unlimited)*

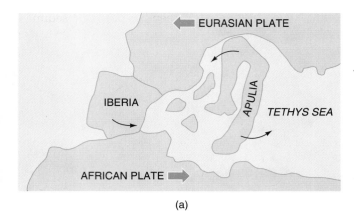

(a)

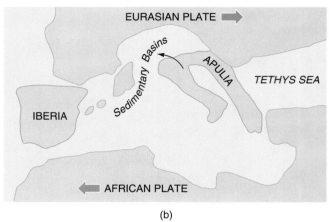

(b)

FIGURE 12.42 As Pangea began to break up, Africa slid eastward relative to Europe. Small lithospheric blocks, including those on which Spain, Italy, Greece, and other Mediterranean islands now lie, broke free and were rolled and crushed between the two large plates. The largest of these was Apulia. *(Readapted with permission from* Nature *from figures 2, 3, and 4 in the article by K. Hsü (Sept. 3, 1971, v. 233, pp. 44-48.) Copyright © 1971 by Macmillan Magazines Limited.)*

them. During the Jurassic period, these collected thick beds of clay, sand, gravel, and carbonate muds. Then about 80 million years ago, during the Cretaceous period, the North Atlantic began to open.

With the opening of the North Atlantic, the European Plate began moving steadily eastward relative to the African Plate. Sediments in the small basins to the west of Apulia were squeezed and folded (Figure 12.43). A subduction zone developed on the southern margin of Europe as Apulia, trapped between Europe and Africa, was now forced northward. An island arc and a back-arc basin thrived briefly and were crushed against Europe as Apulia finally collided at the close of the Cretaceous period, 60 million years ago.

The suture between Apulia and Europe lies in a broad trough between the Pennine Alps and the Bernese Alps in southern Switzerland, where both the Rhône River and the Rhine have their headwaters. As the plates collided, both were compressed and draped to the northwest in massive nappes (Figure 12.44). The

thick layers of Mesozoic sediment that were once in marine basins to the west and north of Apulia are now high in the Alps. Near the border between Italy and

FIGURE 12.43 As the North Atlantic began to open, movements on the boundary between Europe and Africa changed direction and Apulia was pushed to the northwest. The shallow ocean floor ahead of it was subducted below Europe and finally consumed as Apulia collided. The collision produced the Alps. *(Readapted with permission from* Nature *from figures 2, 3, and 4 in the article by K. Hsü (Sept. 3, 1971, v. 233, pp. 44-48.) Copyright © 1971 by Macmillan Magazines Limited.)*

FIGURE 12.44 Massive folds created during the collision between Apulia and the European Plate are now exposed by erosion in the high Alps near the Swiss-Italian border. *(Martin G. Miller, Visuals Unlimited)*

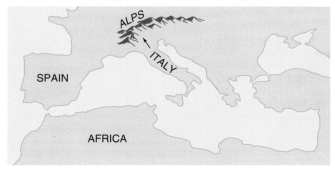

Switzerland, they can be seen in stacks of sediment that have been folded over each other eight to twelve times. By some estimates, if these nappes could be flattened out today, Switzerland would be ten times larger.

Apulia finally stopped moving northward by the mid-Oligocene epoch, about 30 million years ago. The majestic peaks of the Alps have been shaped by isostatic adjustments and erosion since then. Apulia and the other terranes are now firmly attached to the Eurasian Plate along a complex accretionary margin that is one of the most intensely studied and most confusing regions in the world.

The Wilson Cycle

No continental margin behaves the same way forever. As plates change direction and velocity, ocean basins that were once expanding may shrink or be distorted into new shapes. Where plates once converged, they may diverge or slide past each other. J. Tuzo Wilson, the Canadian geologist who postulated the existence of transform faults in the late 1960s, saw a regular pattern of continental evolution in these seemingly random events. This historical pattern, which geologists now refer to as a **Wilson Cycle,** is attractively simple and accounts for the roughly parallel arrangement of orogens on several continents.

Wilson, like many geologists before him, was fascinated by the Appalachian Mountains in the eastern United States and became even more intrigued as the history of seafloor spreading in the North Atlantic was revealed in the 1960s. If the North Atlantic margin is passive as it appears to be, he wondered, why is there a major orogen parallel to the shoreline? More puzzling still, how could geologists explain the older and more subtle mountains of the Grenville orogen that parallels the Appalachians farther inland?

Wilson's solution was to suggest that the Atlantic Ocean basin has opened and closed several times, leaving successive parallel orogens during each closing. Isotopic dating and paleomagnetic studies have confirmed that the Grenville orogen formed on a collisional margin about 1.0 billion years ago. The suture did not last for long, however. A rift developed roughly parallel to the suture during the late Proterozoic eon, and by about 600 million years ago a new ocean basin opened, with passive margins on each side (Figure 12.45a). Geologists have named this the Iapetus Ocean. Sediments began to accumulate along the new coasts. Then, by the late Cambrian period, 500 million years ago, the plates reversed direction again. The continental margin became first a subduction margin and then a collisional margin as Pangea was

Late Precambrian Period
600 million years ago

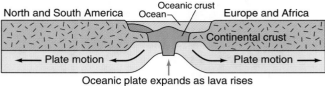

Expansion of "Appalachian Atlantic Ocean" begins

(a)

Mid-Ordovician Period
450 million years ago

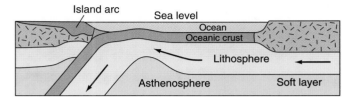

Ocean shrinks. Oceanic crust slips under American continental margin. Some island arcs form along coast.

(b)

Late Devonian Period
350 million years ago

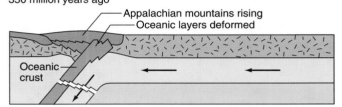

Ocean vanishes. Continents collide, causing further folding of Appalachian mountain system.

(c)

FIGURE 12.45   This sequence of drawings illustrates one interpretation of the events that opened and then closed the Iapetus Ocean. (a) Rifting of a supercontinent in the late Proterozoic eon separated America from Europe and Africa, creating the Iapetus basin. Passive margins developed on both sides of the ocean. (b) During the Ordovician period, the basin became smaller owing to subduction under North America. Island arcs and other subduction features were common along the coast of North America at that time. (c) At the close of the Devonian period, subduction was complete. The Iapetus Ocean disappeared, and the collision between continents on its two shores produced a mountain range that extended from what is now the southern United States to northern Norway. *(The above illustration is based on the diagram accompanying the article "Mountain Birth Linked to Oceans" by Walter Sullivan (Sept. 12, 1970, p. 13). Copyright © 1970 by The New York Times Company. Reprinted with permission.)*

assembled (Figure 12.45b, c). The Appalachian orogen, created during the Ordovician and Devonian periods, is a fold-and-thrust belt that bears many similarities to the younger Alps and Himalayas (Figure 12.46). It is approximately parallel to the old Grenville orogen.

The early Mesozoic rifting that broke Pangea apart again was roughly parallel to the old collisional mar-

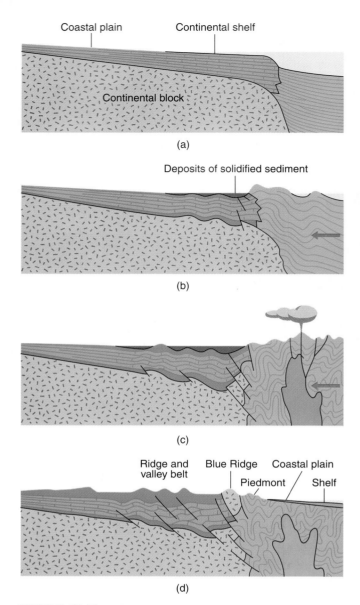

Coastal plain   Continental shelf

Continental block

(a)

Deposits of solidified sediment

(b)

(c)

Ridge and   Blue Ridge   Coastal plain
valley belt   Piedmont   Shelf

(d)

FIGURE 12.46   The Appalachian Mountains were formed as continental shelf sediments, originally deposited on the passive margin of North America, were trapped between colliding continents. As shown in this sequence of illustrations, faulting and folding of the shallow-water sediments produced the major fold belt that is known as the Valley and Ridge. From the Blue Ridge eastward, deep-water sediments were severely deformed and metamorphosed. High mountains that once stood east of the present Appalachians have been eroded away, and their roots are now below the Atlantic coastal plain, which extends out under the continental shelf. *(Readapted from the figure on p. 36 of R.S. Dietz's "Geosynclines, Mountains, and Continental-building" (March 1972, v. 226 (3)). Copyright © 1972 by Scientific American. Used with permission.)*

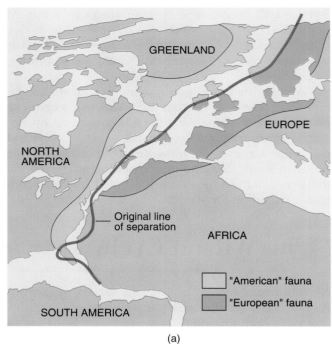

GREENLAND

NORTH AMERICA

EUROPE

Original line of separation

AFRICA

☐ "American" fauna
■ "European" fauna

SOUTH AMERICA

(a)

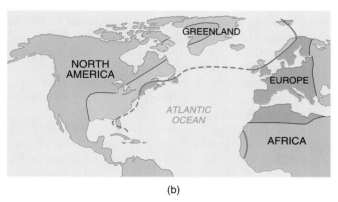

GREENLAND

NORTH AMERICA

EUROPE

ATLANTIC OCEAN

AFRICA

(b)

FIGURE 12.47   When the Atlantic Ocean opened at the end of the Mesozoic era, rifting was parallel to the old suture that marked the close of the Iapetus basin, but did not coincide with it perfectly. Paleozoic North American fossils are now found in western Norway and the northern British Isles, whereas European fossils are found in parts of New England and eastern Canada. *(Readapted with permission from* Nature *from figures 1 and 2 in the article by J.T. Wilson (1966, v. 211, pp. 676-81.) Copyright © 1966 by Macmillan Magazines Limited.)*

gin, although not in precisely the same place. As Figure 12.47 shows, thin slices of pre-Atlantic Europe and Africa can now be found in North America, and pieces of North America (recognizable by fossils) can be found in Scotland, Ireland, and parts of Scandinavia.

As we discussed earlier, the continental margins of the modern Atlantic Ocean have been passive since the mid-Mesozoic era. In some future era a subduction zone may again develop in the western Atlantic basin, and sediments of the continental shelf may be deformed into a new orogen as the ocean closes once again.

The idea of a Wilson Cycle has helped geologists to resolve several of the recurring questions that we have encountered since Chapter 8. It suggests, for example, that a suture is a weak zone in the lithosphere and a potential site for later divergence. From the recurring pattern of tectonic events, geologists also infer that some convection cells remain vigorous at nearly the same place in the mantle for hundreds of millions of years. This may be a useful clue to many unresolved questions about how convection currents work. Perhaps it will help geologists to decide, for example, which one of the several hypotheses about seafloor spreading and subduction suggested in this chapter is the correct one.

The concept of supercontinents dispersing and reuniting also sheds light on the century-old debate about the relationship between sediment accumulation and mountain building. You recall from Chapter 11 that Hall and Dana, studying the Appalachians, deduced correctly that the sedimentary rocks in them were once deposited offshore as a thick sequence of sedimentary beds. Dana even hypothesized that the sediments were crumpled against the continent, but he guessed incorrectly that the squeezing was due to global contraction. In the Wilson Cycle, we see that Hall's Appalachian geosyncline was a continental shelf on the passive margin of North America, much like the one that is there today. It became a mountain range, as shown in Figure 12.46, when the sediments were trapped in a collisional margin.

As geologists have studied each type of continental margin and learned how one type may evolve into another, they have begun to answer the questions with which we opened this chapter, questions regarding how continents may grow through the ages. Figure 12.48 is a generalized geologic cross section of North America from California to Maryland. Flanking the stable platform that is the underpinning of the Great Plains from Colorado to the Ohio River Valley, we see the fold-and-thrust ranges of the Appalachians, the Rockies, and the Coastal Ranges of California. We see also the batholith of the Sierra Nevada range and the horsts and grabens of the Wasatch Range and the Great Basin of Utah and Nevada. Each of these marks an episode in the evolution of North America as it was shaped by collision, subduction, transform faulting, and accretion. The continent has gained size by accumulating pieces of other plates and new volcanic rocks.

The geologic histories of the continents, as we have tried to show, are much different from those of the ocean basins. For all of their dynamic spreading and subduction, the seafloors are young and com-

FIGURE 12.48   A generalized geologic cross section of North America from California to Maryland. The insets highlight geologic structures that formed at different periods as the continent was assembled. (*This illustration was derived from Figure 16.25 in W.M. Marsh's EARTHSCAPE: A PHYSICAL GEOGRAPHY, © 1987 John Wiley & Sons, Inc.*)

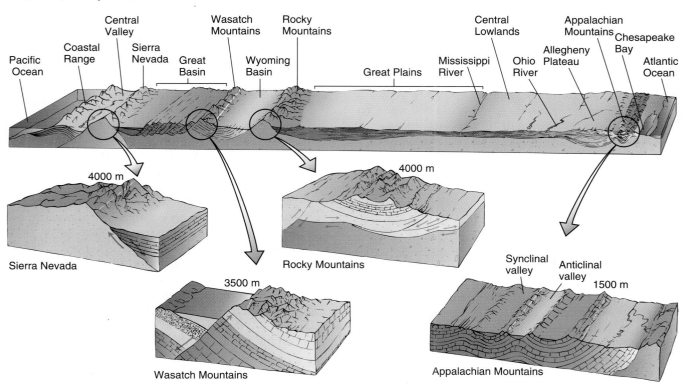

paratively simple. What geologists have learned is that the boundaries between plates, often those boundaries associated with continental margins where the oceanic and continental lithosphere meet, are the places where the greatest changes occur in the Earth.

Summary

1. The total amount of continental crust is greater today than it was during the Hadean eon. Geologists cannot agree, however, about whether the continental crust grew rapidly during the Archean eon, or whether it has been growing continuously throughout Earth's history. Geologists offer two hypotheses about the origin of the oldest pieces of continental crust, the cratons. One hypothesis suggests that cratons formed at hot spots, as shield volcanoes do, without the aid of plate tectonics. The other proposes that cratons were formed by igneous differentiation and andesitic volcanism at subduction zones. There is insufficient evidence to decide which hypothesis is correct. The oldest cratons, formed 4.2 billion years ago, seem to have been affected by both igneous and sedimentary processes. The early relationship between the rock system and tectonic processes, however, is not clear.

2. Geologists recognize two subdivisions in the continental crust. Cratons are low, featureless portions of the crust, composed largely of granite and metamorphic rocks. These regions formed during the Archean and Proterozoic eons and since then have been in nearly perfect isostatic balance with the mantle. Orogens are younger regions of highly deformed marine sediment, commonly metamorphosed or melted during mountain-building episodes. They are typically elongated regions surrounding the cratons. Orogens form as the result of tectonic processes at continental margins.

3. Continental margins that coincide with plate boundaries are called *active margins,* of which there are four kinds (subduction, collisional, transform, and accretionary). Even where the plate boundary is far from the continental margin (a *passive margin*), its development controls the geologic character of the margin. A *passive* margin evolves from a rift on a continent. Widening of the rift produces new seafloor and leaves continental margins that are not volcanically or seismically active. A *subduction* margin is associated with a trench on the seafloor, characterized by blueschist metamorphism and the accumulation of jumbled sediments called *mélange.* Severe, deep-seated earthquakes are common. If oceanic lithosphere is subducted below a continent that is rapidly overtaking it, the result is a margin with fold-and-thrust mountains, a continental arc of andesitic stratovolcanoes, and granitic plutons. If the oceanic lithosphere subducts quickly, island arcs evolve along the continental margin. Each is a chain of andesitic stratovolcanoes separated from the continental mainland by a back-arc basin, which may have a basaltic seafloor. A *collisional* margin forms where the seafloor between two blocks of continental lithosphere has been completely subducted. The lithosphere is thickened. Only shallow earthquakes are common, and no volcanism is observed. A *transform* margin develops from a transform fault that was originally associated with a spreading center on the seafloor. If the spreading center is subducted beneath a continent, then the transform fault may gradually replace the subduction zone at the continental margin. *Accretionary* margins develop where small terranes, usually unrelated to each other, accumulate against larger plates in a patchwork fashion. Each of the individual terranes may have a collisional, subduction, or transform margin with adjacent plates. Taken as a collection of terranes, however, they comprise a complex margin.

4. The Wilson Cycle is a recurring pattern of supercontinent formation and breakup, most evident in the rocks of eastern North America. Geologists see evidence that sediments that accumulated on a passive margin of the continent were deformed by later subduction and collision between the European and North American plates. Divergent plate motion later separated Europe and North America, opening an ocean basin between them and establishing new passive margins on their shores. This pattern of convergence, collision, and rifting was repeated at least twice, leaving behind the folded rocks of the Grenville orogen and the Appalachian mountain range. Geologists may learn more about convection in the mantle by studying this cycle of assembling and disassembling supercontinents.

KEY WORDS AND CONCEPTS

back-arc basin 298
boundary 287
continental arc 296
continental rise 290
continental shelf 290
continental slope 290
craton 284
failed rift 288
fold-and-thrust mountains 296
granitic belt 285

greenstone 284
greenstone belt 284
island arc 286
margin (active: subduction, collisional, accretionary, transform) 287
margin (passive: trailing) 289
mélange (clastic wedge) 295
nappe 296
ophiolite 303

orogen 284
pelagic ooze 290
plate triple junction 288
spreading pole 303
suture 301
terrane (microplate) 306
thrust sheet 296
Wilson Cycle 309

QUESTIONS FOR REVIEW AND THOUGHT

12.1 When did the total continental area of the Earth become as great as it is today? What alternative hypotheses have geologists suggested?

12.2 What distinctive types of rocks are typical of Archean continental crust? Which subdivision of the Archean crust is older?

12.3 What alternative hypotheses have geologists suggested to explain how the cratons began to form? What does each theory imply about the history of plate movement on the Earth?

12.4 What evidence do the rocks at Isua in Greenland offer about the nature of the Earth's surface during the Archean eon? What evidence do geologists have about the nature of an even earlier crust?

12.5 What is the difference between the terms "plate boundary" and "continental margin," as they are used in this chapter?

12.6 Describe how rifts are thought to develop at a plate triple junction. What pattern of igneous and sedimentary events commonly occurs? What kind of deformation is associated with rifting?

12.7 How is the appearance of a rapidly spreading mid-ocean rift different from the appearance of one that is spreading slowly?

12.8 Name the regions of the seafloor that you would encounter along a passive continental margin. What kinds of sediment are normally found on each? Where do the continental crust and the oceanic crust meet?

12.9 What hypotheses have geologists suggested to explain what makes oceanic lithosphere move from a spreading center to a subduction zone? What objections have been raised to these theories?

12.10 What distinctive sedimentary, igneous, and metamorphic rocks are associated with subduction margins?

12.11 In what ways are subduction margins such as the one along the west coast of South America similar to those like the one in the Japanese Islands? How are they different?

12.12 What styles of volcanic and seismic activity are associated with subduction margins?

12.13 How are blocks of continental lithosphere deformed where they meet at a collisional margin?

12.14 What kind of seismic activity is usually associated with a collisional margin?

12.15 Discuss the reasons why parts of a mid-ocean ridge spread at different rates. How does this produce transform faults? How do transform faults come to be associated with continental margins?

12.16 What is a terrane, and what does it have to do with accretionary margins?

12.17 How are accretionary margins similar to other kinds of active margins? How are they different?

12.18 What is the Wilson Cycle? How does it explain the existence of parallel orogens in eastern North America? How might it account for the presence of the Ural Mountains between Europe and Asia?

Critical Thinking

12.19 During its initial reconnaissance from high above the planet Fango, your scientific survey team has noticed several large rifts. Several members of the team believe that these are evidence that plate tectonics is active on Fango; others suggest that the rifts were caused by local vertical movements in the crust. You cannot land on the planet without alarming its inhabitants, so you will have to complete your survey from orbit. What could you look for in order to determine whether there has been plate movement on Fango?

12.20 While you are interviewing for a job with a company in Los Angeles, the personnel manager reads your college transcript and notices that you have taken a geology course. "I understand that California is sliding into the ocean," the manager says. "Is that true?" The manager sounds genuinely curious. This is a good opportunity for you to show how easily you can provide a simple but informative answer to an impromptu question. What could you say?

CHANGES AT THE EARTH'S SURFACE

So far we have examined the formation of rocks and the Earth's internal processes. In Chapters 2–7 we saw how rocks are made and how they change as they progress through the rock system, and in Chapters 8–12 we saw how the internal forces of the Earth operate and particularly how the global system of plate tectonics moves the pieces of the fragmented lithosphere from one place to another, pushing the landmasses above sea level. Yet just as surely as the internal forces, as expressed by plate tectonics, create the continents, so do the Earth's surface processes work to wear the continents away. We devote the next six chapters (Chapters 13–18) to an examination of these surficial processes and their work (Figure T3.1).

We have already discussed one surficial process, weathering, in Chapter 4. There we saw that weathering plays the very important role in the rock cycle of breaking down rocks and turning old minerals into new ones. We now begin to examine the other surficial processes that both erode the weathered material and transport it to new resting places and ultimately to the oceans. These processes include a gravity system that continuously moves masses of material down slopes; a vast surface system of rivers and streams; an enormous supply of underground water moving slowly

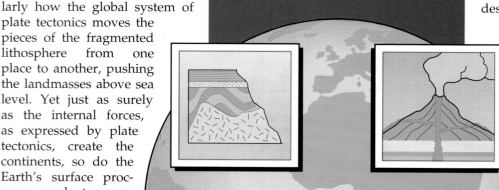

just beneath the surface; glaciers that crawl down the sides of high mountains and squeeze outward from the ice caps of Antarctica and Greenland; ocean waves that chip at the margins of the lands; and wind that drifts large volumes of loose silt and sand across dry desert floors.

Two sources of energy drive the surficial processes. One is gravity and the other is the Sun. Weathering and erosion make slopes of rock and soil increasingly unstable to the point that they are unable to resist gravity, which then moves them to lower elevations. Gravity also pulls water down slopes, and the energy of running water is, in part, used up by transporting materials through the stream channels. Moreover, gravity moves glaciers and underground water. The Sun provides its share of energy for the surficial processes through heat. The land and ocean absorb heat energy from the Sun and reradiate it to heat the atmosphere. It is heat from the Sun that pumps moisture into the atmosphere through evaporation, most of it from the oceans. The Sun's heat even effects the flow of air. Differential heating of the atmosphere—low temperatures at the poles and high temperatures at the equator—produces the circulation that we know as winds. These move moisture

FIGURE T3.1   Clouds represent moisture moving through the atmosphere. In this picture rain descends as a gray veil from the dark cloud at the top of the view.

through the atmosphere and connect land and ocean in the hydrologic system (Figure T3.1). Winds form the ocean waves that erode the shorelines and move sand and dust in the desert environment.

In Chapter 13, which follows, we discuss the way that gravity moves masses of Earth material down slopes from higher to lower levels. Gravity is responsible for the movement of vast amounts of rock and soil. Some of this movement is sudden and catastrophic. More often, however, it is slow, and much is almost imperceptible. Water eases the process of mass movement by saturating the pores and cracks in rock and soil, making it easier for gravity to move materials. The net result of all this movement is to transport the weathered debris of the continents to the rivers that then take over the task of moving it onward toward the ocean.

While gravity works on all slopes, the slopes themselves are quite varied, some long and gentle, others short and steep. Most of them lead to the banks of streams and rivers. These thin threads of water, the subject of Chapter 14, are critical to the creation of most of the Earth's valleys. As the extensive network of rivers drains the continents they also pick up the continents' weathered debris that gravity delivers to them (Figure T3.2). These are sediments destined to become, someday, new sedimentary rocks.

The action of rivers occurs at the surface and can be easily seen. Chapter 15, however, deals with water that flows unseen beneath the ground. This underground water, many times vaster than that in the rivers of the world at any one instant, is the result of water from rain and melting snow seeping through the soil into the underlying rock. There it flows, generally very slowly, through the pores and cracks in the rock, dissolving some minerals as it moves. If rock is

FIGURE T3.2   Gravity has moved rock from this sandstone cliff down to the edge of the San Juan River, Utah. There the stream will pick up the debris and move it on downstream toward the ocean. *(William C. Bradley)*

soluble enough, caves may form (Figure T3.3). More importantly, however, groundwater returns to the surface and serves as a major source of river water.

Rivers contain a significant amount of the Earth's freshwater, but more freshwater is stored in glaciers than in any place else on Earth. Chapter 16 examines

FIGURE T3.3   Underground water dissolved Carlsbad Caverns, New Mexico, out of beds of limestone. Subsequently, deposition of calcite from water dripping into the cave has formed these decorations of stone icicles (stalactites) and columns (stalagmites).

modern glaciers, the conditions that govern their advance and retreat, and how they modify the landscape by erosion and deposition. Glaciers were once more extensive than they are today, and at other times more widespread. We therefore close the chapter with a search for the causes of glaciations and deglaciations.

Glaciation and deglaciation have affected the levels of the oceans. These oceans, where most of the Earth's water has collected, are constantly modifying the shorelines of the world. In Chapter 17 we examine the way the energy of the wind is converted into the energy that changes shorelines, with how headlands are worn back, and with how waves set up currents that move sediments and build beaches. In this chapter we also examine the causes of rise and fall of sea level and look into the conflicts that often arise between people and the shoreline processes.

Deserts and the surrounding dry areas exist because, for a number of reasons, moisture-bearing winds fail to reach them. As a result, as we see in Chapter 18, wind is more important as a process here than any place else on Earth except in a narrow band along the shore. It moves large amounts of sand into dunes and distributes dust well beyond the desert edge. We shall learn also that people bring about desert conditions and initiate, particularly along the margins of the desert, a process of land degradation known as **desertification.**

Together the surficial processes wear down the continents, constantly competing with the process of plate tectonics operating to build them up. Before examining the surficial processes in detail, however, we should consider some more general aspects of continental erosion. We have learned that the great mountain belts rise as lithospheric plates collide with each other along converging borders. Erosion starts immediately as the mountains begin to rise. During the active mountain-building stage, erosion is unable to keep pace with uplift. Only after mountain building ceases does erosion bring a net lowering of the mountains. This process is a lengthy one and can take hundreds of millions of years, even if not interrupted. For example, the final compressive events in the construction of the Appalachian Mountains of the eastern United States occurred about 250 million years ago. Erosion has since removed kilometers of rock from their crests. Now, although the mountains are much lower than they were a quarter of a billion years ago, their erosion continues, but we still class them as mountains (Figure T3.4).

There are places, however, where erosion has had time to eat away at ancient mountain ranges. We find them on the cratons forming the interiors of the modern continents. Here, unhindered by the uplift associ-

ated with mountain building, erosion has been able to attack the continents for hundreds of millions of years. For instance, the Proterozoic and Archaeozoic rocks that form the Canadian shield are, in large part, metamorphic and igneous rocks that record the roots of mountains now vanished. Rocks of similar age extend southward into the United States between the Rocky Mountains and the Appalachian Mountains. They lie beneath a veneer of younger sedimentary rocks that obscures the low, eroded stumps of ancient mountains. On the cratons of continents, therefore, we find that erosion has reduced large expanses of these landmasses to a low, gently rolling landscape. Such surfaces of erosion are called **peneplains** from the Greek *pene* for "nearly" plus "plain."

The contest between erosion and plate tectonics is one in which tectonics is clearly dominant along the margins of converging plates. Within the interior of continents, however, the continental crust is generally buffered from the stresses of tectonic processes, which allows erosion to prevail and to reduce large segments of the landmasses to peneplains. Mountain ranges, although dramatic with their kilometers-high peaks, are restricted in distribution. In fact, if one looks at a relief map of the globe, it is remarkable how much of the land surface is less than a kilometer in elevation (Figure T3.5). Erosion has been able to reduce the average height of the continents to 840 meters.

Although erosion is more active than tectonics in continental cratons, plate tectonics is important in determining the environments in which the surficial processes operate. For instance, the rise of the Sierra Nevada a few million years ago changed a large part of the western United States from a well-watered landscape to arid desert with accompanying changes in the surficial processes. The Sierra Nevada and Cascade Mountains now block moisture-bearing winds from the land directly to the east. Such vertical movements of the crust help determine environmental

FIGURE T3.4   Sunrise in the Blue Ridge Mountains, near Boone, North Carolina. (*Frank J. Miller*)

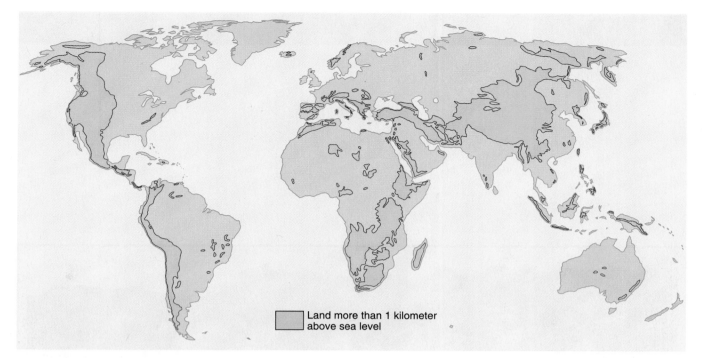

FIGURE T3.5 Area of the landmasses with elevations less than 1 km above sea level.

conditions. However, so too do the horizontal movements of plates. These movements transport crustal fragments thousands of kilometers. In doing so they move landmasses from one environmental zone to another. For example, in the last 300 million years, what is now India has moved, from a climate that buried part of it beneath kilometers of glacier ice, through temperate climes to its present location. Here, India has environments as diverse as glaciated mountains, hot deserts, and seasonally flooded plains. Conversely, lithospheric plates have moved out of nonglacial environments into polar climates. In this new environment glaciers have been able to form and spread over most of Antarctica and Greenland and, at times, over large areas of Canada and the United States, and across northern Europe.

As we study the various surficial processes in the coming chapters, we should remember that, according to the principle of uniformitarianism, these processes have been operating throughout millions of years in much the same way they do today. Plate tectonics, however, has also long been operating as it does today and has thus moved lands into new locations and different climates. As a result, a process of erosion that dominates a particular landmass for a particular span of time may change to another as its plate moves slowly into a new environment. More important, however, is recognizing that the erosion of the continents is an integral part of the rock system and connects the internal with the surficial processes of the Earth through the global system of plate tectonics.

Changes at the Earth's Surface   **317**

13

Ancient debris flow, east side of Lost River Range, south central Idaho (Dr. John S. Shelton)

Mass Movement

OBJECTIVES

▲▲

*As you read through this chapter, it may help if you focus
on the following questions.*

1. What is the relationship between gravity and mass movement?
2. What are the various factors that promote mass movement?
3. How do Earth materials behave during mass movement?
4. What are the various types of mass movement, and how are they separated from each other?
5. How can people protect against the hazards of mass movement?

OVERVIEW

▲▲

The Earth is constantly shedding its surface material, or molting, if you will. This is the continuous and ever-present process of mass movement. Sometimes thick slabs of rock slip rapidly away. Most of the time, however, the movement is slow, almost imperceptible. So slowly and quietly does it operate that soil and weathered rock are replaced, even as the process of removal proceeds. Within the rock system, mass movement is the process transporting Earth materials down slopes to the river systems of the world. Here materials are carried to the ocean as particles of clay, silt, and gravel. Beneath the sea, gravity continues to pull material down seafloor slopes to deeper basins of the world ocean where it settles out eventually to become sedimentary rock. Mass movement is expected in the normal course of surface change. Of immediate human interest is the effect that even the slowest movements can have on people and their activities, and how damaging effects can be avoided or minimized.

Gravity and Mass Movement

The shape of the Earth's surface—its landscape—is made of slopes. Soil and rock move down these slopes under the influence of gravity. The process is called **mass movement**—"mass" because large volumes of Earth material are generally involved. An individual movement may be extremely slow or incredibly fast—from millimeters per year to over 150 km per hour. Whether they are fast or slow, it is helpful to consider that movements are not isolated individual events but are a part of a continuing process. Let us take a rockslide as an example.

Long before a rockslide happens, numerous events occur to weaken the material beneath the slope making it more and more susceptible to gravity. An earthquake may loosen the rock in the slope, or a river may remove some support for a slope by undercutting its base. Even a sonic boom can shake the slope a bit and make it a little weaker. During all this time the slope remains stable but continues to approach a threshold of movement between stability and instability. Finally, a point arrives when the strength of the slope is so reduced that some specific event weakens it enough to push it across the threshold. Material on the slope can no longer resist gravity, and movement begins. The rockslide is triggered by some discrete event such as another earthquake, an exceptionally wet season, or the vibration of heavy traffic. Whatever the initiating event, it is only the last of many that have contributed, bit by bit, to the weakening of the slope. That last event—the trigger—simply pushes the slope across the threshold from stability to instability (see Figure 13.1).

We can express this concept of a threshold more precisely by examining the two forces at work on a hillside. The first is the weight of an object, such as a slab of rock, which is the force caused by gravity and produces **shear stress,** which causes movement of material downward parallel to the slope. The other is the strength of the slope, its **shear strength,** which is the slope's strength to resist movement. Figure 13.2 illustrates the changing relation of these forces.

When a rock rests on a horizontal surface its weight is directed vertically downward perpendicular to the surface, and shear stress between the rock and the surface is zero (Figure 13.2a). On an inclined surface, such as a hill slope, however, gravity can be divided into two components (Figure 13.2b, c). One is directed perpendicular to the slope and downward, and the other is parallel to the slope and downward. The component of gravity perpendicular to the slope holds the rock in place. The component of gravity par-

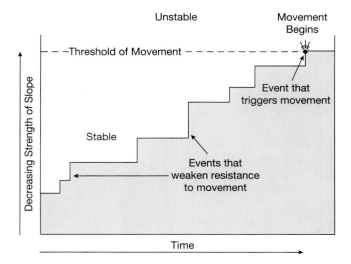

FIGURE 13.1 A series of events can continue to weaken the strength of a slope to a point at which its resistance crosses a threshold of resistance to gravity, and movement begins.

allel to the slope is the shear stress, and it becomes greater as the slope increases. The rock will not move as long as the shear stress is less than the shear strength, which in Figure 13.2b, c is the friction between the rock and the slope. As the slope and the shear stress increase, the rock approaches the threshold of movement. When the slope is steep enough the shear stress becomes greater than shear strength, the threshold is passed, and the rock moves downslope (Figure 13.2d).

The movement shown in Figure 13.2d begins because of the increase in the angle of the slope. Reducing the strength of the slope can also trigger movement. To understand the process of mass movement, therefore, one needs to understand the factors that increase the shear stress (slope angle) and reduce the shear strength (slope strength). Some of these are discussed below, and additional ones are listed in Table 13.1.

Factors Contributing to Mass Movements

Water

Water plays several roles in mass movement. An example from the sandbox or beach illustrates one. If you pour dry sand from a pail, it forms a conical heap. The more sand you add, the higher the cone becomes. No matter how much dry sand you add, however, the angle of the cone gets no steeper. This is the maximum angle of slope that the dry sand can

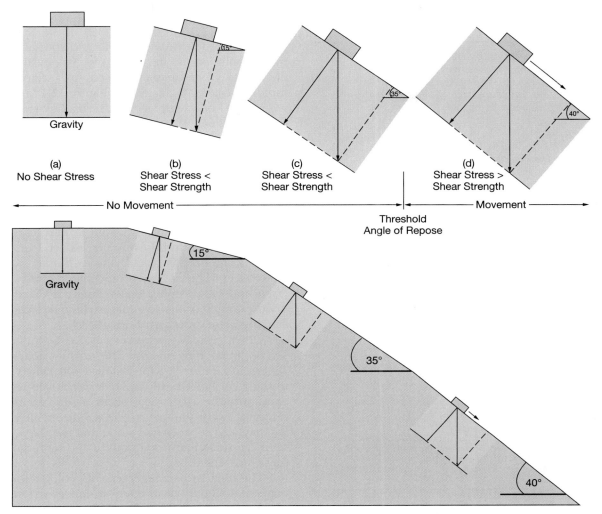

FIGURE 13.2 A diagram to show the relationships among gravity, shear stress, and shear strength on a slope of increasing steepness. See text for discussion.

maintain, and this is called the **angle of repose** (Figure 13.3). In Figure 13.3a, the angle of repose is about 34° at the threshold of movement.

As most people know, to build a good sand castle you need to moisten the sand. This increases the angle of repose by partially filling the pore spaces between the grains of sand with water. Under these conditions the surface tension of the water holds the grains together to withstand gravity, and you can build steep, even vertical slopes that will stand until the sand dries out, at which point the material slumps back to the gentler slopes of dry sand. If you saturate the sand, however, and fill the pore spaces completely with water, then the surface tension of water between the grains is broken and the friction between the sand grains is reduced. The entire mass collapses and flows.

In each of these examples the angle of repose of the sand is different; it is least for the saturated sand, most for the moist sand, and intermediate for the dry sand. These examples also provide another helpful way to think about slope stability. In each, the friction between the sand grains is different. It is least in the saturated sand and most in the moist sand. This internal friction of the slope material determines the angle of repose. The greater the friction, the greater the angle of repose, and the steeper the slope.

Water can also promote mass movement in other ways. As more and more water is added to a slope it adds weight to the slope and may bring the slope material to the threshold between stability and instability. Water from rain or melting snow seeping into the slope contributes to mass movement in a more fundamental way, however. As more and more water seeps into the pore spaces of the slope material the water pressure increases there. The water then acts as a hydraulic jack, pushing apart individual grains or even entire rock units. This decreases the internal friction of the slope material. If it is reduced far enough, then the force of gravity will exceed the strength of the slope material and mass movement begins.

**TABLE 13.1 Some Factors That Reduce the Capacity
of Slope Materials to Withstand Movement**

Removal of lateral support by steepening slope
   Erosion by streams, glaciers, waves
   Road construction, quarrying, terracing
Removal of underlying support
   Solution of rock beneath surface
   Mining
Addition of mass
   Buildings, artificial fills, dumps
   Rain, snow, ice, vegetation
Lateral pressure
   Freezing of water in slope materials
   Swelling of clay minerals
Weathering
   Mechanical disintegration of granular rocks
   Removal of cementing material in granular rocks
   Drying of clays
Pore water
   Increased pressure of fluids in pore spaces of slope
     material
   Increase mass
Fracturing of rock
   Rock expansion due to relief of overlying load
   Tectonic movements
Organic activity
   Burrowing by animals
   Wedging or prying by plants
   Decay of root system
Vibrations
   Earthquakes
   Sonic booms, traffic, blasting

Modified from D.J. Varnes, "Slope Movement: Types and Processes," in
R.L. Schuster and R.J. Krizek (eds.), *Landslide Analysis and Control*,
National Academy of Sciences Special Report 17611-33, Washington, DC,
1978.

Geologic Setting

It is obvious that there must be slopes down which
material can move if mass movement is to occur at all.
The tectonic dynamism of the Earth provides these
slopes through sporadic changes in the elevation of
the land masses and ocean floors. Without this rejuve-
nation of elevations, the system that moves material to
lower and lower elevations would slow and eventually cease.

Most of the extremely rapid and spectacular mass
movements occur in areas of geologically young
mountains. Landmasses raised by tectonism are
rapidly carved by glacier ice (Figure 13.4) and torrential streams into regions characterized by steep, rocky,
unstable slopes. It is here that immense and destructive landslides most often occur. When tectonism
fades and ceases in that area, erosion is able to reduce
the land to lower and lower levels. Thus, as a landscape ages, extremely large and rapid mass movements give way to smaller, less dramatic movements
(Figure 13.5).

We discovered in Chapter 9 that earthquakes
occur on and around the boundaries between adjacent
plates of the Earth's crust. Here earthquakes are very
effective in triggering mass movement, often on a very
large and catastrophic scale. Figure 13.6 presents a
vivid example of the role that earthquakes can play in
shaking loose great quantities of rock.

Other geologic processes besides earthquakes can
set off mass movements. The erosive action of the
stream in Figure 13.7, for instance, weakened the base
of the river bank until the shaley rock slipped downward into the river. Erosion of a sea cliff by waves
(Figure 13.8) or the scouring of a valley wall by glacier
ice may also so steepen a slope that the Earth material
on it can no longer resist gravity and undergoes rapid
mass movement.

The type and arrangement of geologic formations
beneath a slope will in many instances determine the
nature of mass wasting. For instance, where the bedding planes of a series of sedimentary formations lie
parallel to the angle of the surface slope, the layers
may be prone to slippage along these planes (Figure
13.9). Also, zones of clay may provide a slippery surface on which material above it may move easily. In
other places rocks resistant to erosion overlie more
easily eroded rock. In such situations the less resistant
rocks are eroded most rapidly. This undermines the
resistant rock above, and it falls as it has in Figure
13.10.

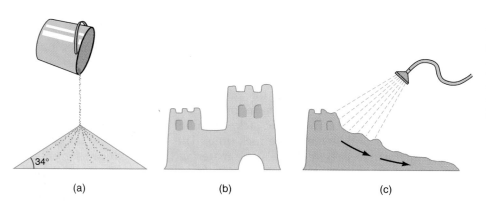

(a)   (b)   (c)

FIGURE 13.3
Sand will stand at slopes of different
angles depending upon the amount of
water in the sand. (a) Dry sand will hold
a constant angle of about 34°. (b) Moist-
ened sand will hold a vertical slope. (c)
Sand saturated with water will flow.

FIGURE 13.4   The steep slopes in high mountains and rigorous climate, as in the Himalayas, are the most susceptible to large and rapid mass movements. *(Roger Chaffin)*

Climate

In Chapter 4 we discussed the roles of alternate freezing and thawing in weathering and moving fragments of rock. In some regions permanently frozen ground called **permafrost** underlies vast areas. A seasonally thawed zone at the surface produces a water-saturated soil that is extremely effective in moving material

FIGURE 13.5   Mass movement continues on gentler slopes and in milder climates. Here, near Gays Mills, Wisconsin, wet spring soil has slipped downhill.

FIGURE 13.6   On August 26, 1963, seven months before the Good Friday earthquake in southern Alaska, the Sherman glacier (a) appeared like this in the camera of Austin Post of the U. S. Geological Survey. (b) The earthquake of March 14, 1964, set off a number of major rockslides, one of which occurred across the Sherman glacier. Post's photograph of August 24, 1964, shows the slide as a 4-km-long tongue of dark debris against the bright background of the glacier. *(Austin Post, U. S. Geological Survey)*

(a)   (b)

FIGURE 13.7 The Wind River has undercut its bank in the Wind River Canyon, Wyoming, and caused this landslip of shaley rock. In addition, rocks, pried from the cliffs above, have moved downslope toward the river.

down a very gentle slope. Similarly, tropical lands with heavy rainfall will have high rates of mass movement, particularly in areas with high elevations.

Plants and Animals

The burrowing of an animal, whether it be a worm or a fox, helps to urge material downslope as the burrow collapses downward under the pull of gravity. Further, the tread of animals on the surface pushes unconsolidated materials lower on the slope (Figure 13.11). The root systems of plants of whatever type, however, tend to slow downward movement of surfi-

cial materials. But when the vegetative cover is broken, as by overgrazing, the strength of the slope is reduced and gravity is more effective. Even when alive, however, the roots of plants can sometimes help move material downslope. For example, winds in the branches of a tree translate a gentle, prying motion to the root system. In addition, when a tree dies the voids left by decaying roots fill downslope. All of these are very slow processes but, in time, are very effective as well.

Humans

The development of humans a brief geologic moment ago brought a new geologic process to the Earth—human activity. People have become extremely effective in modifying old slopes and creating entirely new ones. In the process, mass movement has sometimes slowed, but more often has quickened. One of the most powerful human tools available for slope modification has been the plow. Pulled first by people, then by animals, and now by tractor, the plow has been in use for an estimated 8,000 years. Breaking the protective vegetative cover with a plow renders the slope very susceptible to rapid erosion (Figure 13.12) and increases enormously the rate and amount of material moving down the slope (Figure 13.13).

The pick and shovel, more recently joined by the bulldozer and explosives, have also produced new and artificial slopes. They may be the steep slopes of a quarry (Figure 13.14) or an open-pit mine or the less dramatic slopes of a highway cut or embankment. Some are unstable from the beginning. Other slopes may be stable initially but will fail after a short time. Still others may stand for centuries.

FIGURE 13.8 La Roche Percé (the "pierced rock") stands just offshore from the town of Percé on the Gaspé Peninsula of Quebec, Canada. Its vertical walls of limestone are exposed to wave erosion. Wave action undermines the cliffs, and from time to time large pieces of La Roche Percé crash into the sea. A stage in a collapse is shown in (a). In (b) the debris from the rockfall provides protection against further erosion for a while. When the sea has removed the rubble from the base of the cliff, direct attack on La Roche Percé will continue. (Photographer unknown)

(a)

(b)

Behavior of Material During Mass Movement

Materials involved in mass movement can move in four different ways: They can **slide, fall, flow,** and **heave** (Figure 13.15).

During the process of sliding, material maintains continuous contact with the surface on which it moves, much like a child does on a playground slide. The speed of the movement can range from slow to fast, and the material sliding may preserve its original form or be extensively deformed (Figure 13.15a).

Flowing, like sliding, can be slow or fast. It involves continuous turbulent movement. The consistency of the material involved can vary from a very liquid slurry to a thick paste. Slow flowing can deform material, whereas rapid flowing can thoroughly mix material (Figure 13.15b).

Freezing groundwater or swelling clay, as well as similar processes, can heave material at or near the surface upward. As the ground swells, the upward push is perpendicular to the slope. When the pressure is released, by the melting of ice or the drying of the slope, the material falls vertically downward. Because the particle heaves upward at an angle but then comes straight down, there is a net downslope motion with each up and down cycle (Figure 13.15c).

Fall refers to the free-fall of material. A rock pried from a cliff face, for instance, travels downward with no contact with the surface. It may bounce from one spot to another but in-between bounces it is considered to be in free-fall (Figure 13.15d).

Most movements are not pure examples of any one of these four processes but rather a combination of two or more, as the diagram in Figure 13.16 illustrates. For example, there is a gradation from almost pure sliding to pure flow in the progression from rockslides to river flow. The same diagram suggests the increase in velocities of movement from the slow process of heaving to the more rapid ones of flow and slide/fall.

Types of Mass Movements

Geologists recognize a great number of different mass movements. The following section describes some of them, beginning with the most rapid and potentially most catastrophic to life and property.

FIGURE 13.10   Here, in Chaco Culture National Historical Park, New Mexico, a vertically jointed, thick sandstone unit overlies more easily eroded thin beds of sandstone, shale, mudstone, and coal. Differential erosion of these less resistant units undermines the overlying sandstone. The sandstone breaks away along the vertical joint planes that define a new cliff face. The rubble from the fallen rock accumulates at the base of the cliff. *(Pamela Hemphill)*

Massive, jointed sandstone

Mudstone, thin sandstone beds

Coal

Rubble from rockfall

FIGURE 13.9   At Zumaya, on the north coast of Spain, sedimentary rocks dip at a high angle down toward the beach. There they are undermined by wave erosion, and from time to time slabs slide off into the sea. *(Franklyn B. Van Houton)*

Perspective *13.1*

A Woozy Mountain

The Gros Ventre slide was predicted by a local resident, "Uncle Billy" Bierer, who maintained that

Some of these times, these earthquake tremors that are coming so often are going to hit at about the right time when the mountain is wooziest and down she'll come.

Bierer's prediction is characterized by accuracy and understanding as well as by an absence of scientific jargon. Here is his story.

Albert Nelson, a local rancher along the Gros Ventre Valley in Wyoming came upon some springs high up on the valley wall. This was some time before his neighbor

Bierer sold his cabin in 1920, a cabin that lay downslope from the springs and not far from the river. When Nelson spoke with Bierer about the springs, Bierer is reported to have responded:

Yes, I have noticed that and I cannot see where the water can be going unless it is following the formation between the two different stratifications and coming to the surface at some other water-level point. If not, this mountain would be a mushy, woozy boil and the time will come when the entire mountain will slip down into the canyon below.

Another member of the Nelson family remembers hearing additional Bierer remarks:

Anywhere on that slope, if I lay my ear to the ground, I can hear the water tricklin' and runnin' underneath. It's runnin' between strata, and some day, if we have a wet enough spring, that whole mountain is gonna let loose and slide. Give it a wet enough year and all that rock strata will just slide right down on the gumbo like a beaver's slippery slide.

The cabin that "Uncle Billy" sold in 1920 was buried by the slide five years later.

Source: Barry Voight, "Lower Gros Ventre Slide, Wyoming, USA," in Barry Voight (ed.), *Rockslides and Avalanches*, Vol. 1, p. 116, Elsevier Publishing Co., Amsterdam, 1978.

Rockslides

Of all the mass movements, the most terrifying and catastrophic are **rockslides,** also called **avalanches.** These are masses of broken rock that start suddenly and move extremely rapidly down a mountain slope. Sometimes moving at speeds in excess of 150 km/hour, rockslides can transport millions of cubic meters of earth and broken rock over distances measured in kilometers. This is the type of movement described in Box 13.1. That particular slide is known as the Gros Ventre slide, which occurred in west-central Wyoming in 1925. Several factors brought about the slide. For example, the beds of weak mudstone and weathered sandstone dipping into the valley parallel to the slope of the south valley wall were not very stable (Figure 13.17). Also, minor earthquakes had occurred in the area not long before the slide. The trigger, however, was heavy precipitation and snow melt in the months before the event. Water seeping into the sandstone concentrated above the shale bed and

apparently broke the bond between it and the sandstone, thus initiating the slide.

The slide moved an estimated 40 million cubic meters of rock and debris across the Gros Ventre Valley and dammed the Gros Ventre River to form a lake 5 km long and some 70 m deep. Two years later erosion of the dam along the outlet channel of the lake caused heavy flooding downstream. A wall of water 5 m high destroyed the village of Kelly 6 km downstream and killed six people.

An example of a rockslide triggered by an earthquake occurred in southwestern Montana. In the early hours of August 16, 1959, an earthquake of magnitude 7.1 with an epicenter at the small town of West Yellowstone started the slide at the mouth of the Madison Canyon, 32 miles to the west (Figure 13.18). A mass of rock estimated to be nearly 30 million metric tons fell from the south wall of the canyon. It climbed over 100 m up the opposite valley wall and dammed a lake 8 km long and over 30 m deep. More than 20 people lost their lives in the Madison River campground area,

FIGURE 13.11   Cattle move along a hillside more or less parallel with the contours of the slope. They often break the vegetative cover and push soil downslope to produce cattle trails as along this hillside in southern Wisconsin.

most buried below the slide. Survivors reported extremely powerful winds along the margins of the slide. In fact, some people were blown away, never to be seen again, and wind carried an automobile over 10 m before smashing it against a row of trees. The wind formed as the mass of rock fell to the valley bottom and trapped and compressed large quantities of air.

The geologic conditions at the site of the Madison Canyon slide are shown in Figure 13.19. The south wall of the canyon is underlain by a dolostone and by gneiss and schist. The dolostone helped to buttress the weaker units of strongly foliated metamorphic rocks. Over the years the thinning of the supporting wedge of dolostone by the Madison River's continued canyon cutting brought the slope ever closer to the threshold between stability and instability; thus the earthquake was only the immediate cause of the slide. Under the stress of seismic wave motion the barrier of dolostone gave way, and large masses of dolostone, schist, and gneiss swept down into Madison Canyon, buried the

FIGURE 13.13   Plowing has allowed accelerated erosion of this recently sprouted wheat field.

FIGURE 13.12   For thousands of years the plow has made slopes vulnerable to erosion and mass movement.

campsite, and formed a dam behind which is Earthquake Lake.

One of the first major rockslides to be studied in any detail occurred nearly a century ago and killed over 70 people in the small coal mining town of Frank,

FIGURE 13.14   Human activity plays an increasingly important role in creating new slopes, as in this quarry of loosely consolidated volcanic ash and cinders. These slopes, fashioned by mechanical shovels, are vertical in places and very much steeper than naturally developed slopes of the same materials. Inevitably, mass movement will reduce the quarry walls to the gentler angles of natural slopes. (Pamela Hemphill)

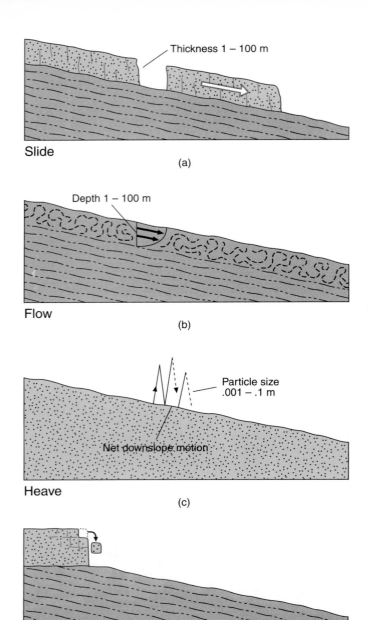

FIGURE 13.15 Material can move down slopes by (a) sliding, (b) flowing, (c) heaving, and (d) falling.

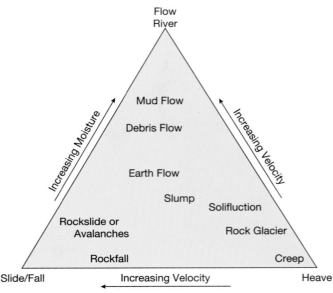

FIGURE 13.16 This diagram shows how most mass movements involve a combination of slide/fall, flow, or heave. The diagram also relates these movements to their relative velocities (base and right arm of the triangle) and to the moisture content of the material involved (left arm of the triangle).

Alberta. Here some 32 million cubic meters of rock crashed down from the crest of Turtle Mountain, which stands over 900 m above the valley (Figure 13.20). Coal mining activities may have triggered the slide, but natural conditions brought it to the threshold of movement. As Figure 13.21 shows, Turtle Mountain has been sculpted from a series of limestone, sandstone, and shale beds, which have been folded and faulted. It also shows four factors that contributed to the slide: The steepness of the mountain; the bedding planes of the limestone dipping parallel to the mountain face; two thrust planes that break the strata part way down the slope; and the weak shale and coal beds in the valley.

The steep valley wall enhanced the effectiveness of gravity, and the bedding planes, along with the fault planes, served as potential planes of movement. The weak shale beds at the base of the mountain probably underwent slow plastic deformation under the weight of the overlying limestone, and as the shale was deformed, the limestone settled lower and lower. The settling action may have been helped along by the coal mining operations in the valley as well as by frost action, rain, melting snows, and earthquake tremors that had shaken the area two years earlier.

Regardless of what finally triggered it, material on the slope finally passed the threshold of movement. The limestone beds gave way along their bedding planes and pushed outward toward the valley along the uppermost fault, and the great mass of rock hurtled down into the valley. The rock material behaved in different ways at different times. First, the shales underwent plastic deformation, producing a condition of extreme instability on the mountain slope. Next, when the strata that still held the limestone mass on the slope sheared, the slide actually began. Once the slide was under way, the rock slid down the mountain slope until it was literally "launched" into the air by a ledge of rock. Thereafter it arched outward and downward to the valley below. Finally, when the rock reached the valley floor, it moved at a high speed and

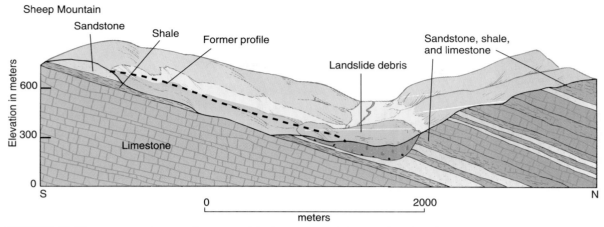

FIGURE 13.17   Diagram showing the nature of the Gros Ventre slide. Note that the sedimentary beds dip into the valley from the south. The large section of sandstone slid downward along the shale bed. *(Redrawn from William C. Alden, "Landslide and Flood at Gros Ventre, Wyoming," Trans. AIME, Vol. 76, p. 348, 1928.)*

traveled up the hills on the far side of the river. In this phase it moved like a fluid with a series of waves spreading out along its front.

The Earth's largest recorded rockslide by volume, modern or prehistoric, took place some 300,000 years ago on the slopes of a volcano that is now Mt. Shasta in northern California. The avalanche extended 43 km northwestward from the volcano and involved 17×10^{12} m^3 of volcanic material. If that amount of material were spread evenly over Manhattan Island in New York City it would make a layer 200 m deep.

Even larger slides take place below the sea. Subsea slides often occur around volcanic islands. Around the flanks of the Hawaiian Islands, for example, submarine mapping has revealed slides up to 200 km in length and over a kilometer in thickness.

Various investigators have reported that the shattered and jumbled materials that result from rockslides maintain the same general relationships that they had before the slide took place. Material toward the back of the rockslide does not overtake that in front. Researchers deduce that the material in the back part of the slide collides with material immediately in front of it, urging that material on faster. This process is repeated throughout the slide; thus material behind rarely overtakes the debris ahead of it. This transfer of energy through the debris of the slide, called **momentum transfer,** makes possible the progressively more rapid movement of material in the downslope positions within the sliding mass.

Mudflows, Debris Flows, and Earthflows

Similar in some respects to rockslides are **mudflows** and closely related movements called **debris flows.** Like rockslides, they move very rapidly and can be very destructive. Unlike rockslides, they have a high water content and flow as though they were rivers. Debris slides, though, have a higher content of rock debris than do mudflows. **Earthflows** resemble mudflows and debris flows in that movement is turbulent within the moving mass. They differ in that they move slowly, although perceptibly, and are generally smaller in terms of the area affected. They have proportionately less water and more debris than do mudflows, and about the same mix as a debris flow.

FIGURE 13.18   An aerial view west across the Madison Canyon rockslide. The scar of the slide is seen on the mountain to the south (left). The slide debris fills the valley bottom and has dammed up a new lake. An outlet channel for the lake has been cut across the rockslide dam. The former Madison Canyon campsite lies beneath the slide material on the north (right) side of the valley. *(William C. Bradley)*

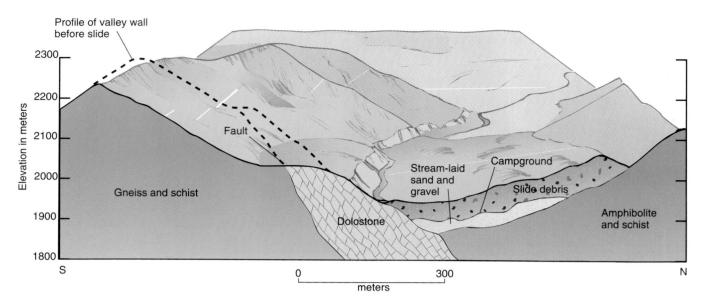

FIGURE 13.19 Geologic conditions at the site of the Madison Canyon slide. See text for discussion. *(From Jarvis B. Hadley,* Landslides and Related Phenomena Accompanying the Hebgen Lake Earthquake, *U.S. Geological Survey Professional Paper 435-K, p. 115, Washington, DC, 1964.)*

There are two types of mud or debris flows. One type originates on the slopes of a small steep-sided canyon or gulch where the slopes have become covered by loose, unconsolidated, and unstable material. A sudden rainstorm in a dry area or water from melting snows in high mountains loosens the unconsolidated materials from the slopes and carries them to a usually dry stream channel at the bottom of the canyon. Water and debris then mix together to form mud or a mudlike substance with the consistency of wet concrete. The advance down the channel may be intermittent, at times slowed by a narrowing of the canyon or an accumulation of old debris; at other times it surges forward, pushing obstacles aside or

FIGURE 13.20 The Frank, Alberta, rockslide left this scar on Turtle Mountain and the sheet of debris across the plain below. *(Photographer unknown.)*

carrying them along with it. Eventually the flow leaves the canyon mouth for gentler slopes and less confined channels. Here it may splay out, its water soaking into the ground. The thinning mass then dries and turns into a hardened sheet of rubble.

Another and more dramatic type of this mass movement is associated with volcanic eruptions. Extensive and unstable deposits of volcanic ash become mobilized by heavy rainfalls. The 1991 eruption and associated mudflows of Mt. Pinatubo in Luzon, Philippines, forced the abandonment of Clark Air Force base and caused the destruction of numerous settlements in the area (Figure 13.22). Also, in 1985 a small volcanic eruption on the slopes of the Nevada del Ruiz volcano in Columbia melted enough snow and glacier ice to create a flow that killed an estimated 25,000 people in the lowlands below.

A mud or debris flow can buoy up and carry boulders weighing 100 metric tons or more and float houses, cars, and giant trees along as though they were toys. Therefore, mud and debris flows in inhabited areas, such as along the lower slopes of the San Gabriel Mountains along the northern edge of the Los Angeles metropolitan area, are a continuing threat to life and property.

An earthflow is marked by multiple pull-away scars at its head. Its sides are well defined, and downslope it ends in a bulging lobelike form (Figure 13.23). It may develop on a hillside, as in Figure 13.23, or at the head of a gully marking the beginnings of a small stream. The movement may involve from a few to several million cubic meters of material.

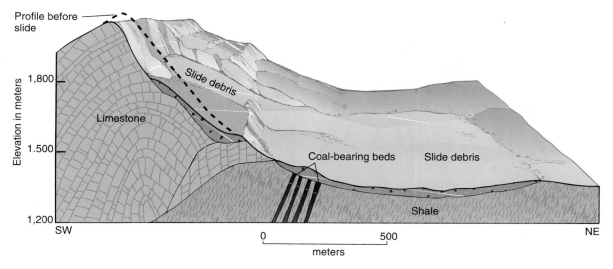

FIGURE 13.21   A cross section showing the geologic setting of the rockslide at Turtle Mountain, Frank, Alberta. *(Modified from D.M. Crudent and J. Krahn, "A Reexamination of the Geology of the Frank slide," Canadian Geotechnical Journal, Vol. 10, (no. 4,) p. 536, 1973.)*

Rockfall and Talus

A **rockfall** is the sudden fall of a large fragment of rock from a steep, usually vertical cliff. It differs from a rockslide in that it is usually a free-fall of a single large block that descends to the bottom of the cliff where it may shatter and spread its pieces over the nearby lower slopes. It is a common occurrence along steep cliffs of massive, vertically jointed or layered rock. When the rock is sufficiently undermined it breaks off along a joint or bedding or cleavage plane. We have seen examples in Figures 13.8 and 13.10. Figure 13.24 provides an example of an incipient rock fall.

FIGURE 13.22   Residents salvage the metal roofing from their home, which was almost completely buried by the mudflow following the eruption of Mt Pinatuba, North of Manila, the Philippines, in June 1991. *(AP/Wide World Photo)*

A **talus** is a slope built up by an accumulation of rock fragments at the foot of a cliff or a ridge. The rock fragments are sometimes referred to as **rock waste** or **slide rock.** Often, however, talus is used as a synonym for the rock waste itself.

Mechanical weathering produces the rock fragments of a talus. Very commonly the process of freezing and thawing is responsible for breaking up the bedrock, but the transfer of the fragments down a slope to form a talus is in the realm of mass movement. A talus is formed as rock fragments are loosened from the cliff and clatter downward in a series of free-falls, bounces, and slides. Eventually the rock waste builds up in a heap or sheet. An individual talus resembles a half-cone with its apex resting against the

FIGURE 13.23   An earthflow down a hillside in California has the typical multiple scarps at its top and the zone of soil flow downslope. A second, smaller earthflow has begun to the left of the main flow. *(Martin Miller)*

Types of Mass Movements   **331**

FIGURE 13.24   This sandstone cliff is vertically jointed and a part of it is very close to pulling away from the cliff and falling to the valley floor. Daylight already shows between the main cliff and the slab of isolated rock. (Chaco Culture National Historical Park.)

(a)

(b)

FIGURE 13.25   (a) This talus near Challis, Idaho, has the near-perfect shape of a half cone. (b). Several taluses merge to form an apron of rock debris along the north valley of Line Creek, southern Montana.

cliff face in a gulch along the cliff (Figure 13.25a). A series of these half-cones often forms an apron at the base of steep mountains, completely obscuring their lower portions (Figure 13.25b). Eventually, if the rock waste accumulates more rapidly than it can be destroyed or removed, even the upper cliffs become buried in their own rubble, and the growth of the talus stops. The angle of repose of the talus varies with the size and shape of the rock fragments. Although some angular material can maintain slopes up to 50°, a talus rarely exceeds angles of 40°. A talus is subject to the normal process of chemical weathering, particularly in a moist climate. The rock waste is decomposed, especially toward its lower limit, or toe, which may grade imperceptibly into a soil.

Slump

Slump is the downward and outward rotational movement of Earth materials traveling as a unit or as a series of units. Slump often occurs where the foot of the original slope has been sharply steepened by, for instance, stream erosion or a cutting for a road. Large blocks of the slope rotate downward and outward along curved surfaces, and the upper surface of a block is tilted backward as it moves. Movement may be fairly sudden, but material does not travel far, usually a few meters or tens of meters. After a slump has started, it is often helped along by rainwater collecting in basins between the tilted blocks and the plane of movement. The water seeps down along the spoon-shaped surface on which the block is sliding and promotes further movement. A slump may involve a single block, as shown in Figure 13.26a, or a series of slump blocks (Figure 13.26b). Two different slumps are shown in Figure 13.27.

Creep

Creep is the very slow, generally continuous, down-slope movement of soil and rock fragments under the influence of gravity. Creep operates even on gentle slopes with a protective cover of grass and trees. It is hard to realize that this movement is actually taking place because the observer sees no break in the vegetative mat and no large scars, and has no reason to suspect that the soil is in motion beneath his or her feet. Yet this movement can be demonstrated in exposures

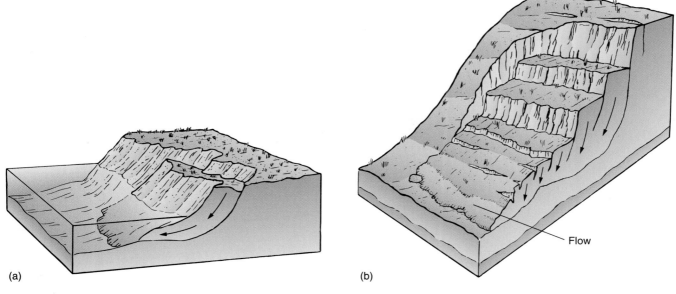

FIGURE 13.26   (a) This diagram shows the type of movement found in slumps. In this case the slump is pictured as occurring along an over-steepened slope along a shoreline. (b) Many slump movements contain several discrete units, as suggested in this diagram.

of downslope bending in beds of layered rocks; by the drift of boulders down a gentle slope; by the bowing of trees; by the tilting of fences, telephone poles, and cemetery markers; and by the bulging of retaining walls (Figure 13.28).

Solifluction

Solifluction, from the Latin *solum*, "soil" and *fluere*, "to flow," involves the turbulent flow of soils or surficial debris. It occurs in a zone that aver-

ages a meter or so in thickness and lies immediately beneath the surface. Movement may be as fast as a few millimeters or centimeters per day or as slow as a few millimeters or centimeters per year. It occurs because the zone overlies impermeable material through which water cannot drain. As a result, the soil becomes saturated and the sodden mass flows slowly downslope.

The process occurs in two widely different climate zones. It is common in tropical rain forests where slope materials may be saturated with rainwater that cannot escape downward. It is equally common in

FIGURE 13.27   (a) This small slump in the Austrian Alps has the typical pull-away scarp at its upper limit. Just to the left of the slump are the vestiges of two older, now partially healed slumps. (b) Slump with multiple blocks along a recent cut for a railroad.

(a)

(b)

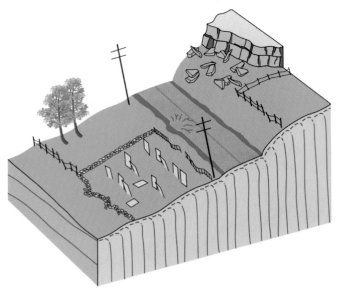

FIGURE 13.28 Creep on slopes can be detected by a number of different types of evidence as suggested in this diagram. All of the features shown here will not necessarily be present in any one locality.

Arctic climates where frozen ground blocks the downward percolation of water accumulated in the soil during the periods of seasonal thaw.

Solifluction in high latitudes occurs where the mean annual temperature is 0°C or less. A thin zone of summer thaw, the **active layer,** overlies permafrost, ground that is frozen throughout the year. Such regions are commonly called **periglacial** for their near-glacial climate (Figure 13.29). The vertical temperature

FIGURE 13.30 Vertical temperature distribution in an area of permanently frozen ground.

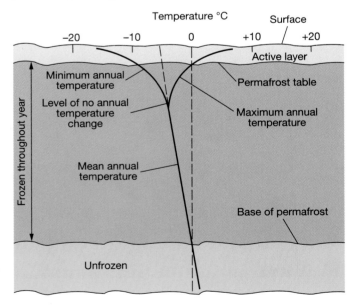

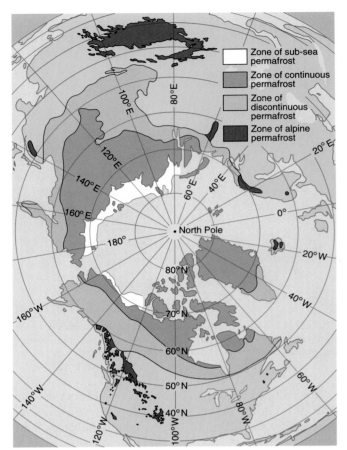

FIGURE 13.29 Distribution of permafrost in the Northern Hemisphere. *(Troy L. Péwé)*

distribution in zones of permanently frozen ground (Figure 13.30) helps to explain the cause of mass movements in such regions. The zone above the **permafrost table** (that depth at which the maximum annual temperature reaches 0°C) thaws each summer, while the ground below remains frozen. The permafrost acts as a barrier to the infiltration of meltwater from the seasonally thawed zone. As a result, the material above the permafrost can become saturated with water. If the material is unconsolidated, great sheets of debris move slowly down even the gentlest slopes by the process of solifluction (Figure 13.31).

Rock Glaciers

Rock glaciers are tongue-shaped masses of rock rubble found in many high, cold mountainous areas (Figure 13.32). The rubble is cemented together by ice that forms in the pore spaces between rock fragments. Sometimes bodies of clear ice accumulate in large lenses within the rubble mass. Rock glaciers move under the influence of gravity, the ice within them deforming as they move. The forward rates of motion

FIGURE 13.31 Soil drapes in solifluction lobes across the slopes of this hillside near Mt. Fairplay in eastern Alaska. *(Steve McCutcheon, Visuals Unlimited.)*

of rock glaciers are as slow as 1 mm/year (slower than the rate at which lithospheric plates drift apart) to as fast as 1.7 m/year. The average is probably between 30 and 40 cm/year.

The surfaces of rock glaciers are characteristically marked by furrows and ridges that occur perpendicular to their direction of movement. These are clearly visible from the air and give the impression of great waves in the rubble. These furrows and ridges are obscured, however, to the ground observer by the huge boulders that litter the surfaces of most rock glaciers. The leading edges of rock glaciers are steep-faced (30 to 60 m high) as are the glaciers' sides.

Active rock glaciers are found at elevations above the tree line, the upper limit of tree growth, and at the lower limit of the zone of permanently frozen ground. They carry rock debris along from the bases of steep, rock-rubble slopes or from the rock-littered margins of stagnant and melting mountain glaciers.

FIGURE 13.33 It would take a foolhardy person, or a very brave one, to build at the base of this magnificent cliff of limestone in the Big Horn Mountains of Wyoming.

FIGURE 13.32 This rock glacier is in the Wrangell Mountains, near McCarthy, Alaska. The ridges and furrows on its surface are characteristic and are at right angles to the direction of flow. It is composed of rock fragments from the talus along the cliffs in the background. This rock rubble is cemented by ice. *(Noel Potter, Jr.)*

Protecting Against Hazards of Mass Movements

The more rapid mass movements can be catastrophic. Millions of tons of rock hurtling down slopes at speeds in excess of 100 km an hour have killed tens of thousands of people and wiped out entire communities. The slower and less disastrous movements can disrupt public utilities, roads, and railroads and damage building foundations. Given the fact that mass movement is inevitable what precautions and defensive measures can we take against the potentially dangerous movement?

The first measure to take against damage from movement on unstable slopes is an examination of the geologic conditions of a given area. At one level, it does not take much geologic knowledge to realize that

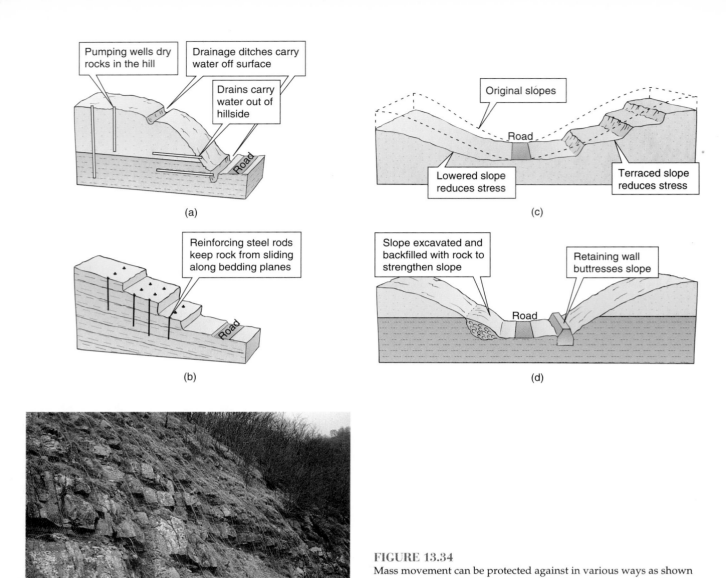

Pumping wells dry rocks in the hill

Drainage ditches carry water off surface

Drains carry water out of hillside

Road

(a)

Reinforcing steel rods keep rock from sliding along bedding planes

Road

(b)

Original slopes

Road

Lowered slope reduces stress

Terraced slope reduces stress

(c)

Slope excavated and backfilled with rock to strengthen slope

Retaining wall buttresses slope

Road

(d)

(e)

FIGURE 13.34

Mass movement can be protected against in various ways as shown here. (a) Drying out the slope; (b) anchoring of rock susceptible to movement; (c) lowering the slope or terracing the slope; (d) buttressing the slope with rockfill or a retaining wall; (e) use of steel netting.

there is little sense in building at the base of a cliff from which large blocks periodically fall (Figure 13.33). At another level, local lore, combined with a careful inspection, may expose an area of unstable slopes. The presence of water seeping from the slope might signal future movement as would open fractures in the soil. Bowed walls, tilted utility poles, and offset fence lines demonstrate creep. Bumpy, irregular topography, combined with break-away scars of old slumps or rockslides, indicate rapid slope movement in the past and the possibility of similar movement in the future. In the case of the Gros Ventre slide described earlier, an extensive field survey along the valley would have revealed irregular terrain below old break-away scars, characteristic of rockslides. Further

field study would have shown that the geologic conditions were favorable to rockslides. On this basis it would have been reasonable to predict future rockslides along the valley.

The best way to protect against damage on dangerously unstable slopes is, of course, to avoid construction on them altogether. Where this is impossible, or, for good reasons, undesirable, there are various methods of strengthening slopes (Figure 13.34). Because water is so important in promoting mass movement, one of the first considerations is to keep the slope dry. This may involve the diversion of surface water to keep it from infiltrating the slope materials, draining subsurface water from the slope material, or both. Other methods of stabilizing slopes include

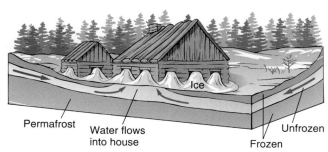

FIGURE 13.35   One of the hazards of construction in areas of permafrost is seen here in two buildings in Siberia. See text for discussion.

FIGURE 13.36   This small village of Civita di Bagnoregio in central Italy has occupied this isolated butte for centuries. The edges of the village continue to disappear as rockfalls push back the cliff. *(Pamela Hemphill)*

installing retaining walls, anchoring dangerous rock units to firm rock beneath, and reducing the steepness of a slope by leveling it back or by stepping it back in a series of terraces. Protection against falling rock can be provided by heavy steel netting, or by coating a crumbly rock face with concrete. How long these slope-strengthening measures remain effective will always depend upon how well they are engineered in the first place, and how well they are maintained thereafter.

Permafrost raises serious obstacles to engineering and thus to extensive settlement of Arctic regions. The presence of an active solifluction layer above the permafrost makes building foundations difficult to construct. Furthermore, permafrost itself can be melted by the heat from an overlying structure, and this causes the building to settle, often unevenly. Another situation may develop as illustrated in Figure 13.35. Here the upper part of the active, thawed zone froze downward trapping still unfrozen material between it and the permafrost below. The still unfrozen zone acted as a conduit to lead water up into the buildings. There it filled the buildings and flowed out through the windows. The final results were ice-filled buildings and cascades of ice pouring out from the buildings and across the ground. Answers to these and associated problems depend on the proper insulation between the permafrost and whatever structures are built.

In some cases undertaking remedial measures may be possible, but not practical. Figure 13.36 shows a small Italian village that has stood for centuries atop a narrow butte. It refers to itself as "the town that waits to die." Its steep-sided cliffs are continuously retreating as blocks slip downward along vertical joints. Each rockfall takes a bit of the village with it. Protective engineering measures are possible, but they would be expensive, and how long they would be effective could not be guaranteed. Relocation might be a more practical approach.

Certain engineering structures can exacerbate pre-existing dangers. The Vaiont dam in the Italian Alps is

an example. There geologic conditions provided very unstable slopes, the rising lake saturated the base of a valley wall, and heavy rains added more water. Finally, a torrential downpour helped trigger the movement of the south wall of the valley into the lake. As a result a wave of water 100 m high sloshed over the dam and rushed down the valley. Nearly 3,000 people died, and property damage was in the millions. Site investigations before dam construction had shown that the south wall of the proposed lake had already had some rockslides in the past and that the geologic conditions favored more. Even as the lake filled a small slide occurred. In the months before the slide, monitoring devices recorded slow movements of the slope. Here, then, was a situation in which knowledge from a preconstruction survey and postconstruction monitoring was not acted upon, and the tragedy became inevitable.

Mass Movement on Other Planetary Bodies

▲▲

Because gravity is operating on other planetary bodies, geologists would expect to find evidence of mass movement on some of them as well as on Earth, and indeed we do. On Mars gigantic mass movements have taken place along the walls of the vast Valle Marineris. Figure 13.37 shows the result of mass movement at a place where the canyon is 2 km deep and 130 km wide.

Craters are common features of many of the solid planetary bodies. Around the inner rims of craters large slump features often occur, as shown in Figure 13.38, an oblique view of a portion of the crater Coper-

FIGURE 13.37   This view shows a section of the Capri Chasm, a portion of a vast section of interconnected canyons on Mars called the Valle Marinris. The canyon here is about 2 km deep and 130 km wide. A series of slump blocks lie well below the lip of the cliff. From the base of the slumps, aprons of material extend striated zones that have the pattern of gigantic flows of dry sand. These end in overlapping, lobate zones toward the bottom of the canyon. *(NASA)*

FIGURE 13.38   An oblique view of the crater Copernicus shows the results of mass movement on the Moon. Forty-five kilometers beyond peaks in the middle foreground are the slumped blocks along the inner wall of the crater. The lip of the crater at this point is about 3 km above the crater floor. *(NASA)*

nicus on the Moon. Slumps such as these occur immediately after excavation of a crater by meteoritic impact. The newly formed crater walls are steep and unstable and rapidly adjust by slumping.

▶▶ EPILOGUE ◀◀

No slope is completely stable over long periods of time. Gravity, aided by water, climate, geologic condi-tions, and living things will, over time, move slope material downward. "As everlasting as the hills" suggests a permanence that does not exist. Mass movement continuously transports Earth materials to lower and lower elevations. Movements as widely different as catastrophic rockslides and slow, almost imperceptible soil creep, are all part of this segment of the rock system. Most of this material moved by mass movement is delivered eventually to the rivers of the world, which is the subject of the next chapter.

SUMMARY

1.   Gravity is always acting on Earth materials and is thus responsible for the their mass movement. When the strength of slope material (its shear strength) is greater than the shear stress generated on it by gravity, the slope is stable. As material approaches a threshold between stability and instability the shear stress becomes greater and greater in relation to the shear strength. When shear stress exceeds shear strength movement begins.

2.   Increasing the shear stress or decreasing the shear strength can promote and initiate mass movement. There are various ways that this hap-pens, and they include steepening the slope by erosion and road construction; removing support by mining; adding mass to slope with buildings, snow, and ice; weathering of slope material; increasing the water pressure in pores of slope; fracturing rock; burrowing by animals and prying by plants; and vibrating caused by earthquakes, traffic, and blasting.

3.   Mass movement of material may involve sliding, flowing, falling, or heaving. Sliding material maintains continuous contact with the surface on which it moves. Flowing involves turbulence, and falling material is airborne. Heaving occurs

as freezing soil raises material at right angles to a slope and, after thawing, gravity brings it back perpendicularly.

4. Mass movements can be separated on the basis of velocity, amount of water involved, and the type of movement. The fastest movements are rockslides or avalanches. Mudflows and debris flows are also rapid. Mudflows contain much more water and less Earth debris than do debris flows. A talus forms at the foot of a steep slope as loose rock fragments from above accumulate in the shape of a half cone. A slump is the downward and outward rotation of a block of Earth material. Creep is the very slow, almost imperceptible movement of soil and slope debris. Solifluction takes place in saturated soils of both tropical rain forests and periglacial areas. Rock glaciers are thick accumulations of ice-cemented rubble that occur in high mountains.

5. Protection against mass movement begins with a search in the field for indications of potential mass movement. If possible, construction should not be attempted in areas found to be prone to mass movement. If construction is carried out in such areas, the first step is to make sure the slope is well-drained. Beyond this, protection can be obtained by various engineering projects, including reduction of slope angle and installation of retaining walls and protective netting. Sometimes relocation of people and their structures is a sensible way of protecting against the hazard of mass movement.

KEY WORDS AND CONCEPTS

active layer 334	momentum transfer 329	rock waste 331
angle of repose 321	mudflow 329	shear strength 320
creep 332	periglacial 334	shear stress 320
debris flow 329	permafrost 323	slide 325
earthflow 329	permafrost table 334	slide rock 331
fall 325	rock avalanche 326	slump 332
flow 325	rockfall 331	solifluction 333
heave 325	rock glacier 334	talus 331
mass movement 320	rockslide 326	threshold of movement 320

QUESTIONS FOR REVIEW AND THOUGHT

13.1 List some factors that affect the movement of Earth materials down slopes.

13.2 From your own experience and observations cite one or more examples of downslope movement of material. Describe one in some detail.

13.3 Assume that you are responsible for assuring the stability of a road embankment cut in layered rock that dips 35° toward the road. What measures might you take to make sure that the embankment does not fail?

13.4 Explain how gravity operates in mass movement.

13.5 What is the "angle of repose"?

13.6 Give an example of each of the following types of movement: slide, fall, flow, and heave.

Critical Thinking

13.7 What would you expect the dominant mass movements to be along each of the three major plate boundaries and why?

13.8 You plan to build a home on a hill underlain by sandstone, shale, and limestone. What will you look for in order to avoid building on an unstable slope?

14

The Merced River, Yosemite National Park, California (Ansel Adams, Copyright 1994 by the Trustees

Streams and Stream Systems

OBJECTIVES

▲▲

A focus on the following questions will help in your study of streams and stream systems.

1. How does water flow through the hydrologic system? How does the volume of water vary in different parts of the system?
2. What are the types of flow of water? What is the relationship between stream discharge, velocity, and the width and depth of the stream channel?
3. What is base level, temporary and ultimate, and how does it function?
4. How does a stream erode, transport, and deposit sediments?
5. What are drainage basins and stream networks? What are some features of narrow valleys? Of broad valleys?
6. What relation exists between drainage systems and plate tectonics?

OVERVIEW

▲▲

Earth is unique among the planets of our Solar System in having presently active streams. They clear the continents of most of their debris, carrying, like endless conveyor belts, the Earth's soil and weathered rock to the oceans. In so doing, they play an integral part in the rock system and form the major link between land and sea in the hydrologic system. Streams flow in an orderly system of channels that change in response to changes, either natural or caused by human intervention, in the water and material delivered to them. At times and places they erode Earth materials; at other times and places they deposit them and are thus responsible for much of the planet's landscape. Movement of the Earth's lithospheric plates has helped to form many of the planet's major stream systems as converging and diverging plates have built long-lasting drainage patterns on all the continents.

341

Streams and the Hydrologic System

For centuries people believed that ocean waters cycled back to the land through a complex subterranean plumbing system. In the process, the salts of the sea-water were somehow distilled out and fresh water broke forth as springs to continuously feed the streams. People who believed this assumed that rain-fall was inadequate to account for the flow observed in streams. After all, streams ran continuously, even though rain fell only occasionally. Furthermore, people generally believed that rainwater could not soak into the ground to replenish springs.

The true source of stream water has been identified only in relatively recent times. In 1674 Pierre Perrault (1608–1680), a French lawyer and sometime hydrologist, presented the results of his measurements in the upper drainage basin of the Seine River. Over a 3-year span he collected data on the amount of precipitation in this portion of the basin. At the same time he kept track of the amount of water discharged farther downstream. The results showed that the flow of the river was surprisingly little compared to the total amount of water available from precipitation. In fact, the annual precipitation was six times the total volume of river flow. Measurements by other researchers soon showed that water could infiltrate into the underground, and still others demonstrated that a tremendous amount of moisture was evaporated back into the atmosphere. Such data laid the basis for the hydrologic system introduced in Chapter 1.

Scientists now accept that water circulates from ocean to atmosphere to land and back to the ocean. Also, scientists now know that the major supply of water for rivers does not come from direct rainfall or surface runoff. Instead, it comes from water that has seeped into the underground to be stored temporarily as ground water and returned slowly to the streams in both clear and rainy weather.

The hydrologic system involves an enormous amount of water. Figure 14.1 shows how many million cubic kilometers of water are stored in the atmosphere, in the oceans, and on the continents. These are the reservoirs for the Earth's water. Figure 14.1 also shows how much water circulates per year among these reservoirs. Stream flow, for instance, carries 37,000 cubic kilometers of water to the oceans annually and is balanced by the same amount of atmospheric moisture moving in the opposite direction. These figures can answer a number of questions. For example: How long would it take streams to fill the ocean basins? The volume of the oceans divided by annual stream flow gives the answer: 37,000 years. This can also be

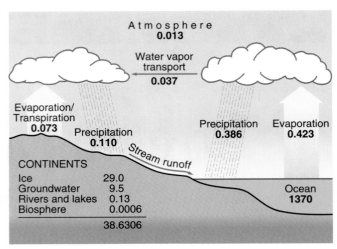

FIGURE 14.1 This version of the hydrologic system shows the volume of water, in millions of cubic kilometers, contained at any one time in the atmosphere, the oceans, and on the land. It also shows how many millions of cubic kilometers of water move in and out of the three reservoirs each year.

thought of as the **residence time** of a molecule of water in the ocean, the average amount of time it spends in the reservoir.

Stream Flow

Stream Channels

Streams flow in channels of varying widths and depths down slopes called **gradients.** A stream's gradient in any one section is the vertical distance a stream descends over a fixed distance of horizontal flow. In general, a stream's gradient decreases downstream. As a result, a stream channel's overall gradient from its headwaters (the "start" of a stream) to its mouth (the "end" of a stream) is concave toward the sky (Figure 14.2). The Mississippi River from Cairo, Illinois, to the mouth of the Red River in Louisiana has a low gradient: Along this stretch the drop varies between 2 and 10 cm/km. Conversely, the Arkansas River, in its upper reaches through the Rocky Mountains in central Colorado, has a high gradient. There the drop averages 7.5 m/km. The gradients of other rivers are even higher. The upper 20 km of the Yuba River in California, for example, has an average gradient of 42 m/km; and in the upper 6.5 km of the Uncomphagre River in Colorado, the gradient averages 66 m/km.

A stream channel is wider than it is deep. The proportion between width and depth, however, generally changes with the amount of water carried by the stream. In a small channel, a few meters wide, the width may be about five to six times its average depth.

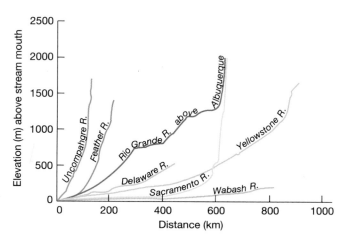

FIGURE 14.2 The gradients of streams always have some irregularities. In general, however, they are concave upward and decrease downstream toward their mouths.

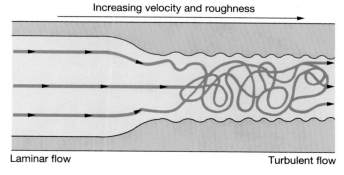

FIGURE 14.3 Diagram to illustrate laminar and turbulent flow of water through a section of pipe. Individual water particles follow paths indicated by the colored lines. In laminar flow the particles follow paths parallel to the walls of the pipe. With increasing velocity, and with increasing roughness of the confining walls, laminar flow changes to turbulent flow. The water particles no longer follow straight lines but are deflected into eddies and swirls. Most stream flow is turbulent.

In contrast, the Delaware River which is 250 to 300 m wide in its middle reaches roughly 50 to 60 times its average depth.

Types of Flow

The type of flow of water in a stream differs according to its velocity. When water moves at low velocities and particles in it follow parallel paths without mixing, it is said to move in **laminar flow** (Figure 14.3). Laminar flow occurs very rarely in streams, and when it does it occurs only in quiet water immediately adjacent to the stream wall. As the velocity of the water increases or the stream walls become rougher, **turbulent flow** (from the Latin for "full of turmoil") begins. This type of flow is marked by swirls and eddies and characterizes most stream flow (Figure 14.3). Finally, **jet flow,** which is characterized by water moving in jet-like surges, occurs at high velocities reached along a very steep stream bed or in the free-fall of water over waterfalls.

Velocity, Turbulence, and Discharge

The **velocity of a stream** is measured in terms of the distance its water travels in a unit of time. A velocity of 5 to 10 cm/s is relatively low, and a velocity of about 625 to 750 cm/s is relatively high. Velocity varies from stream to stream and in any one stream it varies from time to time depending on the amount of water the stream carries. Velocity also varies from place to place in a stream, as shown in Figure 14.4. Flow slows near the margins of the stream because of friction between the water and the stream bed. Along a straight stretch of a channel the greatest velocity lies toward the center of the stream at, or just below, the surface of the water. Around a curve in the channel, however, the zone of greatest velocity is pushed to the

outside of the bend, just as a speeding vehicle is pushed toward the outside of a curve in the road.

Zones of maximum turbulence occur where the fastest-moving sections of the stream come closest to the channel walls of the stream. Figure 14.5 shows the relation of maximum velocity to maximum turbulence in different sections of the stream. Turbulence increases across a stream from the inside of a bend to

FIGURE 14.4 Velocity variations in a stream. Both in plan view and in cross section the velocity is slowest along the stream channel, where the water is slowed by friction. On the surface it is most rapid at the center in straight stretches and toward the outside of a bed in the channel. Velocity increases upward from the river bottom.

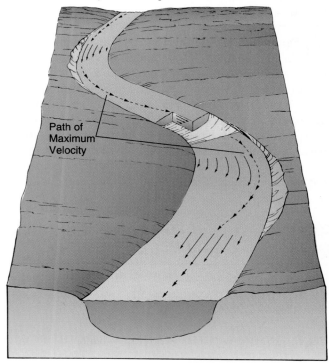

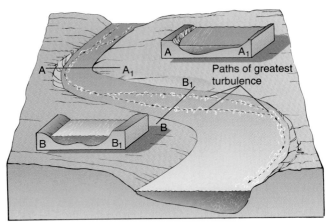

FIGURE 14.5   Zones of maximum turbulence in a stream are shown in cut-away sections through a river channel. They occur where the change between the two opposing forces—the forward flow and the friction of the stream channel—is most marked. Note that the maximum turbulence along straight stretches of the river is located where the stream banks join the stream floor. On bends the two zones have unequal intensity; the greater turbulence is located on the outside of a curve. It is here, on the outside of the bend, that turbulence causes the greatest erosion. By contrast, velocity and turbulence are lowest on the inside of a bend; erosion is least and deposition occurs most often.

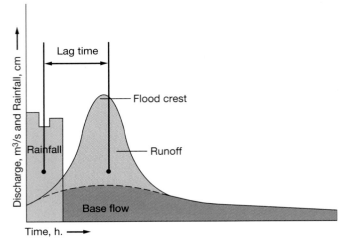

FIGURE 14.6   As rainfall changes so do runoff, base flow, and discharge of a river, as shown in this hydrograph.

the outside, following the increase in maximum velocity. The inside of the bend has low turbulence and quiet water. The zone of maximum turbulence moves across the river and to the outside of the next bend downstream.

Discharge is the amount of water passing through a stream channel in a unit of time and usually is measured in m³/s. Water in a stream channel has two sources: **runoff**, precipitation that flows directly off the surface of the land to the channel, and, except in very dry climates, **base flow**, the ground water that seeps into the channel and keeps the stream flowing even in dry weather. As precipitation increases, runoff into the streams rises rapidly while base flow increases more slowly. Discharge also increases but does not reach a peak until some time after rain has tapered off or totally ceased. Thereafter discharge decreases to more normal flow. The delay in response of the stream between precipitation and peak discharge is referred to as the **lag time.** These aspects of discharge can be recorded in a **hydrograph,** as shown in Figure 14.6.

Adjustments of Discharge, Channel, and Velocity

The discharge of a stream relates directly to the velocity of a stream and to its depth and width. We can state this as follows.

Discharge (m³/s) = channel width (m) × channel depth (m) × water velocity (m/s)

This can be more quickly and simply written as

$$Q = W \times D \times V$$

where Q = discharge; W = width; D = depth; and V = velocity.

This equation tells us that if stream discharge changes, then changes must also occur on the other side of the equation. Thus, width, depth, velocity, or some combination of the three must change to balance the equation.

An example of how increased discharge is balanced by adjustments in width, depth, and velocity is given in Figure 14.7, which records changes during a flood near Bluff, Utah, on the San Juan River. As discharge increased so did the depth. Increase in width of the stream was negligible because the river was confined by bedrock walls. Velocity, however, increased from about 0.9 m/s on September 9 to a maximum of about 3.4 m/s and then dropped to 1.5 m/s on December 9 as discharge fell.

It is not surprising that if the discharge at a given point along a river increases, then the width, depth, velocity, or some combination of these factors must also increase. Many people are surprised, however, to find that the adjustments of width, depth, and velocity to changing discharge are orderly and predictable. In most streams, if the discharge increases, then the width, depth, and velocity each increase at a definite rate.

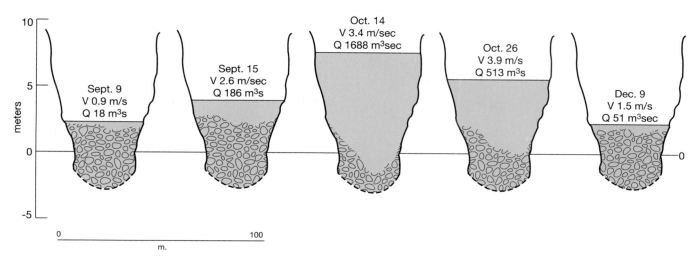

FIGURE 14.7 Changes in stream discharge during a flood on the San Juan River at Bluff, Utah, in 1941 caused changes in the stream channel and in velocity. As discharge increased from September 9 to 15, velocity also increased. In the same interval the channel floor rose, but because the stream level rose faster a deeper channel resulted. At peak discharge velocity increased to about 3.4 m/s, and the current scoured the channel sediments below their level on September 9. As discharge and velocity fell, the sediments built up in the channel and the stream shallowed. Changes in width of channel were minimal because, particularly at high flows, the channel was constrained by bedrock.

Figure 14.8 is a record of how water depth, width, and velocity varied at one spot on a stream in Wyoming as discharge changed during the year. The figure also illustrates the effect of a bridge at higher discharge rates. Channel width and depth and water velocity increase consistently up to a discharge of about 30 m^3/s. At this discharge, however, the bridge begins to modify the depth and the width of the stream. As the discharge increases, the depth increases more rapidly but the width remains nearly constant, controlled as it is by the restricted opening beneath the bridge. The velocity continues to increase at the same rate as it did at lower flows, unaffected by the constriction at the bridge.

Bridges are not the only human structures that can influence stream characteristics. At the town of North Platte, Nebraska, the width of the channel of the North Platte River has shrunk from an estimated average of 790 m in 1865 to an average of 90 m during 1965–1969. Additionally, the average peak flow of the river dropped from 511 m^3/s before 1909 to about 72 m^3/s since 1957 (Figure 14.9). Similar changes have occurred along the Platte River downstream from the junction of the South and North Platte rivers. What caused such drastic changes?

You need not look far for the answer. The North Platte is one of the most intensively developed rivers in the United States. Beginning in 1909 a series of dams have been constructed along the river. These dams capture the spring snowmelt from the headwaters of the North Platte in Wyoming and Colorado.

The water is then released slowly for irrigation during the growing season. As you would guess, this has decreased the size of peak flows each year along the North Platte. This decrease in discharge has also led to the decrease in channel width. There are no data from our diagram about the effect of decreased discharge on depth and velocity. Geologists would predict, on the basis of experience, that both depth and velocity of the North Platte also decreased.

There have been secondary results caused by stream narrowing. For example, a great increase of vegetation has taken place along the river banks and on some of the sand islands in the river as well. Both the reduction in the width of the river and the increase in vegetation on sand islands have severely curtailed the use of the Platte and its tributary, the North Platte, as a stopover point by sandhill cranes in their annual migrations from the northwestern United States and western Canada to Texas and back.

A great deal of change also takes place along the entire length of a stream. For example, the discharge of a stream generally increases downstream as more and more smaller streams, known as tributaries, contribute water to its main channel. The width and depth also increase downstream. When we gather accurate data on the width, depth, velocity, and discharge of a stream from its headwaters to its mouth for a particular stage of flow we again find that the changes follow a definite pattern (Figure 14.10). Not only do the depth and width increase downstream as discharge increases, but velocity also increases toward the river's

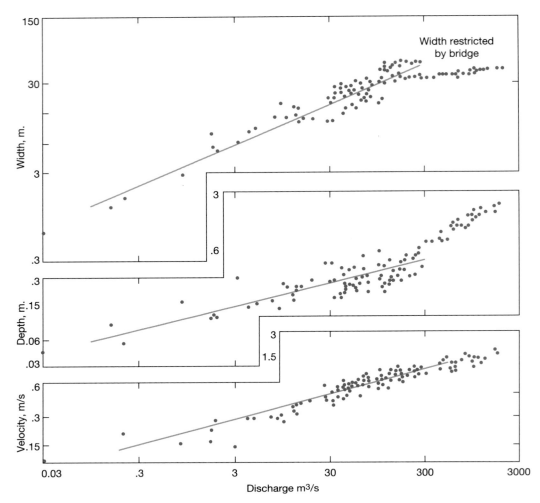

FIGURE 14.8 As the discharge of a stream increases at a gauging station, so do its velocity, width, and depth. The increase is in an orderly fashion, as shown by these graphs based on data from a gauging station on the Powder River at Arvada, Wyoming. Variations in the direction of adjustments at flows greater than the 28 m³/s are due to the effect of a bridge at the gauging station. See text for discussion.

FIGURE 14.9 As dams were built along the North Platte River, the annual average peak flow was reduced. This was accompanied by a decrease in channel width. The flow is given in 5-year averages of annual peak flow.

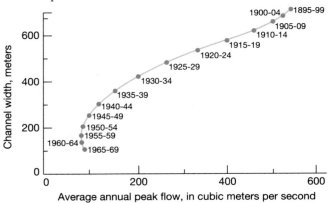

mouth. This is because a stream deepens and widens its channel to handle the greater discharge downstream, but not rapidly enough to handle all of the added water. Some of the increased discharge, therefore, must be accommodated by an increase in stream velocity.

Floods

When the discharge of a stream exceeds the capacity of its channel, water must rise over the banks of the channel and form a **flood** that covers the adjacent low-lying land. A flood is any peak flow that tops the channel banks (Figure 14.11). Floods can be catastrophic in human terms, as were the 1993 floods that ravaged the midwestern United States. They are, however, simply part of the natural behavior

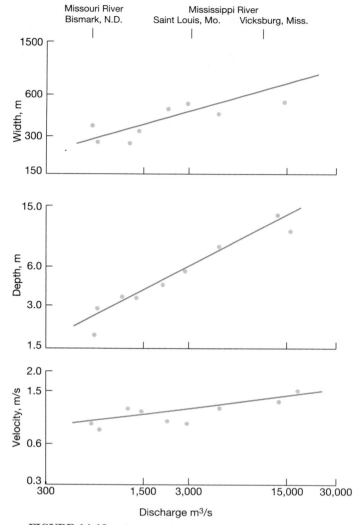

Missouri River
Bismark, N.D.

Mississippi River
Saint Louis, Mo.   Vicksburg, Miss.

FIGURE 14.10   Stream velocity, depth, and width of a channel increase as the discharge of a stream increases downstream. Measurements in this example were taken at mean annual discharge along a section of the Mississippi-Missouri river system.

FIGURE 14.11   Topping and breaching of levees and dikes, letting water into the lower lands behind them, was typical of the 1993 flood along the Mississippi and Missouri Rivers. This low aerial view shows muddy waters rushing across a farm near St. Louis. (*B. Gillette, Gamma-Liaison, Inc.*)

of streams. Geologists have found that floods are actually fairly predictable. For instance, a discharge that just fills a stream channel, so that it is at the **bankfull stage,** occurs every 1.5 to 2 years. The frequency of peak flows or floods, which exceed the bankfull stage, can also be anticipated statistically, as discussed in Perspective 14.1 and shown in Figure P14.1.1.

Floodplains
Floods occur when the stream level rises above the bankfull stage and waters spread across adjacent, low-lying lands. These lands are usually marked by sands and silts left by the subsiding waters of previous floods. The extent of floodwaters defines the **floodplain.** Because floods are of different magnitudes, geologists speak, for example, of the floodplain of a 10-year flood, which will be less extensive than, for instance, the floodplain of a 50-year flood.

People find the floodplain useful for many different purposes. They use it for agriculture, road, railroad, industry, and town sites. The floodplain, however, is really an extension of the stream channel. The stream uses it when the channel is too small to hold the water brought to it. When this happens the stream floods, spilling out across the floodplain. As it does so, it can cause loss of lives and property. Therefore, we should view the floodplain as part of the stream's domain, useful to it in flood times and certain to be claimed by it from time to time. Human uses of the floodplain should be those whose disruption will be the least costly when floods come.

Engineering efforts to control flooding focus on reducing the peak flow of a stream or confining the peak flow to the area of the channel. Reduction of the peak flow depends upon flood-control dams. These are built to store floodwater in the collecting basin behind the dam and then to let it out slowly, thus lowering the flood crest by spreading it out over a period of time (Figure 14.12). Containing the flood crest within the stream channel relies primarily on dredging a stream channel deeper or raising its banks with dikes or embankments.

Urbanization and Suburbanization
The building of cities and their suburbs affects stream flow and changes stream discharge, although in the opposite direction from that of flood-control dams. Large clusters of buildings with their associated roads, streets, sidewalks, and paved parking areas are the cause of changes in stream flow. Precipitation runs rapidly off developed areas directly to the stream channels. At the same time less water infiltrates into the underground, and base flow of the streams is reduced. This has several results. The time lag between precipitation and flood peak is shortened

Estimating Flood Recurrence

We can estimate how often a flood of a given size might occur by using a simple statistical method. It involves a record of stream discharge for as many years as possible. The first step is to list the maximum peak discharge for each year, whether or not it resulted in topping the stream banks. The second is to rank each maximum annual discharge by size from largest to smallest, designating the largest as 1. Next we compute the flood-recurrence interval by using the following formula:

$$R = \frac{N+1}{M}$$

where R equals the **flood recurrence interval** in years, N equals the number of years of the record, and M equals the magnitude or rank of each maximum annual flood.

The results from such an analysis for a 93-year record on the Cumberland River at Nashville, Tennessee, are plotted in Figure P14.1.1. The largest flow, 5,750 m^3/s, is entered at 94 years, (93 + 1)/1 in our formula. The size of the second-largest flood, 5,570 m^3/s, is plotted at 47 years, (93 + 1)/2, because two floods of this size or larger have occurred in the last 93 years. The third flood is shown as 94/3, or 31 years plus, and so on.

Graphs such as the one pictured allow geologists to estimate the size of 10-year or 50-year floods (floods that recur every 10 or 50 years). Because the graph shows points on a relatively straight line, geologists may extend the line to the upper right to estimate the size of floods larger than those on record. For example, it might be justified to predict the size of a flood to be expected once in 200 years. It would not, however, be justified to predict the size of a 1,000-year flood on the basis of the data available.

Flood frequency data have obvious uses for engineers and regional planners. For example,

because water is not slowed on its way to the stream channel by a vegetative cover, nor does any appreciable amount sink into the ground. The flood peak is higher because the stream must carry more water in a shorter period of time. The base flow is lower because of a decreased supply of ground water on which the base flow depends. The net result of these changes is a **flashy stream,** one that has a low base flow and a high, short flood peak as shown in the hydrograph in Figure 14.13.

Changes in Vegetative Cover   Changes in vegetative cover can cause changes in the stream flow. In a general way, a decrease in the vegetative cover affects stream flow in the same way that an increase in urbanization does.

Studies of the effects on stream flow of cutting different percentages of forest cover have been carried out by the United States Forest Service in the Fernow Experimental Forest near Parsons, West Virginia. The results are not entirely unexpected. Removal of the trees increases the flow of streams during the growing season chiefly because the water ordinarily transpired into the atmosphere by trees makes its way to the streams instead. The increase in stream flow is approximately proportional to the percentage of tree cover removed, the greatest increase—over 100 percent—occurring with complete cutting of the forest cover. Concurrently with the increase of stream flow, the flashiness of streams increases. As the tree cover replaces itself, the flow decreases and the flashiness of streams declines. The return to stream-flow conditions similar to those before cutting is about 35 years.

Base Level of a Stream

Base level of a stream is defined as the lowest point to which the stream can erode its channel. Thus, anything that prohibits the stream from lowering its channel creates a base level. For example, the velocity of a stream is slowed to a halt when it enters the standing, quiet waters of a lake. Because its waters are no longer

they tell engineers how great a volume of water a bridge or culvert must accommodate for any given flood size. Regional planners will find it useful to know how high floodwaters can reach for a given flood so they can plan how the land adjacent to the river should be zoned.

Flood-recurrence analysis is a useful technique. But remember, we do not know precisely when a flood of a given size will occur. The historical record merely tells us to expect a flood of a given size sometime within a given span of years.

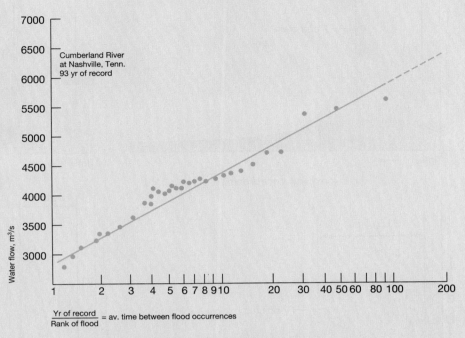

$$\frac{\text{Yr of record}}{\text{Rank of flood}} = \text{av. time between flood occurrences}$$

P14.1.1 Average interval between floods of various sizes on the Cumberland River at Nashville, Tennessee, plotted from data in U.S. Geological Survey, Water Supply Paper 771.

moving the stream loses its ability to erode, and it cannot cut below the level of the lake. The lake's control over the stream is effective along the entire course upstream because no part of the stream can erode beneath the level of the lake. However, the base level created by every lake is temporary. After the lake has been destroyed, the stream will be free to continue its downward erosion. Because of its impermanence, the base level formed by a lake is referred to as a **temporary base level.** Other examples of temporary base levels include layers of resistant rock at the lip of a waterfall or the level of the main stream into which a tributary drains. But even after a stream has been freed from one temporary base level, it will be controlled by others farther downstream. Eventually, though, it makes its way to the ocean, which is the **ultimate base level.** Yet the ocean itself is subject to changes in level, so even the so-called ultimate base level is not fixed (Figure 14.14). But what, you may ask, is the base level for a stream that runs into a basin below sea level such as Death Valley or the Dead Sea? Such a basin can be classed as a temporary base level.

Eventually the basin fills with sediments brought in by the stream, and water spills over to another basin or the sea.

Work of Streams

The water flowing in streams performs three main jobs: It erodes its channel's floor and banks, it transports material in both solid and dissolved form, and it deposits sediments at various points along the valley or delivers them to lakes or to the ocean.

Running water may help to create a chasm like that of the Grand Canyon of the Colorado. During floods it may spread mud and sand across vast expanses of valley flats. Running water may build deltas, such as those at the mouths of the Nile and Mississippi rivers. The type and the extent of these activities depend on the energy being expended by the stream, and this in turn depends on the amount of water in the stream and the gradient of the channel. A

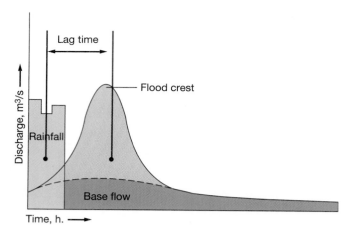

(a) Before flood control dam.

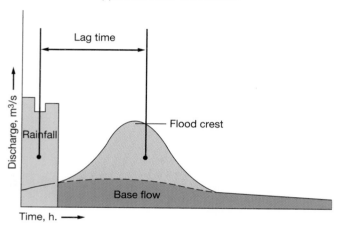

(b) After flood control dam.

FIGURE 14.12 Hydrograph showing the effect on stream flow of a flood-control dam upstream from the hydrograph station. Water is stored in the reservoir behind the dam and then let out slowly. This spreads the flood out over a longer period of time, lowers the crest, and increases the lag time between flood crest and precipitation maximum.

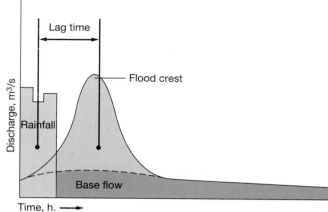

(a) Before urbanization

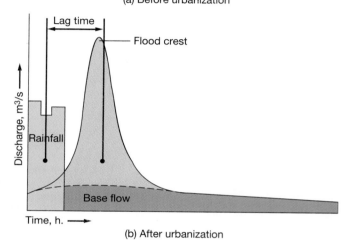

(b) After urbanization

FIGURE 14.13 Hydrograph showing the effect of urbanization on stream flow. See text for discussion.

stream expends its energy in several ways. By far the greatest part is used up in the friction of the water with the stream channel and in the friction of water with water in the turbulent eddies we discussed earlier. Relatively little of the stream's energy remains to erode and transport material. When available energy decreases far enough and the stream can no longer move the material it has been carrying, then deposition takes place.

Erosion

Direct Lifting In turbulent flow, water travels along paths that do not parallel the channel. The water eddies and whirls, and if an eddy is powerful enough, it dislodges particles from the stream bed and lifts them into the stream. Whether this will happen or not

in a given situation depends on a number of variables that are difficult to measure, but if we assume that the bed of a stream is composed of particles of uniform size, then the diagram in Figure 14.15 tells us the approximate stream velocities that are needed to erode clay, silt, sand, granules, and pebbles. A stream bed composed of fine-sized sand grains, for example, can be eroded by a stream with a velocity of a little over 10 cm/s. As the fragments become larger and larger, ranging from coarse sand to granules to pebbles, increasingly higher velocities are required to move them, as we should expect.

What we might not expect is that as the particles become smaller than sand size they are harder for the stream to pick up. In fact, stream velocity must increase to erode silt- and clay-sized particles. The reason for this is that the smaller the particles, the more firmly packed the deposit is and thus the more resistant it is to erosion. Moreover, the individual particles may be so small that they do not project sufficiently high into the stream to be swept up by the turbulent water.

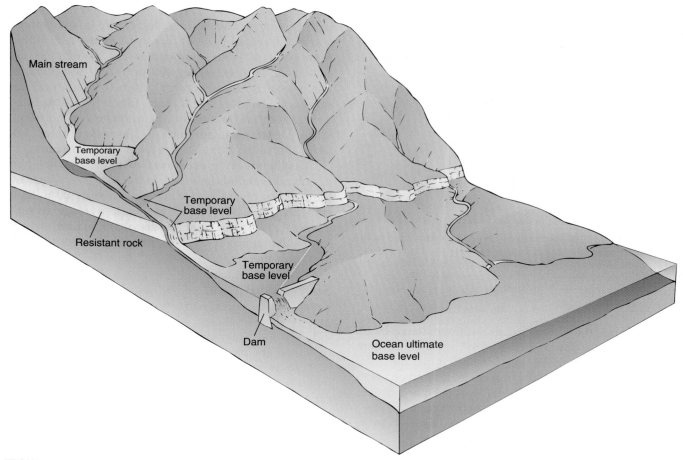

FIGURE 14.14   A temporary base level for a stream may be determined by natural and artificial lakes, by a resistant rock stratum, and by the point at which a tributary stream enters a main stream. The ocean is the ultimate base level for a stream.

Abrasion, Impact, and Solution   The solid particles carried by a stream may themselves act as erosive agents because they are capable of abrading

FIGURE 14.15   The diagram shows the relation between velocity and the erosion, transportation and deposition of particles of different sizes.

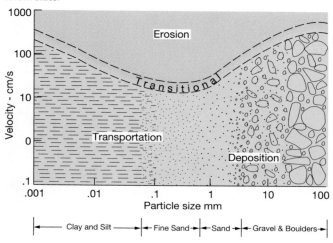

(wearing down) larger fragments in the bed of the stream or even the bedrock itself. When the bedrock is worn by abrasion, it usually develops a series of smooth, curving surfaces, either convex or concave, or a series of holes—**potholes**—drilled by pebbles swirled in the turbulent stream (Figure 14.16). Individual pebbles on a stream bottom are moved and rolled about by the force of the current, and as they rub together they become both rounder and smoother. In addition, the impact of large particles against the bedrock or against other particles knocks off more fragments, which are added to the load of the stream.

Some erosion is also caused by stream waters dissolving some channel debris or bedrock. This accounts for very little erosion, and most of the dissolved matter carried by a stream is contributed by the underground water that drains into the stream.

Transportation

The material that a stream picks up directly from its own channel—or that is supplied to it by tributaries or mass wasting—is moved downstream toward its

FIGURE 14.16   The solid material carried by a stream can erode bedrock of a stream channel. Here, in a low water stage of a stream, are the abraided pits (potholes) and smoothly curved surfaces in the channel and the pebbles and cobbles that carved them. The large stone in the pit in the center-left of the photo is 10 cm in maximum dimension. *(Pamela Hemphill)*

eventual goal, the ocean. The amount of material that a stream carries at any one time, which is called its **load,** is usually less than its **capacity**—that is, the total amount it is capable of carrying under any given set of conditions. The maximum size of particle that a stream can move is a measure of the **competence** of a stream.

A stream transports material as (1) **dissolved load,** (2) **suspended load,** and (3) **bed load.**

Dissolved Load

In nature no water is completely pure. We have already seen in Chapter 6 that when water falls and filters down into the ground it dissolves some of the soil's compounds. Then the water may seep down through openings, pores, and crevices in the bedrock and dissolve additional matter as it moves along. Most of this water eventually finds its way to streams. The amount of dissolved load varies with climate, season, and geologic setting, and is measured in terms of *parts* of dissolved matter *per million* **ppm** parts of water. Sometimes the amount of dissolved material exceeds 1,000 ppm, but usually it is much less. By far the most common compounds in solution in running water, particularly in arid regions, are calcium and magnesium carbonates. In addition, streams carry small amounts of chlorides, nitrates, sulfates, and silica, with perhaps a trace of potassium.

Geologists have estimated that the total load of dissolved material delivered to the seas every year by the streams of the United States is about 100 million cubic meters. The rivers of the world average an estimated 115 to 120 ppm of dissolved matter, which means that annually they carry to the sea about 1.6 billion cubic meters.

Suspended Load

Solid particles will settle in water unless some force counters the influence of gravity. The swirls and eddies of turbulent water provide this force in a stream. The greater the turbulence, the larger the particle that the water can buoy up, that is, hold in **suspension.** Because turbulence increases when the velocity of stream flow increases, the greatest amount of material is moved during flood time, when velocities and turbulence are highest (Figure 14.17a,b). In just a few hours or a few days during flood time, a stream transports more suspended material than it does during the much longer periods of low or normal flow. Observations of the area drained by Coon Creek, at Coon Valley, Wisconsin, over a period of 450 days showed that 90 percent of the stream's total suspended load was carried during an interval of 10 days, slightly over 2 percent of total time.

FIGURE 14.17   (a) The Virgin River, Utah, is muddy with suspended load during high velocity in flood. (b) In contrast, the gravel bed of the Cincinnati Creek at Barneveld, N.Y., is visible through the clear water during a period of low flow.

(a)

(b)

Bed Load Materials in motion along a stream bottom constitute the stream's bed load. These particles move by means of saltation, rolling, and sliding.

Saltation has nothing to do with salt. It is derived from the Latin *saltare* "to jump." Thus a particle moving by saltation jumps from one point on the stream bed to another. First the particle is shaken loose from the bed of the stream by a surge of turbulent water or the impact of another particle and carried upward into the forward-moving current. When the upward motion of the turbulent water is no longer strong enough, the particle drops back to the bed of the stream. In this process of rising and falling, the particle is carried forward by the stream to a new resting place downstream from its original position. Movement by saltation, then, is transitional between movement by continuous suspension and movement by continuous contact between the particle and the floor of the channel. Particles that maintain continuous contact with the floor of the channel are too heavy to be picked up even momentarily by water currents. These particles, however, may be pushed along the stream bed and, depending on their shape, move forward by either rolling or sliding.

Deposition

As soon as the velocity of a stream falls below the level necessary to hold material in suspension, the stream begins to deposit its suspended load. Deposition is a selective process. First, the coarsest material is dropped in waters still swift enough to carry finer sediments. Then, as the velocity (and hence the turbulence) continues to slacken, finer and finer material settles out. Reexamination of Figure 14.15 shows at what flow velocities particles of differing sizes begin to deposit.

Distribution of turbulence in a winding stream provides an example of the differing environments of deposition within a stream. You should recall that zones of greatest velocity and turbulence in a stream swing to the outside of a bend, and the zones of least turbulence and lowest velocity are found on the inside of the bend. As the turbulent current swings up against the outside of the bend it erodes the bank. Any sediments too coarse for the stream to carry are deposited along the outside of the bend. The current sweeps finer material downstream and some of this is deposited in the quieter waters on the inside of the next bend downstream (Figure 14.18). When a stream floods, waters, muddy with fine-grained material in suspension spread across the floodplain. There, unconfined by the normal channel, velocity is quickly stemmed and the fine sediments settle out onto the floodplain.

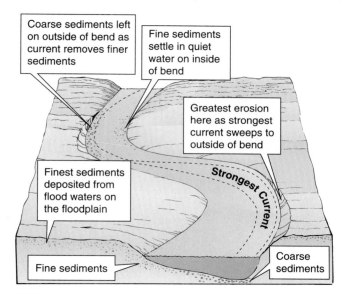

FIGURE 14.18 Deposition of coarsest sediments in a winding stream occurs on the outside of a bend and the finest occurs on the inside.

Features of Stream Valleys

Drainage Basins, Stream Networks, and Patterns

A **drainage basin** is the entire area from which a main stream and its tributaries receive their water. Each drainage basin is separated from its neighbor by a **drainage divide** (Figure 14.19). The Mississippi River and its tributaries drain a tremendous section of the North American continent, reaching from the Rockies to the Appalachian Mountains and north across the Canadian border. Its drainage divide on the west is part of the **continental divide,** which separates drainage to the Pacific Ocean from that to the Atlantic Ocean via the Gulf of Mexico. The Columbia River and the Rio Grande also have drainage basins that extend across national boundaries. In each, tributaries to the main stream have their own drainage basins, and these in turn can have still smaller basins. The drainage basins continue to subdivide into smaller and smaller basins until the unbranched tributaries are reached.

Individual streams and their valleys are joined together into networks. In each network the streams prove to have a definite geometric relationship. We can demonstrate this relationship by assigning a **stream order** to each stream in a network. As shown in Figure 14.20, we label small headwater streams without tributaries as belonging to the first order in the

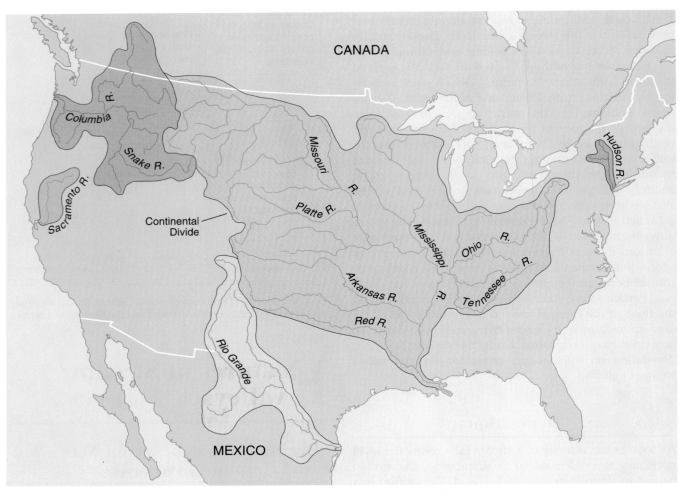

FIGURE 14.19 The stream's drainage basin is the area from which the stream and its tributaries receive water. Some well-known drainage basins in North America are shown here.

hierarchy of the network. Two or more first-order streams join to form a second-order stream; then two or more second-order streams join to form a third-order stream; and so on. In other words, a stream segment of any given order is formed by the junction of at least two stream segments of the next lower order. The main segment of the system always has the highest order number in the network, and the number of this stream is assigned to describe the drainage basin. Thus the main stream in the basin shown in Figure 14.20 is a fourth-order stream, and the basin is a fourth-order basin.

It is apparent from Figure 14.20 that the number of streams of smaller order is much greater than the number of those of higher order. The same is true of drainage basins. There are many more small drainage basins than there are large ones. These relationships are shown in Table 14.1, a close inspection of which shows that the variations among streams and drainage basins of different orders occur in an orderly fashion.

The overall pattern developed by a system of streams and tributaries depends partly on the nature

FIGURE 14.20 Method of designating stream orders.

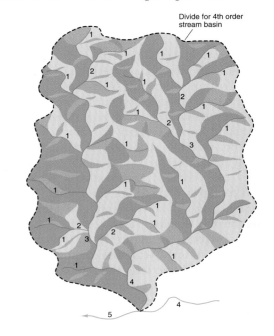

TABLE 14.1 Number and Length of River Channels of Various Sizes in the United States[a]

ORDER	NUMBER	AVERAGE LENGTH (KM)	TOTAL LENGTH (KM)	MEAN DRAINAGE AREA. INCL. TRIBUTARIES (KM²)	RIVER REPRESENTATIVE OF SIZE
1[b]	1,570,000	1.6	2,512,000	2.6	
2	350,000	3.7	1,295,000	12	
3	80,000	8.5	680,000	59	
4	18,000	19	342,000	283	
5	4,200	45	189,000	1,348	Charles
6	950	102	97,000	6,400	Raritan
7	200	219	44,000	30,300	Allegheny
8	41	541	22,000	144,000	Gila
9	8	1,243	10,000	683,000	Columbia
10	1	2,880	2,880	3,238,000	Mississippi

[a]From Luna B. Leopold, "Rivers," *Am. Sci.*, vol. 50, p. 512, 1962.
[b]The size of the order 1 channel depends on the scale of the maps used; these order numbers are based on the determination of the smallest order using maps of scale 1:62,500.

of the underlying rocks and partly on the history of the streams. However, almost all streams follow a branching pattern. They receive tributaries, and the tributaries, in turn, are joined by still smaller tributaries. The manner of branching, however, varies widely.

A stream that resembles the branching habit of a maple or similarly branched, leafy, or deciduous tree is called **dendritic,** "treelike" (Figures 14.21 and 14.22a). A dendritic pattern develops when the under-

FIGURE 14.21 This photo, taken from an orbiting space craft, shows the branching, treelike pattern of a dendritic stream system in Yemen. Width of view is about 75 km. *(NASA)*

lying bedrock is uniform in its resistance to erosion and exercises little or no control over the direction of valley growth. This situation occurs when the bedrock is composed either of flat-lying sedimentary rocks or of massive igneous or metamorphic rocks. The streams can cut as easily in one place as another; thus, the dendritic pattern is, in a sense, the result of the random orientation of the streams.

Another type of stream pattern is called **radial** and is characterized by streams radiating outward in all directions from a high central zone. Such a pattern is likely to develop on the flanks of a newly formed volcano where the streams and their valleys radiate outward and downward from various points around the cone (Figure 14.22b).

A **rectangular** pattern occurs when the underlying bedrock is crisscrossed by fractures that form zones of weakness particularly vulnerable to erosion. The master stream and its tributaries then follow courses marked by nearly right-angle bends (Figure 14.22c).

In the United States, particularly in a belt of the Appalachian Mountains running from New York to Alabama, some streams follow what is known as a **trellis** pattern, after its resemblance to the stems of a vine on a trellis (Figure 14.22d). This pattern, like the rectangular one, is caused by zones in the bedrock that differ in their resistance to erosion. The trellis pattern usually, though not always, indicates that the region is underlain by alternate bands of resistant and nonresistant rock, inclined as the result of folding.

Water Gaps Particularly intriguing features of some valleys are short, narrow segments walled by steep, rocky slopes or cliffs. These are called **water gaps.** The stream, in effect, flows through a narrow notch in a ridge or mountain that lies across its course, as shown in Figure 14.23. During the period of American expansion westward in the eighteenth and nineteenth centuries, water gaps became major routes

Features of Stream Valleys **355**

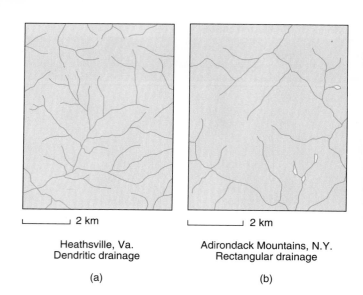

Heathsville, Va.
Dendritic drainage

(a)

Adirondack Mountains, N.Y.
Rectangular drainage

(b)

Mount Hood, Ore.
Radial drainage

(c)

Saypo, Mont.
Trellis drainage

(d)

FIGURE 14.22 The overall pattern developed by a stream system depends in part on the nature of the bedrock and in part on the history of the area. See text for discussion.

FIGURE 14.23 The notch in the ridge on the skyline is the Delaware Water Gap. It was cut by the Delaware River, which here flows along the New Jersey–Pennsylvania border.

FIGURE 14.24 One way that water gaps are created. An ancient landscape of hills and valleys, (a), is buried by younger sediments across which streams flow, (b). Streams cut through the younger sediments, and encounter the old topography. The main stream has enough energy to cut through the old ridges, thus creating water gaps. See text for further discussion.

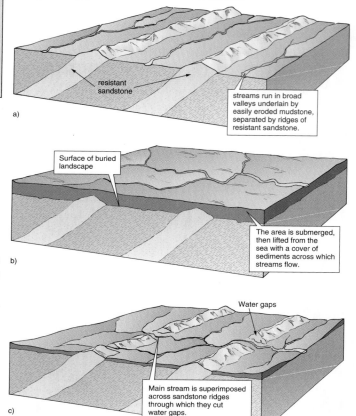

resistant sandstone

streams run in broad valleys underlain by easily eroded mudstone, separated by ridges of resistant sandstone.

a)

Surface of buried landscape

The area is submerged, then lifted from the sea with a cover of sediments across which streams flow.

b)

Water gaps

Main stream is superimposed across sandstone ridges through which they cut water gaps.

c)

across the Appalachians onto the fertile lands of the mid-continent. They remain today important routes for highway and railroad systems.

How can a stream cut through mountains? Figure 14.24 will help answer the question. An area of hills and valleys underlain by different rock types (Figure 14.24a) is buried in younger sediments. One way that this could have happened was by submergence of the area beneath the sea. When the area emerges from the sea, streams begin to flow across the newly exposed marine sediments (Figure 14.24b). As the streams erode down through this cover they are superimposed across the hills of the old landscape. If they have enough energy they will cut down through the old ridges. In Figure 14.24c, the main stream is doing just this. The stream is called a **superimposed stream,** and when it cuts through the hill it forms a water gap. Then other surficial processes, including erosion by the tributary streams in Figure 14.24b, continue to remove the cover of younger sediments and create a new landscape. One example of such a gap was carved by the Big Horn River through the northern end of the Big Horn Mountains in Montana; another is that followed by the Wind River across the Owl Creek Mountains in Wyoming (Figure 14.25).

Stream Piracy An unusual history recorded by some streams results from the enlargement of one stream valley at the expense of another. An example of this, called **stream piracy,** is given in Perspective 14.2.

Enlargement of Valleys

Geologists do not know how running water first created the great valleys and drainage basins of the continents because the record has been lost in time. But we do know that certain processes are now at work in widening and deepening valleys, and, using the principle of uniformitarianism, it seems safe to assume that these processes also operated in the past.

If a stream were left to itself in its attempt to reach base level, it would erode its bed straight downward, forming a vertically walled chasm in the process (Figure 14.26). But because the stream is not the only agent at work in valley formation, the walls of most valleys slope upward and outward from the valley floor. In time, even the steepest cliffs will be angled away from the valley floor.

As a stream cuts downward and lowers its channel into the land surface, both weathering and mass

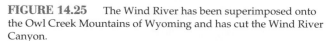
FIGURE 14.25 The Wind River has been superimposed onto the Owl Creek Mountains of Wyoming and has cut the Wind River Canyon.

FIGURE 14.26 Erosion by the Aude River in southwestern France has fashioned this limestone-walled gorge. Slope processes have only just begun to lower the angle of the valley walls.

movement come into play, constantly wearing away the valley walls and pushing them farther back. Under the influence of gravity, material is carried down from the valley walls and dumped into the stream (Figure 14.27), to be moved toward the sea. The result is a valley whose walls flare outward and upward

FIGURE 14.27 Mass wasting is a major process in widening valleys. Here debris has slipped from the valley wall into an Italian alpine stream.

Perspective *14.2*

Captured and Beheaded by Pirate

Though our box title sounds like a tabloid newspaper headline, it describes a geological process known as *stream piracy*, in which one stream captures a part of an adjacent stream and its drainage basin. Here is an example from the Appalachian Mountains of Virginia.

The Potomac River cuts through the Blue Ridge Mountains in a water gap where Virginia, West Virginia, and Maryland meet near Harpers Ferry. Before it does, the Potomac is joined by a large tributary, the Shenandoah River, flowing northeastward. At some time in the past the Shenandoah was a much shorter river. At that time, other streams to the south of Harpers Ferry also crossed the Blue Ridge in their water gaps, as did the Potomac River of the time (see Figure P14.2.1). The Shenandoah flowed northeast parallel to the linear arrangement of the rocks within a belt of nonresistant limestone, dolostone, and shale. At the same time the other streams flowed southeast across the general trend of the rocks, which included the resistant rocks of the Blue Ridge in which they had fashioned their water gaps. This set the stage for piracy by the Shenandoah.

The Shenandoah's channel lay in the easily eroded belt of nonresistant rock; the channels of the streams destined for capture cut

FIGURE P14.2.1 The Shenandoah River has expanded its drainage basin at the expense of several other streams through a process of stream piracy. Along the way some old water gaps have become wind gaps.

(a)

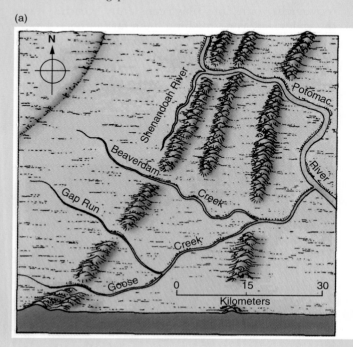

(b)

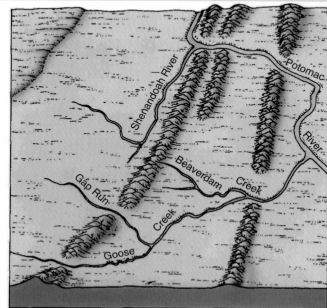

through resistant rocks of the Blue Ridge. These rocks served as a base level for the upper parts of the drainages of Beaverdam Creek, Gap Run, and Goose Creek. The ancestral Shenandoah was able first to lower its valley floor below that of the upper Beaverdam, and then to erode headward until it intercepted the headwaters of the Beaverdam and diverted them into its own drainage. We can say, therefore, that the **pirate stream,** the Shenandoah, has captured the upper drainage of the Beaverdam. The remaining drainage of the Beaverdam downstream from the point of capture is now a **beheaded stream.** This process was repeated as the Shenandoah captured, in succession, the headwaters of Gap Run and Goose Creek.

The water gaps that originally carried the channels of the three drainages across the Blue Ridge were abandoned during this process of piracy. Now they stand as gaps along the Blue Ridge and are known locally, from northeast to southwest, as Snicker's Gap, Ashby Gap, and Manassas Gap. These had previously been the water gaps for the Beaverdam, Gap Run, and Goose Creek rivers. Because today only wind flows through these abandoned water gaps, they are now known as **wind gaps.**

From the human point of view, the three gaps carry more than just wind. Several four-lane highways convey traffic across the Blue Ridge from and to the Shenandoah valley through these gaps: Virginia State Highway 7 (Snicker's Gap), U.S. Highways 17 and 50 (Ashby Gap), and Interstate 66 (Manassas Gap).

(c)

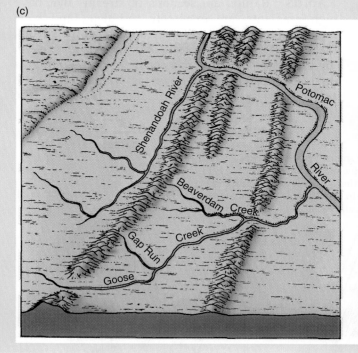

(d)

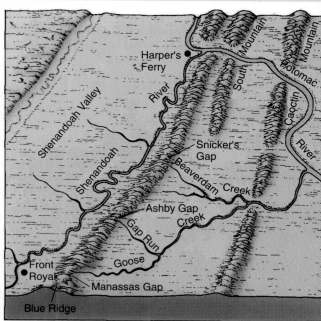

from the stream (Figure 14.28a,b). The stream, of course, contributes to mass wasting as it removes debris and weakens the slopes by undercutting the valley walls.

In addition to cutting downward into its channel, a stream also cuts from side to side, or laterally, into its banks. In the early stages of valley enlargement, when the stream is still far above its base level, downward erosion is dominant. Later, as the stream cuts its channel closer and closer to base level, downward erosion becomes progressively less important and a larger percentage of the stream's energy is directed toward eroding its banks and valley walls. As the stream swings back and forth, it forms an ever-widening floodplain on the valley floor, and the valley itself broadens (Figure 14.28c).

How rapidly do these processes of valley formation occur? In some instances geologists can measure the ages of valley formation in years and even months, but for large valleys we must be satisfied with rough approximations—and usually with plain guesses. For instance, one place where we can estimate the rate of valley formation is the Grand Canyon of the Colorado River in Arizona. Several lines of evidence suggest that it has taken roughly 10 million years for the Colorado to cut its channel downward about 1,800 m and push back the margins of the canyon walls between 6 and 20 km.

Features of Narrow Valleys

Waterfalls Waterfalls are among the most fascinating spectacles of the landscape. Thunderous and powerful as they sometimes are, however, they are actually short-lived features in the history of a stream. They owe their existence to some sudden drop in the river's gradient—a drop that will be eliminated with the passage of time.

Waterfalls are caused by many different conditions. Niagara Falls, for instance, is held up by a relatively resistant bed of dolostone underlain by beds of nonresistant shale (Figure 14.29a). The shale is easily undermined by the swirling waters of the Niagara River as they plunge over the lip of the falls. When the undermining has progressed far enough, blocks of the dolostone collapse and tumble to the base of the falls (Figure 14.29b). The same process is repeated over and over again as time passes, and the falls slowly retreat upstream. Historical records suggest that the Horseshoe or Canadian falls (by far the larger of the two falls at Niagara) have been retreating at a rate of 1.2 to 1.5 m/year, whereas the smaller American falls have been eroding away 5 to 6 cm/year. The 11 km gorge of the Niagara River between the foot of the falls and the Lake Ontario plain are evidence of the headward retreat of the falls through time (Figure 14.29c).

The waters of Yosemite Falls in Yosemite National Park, California, plunge 770 m over the Upper Falls, down an intermediate zone of cascades, and then over the Lower Falls. The falls leap from the mouth of a small valley high above the main valley of the Yosemite. The Upper Falls alone measure 430 m, nine times the height of Niagara. Several times in the past, glaciers scoured the main valley much deeper than they did the side valleys. Then, when the glacier ice melted, the main valley was left far below its tributary, which now joins it after a drop of nearly three quarters of a kilometer.

Rapids Rapids are sections of streams that have high gradients, are extremely turbulent, and have higher velocities than adjoining sections of the stream. Like waterfalls, rapids occur at a sudden increase in the gradient of the stream's channel. Although rapids do not plunge straight down as waterfalls do, the underlying cause of their formation may often be the

FIGURE 14.28 If a stream of water were the only agent in valley formation, we might expect a vertically walled valley no wider than the stream channel, as suggested by the rectangular outlines in (a), (b), and (c). Mass movement and slope wash, however, are constantly wearing away the valley walls, carving slopes that flare upward and away from the stream channel, as shown in the diagrams. In (c) the stream is shown swinging back and forth as it erodes laterally, pushes back the valley wall, and widens the valley.

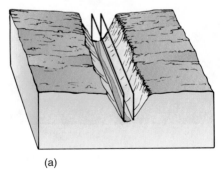
(a)

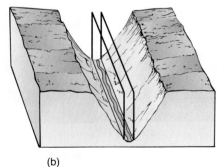
(b)

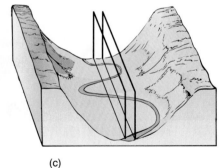
(c)

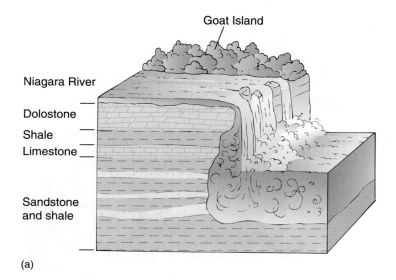

(a)

(b)

Grand Island   Goat Island

American Falls

Canadian Falls

Niagara

Niagara Gorge

Escarpment

Lewiston

Niagara River

Lake Ontario

(c)

FIGURE 14.29

(a) The waters of Niagara Falls tumble over a resistant bed of dolostone, underlain chiefly by shale. As the less-resistant bed of shale erodes, the undermined ledge of dolostone breaks off, and the lip of the falls retreats. (b) Niagara Falls. Rock from an old collapse of the lip is partially obscured in mist and foam at the base of the American falls. The Canadian or Horseshoe falls are in the distance. (c) The falls of Niagara have been retreating headward from the lip of Niagara escarpment ever since the last glacier left the area about 12,000 years ago. The retreating falls have formed the gorge of the Niagara River.

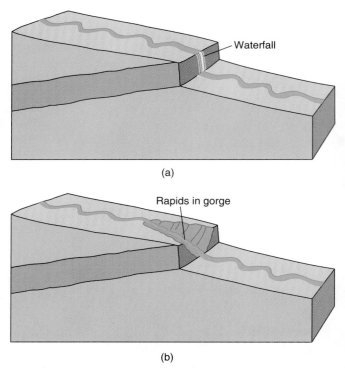

(a)

Waterfall

Rapids in gorge

(b)

FIGURE 14.30 Rapids may represent a stage in the destruction of waterfalls, as suggested in this diagram.

same. In fact, many rapids have developed directly from preexisting waterfalls (Figure 14.30). Others can develop because of special circumstances. For instance, most of the rapids in the Grand Canyon of the Colorado result from bouldery debris dumped across the channel during rare, torrential discharges from side canyons. Even without waterfalls and rapids the streams that run through narrow valleys usually have steep gradients as compared with those that run through broad valleys.

Features of Broad Valleys

The various agents working to enlarge a valley will eventually produce a broad valley with a wide, relatively level floor. During periods of normal or low water the stream is confined to its channel, but during high water, or flood, it overflows its banks and spreads out across the floodplain.

Meanders The hairpin bends of a meandering stream (Figure 14.31) provide an extreme example of the normal process of stream erosion on the outside of a bend and deposition on the inside, a process discussed earlier. The term **meander** comes from the Menderes River in western Turkey, which bends back on itself in a series of broad loops. The erosion has both a lateral, cross-valley component and a down-valley component. The result is that the bends of a meandering stream move both across and down the valley.

FIGURE 14.31 The Kobuk River, Alaska, flows across its flood plain in well-developed meanders. Scars of old meanders, some containing oxbow lakes, testify to the shifting nature of the river's channel. The Brooks Range is in the background. *(Michael Male, Photo Researchers, Inc.)*

As the stream moves down-valley, different meanders as well as different parts of the same meander may move at different rates. The result is that the narrow neck of a meander may be cut through, and the stream, taking a new course, will abandon the old bend. The new channel is called a **neck cutoff.** The abandoned meander—the **meander scar**—is called an **oxbow** because it has a shape similar to the U-shape of an ox's collar. Usually both ends of the oxbow are gradually filled with sediment, and the old meander becomes completely isolated from the new channel. If the abandoned meander fills up with water, an **oxbow lake** results. Although a cutoff will eliminate a particular meander, the stream's tendency toward meandering still exists, and before long the entire process begins to repeat itself.

A meander grows and migrates by erosion on the outside and downstream side of the bend and by deposition on the inside. This deposition on the inside leaves behind a series of low ridges and troughs, col-

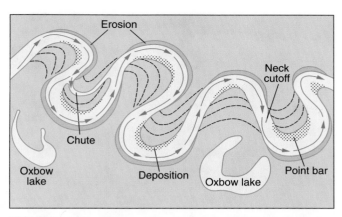

FIGURE 14.32 Erosion takes place on the outside of a meander bend, whereas deposition is most marked on the inside. The deposition produces point bars. If the neck of a meander is eroded through, the cutoff bend forms an oxbow or oxbow lake. A chute originates in a trough between low ridges created where irregular deposition takes place on the inside of a meander as the meander migrates.

lectively known as **point bars.** These features are made of material deposited in the slack water on the inside of the bends of a winding or meandering river. Swamps often form in the troughs, and during flood time the river may develop an alternate channel through one of the troughs. Such a channel is called a **chute cutoff,** or simply a **chute** (Figure 14.32).

Floodplains and Their Deposits Most streams are bordered by floodplains. Floodplains may range from very narrow (a few meters wide) to those whose widths are measured in kilometers, as is that of the lower Mississippi River. Floodplains appear very flat, particularly when viewed in relation to the slopes of the valley walls that flank them. But plains are not without their ups and downs. A large floodplain may have differences of relief of several meters and be marked by such features as natural levees, meander scars, and oxbow lakes.

Floodplains are made up of stream-carried sediments. Two general categories of deposit are common. One is made up of fine silts, clays, and sands, which may be spread across the plain by a river that overflows its banks during a flood. These are called **overbank deposits.** The other one is the point bars discussed earlier. These are made of coarse material, gravel and sand, and are related directly to the channel of the stream.

In discussing meanders, we stated that the loops of a channel migrate laterally so that the river swings both across the valley and down-valley. While erosion takes place on one bank of the river, deposition of sediments from upstream takes place on the point bars on the opposite side. The river thus builds its floodplain in some places and simultaneously destroys it in oth-

ers, and the floodplain then becomes a temporary storage place for sediments.

We can view the floodplain, then, as a depositional feature more or less in equilibrium. This equilibrium may be disturbed in several ways. The stream may lose some of its ability to erode and carry sediment so that a net gain of deposits occurs and the level of the floodplain rises. On the other hand, the flow of water or the gradient may increase or the supply of sediments may decrease, and there will be a net loss of sediments as erosion begins to destroy the previously developed floodplain.

Braided Streams On some floodplains, particularly those where large amounts of material are dropped rapidly, a stream may build up a complex tangle of converging and diverging channels separated by sandbars or islands (Figure 14.33). A stream of this sort is called **braided.** This braiding generally occurs when a stream's discharge is highly variable and its banks are easily eroded to supply a heavy load. In general, the gradient of a braided stream is higher than that for a meandering stream of the same discharge. This increase in gradient gives the braided stream the extra energy needed to transport the greater load.

Levees In many floodplains the water surface of the stream is held above the level of the valley floor by banks of sand and silt known as **levees,** a name derived from the French verb *lever,* "to raise." These banks slope gently, almost imperceptibly, away from their crest along the river toward the valley wall. Levees are built up when water spills over the river banks onto the floodplain during floods. Because the muddy water rising over the stream bank is no longer confined by the channel, its velocity and turbulence drop immediately, and much of the suspended load is

FIGURE 14.33 This low aerial photograph shows the complex braided pattern of the Platte River near Grand Island, Nebraska. *(Kevin Crowley)*

Features of Stream Valleys **363**

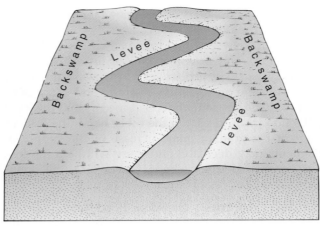

FIGURE 14.34 Natural levees characterize many aggrading streams. They build up during periods of flood as coarser material is deposited closest to the stream channel to form the levee and finer material is deposited in the backswamps. As the banks build up, the floor of the channel also rises.

FIGURE 14.35 In this picture taken from Gemini 4 the Nile Delta is prominent because its well-vegetated surface stands out in contrast to the neighboring desert. *(NASA)*

deposited close to the river. The deposit of one flood is a thin wedge tapering away from the river, but over many years the cumulative effect produces a natural levee that is considerably higher beside the river bank than away from it (Figure 14.34). The low-lying floodplain away from the levee may contain marshy areas known as **backswamps.** On the Mississippi Delta, for instance, the levees stand 5 to 6 m above the backswamps. Although natural levees tend to confine a stream within its channel, each time the levees are raised slightly, the bed of the river also rises. In time, the level of the bed is raised above the level of the surrounding floodplain. A stream may then cut a new channel (called a **crevasse**) through a confining levee and assume a new course across the lowest parts of the floodplain toward the backswamps.

A tributary stream entering a river valley where the main stream has high levees may be unable to find its way directly into the main channel. Therefore it will flow down the backswamp zone and may run parallel to the main stream for many kilometers before finding an entrance. Because the Yazoo River typifies this situation by running 320 km parallel to the Mississippi, all rivers following similar courses are known as **yazoo-type** rivers.

Deltas and Alluvial Fans For thousands of years the Nile River has deposited sediments as it empties into the Mediterranean Sea, thereby forming a great triangular plain that spreads out downstream. In the fifth century B.C. the historian Herodotus introduced the term **delta** for this plain because of its similarity to the outline of the Greek letter Δ (Figure 14.35). Very few deltas, however, show the perfect delta shape. Many factors, including shore currents, varying

rates of deposition, the compaction of sediments, and the downwarping of the Earth's crust, act to modify the typical deltaic form and sequence.

Whenever a stream flows into a body of standing water, it quickly loses its velocity and transporting power. If the stream carries enough debris and if conditions in the body of standing water are favorable, a delta will gradually form. A series of channels that fan out from the delta's upstream point provide the area's drainage. These branching channels, or **distributary channels,** build the delta. As sediment is deposited along a distributary channel, its bed and levees build up above the surrounding area. In time, the distributary stream escapes its old channel and forms a new one. The process then continues to repeat itself as the delta grows (Figure 14.36).

At the outer end of a distributary channel sediments may settle out in a definite pattern. The coarse material is dumped first, forming a series of dipping beds called **foreset beds,** but the finer material is swept farther along to settle across the sea or lake floor as **bottomset beds.** As the delta extends farther and farther out into the water body, the stream must extend its channel to the edge of the delta. As it does so, it covers the delta with **topset beds,** which lie across the top of the foreset beds (Figure 14.37).

Deltas are characteristic of many of the larger rivers of the world, including the Nile, Mississippi, Ganges, Rhine, and Rhône. On the other hand, many rivers have no deltas, either because the deposited material is swept away as soon as it is dumped or, as

FIGURE 14.36   The Mississippi delta is called a "birdfoot delta" for its similarity to the clawed foot of a bird. Here, at the outermost part of the delta, the main distributary channel, as well as the smaller ones leading from it, are outlined by natural levees. Adjacent to the delta the muddy waters record the suspended load of the Mississippi and the way it is distributed by the local currents. Width of view about 75 km. Photo taken April, 1984. *(NASA)*

in the case of the St. Lawrence River, because the streams do not carry sufficient detrital material to build up a delta.

The counterpart of a delta on land is an **alluvial fan** (Figures 14.38 and 14.39). These fans are typical of arid and semiarid climates, but they may form in almost any climate if conditions are right. A fan marks a sudden decrease in the carrying power of a stream as it descends from a steep gradient to a flatter one—for

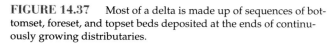

FIGURE 14.37   Most of a delta is made up of sequences of bottomset, foreset, and topset beds deposited at the ends of continuously growing distributaries.

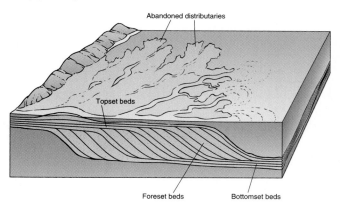

FIGURE 14.38   Alluvial fans in Death Valley, California. The varying sizes of fans relate to the varying sizes of the drainage basins providing material to the fans. The larger the drainage basin, the larger the fan. The biggest fan shown here has a radius of about 2.5 km. *(Fairchild Aerial Photograph Collection, Whittier College)*

example, when the stream flows down a steep mountain slope onto a plain. As the velocity is checked, the stream rapidly begins to deposit the sediments it is carrying. Eventually the fan is established, and its gradient is close to that of the stream as it flows from the

FIGURE 14.39   The apexes of alluvial fans lie at the mouths of drainages emerging abruptly from higher to lower land, as they do in these fans in Glacier National Park, Montana. *(William E. Ferguson)*

mountain. Deposition on the fan continues, however, for two reasons. First, as the stream emerges from the mountain, it loses water into the permeable sediments of the fan and thereby loses its ability to transport material. Second, as the stream encounters the top of the fan, it is able to widen its channel in the unconsolidated sediments of the fan, and this decreases the velocity and leads to deposition. As the stream deposits material, it builds up its channel, often constructing small levees along its banks. Eventually, as the levees continue to grow, the stream may flow above the general level of the surrounding fan. Then during a time of flood it seeks a lower level and shifts its channel to begin deposition elsewhere. As this process of shifting continues, the alluvial fan continues to grow.

Stream Terraces

A **stream terrace** is a relatively flat surface running along a valley, with a steep bank separating it either from the floodplain or from a lower terrace. It is a remnant of an old, higher floodplain of a stream that has now cut down to a lower level (Figure 14.40).

Stream terraces form in different ways. A common type of terrace formation occurs when a stream first clogs its valley with sediments and then cuts its way down to a lower level (Figures 14.41 and 14.42). The initial increase in sedimentation may be caused by a change in climate that leads either to an increase in the stream's load or to a decrease in its discharge. Deposition can also occur if the base level of the stream rises, reducing the gradient and causing the stream to drop its load. Whatever the reason for this deposition, the stream fills the valley with sediment, causing the floodplain to rise gradually. If the stream then erodes its channel, it will cut a new channel down through

FIGURE 14.41 Terrace levels are matched in elevation across this intermittent stream, La Plaza, 15 km north of Gela in southern Sicily. The top of the terrace marks the elevation of a past floodplain. Since then the stream has cut downward to leave behind a terrace on either side of the new floodplain.

the deposits it has already laid down. The level of flow will be lower than the old floodplain, and at this lower level the stream will begin to fashion a new floodplain. As time passes, remnants of the old floodplain may be left standing as terraces on either side of the new one.

FIGURE 14.42 One example of the formation of terraces. (a) The stream has partially filled its valley and has created a broad floodplain. (b) Some change in conditions has caused the stream to erode into its own deposits; the remnants of the old floodplain stand above the new river level as terraces similar to those in Figure 14.41.

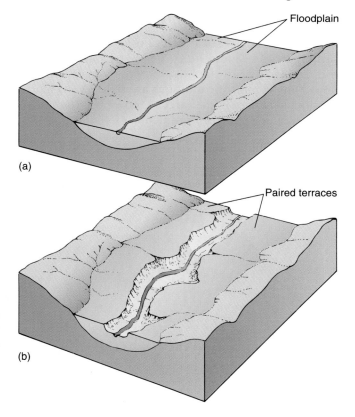

FIGURE 14.40 In the past the Madison River, Montana, deposited sand and gravel washed from now-vanished glaciers. The modern river has cut through the old deposits and has left well-developed terraces. *(William C. Bradley)*

Rivers and Continents

Major Drainage Divides and Drainage Basins

If you look at a world map you will see that many of the major drainage divides are related to plate boundaries, both present and past. Some interesting examples are found in North and South America and along the Nile River in Africa.

North and South America In western North America a vast mountain region, the North American cordillera, stretches from Alaska to central America. This zone of plate collisions, present-day and past, blocks most of the continental drainage from flowing to the Pacific Ocean. Instead, runoff from the continent is diverted to the Arctic Ocean and the Atlantic Ocean, directly or via Hudson Bay or the Gulf of Mexico. The largest drainage basin on the North American continent is that of the Mississippi-Missouri system. Its western headwaters lie along the continental divide in the Rocky Mountains. Its southeastern headwaters, however, reach to the western flanks of the Appalachian Mountains, the site of converging plates throughout most of the Paleozoic Era (Figure 14.43).

In South America, the Andes, raised by the continuing collision of the Nazca and American plates, form the continental divide. This mountain range separates river drainage to the Pacific Ocean from that to the Atlantic. Over 90 percent of the continent drains eastward to the Atlantic and adjacent Caribbean Sea. Three of the world's largest rivers, the Amazon, Paraná, and Orinoco, rise on the eastern slopes of the Andes and drain 63 percent of the continent (Figure 14.43). On a world scale, the Atlantic receives more continental drainage than all the other oceans combined, nearly 60 percent. By contrast, the world's largest ocean, the Pacific, receives about 17 percent, the Arctic Ocean about 14 percent, and the Indian Ocean about 10 percent.

The Nile The course of the world's longest river has been determined by the diverging plate boundaries along three rifts focused on Djibouti at the southern end of the Red Sea (Figure 14.44a). The Red Sea is one spreading axis, the Gulf of Aden is a second, and the third, less well-developed, is the East African rift.

To trace the relation between the present course of the Nile and the diverging plate boundaries let us follow the river from its mouth at Cairo and the delta region southward to its headwaters in Lake Victoria (Figure 14.44b). Between its mouth and the junction of the Blue Nile with the White Nile at Khartoum in Sudan, the Nile has been forced to flow parallel to and to

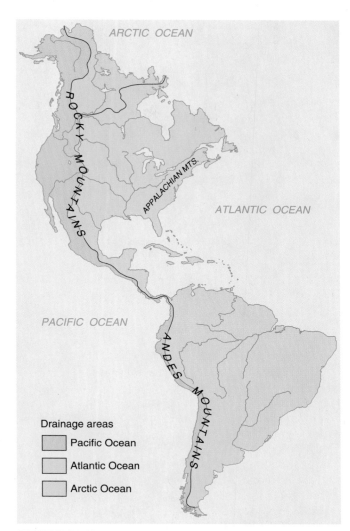

FIGURE 14.43   In the Americas most of the drainage from the continents is directed to the Atlantic and Arctic oceans, and very little to the Pacific Ocean.

the west of the uplift associated with the Red Sea spreading axis. The Blue Nile brings waters from the Abyssinian Highlands, an uplift related to the triple junction defined by the three spreading axes (see Chapter 11). The channel of the White Nile lies in the lowlands to the west of the uplift. At Nimule, in extreme southern Sudan, the river has found a low spot in the western wall of the East African rift. It has been able to acquire—probably by capture—water from the rift, including that from the drainage basins of Lake Victoria and Lake Albert. In summary, then, the headwaters of the Nile escape from the East African rift. Thereafter the river is constrained to flow northward through Sudan by the triple junction uplift to the east, and is kept on its northward journey through Egypt by the uplift between it and the Red Sea spreading axis.

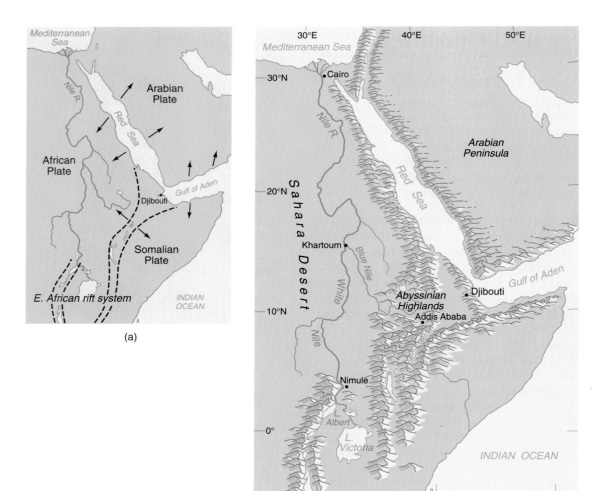

FIGURE 14.44 (a) Three sets of diverging plate boundaries radiate out from a triple junction focused on Djibouti, at the southern end of the Red Sea. They separate the Arabian, African, and Somalian plates. (b) One of the results of this recent rifting of the continental crust has been to determine the course of the Nile River. See text for discussion.

Plate Tectonics, Climate, and Rivers

Those areas where large river systems exist must, of course, have sufficient rainfall to nourish them. The coincidence of plentiful rainfall and large rivers, therefore, is due to a continent's position within the world's climatic system. This position is determined by plate tectonics, the mechanism responsible for moving the continents. If continents, or some portions of them, happen to end up in a rainy climate, then well-developed drainage networks and basins develop. The Amazon has the largest flow of all the world's rivers, and the Zaire-Lualaba, the second largest flow. Both are located in the zone of heavy tropical rainfall. In contrast, areas that are in a dry climatic zone, owing to tectonic movement, have poorly developed river systems, if any at all.

It follows that major river systems may appear and disappear depending upon the positions of the continents. Consider Australia, for example (Figure 14.45). Today the island continent lies almost completely within a belt of dry, desert climates. About 70 million to 80 million years ago it lay to the south, largely in a well-watered, temperate climatic zone. The rivers of that time have all but disappeared. If present-day Australia continues to move northward, it will eventually edge into a zone of tropical rainfall similar to that of the Amazon and the Zaire-Lualaba, and extensive, well-integrated river systems will reappear.

India has had a longer trip through more climatic zones than has Australia. A hundred million years ago it separated from Gondwana and began its northern journey, which has carried it nearly a quarter of the way around the globe. Its southern margin probably

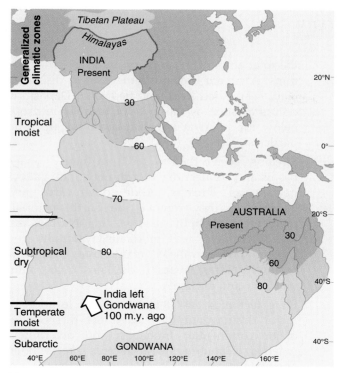

FIGURE 14.45   Australia and India split off from the ancient supercontinent of Gondwana and moved northward to their present positions. In their journeys they encountered different climates which in turn affected stream flow. See text for discussion.

then lay in a subarctic zone. Since then India has moved through temperate moist climates, through subtropical deserts and savannahs, and into the tropical rain forest of the tropical moist climates. On its way, rivers ran when India lay in the moist climatic zones. They stopped when it was in the tropical desert climate. Today, India's climate ranges from rain forest to desert and is complicated by the presence of the Himalayas and the Tibetan desert, themselves the products of plate tectonics.

Water on Mars

Water exists as ice on several of the moons of the outer planets of our Solar System. No place in this system but on Earth, however, is there presently water in fluid form. This was not always the case, though. Water once flowed on Mars.

The evidence for Martian water lies in the presence there of stream channels (Figure 14.46), which are now dry. There is, however, no explanation for their presence but that running water was once active on the planet at some time in the past. Geophysical investigations of the planet and its atmosphere suggest that

FIGURE 14.46   Water-formed channels on Mars. *(NASA)*

the larger channels held running water 3.5 billion years ago. As tectonic activity and volcanism gradually ceased, the global climate changed as well. The surface temperature on Mars plunged to −25°C. Water in the soil became frozen in a planet-wide zone of permafrost, where much of it probably remains today. Some of the last valleys to form may have been fed by melting permafrost during the final stages of volcanism. When water evaporated into the thin atmosphere, it was dissociated into hydrogen and oxygen. The hydrogen escaped to space and the oxygen combined with iron to give the soil of Mars its red color.

Is this what will happen on Earth? Probably not. Our planet is much larger than Mars and closer to the Sun; thus it should retain its massive hydrosphere as long as the Sun shines.

▶▶ EPILOGUE ◀◀

Water running in streams is a major agent of change at the Earth's surface. Streams form well-defined channels arranged in intricate networks, which characterize the land surfaces of the planet. Through these channels turbulent water moves sediments toward the ocean. In the process, streams play a critical role in the hydrologic system and help to fashion many of the world's landforms. Within the rock system streams gather the products of weathering and transport them to depositional areas where they accumulate to form members of the sedimentary rock family. Streams relate to the global tectonic system as the Earth's internal forces rearrange the lithosphere, construct major drainage divides, and move continents in and out of the world's climatic zones.

SUMMARY

1. Water circulates from ocean to atmosphere, to land to streams, and back to the ocean. Oceans hold over 97 percent of Earth's water, whereas the atmosphere holds only 0.001 percent at any one moment, and the streams hold less than 0.0001 percent.

2. Flow of water can be laminar, turbulent, or jet. Most stream flow is turbulent and varies with velocity. Velocity is also related to stream discharge, or the amount of water passing through a stream channel in a fixed unit of time. Changes in discharge can in turn affect velocity, as well as the width and depth of the stream channel. (Stream discharge = velocity of stream × width of stream × depth of stream).

3. Base level is the elevation below which a stream is unable to erode. It may be temporary, as it is at a dam or a waterfall. When base level falls erosion takes place. When base level rises deposition occurs. Sea level is the ultimate base level.

4. A stream erodes by direct lifting, abrasion, impact, and solution. The amount of material carried by a stream is its load. Stream capacity is the total amount it can carry under given conditions. A stream transports material as dissolved load, suspended load, and bed load. Deposition occurs as soon as the level of turbulence drops below the point at which it can buoy up solid particles.

5. Streams receive their water from their drainage basins and connect together in networks. Valleys deepen as streams erode downward, and broaden as they remove the Earth materials delivered to them by mass wasting. Narrow valleys often have steep gradients, waterfalls, and rapids. Broad valleys may have meanders, braided streams, levees, floodplains, deltas, alluvial fans, and terraces.

6. Plate tectonics has determined the location of major drainage divides, fixed the direction of major continental drainages, and determined whether a landmass will receive precipitation adequate to maintain large stream systems. More continental drainage is directed to the Atlantic Ocean than to all the other oceans combined.

KEY WORDS AND CONCEPTS

alluvial fan 365
backswamp 364
bankfull stage 346–347
base flow 344
base level of a stream, temporary, ultimate 348, 349
bed load 352
beheaded stream 358
bottomset bed 364
braided stream 363
capacity 352
chute, chute cutoff 363
competence 352
continental divide 353
crevasse 364
delta 364
dendritic drainage 355
Discharge = Depth × Width × Velocity 344
discharge 344

dissolved load 352
distributary channel 364
drainage basin 353
drainage divide 353
flashy stream 348
flood 346
floodplain 347
flood-recurrence interval 348
flow, laminar, turbulent, jet 343
foreset bed 364
gradient 342
hydrograph 344
lag time 344
levee 363
load 352
meander, meander scar 362
neck cutoff 362
overbank deposits 363
oxbow, oxbow lake 362
pirate stream 358

point bar 367
pothole 351
radial drainage 355
rapids 360
rectangular drainage 355
runoff 344
saltation 353
stream order 353
stream piracy 357
stream terrace 366
suspended load 352
suspension 352
topset bed 364
trellis drainage 355
turbulent flow 343
velocity 343
water gap 355
waterfall 360
wind gap 358
yazoo-type river 364

QUESTIONS FOR REVIEW AND THOUGHT

14.1 At one time it was thought that rivers were fed by waters that cycled from the oceans through the underground. What is the present view and what types of data brought about the change in thinking?

14.2 How would you determine how much water was being returned to the atmosphere from an area by evaporation and transpiration if you knew the runoff and infiltration rates in the area?

14.3 What are three types of water flow and how are they different?

14.4 What are the relationships among discharge, turbulence, and velocity?

14.5 Where is the greatest velocity in a sinuous stream found? Where is the greatest turbulence?

14.6 What do Q, W, D, and V mean in the formula $Q = W \times D \times V$?

14.7 What is the base level of a stream? What is the ultimate base level? Is it constant or changing? Give a reason for your answer. Give two examples of temporary base level. (Here is a place where a simple sketch can sharpen your answer.)

14.8 What happens to the activity of a stream when the base level rises? When it falls?

14.9 Referring to Figure P14.1.1, what is a reasonable suggestion for the water flow of a flood with a recurrence interval of 10 years?

14.10 Distinguish among load, capacity, and competence in a stream.

14.11 What are the ways by which material is carried by a stream?

14.12 How does land use affect the availability of sediments to a stream?

14.13 Nature provides a number of different stream patterns that can be seen on a map or from the air. List two different patterns and discuss what each means in terms of the geology of an area.

14.14 Many different features distinguish streams and valleys. You should be familiar enough with them to define and know the origin of each of the following: waterfalls, rapids, floodplain, meander, oxbow, cutoff, braided stream, natural levee, point bar, delta, alluvial fan, stream terrace.

14.15 Describe how plate tectonics can affect the courses of rivers.

Critical Thinking

14.16 Suppose that the western coast of Africa were to become the location of a collision boundary between the continent and the Atlantic ocean. What could be expected to happen to the drainage of Africa?

14.17 How does Mars provide us with an example of uniformitarianism?

15

Detail, Papoose Room, Carlsbad Caverns National Park, New Mexico

Ground Water

OBJECTIVES

▲▲▲

As you study this chapter, consideration of the following questions will help your understanding of ground water.

1. What are the two basic zones of ground water? What is their relation to the water table? What accounts for changes in the shape of the water table?
2. How do porosity, permeability, and Darcy's law relate to the movement of ground water?
3. How do simple wells and springs differ from artesian wells and springs?
4. Where and how do caves form? What are some of the features of topography that form above caves?
5. How do humans affect the quality and behavior of ground water?

OVERVIEW

▲▲▲

It is easy to describe the movement of rivers. They flow downhill and occupy definite channels, which they sometimes overflow. Ground water, however, is completely different. Once rainwater falls and soaks into the ground, it generally remains out of our sight until it returns to the surface. It may take a few days, weeks, months, or decades, or even centuries or millennia, but eventually ground water comes back to the surface as springs, swamps, or the base flow of streams.

When ground water circulates through soluble rocks it may carve caves, those exciting and sometimes dangerous geologic attractions. As a by-product of this underground activity, it produces a landscape, called *karst,* marked by collapse depressions at the surface and by disappearing streams.

Humans have long benefitted from the supply of underground water. Properly managed, it is a reliable, renewable natural resource. In too many situations, however, human intervention in the natural system of ground water flow produces long-term pollution of the resource. In other places excessive pumping of ground water causes threatening and expensive subsidence of fields and cities.

Basic Distribution of Ground Water

Beneath our feet, hidden from our view, is an enormous supply of water, many times the amount stored in rivers, lakes, and reservoirs. This water, **ground water,** circulates slowly through the pores and crevices of Earth materials. It provides water for home, farm (Figure 15.1), and industrial use. It nourishes plant growth, forms caves, sustains the flow of rivers, furnishes geothermal power, and has helped to create many of our commonly mined mineral deposits.

The Two Zones of Ground Water

Underground water occurs in two distinct zones. Just below the planet's surface lies a thin, upper zone known as the **zone of aeration,** in which the pore spaces in the Earth materials are filled partially with air and partially with water. Below this zone lies the zone of saturation where the ground is saturated and all available pore spaces are filled with water (Figure 15.2).

Soil Moisture

Water seeps down into the zone of aeration from the surface. Some of it keeps on going down to the zone of saturation. The water that remains in the zone of aeration is used by plants, evaporated back to the surface, or held suspended by molecular attraction between water particles and between water and soil particles.

FIGURE 15.1   Ground water flows to the surface as a spring that has been developed to supply a watering trough for cattle. Water, flowing throughout the year, collects in a spring box and is led by a conduit to the trough.

FIGURE 15.2   Water infiltrates through the zone of aeration to the zone of saturation, the upper surface of which is the water table. Some water remains in the zone of aeration, partly filling pore spaces. There it is held by molecular attraction among water molecules and with soil particles. Plants use some of this soil moisture and also transpire some of it back into the atmosphere. Additional moisture is evaporated directly from the soil to the air. Unless the soil moisture is periodically renewed by precipitation, the soil dries out and plants wither and soon die. Some water may rise from the water table in narrow passages by capillary attraction.

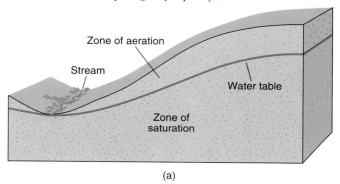

(a)

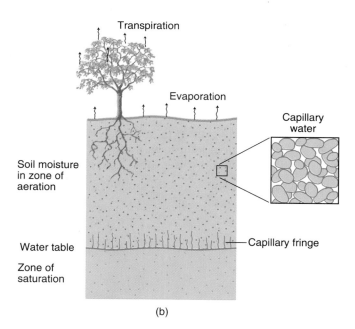

(b)

We call this water **soil moisture.** Because of it crops thrive, flowers bloom, and forests grow. When it is not available, plants wilt and die. Soil moisture, however, does more than support plant life. The very presence of water in the soil facilitates the weathering processes we discussed in Chapter 4. Water, able to reach the zone of saturation, carries dissolved material out of the soil zone. Additionally, zonation of soils, also described in Chapter 4, develops as soil waters move materials downward and deposit some of them lower in the soil profile.

Soil moisture assumes several forms. During a period of precipitation water seeping through the pore space in the soil is drawn downward to the water table. This is called **gravitational water,** and as precipitation slows and ceases it is drained to the saturated zone. The moisture left over is **capillary water** and is held close to the soil particles by the surface tension of the water molecules for each other and for the soil particles. The amount of capillary water retained after the drainage of the gravitational water is a measure of the soil's **specific retention,** also called the **field capacity.** Gravity is unable to move capillary water, but plants rely on it for almost all their water. When virtually all the capillary water is either used by plants or evaporated to the atmosphere, the **wilting point** of plants is reached. The only moisture left is a very thin film that is bound so tightly to soil particles that the plant cannot use it, although it may evaporate. Once the wilting point is reached plants will survive only if soil moisture is quickly renewed. In fine-grained materials a capillary fringe may draw water a few centimeters to a meter up into the zone of aeration from the zone of saturation.

The Water Table

The top of the zone of saturation is known as the **ground-water table** or, simply, the **water table.** In arid and desert areas the water table lies deep beneath the surface, whereas in areas of moist climate the water table is generally close to the surface—deeper beneath the hills and shallower beneath the valleys. Here the water table can be thought of as mirroring the surface topography (Figure 15.3).

In humid climates, water beneath the water table drains to the surface through areas of low elevation, such as stream channels and lakes. An increase or decrease in precipitation will raise or lower the water table in these humid areas. Even in a desert environment, however, the much deeper water table will also fluctuate. The rare precipitation that reaches the zone of saturation under deserts will raise the water table slightly. The water table in desert areas also falls, primarily as ground water is lost by leakage to some distant outlet.

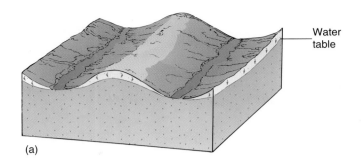

(a)

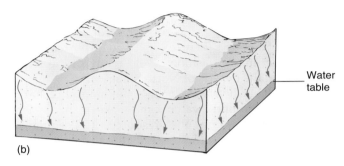

(b)

FIGURE 15.3 In a humid climate (a) the water table will develop a shape that mimics the surface topography and drains to the low spots, here the stream channels in the valley bottoms. The water table in a desert climate may be very flat as in (b).

Recharge of Ground Water

As we have already stated, the ultimate source of ground water is precipitation that finds its way below the land surface. Some of this precipitation seeps into the ground, reaches the zone of saturation, and raises the water table. Measurements show an intimate connection between the level of the water table and rainfall (Figure 15.4). Because water moves relatively slowly in the zone of aeration and the zone of saturation, fluctuations in the water table usually lag a little behind fluctuations in rainfall. These, in turn, lead to the lag time in stream discharge, which we discussed in Chapter 14 and illustrated in Figure 14.6.

Several factors control the amount of precipitation that actually reaches the zone of saturation. For example, rain that falls during the growing season must first replenish moisture being used up by plants or passed off through evaporation. This may use up a great deal of water, leaving very little water to find its way down to recharge the zone of saturation. Also, a slow, steady rain is much more effective than a heavy, violent rain in replenishing the supply of ground water. Melting snow contributes to the recharge of ground water where the ground is unfrozen. High slopes, lack of vegetation, or the presence of impermeable rock near the surface may promote runoff of

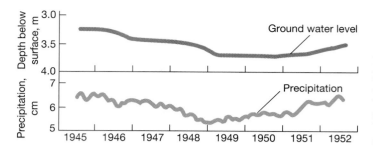

FIGURE 15.4
Relationship between the water level in an observation well near Antigo, Wisconsin, and precipitation from 1945 to 1952. The water table reflects the changes in precipitation. Data are given as 3 year running averages. (*After A.H. Harder and William J. Drescher, "Ground Water Conditions in southwestern Langlade County, Wisconsin," U.S. Geological Survey, Water Supply Paper 1294-c, 1954.*)

heavy rains into streams and reduce the amount of water that reaches the zone of saturation. In dry climates, however, some streams are themselves sources for the recharge of underground water. Water from the streams leaks into the zone of saturation through the zone of aeration (Figure 15.5). The lower reaches of the Nile act in this fashion, as do many of the rivers that flow in the deserts of North America.

In areas of heavy water use, the natural recharge of the ground water cannot keep pace with human demands on the resource. Consequently, arrangements are often made to recharge these supplies artificially. On Long Island, New York, for example, water that has been pumped out for air-conditioning purposes is returned to the ground through special recharging wells or, in winter, through the same wells that are used in summer to pump the water out for air conditioning. Recharge is also sometimes carried out by intentional flooding of the surface. In the San Fernando Valley, California, water from the Owens Valley aqueduct is fed into the ground in an attempt to keep the local water table at a high level.

Movement of Ground Water

▲▲▲

There is a limit to how much water can filter into the underground. Yet there is clearly no backup of water

that is unable to soak into the ground, otherwise the Earth's surface would be a swamp. Therefore the water that soaks into the ground must escape someplace. In other words, ground water must move.

Ground water begins its journey through the underground as water moves from the surface directly downward through the zone of aeration to the zone of saturation. Below the water table the direction of flow changes. Overall, ground water flows from areas of high water table to areas of low water table. Two factors influence this flow. The first is elevation, called **head** or **hydraulic head.** This is the level to which water will rise at a given point. The slope of the water table between two points represents the loss of head along the water table. Second, ground water is constantly under pressure of the overlying ground water. Water beneath a ground water "hill" is under greater pressure than ground water beneath a ground water "valley." Ground water therefore moves down beneath the hill and up beneath the valley. The combined effect of pressure and head is that ground water flows in broad looping curves toward low spots in the ground-water table such as a stream valley, as suggested in Figure 15.6.

FIGURE 15.6   Flow lines in ground water through a homogeneous material above an impermeable shale with low permeability. See text for discussion. Points *A* and *B* refer to Question 15.13 at end of chapter.

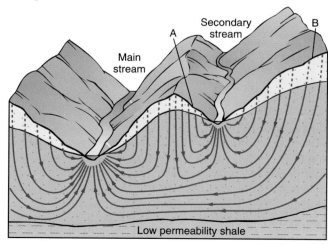

FIGURE 15.5   Ground water may be recharged by water from a surface stream leaking into the underground.

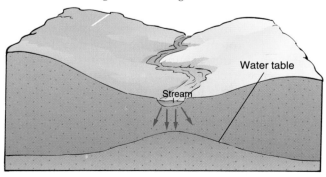

Because the flow of ground water is usually very slow and through narrow passages, it is largely laminar, whereas stream flow is rapid and turbulent. The exception to slow flow in the underground is found in the large passages of caves where flow may reach velocities of meters per second and become a danger to cavers. In laminar flow the water near the walls of small openings is presumably held motionless by the molecular attraction of water particles among themselves, as well as their attraction to the passage walls. Water farther away from the walls moves in smooth, threadlike patterns. In most Earth materials the rate of water flow is seldom faster than 1.5 m/day or slower than 1.5 m/year, although rates of over 120 m/day and as low as a few cm/year have been recorded. These velocities are in large part controlled by the size and shape of the pore spaces in rock. This leads us to two characteristics of Earth materials, **porosity** and **permeability.**

Porosity

The porosity of any Earth material is the percentage of its total volume occupied by void space. It follows, then, that the greater the porosity of the rock the more water it can hold. In a rock made up of individual particles, such as sand grains, the porosity is the space between individual grains. This *intergranular porosity* can vary widely, as suggested by the three diagrams in Figure 15.7 in which the size and shape of the larger sand grains are similar. In Figure 15.7a the space between grains is open and the porosity is high, 45 percent, more or less. When a similar collection of sand grains is mixed with smaller particles, the smaller grains partially fill the spaces between the larger grains, and the pore space is reduced, perhaps to about 25 percent as in Figure 15.7b. Intergranular pore space also generally decreases during diagenesis, as shown in Figure 15.7c. In this example a cementing material has been deposited around the original grains, in part filling the pore spaces and reducing the

porosity to approximately 10 percent. Even well-lithified sediment, however, retains some of its initial porosity.

Porosity is also provided by the bedding plane partings and fractures in sedimentary rocks (Figure 15.8a). Massive igneous rocks such as granite acquire porosity by fracturing (Figure 15.8b), and many lavas have porosity due to cooling cracks (Figure 15.8c). Cleavage and foliation planes may give porosity to some metamorphic rocks (Figure 15.8d). Finally, solution along bedding and fracture planes in limestone and dolostone gives added porosity to these carbonate rocks (Figure 15.8e).

The range in porosity is very large. Recently deposited muds, called *slurries,* can hold up to 90 percent by volume of water, whereas unweathered igneous rocks such as granite or gabbro may hold only a fraction of 1 percent. Porosity of unconsolidated deposits of clay, silt, sand, and gravel varies from about 20 percent to as much as 50 percent. When these deposits are turned into rock by cementation and compaction, however, their porosity is sharply reduced. Average porosity values for individual rock types have little meaning because of the extreme variation within each type. In general, however, a porosity of less than 5 percent is considered low; from 5 to 15 percent represents medium porosity; and over 15 percent is considered high. Table 15.1 lists the types of porosity in some common Earth materials.

FIGURE 15.7   The intergranular porosity of a sandstone is the pore space between individual sand grains, here shown magnified 10 times. In (a) the original pore space is completely open. In (b) it is partially reduced by smaller particles within the original pore spaces. In (c) the pore space is further reduced, in this instance by the introduction of cementing material between grains.

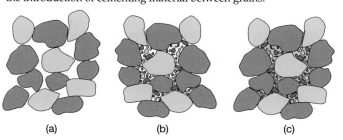

(a)   (b)   (c)

TABLE 15.1   Types and Ranges of Porosity in Some Common Earth Materials

MATERIAL	TYPE OF POROSITY (POROSITY RANGE %)
Mud, silt, sand, and gravel	Intergranular pores (35–65)
Mudstone, shale, siltstone, sandstone, conglomerate	Intergranular pores, bedding plane partings, fracture planes (3–30)
Limestone, dolostone	Fracture planes, bedding planes, and intergranular pores. All may be enlarged by solution (3–30)
Massive igneous rocks such as granite	Fracture planes (<1–2)
Basalt	Cooling joints, vesicular openings (<1–15)
Volcanic ash and cinders	Intergranular pores (5–60)
Slate	Cleavage and fracture planes (1–4)
Schists and gneisses	Foliation and fracture planes (1–4)
Marble	Fracture planes, commonly enlarged by solution (1–5)

(a)

(b)

(c)

(d)

(e)

FIGURE 15.8
Porosity is due to bedding planes, fractures, rock cleavage, and solution. (a) Bedding and fractures in sedimentary rock, Spring Dale, West Virginia; (b) Fractures in granite, Branford, Connecticut; (c) Cooling fractures in basalt, Devils Post Pile, California; (d) Rock cleavage, slate quarry, Pen Argyl, Pennsylvania. Ground-water seepage darkens large areas of the quarry walls; (e) Solution-enlarged joints in limestone, Gait Barrows, Arnside, England. *((c) Pamela Hemphill)*

Permeability

Just because the Earth materials in a given area are porous does not mean we will find an abundant supply of ground water there. The material must also allow water to move through it. We call this capacity to transmit water *permeability.*

The rate at which a material transmits water depends less on its porosity and more on the size and arrangement of the interconnections between its openings. For example, although a clay may be more porous than a sand, the particles that make up the clay are minute flakes and the spaces between them are very small. Water passes more readily through the sand than through the more porous clay because the molecular attraction on the water is much stronger in the tiny openings of the clay. The water moves freely through the sand because the passageways between particles are relatively large and the molecular attraction on the water is relatively low. Of course, no matter how large the spaces in a material are, there must be connections between them if water is to pass through. If they are largely unconnected, as in pumice or scoria, the material has a very low permeability.

A permeable Earth material that yields a strong flow of water is called an **aquifer,** from the Latin for "water" and "to bear." Among the most effective aquifers are unconsolidated beds of sand and gravel, sandstone, and some limestones. Limestone usually owes its high permeability to solution that has enlarged the fractures and bedding planes into open passageways. The fractured zones of some of the more compact rocks such as granite, basalt, and gabbro also act as aquifers, although the permeability of such zones decreases rapidly with depth as fracture zones are closed by the pressure of overlying rock.

Clay, shale, and most metamorphic and igneous crystalline rocks transmit very little if any water. These rocks of low permeability are called **aquitards,** from the Latin words for "water" and "slow" in reference to their ability to retard the passage of water. Table 15.2 compares the permeabilities of some common rock types.

Darcy's Law

The groundwork for our understanding of ground water flow began in the first half of the nineteenth century with the laboratory experiments of the French hydrologist Henri Darcy. His measurements led to the law that now bears his name and allows us to calculate the discharge of ground water through various types of aquifers.

Darcy started with experiments that showed that the velocity of ground water flow was proportional to the slope of the ground water table. The slope of the water table is measured by the difference in elevation of two points along the slope, divided by the distance in flow between them, and is known as the **hydraulic gradient** (Figure 15.9). The velocity of flow then is proportional to the hydraulic gradient

$$V \propto \frac{(h_1 - h_2)}{d}$$

where V is velocity, $(h_1 - h_2)$ is the loss of elevation between two points on the slope of the water table, and d is the distance of flow between the two points.

Darcy experimented with different Earth materials, such as coarse sand and fine sand, and found that the velocity of flow varied with their porosity. There-

FIGURE 15.9 The hydraulic gradient $(h_1 - h_2/d)$ is measured by the difference in the elevation of the water table at two points, $(h_1 - h_2)$, divided by the distance between the two, d.

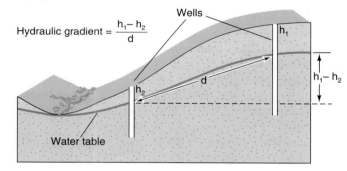

TABLE 15.2 **Relative Permeability of Some Common Rock Types**

MOST PERMEABLE		LEAST PERMEABLE
Cavernous limestone		
Gravel, sand, silt		Clay
Volcanic ash and cinders	Sandstone, siltstone	Shale, mudstone
	Noncavernous limestone	
Prismatically fractured basalt		Slate
		Schist, gneiss
		Massive igneous rocks such as granite

fore, he introduced the factor K, a measure of permeability, known as the **hydraulic conductivity**. This coefficient takes into consideration the nature of the aquifer and the fluid. The velocity equation becomes

$$V = \frac{K(h_1 - h_2)}{d}.$$

To determine the discharge, Q, through an aquifer one must know the cross-sectional area A through which the water flows. Because the cross section is not a nice open channel, as with a river, the void space in the aquifer—that is, its porosity—must be a factor in determining A. **Darcy's law** then is expressed

$$Q = \frac{AK(h_1 - h_2)}{d}.$$

Wells, Springs, and Geysers

▲▲

Simple Springs and Wells

A **spring** occurs where the water table naturally emerges from the underground, causing ground water to flow to the surface (Figure 15.10). Springs have attracted human attention throughout history. Hundreds and even thousands of years ago they were regarded with superstitious awe and were sometimes selected as sites for temples and oracles. To this day many people feel that spring water possesses special medicinal and therapeutic values. Water from "mineral springs" contains salts in solution that were picked up by the water as it percolated through different Earth materials. Paradoxically, the same water pumped up out of a well might be regarded as tainted and undesirable for general purposes.

Springs vary greatly in discharge. Some are small, intermittent trickles that disappear when the water table recedes during a dry season. Others flow at

FIGURE 15.10 Multiple springs flow from limestone in the Grand Canyon of the Colorado just above river level at Vesey's Paradise. *(Pamela Hemphill)*

tremendous rates, as do the springs along a 16-km stretch of Fall River, California, where the daily discharge is 3.8 billion liters.

Various subterranean conditions account for springs, some of which are illustrated in Figure 15.11. Wherever ground water flows out at the surface a spring develops. In Figure 15.11 the permeable sandstone transmits water to the surface where low-permeability granite interferes with the flow. Spring water feeds the swamp in the low spot between the two sandstone hills. In the larger of these hills is a lens of

FIGURE 15.11
Ground water comes to the surface to feed swamps, springs, and rivers wherever a perched water table or a main water table intersects the surface. Some typical situations are shown here.

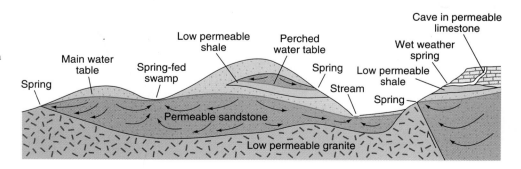

low-permeability shale. A water table forms on top of it above the main water table. This is a **perched water table.** When there is enough water, some of it will seep out above the shale along the hillside as a spring. Another lens of shale occurs beneath limestone on the right side of the diagram. A cave has formed in the limestone above the shale, and during wet weather water drains through the cave and emerges from it as a spring. Finally, the stream is itself fed by discharge of ground water to springs in its channel.

Even in low-permeability rocks permeable zones may develop as a result of fractures or solution channels. If these openings fill with water and their lower parts are intersected by the ground surface, the water will emerge as a spring. Because water may flow through these *fissure springs* rapidly, they often turn out to have highly variable flow rates.

In contrast with a spring, a **well** is an artificial intersection of the surface and the water table. To be productive it must penetrate permeable material below the water table. If it is not sunk deep enough, the water table may fall below the bottom of the well during dry weather, and the well will go dry until the next spell of rainy weather raises the water table. When a well is pumped, ground water immediately adjacent to the well flows into the well and is taken to the surface. Because the water is removed more rapidly than it can flow through the rock, the water table is distorted and drawn down in a **cone of depression.** The diameter of the cone of depression depends on the permeability of the aquifer and the rate of pumping. The higher the permeability, the wider and shallower the cone. If the well has not been drilled deep enough, or is overpumped, the cone of depression may reach the bottom of the well and render it unproductive until the water table recovers (Figure 15.12).

FIGURE 15.12   To provide a reliable source of water a well must penetrate deep into the zone of saturation. In this diagram Well 1 reaches only deep enough to tap the groundwater during periods of high water table; a seasonal drop of the water table surface will dry up this well. Well 2 reaches to the low water table but continued pumping may produce a cone of depression that will reduce effective flow. Well 3 is deep enough to produce reliable amounts of water, even with continued pumping during low water-table stages.

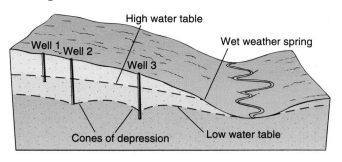

The types of bedrock that make up aquifers will affect the yield of ground water. For instance, wells drilled into an aquifer of fractured crystalline rock, such as granite, may produce an adequate supply of water at relatively shallow depths, but we cannot increase the yield of such wells much by deepening them because the number and size of the fractures commonly decrease the farther down into the Earth we go. It is clear, also, that a highly permeable aquifer, such as the unconsolidated gravel of an old stream bed, will usually yield larger supplies of water than a tightly cemented conglomerate.

The different types of rocks that serve as aquifers also affect the quality of ground water. Water flowing through rock like limestone, or through limestone-rich sand and gravels, will acquire calcium ions in solution. Such water does not lather readily with soap and forms a scale when evaporated in a tea kettle or shower head. It is familiar to us as *hard water.* In contrast, water flowing through aquifers made up of relatively insoluble silicate minerals carries very little material in solution, and essentially no calcium. This water lathers well with soap and leaves no scale when it evaporates; we call it *soft water.* Should an aquifer contain minerals such as halite or gypsum, the ground water in it may carry so much of the minerals in solution as to be undrinkable. Halite will make the water salty, and the sulfate from gypsum will render it bitter. Water containing iron or sulfur compounds may have an unpleasant taste or smell but may still be drinkable.

Artesian Water

Most artesian systems are similar to the one shown in Figure 15.13. The water is contained in an aquifer inclined so that it crops out at the surface where it receives water. Beneath the surface the aquifer is confined by low permeability rock that keeps water from escaping. If the overlying aquitard is penetrated by wells the water rises above the top of the aquifer, and in some instances may rise above the land surface. The cause of the rise is the difference in elevation, or head, between the recharge area and the intake point of the well and the pressure of the water all along the aquifer between these points. As soon as the overlying, low permeability rock is pierced, water will rise to a level called the **potentiometric surface.**

The name for **artesian wells** comes from the name for the northern French region of Artois (originally called Artesium by the Romans) where this type of well was first studied. Since the 1880s, wells of this type have provided ground water across much of the Great Plains of North America from Texas northward to Manitoba.

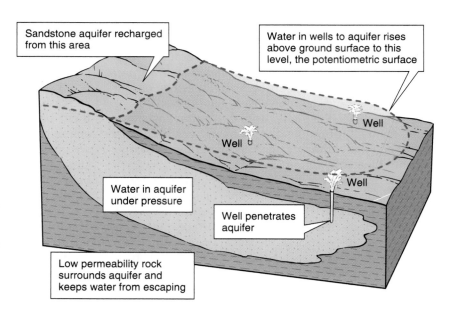

FIGURE 15.13
The wells in the diagram meet the conditions that characterize an artesian system: (1) an inclined aquifer, (2) confined by low permeability layers that prevent water from escaping vertically or laterally, and (3) sufficient pressure to force the water above the aquifer wherever it is tapped. Water in the wells shown rises to the level of the potentiometric surface.

Thermal Springs

Springs that bring warm or hot water to the surface are called **thermal springs,** hot springs, or warm springs (Figures 15.14 and 15.15). A spring is usually regarded as a thermal spring if the temperature of the water is about 6.5°C higher than the mean temperature of the surrounding air. There are over 1,000 thermal springs in the western mountain regions of the United States, 46 in the Appalachian Highlands of the East, 6 in the Ouachita area in Arkansas, and 3 in the Black Hills of South Dakota. As we noted in Chapter 5, spring waters coming to the surface may deposit mineral matter. Water, particularly warm or hot water, dissolves mineral matter in its passage beneath the

FIGURE 15.14   Boiling Springs, California, is a hot spring in the Long Valley caldera at the foot of the eastern escarpment of the Sierra Nevada. This is an area being closely monitored for warnings of possible explosive volcanic activity. Heat is supplied indirectly from a magma chamber below the caldera. *(Pamela Hemphill)*

surface and then deposits the dissolved material as the springs flow out into the changing conditions at the surface (Figure 15.15).

Most of the western thermal springs derive their heat from masses of magma that have pushed their way into the crust almost to the surface and are now cooling. In the eastern group, however, the circulation of the ground water carries it so deep that it is warmed by the normal geothermal gradient. The well-known spring at Warm Springs, Georgia, is heated in just this way. Rain falling on Pine Mountain, about 3 km south of Warm Springs, enters a formation known as the Hollis quartzite (Figure 15.16). The Hollis quartzite is a good aquifer in both its top and bottom zones. Through the middle of the Hollis, however, is a zone of low permeability separating the formation into upper and lower aquifers. Recharge of the lower aquifer is in the Pine Mountain region where erosion has removed the low permeability zone. At the start of its journey downward, the average temperature of the water is about 16.5°C. It percolates through the lower Hollis formation northward under Warm Springs at a depth of about 100 m, and then follows the rock downward as it plunges into the Earth to a depth of 1,140 m, 1.6 km farther north. Normal rock temperatures in the region increase about 1.8°C/100 m of depth, and the water is warmed as it descends along the bottom of the Hollis bed. By a depth of 1,140 m the water has warmed to about 37°C. Here the bed has been broken and shoved against a low-permeability gneiss that turns the water upward. The heated water escapes into the upper aquifer of the Hollis. Here, under artesian pressure, it moves upward along the top of the Hollis quartzite, cooling somewhat as it rises. Finally, it emerges at Warm Springs at a temperature of 31°C.

FIGURE 15.15 Mammoth Hot Springs, Yellowstone National Park, Wyoming, are thermal springs that have built this flight of terraces by depositing travertine, a variety of limestone. *(Jane Grigger)*

Less than 1 km away is Cold Spring, whose water comes from the same rainfall on Pine Mountain. A freak of circulation, however, causes the water at Cold Spring to emerge before it can be conducted too far downward and be warmed. Its temperature is about 16.5°C.

Geysers

A **geyser** is a special type of thermal spring that ejects water intermittently and with considerable force (Figure 15.17). The word comes from the name of a certain spring of this type in Iceland, *geysir;* probably based on the Icelandic verb *gjosa*, "to rush forth."

FIGURE 15.16 Warm Springs, Georgia, flows from an artesian system in which the water, derived from normal precipitation, is heated by the normal geothermal gradient before being driven back to the surface. *(After Hewett, D. F. and G. W. Crickmay, "The Warm Springs of Georgia," U.S. Geological Survey, Water Supply Paper 319, 1937.)*

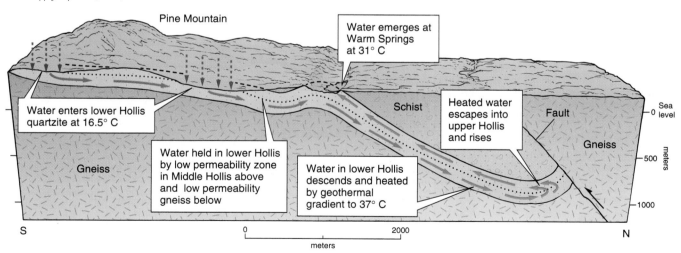

Pine Mountain

Water emerges at Warm Springs at 31° C

Water enters lower Hollis quartzite at 16.5° C

Schist

Heated water escapes into upper Hollis and rises

Fault

Gneiss

Water held in lower Hollis by low permeability zone in Middle Hollis above and low permeability gneiss below

Water in lower Hollis descends and heated by geothermal gradient to 37° C

Sea level

meters

500

1000

S

0   2000

meters

N

FIGURE 15.17 A geyser on the Volcanic island of Iceland. *(Terraphotographics/BPS)*

Scientists do not fully understand how geysers work, but here is what they think happens: Ground water fills a natural pipe, or conduit, that opens to the surface. Hot igneous rocks, or gases emitted by them, heat the water column in the pipe, most rapidly toward its base, and raise its temperature toward the boiling point. Deep in the water column, however, water pressure is higher and so, therefore, is the boiling point. The conduit is much like a pressure cooker. Eventually the column of water becomes so hot that it is close to boiling. At this stage a small increase in temperature or decrease in pressure will bring the water to a boil. Probably it is a small increase in temperature at some point in the pipe that starts the boiling. This, in turn, causes expansion, which pushes the overlying column of water upward until some of the water spills out at the surface. The consequent reduction in the amount of water in the conduit reduces pressure all along the water column, and the result is much like accidentally removing the weight from the top of a pressure cooker. The heated column boils explosively and steam and water gush upward to the surface. The process continues until the energy is spent. Then the pipe fills again with water and the process begins over again (Figure 15.18).

FIGURE 15.18 Stages in the development of an eruption of a geyser.

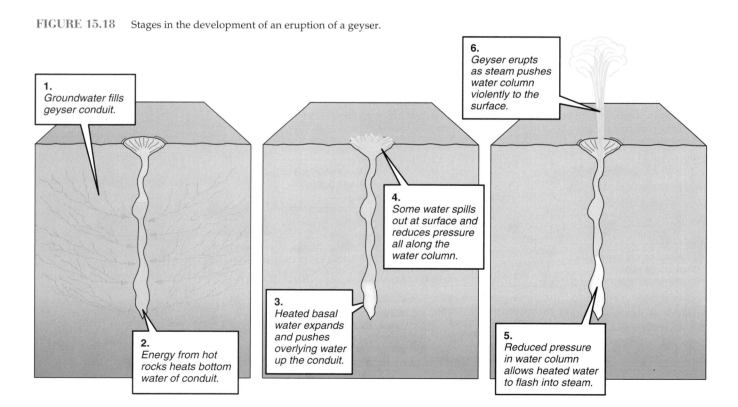

Karst
▲▲

Karst is a landscape that develops from the action of ground water in areas of easily soluble rocks such as the carbonates, limestone, dolomite, and marble, and, less often, gypsum and halite. It is characterized by caves, underground drainage, and sinkholes, and it takes its name from a region in Slovenia and northeastern Italy where this type of topography is well developed.

Caves are the most spectacular feature of karst. Certainly the exploration of caves can provide a great experience. Most caves form in limestone—less commonly in dolomite—as slightly acidic water circulates through the rock. The acid-bearing water slowly dissolves the rock and brings it in solution to the surface to be carried away by springs and streams. Caves, in turn, tend to cause other features of karst, such as depressions that pock the ground surface and serve as funnels draining water into the underground.

By far the greatest number of caves form by water entering from the surface and moving through the zone of aeration to the zone of saturation and through it to emerge as springs in nearby valleys. A much smaller number, perhaps less than 10 percent, are created by water rising from a source deep beneath the surface in the zone of saturation.

Caves Formed by Water From the Surface

All rainwater brings with it a small amount of carbon dioxide dissolved from the atmosphere. As this water soaks into the ground it becomes more acidic as it acquires more carbon dioxide, the product of organic activity in the soil. The carbon dioxide combines with the water to form a weak acid, carbonic acid. It is this carbonic acid that dissolves the limestone as it reacts with calcite, the chief constituent of limestone, as we discussed in Chapter 4.

We stated earlier that the porosity of limestone is along bedding plane partings and fractures and in intergranular pores. Most important in cave formation are bedding planes and fractures. Water seeps into these partings of the rock and enlarges them (Figure 15.8e). The enlarged passages allow water to move easily from the surface downward. At first the flow of water through these partings is laminar, and dissolution of the limestone is very slow. Eventually the passages are enlarged enough to allow turbulent flow. At this stage the cave widens at a rate between 0.01 to 0.1 mm/yr. Eventually passages can enlarge enough to allow humans to move through them (Figure 15.19). This may take 100,000 years or more. Once formed, few caves survive for more than a million years before

FIGURE 15.19   These irregular vertical openings represent joint planes widened by ground water solution. Wind Cave, Wind Cave National Park, South Dakota. *(Arthur N. Palmer)*

being destroyed by erosion from the surface downward.

The patterns of caves formed by water passing from the surface through the zone of aeration to the zone of saturation are of two basic types. Most common is a **branchwork** arrangement. It forms primarily along bedding planes and is reminiscent of the network of many surface streams (Figure 15.20). During formation each small branch is fed by an individual source of water. These branches converge to form fewer but larger conduits that eventually feed into a single passage draining through a lower outlet to the surface. In many branchwork caves, fractures will impose an angularity to the cave configuration (Figure 15.21a).

In contrast with the branchwork caves are those with interconnecting loops that form mazelike patterns. In one form of **maze caves,** fractures in the limestone determine the arrangement of interconnecting passages (Figure 15.21b). A second form resembles that of a braided surface stream (Figure 15.21c).

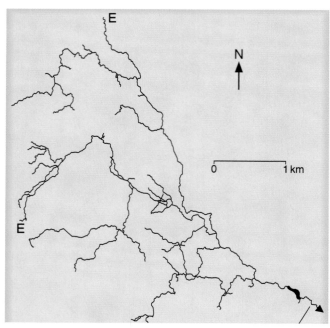

FIGURE 15.20 This map of Crevice Cave, Missouri, illustrates the pattern of a branchwork cave. Points marked with an *E* indicate entrances. *Courbon, Paul, et. al, Atlas of the Great Caves of the World, p.101, Cave Books, St. Louis, 1989.)*

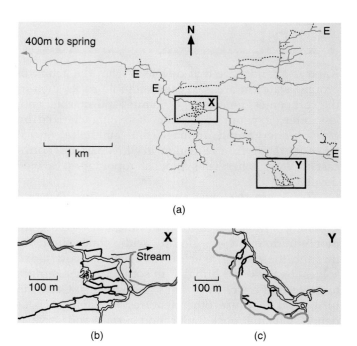

FIGURE 15.21 Map of Blue Spring Cave, Indiana. (a) A strongly jointed limestone has given the branchwork pattern of this cave a distinct angularity. An enlarged portion of this same cave (b) shows a maze pattern with a rectilinear network. Another portion (c) has developed into a maze with a braided or anastomosing pattern. Points marked *E* indicate entrances. Blue indicates active stream passages. Dotted, solid or open symbols indicate relict upper level passages. *(After Palmer, Arthur N. "Origin and Morphology of Limestone Caves," Geological Society of America Bulletin, Vol. 103, Figure 16, p.12, 1991.)*

Figure 15.22 illustrates the development of a branchwork cave. Before the cave begins to develop, a normal, stream-formed landscape has formed on shale overlying limestone (Figure 15.22a). The shale has very low permeability and little water can get through it to the limestone below. At this stage erosion has yet to expose the limestone, even along the main stream, and caves have yet to develop.

Figure 15.22b shows that the main stream has cut a canyon down into the limestone, and erosion has stripped back some of the shale, exposing limestone at the surface. With the exposure of the limestone and the downward incision of the master stream, caves have developed. Two distinct zones characterize the caves in Figure 15.22b. The first one lies in the zone of aeration between the surface and the water table. Here, water passing from inlets at the surface and down through the zone of aeration to the water table dissolve a series of ever-descending passages, some of them vertical (Figure 15.23). The second and lower zone contains nearly horizontal passages that lead along the water table (Figure 15.24), or just below it, to an outlet spring near river level low in the canyon wall.

In addition to the large cave system shown in Figure 15.22b, a smaller cave is shown on the right-hand side of the diagram. It formed above a perched water table held up by a lens of shale.

As caves develop in the underground, **sinkholes,** so characteristic of karst topography, begin to pock the surface (Figure 15.25). These depressions result from collapse into solution-formed caves below. Surface streams drain through these depressions and are called **sinking streams** (Figure 15.26), and the valleys in which they disappear are known as **blind valleys.** In some places sinkholes form lake basins as they do over large portions of Florida. In such a situation a rising water table floods an old cave network and reaches close enough to the surface to turn sinkholes into lakes.

A later stage of cave development appears in Figure 15.22c. At this stage the river has deepened its canyon. As a result the water table has fallen and forms a new, lower base for cave development. Water passing through the zone of aeration now flows downward to the new water table. It forms a new set of downward-trending conduits below the original cave system. New, horizontally oriented passages develop

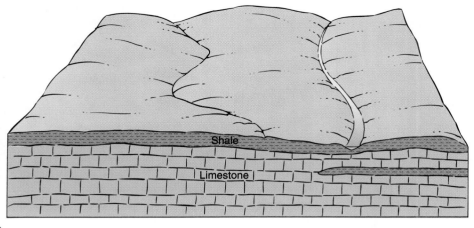

(a)

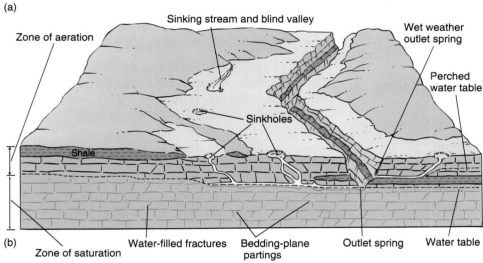

(b)

Zone of aeration

Sinking stream and blind valley

Wet weather outlet spring

Perched water table

Sinkholes

Shale

Zone of saturation

Water-filled fractures

Bedding-plane partings

Outlet spring

Water table

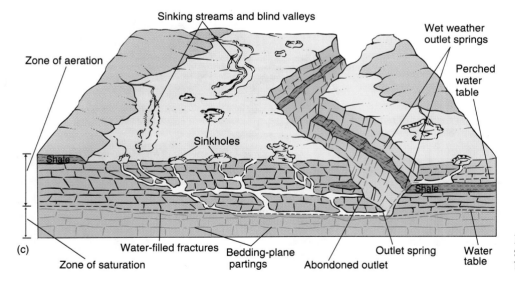

(c)

Zone of aeration

Sinking streams and blind valleys

Wet weather outlet springs

Perched water table

Sinkholes

Shale

Zone of saturation

Water-filled fractures

Bedding-plane partings

Abondoned outlet

Outlet spring

Water table

FIGURE 15.22
Sequence of development of a branchwork cave.

FIGURE 15.23   This vertical shaft in Mammoth Cave, Kentucky, was formed by water descending through the zone of aeration to the water table below. Distance to water at the bottom of the shaft is about 18 m and the diameter of the shaft, about 2.5 m. (*Arthur N. Palmer*)

along the lowered water table. A new and lower outlet forms along the canyon wall near the new stream level. The old outlet has been abandoned.

As cave development progresses, so too does the development of karst topography on the ground sur-

FIGURE 15.24   At the water table horizontal passages form that lead to an outlet at the surface. This passage is in Blue Spring Cave, Indiana. (*Arthur N. Palmer*)

FIGURE 15.25   The depression between the viewer and the farm house is a sinkhole. It has formed in part by the collapse into McClung's cave, West Virginia, to which it is the entrance, and in part by the subsidence of soil into the opening. (*Arthur N. Palmer*)

face. The sinkholes increase in size and number as do the blind valleys. If this process of limestone solution is carried on long enough a weird landscape can develop in which certain formations of limestone, less susceptible to erosion because of position, structure, or composition, remain as steep-sided residuals that may persist for long periods (Figure 15.27).

FIGURE 15.26   This sinking stream is in Yorkshire, England. The stream runs across low permeability peat on sandstone and disappears here as it encounters an exposure of limestone. The water drops vertically about 70 m through a solutionally enlarged joint in the limestone. (*Arthur N. Palmer*)

FIGURE 15.27   Solution of limestone has produced these steep-sided hills along the Li River in southern China. They are what remains of a once-continuous layer of limestone. *(Richard J. Cross)*

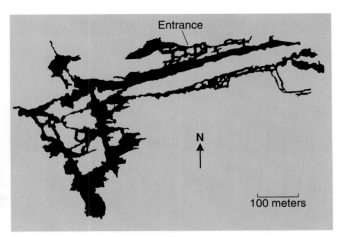

FIGURE 15.28   This plan of the main level of Carlsbad Cavern, New Mexico, outlines the large rooms and galleries with smaller passages extending out from them, characteristic of a cave formed by waters rising from depth. *(Based on a map by the Cave Research Foundation.)*

Caves Formed by Water Rising From Depth

Caves formed by water rising from depth toward the water table are much less numerous than those formed by waters sinking downward from the surface. They include, however, some of the largest known caves, such as Carlsbad Caverns in New Mexico. Acid hydrothermal waters are the cause of this type of cave formation. Being warm, the waters rise. The acid, most commonly sulfuric, comes from the reduction of sulfate minerals such as gypsum in the presence of underground organic material such as petroleum.

The patterns of caves formed by rising waters differ from those of the more common branchwork caves formed by surface waters. They are characterized by large rooms and galleries with smaller passages extending laterally outward from them (Figures 15.28 and 15.29). Passages are not a series of convergent conduits fed by distinct sources of surface water as in the more common branchwork caves.

Stalactites, Stalagmites, Columns, and Flowstone

With time, a cave formed at or below the water table will be drained as its outlet is lowered. Deposition of calcite in the caves then creates the remarkable rock forms so characteristic of caves (Figure 15.30). Their formation is the result of reversing the chemical reaction that dissolves calcite. Remember from Chapter 4 that the dissolution of calcite left soluble calcium and

FIGURE 15.29
This photograph of Carlsbad Caverns, New Mexico, illustrates the large interconnecting spaces or rooms found in caves formed by ascending water. *(Arthur N. Palmer)*

Mushrooms, Wine, and Roquefort Cheese

We don't think of the dark, cool, dank environment of a cave as one that favors the creation of gastronomic delicacies. Consider, however, some of the conditions under which some gourmet delights begin their journey to the table.

Mushrooms grow in woods and open fields. Commercial production, however, calls for cool, constant temperature, high humidity, and darkness. This can be provided by caves. In fact, caves (and abandoned mines) gave commercial mushroom culture its start in the United States.

Of longer standing has been the use of caves for the aging of wine. Once fermentation of the grape juice is completed, the wine is drawn off into barrels and stored in a place where there is a cool, constant temperature of 10° to 12°C to begin the aging process. Here, again, a cave is ideal. Even today's fabricated storage units may be called "caves."

The small town of Roquefort-sur-Soulzon in southern France

---

bicarbonate ions. When water charged with these ions drips into the cave, calcite (calcium carbonate) forms as follows:

$$Ca^{2+} + 2HCO_3^- \rightarrow H_2CO_3 + CaCO_3$$

The deposition of calcite takes place in two major forms. **Stalactites** (from the Greek *stalactos,* "oozing out in drops") look like stone icicles hanging from the cave roof (Figure 15.30a). **Stalagmites** (from the Greek *stalagmos,* "dripping") resemble posts growing up from the cave floor (Figure 15.30a).

A stalactite forms as water containing calcium and bicarbonate in solution seeps through a crack in the cave roof. Drop after drop forms and then falls to the floor. But during the few moments before each drop falls some carbon dioxide is lost, and a small amount of calcium carbonate is deposited (Figure 15.30b). Over the centuries enough calcium carbonate may be deposited that a large stalactite gradually develops. When water falls from the stalactite to the cave floor more carbon dioxide is lost, more calcite is deposited, and a stalagmite grows upward toward the stalactite. When a stalactite and a stalagmite meet a **column** forms (Figure 15.30c). **Flowstone** is a general term for deposits formed from dripping and flowing water on the walls and floors of a cave (Figure 15.30e).

Ground Water and Human Affairs

▲▲

Ground water is an extensive and extremely important natural resource. We have learned how to find and develop it and to maintain it as support for human activities. But humans can affect the ground water system, and not always in beneficial ways. We can change the quality of ground water, diminish its availability, and interfere in its role of maintaining land stability.

Ground Water Pollution

The estimated amount of available ground water in the United States is enormous: Over 750 quadrillion liters. The good news is that much less than 1 percent of this is polluted. The bad news is that each year in this country we trickle over 11 trillion liters of contaminated fluid into the underground. Much of this enters facilities, properly designed and controlled to contain the potential pollutants and allow them to decompose without affecting nearby environments. In the absence of responsible waste management, however, ground water can be degraded beyond human use. It is then of little solace to people, originally dependent upon that

gives its name to one of the world's most famous blue cheeses. Nearly 2,000 years ago the Roman naturalist Pliny the Elder noted the high quality of Roquefort cheese. In the Middle Ages, the Emperor Charlemagne is said to have had Roquefort delivered to him regularly at his capital in Aachen nearly 1,000 miles to the north. But the real story behind this cheese involves mass movement of limestone blocks, caves, and a particular type of penicillin.

The River Soulzon has cut down through a limestone plateau and into the underlying, weak and slippery, limey shales. As a result, blocks of limestone from the plateau have slumped down into the valley. At the foot of these blocks is the town of Roquefort-sur-Soulzon, and within the jumble of rocks caves have formed. It is this situation that has favored the production of Roquefort cheese.

Roquefort cheese is made of sheep's milk. The ewes, which number nearly 800,000, graze on the limestone plateau near the town. Their milk is gathered and curdled, and the curds, combined with mouldy bread crumbs, are pressed into forms 20 cm in diameter and 15 cm in depth. The forms are then stored in the caves near the town. These have been adapted as storage rooms for curing the small drums of Roquefort cheese. Here the drums are punctured to let air into them. Cool, moist air circulates through the fissures in the limestone (and into the drums of cheese) and maintains a constant humidity and a temperature between 6° and 8°C. The cool and moist conditions of the cave are also perfectly suited for the growth of a very specific mould, *Penicillium roqueforti*. This is the chief ripening agent in the curing of the cheese. It grows in the punctures made in the cheese when the drums were first stored in the caves and gives the cheese its blue veination and mottling.

now polluted source, that much usable ground water is available elsewhere.

Sources of ground water pollution are many and fall into one of two general groups: those from point sources and those from dispersed sources. Point sources include leakage from landfills and dumps; leaking sewage, petroleum and chemical lines; underground petroleum and chemical tanks; septic tanks; injection wells; mine tailings; inadvertent spills; illegal dumping; and radioactive waste. Dispersed sources of contamination include agricultural chemicals, highway de-icers, and other materials that are applied to extensive land areas and may gradually concentrate in ground water.

Polluted ground water presents us with a different situation than does polluted surface water. Sources of surface water contamination, precisely because they are on the surface, are usually more easily identified than those for underground water. Furthermore, once the origin of surface water pollution is eliminated, the stream cleanses itself fairly rapidly because of its rapid, turbulent flow. Underground water, by contrast, moves very slowly in laminar flow. In one example, in 13 years ground water polluted by chromium moved 1,200 meters at a rate of 25 cm/day. Turbulent stream water would have moved at velocities measured in centimeters or meters per second. Once a source of contamination for ground water is identified and eliminated, therefore, it takes years for the aquifer to cleanse itself. Cleanup can be hastened by pumping the contaminated water out of the ground, although this can be a slow and expensive process.

Ground water flow presents another problem in terms of pollution. Because ground water moves along well-defined flow lines, polluted ground water from a point source will follow these lines and form a **plume** extending from the source of contamination toward whatever point it will eventually appear at the surface. In Figure 15.31, the pollution will eventually reach the stream and thus pollute surface water. Figure 15.31 also demonstrates how important it is to know the direction of ground water flow so that well drilling can avoid the polluted zone. Figure 15.32 illustrates some of the details of the formation of a pollution plume. Contaminants from a dispersed source, of course, will not travel in a plume and therefore may be much harder to trace and eliminate.

A particular type of contamination to which underground freshwater supplies are subject is **saltwater invasion.** This may be salt water of an adjacent ocean environment or it may be salt water trapped in rocks lying beneath the freshwater aquifers.

Fresh water has a lower specific gravity (1.000) than normal salt water (about 1.025). Therefore, fresh water will float on top of salt water, and if there is little or no subsurface flow to mix them a body of

(a)

(b)

(c)

(d)

FIGURE 15.30 (a) Stalactites grow down from the roof of a cave and stalagmites up from the floor. These are in a cave in the Bob Marshall Wilderness, Montana. Impurities of organic acid have colored the calcite red. (b) A needlelike crystal of calcite forms in the drop of water at the end of a growing stalactite. (c) Columns form as stalactites meet stalagmites. Here, in Lechuguilla Cave, Carlsbad Caverns National Park, New Mexico, a pair of columns stand behind the figure in the middle distance. To the left are several stalagmites. (d) A cascade of flowstone in Lechuguilla Cave. Red and yellow colors are due to impurities of iron oxide in the calcite. *((a, c, and d) Arthur N. Palmer. (b) Chris Anderson)*

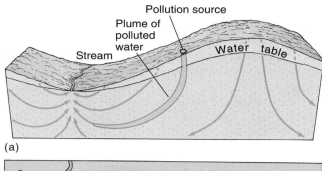

(a)

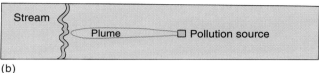

(b)

FIGURE 15.31 Contamination of ground water follows flow lines from the water table toward the point of emergence at the surface. The pollution could be from any one of several sources including an injection well, a leaking landfill, oil storage tank, or a spill. (a) The direction and distance of flow is shown in section. (b) The apparent length of flow in map view.

fresh water may be buoyed on salt water. Where fresh and salt ground water are juxtaposed, pumping can disturb the natural equilibrium of the two, thereby mixing salt water with fresh water. Figure 15.33a illustrates an island (or a peninsula) underlain by a homogeneous aquifer that extends under the adjacent ocean. Rainwater falling on the land has built up a lens of nonsalty ground water that is buoyed up by the salt water that surrounds it. Creation of a cone of depression in the water table by pumping raises the elevation of the saltwater–freshwater interface. For each meter of lowering of the ground-water table, the saltwater–freshwater contact rises about 40 feet. It is very possible, then, that a well that originally bottomed in fresh water could, by lowering of the water table, turn into a saltwater well, as suggested in Figure 15.33b.

Radioactive waste is a particularly worrisome type of ground water contaminant both because of its danger to living things and because many radioactive species last for very long periods of time. The very fact that radioactive contamination can be so dangerous demands that we be careful to distinguish between two types of radioactive waste: **high level nuclear waste** (referred to as *HLW*) and **low level nuclear waste** (referred to as *LLW*).

In the United States, HLW comes from two sources. One is the supply of spent fuel rods from nuclear reactors. The other source is the radioactive waste from the defense programs of the United States government. The Congress, the Environmental Protection Agency, and the Department of Energy have all been involved in determining where and how these contaminants are to be stored. The decision was long ago reached that they should be buried. One of the most important requirements to satisfy in burying high level waste is that there be no migration of the material for 10,000 years. An obvious question arises: Will earthquakes disrupt the storage and allow escape of radioactive waste into the ground water?

The storage of HLW has been a very difficult matter. Completion of an underground waste depository was originally set for 1998, a date that has been twice extended, first to 2003 and then to 2010. The final site chosen by the Department of Energy is the Yucca Mountain Site in southwestern Nevada. Progress there has become stalled in scientific, engineering, and political arguments. In the meantime the cannisters of HLW are stored in pools of water on the surface. By the year 2000 it is estimated that there will be over 40,000 metric tons of spent fuel. Another 8,000 metric tons of HLW from defense programs of the United States government will have accumulated at the surface awaiting an approved, permanent geologic depository.

Low level nuclear waste is now being stored, on a test basis, in the Waste Isolation Pilot Plant, known as *WIPP*, near Carlsbad, New Mexico. The facility was neither designed nor licensed to receive high level waste. The waste consists largely of material used by people working at national defense utilities. It includes lab coats, gloves, glass and metalware, as well as some

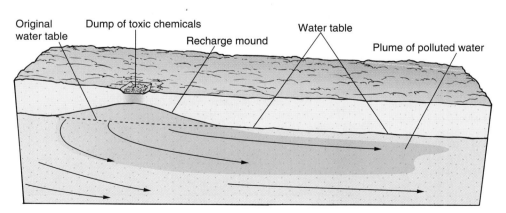

FIGURE 15.32
The water table is distorted by the buildup of a recharge mound from the contaminant source, here a dump for toxic chemicals. The pollution plume then follows the ground-water flow lines.

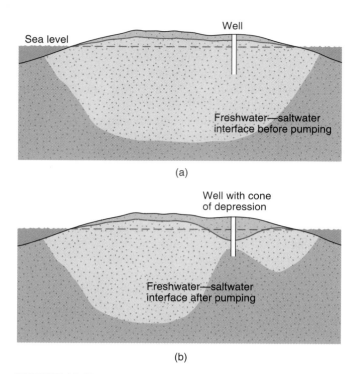

(a)

(b)

FIGURE 15.33 A cross section through an oceanic island (or a peninsula) underlain by a homogeneous permeable material. The fresh ground water forms a lens floating in hydrostatic balance on the neighboring salt ground water. Lowering the water table by pumping will cause a change in the saltwater–freshwater interface, as shown in (a), to that after pumping, as shown in (b). See text for discussion.

discarded machine tools. The waste is officially known as *TRU,* transuranic waste, in reference to radioactive elements heavier than uranium.

The waste is placed 2,150 feet underground in vaults carved from bedded salt deposits. The presence of these salt beds was critical in the selection of the site. The salt indicates that there is very little ground water circulation, or the salt long ago have would been dissolved away. Salt has another advantage. The salt, under the weight of overlying rock, will flow to heal a fracture. In addition, when a waste storage vault is filled its walls can be allowed to flow and seal off the radioactive waste containers. Furthermore, there has been little tectonic activity here since the beds were deposited 225 million years ago, and there is little or no earthquake activity today.

FIGURE 15.34

Three factors have been involved in the relative rise of sea level in respect to Venice, Italy. See text for discussion. *(Modified from Carbognin, Laura, and Paolo Gotti "An Overview of the Subsidence of Venice," in Land Subsidence, A. I. Johnson, Laura Carbognin, and L. Ubertini, eds. Third, International Symposium on Land Subsidence, International Association, of Hydrological Sciences Publication 151, Wallingford, England, 1986, p. 326, Figure 8.)*

Land Subsidence

The withdrawal of ground water can cause subsidence of the overlying surface. There are numerous examples, but one of the most widely known occurred in Venice, Italy.

Venice was first settled in the fifth century A.D. by people fleeing barbarians invading across the Alps. Their settlements occupied low, sandy islands in a lagoon near the head of the Adriatic Sea. Through its history the life of Venice has continued to focus on the sea. The city has long been subject to the *alte aqua,* the high water that floods into the lowest parts of the city. In the twentieth century these floods have become more and more frequent and more and more disastrous. A particularly high water in 1966—caused by torrential rains, high tides, and onshore winds—damaged much of the historic city, its architecture, and its art, and elicited help from around the world. It seemed that Venice was sinking into the sea that had nurtured it for so long.

The immediate cause of the increase in flooding was land subsidence caused by overpumping artesian aquifers. After World War II heavy pumping of ground water began in the industrial zone on the mainland and to a lesser extent in the historic center of the city to the east. The aquifers beneath Venice are unconsolidated, fairly fine-grained sediments. Water removed from the intergranular pore spaces allowed compaction of the aquifer and consequent subsidence of the surface. By 1969 the subsidence rate due to exploitation of ground water had reached 1.7 cm/yr in the industrial zone and 1.4 cm/yr in the old city. The remedy to halt this subsidence was to import surface water by aqueduct from nearby rivers, close down the wells, and forbid new ones. By 1975 rapid subsidence had stopped and a ground recovery of 2 cm had been recorded. Subsidence due to ground water removal has now stabilized in Venice.

Although Venice is no longer sinking because of withdrawal of ground water by pumping, that is not the end of the story for the city. Subsidence still continues because of compaction of sediments, some of it due to the weight of the city itself. Add to this a general worldwide rise of sea level, and periodic flooding of low parts of Venice should increase in the years to come (Figure 15.34).

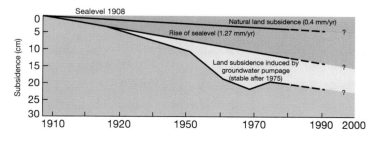

FIGURE 15.35
The Leaning Tower of Pisa, begun in 1173, started to subside during the construction of the first three floors, at which time the builders made an adjustment and the next four floors tilt at a slightly smaller angle. Tilting continued and another adjustment was made for the topmost floor. By the late 1980s the tower was over 5 m out of line and continued to tilt at over 1 mm/yr. It was then closed to the public. By 1993, engineers had managed to halt the tilting and to actually reverse the tilt by a few millimeters.

The Leaning Tower of Pisa (Figure 15.35) provides a classic example of subsidence. The tower, a campanile or bell tower, was begun in 1173 and completed in 1372. It stands on deltaic sediments near the mouth of the River Arno. Weight of the structure has caused differential compaction of the sediments and uneven settling.

►►► EPILOGUE ◄◄◄

In the hydrologic system, some precipitation falling on the Earth's surface seeps into the ground. Some of this is used by plants, some is evaporated back into the atmosphere, and some constitutes a vast underground supply of ground water that moves slowly through the pores and crevices of Earth materials. Ground water eventually reappears at the surface and there nourishes streams and springs. It is responsible for the formation of caves in soluble rocks. Associated with cave formation is the development of a landscape characterized by sink holes, blind valleys, and sinking streams. In many areas people rely on ground water for their domestic supply of fresh water, and it is also widely used in agriculture and industry. Ground water is, however, easily polluted, a situation that its slow flow makes difficult to remedy.

SUMMARY

1. The first zone of ground water is called the *zone of aeration,* in which the pore spaces between rocks and grains are partially filled with water. Directly below this lies the *zone of saturation,* in which all of the pore spaces are entirely filled with water. The water table marks the transition between these two zones. In humid regions the ground water is a subdued reflection of the surface topography, and in dry regions it is very flat. The shape of the water table varies with the recharge of ground water through the zone of aeration.

2. Porosity is the amount of total void space in Earth materials. Permeability is the ability of a material to transmit water. Darcy's law describes the amount of water a rock can transmit in terms of the hydraulic gradient, the area through which the water flows, and a permeability factor.

3. Simple springs are the natural intersection of the ground water table with the surface, whereas a well is an artificial penetration through the water table into the zone of saturation. Artesian wells and springs flow under pressure in aquifers confined by low permeability rock.

4. Most caves are formed by waters passing down from the surface and dissolving passageways in limestone. A few are created by acid-bearing water rising from depth toward the surface. Caves produce a distinctive surface topography of sinkholes, sinking streams, and blind valleys. Stalactites grow downward from cave ceilings, stalagmites up from cave floors, and form columns when they meet.

5. Ground water pollution can be from point or diffused sources. Contamination moves in distinct plumes, from point sources. From diffused sources the extent of pollution is less well defined. Pumping ground water may cause subsidence of the land surface. Disposal of radioactive wastes demands that they do not come in contact with ground water.

KEY WORDS AND CONCEPTS

aquifer 379	head, hydraulic head 376	sinkhole 386
aquitard 379	high level nuclear waste (HLW) 393	sinking stream 386
artesian well 381		soil moisture 375
blind valley 386	hydraulic conductivity 380	specific retention 375
branchwork cave 385	hydraulic gradient 379	spring 380
capillary water 375	karst 385	stalactite 390
cave 385	low level nuclear waste (LLW) 393	stalagmite 390
column 390		thermal spring 382
cone of depression 381	maze cave 385	water table, ground water table 375
Darcy's law 380	perched water table 381	
field capacity 375	permeability 377	well 381
flowstone 390	plume 391	wilting point 375
geyser 383	porosity 377	zone of aeration 374
gravitational water 375	potentiometric surface 381	zone of saturation 374
ground water 374	saltwater invasion 391	

QUESTIONS FOR REVIEW AND THOUGHT

15.1 What are the two basic zones of ground water distribution?

15.2 How is the water table defined? Draw a diagram showing a cross section of two hills separated by a valley. On it draw and label the zones of aeration and saturation and the water table.

15.3 Distinguish between porosity and permeability.

15.4 Draw a labeled diagram to show the nature of flow of ground water through a uniformly permeable material. The diagram you used in question 2 would be a good place to start.

15.5 What is the hydraulic gradient? How is it measured?

15.6 What is Darcy's law? What does it measure? What do its several terms represent?

15.7 What is a perched water table? Draw and label a diagram to illustrate it.

15.8 How may a producing well become a dry well?

15.9 How does an artesian well work?

15.10 How does a geyser work?

15.11 How does a simple branchwork cave form?

15.12 How does the pattern of a cave formed by waters rising from depth differ from that of a cave formed by water descending from the surface?

15.13 Referring to Figure 15.6 you will find two points on the surface labeled *A* and *B*. At each point a surface pollutant filters down to the water table. Trace the path of the resultant plume in each instance.

Critical Thinking

15.14   You have inherited rural property in the eastern United States on which you plan to build a home. It is underlain by granite, and you need a well for domestic water supply. Having studied the flow of ground water and the structure and composition of granite, will you drill a deep well or a shallow well and why? Can you expect a good quality of water and why? What elements can you expect to be in solution in the well water?

15.15   A small stream valley separates a town's water supply well from a landfill. If the landfill leaks will the leakage affect the town's water supply? Why? Draw a cross section through the valley and through the town well and the landfill to help explain your answer.

15.16   Within the framework of plate tectonics where would you expect to find geysers and hot springs?

16

Two Alaskan glaciers flow down their valleys. (Bradford Washburn)

Glaciers and Glaciation

OBJECTIVES

▲▲▲

As you read this chapter, attention to the following questions will help in your study of glaciation.

1. How does glacier ice form, and why can the process be compared with metamorphism? What are the different types of glaciers, how do their budgets work, and how do they move?
2. How do glaciers erode and what features are the major results? What are the main types of glacial deposits, and what topographic forms do they produce?
3. What are some of the indirect results of glaciation?
4. How have geologists determined that glaciers were much more extensive in the past than they are today?
5. What are considered to be the causes of glaciation?

OVERVIEW

▲▲▲

Glaciers form when snow, persisting from year to year, is eventually transformed into ice. When enough ice forms, it begins to flow under the influence of gravity. Glacier ice today covers a tenth of the globe's land surface. In the recent geologic past it spread across a third of the land, carved the world's spectacular high mountain scenery, and left its mark on lower lands, particularly the continents of the Northern Hemisphere. The effect of glaciers, however, has been felt far beyond their margins.

The advance and retreat of glaciers many times in Earth history has been accepted for over a century. Much of the proof of the movement and extent of past glaciations has come from the application of the principle of uniformitarianism. Glaciation and deglaciation are now explained by the slow movement of landmasses in and out of polar latitudes combined with short-term fluctuations of the Earth's climate, which result from variations of the Earth's position in relation to the Sun.

Glaciers
▲▲▲

A **glacier** is a mass of ice formed by the recrystallization of snow. Under the influence of gravity it deforms and flows. This definition eliminates the sea ice that forms from frozen seawater in polar latitudes and, by convention, icebergs, even though they are large fragments broken from the seaward margins of glaciers. **Glaciation** refers to the formation, advance, and retreat of glaciers and the results of these activities.

Modern glaciers blanket about 10 percent of the land area of the Earth and contain most of the freshwater on the planet. Most of this glacier ice (about 96 percent) is in Antarctica and Greenland. The Antarctic ice sheet covers about 15.3 million km², and the Greenland sheet covers approximately 1.9 million km². Smaller glaciers are found in widely scattered locations in North and South America, Europe, Asia, and Africa. They occur on many of the islands near the North Pole and on the Pacific Islands of New Guinea and New Zealand. A few small glaciers are located almost on the equator. Mount Kirinyaga in Kenya, for instance, is on the equator and supports at least 10 small glaciers.

How Glacier Ice Forms

Like rivers on the surface and water underground, glaciers are a part of the hydrologic cycle. They depend, primarily, upon the oceans for their nourishment, which comes to them as snow. If the climate is cold enough, some of this snow will not melt during the summer months. The elevation at which snow persists the year round is the **snow line.** Above the snow line lie **snowfields.** In places, where the snowfields become deep enough, ice may collect and glaciers form.

The first step in the creation of a glacier, therefore, is newly fallen snow. It falls as a feathery aggregate of complex and beautiful ice crystals displaying an endless variety of patterns. Despite the variety, however, each snowflake reflects the orderly arrangement of its hydrogen and oxygen atoms in the hexagonal crystal system (Figure 16.1). Snow is not frozen rain; rather, it forms from the crystallization of water vapor at temperatures below the freezing point. This is called **sublimation,** the process by which material in the gaseous state passes directly to the solid state without first becoming liquid.

Soon after snow has fallen it begins to change. Because it is very porous, air circulates through it. Here, again, sublimation begins to play a role, but this time solid material passes to the gaseous state rather than in the opposite direction, as in the formation of snow. Molecules of water vapor escape from the snow, particularly from the edges of the flakes. Some move off into the atmosphere, but other molecules attach themselves to the center of flakes, where they adapt themselves to the crystal structure of snow. This process makes the original snowflakes smaller, and they begin to be pressed closer and closer together under the weight of successive snowfalls. The originally light, fluffy snow becomes granular, eventually changing into an accumulation of ice granules called **firn** or **névé** (Figure 16.2). *Firn* derives from a German adjective meaning "of last year," and *névé* is a French word from the Latin for "snow." Solid remnants of snowbanks that linger on after winter are largely firn or névé.

As one snowfall follows another, firn, which has already begun to grow as a result of sublimation, is packed tighter and tighter under the pressure of the overlying snow. This pressure causes melting, particularly at the points of contact between granules. The process, known as **pressure melting,** is a familiar one in ice skating. As the skate's blade presses down on the ice, a small amount of ice melts, and the resulting film of water lets the skate glide smoothly over the ice. As soon as the pressure is released, freezing recurs. In

FIGURE 16.1   (a) Snowflakes exhibit a wide variety of patterns, all hexagonal and all reflecting the internal arrangement of the hydrogen and oxygen atoms. (b) Crystal structure of ice showing six-sided rings formed by 24 water molecules. *((b) Gross, M. Grant, Oceanography, 6th ed., Englewood Cliffs, N.J.: Prentice Hall, Fig. 4–3a, p. 75, 1993.)*

(a)

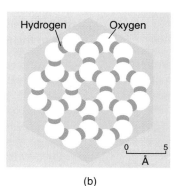

(b)

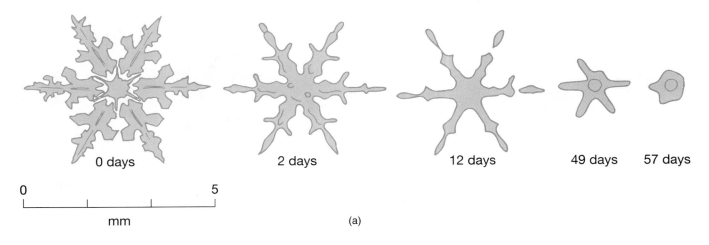

0 days   2 days   12 days   49 days   57 days

0 ⊢——⊥——⊥——⊣ 5

mm   (a)

(b)

FIGURE 16.2

(a) Over 50 years ago Swiss geologists followed the transformation of snowflakes with time. This sequence shows the modification of an original snowflake on the far left to a particle of granular snow on the right, 57 days later. With more time the granular snow changes to pellets of firn as shown in (b). Firn eventually changes to glacier ice. (*(b) Pamela Hemphill*)

firn, melting occurs at points of pressure between the growing granules of ice. The meltwater then moves to places of lower pressure in the still-porous mass where it freezes on portions of granules that are not yet in contact with their neighbors. This new ice continues the hexagonal crystal structure of the original snowflake and the granule that succeeded it.

Gradually, then, a layer of individual firn granules, each a fraction of a millimeter to approximately 3 or 4 mm in diameter, is built up. The firn itself undergoes further change as continued pressure forces out most of the air between granules, reduces the space between them, and finally transforms the firn into **glacier ice,** a solid composed of interlocking crystals (Figure 16.3). In large blocks the ice may be opaque because of the air and fine dirt trapped in it. Its color is blue and is due to the manner that light is absorbed as it travels through the crystal structure of ice.

The time it takes for a snowflake to turn to glacier ice and the depth at which the transition from firn to ice occurs both vary with the climate. On the Seward Glacier in Alaska, located in a moist, maritime climate, the firn layer is about 13 m thick, and it takes between 3 and 5 years to turn snow to ice. In contrast, in the extremely cold central plateau of the Antarctic ice sheet the firn-ice transition is 168 m below the surface, and it takes about 3,500 years to turn snow and firn to glacier ice.

Glacier ice, then, is the end product of several processes. It starts off as snow, which we can consider the equivalent of the sediment in a sedimentary rock. Diagenesis converts the snow to firn and from firn to glacier ice. This final product is really a metamorphic rock, made up of interlocking crystals of ice. As this ice moves in and with the glacier, it undergoes even more metamorphic transformations. The ice crystals, originally a few millimeters in size, grow to several centimeters in maximum dimension, and their original, generally equidimensional texture changes to one in which the crystals are elongated. This traveled ice also becomes marked by bands that reflect planes of shear, and recrystallization of the ice, in fact, takes place along these planes of shearing. Thus, glacier ice represents several stages of metamorphism, which resemble those of the metamorphic rocks we studied in Chapter 6. Of course, in the case of glacier ice, the temperatures and pressures of metamorphism are much lower than they are for most rocks.

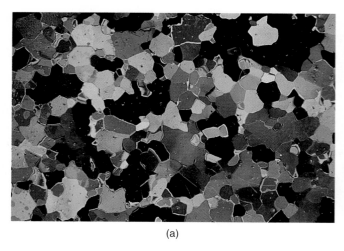

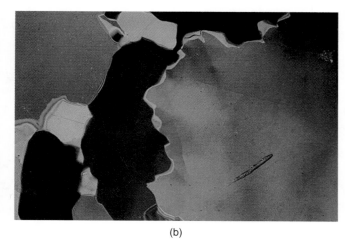

(a) (b)

FIGURE 16.3 Two photomicrographs of thin sections of glacier ice are shown in (a) and (b). Both have been photographed in cross polarized light to show the outline of individual crystals. The sample for each thin section was obtained from a core of ice taken from the Greenland icecap. (a) This sample of glacier ice came from a depth of 314 m below the surface. The interlocking, individual ice crystals range between 1 and 8 mm in maximum dimension and average about 3.5 mm. (b). This sample came from the same ice core as did the specimen in (a) but at a depth of 2991 m. The two thin sections are shown at the same scale, showing that larger individual ice crystals are found deeper in the glacier. The Greenland ice is 3053 m thick at the site where these samples were obtained. *(A. J. Gow, CRREL)*

Types of Glaciers

Geologists classify the glaciers of the world by four principal types: valley glaciers, piedmont glaciers, ice sheets (continental glaciers), and ice shelves.

Valley glaciers—sometimes called **mountain glaciers** or **Alpine glaciers** (Figure 16.4)—are streams of ice that flow down the valleys of mountainous areas. Like streams of running water, they vary in width, depth, and length. A branch of the Hubbard glacier in Alaska is 120 km long, whereas some of the valley glaciers that dot the higher reaches of the Canadian and U.S. mountains are only a few hundred meters in length. A glacier at the foot of a mountain range, a **piedmont glacier,** forms when two or more valley glaciers emerge from their valleys and coalesce to form an apron of moving ice on the flatter lands beyond (Figure 16.5).

Ice sheets are broad, mound-like masses of glacier ice that tend to spread radially outward from a central zone under their own weight. The Vatna glacier of Iceland is a small ice sheet measuring about 120 by 160 km and 225 m in thickness (Figure 16.6). A localized ice sheet of this sort is sometimes called an **ice cap.** The term **continental glacier** is usually reserved for

FIGURE 16.4 Valley glaciers in the Alaska range. Muldrow glacier in the foreground, flowing away from the viewer, is joined by the Traleika glacier on the right. In the distance additional valley glaciers enter the main valley from the right. *(Peter Kresan Photography)*

FIGURE 16.5 The Malaspina glacier is a piedmont glacier that is fed by valley glaciers of the St. Elias Range, southern Alaska. It spreads out in front of the mountains over a width of approximately 50 km and drains to the Gulf of Alaska. *(Steve McCutcheon, Visuals Unlimited)*

FIGURE 16.6 The Vatnajökull ice cap in southeastern Iceland (*NASA*)

great ice sheets that obscure all but the highest peaks of large sections of a continent, such as those of Greenland and Antarctica. On Greenland, ice exceeds 3,000 m in thickness near the center of the island (Figure 16.7a). Ice in Antarctica averages about 2,300 m in thickness, and in some places is over 4,000 m thick (Figure 16.7b). Such ice sheets are so large that they isostatically depress the crust beneath them. An **ice shelf** is essentially a floating ice sheet extending out from a land-based glacier. The largest of these ice shelves are extensions of the Antarctic ice sheet and are known as the Ross Ice Shelf and Filchner-Ronne Ice Shelf (Figure 16.7b).

The Glacier Budget

When the weight of a mass of snow, firn, and ice becomes great enough, the ice begins to move under the influence of gravity, and a glacier is born. Its life thereafter is described in terms of its *budget,* the rate of formation and destruction of ice.

Mountain glaciers provide a good example of the glacier budget (Figure 16.8). A mountain glacier is fed directly by snow above the snow line as well as from tributary glaciers and from avalanches of snow and ice off slopes that rise above it along its route. As the glacier descends to lower elevations it begins to waste by the general process called **ablation,** from the Latin *ab,* "from," and *latus,* "carried." It includes melting, evaporation, and **calving,** the breaking away of ice from the front of the glacier when it ends in the ocean or a lake.

The area where the glacier is fed is called the **zone of accumulation,** and the area of wastage is called the **zone of ablation.** If the supply of ice from the zone of accumulation exceeds the ablation, the glacier will continue to advance. If the rate of ablation exceeds the supply of ice, the front of the glacier will retreat and continue to do so until the supply of ice reaching it equals the amount of wastage. At this point the ice front is stationary, and the glacier's budget is in balance. Because ice moves so slowly there is a time lag in the effect of an increase or decrease of activity in either the zone of accumulation or the zone of ablation. Thus, for example, the amount of ice in the zone of accumulation may increase markedly, but it can be years before this is recorded in an advance of the glacier front. The budget of an ice sheet operates in the same way as does that of a mountain glacier. The only difference is that more ice is involved in the budget of an ice sheet, and its leading edge responds more slowly to a change in its budget than does that of a mountain glacier.

The line separating the zones of ablation and accumulation is the **equilibrium line.** The elevation of the equilibrium line varies according to climatic region. In polar regions, for example, the equilibrium line reaches down to sea level, but near the equator it recedes to the mountaintops. In the high mountains of East Africa it ranges between elevations of 4,500 to 5,400 m. The equilibrium line reaches its greatest elevation in the dry latitudes known as the "horse latitudes" between 20° and 30° north and south of the equator. Here it reaches over 6,000 m. During the last glacial maximum the elevation of the line of equilibrium was, on average, about 1.5 km lower than today (Figure 16.9).

Today most of the mountain glaciers of the world are receding. They operate with deficit budgets, and so their lines of equilibrium are rising in altitude. With only a few exceptions this recession has been going on since the latter part of the nineteenth century, although at varying rates. In contrast, the great ice sheets of Antarctica and Greenland appear to be almost in balance.

Glaciers, Warm and Cold

All glaciers are cold, but some are colder than others. Geologists studying glaciers, therefore, classify them as **polar glaciers** or **temperate glaciers.** A polar glacier is one in which no surface melting occurs even during the summer months; its temperature throughout is always below freezing. A temperate glacier is one in which the temperature throughout is at, or close to, the pressure melting point of ice, except during winter when its upper part is frozen for several meters down from the surface.

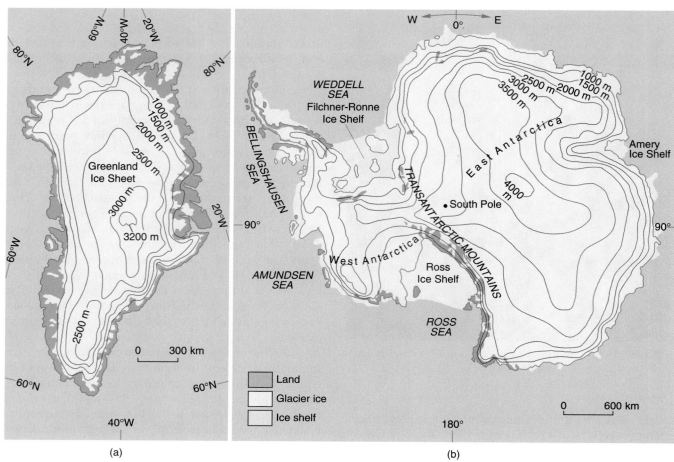

FIGURE 16.7   (a) The Greenland ice cap. (b) The Antarctic ice cap.

FIGURE 16.8   In a glacier a zone of accumulation is separated from a zone of ablation by a line of equilibrium. When accumulation equals ablation the glacier is in balance, and the equilibrium line and the glacier's front are stationary. When accumulation exceeds ablation the front advances and the elevation of the equilibrium line falls. Conversely, when ablation is greater than accumulation the glacier's front retreats and the elevation of the equilibrium line rises.

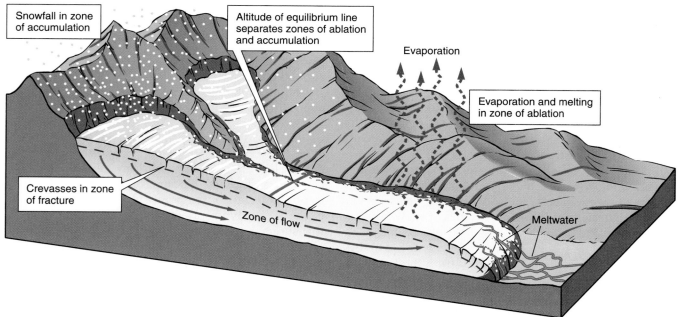

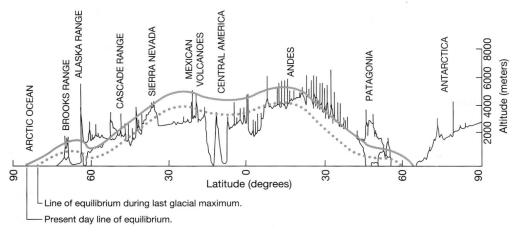

Line of equilibrium during last glacial maximum.

Present day line of equilibrium.

FIGURE 16.9 Elevation of glacier equilibrium line today and during the last major glaciation. *(Broecker, W. S. and Denton, G. H., "The role of ocean-atmosphere reorganizations in glacial cycles," Geochemica et Cosmochemica Acta, vol. 53, p.2470, Fig. 6, 1989.)*

All glaciers must form at subfreezing temperatures, and obviously no glacier can exist above the melting point. Yet there are glaciers that hover near this temperature. How, then, does a glacier become warmed close to its melting point? The Sun's heat cannot penetrate more than a few meters into the poorly conducting ice; therefore, some other mechanisms must operate to warm a glacier. One such process involves the downward movement of water from the surface of the glacier. During the summer season, the Sun melts some of the snow and firn near the surface of the glacier, and this meltwater trickles downward into the glacier. There, in contact with the ice, it begins to freeze, and as it freezes it gives off a small amount of heat. This is the major source of warming in temperate glaciers. In addition to this source of heat, however, there is the normal geothermal flow of heat from the Earth, which helps to raise glacier temperature a bit, as does the frictional heat generated by glacier movement.

How Glaciers Move

Except in rare cases glaciers move only a few centimeters or at most a few meters per day. Geologists can demonstrate and measure the movement of glaciers in several ways.

A very conclusive test is to measure the changing position of a row of stakes driven into the ice across the surface of a glacier. Successive measurements show that, on a mountain glacier, the stakes move down the valley most rapidly along the axis of the glacier and more slowly near the valley walls. This velocity distribution is reminiscent of the movement of water in a stream. Distribution of rock debris on the surface of a glacier provides geologists with more evidence of ice movement. When we examine these rocks and cobbles we find that many of them could not have fallen from the rock walls immediately above, so their sources must lie farther up the valley. We can infer, then, that the glacier carried the debris to its present position. A melting glacier may give us an added indication of glacier movement. As the glacier snout melts and shrinks back, it may expose a rock floor that has been polished, scratched, and grooved. The simplest explanation of these phenomena is to assume that the glacier actually moved across the rock, using embedded debris to polish, scratch, and groove it. Additionally, holes and tunnels cut into glaciers are deformed and offset in a down-valley direction.

Demonstration of movement in an ice sheet can be more difficult if only because of the difficulty of observation. One very obvious proof of movement of ice sheets, however, is that they produce icebergs as their fronts push out into the sea and calve off in deep water.

Once it was established that glaciers move, geologists set about determining *how* they move. Two zones of movement are distinguished: (1) an upper zone that reaches up to 35 m in thickness where ice reacts like a brittle substance—that is, it breaks sharply like the near-surface rock it is; and (2) a lower zone where, because of the pressure of the overlying snow, firn, and ice, the glacier ice deforms by slippage along planes of weakness in the ice crystals. The upper zone is the **zone of fracture;** the lower zone is the **zone of flow.**

The general movement of ice within the zone of flow is downward in the zone of accumulation and upward in the zone of ablation. The forward flow of the glacier is made up of three major elements. First, there is the internal deformation of the ice in the zone of flow. Another important source of movement is slippage of the glacier across a bedrock floor, but this

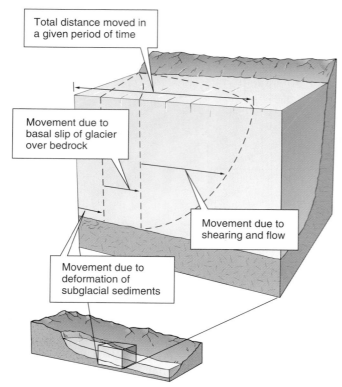

FIGURE 16.10   Movement of a glacier has three components: plastic flow and shearing within the ice; slippage along glacier's contact with the ground; and deformation by the glacier of soft sediments beneath it.

Labels in figure:
Total distance moved in a given period of time
Movement due to basal slip of glacier over bedrock
Movement due to shearing and flow
Movement due to deformation of subglacial sediments

FIGURE 16.11   This glacier in the Himalaya Mountains of Nepal is a chaos of crevasses as it falls down a very steep slope. (Roger Chaffin and Mary Crawford)

occurs only in temperate glaciers, for here water at the glacier's base provides a lubricant. Finally there is the movement caused by the crumpling and deformation of unconsolidated material beneath the ice as the glacier moves over it. Glacier flow is slowed by friction with valley walls and floor (Figure 16.10).

As movement takes place in the zone of flow, the brittle ice above is carried on top of it. The zone of flow, however, moves at different rates in different places—faster in some parts, more slowly in others—and the rigid ice in the zone of fracture above adjusts by fracturing. Consequently, the upper part of the glacier cracks and shatters, producing a series of deep **crevasses** (Figure 16.10 and Figure 16.11).

It is possible to draw an analogy between glacier flow and plate tectonics. The glacier's zone of fracture, where crevasses develop, compares to the Earth's lithosphere. The zone of flow in a glacier flows, and so does the asthenosphere. As the ice in the zone of flow moves, it carries along the zone of brittle ice, just as the lithosphere rides on the back of the underlying asthenosphere.

As we have noted, glacier movement is generally slow. It varies from 3 to 5 m/y for some Antarctic ice to over 300 m/y for the Franz Josef glacier in New Zealand. Exceptional speeds, 10 km/y or more, are reached by some ice on Greenland as inland ice squeezes through mountain valleys to the sea. Surging glaciers, as described in Perspective 16.1, provide another example of the rapid flow of glacier ice.

Results of Glaciation

The glaciers of the last glacial maximum were three times as extensive as they are today. Spectacular mountain scenery and vast plains of glacial debris now record their former presence. This legacy of a past world includes landscapes fashioned by both glacial erosion and deposition. In addition, the vast bodies of glacier ice indirectly affected the Earth well beyond the glacial margin.

Erosion by Glaciers

It is easy to visualize glaciers as effective agents of erosion. The forces generated by masses of moving ice are more than strong enough to tear large fragments from the bedrock as well as to move accumulations of unconsolidated sand and gravel.

When a glacier moves across a stretch of previously fractured bedrock, it can lift large blocks and move them off. This process is known as **plucking,** or **quarrying.** It involves the freezing of water along fracture and bedding planes allowing the glacier to move them as it advances. As it proceeds, a glacier also picks up loose boulders, pebbles, sand, and silt and drags them across the glacier floor. These sediments then cut away the bedrock as would a great rasp or file. On a small scale this process of abrasion produces scratches, or **striations,** and **polish** on the bedrock floor (Figure 16.12). On a somewhat larger scale this abrasion causes **glacial grooves** (Figure 16.13). As the sediments under a glacier abrade the underlying bedrock, they, in turn, are ground down into finer and finer particles. One result of this is that the streams that drain from the front of a melting glacier are charged with **rock flour,** very fine particles of pulverized rock. So great is the volume of this material that it gives the water a characteristically grayish-blue color similar to that of skim milk, testimony to the grinding power of glaciers.

Features of Glacial Erosion

Colliding plates have created the great mountain systems of the world. It has been glaciers, however, that have carved the mountains' towering spires and breathtaking cliffs, created their hidden lakes, gouged their deep valleys, and generally been responsible for the landscapes we typically associate with high mountains. Across the lower sections of the continents ice sheets have ground away at the hills and knobs of the underlying rock and deepened preexisting valleys.

FIGURE 16.12   Glacial polish and striations on basalt at the Devil's Post Pile, California. The striations are oriented from left to right, parallel to the direction of ice movement. The rough, dull areas represent postglacial weathering of the polished surface. Diameter of an individual basalt prism is about 40 cm. *(William C. Bradley)*

FIGURE 16.13   In what is now Central Park, New York City, the last continental glacier carved these grooves in the Manhattan gneiss about 18,000 years ago. *(Jack Burger)*

Glaciated Valleys   Glaciers seldom create their own valleys from the beginning. Rather, they take advantage of preexisting valleys, almost invariably those eroded by streams. In moving through these older valleys, glaciers modify them in a variety of ways. The resulting valleys are broad and **U-shaped** in cross profile (Figure 16.14) whereas mountain valleys created dominantly by streams have narrow, *V-shaped* cross profiles. The advancing glaciers tend to broaden and deepen the **V-shaped** stream valleys, transforming them into wide, open troughs. The deepening takes place by abrasion and plucking, and is hastened because the ice accumulating in a V-shaped valley is thickest along the axis of the valley, making it the site of the most effective erosion. The valleys broaden because the ice has difficulty getting around the bends in the stream valley and tends to straighten and simplify its course. In this process of straightening, the ice snubs off the spurs of mountains extending into the valley. The cliffs thus formed are shaped like large triangles with their apices pointing skyward and are called **truncated spurs.** Glaciers can also erode depres-

Surging Glaciers

Most glaciers move fairly slowly, a few meters to 300 m/y, and at a fairly constant rate. But geologists now know that some glaciers, which have for many years flowed with a certain slow sedateness, suddenly move very rapidly. Such glaciers are called **surging glaciers.**

The largest reported surge occurred on the Bråselbreen ice cap in Spitzbergen, north of Norway. The ice advanced 20 km along a 21 km front sometime between 1935 and 1938, as recorded in aerial photographs. In the summer of 1986 the

Hubbard glacier in Alaska surged at a rate of 12 m/day, cutting off the Russell fjord from the sea and turning it into a lake, at least temporarily (Figure P16.1.1). The most rapid surge known was 110 m/day on the Kutiah glacier, northern India.

Glaciers surge because of an instability that allows them to tear loose from the ground surface on which they rest. The sequence of events involved in the surging of a valley glacier appears to be as follows. More and more ice builds up in the zone of accumulation. If the

P16.1.1   In the summer of 1986 the Hubbard glacier surged into Disenchantment Bay, Alaska, seen in the foreground. The ice advance cut off Russell fjord, on the right, from the sea. The newly formed lake rose about 25 m before the ice dam broke in October and the Russell fjord again became an arm of the sea. *(Steve McCutcheon, Visuals Unlimited)*

sions into the valley floor. These depressions may then fill up with glacial meltwater, creating numerous small lakes. Other lakes may be formed when debris dumped by a retreating ice front forms a natural dam behind which water collects.

In some areas of glaciated, mountainous coasts, deep, narrow arms of the sea snake between steep cliffs and cut far into the coastline. These inlets—known as **fjords,** a name derived from the Norwegian term for this feature—are stream valleys, generally in mountainous areas, that were modified by glacier erosion and then partially filled by the rising sea (Figure 16.15). The deepest known fjord, Vanderford in Vincennes Bay, Antarctica, has a maximum water depth of 2,287 m. In North America the southernmost fjord is the lower Hudson River and New York harbor.

Hanging valleys are another common feature of glaciated mountains. The mouth of a hanging valley is left stranded high above the main valley, through which a glacier has passed. As a result, streams in hanging valleys plummet into the main valley by a series of falls and rapids (Figure 16.16). Hanging valleys are formed as glaciers deepen and broaden existing stream valleys. In the mountains the greatest

amount of ice moves down the main stream valleys. Consequently, erosion there is greater than in the valleys made by tributary streams, which tend to have smaller glaciers in them. Thus the main valley floor becomes correspondingly deeper. Additionally, some hanging valleys have been accentuated by the straightening and widening action of glaciers in the main valleys. The difference in level between the tributary valleys and the main valleys does not become apparent until the ice has melted. Hanging valleys may be formed by any process that deepens a main valley more rapidly than its tributary valleys, but hanging valleys are almost always present in mountainous regions that have been glaciated and are very characteristic of past valley glaciation.

Not all glaciated valleys have been modified by mountain glaciers. Some have instead been changed by continental glaciers. The valleys occupied by the Finger Lakes in central New York State are good examples. As successive continental ice sheets moved southward they encountered the steep, northward-facing edge of the Appalachian Plateau. When the ice rode over this edge and onto the plateau, glacial flow was concentrated in old valleys draining the plateau.

ice here is close to its pressure melting point throughout its mass it can generate water within itself. Geologists believe that some of this water gets trapped beneath the glacier. As more ice accumulates, this section of the glacier becomes more and more unstable. Eventually it may break loose and surge rapidly down valley on a layer of water.

There is evidence that surges may occur in large ice sheets as well as in valley glaciers. Geologists now think that periodically, during the last glacial epoch, rapid flow of portions of the North American ice sheet dumped unusually large amounts of glacier ice into the North Atlantic Ocean from Canada to Europe. As the icebergs melted they dropped the fine stones eroded from the rocks of eastern Canada. Today these pebbles, concentrated at least at six different levels, record these "armadas" of icebergs and testify to the possibility of surges in the Pleistocene ice sheet in North America.

Surges are yet to be recorded from the Antarctic ice sheet. But drilling to the base of the ice has shown that in some places water exists between the ice and the bedrock surface, a condition associated with glacier surges. Scientists have estimated that a surge of the Antarctic ice sheet might well dump enough ice into the Antarctic Ocean to raise the world's sea level between approximately 10 and 30 m during a period of a few years or dozens of years. No imagination is needed to figure out what would happen to the low-lying coasts of the world. New Orleans would be submerged and New York devastated. Coastal nations from the Netherlands to Bangladesh would be drowned, and maritime settlements around the world would be lost. So, anyone interested in a plot for a terrorist science-fiction novel need only to devise a plausible way to start the Antarctic ice sheet surging.

The energy focused along these valleys was great enough to scoop out basins, some of them to depths below present sea level. These have since filled with water to form the Finger Lakes (Figure 16.17).

The Great Lakes offer an even larger-scale example of erosion by continental glaciers. Prior to glaciation, streams in this area flowed toward the north and east. A succession of glaciers deepened and disrupted these valleys, converting them into a series of interconnected basins that now contain the Great Lakes.

Cirques, Horns, Arêtes, and Cols A **cirque** (from the French for "circle") is the basin from which a mountain glacier flows. Cirques are found high in the mountains at the heads of valley glaciers. When the glacier has melted, the cirque is revealed as a great amphitheater, or bowl, with one side partially cut away (Figure 16.18). The headwall of a cirque can be anywhere from less than 100 m to over 900 m above the floor, in many places as an almost vertical cliff. The floor of the cirque lies somewhat below the level of a low ridge or sill separating it from the glacial valley below. A lake, called a **tarn**, may fill the cirque's bedrock basin.

A cirque begins with an irregularity in the mountain side formed either by preglacial erosion or by a process called **nivation**, a term that refers to erosion beneath and around the edges of a snow bank. Frost action loosens and pries free rock debris. Solifluction carries it away, thus creating a shallow basin. Snow that accumulates here eventually turns into a glacier that helps develop the basin into a cirque. The newly formed cirque is deepened and expanded by frost action, plucking, and abrasion.

As several cirques erode around a single mountain, they carve out a sharp spire of rock called a **horn.** The classic example of a horn is the famous Matterhorn of Switzerland (Figure 16.19). An **arête** (French for "ridge," or "sharp edge") forms when cirques expand into a mountain ridge from opposite sides. The ridge, with the additional help of frost action, eventually becomes serrated. If the cirques that form arêtes continue to grow through a ridge until they meet, they form a **col** (from the Latin *collum*, "neck") or pass.

Asymmetric Rock Knobs and Hills In many places glacial erosion of bedrock produces small, rounded, **asymmetric knobs** of rock with gentle, stri-

FIGURE 16.14   The valley of Rock Creek on the northern edge of the Big Horn Mountains in southern Montana has the open U-shape typical of a glaciated valley.

ated, and polished slopes on one side and steep, often rough slopes on the opposite side. The now-gentle slopes faced the advancing glacier and were eroded by abrasion. The opposite slopes were steepened by the plucking action of the ice as it rode over the knobs. Some larger hills have been similarly shaped by glaciers (Figure 16.20).

Deposition by Glaciers

The debris a glacier acquires by erosion is eventually deposited either because the ice that holds it melts or,

FIGURE 16.16   Bridal Veil Falls tumbles from a hanging valley into U-shaped Yosemite valley. *(Pamela Hemphill)*

FIGURE 16.15   The long, narrow, finger-like bays of the sea show up dark against the partially snow-covered land. These are fjords, which are glaciated valleys flooded by the sea after the disappearance of the glaciers. The swirls of light blue represent sea ice. Width of view about 120 km. *(NASA)*

less commonly, because moving ice plasters it on the surface beneath the glacier. Deposits that are laid down directly by glaciers or that are laid down in lakes, ocean, or streams as a result of glacial activity are referred to by geologists as **drift,** a term surviving from an early nineteenth-century belief that such material was deposited from icebergs drifting across the land. Geologists divide drift into two general categories: **unstratified** and **stratified.**

FIGURE 16.17   The basins for the Finger Lakes of central New York State were formed by glacial deepening of stream valleys. Here the lakes appear as dark, narrow "fingers" oriented generally north-south. Lake Ontario is the large lake at the top of the image. The largest of the Finger Lakes, Cayuga Lake, is about 60 km long. *(NASA)*

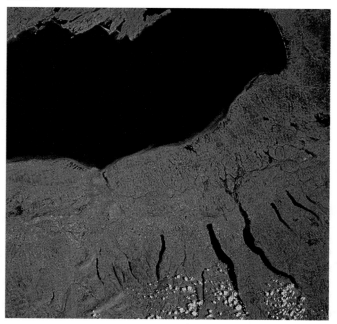

FIGURE 16.18   Cirques with tarns in their basins mark the crest of the Wind River Mountains, Wyoming. *(Austin Post, United States Geological Survey)*

Deposits of Unstratified Drift   Unstratified drift, also called **till,** is made up of rock fragments ranging in size from huge boulders to tiny clay-sized particles that are completely stirred together. Till is made up of a variety of different rock types that vary

FIGURE 16.20   The smooth, gentle slope of Lembert Dome in the Tuolumne Meadows area of Yosemite National Park is the result of glacial abrasion. Plucking has formed the steep slope to the left. Ice moved from right to left. *(Pamela Hemphill)*

FIGURE 16.19   The Matterhorn in the Swiss Alps, like all glacial horns, was formed by the intersecting walls of cirques eroded by glaciers into a central mountain mass. The result is a steep, rocky spire. *(Warren Stone, Visuals Unlimited)*

according to the kind of bedrock across which the glacier moved. This type of drift is termed unstratified because it is without apparent layering and is not sorted into beds of sand, or silt, or gravel (Figure 16.21). Some till lodges in well-consolidated patches as the ice moves across the glacier floor. Most of it, however, is dropped as the glacier melts away. Characteristic of till is that many of the large pieces of the deposit are striated, polished, broken, and faceted as a result of the wear undergone during glacier transport (Figure 16.22).

Moraine is a general term used to designate a landform made largely of till, which can vary widely in size. A **terminal moraine** or **end moraine** is a ridge of till marking the farthest limit of a glacier's advance. It forms as the glacier bulldozes loose material into a ridge, and as melting ice releases additional debris along the glacier's margin. It is broadly convex away from the ice, reflecting the lobate form of the ice front. The end moraines of a continental glacier are broad loops or series of loops traceable for many kilometers

FIGURE 16.21   The range in particle size in a till is very large. In this exposure, in the eastern Pyrenees Mountains of France, boulders of varying size are mixed with finer particles, that, when examined carefully, turn out to range downward in size from boulders, through pebbles, grains of sand and silt, to clay-sized particles. The lack of layering is also characteristic of till. The reddish-brown zone at the top of the exposure is the soil formed in postglacial time.

FIGURE 16.22   Shapes and surfaces of many glacially transported stones are distinctive. These stones have been striated while being carried in the ice. The upper black pebble has been polished as well. The two lower stones were faceted and their margins broken. Largest pebble has a maximum dimension of about 10 cm. *(John Simpson)*

across the countryside (Figure 16.23). A series of similar ridges, known as **recessional moraines,** may also build up as the glacier retreats. These moraines mark positions at which the glacier front readvanced briefly during the general retreat of the glacier. The end moraine of a mountain or valley glacier is much smaller, as befits the glacier's size (Figure 16.24).

FIGURE 16.23   The flow of a continental glacier reflects variations in the topography beneath the ice. Here the outer margin of glacier advance during the last major glaciation has the characteristic lobate form of continental glaciers. Later moraines show this same pattern. *(Charles Denny, United States Geological Survey)*

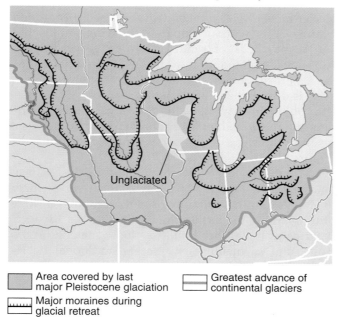

Unglaciated

☐ Area covered by last major Pleistocene glaciation
☐ Major moraines during glacial retreat
☐ Greatest advance of continental glaciers

Not all the rock debris carried by a glacier finds its way into the terminal and recessional moraines. As the main body of the glacier melts, a great deal of debris is released and forms gently rolling topography. Till in this form, called **ground moraine,** may be a thin veneer lying on bedrock, or it may be tens of meters thick, partly or completely clogging preglacial valleys. In addition an **interlobate moraine** may build along the junction of adjacent glacier lobes.

Finally, valley glaciers produce two special types of moraine. While a valley glacier is still active, it col-

FIGURE 16.24   The semicircular ridge around the mouth of U-shaped Little Cottonwood Canyon, Utah, is a terminal moraine of the last major ice advance in the Wasatch Mountains. Cottonwood Canyon, to the left, is larger and was more severely glaciated. The cliff across the moraine of Little Cottonwood Canyon is a fault scarp related to movement along the major fault defining the western margin of the Wasatch Mountains. Little Cottonwood Canyon has the typical U-shaped profile of a glaciated valley. Truncated spurs are present as well.

—Truncated spurs

—U-Shaped valley

—Terminal moraine

—Fault scarp

FIGURE 16.25   A lateral moraine has dammed lakes in side valleys. Dead ice lies behind the lateral moraine. Himalayan Mountains, Nepal. *(Roger Chaffin and Mary Crawford)*

FIGURE 16.26   A drumlin near Elbridge, west of Syracuse, in central New York State. The blunt end of the asymmetric long profile faces north, the direction from which the glacier advanced.

lects large amounts of rubble from the valley walls along its sides. When the ice melts, this debris is stranded along the side of the valley as ridges called **lateral moraines** (Figure 16.25). At its down-valley end a lateral moraine may grade into a terminal moraine. A **medial moraine** is created whenever two valley glaciers merge to form a single stream of ice, and their adjacent lateral moraines merge to form a single, central moraine (see opening photograph). Although medial moraines are very characteristic of existing glaciers, they are seldom preserved as topographic features after the disappearance of the ice.

Drumlins are streamlined hills composed largely of till. Their profile resembles a whale's back or the inverted bowl of a spoon (Figure 16.26), with the blunt end pointing in the direction from which the vanished glacier advanced. Drumlins occur in clusters, called **drumlin fields.** In the United States they are well developed in the Boston area; in eastern Wisconsin; in New York State, particularly near Syracuse; in Michigan; in Minnesota; and in parts of southern Canada. Glacial geologists hypothesize that drumlins were formed as water-rich till was molded into streamlined forms as ice flowed over it.

An **erratic** is a stone or boulder carried from its place of origin and left stranded by the ice in areas of different bedrock composition. The term applies whether the stone is embedded in till or rests directly on the bedrock. **Boulder trains** consist of clusters of erratics from the same source, and with some distinctive characteristic that makes their common source easily recognizable. They appear either as a line of erratics stretching down-valley from their source or in a fan-shaped pattern with the apex near the place of origin. By mapping boulder trains geologists can gain an excellent indication of the direction of ice flow (Figure 16.27).

Deposits of Stratified Drift   Debris washed away by water from a melting glacier is laid down in well-defined layers of sand, gravel, silt, and clay. These deposits are called **stratified drift** because they are layered rather than randomly mixed as is the material in a till. The sand and gravel washed out by meltwater from the glacier is called, appropriately enough, **outwash** (Figure 16.28). Extensive beds of sand and gravel are deposited out beyond the glacier and underlie smooth plains known as **outwash plains,** or, where restricted by valley walls, as **valley trains** (Figure 16.29).

An **esker** is a winding, steep-sided ridge of stratified sand and gravel, sometimes branching and often discontinuous. Eskers vary in height from about 3 to 15 m, although a few are over 30 m high. They range from less than a kilometer in length to over 160 km. They result from deposition from streams flowing in ice tunnels beneath stagnant glacial ice. When the ice finally disappears, the old stream deposits are left standing as sinuous ridges (Figure 16.30).

In some areas, stratified sand and gravel form low, steep-sided hills, called **kames,** that can occur either by themselves or in clusters. Unlike drumlins, kames are irregular in shape and are composed of stratified deposits rather than till. They are associated with stagnant ice in which melting produces large cavities. These fill with outwash sand and gravel, and when the retaining walls of ice melt an irregularly shaped mound replaces the former cavity. A **kame terrace** forms from stratified sand and gravel deposited between a wasting glacier and an adjacent valley wall. When the glacier disappears, the deposit stands as a terrace along the side of the valley. Deposits in kames and kame terraces may be distorted as their ice walls collapse or as a readvancing glacier shoves into them.

Varves, which we discussed in connection with geologic time in Chapter 7, constitute another type of

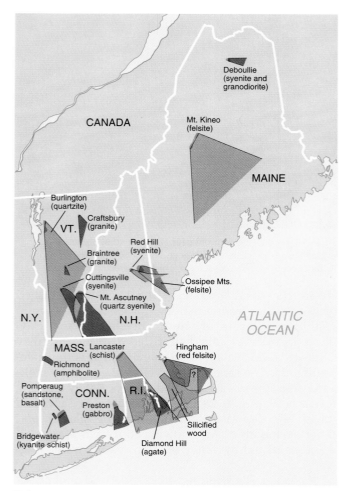

FIGURE 16.27 The boulder trains plotted on this map indicate the general direction of the last glacier movement across New England. The apex of each fan indicates the area from which its boulders came. The fan itself covers the area across which the boulders were deposited. *(In part after J. W. Goldthwait in R. F. Flint, "Glacial Map of North America," Geol. Soc. Am. Spec. Paper 60, 1945.)*

FIGURE 16.28 Layered outwash deposits near Watertown in northern New York State. The beds have been slightly deformed by a readvance of the ice.

FIGURE 16.29 Sand and gravel carried from melting alpine glaciers have clogged the Inn Valley, Austria, and built up this very flat surface typical of a valley outwash plain or valley train.

stratified drift. Their rhythmic banding, in which each bed grades upward from coarser sediments to finer, is a type of graded bedding discussed in Chapter 5 on sedimentary rocks.

Kettles are pits or depressions that occur in areas of both till and outwash. In a way they are the counterparts of kames. When blocks of stagnant ice, partially or completely covered by debris, melt they leave behind kettles (Figure 16.31). These kettles range from a few meters to several kilometers in diameter and from a few meters to over 30 m in depth. In many places water fills the kettles to form lakes, ponds, and swamps, features widespread through Canada and the northern United States (Figure 16.32).

Some of the glacial features we have been discussing are more common in areas of mountain glaciation; others are more characteristic of areas of continental glaciation. Many of the features are found in both areas. They are summarized in Figures 16.33, 16.34, and 16.35, and Table 16.1.

FIGURE 16.30 View along the top of an esker at Belgrade, Maine *(Stephen J. Robinson)*

FIGURE 16.31   Kettle holes forming along the wasting margin of the eastern lobe of Woodworth Glacier Tosauma Valley, Alaska. (*Bradford Washburn*)

FIGURE 16.32   Ponds partially filling kettle holes in a recessional moraine in the Chenango Valley, north of Shelbourne, N.Y.

Indirect Effects of Glaciation

Modern society finds that one of its most valuable resources is common sand and gravel, which is the basis of a $4 billion a year business. This work-a-day material is critical to the construction industry and finds its way into roads, bridges, and buildings. If you live in a glaciated area one very good source of sand and gravel is glacial outwash. It has another important use, namely as an aquifer providing us with supplies of ground water for home, industry and agriculture. Clays from old glacial lakes have been the base of another industry—the manufacture of bricks.

Glacial debris, whether till, outwash, or associated wind-borne dust, has provided the parent material for the development of soils in many of the richest agricultural regions of the world. Glacial deposits, particularly the more recent ones, are fertile because there has not been enough time since their deposition for deep chemical weathering to remove many of the essential minerals of plant growth.

Low ocean levels during glaciations have permitted river erosion of the continental margins. These deepened valleys later became flooded as sea level rose during deglaciation. The result has been long estuaries reaching deep into the continents and pro-

FIGURE 16.33

Glaciated landscape in the Himalaya Mountains, Bhutan. The U-shaped valley in the foreground lies at an elevation of about 4,100 m. About 3 km distance from the observer it is joined from the left by a valley in which the snout of a valley glacier is visible. In front of the glacier is a lake held in by a moraine. The dam was partially breached in the 1950s and the outlet channel is visible as a white, V-shaped notch across the moraine. The lake level fell rapidly to its present level and floodwater caused widespread loss of life and damage downstream at the religious center of Punaka. The gray strip along the river is a "trim line" below which vegetation was swept away during the flood and has yet to recover. Up the main valley is the nose of another glacier, out in front of which are moraines. The unvegetated sides of the moraines, facing toward the axis of the valley, are white because they are made up of debris weathered from a white granite, the dominant rock type in the region. (*Lincoln Hollister*)

(a)

(b)

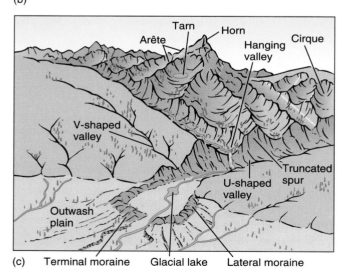

(c)

FIGURE 16.34 The development of the major features of mountain glaciation. (a) A mountainous area before glaciation. (b) The same area during glaciation. (c) The area after disappearance of the glaciers.

viding sites for settlement as well as wetlands for wildlife. The Delaware estuary in the eastern United States is an example. Here the inland cities of Wilmington, Philadelphia, and Trenton have direct access to the sea. More spectacular is the St. Lawrence–Great Lakes waterway. Both the lake basins and the St. Lawrence River estuary have been fashioned chiefly by glacial erosion. With the help of canals and locks, oceangoing vessels have access to the heartland of the continent at Duluth, Minnesota, over 4,500 km from the open Atlantic Ocean.

Glaciers of the Past

In the first half of the nineteenth century, scientists learned that glaciers were of much greater extent in the past than they are today. The discovery was made during investigations in the Swiss Alps by several scientists, foremost among them a young zoologist named Louis Agassiz. He and his colleagues found that certain features produced by glacier ice are produced by no other known process. They saw that boulders were being transported from their bedrock source by modern glaciers. Beyond the glaciers they found isolated stones and boulders quite alien to their present surroundings. Some of the boulders, they observed, were so large that rivers could not possibly have moved them, and others were perched on high places that rivers could reach only by flowing uphill. They also noted that when modern ice melted, it revealed a polished and striated pavement unlike the surface fashioned by any other known process. They found, in addition, that the glaciers in many places had piled up ridges of debris at their snouts. When they found similar ridges, called moraines by the local farmers, now located well beyond the margin of the present ice, they reasoned that the glacier once had a greater extent than it currently had. By accumulating a large mass of field observations and correlating them carefully, Agassiz was able to document that the Swiss glaciers once extended far beyond their present positions. Agassiz was soon to propose that there had once been a "Great Ice Age" that submerged Switzerland and much of northern Europe in a "sea of ice."

The proof of Agassiz's Ice Age lay in applying the principle of uniformitarianism introduced in Chapter 1. Agassiz observed modern glacial activity directly and compared the results of this activity with similar features and deposits having no association with modern glaciers. He wrote that "only a glacier produces all these features at once."

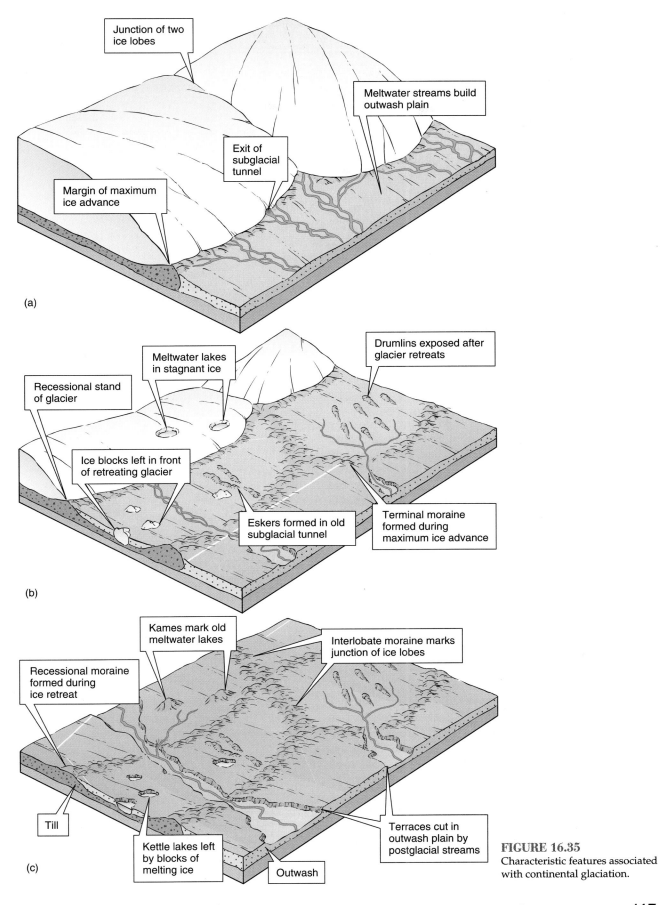

(a)

Junction of two ice lobes

Meltwater streams build outwash plain

Exit of subglacial tunnel

Margin of maximum ice advance

(b)

Meltwater lakes in stagnant ice

Drumlins exposed after glacier retreats

Recessional stand of glacier

Ice blocks left in front of retreating glacier

Eskers formed in old subglacial tunnel

Terminal moraine formed during maximum ice advance

(c)

Kames mark old meltwater lakes

Interlobate moraine marks junction of ice lobes

Recessional moraine formed during ice retreat

Till

Kettle lakes left by blocks of melting ice

Outwash

Terraces cut in outwash plain by postglacial streams

FIGURE 16.35
Characteristic features associated with continental glaciation.

TABLE 16.1 Features of Mountain and Continental Glaciation Compared

FEATURES	MOUNTAIN GLACIER	CONTINENTAL ICE SHEET
Striations, polish, etc.	Common	Common
Cirques	Common	Absent
Horns, arêtes, cols	Common	Absent
U-shaped valleys, hanging valleys	Common	Rare
Truncated spurs	Common	Rare
Fjords	Common	Rare
Till	Common	Common
Terminal moraine	Common	Common
Recessional moraine	Common	Common
Ground moraine	Common	Common
Lateral moraine	Common	Rare
Interlobate moraine	Rare	Locally common
Medial moraine	Common, easily destroyed	Absent
Drumlins	Rare, or absent	Locally common
Erratics	Common	Common
Eskers	Rare	Locally common
Kames	Common	Common
Kame terraces	Common	Common

Pleistocene Glaciation

Approximately 20,000 years ago, during the last major advance of continental ice sheets, 39 million km² of the Earth's surface—about 27 percent of the present land area—were buried by ice. Ice covered approximately 15 million km² of North America. Greenland and Antarctica were covered by ice sheets, as they are today. In Europe an ice sheet spread from Scandinavia across the Baltic Sea and into Germany and Poland. The Alps and the British Isles supported their own ice caps. Eastward, the northern plains of Russia were covered by glaciers. On the north they extended out across the continental shelf and covered Spitzbergen and the islands of Franz Josef Land. Ice was continuous across northern Siberia and its continental shelf to the Bering Strait separating Asia from North America. South of the Eastern Siberian ice sheet were regions of mountain glaciation on the Kamchatka Peninsula and in the Cherskiy Mountains as well as in the high plateaus of central Asia (Figure 16.36).

In North America ice covered Canada, crunched southward to New Jersey in the East, and reached as far south as St. Louis in the Midwest. From there the ice margin extended northwestward to the Canadian border at the foot of the Rocky Mountains. Ice caps and mountain glaciers characterized the high mountains of the western states and Canada. Southernmost glaciation in the United States occurred in the Sierra Blanca of south-central New Mexico. The high peaks of central Mexico were glaciated, and some small glaciers still cling to them.

FIGURE 16.36   Extent of glaciation in the Northern Hemisphere during the last major glaciation.

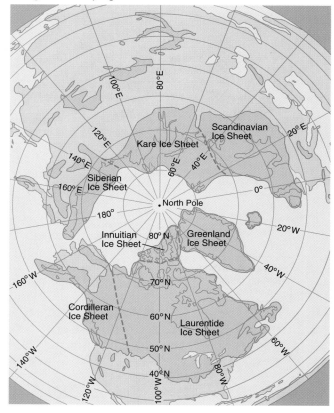

☐ Principal areas covered by glacier ice (Very small areas not shown. In central and northeastern Asia, includes areas of more extensive earlier glaciation.)

Even before most geologists had yet to accept the theory of a single great Ice Age, some researchers were finding evidence that glacier ice had advanced and retreated not just once but several times in the recent geological past. By the early twentieth century a broad, fourfold division of the Ice Age, or the Pleistocene as it is more formally called, had been demonstrated in North America and Europe. Early in this century four stages of continental glaciation were discovered in the United States. Originally they were called, from oldest to youngest, *Nebraskan, Kansan, Illinoian,* and *Wisconsin* for the midwestern states where deposits of a particular age were first discovered or where they are well exposed. This fourfold subdivision eventually proved to be an oversimplification. Glacial and interglacial times (periods of glacial advance and retreat, respectively) alternated with each other throughout the Pleistocene. Evidence now indicates that there have been as many as 30 cycles of glaciation and deglaciation in the last 2 million years (Perspective 16.2).

Pre-Pleistocene Glaciations

Geologists have found evidence that glaciers appeared and disappeared in other periods before the Pleistocene, as well. The record is fragmentary as we would expect, for time tends to conceal, jumble, and destroy the effects of glaciation. Recent evidence indicates that glaciation and deglaciation reach back into the late Miocene in Alaska. Glacial deposits in Antarctica have been dated by the potassium-argon method as 10 million years in age. Further back in geologic time there were other extensive glaciations. From about 230 million to 290 million years ago (toward the end of the Paleozoic Era) glaciation was widespread in what is now South America, South Africa, India, Australia, and Antarctica. As we found in Chapter 11, the evidence of this glaciation constitutes one of the proofs of plate tectonics and the drift of continents. There is evidence of glaciation during the Silurian and Devonian in South America; and earlier, in the Ordovician, ice spread across what is now the Sahara Desert. In the late Proterozoic, a little over 600 million years ago, glaciation affected landmasses now in the Northern Hemisphere. Still earlier, perhaps 850 million years ago, ice covered large parts of the ancient continent of Gondwana. Some Russian geologists believe that sections of northwestern Siberia were glaciated some 1.2 billion years ago. Extensive glaciation, dating about 2.2 billion years ago, is recorded in what is now south-central Ontario, Canada. Most of these anciently glaciated lands have since drifted into areas where glaciers cannot form today.

What Causes Glaciation?

To begin the search for causes of glaciation we can start with the knowledge that glacier ice forms in a climate in which snow survives the summer months and accumulates from year to year. Such climates are most widespread in polar latitudes. Also, glaciers need a landmass on which to form. They will not form from the accumulation of sea ice. Thus geologists know that continental glaciation needs extensive land surfaces in high latitudes where snow can last through the high sun months and firn and ice can continue to form as in Antarctica and Greenland today.

Plate Tectonics

The process of plate tectonics is constantly moving landmasses to various places on the globe. At times continents are moved into the polar positions that favor glaciation. For instance, geologists know that with the breakup of Gondwana some fragments of that ancient megacontinent moved into polar regions, arriving there in the late Miocene after voyages lasting tens of millions of years. Their arrival corresponded to the beginning of late Cenozoic glaciations, the most extensive of which mark the Pleistocene Epoch.

Necessary as plate tectonics is to shove landmasses into the proper environment for widespread continental glaciation, it is not sufficient to explain the multiple glaciations of the Pleistocene. These glacial-interglacial episodes occurred a number of times at intervals varying between about 80,000 and 130,000 years. It is impossible that the process of plate tectonics moved landmasses so rapidly in and out of polar regions. What plate tectonics *does* do is move continents into position where ice sheets may form, advance, and retreat repeatedly. What we now need to explain, therefore, is the reason, or reasons, for these pulses of glaciation and deglaciation.

The Milankovitch Astronomic Theory

The Earth's motions are affected by the gravitational attraction of the Sun and other planetary bodies in our Solar System. Astronomers have long recognized that, as a result, the position of the Earth in relation to the sun fluctuates slightly. In 1875 Scottish geologist James Croll suggested that variations in the Earth's orbital motion cause long-term differences in the Earth's exposure to the Sun's rays, and these differences

Deep-sea Cores Monitor Glacial and Interglacial Times

Fine sediment and the tests (shells) of one-celled animals are constantly settling slowly to the deep seafloor. These microfossils can reflect the climate at and near the ocean surface. For example, in core samples taken from the ocean floor researchers have found successive zones of cold water and warm water fossil forms. Beginning at the very top of the core are fossils of life-forms that exist in the warm climate of the present. Deeper in the core, cold water forms replace warm water forms. These zones of cold and warm water forms keep alternating down the core. Because

the cooling that produced glaciers on land also produced colder ocean water, this zonation of fossils is interpreted as the result of repeated glaciation and deglaciation.

The chemical composition of these microfossils provides scientists with yet another and intriguing way to determine what past climates were like. Geologists use a method based on two stable (nonradioactive) isotopes, both of which are present in any environment. Water molecules containing only ^{16}O are a little lighter than those containing ^{18}O and are therefore a little easier to evaporate. For this

reason clouds, and the snow that falls from them, have proportionately less ^{18}O than does seawater. During a glacial period, when snow accumulates on the land rather than returning to the ocean, progressive evaporation leaves seawater richer in ^{18}O than in interglacial time. The $^{18}O/^{16}O$ ratio of seawater, therefore, is an indirect measure of how much glacial ice there is in the world. Of course, scientists cannot sample ancient seawater directly, but we can analyze the shells of microfauna (foraminifera) that lived in it. When these one-celled animals die and settle to the bottom of the

might cause changes in the planet's climate great enough to cause glaciation and deglaciation. This theory enjoyed a brief popularity and then was forgotten until early in the twentieth century. Then Milutin Milankovitch, a Serbian astronomer, greatly extended and refined the idea. Using three different orbital motions of the Earth, Milankovitch offered an explanation for the goings and comings of Pleistocene glaciers over the last 600,000 years. By the last quarter of the twentieth century, his theory was achieving a widespread acceptance. It states that when the orbital motions minimize the Earth's receipt of solar energy in high northern latitudes, then glaciation occurs.

The **astronomic theory** rests on three factors that enter into the changing position of the Earth in its orbit

around the Sun. These are shown in Figure 16.37 and described as follows:

- The Earth's path around the Sun is an ellipse with the Sun as one focus. This is called the **eccentricity** (measure of circularity) **of the Earth's orbit,** and it varies with time from an almost circular path to a path of greater eccentricity than today. This means that the Earth's distance to the Sun changes as the eccentricity changes. The shifts back and forth between greater and smaller eccentricities take place in cycles of about 100,000 and 400,000 years (Figure 16.37a).

- A second factor is known as the **obliquity of the ecliptic.** This is the tilt of the Earth's axis of rotation in relation to the plane (the ecliptic) in which the Earth circles the Sun. If the Earth's axis of rotation were perpendicular to the plane of the Earth's plane of rotation around the Sun then the radiation received from the Sun at a given latitude

ocean, they carry a $^{18}O/^{16}O$ climatic record with them. By studying the isotopic composition of the foraminifera in the deep-sea sediments, therefore, geologists have been able to "read" the history of the Pleistocene glacial advances and retreats (Figure P16.2.1).

How are the dates of the warm and cold periods determined? In Chapter 7 we studied the determination of age of samples by both radiocarbon and magnetic reversals. From the latter part of the last glaciation toward the present, radiocarbon can be used to date the fossils in deep-sea sediments. Further back in time the known ages of microfossils and the record of magnetic reversals in the sediments form the basis for establishing ages.

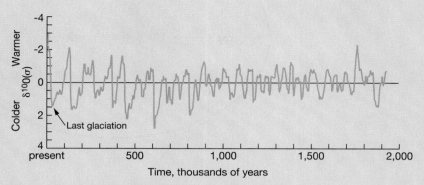

P16.2.1 The variations in temperature in the ocean from the present back almost 2 million years are shown in the composite record of oxygen-18 isotopes in deep-sea cores. The isotope values are shown as differing from a standard value ("0' line). Increase in the $\delta\,^{18}O\,\sigma$ value indicates colder water, and a decrease signals warmer water. Colder water is interpreted as contemporary with glaciations and warmer water with deglaciation. (The symbol δ stands for "difference" or "departure," and the symbol σ is a statistical unit called "standard deviation." *(Williams, Douglas F., et al, "Chronology of the Pleistocene Oxygen Isotope Record of 0–1.88 my. B. P., "Palaeogeography, Palaeoclimatology, Palaeoecology, vol. 64, 1988, Fig. 15.)*

would be constant the year around. The Earth's axis, however, is tilted and this accounts for the annual march of temperatures at all latitudes—in short, for our seasons. Today the tilt is approximately 23.5°. It varies, however, over a period of 41,000 years, through about 3° between a minimum of 21.5° and a maximum of 24.5°. As the tilt increases the contrast between seasons increases (Figure 16.37b).

- The third element in the Milankovitch theory is the **precession of the equinoxes.** The rotating Earth behaves like a giant, wobbling top. Today its axis of rotation points toward the North Star. Over time, however, the wobble causes the axis to point to successively different stars. Three thousand years ago it pointed to the star Thuban, and 12,000 years from now the star Vega will be the pole star. One wobble of the Earth is completed about every 23,000 years. Because of this wobble the times of the equinoxes, when days and nights of equal length usher in

spring and fall, shift or *precess* along the path of the Earth's orbit (Figure 16.37c).

Each of these three periodic motions affects the way the Sun's energy is distributed at the Earth's surface. Together, these three factors can cause long-term changes in the amount of insolation at any one latitude on the Earth's surface. Graphically representing the effects of these three factors on the Earth's climate, scientists have derived a chart known as the **ETP curve** (for eccentricity, tilt, and precession). The curve, often referred to as the **Milankovitch curve,** traces a succession of cold and warm times, and these are seen as corresponding to glacial and interglacial periods (Figure 16.38). Scientists can determine the ages of the various

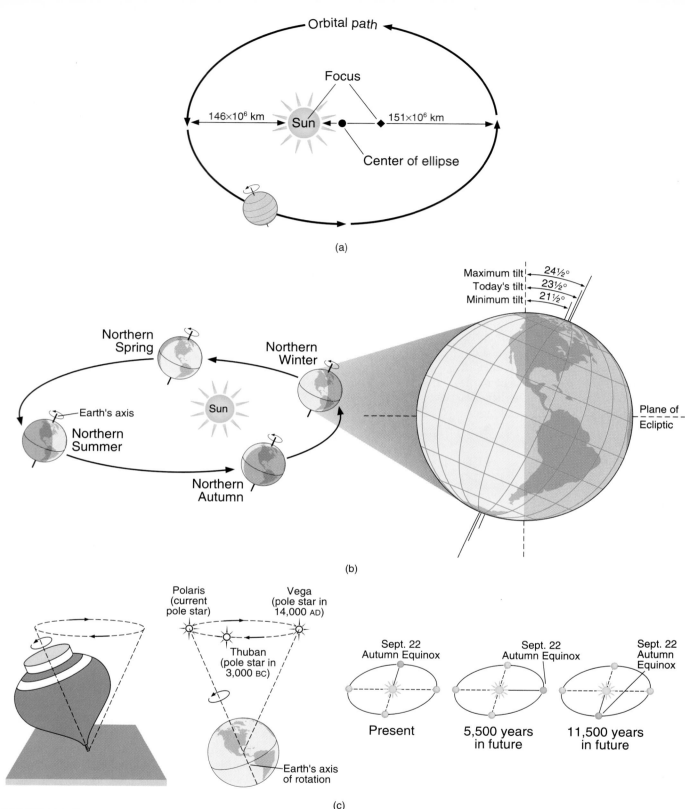

FIGURE 16.37 The orbital motions of the Earth are threefold and are diagrammed here. (a) The Earth's orbital path is today an ellipse with the Sun at one focus. The shape of the path varies with time from almost circular to one of greater eccentricity than today's path. This eccentricity changes in cycles that repeat every 100,000 and 400,000 years. As a result the Earth is closer to the sun at some times than at others. (b) The axis of rotation of the Earth is tilted today at about 23.5° to the plane in which the Earth moves around the Sun. This tilt changes and varies between a maximum of 24.5° and a minimum of 21.5°, completing one cycle every 41,000 years. (c) The axis of Earth's rotation wobbles like that of a spinning top. Consequently, the axis points to different spots in the heavens through a cycle of about 26,000 years. This cycle is reduced by a slower wobble of the Earth's elliptical orbit in the opposite direction. These motions together produce a shift, or precession, of the spring and autumn equinoxes every 23,000 years. See text for discussion. *((c) In part after Chaisson, Eric, and Steve McMillan,* Astronomy Today. *Englewood Cliffs, N.J. Prentice Hall, Fig. 1.21, p. 21, 1993.)*

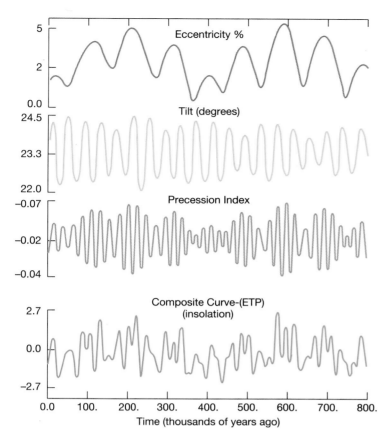

FIGURE 16.38

The variations through time of the Earth's orbital motions can be plotted independently as they are in the upper three curves in this figure. When these variations are added together a fourth curve is obtained and is shown in the lowermost graph. This composite curve (the ETP or Milankovitch curve) represents the variation in radiation received and hence the succession of colder and warmer periods. *(After Imbrie, J. et al, "The orbital theory of Pleistocene Climate: Support from a revised chronology of the marine δ18 record" in* Milankovitch and Climate, Part 1, *A. Berger et al, eds. Dordrecht: D. Reidel Publishing Co., Netherlands, 1984, p. 276, Fig. 2.)*

cool and warm periods because the orbital motions of the Earth that go into the ETP curve are predictable.

▶▶ EPILOGUE ◀◀

Plate tectonics sets the stage for glaciation by clustering landmasses in polar latitudes. Changes in the orbital motion of the Earth in relation to the Sun has accounted for alternating glacial and interglacial periods. Where snow persists through the summer it turns first to firn and then to ice and, when enough ice has accumulated, it begins to flow, and a glacier is born. During the last 2 million years successive advances of ice sheets and mountain glaciers have molded the landscape of a third of the Earth's land surface. As the ice advanced and retreated the climates of the world swung back and forth, drastically changing the habitats for flora and fauna, including humans. At the same time sea level changed, generally rising with deglaciation and falling with glaciation, affecting the shorelines of the world, the general subject of the next chapter.

SUMMARY

1. Glacier ice forms by the transformation of snow into firn and then into glacier ice, a process analogous to metamorphism at low temperatures and pressures. Glacier ice forms ice sheets, ice caps, piedmont glaciers, and mountain glaciers. Every glacier has a zone of accumulation and a zone of ablation. When accumulation exceeds ablation the glacier advances, and when it is less than ablation the glacier retreats. This process is known as the glacier budget. Glaciers move by internal deformation of the ice, basal slippage, by deforming underlying sediments, and by surging.

2. Glaciers erode by abrasion and plucking. Erosional features include U-shaped valleys, truncated spurs, arêtes, cirques, horns, cols, hanging valleys, polished and striated bedrock, and asymmetric rock knobs and hills. Unstratified

glacial deposits are called *till.* They make up drumlins and various kinds of moraines. Stratified deposits are called *outwash* and make up outwash plains, kames, eskers, and varves.

3. Indirect effects of glaciation include falling sea level, deposition of material that forms the basis for modern soils, deposition of valuable deposits of sand, gravel, and clay, and creation of harbors and waterways.

4. By applying the principle of uniformitarianism, geologists have determined that glaciers were of much greater extent at times in the past than they are today. Researchers discovered that present-day glaciers produced geologic features created by no other known processes. When these researchers then found similar geologic features that were unassociated with modern glaciers, they reasoned that glaciers must have been present and created these features in the past.

5. To form, glaciers require a landmass with a climate in which snow lasts through the summer months. Plate tectonics has moved continents into polar positions and set the stage for glaciation. Variations in the orbital position of the Earth in relation to the Sun have caused variation in the amount of insolation received by the Earth. These insolation variations have triggered glaciation and deglaciation.

KEY WORDS AND CONCEPTS

ablation 403
arête 409
astronomic theory 420
asymmetric rock knob, hill 409
boulder train 413
calving 403
cirque 409
col 409
crevasses 406
drift, unstratified, stratified 410
drumlin, drumlin field 413
eccentricity of Earth's orbit 420
equilibrium line 403
erratic 413
esker 413
ETP curve 421
firn 400
fjord 408
glacial grooves 407

glaciation 400
glacier ice 401
glacier, valley, mountain, Alpine, piedmont, continental, surging, polar, temperate 400, 402, 403
hanging valley 408
horn 409
ice sheet, ice shelf, ice cap 402, 403
kame, kame terrace 413
kettle 414
Milankovitch curve 421
moraine, end, terminal, ground, interlobate, medial, lateral, recessional 411, 412, 413
névé 406
nivation 409
obliquity of the ecliptic 420
outwash, outwash plain 413
plucking 407

polish 407
precession of the equinoxes 421
pressure melting 000
quarrying 407
rock flour 407
snow line 400
snowfield 400
striation 407
sublimation 400
tarn 409
till 411
truncated spur 407
U-shaped valley 407
valley train 413
varve 413
zone of ablation, accumulation, flow, fracture 403, 405

QUESTIONS FOR REVIEW AND THOUGHT

16.1 How does glacier ice form?
16.2 Why can glacier ice be considered to be a metamorphic rock?
16.3 What are the different types of glaciers?
16.4 What is the difference between a polar and a temperate glacier?
16.5 How does a glacier's budget work?
16.6 How do glaciers move?
16.7 How do glaciers erode?

16.8 What are the features of erosion caused by glaciers? How do they form?
16.9 What are the two main types of glacial deposits?
16.10 What are moraines? How do they form?
16.11 What are drumlins? Boulder trains? How do they tell the direction of ice movement?
16.12 What are some features made of glacier outwash? How do they form?
16.13 What are kettles? How do they form?

16.14 What are some indirect results of glaciation?

16.15 What role does plate tectonics play in glaciation?

16.16 What is the astronomic theory of glaciation?

Critical Thinking

16.17 You are assigned the task of mapping the glacial geology of an area. You know that some features will have been formed by water. How will you distinguish between these and features formed by deposition directly from the ice?

16.18 You have been asked to locate sand and gravel deposits that could be used in a major highway project across southern Wisconsin. What would you look for? What factors, other than location of the deposits, will have to be considered before the resources can be used?

16.19 You find that your friend comes from the state of Michigan. You remark that the entire state was once covered by a glacier. He doesn't believe you. How would you go about proving that Michigan was indeed covered by an ice sheet? Assume you have plenty of time and an unlimited travel budget for two.

17

Storm Surf, Timber Cove, California

Coasts and Coastal Processes

OBJECTIVES

▲▲

*As you read through this chapter, it may help if you focus
on the following questions.*

1. How do wind-formed waves move toward the shore and how is their energy distributed along it? What are the ways that materials are transported along the shore? What roles do wind, animals, and plants have along the shore?
2. What are the different types of coasts, how are they formed, and what are some of their characteristics?
3. How do people protect against potentially harmful coastal processes?

OVERVIEW

▲▲

The shore is shaped primarily by wind-driven waves. The energy of these waves carves cliffs and transports material to form beaches. The work of waves is aided by local winds moving sand along the beach, and by plants and animals interacting with the coast. The change in position of sea level determines where the coastal processes will be focused.

Different types of coasts define the margins of the continents. Some coasts have been heavily glaciated in the past. Others characterize the trailing edge of a continental margin. Still others develop where plate boundaries affect the continental edge. These different types of coasts all have their own unique features and forms.

Hundreds of millions of people live along, or close to, the world's shores and can be affected by the coastal processes. Some inhabitants choose engineering works to protect themselves and their property from the ravages of storm and flood. Others retreat to safer places.

Coastal Processes

The **coast** is a narrow strip of land along the margin of the ocean and extends inland for a variable distance from low water mark. The **shore** is the seaward edge of the coast, is alternately covered and exposed by the tides and waves, and thus lies between low tide and the extent of effective wave action. The **shoreline** is the line separating land and water and its position fluctuates as water rises and falls. Lakes as well as oceans have shores, but the term *coast* refers to the zone that includes an oceanic shore.

Wind-Formed Waves

Wind-formed waves (Figure 17.1) dominate the processes shaping the shoreline. They are produced purely by the stress and friction created as wind moves across the water. The harder and longer the wind blows, and the greater the distance over which it blows (called the **fetch**), the higher the water piles up into waves. The distance between two successive waves is the **wavelength,** and the vertical distance between the **wave crest** and the bottom of an adjacent **wave trough** is the **wave height** (Figure 17.2). These wind-formed waves persist even after the wind that formed them dies. Such waves, broadening and flattening as they move, are **swells,** and they may travel for thousands of kilometers from their zone of origin.

Although waves can travel great distances in the open ocean, it is the wave form that actually moves. In deep water, the water in the waves is not really transported. Instead, water within the wave has a circular motion. Water in the wave moves forward on the crest of the wave, slips backward as the following trough nears, and moves forward as it rises with the next

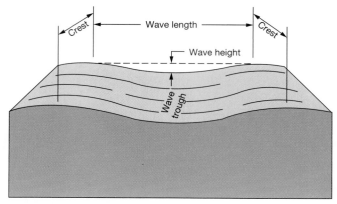

FIGURE 17.2   Diagram to explain terms used in describing water waves.

approaching crest. The diameter of this orbit is equal to the wave height. The diameter of the water's circular orbit decreases beneath the surface. At a depth equal to one-ninth the wavelength the orbit has decreased to a diameter of one-half the wave height. By a depth equal to about one-half the wavelength, the motion is negligible. This depth is the **wave base** (Figure 17.3). Below this level wave motion has practically disappeared, which explains why a submarine can ride out the worst surface storms by submerging to the quiet waters below wave base.

As the wave leaves deep water and advances toward the shore, several important changes take

FIGURE 17.3   Water particles at the surface in deep water move in virtually circular orbits whose diameter approximates the wave height. Water particles move forward under the crest and backward under the trough. Their paths describe orbits with a diameter equal to the wave height. These orbits decrease in diameter rapidly downward until, at a depth of approximately one half the wavelength, motion is negligible. This depth is known as *wave base*. Only a few water particle orbits are shown

FIGURE 17.1   Waves, the most important agents in fashioning the shoreline, are caused by the friction of the wind moving across the surface of the water. Here a series of waves break along the shore.

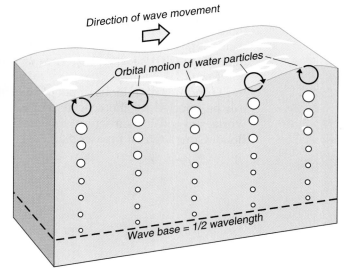

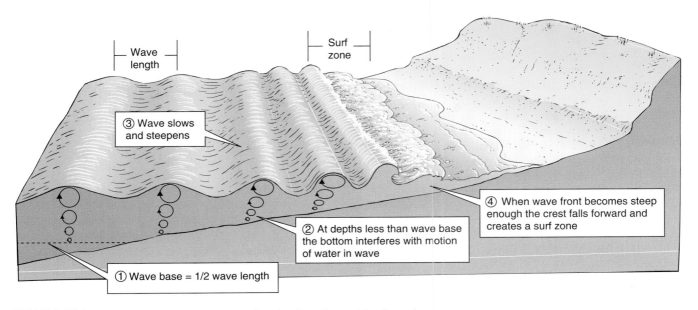

③ Wave slows and steepens

Wave length

Surf zone

② At depths less than wave base the bottom interferes with motion of water in wave

④ When wave front becomes steep enough the crest falls forward and creates a surf zone

① Wave base = 1/2 wave length

FIGURE 17.4　Changes in a wave as it approaches the shore. See text for discussion.

place in the form and velocity of the wave, which affect the water in the wave and are important to shoreline processes. When the overall depth of the water shallows to wave base, the ocean bottom begins to interfere with the circular motion of the water in the wave. As water depth decreases further, the motion of the water is forced to become increasingly elliptical. The velocity of the wave decreases, as does the wavelength, and the front of the wave becomes steeper. When the wave front becomes steep enough the crest falls forward and breaks, producing **surf** in a zone usually just outside the shoreline (Figures 17.4 and 17.5). At this moment water within the wave is thrown forward against the shore. The energy thus released is

then available to erode the shore and to set up currents that move water and sediments along the shore.

As waves break against the shore they erode by the hydraulic pressure and turbulence they generate. As waves strike a headland, for example, a cliff forms. Continued wave action cuts away at the cliff at its base, and, as a result, mass movements dump debris at the water's edge. The cliff continues to retreat, and waves continue to sweep away the debris eroded from the cliff. The result is an erosional platform called the **shore platform** (Figure 17.6) which slopes gently seaward to the low-tide mark. The platform is usually covered with a thin veneer of debris, which is being transported out to sea or laterally along the shore. A

FIGURE 17.5　Wind-driven waves are the major source of energy for erosion of the shore and for transportation of material along it. Here waves break against the Tyrrhenian shore near Ansedonia, Italy. *(Pamela Hemphill)*

FIGURE 17.6　Wave erosion on a headland erodes a shore platform, an erosional feature, which slopes gently seaward. At its shoreward edge it terminates at the base of a sea cliff. This example is from Australia. *(William C. Bradley)*

fully developed shore platform may have a breadth of several hundred meters.

Most waves approach the shoreline at an angle, but they tend to bend as they move into shallower and shallower water. This bending is called **refraction,** and it causes waves to strike the shore nearly head-on. The effect of refraction can best be seen on a relatively straight stretch of shore that waves approach at an angle over a bottom that grows shallower at a constant rate (Figure 17.7). As a wave crest nears the shore, the section in the shallower water close to shore first feels the effect of the upwardly sloping bottom. This section of the wave thus slows while the seaward part continues at its original speed. The effect is to swing the wave sharply toward the shore and focus its energy directly onto the shore.

Refraction also helps to explain why, on an irregular coast, the greatest energy is concentrated on headlands and the least in the bays, and why the wave crests generally conform to the shape of the bay. In the diagram in Figure 17.8 a wave advances directly on an irregular shoreline. A portion of the wave is divided into two equal segments, A-B opposite a bay, and B-C opposite a headland. Because the bottom is shallower off the headland than off the bay the wave segment B-C is bent more and more toward the headland. At the same time the segment A-B is stretched out along the bay. Each segment, however, starts out with the same amount of potential energy. By the time they reach the shore, however, more energy is concentrated per unit length of shore on the headland. Around the bay the shore receives the same total amount of energy but it is less per unit length of shoreline. The energy is dis-persed around the bay and focused on the headland. The headland, therefore, is a zone of high erosion, and the bay becomes a place for deposition, here shown as a beach. Figure 17.9 is an example of the way a wave front spreads out around a bay, conforming to its general outline.

Despite refraction, most waves tend to hit the shore at a slightly oblique angle. Excess water thus piles up in the surf zone and must escape in some way. To do this, the water flows in a current parallel to the shore, just inside the line of breaking waves. This is called a **longshore current** or **littoral current.** Breaking waves add more and more water to the current until it must escape seaward. It then breaks through the surf zone as a **rip current,** which carries the excess water away from shore (Figure 17.10). Velocity in a rip current may be high (1–2 m/s) and dangerous to swimmers.

Longshore currents move sediments parallel to the shore over long distances. In places where wave and current energy is low enough, sediments are deposited along the shore to form **beaches.** These are temporary accumulations of usually well-sorted sediments ranging in size from clay-sized particles to boulders that collect between the low- and high-tide marks. Rip currents carry beach material seaward where it is picked up by incoming waves and brought back to the beach or lost to deeper water.

In addition to longshore and rip currents, material along the shore is also moved by a process known as **swash and backwash.** As a wave breaks at an angle on the shore, water rushes, or swashes, up the beach and then retreats as it washes back seaward. If the wave

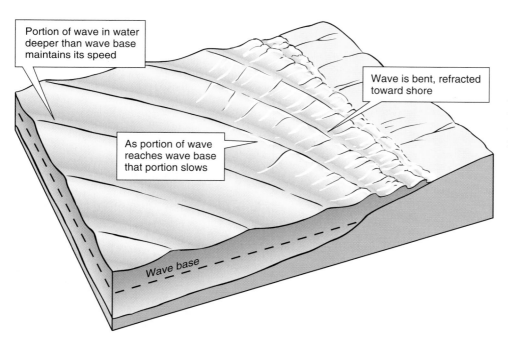

Portion of wave in water deeper than wave base maintains its speed

As portion of wave reaches wave base that portion slows

Wave is bent, refracted toward shore

Wave base

FIGURE 17.7
Wave crests, which advance at an angle on a straight shoreline and across a uniformly shallowing bottom, bend shoreward, as suggested in this diagram. Refraction is caused by the increasing interference of the bottom with the orbital motion of water particles within the wave as it nears the shore.

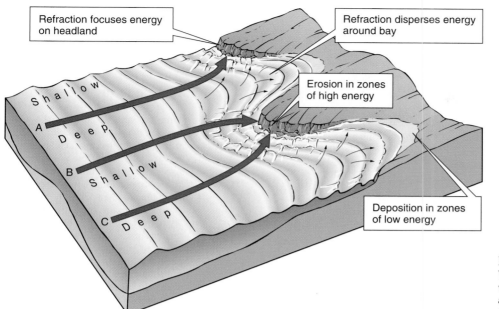

Refraction focuses energy on headland

Refraction disperses energy around bay

Erosion in zones of high energy

Deposition in zones of low energy

Shallow

Deep

Shallow

Deep

A

B

C

FIGURE 17.8
Wave refraction helps explain why wave energy is high on headlands and low in adjoining bays. See text for discussion.

breaks on the beach at an acute angle then the swash will wash over the beach at a similar angle. Water then flows back down the beach at right angles. Therefore, sand and pebbles caught in the swash and in its backwash travel along the beach in a saw-toothed path.

Most beaches have the features diagrammed in Figure 17.11. They are concave from the high-tide mark down to a depth between 5 and 20 m. This is called the **shoreface,** and at its lower extent it grades into a gently seaward-dipping planar surface known as the **ramp.** Toward the base of the shoreface linear **troughs** and **bars** form as a result of breaking waves. The **foreshore** lies between the high- and low-tide marks, and the **backshore** from the high-tide mark to the base of a sand dune. In the backshore area are one

FIGURE 17.9   This bay, Porth Oer, in northwest Wales, lies between two headlands. Refraction has turned these waves almost parallel to the shore.

or more **berms,** resembling small terraces with low risers on their seaward sides.

Wind, Plants, and Animals

In addition to generating waves, wind plays another role along the shore. When the beach is dry, wind moves sand grains along it. Many of these grains move within reach of swash, long shore currents, and surf. Many others, however, are blown inland off the beach, where they are trapped by vegetation to form dunes. In time these shore dunes, covered with a sparse but hardy vegetation, can build into formidable barriers between the sea and the land beyond (Figure 17.12). They become a characteristic feature of the beachscape and provide an effective protection to the beach and to the land immediately behind it.

Plants also play other roles in the coastal process. In quiet waters, saltwater plants may establish themselves, trap sediments, and help to form a saltwater marsh or swamp. Plants along the shore are characteristically grasses, reeds, and rushes, but in many tropical and subtropical areas mangrove swamps thrive along the shores.

Animals also play a part in building shores. For instance, the white and pink sands of the warmer ocean shores are usually composed of the carbonate fragments of shells and corals.

Changing Level of the Sea

Many people think of sea level as being a constant level. Actually, in addition to the daily tidal changes, it may rise or fall, slowly or suddenly. As it does so it

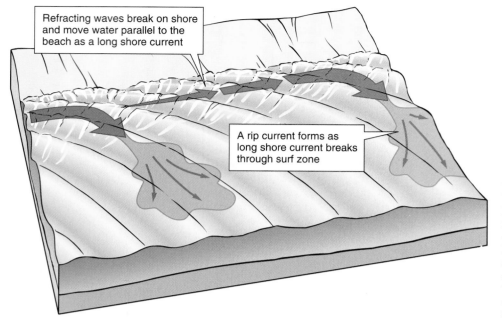

Refracting waves break on shore and move water parallel to the beach as a long shore current

A rip current forms as long shore current breaks through surf zone

FIGURE 17.10
A diagram to illustrate one origin of longshore and rip currents. Waves striking the shore at an oblique angle pile up water that must escape back toward the sea in some way. The excess water first moves parallel to the shore as a longshore current. Each breaking wave down the shoreline increases the volume and velocity of the current. Eventually this current breaks out to sea through a low spot in the shore or a weaker section of the surf zone. This is a rip current, a narrow, rapid stream of water that eventually spreads and dissipates at some distance from shore.

displaces the position of the shore. It is quite clear that if the net motion of sea level is upward then the position of the shore will not only be displaced upward but also will migrate landward. Conversely, a drop in the sea relative to the land will have the opposite effect: The shore will move downward and seaward. The amount of shoreline displacement is generally much greater horizontally than vertically. For example, if a shore platform has a seaward dip of about 1°, then a 1-meter drop of sea level will displace the shore out across the platform about 60 m seaward of its previous position. When the shore is displaced the coastal processes of erosion, transportation, and deposition begin to work on the new shore. Shore platforms are eroded, cliffs cut, sediments transported, and beaches built.

There are several ways by which the sea level changes and the shore is displaced. A **eustatic** change of sea level is worldwide. One cause of eustatic change of sea level has been the repeated waxing and waning of ice sheets during the Pleistocene. During cold peri-

ods, as water evaporated from the ocean and helped to form the ice sheet, sea level fell between 100 and 125 m. The result was the withdrawal of the sea from a large percentage of the continental shelves, shifting the shore seaward more than 200 km along some coasts. When climate warmed, glaciers melted and returned water to the ocean, and sea level rose, pushing the shore back across the margins of the continents.

Changes in the depth of the world's oceans, which also cause eustatic changes in sea level, are most commonly due to changes in the rates of seafloor spreading. With rapid seafloor spreading more lava and heat raise the spreading ridges, shallowing the deep ocean a bit and forcing water across unflooded continental margins. A decrease in the rate of seafloor spreading allows cooling of the oceanic crust and its concurrent contraction and subsidence. The result is to deepen the ocean basin and drain water from the continental margins.

Isostatic changes in sea level occur in response to changes in load on the Earth's crust as discussed in

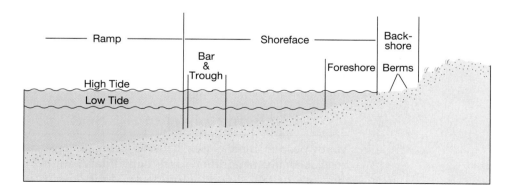

FIGURE 17.11
Terms used in description of beaches.

Ramp — Shoreface — Back-shore

Bar & Trough

Foreshore | Berms

High Tide

Low Tide

FIGURE 17.12 Many of the long beaches along the eastern and gulf coasts of the United States are backed by sand dunes. This dune along the beach on the eastern coast of Cape Cod is among the most impressive of these features. Storm waves have eroded its seaward side. It is replenished by sand blown off the beach and trapped by sparse dune vegetation. The dune stands over 12 m above the beach. (Rick A. Ross)

Chapter 11. There the example used was the rise of land around Hudson's Bay as the last ice sheet melted.

Isostatic and eustatic changes in sea level, and hence the changes of position of the shore, are relatively slow compared with the sudden movements along tectonically active coasts where they can be almost instantaneous. For example, during the 1964 Good Friday earthquake in Alaska, the shoreline at Whittier, Alaska, subsided 1.8 m, and on Montague Island in places the shoreline emerged nearly 10 m.

Predicting future changes in sea level, and thus the changing position of the shore, is full of uncertainty. There are, however, some guides as to what to expect about sea-level changes. For example, the odds are high that major glaciations will follow in much the same pattern as those of the last million years or so. We can expect, therefore, an ice advance with an accompanying fall of sea level of 100 m or more 5,000 to 10,000 years in the future. That change should be followed 90,000 to 100,000 years later by glacier melting and a return to current levels. Present generations, naturally, won't be around for either of these events. However, if the predictions of many scientists are correct, some current generations will see a significant rise in sea level during their lifetimes.

During the last hundred years the global rise in sea level has been between 10 and 30 cm. The rise has been matched by the increase in global temperatures during this time. Concurrent with these changes has been an increase in **greenhouse gases** in the atmosphere. These gases, products of increasing industrialization, trap heat energy in the atmosphere. Many scientists see them as responsible for the increase in worldwide temperature. They feel that the predicted increase of these gases will lead to **global warming** and to an accompanying rise in sea level. Some of the expected consequences of a rise in sea level, whatever its cause, include:

- increased beach erosion;
- decrease in wetlands area, particularly if sea-level rise is rapid;
- damage to structures on low-lying coasts;
- saltwater invasion of groundwater supplies (see Figure 15.33); and
- saltwater invasion of coastal rivers and pollution of the freshwater supplies of adjacent cities.

Types of Coasts

Barrier-Beach Coasts

From Cape Cod, southward to Florida and on around the Gulf coast to the U.S.-Mexico border and beyond, the North American coast is backed by a low-lying coastal plain underlain by easily eroded, gently seaward-dipping, sedimentary deposits sloping out beneath the sea onto a wide continental shelf. This is the passive margin, or trailing edge, of the North American continent. It is characterized by a series of sandy beaches, separated in many places from the mainland by shallow lagoons. In these locations they serve as barriers against storm waves and are called, appropriately, **barrier beaches.** These beaches bear the full brunt of major storms including hurricanes, tropical storms, and in the eastern United States, damaging nor'easters. They are the most dynamic sections of the eastern North American shore, and the continuous changes that occur along these coasts are often sudden and catastrophic. Yet the beaches are magnificent, and they constitute a major recreational resource and protect a vast area of wetlands. In addition, many sections, such as Miami Beach, Florida, and Ocean City, Maryland, are essentially urban.

These barrier beaches, sometimes called **barrier islands,** are relatively narrow, a few hundred meters to a kilometer or two in width, but are several kilometers to tens of kilometers long (Figure 17.13). Commonly, a barrier beach that is not fully attached to the mainland may be attached at one of its ends. On its landward side a shallow lagoon stands between it and the mainland. The lagoons have access to the sea through tidal inlets.

The coastline of northeastern New Jersey illustrates some of the features of barrier beaches along the eastern shores of the United States (Figure 17.14). The Asbury Park–Long Branch section of the coastline is a zone of erosion formed by the destruction of the broad

FIGURE 17.13   A barrier beach along the Delaware shore stands out as a bright narrow ribbon. Shallow lagoons separate the beaches from the mainland. They have access to the ocean through tidal inlets, as does Indian River lagoon in the foreground. A plume of sediment marks the entry of the muddy water of the lagoon into the Atlantic Ocean. *(Terraphotographics/BPS)*

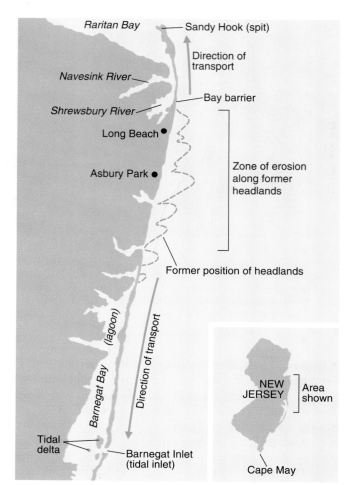

FIGURE 17.14   This map of northeastern New Jersey shows some of the features along a barrier beach coast and its adjacent headland. See text for discussion. *(After an unpublished map by Paul Mac-Clintock)*

headland along this part of the coast, where the sedimentary rocks are easily cut by the Atlantic waves. The material eroded from this section is moved both north and south along the coastline by longshore currents. Much additional sediment is contributed to these currents from just off shore as storm waves move sediment shoreward. Sand swept northward is deposited in Raritan Bay and forms a long sandy beach projecting northward, a **spit,** known as Sandy Hook. Just south of Sandy Hook the flooded valleys of the Navesink and Shrewsbury rivers are bays that have been almost completely cut off from the open ocean by sandy beaches called **bay barriers.**

Sand moved southward from the zone of erosion has built up another sandy beach. Behind it lies the shallow lagoon of Barnegat Bay that receives water from the sea through Barnegat Inlet, a **tidal inlet.** This passage through the beach was probably first opened by a hurricane. Just inside the inlet a delta has been formed of sediment, partly during the original breakthrough of the bar and partly by continued tidal currents entering the lagoon. Geologists call it a **tidal delta.** Longshore currents continue southward along the shore to Cape May and Delaware Bay.

Hurricane Hugo, which struck the South Carolina coast just north of Charleston on September 22, 1989, demonstrated the destructive nature of Atlantic and Gulf Coast storms. Hugo made landfall on the Carolina coast with sustained winds of over 200 km/hr. The barrier islands were submerged by high waters, buildings were swept away or destroyed, and flooding extended several kilometers inland. Erosion pushed back the beach front an average of 30 m and in some places as much as 45 m. As experience with earlier

hurricanes had shown, the greatest damage along the shore occurred where the barrier island was narrow and protective dunes were low or lacking.

Smith Island, Virginia, off the southern tip of the Delmarva Peninsula, is an excellent example of what can happen over time to a barrier island. Between 1852 and 1980 the entire island—beach, dune, and lagoon side—apparently has moved 700m westward toward the mainland. Actually, the island did not move as a body. As hydrographic surveys in 1852, 1921, and 1954 showed, the shoreface of the beach retreated, the ramp followed it, extending the old ramp section shoreward. Erosion on the seaward side of the island was made up for by deposition on the side toward the mainland. Tidal gauges and releveling surveys suggest that the rate of sea-level rise is large in the area and that sea level rose at least 30 cm from 1852 to 1980. Taking this, and other examples, geologists can suggest that, with time and a rising sea level, a barrier beach and its

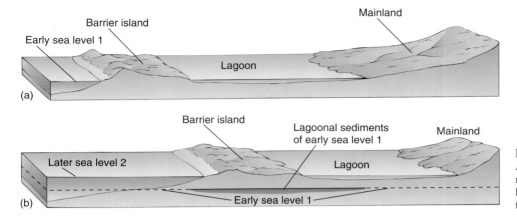

FIGURE 17.15
A diagram to suggest the landward migration of a barrier beach and its lagoon with a rise of sea level. See text for discussion.

lagoon can follow a pattern of shoreward migration, as shown in Figure 17.15.

Glaciated Coasts

Glaciated shorelines, particularly those underlain by resistant rocks, have irregular outlines, often with steep headlands with protected bays and beaches. Also, numerous rocky islands tend to lie just off the main shorelines of these coasts. Many glaciated coasts also have long, narrow fjords, as discussed in the last chapter. Repeated glaciations of the United States and Canada, from New York northward, have fashioned North America's northeastern coast, which is charac-

terized by long estuaries, bays and inlets, many rocky islands, cliffed headlands, and gravelly beaches (Figures 17.16 and 17.17).

Small beaches, called **pocket beaches,** are cradled at the head of many of the coves and bays. The sediment on these beaches, which is frequently gravel and boulders, comes from glacial till and outwash that have been reworked by waves and from the slow erosion of resistant cliffs. Saltwater marshes are found along small inlets, the mouths of streams, and in other protected spots. Unlike the long, sandy barrier islands farther south on the North American coast, the northeast islands are the unflooded high spots of an irregular, preexisting topography. The resistance of the rock that forms them ensures that wave erosion is slow.

Deltaic Coasts

The location of the major deltaic coasts are shown in Figure 17.18. Some, such as those of the Nile, Rhine, Ganges, Mekong, and Mississippi are among the most

FIGURE 17.16   A satellite view of the New England Coast from Penobscot Bay to the Canadian–U.S. border illustrates the irregularity of a previously glaciated shoreline now flooded by the ocean. In this infrared imagery, vegetated areas show up in various tones of yellow and red. *(NASA)*

FIGURE 17.17   The New England coast along the Penobscot Bay, Maine.

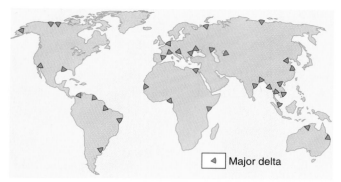

FIGURE 17.18   Major deltas are significant features of the world's shoreline.

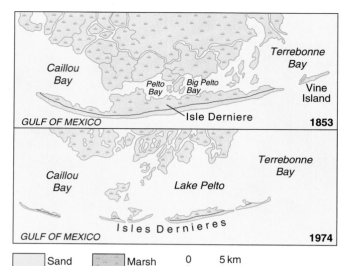

Sand   Marsh   0   5 km

FIGURE 17.19   Between 1853 and 1978 Terrebonne Parish, Louisiana, has lost a large amount of its wetlands to the sea, as shown in these two maps. *(Penland, Shea, Ron Boyd, Dag Nummedal and Harry Roberts, "Deltaic Barrier Development on the Louisiana Coast," Transactions, Gulf Coast Association of Geological Societies, vol. 31 supplement, p.475, Fig. 11, 1981.)*

heavily populated regions in the world. All are susceptible to flood and wave damage. For example, the coast of Louisiana has a higher rate of shoreline erosion and land loss than any other section of the North American coast. Scientists estimate that each year the sea claims 130 km². The greatest annual loss is currently along the shore of Terrebonne Parish in the western delta (Figure 17.19). A sobering prediction is that, in 100 years, at the current rate of loss, the sea will have consumed the entire parish, leaving only outdated maps as its record.

The slow flooding of the Mississippi Delta provides us with an excellent example of how complex sea-level fluctuation can be and the way many different factors enter into the ups or downs of sea level and the position of the shore. The single most important cause of the sea's transgression of the Mississippi Delta is the subsidence of the delta. Over the last 7,000 years the Mississippi River and its distributaries have built the Mississippi Delta out into the Gulf of Mexico. As the fine sand, silt, and clay carried from the continent have spread across the delta their weight has compacted the underlying sediments. In addition, the great weight of the delta has been bowing down the Earth's crust isostatically. In order to keep up with this subsidence new sediments must be added at the surface. In the normal course of events nature would probably provide for this. But humans have interfered with sediment renewal by building dams and levees along the river system thereby trapping sands and muds in artificial lakes and behind artificial river embankments.

Other factors conspire in the loss of land in the Mississippi Delta. In the delta area the annual rise of sea level is over 1 cm on average. Of this, 5 to 10 percent is attributed to the return of water to the ocean by melting glaciers and by the thermal expansion of the ocean due to postglacial warming. This rise in sea level has made it easier for the hurricanes and tropical storms to destroy barrier islands and the wetlands they protect. Human activity on the delta contributes further to the loss of delta land. Canals dredged by petroleum companies disrupt patterns of water flow and erode and enlarge areas of open water. Extraction of petroleum and subsurface brines can also contribute to subsidence.

Tectonically Active Coasts

The western margin of the North American continent is tectonically active, and this activity affects the nature of the shoreline. Convergent, divergent, and transform plate boundaries are all involved (Figure 17.20).

From the Guatemala-Mexico boundary northward for 2,000 km, the North American coast parallels the trace of a convergent plate boundary between the American plate on the east and the Cocos plate on the west. This gives way to the youthful spreading axis of the Gulf of California, shaped like a long finger pointing to the mouth of the Colorado River. Mainland Mexico and the scraggly peninsula of Baja California are moving away from each other on either side of the divergent boundary, which is thus similar to the Red Sea rift. The delta of the Colorado River dominates the north end of the gulf.

From southern to northern California the coast parallels the transform faults of the San Andreas fault

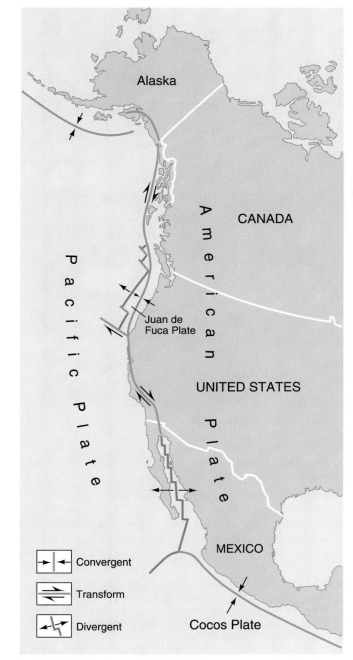

FIGURE 17.20 Plate boundaries of western United States affect the direction and nature of the coast. See text for discussion.

system. In contrast to the low, sandy barrier beaches and lagoons that characterize much of the eastern and gulf coasts of the United States, great lengths of the California coast are marked by bold, tectonically raised cliffs where few beaches or protected bays exist (Figure 17.21). At their feet waves are cutting shore platforms. Associated with these are **arches** (Figure 17.22), cut through headlands by wave erosion and mass movement, and **stacks** (Figure 17.23), rock resid-

FIGURE 17.21 The rugged coast of California, south of Monterey, where the edge of the Santa Lucia Mountains comes down to the sea, provides little protection from Pacific storms. *(Rick A. Ross)*

uals left as cliffs retreat. Many of the cliffs rise to terrace levels, which are the abandoned and uplifted shore platforms that had been cut at lower sea levels (Figure 17.24).

Along the California coast, San Francisco Bay provides sheltered harbors and access from the sea to inland areas. It is the result of the Pacific Ocean flooding through a narrow opening, the Golden Gate, into the lowest portion of a tectonically formed valley. This valley parallels the main coastline and faults of the San Andreas system. Before the modern bay was formed, the ancestral Sacramento River drained this valley and the much larger Central Valley of California at the western foot of the Sierra Nevada. The Sacramento reached the sea via a water gap it had cut as it maintained its course through a rising coastal range. Later, when land sank low enough and the sea rose high enough, marine water flooded through the water gap and into the lower portions of the Sacramento drainage to form what is now San Francisco Bay and the Golden Gate (Figure 17.25). Today, shores of the

FIGURE 17.22   The sea has cut an arch through this promontory along Monterey Bay, California.

bay are heavily modified by human activity, as illustrated, for example, by the San Francisco International Airport, built out into the bay on artificially constructed land.

From northern California to southern British Columbia, the small Juan de Fuca plate is converging with, and thrusting beneath, the North American plate. The bold, cliffed coasts of Oregon and Washington facing the open Pacific are similar to parts of the California coast. They are locally dotted by stacks and actively forming shore platforms and are backed by wave-cut cliffs (Figure 17.22). North of the Juan de

FIGURE 17.23   Rocks and pinnacles, called stacks, are residuals of erosion left behind as the ocean at Bandon Beach, Oregon, pushes back the sea cliff. (R. Kolar, Earth Scenes)

FIGURE 17.24   The California coast is tectonically active and has been rising in relation to sea level. In many places this is recorded by old shore platforms now emerged from the sea to form marine terraces, as seen here in the flat-surfaced features on San Clemente Island off southern California.

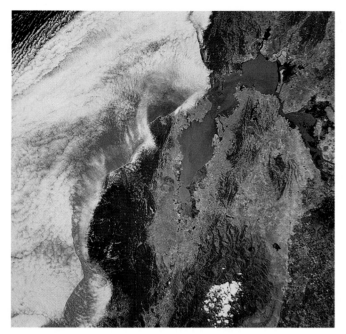

FIGURE 17.25   Coastal California in the San Francisco area. See text for discussion. *(NASA)*

Fuca plate the Queen Charlotte-Denali transform fault system parallels the coast until it converts into the convergent boundary of southern Alaska and the Aleutian Islands. As with the northeastern coast of the continent, repeated glaciations during the Pleistocene have scoured this Canadian-Alaskan coast. Fjords, long straits, narrow channels, and islands, large and small, abound. Unlike the northeastern coast, however, geologically recent tectonic activity has produced a much more rugged coastal zone here than is found in the east. Furthermore, the northwest-southeast orientation of the major coastal landforms reflect the major tectonic pattern of the plate boundary. In addition, in some places, glaciers still come down to the shore.

The Aleutian Islands have their own characteristic coasts. The islands are a still-active chain of volcanoes, a result of the continued thrusting of the Pacific plate beneath the American plate. The coast is dominantly one of volcanic construction. Wave erosion has cut steep cliffs on exposed headlands. Drowned glacial valleys provide sheltered segments and pocket beaches along the coasts.

Coral Reef Coasts

The coral reef coast is organically formed, being a mass of limestone built of both the skeletal remains of corals and the products of calcite-depositing algae. Corals thrive best in tropical and subtropical waters whose winter temperatures do not fall below 18°C, where ocean salinity is normal, and where the water is clear, conditions reflected by the modern distribution of coral reefs. Corals also need a firm base on which to grow and spread. Once they have started to grow and spread, corals will grow to the average level of low tide, and calcareous algae will join them to carry the reef up a meter or more higher into the zone of breaking waves, the surf zone. Large and violent storms can submerge a reef and deposit wave-broken coral and coral sand behind it. In time a low island can develop from this debris, high enough to support vegetation and a lens of fresh water. The major reef types are termed a **fringing reef,** attached directly to a mainland; a **barrier reef,** separated from the mainland by a shallow lagoon; and an **atoll,** a roughly circular reef with an occasional small, low islet, surrounding a shallow lagoon.

The origin of an atoll was suggested by Charles Darwin over 150 years ago. He proposed that the fringing reef bordering a volcanic island developed into a barrier reef and finally into an atoll. His explanation, with some modification, is still valid today. Present-day understanding is illustrated in Figures 17.26 and 17.27 and outlined below.

• A fringing reef establishes itself around a volcanic island.

FIGURE 17.26   In (a) a fringing reef forms around a volcanic island to initiate the process of atoll formation. (b) As the seafloor subsides the coral reef grows upward and forms a barrier reef separated from the sinking volcano by a lagoon. (c) Finally, the volcano disappears below sea level, and as it does the coral continues to grow upward to create a reef surrounding a lagoon.

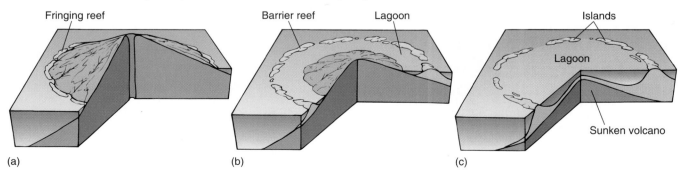

(a)   (b)   (c)

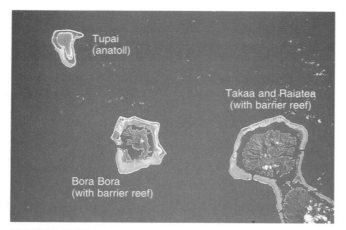

FIGURE 17.27 This northwestern group of the Society Islands is over 3,000 km south of Hawaii in the South Pacific. The atoll in the upper left is Tupai. The small volcanic island surrounded by a barrier reef is Bora Bora. The two larger volcanoes are Tahaa and Raiatea and are encircled by a common barrier reef. See text for discussion. *(NASA)*

- The volcano and its reef subside slowly as seafloor elevation drops off along a hotspot track or as it moves away from a spreading center. As they sink, the coral reef is able to grow fast enough to keep its growing, seaward edge close to sea level. A lagoon forms between the reef and the disappearing volcanic island.

- As subsidence continues the reef continues to grow upward. The volcano, unless it renews activity, eventually sinks below sea level and is buried in the coral and coral debris of a shallow, interior lagoon ringed by the outer reef. An atoll has formed.

There is evidence for subsidence of volcanic islands elsewhere, such as in the evolution of the Hawaiian-Emperor seamount chain and in the wave-beveled, now-submerged, flat-topped seamounts called guyots (Chapter 11). There is even further proof that atolls are attached to sunken volcanoes. A number of geologic drill holes have punched deep below the surface of atolls. At Eniwetok atoll, in the Marshall Islands, the drill hit basalt after going through over 1.4 km of coral rock. It had taken over 50 million years for the slowly subsiding volcano to accumulate its coral cap. On the nearby Bikini atoll the basalt lies beneath more than 1.3 km of coral.

Human Responses to Changing Shorelines

▲▲▲▲▲▲▲▲▲▲▲▲▲▲▲▲▲▲▲▲▲▲▲▲▲▲▲▲▲▲▲▲▲▲▲▲▲▲

Because so many millions of people live, work, and play along or close to the shore, coastal changes understandably invoke concern. Changes causing the most alarm are, not surprisingly, the loss of beaches and the land immediately behind them. Such changes mean loss of property, buildings, recreational beaches, tourist trade, public utility installations, and loss of spawning grounds for shellfish and finfish, and a habitat for bird and beast.

There are only two basic ways of handling the problem of shoreline erosion and the landward push of the shoreline: Either protect the shore with engineering projects, including structures and beach replenishment or admit defeat and retreat.

Groins, Jetties, Breakwaters, and Seawalls

A **groin** is a long wall built out from the shore, usually perpendicular to it. Its purpose is to nourish the beach by trapping sand carried parallel to the shore by longshore currents. Groins are often built in sets so that, in effect, they form a series of cells or giant sandboxes. As sand moves parallel to the beach it is caught on the groin's updrift side, and when it spills around the end it is caught behind the next groin downdrift. A major problem encountered in the use of groins is that by slowing the supply of sand moving along the beach groins can cause erosion of unprotected beaches downdrift (Figure 17.28). **Jetties** are similar to groins and are built out from the shore to protect the entrance to a harbor or inlet. They are often constructed in pairs, one on either side of the entrance. Like groins, however, jetties usually cause beach erosion along the shore downdrift from them.

A **breakwater** is a wall built offshore and generally parallel to it. It affords protection to the water and shore behind by breaking the force of the waves. The breakwater off Santa Monica pier (Figure 17.29a) was built to provide protection for small boats. Figure 17.29b shows the changes this produced. Sand was deposited on the shore behind the breakwater. The breakwater then offered protection for the expanded beach, which reduced the area open to small boats. Deposition was caused by the refraction of waves around the breakwater and the consequent disruption of the longshore drift of sediment. The use of breakwaters is far from new, as illustrated in Figure 17.30, a Roman breakwater along the Costa Brava, northeastern Spain.

FIGURE 17.28 Diagram to show the effect of groins along a beach.

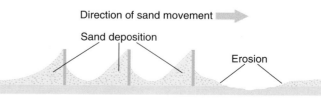

(a)   (b)

FIGURE 17.29   (a) The Santa Monica, California, pier in 1931. (b) In 1933 a breakwater was built
to provide protection for small craft. Waves, refracted by the breakwater, were spread out behind it.
By 1949 extensive deposition had occurred along the beach in the lee of the breakwater. *(Photos,
Fairchild Aerial photography collection, Whittier College.)*

As a beach is narrowed by erosion and provides less and less protection to the buildings, roads, and other installations behind it, a **seawall** may be built. This is a wall built to stop further retreat of the shore and to protect buildings and other structures from damage by waves or water (Figure 17.31). A seawall works as long as it is kept in repair and is not subjected to storms stronger than those for which it was designed. A seawall will save, at least temporarily, the area behind it. However, any beach that remained when the wall was built will not last long. The line of the seawall defines the shoreline.

The Netherlands well illustrates the effective use of engineering projects to protect land, structures, and people from the sea. Over 40 percent of the country once lay beneath sea, swamps, and lakes. Over the centuries it has been drained and the sea kept at bay

FIGURE 17.30   Roman engineers built these breakwaters for the port of the Roman imperial town of Ampurias in what is now north-eastern Spain. Note that the effect they have had on the beach is very similar to that of the Santa Monica breakwater of Figure 17.29. *(John Sutcliffe)*

FIGURE 17.31   A seawall protects land and buildings along the northern N.J. shore at Long Branch. The Army Corps of Engineers has undertaken to build a 42 m wide beach here and along over 50 km of adjoining shoreline north and south. It is estimated that the Long Branch section will cost about $80 million a mile to build and maintain over the next 50 years.

Perspective *17.1*

The Delta Project of the Netherlands

*T*he compound delta built by the Rhine, Scheldt, and Maas rivers includes much of the Netherlands. Higher portions of the delta have long been pro- tected by dikes. Like other deltas, however, it is becoming lower in relation to sea level because of com- paction of the deltaic sediments, the nonrenewal of sediments at the surface, and the eustatic rise of the sea. As a result, disastrous floods had long been common to the delta area and the Nederlanders were in increasing danger of losing land

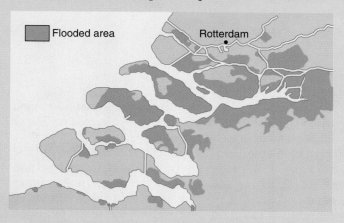

P17.1.1 The extent of flooding in southwestern Netherlands by the storm of February 1, 1953. *(Ministry of Transport and Public Works, The Hague, Netherlands)*

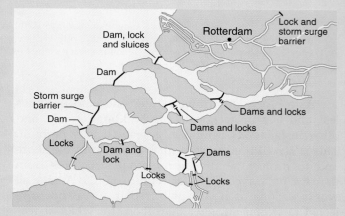

P17.1.2 The system of dams, locks, sluices, and storm barriers installed to protect the southwestern Netherlands from flooding. *(Ministry of Transport and Public Works, The Hague, Netherlands)*

by a complex system of dikes and pumps. The new land has provided sites for the country's largest cities and richest farmlands. The Netherlands also provides a different example of shore protection, as described in Perspective 17.1.

Beach Replenishment and Shore Management

When a beach is gone (or nearly so) one option is to rebuild it. The process is known as **beach replenish- ment.** To rebuild a beach one needs a source of sand, either onshore or off. Sand is dumped on the beach by truck, or is dredged from an offshore site and pumped onto the beach. The success of beach replenishment is variable. Among the most successful ventures have been those along the east coast of Florida, particularly at Miami Beach. Here a replenished beach is expected to last well over 10 years. Most other replenished beaches do not fair nearly as well, however. In fact, half of all replenished beaches have lasted two years or less. A most discouraging example comes from Ocean City, New Jersey, where a beach that cost $5.2 million to replenish was gone again in two and a half

442   **Chapter 17 ·** Coasts and Coastal Processes

even though protected by dikes. Eventually, it was the marine flooding of February 1, 1953, which swept over much of the delta lands of southwestern Netherlands (Figure P17.1.1), that triggered the largest shoreline control project ever undertaken. To protect against the further loss of both citizens and land the Dutch embarked upon what they called the Delta Project. Work began in 1958 and was completed in 1987.

The final water-control system included installation of dams, locks, canals, sluices, and storm barriers, and raising existing dikes. The system has effectively isolated most of the delta islands of the area from flooding and yet made provision for shipping through the area (Figure P17.1.2). The final system reduced the length of sea-dikes from 700 km to 25 km. The centerpiece of the program is the storm barrier at the mouth of the eastern Scheldt River (Figure P17.1.3). This serves to preserve the marine environment of the eastern Scheldt estuary by allowing the passage of tidal waters. The barrier can be closed, however, against a potentially damaging storm or a near-shore pollution incident, such as a major oil spill. With the present level of the sea as a reference the Delta Project has been designed to reduce the land flooding to once in 10,000 years and flooding of the estuaries to once in 4,000 years.

P17.1.3   The storm barrier of the eastern Scheldt. The gates of the barrier are normally open to maintain the marine environment of the estuary. During exceptionally high water in the North Sea they are closed as they are here. (*Ministry of Transport and Public Works, The Hague, Netherlands*)

months. It is true that a beach can be replenished, but before long it will have to be rebuilt. The lesson is that once a community embarks upon a program of beach replenishment it should be prepared to assume a continuing commitment.

An important aspect of shoreline management is controlled withdrawal from a threatened shore. North Carolina, for example, has legislated against the rebuilding of shore structures damaged or destroyed by large storms, such as a hurricane. Another approach is to encourage the demolition or relocation of structures threatened by the most powerful storms.

Many coastal governments have adopted building codes to be applied to coastal structures and have established a minimum distance between structures and the beach.

▶▶ EPILOGUE ◀◀

Over 97 percent of the water in the hydrologic cycle is contained in the ocean basins of the world. At the seaward edge of the coast, the shore, ocean water joins the other processes that are constantly changing the Earth's surface. Here are concentrated the coastal activities of erosion, transportation and deposition. These activities involve the work of plants and animals and wind-driven waves. The most important of these processes are wind-driven water waves. Such waves are responsible for both long sandy beaches and bold cliffs. They also flood into drowned valleys and cut away at the margins of growing volcanoes. Along the narrow coastal strip that makes up 10 percent of the Earth's land surface live over 65 percent of the world's population, which is directly or indirectly affected by the coastal processes.

Summary

1. Wind-formed waves are refracted as they move against the shore. This concentrates their energy on headlands and disperses it around coastal reentrants. The sediments loosened by this wave action are moved along the shore by longshore currents, by swash and backwash action, and by local winds. Winds also drive sand inland to form dunes behind the beach. Corals build coral reef shores, and plants collect sediments to modify areas of quiet water along the shore.

2. Barrier beach coasts form on the trailing edge of a continent and are marked by long barrier beaches, which are separated by lagoons from the mainland. Formerly glaciated coasts are characterized by flooded valleys, offshore islands, and small gravelly beaches. On tectonic coasts, waves have cut shore platforms backed by cliffs and marked by stacks and sometimes by arches. The major outlines of tectonic coasts are determined by the direction of plate boundaries, or by the volcanoes forming along colliding margins. Deltaic coasts form at the mouths of sediment-laden rivers, and coral reef coasts are prominent in tropical and sub-tropical waters.

3. People protect themselves and shore property from waves and storms by building breakwaters, groins, jetties, seawalls, and dikes. They replenish eroded beaches by adding more sand to them. In some places it has become more reasonable to move farther away from the shore.

Key Words and Concepts

arches 437
atoll 439
backshore 431
backwash 430
barrier beach or barrier island 433
barrier reef 439
bars 431
bay barrier 434
beach 430
beach replenishment 442
berm 431
breakwater 440
coast 428
eustatic change of sea level 432
foreshore 431

fringing reef 439
global warming 433
greenhouse gases 433
groin 440
isostatic change in sea level 432
jetty 440
littoral current 430
longshore current 430
pocket beaches 435
ramp 431
refraction 430
rip current 430
seawall 441
shore 428
shore platform 429
shoreface 431

shoreline 428
spit 434
stacks 437
surf 429
swash 430
swells 428
tidal delta 434
tidal inlet 434
troughs (below low-water line) 431
wave base 428
wave crest 428
wave height 428
wave trough 428
wavelength 428

QUESTIONS FOR REVIEW AND THOUGHT

17.1 How does refraction affect wind-driven water waves?

17.2 What is the shore platform? How does it form?

17.3 How does a longshore current form? What does it do?

17.4 How does a rip current form?

17.5 Wind forms water waves, but what else does wind do along the shore?

17.6 What are the characteristics of a barrier beach coast?

17.7 How do atolls form?

17.8 How do groins work in stabilizing a shore?

17.9 What is the function of a breakwater and how does it affect wave patterns and deposition?

17.10 How does a eustatic change in sea level differ from an isostatic change in sea level?

Critical Thinking

17.11 You have a cottage on the seaward side of a barrier island. Over the past few years the beach has become narrower and narrower, and the ocean has continued to move closer and closer to your cottage. You are afraid that the advancing sea will soon remove the rest of the beach and take your cottage with it. What information would you want to have to explain the change along the beach? What options might you have in meeting the problem? Which will be most effective and why?

17.12 Why are some continental margins marked by cliffs and mountains and others by low coasts, long sandy beaches, and shallow embayments? Why are some other coasts very irregular in plan view and have deep embayments and small gravelly beaches? How do these variations relate to plate tectonics and other geologic conditions?

18

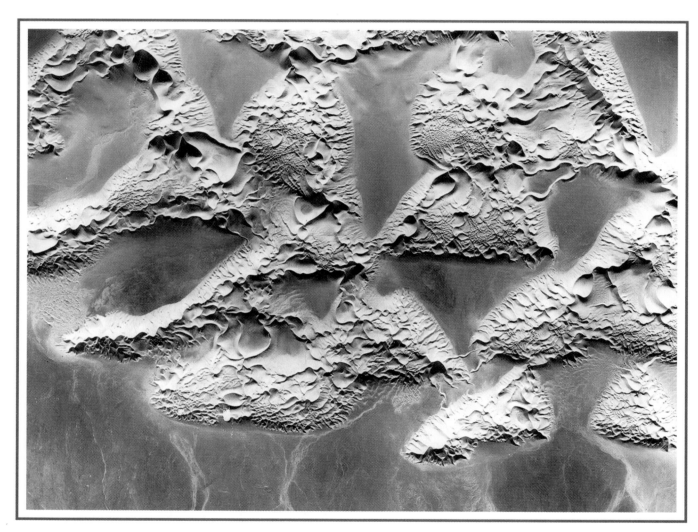

Complex sand dune patterns, North Africa.
(United States Geological Survey)

Wind and Deserts

OBJECTIVES

▲▲

*As you read through this chapter it may help if you focus
on the following questions.*

1. What are the various types of deserts and how are they formed?
2. What are the differences between dust storms and sandstorms? What are the characteristics of dust deposits? Of sand deposits?
3. What is the nature of other desert processes, in addition to wind?
4. What is desertification? How does it start? What are the mechanisms of desertification?

OVERVIEW

▲▲▲

The Sun's energy drives the global circulation of air, and the patterns of this circulation create deserts. Descending dry, warm air spreads vast deserts across the subtropics, and mountain ranges and continental isolation help to create deserts in higher latitudes. It is in these desert regions, and the bordering semiarid zones, that wind is a major surficial agent.

Wind moves millions of cubic meters of dust annually, distributing it over wide areas. Accumulations of blown sand have all but forbidden human activity on millions of square kilometers and threatened, or even overrun, attempts to use the desert. In contrast, wind-deposited dust forms some of the world's most productive soils. Although wind is the most impressive of the desert processes, weathering, running water, mass movement, and ground water are also at work in the deserts.

Many of the Earth's arid and semiarid lands are affected by desertification, better called *land degradation.* Human activity initiates desertification, which can turn potentially arable land into unworkable desert.

Deserts: Causes and Distribution

Deserts cover about 20 percent of the land surface of the Earth and are home to about 4 percent of the population. In general, they receive less than 25 cm of precipitation each year. Bordering the true deserts are belts of near deserts, the semiarid lands, constituting 15 percent of the globe's land surface, where annual precipitation is generally less than 50 cm (Figure 18.1). These regions, transitional from the deserts to more humid lands, share many of the same characteristics of the true desert, although on a less extensive scale. Collectively, deserts and semiarid lands are known as **drylands.**

While aridity is a main characteristic of deserts, so too is windiness. Although wind is present everywhere on Earth, it achieves great effectiveness in the desert, where sparseness or complete lack of vegetative cover makes it a powerful agent of erosion, transportation, and deposition. This is why sandstorms, dust storms, and sand dunes are the hallmarks of deserts and near deserts.

Geologists classify deserts into four main types: **subtropical deserts, rain shadow deserts, continental deserts,** and **polar deserts.**

Subtropical Deserts

Most extensive of the Earth's deserts are the subtropical deserts. They lie in regions of descending air in a pair of climatic zones lying between about 25° and 30° north and south of the equator. As the air descends it also heats up and becomes very dry. Clouds are rare, and thus solar radiation is intense and surface temperatures are the highest on Earth. In the Northern Hemisphere an almost unbroken subtropical desert stretches for over 9,000 km from the Atlantic coast of Africa to northwest India. This includes the Sahara Desert of Africa, the Rub'al Kali Desert of Saudi Arabia, and the deserts of Pakistan and northwest India. In North America the small Sonoran Desert of northwestern Mexico and extreme southwestern United States is a subtropical desert. The Simpson and Great Sandy deserts of Australia lie in the southern belt of high pressure, as do the Namib and Kalahari deserts in southwestern Africa. In South America a small subtropical desert, the Atacama, appears in Chile.

Rain Shadow Deserts

When a mountain range lies across the path of rain-bearing winds it creates desert conditions on the far side of the range. These are the rain shadow

FIGURE 18.1   Distribution of desert and semiarid regions of the world. (*David Turnley, Black Star*)

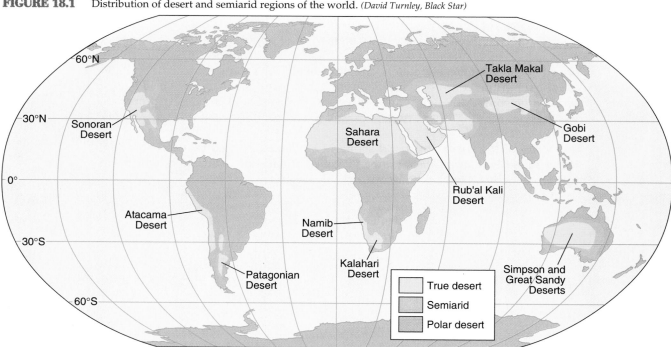

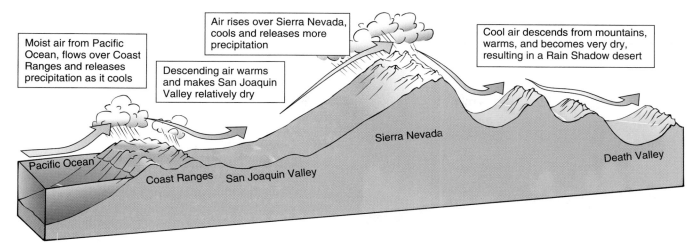

FIGURE 18.2 In the latitude of Fresno, California, the Coast Ranges and the Sierra Nevada each create a rainshadow. Moisture-laden winds from the Pacific Ocean cool and drop moisture as they rise over the low Coast Ranges. Once over the mountains the air descends, warms and keeps the San Joaquin Valley comparatively dry. The same winds continue eastward and cool as they rise up the slopes of the Sierra Nevada. As a result, still more precipitation falls. When it reaches the summit the air is drier than when it started out. Moving down the eastern slope of the Sierra Nevada the air warms and its relative humidity becomes still lower, which accounts for the rainshadow desert of Nevada and eastern California.

deserts, well illustrated by the desert country of most of Nevada. In this case, moisture-bearing winds from the Pacific Ocean move up the western, or windward, slopes of the Sierra Nevada. As the air rises it becomes cooler, is unable to hold as much moisture, and loses most of it as rain and snow. Once over the mountain crests, the now-drier air descends down the lee side of the mountains. As it descends it warms and becomes drier still. Because there are no additional sources of moisture for this area, the country on the lee side, in the "shadow" of the mountains, becomes desert (Figure 18.2). The Andes have created a similar, but smaller, desert, the Patagonia, in western Argentina.

Continental Deserts

Central Asia contains two of the world's greatest deserts, both of them examples of continental deserts. The Takla Makan and the Gobi deserts are located deep in the Asian continent, far from large bodies of water and isolated behind the high Tibetan Plateau and Himalayan Mountains. Because of their distance from any immediate sources of water, these areas are naturally arid. Not only are they extremely arid but they are also characterized by severe winters.

Polar Deserts

Polar deserts exist in northern Canada, Greenland, and Siberia, as well as in the ice-free valleys of Antarctica. Like other deserts, they are arid. Unlike the others, however, their aridity is a function of cold. Ample water exists in these climates, but it exists chiefly as ground ice and is thus generally unavailable to living things. The polar deserts lie on the poleward margins of the regions of permafrost.

Work of the Wind

Air Movement

Air, like water, is a fluid, and flowing air—wind—moves in much the same way as does running water. Wind velocity increases rapidly with height above the ground, just as the velocity of running water increases at levels above its channel. Furthermore, like running water, most air movement is turbulent. Wind velocity, however, increases at a greater rate than does water velocity, and the maximum velocity attained is much higher. Maximum wind velocities are reached in hurricanes, typhoons, and tornadoes where they commonly exceed 120 km/h and sometimes three or four times this. These destructive storms, which can pick up cars and blow them through the air, do not occur in the deserts. A strong desert wind may reach 50 or 60 km/h, but even a wind of a few km/hr can move sand and dust that is unprotected by vegetation.

The greatest physical difference between air and water is density. A cubic centimeter of water weighs over 750 times as much as a cubic centimeter of dry air

FIGURE 18.3 A dust storm darkens the Iowa sky. *(Glennie Murray Wall, Terra Photographics/BPS)*

at sea level. Because of its much greater density, running water is able to move particles the size of gravel or even boulders along its bed. By contrast, even during high desert winds air can move only sand- and dust-sized particles.

Dust Storms and Sandstorms

As it moves, wind sorts sediment into two distinct sizes of particles: sand and dust. The diameter of wind-driven sand grains averages between 0.15 and 0.30 mm but can be as fine as 0.06 mm. All particles smaller than 0.06 mm, whether silt-sized or clay-sized, are classified as dust.

In a true **dust storm** (Figure 18.3), the wind picks up fine particles and sweeps them upward hundreds or even thousands of meters into the air, forming a great cloud that may obscure the Sun and darken the sky. In contrast, a true **sandstorm** is a low-moving blanket of wind-driven sand with an upper surface 1 m or less above the ground. The greatest concentration of moving sand in a sandstorm is usually just a few centimeters above the ground, and individual grains seldom rise even as high as 2 m. Above the blanket of moving sand the air is quite clear, and a person standing in a sandstorm appears to be partially submerged, as though standing in a shallow pond. Often, of course, dust and sand are mixed together in a wind-driven storm (Figure 18.4). However, the wind treats the grains differently. The dust-sized grains are light enough to be soon swept off and buoyed up by the air. In contrast, sand particles are so heavy that the near-surface winds are unable to move them far off the ground.

Movement of Sand Grains Sand grains move forward in a series of jumps, in a process known as **saltation.** The same term was used to describe the

FIGURE 18.4 Wind drives sand and dust across U.S. Highway 160 near Kayenta, Arizona. Vehicles are barely visible in the dust.

motion of particles along a stream bed, but the term is different as it pertains to wind: An eddy of water can actually lift individual particles into the main current, whereas wind, by itself, cannot pick up sand particles from the ground.

Sand particles rise into the air because of the impact of other particles. When the wind's velocity is great enough, grains of sand begin to roll forward along the surface. When a rolling sand grain bumps into another, the impact may lift one or both particles into the air. Once in the air, the sand grain is influenced by gravity, which eventually pulls the grain back down to Earth. Even as the grain falls, however, the horizontal velocity of the wind drives it forward. The resulting path of the sand grain is parabolic from the point where it was first thrown into the air to the point where it finally hits the ground. The angle of impact varies between 10° and 16° (Figure 18.5). Saltation keeps a sand cloud in motion. Countless grains are lifted into the air by impact and are driven along by the wind until they fall back to the ground. Then they either bounce back into the air or push other grains upward by impact. Although the energy that lifts each grain into the air comes from the impact of another grain, it is the wind that provides the initial energy to get the particles in motion and then contributes additional energy to keep those particles mov-

ing once they are airborne. When the wind dies, all the individual particles in the sand cloud settle quickly to earth.

Some sand grains, particularly the large ones, never rise into the air at all, even under the impact of other grains. They inch along the ground, pushed either by the impact of other grains or directly by the wind. This is comparable to the movement of particles in the bed load of a stream. It is as if the surface layer of sand were creeping forward, and, in fact, the term **creep** is applied to the process. Between 20 and 25 percent of sand moved in a sandstorm travels in this manner; the rest moves by means of saltation.

After the wind has initiated saltation by starting the sand grains moving along the surface, it no longer acts to keep them rolling. The cloud of saltating grains actually shields the ground from the direct force of the wind. Thus, as soon as saltation begins, the velocity of near-surface wind drops rapidly. Saltation continues only because the impact of the grains continues. It is a chain-reaction process. The stronger the wind blows during saltation, the more grains are kept airborne, and the denser will be the blanket of sand. Once the wind abates, grains settle back to Earth. They stay there until the wind starts up again and begins moving sand grains across the surface, thus initiating the process of saltation once more.

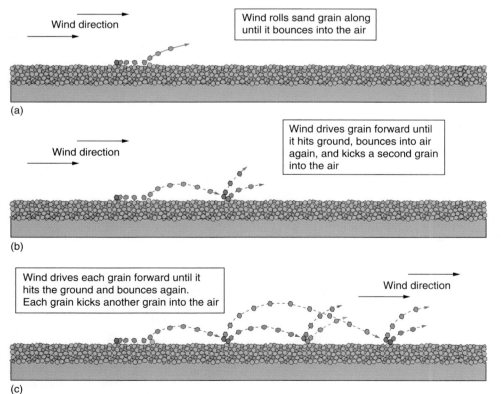

(a) Wind direction — Wind rolls sand grain along until it bounces into the air

(b) Wind direction — Wind drives grain forward until it hits ground, bounces into air again, and kicks a second grain into the air

(c) Wind drives each grain forward until it hits the ground and bounces again. Each grain kicks another grain into the air — Wind direction

FIGURE 18.5
A sand grain is too heavy to be picked up by most winds but can be put into the air by saltation. (a) Here a single grain is rolled by the wind along a sandy surface until it bounces off another grain. Once in the air, the wind drives it forward even as it is simultaneously pulled back to ground by gravity. The grain follows a parabolic path, hitting the ground at an angle between 10° and 16°. (b) Its impact can pop another grain into the air, and the grain itself can bounce back up again as well. In (c) these two grains bounce into the air, and each puts another grain into the air. The process resembles a chain reaction that continues as long as the wind blows.

Movement of Dust Under ordinary conditions, particles smaller than 0.03 mm in diameter cannot be swept up by the wind after they have settled to the ground. In dry country, for example, dust may lie undisturbed on the ground even though a brisk wind is blowing. The reason for this behavior lies in the nature of air movement.

A close look at wind velocity near the ground shows that just along the surface of the ground there is a thin but definite zone where the air moves very little, if at all. Field and laboratory studies have shown that the depth of this zone depends on the size of the particles that cover the surface. On average, the depth of this zone of quiet air is about one-thirtieth the average diameter of the surface grains (Figure 18.6). The small dust grains lie within the thin zone of quiet air just above the surface. They are so small that either the wind passes them by, or larger particles shield them from the action of the wind. Some special conditions must occur to allow the wind to pick up dust particles.

Commonly dust is pushed into the air by the impact of saltating sand grains, and it remains in sus-pension long after the sandstorm has subsided. Wind moving over the irregularities in a newly plowed field or a recently exposed stream bed may become turbulent enough to penetrate through the zone of quiet air and propel dust particles aloft. This tends to cause small dust-bearing whirlwinds called **dust devils** to cross arid lands (Figure 18.7). They originate as air, heated at ground level, becomes unstable and rises rapidly for a few meters or tens of meters, carrying dust and light debris with it. More spectacular, however, are the true dust storms, many involving millions of tons of suspended sediment. They originate with downdrafts of cold air moving at speeds of 40 to 80 km/h, more than strong enough to break through the zone of quiet air and lift vast amounts of dust into suspension. As descending air reaches the ground it spreads forward as a great wall of billowing dust.

Deserts provide the sites for most dust storms, but these storms can also occur in the semiarid regions. The Great Plains of the United States provide a good example. This was the site of the great **Dust Bowl** of the 1930s (Figures 18.8 and 18.9). The area earned the

FIGURE 18.6 In a thin zone close to the ground there is little or no air movement, regardless of the wind velocity immediately above. This zone is approximately one-thirtieth the average size of surface particles. In this diagram the average diameter of surface particles is assumed to be 1 cm. Therefore the zone of no air movement has a depth of 0.03 cm. The diagonal line represents the increase in velocity of a wind of a given intensity blowing over surface. *(Adapted from R.A. Bagnold,* The Physics of Blown Sand and Desert Dunes, *London: Methuen and Co., 1941, p. 54)*

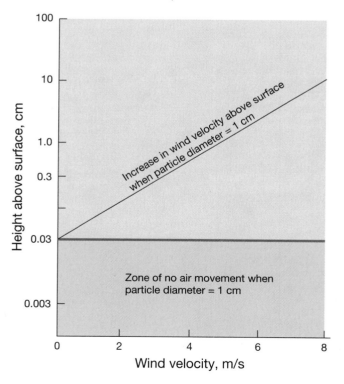

FIGURE 18.7 This dust devil was photographed in Amboseli National Park, Kenya. *(Pat Armstrong, Visuals Unlimited)*

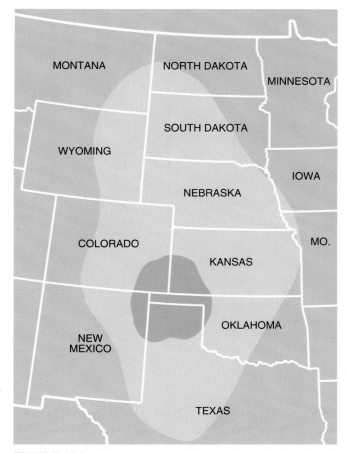

FIGURE 18.8 The greatest wind damage during the Dust Bowl of the 1930s occurred where Colorado, Kansas, Texas, and New Mexico come together, shown here in the darker color. The effects were felt well beyond this area over the greater part of the Great Plains, as suggested by the lighter color.

FIGURE 18.9 A dust storm on a farm in Bent County, Colorado, during the Dust Bowl of the 1930s. *(Ewing Galloway)*

name when strong winds stripped the soils from plowed lands and dumped their dusty burden on parched and struggling young shoots of winter wheat planted in preceding autumns. Towering walls of swirling dust—"black rollers"—turned day to night and sent people scurrying for whatever shelter they could find.

How far does dust travel? Some of the dust, particularly the larger particles, may settle out of the air fairly promptly. The finer particles, however, can remain suspended for long periods before drifting back to the ground, or until precipitation washes them out of the air. Numerous studies show that airborne dust can travel thousands of kilometers from its source. For example, dust from the dust storms of the 1930s dimmed the Sun over the eastern states, and dust from the Sahara Desert has been identified in the air over the Caribbean Sea. Dust is acknowledged as the chief cause of hazy conditions over the oceans, and marine geologists report the presence of wind-borne dust in the sediments of the deep sea (Figure 18.10).

Erosion

Erosion by the wind is less obvious than, for instance, erosion by glaciers or streams. It takes place by one of two processes: **abrasion** and **deflation**. Abrasion is most often the result of wind driven sand grains, commonly made of quartz. Deflation involves both sand grains and dust particles.

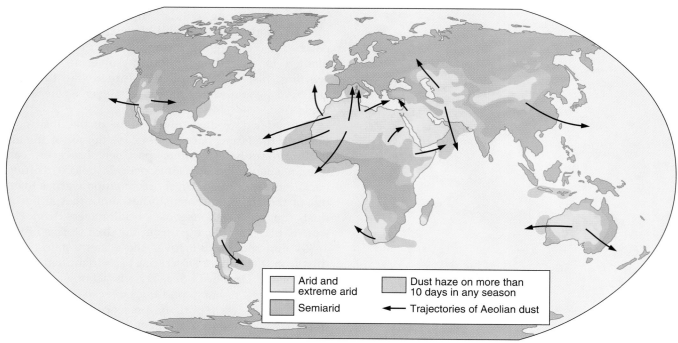

FIGURE 18.10   Trajectories of major dust storms and areas of dust haze over the ocean on 10 days or more in any one season. *(Middleton, Nicholas J., "Desert Dust" in* Arid Zone Geomorphology, *David S.G. Thomas ed, New York: Halsted Press, 1989, Fig. 12.3, p. 267.)*

Abrasion   Like the debris carried in glacier ice or in the turbulent water of a river, saltating grains of wind-driven sand are effective abrasive agents in eroding rock surfaces. As we have seen, wind-borne sand seldom rises more than 1 m above the surface, and measurements show that most of the grains are concentrated in the 0.5 m closest to the ground. This is the zone where abrasion takes place.

Wind-borne sand can be a very effective abrasive. For example, it is not uncommon for sand to cut notches into the bases of bedrock cliffs, as well as to eat into the bottoms of fence posts and telephone poles. In the more arid deserts the evidence of sand abrasion is most impressive, for here the wind-driven sand has in some places cut sharp, irregularly crested ridges called **yardangs.** Yardangs form in soft, easily eroded materials and are oriented in the direction of the dominant wind. They are flanked by concave, even undercut, slopes and may be several meters to kilometers in length. Their crests stand a few centimeters to a known maximum of 80 m above the adjacent troughs.

Among the most common results of sand abrasion are pebbles, cobbles, and even boulders that have been cut by wind-driven sand. These are called **ventifacts,** from the Latin *ventus* for "wind" and *factum* for "made." They are found not only on deserts but also along some modern beaches—in fact, wherever the wind blows sand grains against rock surfaces. Facets (smooth surfaces), pits, gouges, and ridges as well as a

relatively high gloss, or sheen, characterize their surfaces. The surface of an individual ventifact may display one facet, or as many as 20 or more facets. These are sometimes flat but are more commonly slightly curved. Where two facets meet, they form a well-defined ridge, and the intersection of three or more facets gives the ventifact the appearance of a small pyramid (Figures 18.11 and 18.12).

Deflation   Deflation (from the Latin *deflare,* "to blow away") is the process by which wind picks up and carries away unconsolidated sediments. The

FIGURE 18.11   The facets on these desert stones are characteristic of ventifacts. The maximum dimension of the largest ventifact is 10 cm. *(John Simpson)*

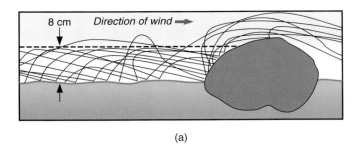

(a)

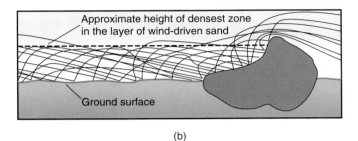

Approximate height of densest zone
in the layer of wind-driven sand

Ground surface

(b)

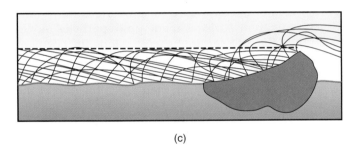

(c)

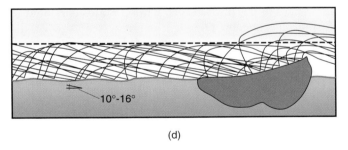

10°-16°

(d)

FIGURE 18.12 Saltating sand grains cut a facet on a desert stone. Most of the abrasion takes place within a few centimeters of the ground. *(After Robert P Sharp, "Pleistocene Ventifacts East of the Big Horn Mountains, Wyoming," Journal of Geology, vol. 57, p. 182, 1949.)*

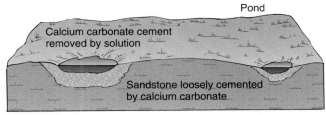

(a) Moist climate

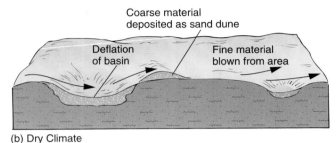

(b) Dry Climate

FIGURE 18.13 Depressions on the High Plains of eastern New Mexico and the western panhandle of Texas are the results of times of deflation alternating with times of solution of calcium carbonate cement in the bedrock. See text for discussion.

process is volumetrically more important than abrasion and creates several recognizable features in the landscape. For example, it often scoops out basins in soft, unconsolidated deposits ranging from a few meters to several kilometers in diameter. Even in relatively consolidated sediment the wind can excavate sizable basins if some other process is at work loosening the material. Such depressions exist in the almost featureless High Plains of eastern New Mexico and western Texas, where the bedrock is loosely cemented by calcium carbonate. Several times during the Pleistocene the climate in this area shifted back and forth between moist and dry. During the moist periods water dissolved some of the calcium carbonate and left the sand particles lying loose on the surface. Then, during the dry periods, the wind was able to remove the loosened sediment, creating numerous basins. Today the larger particles are piled up in sand hills on the leeward sides of the basins excavated by the wind. The smaller dust particles were carried farther along and spread in a blanket across the plains to the east. During moist weather, depressions may contain a shallow lake or pond (Figure 18.13). In Egypt the great Qattara depression is a deflation basin that lies below sea level and is over 300 km in maximum dimension.

Deflation removes only the sand and dust particles from an area; it leaves behind the larger pebble- or cobble-sized particles. As more and more sand and dust are blown away by the wind these larger particles may become concentrated at the surface. With time the large particles form a fairly continuous veneer, a **desert pavement,** that cuts off further deflation (Figure 18.14).

Deposition

Whenever the wind loses its velocity, and hence its ability to transport the sand and dust particles it has

FIGURE 18.14 As more and more dust and sand are moved from a desert surface, the larger particles, unable to be moved by the wind, are concentrated at the surface as in this stone-littered area in Owens Valley, California.

picked up from the surface, it drops them back to the ground. Most surficial deposits and soils have some wind-derived particles within them. In some places, however, dust deposits are abundant enough to form extensive accumulations. Thick deposits of wind-borne dust form **loess** (from the German word *lös*, "loose"). Sand-sized particles, however, pile up in various types of **sand dunes.**

Loess Loess is a buff-colored, unstratified, wind-deposited material composed predominantly of silt-sized mineral particles but with some fine sand and clay-sized particles mixed in as well. Deposits of loess range in thickness from a few centimeters to 10 m or more in the central United States to over 100 m in parts of China. Loess can hold together tightly, and an exposure in loess often holds a vertical face. In some places deposits hold together well enough so that dwellings can be carved out in them, as in western China.

Loess is common in many parts of the world (Figure 18.15) and forms some of the most fertile land on Earth. A large part of the surficial deposits across 0.5 million km² of the Mississippi River basin is made up of loess, and this ancient dust has provided the parent material for the agricultural soils of several midwestern states, particularly Iowa, Illinois, and Missouri. In Washington State productive soils have developed on blankets of loess east of the Cascade Mountains on the Columbia Plateau.

The major deposits of loess in the world have two main sources. Some of it has been blown from major deserts and deposited in moister lands, whereas some of it is associated with glaciation.

The Gobi Desert provided the source material for the vast stretches of yellow loess that blanket much of northern China and whose erosion gives the characteristic color to the Yellow River and the Yellow Sea. Also, much of the land used for growing cotton in the eastern Sudan of Africa is made up of particles blown from the Sahara Desert in the west. Along the northern edge of Africa there are additional thin deposits of loess, yet the loess deposits associated with the Sahara Desert are surprisingly restricted given the size of the desert and the amount of dust that comes from it. Elsewhere, little or no loess deposits of significance are associated with the Australian deserts, the Kalahari Desert of South Africa, or the Sonoran Desert of North America.

Loess in the midwestern United States is related to former ice sheets. During the great Ice Age of the Pleistocene, the rivers of the Midwest carried large amounts of debris-laden meltwater from the glaciers. Consequently, the flood plains of these rivers built up

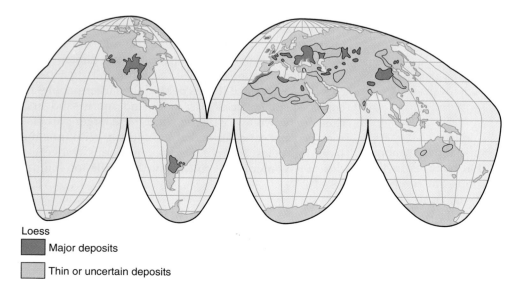

FIGURE 18.15
Major loess deposits of the world
(Thomas, David S.G. "Aeolian Sand Deposits" in David S.G. Thomas, ed. Arid Zone Geomorphology, *New York: Halsted Press, 1989, Fig. 11.1, p. 233.)*

Loess

Major deposits

Thin or uncertain deposits

at a rapid rate and were broader than they are today. During periods of low water the floodplains were wide expanses of gravel, sand, silt, and clay exposed to strong westerly winds. These winds whipped the dust-sized material from the floodplains, moved it eastward, and laid down the thickest and coarsest of it closest to the rivers. Loess deposits in Europe, the Balkan countries, southern Russia, and the Ukraine are also related to the floodplains of major glacial rivers of the Pleistocene.

Sand Dunes Unlike the generally featureless deposits of loess, most deposits of sand assume recognizable and characteristic shapes. These range from sand ripples, which resemble small waves a few centimeters in height and wavelength, to forms a kilometer or more in width, 100 km or more in length, and up to 500 m in height known as **sand dunes** (Figure 18.16). Some deserts support nearly continuous zones of complex sand dunes. These areas are called **sand seas** because the numerous dunes resemble sea waves (Figure 18.17). In other places well-defined forms of simple dunes are the rule.

Wind-driven sand will accumulate wherever an obstruction interrupts the flow of air. For instance, air will flow over a boulder, but in the lee of the boulder there is a small area protected from the full effect of the wind, called the **wind shadow.** Sand settles here in the quieter air. In a true sand dune the wind shadow is formed by the dune itself. Sand is driven up the windward side of the dune. As the sand blows over the dune crest (Figure 18.18) it falls into the wind shadow, which is bounded by a **slip face,** a slope that defines the lee side of the dune. It is also the place where most deposition takes place in a dune. The roles of these two features can best be seen by examining the profile of a dune (Figure 18.19).

FIGURE 18.17   This aerial view in the Namibian Desert illustrates the complexity of form and vastness of extent of dunes in a sand sea. (*Anthony Bannister, Earth Scenes*)

The profile of a dune is typically asymmetric. The windward side is a ramp with a slope of about 10° to 12°. The lee side along the slip face is approximately three times as steep. Lines of air-flow moving up the dune are pushed closer together as they sweep over the dune, thus converging as they cross the crest. Immediately beyond the crest they diverge again. Here the rapidly moving air is separated from the pocket of quieter air in the wind shadow, and the plane separating them is a **surface of discontinuity.**

FIGURE 18.16   These sand dunes are in the Namib Desert of Namibia. The crest of the highest dune stands over 100 meters above the sparsely vegetated plain in the foreground. (*Arthur Gloor, Earth Scenes*)

FIGURE 18.18   Wind drives sand up the windward side of this sand dune in Namib and blows it over the crest of the dune. It falls through the wind shadow and accumulates on the slip face on the lee side of the dune. (*Tad Nichols, Peter Kresan Photography*)

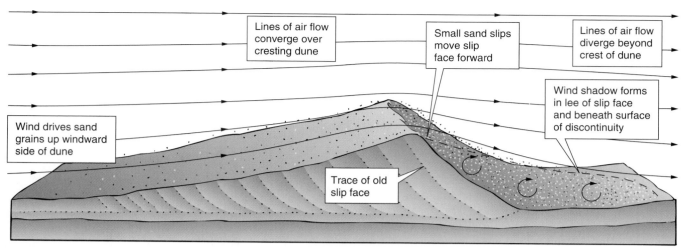

FIGURE 18.19   The wind shadow, slip face, and surface of discontinuity of a sand dune. See text for discussion.

Within the wind shadow air is in motion but only in gentle eddies. On the windward slope, wind sends sand grains saltating up to the crest. Here the grains drop into the wind shadow where, as the wind velocity falls abruptly, they are deposited. Most of them settle immediately and pile up on the upper section of the slip face. The continued deposition steepens this section, and it becomes unstable. Small sand avalanches then occur and return the slope to a stable angle of repose, about 34° (Figure 18.20). As long as deposition continues, small, periodic sand slides occur and the dune moves forward with each small advance of the slip face.

A sand dune begins when some surface obstacle such as a boulder, a bush, or a slight topographic irregularity interferes with the flow of near-surface air, creating eddies of varying velocities. In the slower of these, sand may begin to accumulate. As the mound of sand increases it begins to create its own wind shadow, and a true slip face promptly follows.

A few simple dune types are quite well understood. These different types may overlap and coalesce to form complex dune patterns that defy easy description or classification. The basic dune types are shown diagrammatically in Figure 18.21 and in photographs in Figure 18.22.

Wind blowing consistently in a single direction may form a **transverse dune,** a long, straight dune perpendicular to the direction of the wind and with a single, continuous slip face. It is asymmetric in cross profile. If the supply of sand diminishes downwind the transverse dune may break up into a series of individual, crescent-shaped dunes called **barchans.** The horns of the crescent point downwind and enclose the

slip face. They may reach 30 m in height and 300 m between horns. The larger ones migrate at about 15 m/y, and the smaller ones at half this rate.

Dunes formed by winds of varying directions have two or more slip faces. **Linear dunes** are long straight dunes with slip faces on both sides. They are thus symmetrical in cross profile. Their origin remains controversial, but the paired slip faces indicate that they are formed by winds from both sides. Sand in the dune moves downwind along the long axis of the dune. Winds alternating in several different directions create dunes called **star dunes** with three or more arms that radiate from high central points. These dunes build vertically rather than migrating and growing laterally.

FIGURE 18.20   The steep slope facing the viewer is the slip face on the lee side of a New Mexican sand dune. The dune moves forward into the wind shadow and toward the viewer. Sand is pushed up the windward side to the lip of the slip face. Continuing small slips of sand keep the slip face steep. *(Franklyn B. Van Houten)*

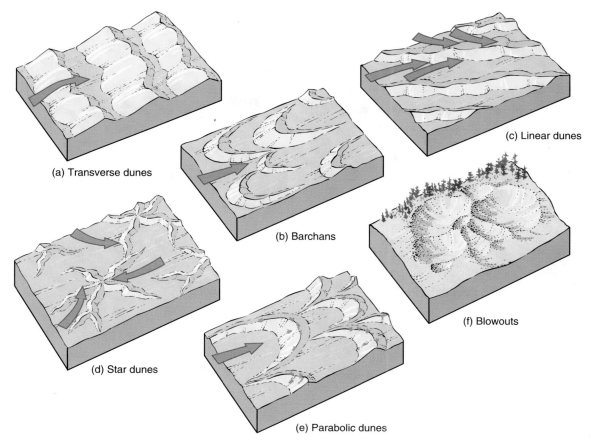

(a) Transverse dunes

(b) Barchans

(c) Linear dunes

(d) Star dunes

(e) Parabolic dunes

(f) Blowouts

FIGURE 18.21 Major dunes types: (a) barchans; (b) transverse dunes; (c) linear dunes; (d) star dunes; (e) parabolic dunes; (f) blowouts. Arrows show directions of dominant winds. (*Modified from McKee, Edwin D., ed, A Study of Global Sand Seas. U.S. Geological Survey Professional Paper 1052, 1979.*)

Dunes are stabilized as vegetation takes over and shields them from the direct attack of the wind. One type of dune forms as wind scoops out a basin in previously deposited, often partially vegetated, sands. A rim usually marks its border. These dunes, called **blowouts,** are often found along shorelines in the sandy ridges that back the beach along many shores. **Parabolic dunes** are a variation of blowouts. As their name implies they are parabolic in shape, but, unlike the barchans, their slip faces are convex downwind. The horns, pointing into the wind, are partially anchored by vegetation while the fronts of the dunes migrate downwind.

Not all dunes are found in deserts. Along the shores of the ocean and of large lakes, ridges of wind-blown sand accumulate even in humid climates. They are well developed along the southern and eastern shores of Lake Michigan, along the Atlantic coast from Massachusetts southward, along the southern coast of California, and at various points along the coasts of Oregon and Washington. In arid climates, as in the Namib Desert of South Africa, windblown sand from the beach is driven inland to merge with the sandy deposits of the desert beyond.

Stratification in Dunes Often sand dunes have an inclined or cross-bedded stratification (Figure 18.23). The inclination is greatest in those beds deposited on the slip face of the dune, about 34°. These beds are analogous to the foreset beds in a delta. In the wind shadow, sand may accumulate in thin and more or less horizontal beds. These can be preserved as the sand deposited along the slip face rides over them, preserving the equivalent of bottomset beds in a delta. Bedding may form along the windward slope of a dune as sand moves up it toward the dune crest and assumes the angle of the windward slope 10° to 12°. Such bedding is not usually long preserved.

Wind Action of Mars and Venus

Both Mars and Venus have atmospheres. The atmosphere of Mars is very thin as compared with that of Earth and is largely carbon dioxide. Venu-

(a)

(b)

(c)

(d)

(e)

(f)

FIGURE 18.22 Photographs of some typical dune forms. (a) These transverse dunes are located on Mesquite Flat, in Death Valley National Monument, California. Their crests are perpendicular to the dune-forming winds. (b) A barchan dune south of the Salton Sea, California. Dune movement is in the direction in which the horns point. Other barchans are visible in the distance. (c) Linear dunes in the Simpson Desert, Nothern Territory, Australia. (d) A low aerial view of a star dune in the Gran Desierto, northwestern Sonora, Mexico. (e) Vegetation covers the Sand Hills region of Nebraska. In places, however, the vegetative cover is broken by the power of the wind, which exposes the underlying windblown sand, as in this blowout. (f) A parabolic dune has a plan similar to that of a barchan, but the horns point into the wind rather than in the direction of the wind. This example is in Death Valley National Monument, California-Nevada. *((a) Peter Kresan Photography, (b) Dr. John S. Shelton, (c) William C. Bradley, (d) Peter Kresan Photography, (e) William E. Ferguson, (f) Pollack, 1993, Biological Photo Services.)*

FIGURE 18.23 These inclined beds of ancient windblown sands are now lithified. The dipping beds record the slip faces of dunes in a Jurassic desert. Kanab Canyon, Utah.

FIGURE 18.24 Windblown particles partially obscure rock debris at hander 1 site on the moon. The large boulder measures about 2 m across. (NASA)

sian atmosphere is also chiefly carbon dioxide. It is, however, much thicker than the atmospheres of Mars and Earth—so thick that it shrouds the surface of the planet to visual observation. The presence of atmosphere on these planets makes both dunes and dust storms possible on them.

Long before spacecraft circled and eventually landed on Mars, telescopic observation had recorded dust storms on the red planet. The first Martian lander was delayed by dust storms that sent particles into the atmosphere where they stayed for several months. The first Martian lander set down in an area where wind deposits had collected in the lee of boulders and smaller rough fragments (Figure 18.24). Orbiting spacecraft have mapped dune fields on the floors of some of the large impact craters and in the north polar area (Figure 18.25).

Not until the Magellan space mission to Venus were scientists able to see through the thick clouds of carbon dioxide that obscure the planet's surface. Side-scanning radar, carried by the spacecraft, has now mapped the entire planet and shows that wind activity

FIGURE 18.25 Sand dunes on Mars are locally extensive. Here, near the edge of the northern polar ice cap, this linear pattern marks strings of individual dunes. The view is over 60 km across. (NASA)

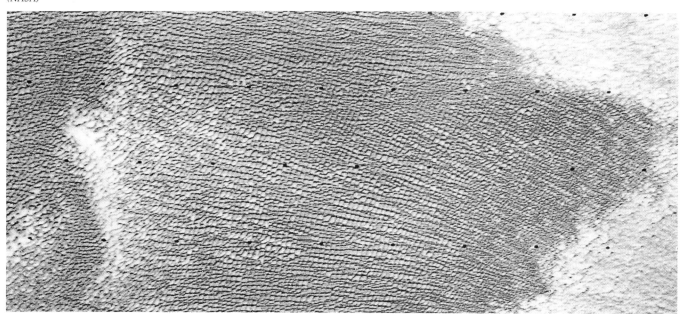

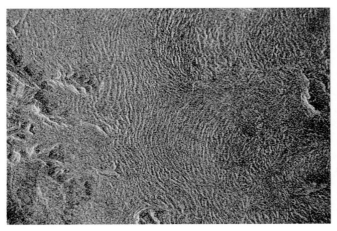

FIGURE 18.26   Wind direction on Venus is generally east to west and these curvilinear ridges, which somewhat resemble whorls of a giant fingerprint, have been interpreted as transverse sand dunes. Width of view is about 55 km and spacing between ridges is about 600 meters. North is to the top. *(NASA)*

FIGURE 18.27   These petroglyphs, associated with the Puerco Ruin in the Petrified Forest National Park, Arizona, were chipped on rock covered with rock varnish. The artist pecked away the varnish and exposed the light-colored rock beneath. The age of the design is unknown, but presumably is less than a thousand years old.

is widespread on Venus. Features include wind streaks (Figure 18.26), dune fields, and, possibly, yardangs.

Other Surficial Processes in the Desert

Weathering

Because of a lack of moisture in the desert, the rate of weathering, both chemical and mechanical, is extremely slow. Mechanical weathering, however, dominates, and most of the weathered material consists of chemically unaltered rock and mineral fragments.

Some mechanical weathering is the result of gravity moving material to lower levels during the process of mass movement. An obvious example is the shattering of rock material when it falls from a cliff. Slower movement, however, usually weathers slope materials in deserts. Abrasion by wind-driven sand is another factor in mechanical weathering. Furthermore, in almost every desert in the world, temperatures fall low enough at some time during the year to allow frost action to function. Here again, however, the deficiency of moisture slows the process. Finally, the wide temperature variations characteristic of deserts cause rock materials to expand and contract and may thus produce some mechanical weathering.

This low rate of weathering is reflected in the soils of the desert. Extensive areas of residual soil are rare in deserts because the lack of protective vegetation permits the winds and occasional floods to strip away the soil-producing minerals before they can develop into true soils. Even so, soils sometimes develop in local areas, although they lack the humus of the soils in moister climates. In addition, they contain concentrations of such soluble substances as calcite, gypsum, and even halite, because there is insufficient water to carry these minerals away in solution.

An intriguing weathering feature of the desert is the accumulation of a thin, shiny veneer on some rock surfaces (Figure 18.27). This **rock varnish** is less than a millimeter thick (Figure 18.28) and is made up chiefly of clay minerals plus 20 to 30 percent of iron and manganese oxides and traces of many other elements. When manganese oxide dominates, the color of the varnish is black, and when iron is the chief oxide the

FIGURE 18.28   Rock varnish is less than a millimeter thick. In this photomicrograph the black zones are manganese rich and the orangish-red layers are low in manganese and higher in iron. *(Ronald I. Dorn)*

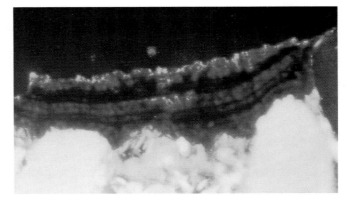

degradation that leads to a reduction or destruction of the land's ability to support plant and animal life, including human life. It can occur in deserts but more often occurs along their margins in the semiarid regions, sometimes even extending into the moister, subhumid lands.

Several different types of land abuse, each involving damage to the vegetative cover, can start desertification. The abuse may take different forms, including, among others, overgrazing, disruption of the vegetative cover by plowing, and damage to it by road construction and mining operations. The results include the appearance of undesirable plants, exposure of the soil to erosion, loss of soil moisture, increase of salts in the soil, and lowered fertility of the land.

One of the mistaken beliefs about desertification is that it spreads from a desert center. Actually, the close presence of a desert has little direct relation to the process. It can start in a semiarid or subhumid climate because of land abuse and spread outward with continued abuse. Several spots of desertification may merge into a larger area, but this is not usual on a large scale. A second misconception about desertification is that it is caused by droughts. Droughts increase the vulnerability of land, but well-tended land will survive a drought and recover. A combination of land abuse during good years and continued abuse during drought years will inevitably lead to desertification.

Extent of Desertification

All the continents, except Antarctica, are affected to a greater or lesser extent by desertification. Although reliable estimates of the total land surface involved are difficult to gather, at least a hundred countries are directly affected. The United Nations estimates that desertification, from slight to severe, affects 45 million km² of the world's land surface, approximately the size of North America and Australia combined. The United Nations also estimates that each year about 210,000 km² of formerly productive land becomes essentially useless, and another 60,000 km² is converted to barren desert. In Mexico, the United States, and Canada approximately 6,360,000 km² of land show desertification to some extent (Figure 18.31).

Mechanisms of Desertification

Wind Erosion Weakening or destroying their vegetative cover exposes soils to wind erosion. Because fine particles make up a large volume of most soils, deflation can be very damaging. The southern part of the San Joaquin Valley of central California, in the neighborhood of Bakersfield, provided a vivid example in 1977. The land was being used for both grazing and agriculture. Rangeland had been over-

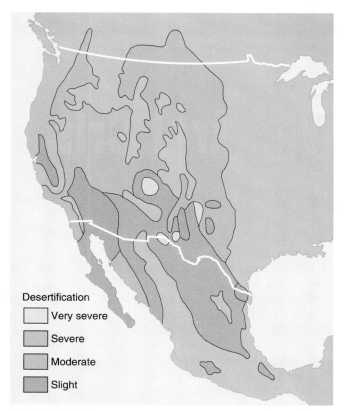

Desertification
- Very severe
- Severe
- Moderate
- Slight

FIGURE 18.31 Extent of desertification in North America. *(From Dregne, H.E., "Desertification of Arid Lands" in* Physics of Desertification. *El-Baz, Farouk and M.H.A. Hassan, eds. Dordrecht: Martinus Nijhoff Publishers, The Netherlands, 1986, Fig. 5, p. 23.)*

grazed, fields had been plowed for planting, and other areas had been stripped of natural vegetation to extend cultivation. In December, unusual meteorological conditions produced winds that averaged 75 km/h and reached gusts of up to 300 km/hr. The result was immediate. In 24 hours wind stripped an estimated 5.5 million m³ of soil from the grazing lands and a similar amount from the agricultural lands. Another consequence of the dust storm was the dramatic increase of "valley fever," a disease caused by fungal spores from the soil. The winds had picked these up along with the dust and spread them widely.

Water Erosion The infrequent rains that fall in dry climates can be torrential when they do come. When human activities weaken the dryland vegetation, infiltration is slowed and runoff is increased. The San Joaquin Valley dust storm of 1977 also provides an example of this. Three months of heavier than usual rains followed in the wake of the dust storm. These caused extensive gullying, mud slides, floods, and clogging of smaller stream channels with sediments.

The Aral Crisis

he Aral Sea lies in Asia's Kyzil Desert about 280 km east of the Caspian Sea. Until recently within the Union of Soviet Socialist Republics, this body of water is now within the boundaries of the newly declared republics of Uzbekistan and Kazakhstan. Thirty-five years ago it was one of the world's largest bodies of inland water spreading over an area of 66,500 km². Today it is half that size, and projections estimate that by the turn of the century it will be down to 25,000 km² and its volume will have decreased from 1,090 km³ in 1960 to 175 km³ (Figure P18.1.1). What has caused this desertification of the Aral Sea and brought on what has been called the "Aral Crisis"?

Two rivers, the Amu and the Syr, empty into the Aral Sea and take their water from the mountains to the southwest. Both rivers are exotic streams, for they receive no tributaries of any importance after leaving the highlands to flow through 800 km of sandy desert to the Aral. The crisis had its beginnings after World War II when the USSR decided to use the waters of the Amu and the Syr for extensive irrigation systems. The plan was to increase the production

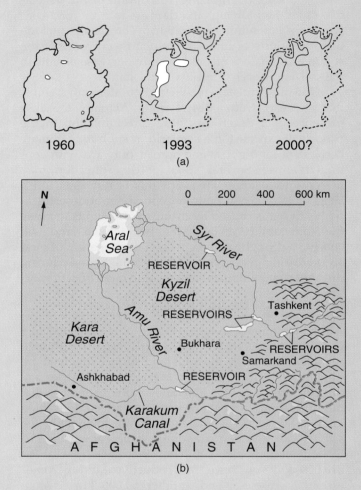

P18.1.1   (a) The Aral Sea lost 50 percent of its area between 1960 and 1993. The prediction is that by the year 2,000 it will have shrunk still further as a result of irrigation water diverted from the Syr and Amu rivers. (b) The Aral sea in 1960 and some of the region's water-development projects. ((a) *Modified from Micklin, Philip P., "The Shrinking Aral Sea." Geotimes, April 1993, p. 16.*)

Salinization of Soils   **Salinization,** the addition of unwanted salts to a soil, is a process that plagues many irrigation projects. Drylands have insufficient water to flush salts through the system and out to the sea. For this reason, the soluble salts of sodium, magnesium, and calcium collect more readily in the dryland environments than in more humid lands. As a result, in drylands, where artificial irrigation is most

of cotton, cotton cloth, and cotton goods; to provide a "national garden" for fruits and vegetables; and to produce meat and rice for an increasing Aral basin population. To achieve this, an extensive irrigation system was planned, which involved the building of numerous irrigation canals and reservoirs. The longest system, the Karakum Canal completed in 1975, draws water from the Amu near the Afghanistan border. The system leads 1,100 km westward along the southern margin of the Kara Desert and includes 450 km of navigable waters. Elsewhere, reservoirs and irrigation canals draw water from the Syr River as well as the Amu.

Because the Aral Sea relies almost completely for its water supply on the Amu and Syr rivers, the Aral received less and less water as more and more water was withdrawn from the rivers by canals and reservoirs. Up until 1960, 56 km^3 of water reached the Aral Sea annually. By the mid 1970s this had fallen to between 7 to 11 km^3. In the 1980s there were several years of no inflow into the Aral. Inevitably, the Aral Sea has shrunk drastically. There are other results besides a shrinking sea. The deltas of the Syr and Amu have lost their wetlands, and wildlife has disappeared. Farms on the drying delta have failed. What is left of the Aral Sea has become so salty that its commercial fish population has disappeared along with the fishing industry that was dependent upon

it. The shrinkage of the sea has exposed large areas of dry seafloor, the source of an increasing number of dust storms. Beyond these effects, human health has declined precipitously as shown by an increased child mortality rate and disease, a marked increase in intestinal cancer, and a generally low level of immunity (Figure P18.1.2).

Meanwhile, the irrigated regions upstream have been having their own problems. From 1959 to 1987 population increased two and a half times, placing new demands on the resources, including water. By 1989 over 50 percent of the irrigated lands in the region were moderately to severely salinized. Estimates by Russian scientists indicate that by 1989 the yields from irrigated lands in the Aral basin had decreased by over 30 percent.

The cure for all these problems is difficult and costly to effect. More efficient irrigation would release some water to the Aral, but would demand funds that are not presently available. Even if funds were available, the amount of water released to the Aral would by no means be enough to restore the sea to health. Elimination of salinized soils will also take a complete redesign of the irrigation systems. The restoration of the Aral Sea and its drainage basin to their earlier conditions would take severe cutbacks of the irrigation networks in the upper Syr and Amu, cutbacks that the political system probably

P18.1.2   Abandoned fishing boats in the dried-up section of the Aral Sea. (David Turnley, Black Star)

cannot bring about, or the social and economic arrangements cannot tolerate. Reversal of the effects of human activity in the Aral basin appear doomed for a long time into the future.

widespread, waters carry a high percentage of salt. Much of this irrigation water is used by plants, as it is intended to be, but some of it evaporates, leaving behind the salts that had been dissolved originally in

the irrigation water. Unless provisions are made to keep the soluble salts moving through the system, they continue to accumulate in the soil to levels that affect plant growth. If this process goes far enough, the irrigated fields become too salty to sustain plant life and must be abandoned.

Demographic Pressure   Because desertification is caused by human misuse of drylands, it follows that the more people demand of these areas, the greater are the probabilities that desertification will result. The pressure may come from an increase of population of pastoral peoples. In this case increased population can mean more herds on limited lands or movement into marginal lands during rainy years. In either situation, overgrazing results and is catastrophic in the drought years. Development of natural resources such as minerals, oil, and gas demands extensive support arrangements from roads to housing and related necessities. These, in turn, can seriously affect the vegetation and bring on desertification. Desertification may also come from a society's long-term plan to supply large amounts of water to an area that under natural conditions has only a limited supply. Perspective 18.1 describes some of the results of such a venture.

▶▶ Epilogue ◀◀

This discussion of winds and deserts ends the group of six chapters devoted to the surface processes of the Earth. These chapters covered the processes that are directly responsible for most of the landscapes of the Earth. At some times and in some places, one process may dominate all others, whether it be mass movement, running water, ground water, glaciation, or wind-driven waves.

Although wind has been the most pronounced process discussed in this chapter, we have seen that other processes have also been operating in the drylands. This should remind us again that Earth processes are interlinked. Geologists separate them out in large part for ease of discussion. This, however, is an entirely artificial division. The various processes go on together, some with greater intensity than others. Glaciers melt to produce rivers. Drying floodplains provide a source of dust for winds, and the loess that they deposit becomes the site of weathering where one set of minerals is exchanged for another.

The interconnectedness of Earth processes does not stop with those of the surface, as we have seen earlier. Without the internal processes of the Earth, lands would not be crumpled, lifted, and offered to the surface processes for continuing change. Without plate tectonics the sediments worn from the continents and taken to the ocean by wind and water would not reappear as sedimentary rocks, or in time as metamorphic rocks and new igneous rocks. We have seen, also, that humans are involved in the Earth processes, sometimes very directly as the next section of this book emphasizes.

Summary

1. Subtropical deserts form in zones of descending air where air warms and dries as it descends. Rain shadow deserts form in the lee of topographic barriers that block rain-bearing winds. Continental deserts are located toward the interior of continents and far from any source of atmospheric moisture. Polar deserts exist because water is locked up in ice and is generally unavailable.

2. In a dust storm fine particles are carried by suspension high into the atmosphere. In a sandstorm sand grains move along close to the ground in a saltating fashion. Dust, or loess, is deposited as a blanket of variable thickness, usually beyond the limits of the driest deserts. Sand is deposited in distinctive forms—sand dunes—in deserts or along some coasts. Sand deposits are generally characterized by slip faces and wind shadows.

3. Chemical weathering is slow in the desert, and thus mechanical weathering dominates. Mass movement is effective but is slowed by lack of water. Streams operate as in humid climates, but less continuously and predictably. Larger streams are often exotic, and smaller streams, intermittent. Ground water usually is deep beneath the surface, is only slowly replenished, and may be quickly depleted. In many places it is high in dissolved mineral matter.

4. Desertification turns productive drylands into unproductive land. Humans start desertification by land use that weakens or destroys the vegetative cover. This includes overgrazing, intensive agriculture, establishment of communities, and road building. Thereafter mechanisms of desertification include wind and water erosion, salinization of soils, and demographic pressures.

KEY WORDS AND CONCEPTS

abrasion 453	exotic rivers 463	saltation 450
barchan 458	flash flood 463	sand dune 456
blowout 459	linear dune 458	sand sea 457
continental desert 448	loess 456	sandstorm 450
creep 451	parabolic dune 459	slip face 457
deflation 453	playa 463	star dune 458
desertification 464	pluvial lake 464	subtropical desert 448
desert pavement 455	pluvial period 464	surface of discontinuity 457
drylands 448	polar desert 448	transverse dune 458
Dust Bowl 452	rain shadow desert 448	ventifact 454
dust devil 452	rock varnish 462	wind shadow 457
dust storm 450	salinization 466	yardang 454

QUESTIONS FOR REVIEW AND THOUGHT

18.1 What characterizes a desert?

18.2 What causes a subtropical desert? A rain shadow desert? A continental desert? A polar desert?

18.3 How does the movement of wind differ from that of water? In what ways are they similar?

18.4 What are the differences between a dust storm and a sandstorm?

18.5 Explain how wind moves sand grains. How it moves dust particles.

18.6 What are some of the results of abrasion in a desert?

18.7 What is deflation, how does it work, and what are some of the results?

18.8 What is loess?

18.9 What are the functions of the slip face and the wind shadow in a sand dune?

18.10 What is the evidence that transverse dunes and barchans form at right angles to the wind? What effect does the wind direction have on their shape?

18.11 How do linear dunes differ from star dunes in shape and origin?

18.12 Why is the stratification in sand dune deposits commonly inclined from the horizontal?

18.13 How do the processes of weathering, mass movement, running water, and ground water in a desert differ from the same processes in humid regions?

18.14 What is desertification? How does it start and what are some of the mechanisms involved?

Critical Thinking

18.15 Wind is an effective agent in the desert because of the scarcity of water. Why is it, then, that wind is so important in fashioning the shores and coasts of the world where water abounds?

18.16 Why is wind more effective as a surficial agent than streams on Mars and Venus? Why is it unimportant on the moon?

HUMANS AND EARTH CHANGE

This book focuses on the changes that characterize the Earth. The constancy of change, in fact, proves to be the hallmark of the Earth. We began by describing how materials of the Earth change from one form to another as they respond to changing environments. Solid rocks of the rock system change as heat and pressure transform them into the new minerals and forms of metamorphic rocks. This process, carried further, melts rocks into the magmas from which form the igneous rocks. In the more familiar realm of the Earth's surface, rocks also change in response to their exposure to air, water, and living things, and are converted into new minerals and fertile soils. We then dealt with internal forces. Heat deep in the Earth provides the energy to move large fragments of the Earth's crust—plates—on global voyages, which change the shapes of continents and oceans, cause volcanic eruptions, set off destructive earthquakes, and raise mountain chains. Next we discussed what happens when the internal forces of the Earth expose the rocky crust to the surface processes of weathering and erosion, which set about changing the Earth's surface (Figure T 4.1). Mass movements, rivers, ground water, glaciers, wind-driven waves, and desert winds transport the products of weathering and erosion to new basins of deposition where these products begin to be changed into sedimentary rocks. Throughout the discussion of these changes runs the thread of the environment and environmental change. We have come to understand that changes take place because of changes in the environment.

Life Appears

▲▲▲▲▲▲▲▲▲▲▲▲▲▲▲▲▲▲▲▲▲

Physical changes on Earth have been continuous ever since the earliest history of the globe. Somewhat later, perhaps 3 billion years or more ago, life developed. However it originated, it soon joined the physical processes as an agent of change. Life has continued to develop, adapting itself to the many different and changing environments the Earth offers, whether in the air, on the land, or in the sea. As life adapted to these environments it also brought about physical changes on Earth. This has been going on ever since the first simple plants began to add oxygen to the primitive atmosphere three billion years ago or longer. This event foreshadowed the development of oxygen-breathing forms—animals, including ourselves. It made possible, also, other changes, such as oxidation, which has created the sim-

FIGURE T4.1   Weathering and erosion have fashioned this landscape within the canyon of the Colorado River in Arizona. The sediments produced move downslope to the river, which in turn moves them downstream until they reach Lake Mead, impounded behind Hoover Dam. Here they settle out and will stay until the dam fails and they are swept down the Colorado to the Gulf of Baja California and the Pacific Ocean. *(Pamela Hemphill)*

ple oxide minerals, among them hematite and goethite, that provide sources of iron ore and color today's soils as well. Other examples of the changes wrought on the environment by living things are fascinating and endless. In the final section of this book, however, we are concerned with some of the ways that biological latecomers—humans—have used the

Earth, how they have changed some Earth environments, and how Earth environments have affected them.

The Rise of Humans

Although the details of human origins are still being worked out, most authorities agree that the first humans had already appeared in Africa by the beginning of the Pleistocene, 1.8 million years ago. From here they spread through the Near East into southern Europe and eastward to China and Southeast Asia. Somewhat less than 100,000 years ago essentially modern types of humans appeared. We have named ourselves (either hopefully or arrogantly) *Homo sapiens sapiens*, from the Latin for "man," and "wise," twice over. By 15,000 years ago we had colonized most of the world's land masses (Figure T4.2). Our numbers by then had reached an estimated five million, 30 percent less than that of New York City today. Our technology included the use of fire, and we had learned that flint, chert, and obsidian were the best materials from which to shape stone tools. We made clothing and constructed crude shelters. We were hunters and gatherers relying on our ability to catch game and find seeds, roots, and fruit. Our effect on our environment was negligible when compared with what was to come.

FIGURE T4.2   Humans are latecomers to the Earth. Since their appearance, however, they have spread rapidly, as indicated on this world map. *(Modified from Roberts, Neil,* The Holocene: An Environmental History, *Oxford: Basil Blackwell, 1989, Fig. 3.7, p. 56.)*

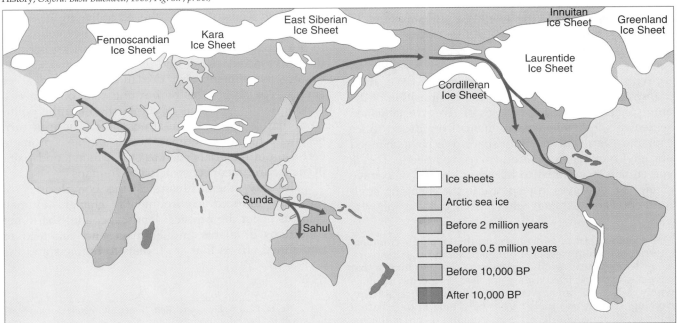

Humans and the Environment

Beginning about 10,000 years ago we began to lay the groundwork for our emergence as a major environmental force and agent of change. In the area we now call the Middle East we introduced agriculture, which included the domestication of both plants and animals. By 7,500 years ago we learned how to find and smelt copper and tin minerals and to convert them into the alloy bronze. In the last 2,000 years people have become increasingly skilled in locating and extracting materials useful to them. These materials have been not only metals such as iron, chromium, and uranium, but have also included nonmetals, such as potassium and sulfur used in the chemical industry. We have sought materials as common as sand, a component of concrete, and as rare as diamond, used both for jewelry and as abrasives for drilling the hardest rock. The materials that probably have had the most impact on people have been the organically formed fuels: coal, oil, and gas. These are the energy sources that drive our cars, planes, trains and ships, run our factories, heat (and cool) our homes, schools, and offices. In Chapter 19, "Energy and Mineral Resources," we examine how the minerals and fuels that underpin our sprawling industrial and technological society have formed, where they are located, how long they may last, and what the alternatives may be when the supply is exhausted.

We can view our manipulation of the environment in different ways. One way, for instance, is to recognize that we have behaved as all species do. We have sought to earn our living from our surroundings. In doing this, however, we have differed from all other species in the extent and nature of our use of the environment. No other species has had such a rapid and far-reaching effect on the environment. We have deforested continents, changed the composition of the atmosphere, moved the courses of rivers, established settlements in hostile environments from desert to ice cap, and even ventured into space. We have changed our environment to meet our perceived needs. This is very much in contrast to all other species who have changed themselves in response to the changing environment. Ecologist Paul Sears in 1957 put it as follows:

Man's unique power to manipulate things and accumulate experience presently enabled him to break through the barriers of temperature, aridity, space, seas, and mountains that have always restricted other species to specific habitats within a limited range. With the cultural devices of fire, clothing, shelter, and tools he has been able to do what no other organism could do without changing its original char-

FIGURE T 4.3 A city, such as New York, is the ultimate transformation of a natural landscape by human beings.

acter. Cultural change was, for the first time, substituted for biological evolution as a means of adapting an organism to new habitats in a widening range that eventually came to include the whole earth.[1]

The changes humans have made in the environment have had numerous unplanned effects. Consider the urban environment, for instance (Figure T4.3). Concrete and asphalt have replaced an original vegetative cover. Surface streams have been severely constrained or buried in artificial tunnels. The urban view bears little resemblance to the landscape that was there a few thousand or a few hundred years ago, or even a few decades ago. In cities, humans have crafted their own environment for their own purposes. Intense human activities that characterize the city trigger another change, though. Cities exclude most species of animals and plants. Those that are allowed to live in the urban environment do so because we think they contribute to our amusement, pleasure, and comfort, or because their presence does not unduly inconvenience us. The urban environment also demands a great deal of energy. We supply this by burning coal, oil, and gas. Among the by-products of combustion are the pollutants that escape into the atmosphere. These in turn become a hazard to human health and also promote corrosion of materials such as stone, metal, and paint in the environment constructed by humans.

The city, then, illustrates the complexities of humanly induced environmental change. Human changes affect the physical landscape, alter the living conditions of plants and animals, and can lead to unplanned effects that are threats to humans and the

[1]Sears, Paul B, *Ecology of Man,* Eugene, Oregon: Oregon State System of Higher Education, 1957, pp. 20–21.

environment they create. Human history provides countless examples of the way we have changed the environment to the detriment of both humans, other living things, and to the environments that people have built. Chapter 20, "Humans as Agents of Environmental Change," the final chapter of this book, explores some of the ways in which human beings affect the environment, and how some changes can have unexpected results.

The Environment and Humans

▲▲

If humans are agents of environmental change there are, nevertheless, a number of natural processes over which we have no control and which can overwhelm us, destroying both life and property. The natural realm still remains our master in many ways. For example, if we build on a fault zone, we must be prepared to deal with the hazard of earthquakes. Currently, we cannot control earthquakes (Figure T4.4), only protect against them. Also, we have yet to contain the largest and most disastrous floods, and even with the most sophisticated equipment we are barely able

FIGURE T 4.4   A number of natural disasters cannot be controlled by humans. Their damage to life and property can be catastrophic. For example, floods up to a certain size can be controlled. When they are too large, however, there is no adequate protection, as this view of Clinton, Iowa, during the 1993 Mississippi flood demonstrates. *(M. Springer, Gamma-Liaison, Inc.)*

to predict volcanic explosions, much less control them. We have, as yet, no way of protecting ourselves against the impact of an unexpected meteorite. Chapter 20 considers natural hazards, as well as other ways the environment impacts humans.

19

Bingham Canyon copper mine, near Tooele, Utah (Dr. John S. Shelton)

Energy and Mineral Resources

OBJECTIVES

▲▲▲

*As you read through this chapter, it may help if you focus
on the following questions:*

1. How do geologists use the terms "resources" and "reserves"?
2. What are the major sources of energy that we derive from the Earth, and how are they related to each other?
3. How did fossil fuels form and how were they preserved in the Earth?
4. How do potentially valuable mineral deposits form?
5. What general criteria do geologists use for categorizing valuable mineral deposits?

OVERVIEW

▲▲▲

The processes by which nature has concentrated useful energy and material resources are the same processes we have been considering since Chapter 1. The gas you filled your car's tank with this morning was derived from marine organisms that died and accumulated in the sediments on some continental shelf. The gold in your ring or bracelet may have concentrated in a quartz vein as hydrothermal fluids percolated through fractures during the final stages of an igneous event. Locating these valuable resources is the job of geologists.

Modern civilization, particularly in industrialized nations, relies on the availability of energy. In the past century our technological society has begun using energy from sources that our ancestors knew nothing about, for purposes they would not have understood. Just as coal once displaced water, wind, and wood as the industrial world's dominant sources for energy, petroleum and nuclear power (in some places) have now displaced coal. Now, at the end of the twentieth century, industrial society seeks ways to maintain reliable and economically feasible supplies of energy.

As the demands of our technological world increase, we also need continually to locate valuable minerals in previously unexploited parts of the crust. Just as with energy sources, some materials that we extract from the Earth were familiar to civilizations thousands of years ago. Others, like the silicon and germanium used in computer chips, have exotic uses undreamed of before the twentieth century. In this chapter we will put to work everything you have learned about the Earth as we look at sources of energy and valuable materials.

General Considerations

Resources and Reserves

Geologists and others who work to extract valuable materials from the Earth describe the supplies of these materials with two terms that are commonly confused by laypersons: **resources** and **reserves**. The word "resource" refers to the total amount of a potentially valuable material (an element, a mineral, or a liquid such as water or petroleum, for example) that exists in the Earth's crust. When geologists consider the resources of coal in North America, for example, they include not only the coal that can currently be mined, but also any coal known or hypothesized to be in the crust, whether it is feasible to extract it today or not. The "reserves" of coal or of any other valuable material, however, include only the portion of the resources that can be mined economically today (Figure 19.1).

Of these two terms, "resources" is a purely geological one. Using many of the principles we have described throughout this book, geologists explore the crust for valuable materials. Geologists' familiarity with the basic processes and the history of the Earth allows them to estimate how much of any given material there is below the surface. Estimating the full extent of a resource is at best a difficult and often imprecise exercise because most Earth resources are hidden. Direct evidence of their abundance comes from surface exposures, existing mines, and exploratory drilling, but most estimates depend heavily on indirect data. Once geologists have found valu-

able materials in a specific place, and once the legal, environmental, and technological costs of extraction have been weighed against the potential market for a refined mineral product, the new discoveries may be termed "reserves." Thus, unlike "resources," the word "reserves" implies profitability and is therefore an economic term.

Renewable and Nonrenewable Resources

The world's mineral and energy resources are distributed unevenly in the crust and are concentrated in relatively small deposits. They have been concentrated by the geologic processes we have been studying, such as weathering, running water, igneous activity, and metamorphism; but only a few geologic environments have favored their formation. Deposits that have been located may soon be exhausted unless they are huge or society's consumption of them is slow. And after they are depleted, resources are gone forever, or at least until geologic processes slowly reconcentrate them. Unlike forests and fish, which are called **renewable resources** because they are replaced approximately as fast as we harvest them, we cannot grow another crop of mineral resources.

To put this another way, most of the natural resources that we exploit are **nonrenewable**. Once they have been used up, they cannot be replenished—at least not in our lifetimes.

Energy Resources

The energy that we use to heat our homes, to power planes, trains, and automobiles, and for thousands of industrial purposes comes from two ultimate sources: nuclear reactions and gravity. Nuclear reactions in the sun release torrents of energy to space, mostly in the form of light and ultraviolet and infrared (heat) waves. A small fraction of this energy, reaching the Earth, heats the atmosphere unevenly and produces **wind power.** Some of it causes water to evaporate from the land and oceans, thus driving the hydrologic cycle. Water returning to the sea under the pull of gravity then produces **water power,** once used to turn waterwheels and generate mechanical power and now tapped to turn electric generators. Green plants use some of the solar energy as they convert water and carbon dioxide to hydrocarbons by the process of photosynthesis, thus storing energy from the sun's nuclear furnace in the form of chemical energy. We can release this stored energy later by burning leaves, twigs, and branches. The **fossil fuels** (coal, petroleum, and nat-

FIGURE 19.1 As this diagram suggests, the reserves of a valuable Earth resource are only a small part of the total resources. To count as part of the reserves, a mineral or energy deposit must be able to yield a profit in today's market. Other deposits, some well explored and others only presumed to exist on the basis of geologic evidence, may be rich enough to become profitable at some future time. Yet others will never be profitable to exploit because they do not contain enough of the valuable material.

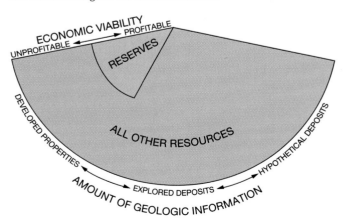

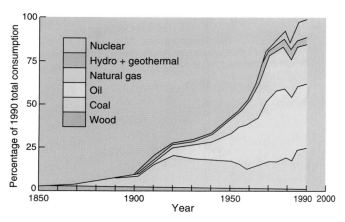

FIGURE 19.2 Energy consumption in the United States has expanded dramatically from 1850 to 1990. During this time the major source of energy has shifted from wood to coal to oil and gas. The slight decline in energy consumption in the early 1980s was due to fuel conservation measures. Since then, the demand for energy has increased again. Notice that the percentages of energy consumption shown in this diagram are relative to the total energy consumption in 1990 and are cumulative at any particular time. That means, for example, that wood, coal, and oil made up 1 percent, 23 percent, and 40 percent of the total energy consumption, respectively, in 1990. The total energy consumption in 1920, by comparison, was only 30 percent of the total in 1990. Relative to the 1990 total, the amount of energy derived from coal stayed nearly constant over the 70-year period, although it was over 60 percent of the total in 1920.

ural gas) represent the remains of plants or animals that gathered their energy indirectly from the sun millions of years ago. These are all traditional sources of energy, used by societies for many thousands of years.

Gravity and nuclear reactions have been exploited in new ways in modern times. For example, solar radiation can now be converted directly to electricity in **solar power** cells. The gravitational pull of the sun and moon on the oceans can be harnessed in the form of **tidal power,** which can now also be converted to electricity. The radioactive and gravitational heat sources within the Earth serve as sources for **geothermal power** for heating homes or driving electrical turbines. **Nuclear power** extracted from the controlled decay of radioactive elements in commercial power plants can also be used to generate heat and electricity. Each of these is used to meet human society's demand for energy. Past and present major sources of energy used in the United States are shown in Figure 19.2.

Fossil Fuels

By far the most important sources of energy are the fossil fuels, so named because they are the organic remains of long-dead animals and plants from which energy can be released by burning. In the United States they account for about 90 percent of the energy consumed.

Coal Coal is a biochemically formed sedimentary rock sometimes considered metamorphic because it is transformed from buried plant matter by the gentle application of heat and pressure. The plants that are now coal once thrived in swampy environments, commonly along an ancient coastline in a tropical or subtropical climate. A deltaic or lagoonal swamp is ideal for preserving plant matter (Figure 19.3) because in such places vegetation grows rapidly and, when dead, is buried underwater quickly by muds and other plant matter. These conditions isolate the organic debris from oxygen in the atmosphere and bathe it in acidic solutions produced by partial decay in the stagnant water. The acidic conditions inhibit bacteria, which would otherwise consume the plant debris. Most plant litter that falls to the floor of a forest or accumulates on a prairie, by contrast, is destroyed completely by bacterial decay and by oxidation.

Dead plant matter begins to experience diagenesis as soon as it is buried. The rising temperature and pressure associated with burial slowly break chemical bonds in the dead tissue, separating large and complex molecules into smaller, simpler ones. With each bond that is broken, some hydrogen and oxygen escape from the changing debris and are lost. Some of the stored solar energy is also lost, but most remains in the carbon-rich residue. The more that diagenesis proceeds, the more concentrated the carbon and the stored energy become. This concentration continues even when pressure and temperature conditions are beyond those of diagenesis and are more properly metamorphic.

Ultimately, the value of coal is determined by the amount of energy that it stores. Generally, the more carbon coal contains, the more energy it contains. Geologists classify coal into three general **grades** or **ranks** on the basis of their energy or carbon content, as shown in Figure 19.4. The lowest grade of coal, **lignite,** has about 70 percent carbon; **bituminous,** or soft coal, has about 80 percent carbon; and the highest rank of coal, **anthracite,** contains from 90 to 95 percent carbon.

As the grade of a coal deposit increases, the coal also becomes denser—more compact. A layer of dead plant matter 2 m thick compacts to produce 1 m of soft coal or 0.5 m of anthracite. The average coal seam is less than a meter thick, and seams greater than 3 m are rare (Figure 19.5).

The geologic occurrence of coal is so well understood that geologists have learned to explore for it by recognizing patterns of sedimentary facies and primary features of bedding associated with deltaic or lagoonal swamps. One such pattern, a repeating sequence of sandstone, shale, and coal beds called a **cyclothem,** typically develops when coastal swamps are periodically buried by marine sediments

FIGURE 19.3
A coal-forming swamp as it might have appeared during the Jurassic period. Trees and other vegetation falling into the oxygen-poor waters of the swamp would have been protected from bacterial decay and might have been buried rapidly by mud. (*Visuals Unlimited*)

as local sea level rises and falls (Figure 19.6). Lignite and bituminous coal are usually found in nearly flat-

FIGURE 19.4   Increasing energy yield of coal with increasing rank. Hydrogen and oxygen are lost from coal as it matures, leaving the remaining organic matter with a higher percentage of carbon and thus a higher energy yield.

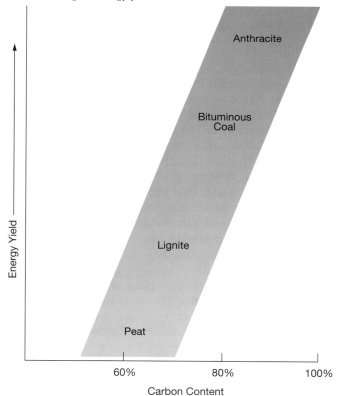

lying beds typical of sedimentation along passive continental margins or the shorelines of inland seas that have at various times flooded the stable platforms of the continents. Anthracite is more commonly mined in the folded rocks of orogens, where heat and pressure associated with mountain building have metamorphosed the coal to its highest grade.

Coal occurs in varying amounts all around the world; the United States has a particularly large supply (Figure 19.7). Most of the major coal-bearing basins have been identified, and geologists have a good idea of both the known reserves and potential resources yet to be studied in detail. Estimates of the total recoverable coal in the world vary, but 7.1×10^{12} (7.1 trillion) metric tons is a rough average. For the United States,

FIGURE 19.5   A coal seam being strip mined. (*R. Ashley, Visuals Unlimited*)

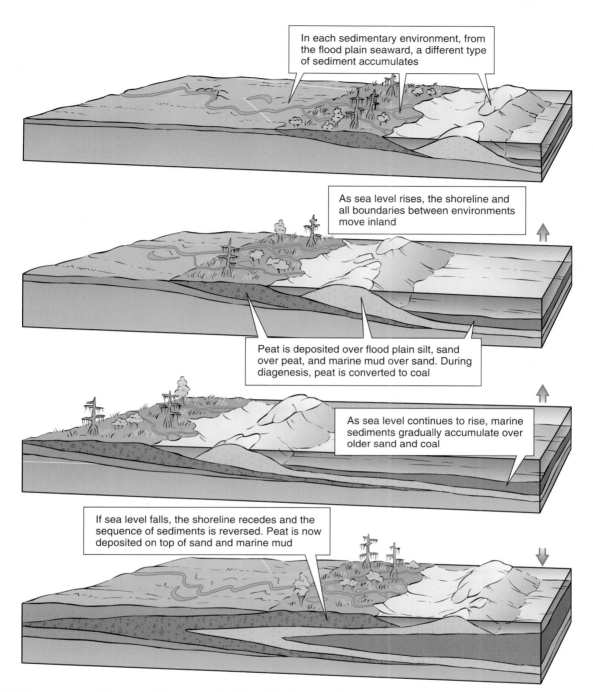

FIGURE 19.6   As sea level rises and falls, or as the land is uplifted or drops because of tectonics, a shoreline may move back and forth across coal-forming swamps. The result, seen here in cross section, is a repeating pattern of sandstone, shale, and coal beds called a *cyclothem*. *(Reprinted with the permission of Macmillan Publishing Company from* EARTH'S DYNAMIC SYSTEMS 6/E *by Kenneth W. Hamblin. Copyright © 1992 by Macmillan Publishing Company.)*

the amount of total recoverable coal is about 1.5×10^{12} metric tons. These are impressive figures and represent a resource that can last for a few hundred years.

A few hundred years, of course, is very short compared to the extremely long time it took for the world's coal resources to accumulate. Plant matter has been collecting in sediments ever since the Silurian period, 450 million years ago, when land plants first appeared. The rate of coal formation, however, has varied because global climate and coastal geography have not stayed the same. The greatest thicknesses of coal were deposited during the late Paleozoic era, while Pangea was a single landmass. Carboniferous and Permian coals are found on every continent

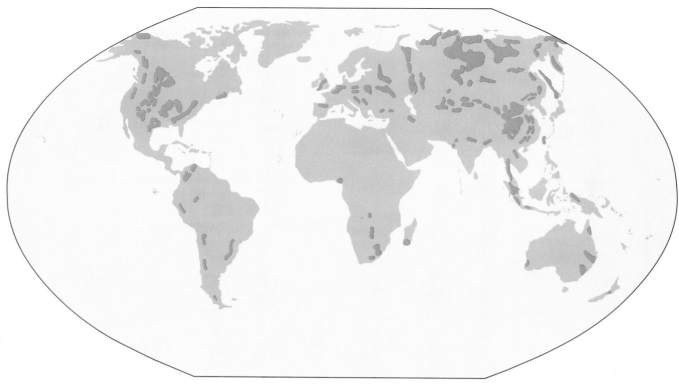

FIGURE 19.7   Known and assumed occurrences of coal in the world. *(After Gunter B. Fettweis,* World Coal Resources, *New York: Elsevier Scientific Publishing Co., 1979, Fig. J, p. 415.)*

including Antarctica. Another great coal-forming age lasted from the mid-Jurassic period until the mid-Tertiary. Most of our coal resources formed during one of these two periods of time—a combined span of roughly 250 million years. There have been times, however, when coal-forming environments were uncommon. In our own epoch, for example, very little coal is accumulating.

Oil and Natural Gas   Oil and natural gas, collectively known as **petroleum,** are mixtures of hydrocarbon compounds (composed largely of hydrogen and carbon) containing minor amounts of sulfur, nitrogen, and oxygen. The liquid and semi-solid compounds in oil are chemically complex, and there are so many of them that no two oils are exactly alike. Natural gas, by contrast, is a fairly simple mixture of methane (CH_4) and minor amounts of other light gaseous hydrocarbons. Like coal, oil and natural gas originated as living tissue. Unlike coal, they were formed from simple marine animals and plants—plankton—that live in the surface waters of the ocean. Oil and natural gas formed as the dead remains of these organisms accumulated in muddy sediments on the seafloor.

The formation of oil and natural gas is not as simple as it sounds, of course. Just as those on the continents, most marine environments contain enough

oxygen and bacteria to destroy the organic remains before they can even settle as far as the seafloor. Those remains that do reach the bottom must be buried quickly to prevent them from degrading further. These conditions can be met only where the water is fairly shallow, deep circulation of oxygen-rich surface water is limited, and sedimentation is rapid. From the Permian to the Tertiary periods, one such environment existed in what is now northern Saudi Arabia, Kuwait, and Iraq and was then a shallow asymmetric synclinal basin (Figure 19.8). Several thousand meters of clastic sediments and shallow marine limestones filled the basin as fast as it began to fold downward, burying large quantities of organic matter derived from plankton that thrived in the warm and shallow waters. The greatest volume of petroleum in the region accumulated during the Tertiary period. Today, the oil fields of these Persian Gulf states are some of the richest in the world. (See Perspective 19.1.)

The sediments destined to provide a source of petroleum start off as muds and silts, and are lithified to form shales, siltstones, and fine-grained limestones during the same diagenetic events that produce petroleum. Their organic content varies between 0.5 and 5 percent, and averages about 1.5 percent. The conversion of hydrocarbons begins at a temperature of about 50° to 60°C, once the sediments are buried to a depth

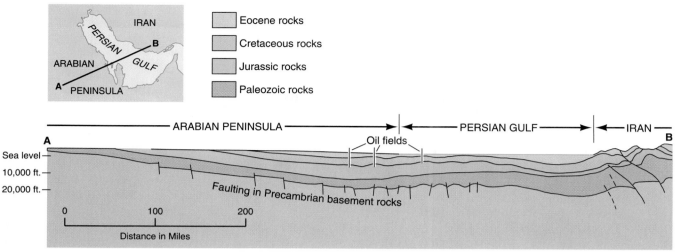

FIGURE 19.8 Generalized geologic cross section of the oil-producing region in the Persian Gulf. The basin in which organic matter and sediments accumulated throughout the Mesozoic era and much of the Tertiary period is a large synclinal fold. The Zagros Mountains in Iran, to the northeast, were formed during the Tertiary period as the Arabian Peninsula was compressed against the Eurasian plate. *(After D.A. Greig, Oil Horizons in the Middle East, American Association of Petroleum Geologists (1958), pp. 1182-1193.)*

of 1 to 2 km. Hydrocarbon formation continues to depths of 6 to 7 km and temperatures of 200° to 250°C. Oil forms largely in the shallower and cooler parts of this range, and gas in the deeper and hotter parts, as suggested in Figure 19.9. Oil and gas mature (that is, they break down to form simpler, lighter hydrocar-bons) as they age, especially as they reach higher temperatures.

A mudstone or fine-grained limestone whose organic content has been converted into petroleum is called a **source rock.** Although such a rock contains a potentially valuable resource, it can rarely be extracted

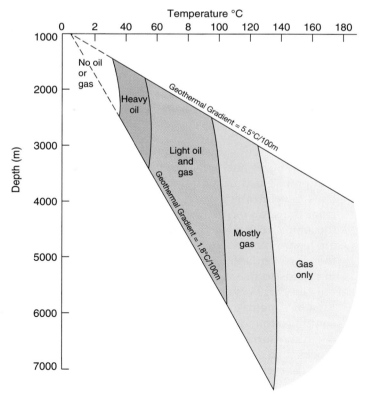

FIGURE 19.9

Organic matter must be heated to at least 50°C before it starts to convert to oil, but must be several hundred meters below the surface at the time or it will escape. With increasing depth of burial and heating, oil matures further. Petroleum heated to much more than 100°C eventually breaks down to form natural gas. In most regions, the probability of this happening increases at depths beyond about 2 km. *(This illustration was derived from Figure 17.14 in B.J. Skinner and S.C. Porter's The Dynamic Earth, 2/E, © John Wiley & Sons, Inc.)*

Drake's Folly

"Drake's Folly" marked the real beginnings of the use of petroleum as a major fuel. The "folly" of "Colonel" Edwin L. Drake (his military title was meaningless) was the drilling of the first oil well. Drake's well struck oil on Sunday, August 28, 1859, near Titusville, Pennsylvania, along Oil Creek, a tributary of the Allegheny River. The site had been chosen on the basis of persuasive evidence. Natural seeps of crude oil had long been occurring along Oil Creek. The oil had previously been collected by immersing blankets in oil-bearing springs along the creek and then wringing out the oil. The product was variously known as "rock oil," "Seneca Oil" (after the Iroquois Indian tribe that had exploited the source long before the white man), and "snake oil." Its chief use was medicinal—it was applied both internally and externally for a wide variety of ailments. (See Figure P19.1.1.)

A number of investors from New York City and New Haven, Connecticut, sponsored Drake's effort. They were convinced, in no small measure by a report written by Yale Professor Benjamin Silliman, Jr., that oil had great economic potential if it could be obtained in large enough quantities.

One of the promoters of the new enterprise, George H. Bissell, had once examined a bottle of "Kier's Petroleum, or Rock Oil" in a New York drugstore and had been struck by the fact that the oil had come from a salt well and was associated with the brine that was pumped. The well had been a drilled well, not a shallow, hand-dug one. If you could drill for water, Bissell thought, why not for oil? It is very possible that Bissell's analogy was decisive in the subsequent operation at Titusville.

Bissell and the other promoters dispatched Drake from New Haven in December 1857 to oversee the venture. On his way west he visited salt wells at Syracuse in central New York State, and when he got to Titusville, he hired an old-time salt-well driller to take charge of the operation. Twenty months later Drake pumped the first oil from his well, much to the amazement of those who had labeled the enterprise a great folly. (See Figure P19.1.2.)

Drake's strike started an oil boom in northwestern Pennsylvania as frantic as the Gold Rush in California a decade earlier. Drilling went on virtually everywhere in the area. In the 20 years after Drake drilled his well the use of rock oil

profitably and thus is not usually included in a survey of petroleum reserves. For one thing, the concentration of petroleum in most source rocks is a few percent or less. More importantly, however, most source rocks have such low permeabilities that oil cannot be pumped from them easily. Geologists do not usually consider source rocks as targets for drilling, therefore. Instead, they look for hydrocarbons that have migrated and are now concentrated in **reservoir rocks**—sedimentary rock units with greater permeability. Sandstone and porous limestone are the most common reservoir rocks.

Several processes move the hydrocarbons in the source bed to a reservoir rock. First, the weight of the overlying rocks tends to squeeze the oil and gas through pores and cracks in the rock. Second, the formation of gas during the breakdown of the hydrocarbons generates a pressure that helps to drive the oil from the source rock. Third, there is usually water in the source rock as well as in adjacent rocks. Petroleum moves upward because it is less dense than water and therefore tends to float, just as the oil in a salad dressing tends to rise and form a separate layer on top of the vinegar.

Once petroleum begins to migrate, of course, it continues to move upward. Eventually the oil and gas reach the surface and are lost there unless something blocks their way and holds them in the ground. A **cap rock,** a rock unit with a low permeability (such as shale), will block or divert oil and gas in their migra-

(petroleum) for illumination and lubrication grew rapidly. A modest commercial venture had introduced the world to a valuable new energy resource.

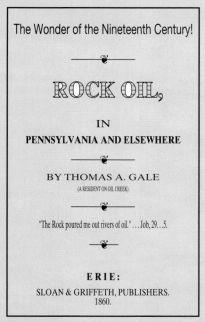

The Wonder of the Nineteenth Century!

ROCK OIL,

IN

PENNSYLVANIA AND ELSEWHERE

BY THOMAS A. GALE

(A RESIDENT ON OIL CREEK)

"The Rock poured me out rivers of oil." ...Job, 29...5.

E R I E :
SLOAN & GRIFFETH, PUBLISHERS.
1860.

P19.1.1 Thomas Gale lived on Oil Creek near the spot where, in the summer of 1859, "Colonel" Edwin Drake drilled the first oil well in the United States. He lost no time in writing a short treatise on petroleum—or, as many then called it, "rock oil."

P19.1.2 A reconstruction of the pump house and derrick housing for Edwin L. Drake's first oil well at Titusville, Pennsylvania, in 1859. *(Jane Grigger)*

tion. If the cap rock is properly positioned, it and adjacent reservoir rocks together form a **trap** and serve as a natural tank. Traps can have several forms, as illustrated in Figure 19.10.

Once the seal provided by the capping bed is broken by a drill, the oil and natural gas move from the pores of the reservoir rock to the drill hole and then can be brought to the surface for processing and distribution. Of course, the trap may be broken by natural means as well. For instance, Earth movements after the formation of the trap may so disturb it that fractures develop, and the petroleum will escape either to another trap or to the surface. Erosion may also eventually reach and destroy the trap. Geologists observe, therefore, that the older the rock, the less chance there

is that it still contains oil. The greatest amount of oil discovered (nearly 60 percent) comes from rocks of Cenozoic age, followed by Mesozoic and then Paleozoic rocks. There is essentially no oil remaining trapped in Precambrian rocks.

Over the past century, geologists have become adept at finding promising combinations of source rocks, reservoir rocks, and natural traps, and have found immense quantities of petroleum. Because petroleum is a nonrenewable resource, however, many people would like to know how much longer we can continue to find untapped sources of oil and gas. Even as exploration methods become more sophisticated, the amount of newly discovered petroleum diminishes each year. It is difficult to estimate with precision how

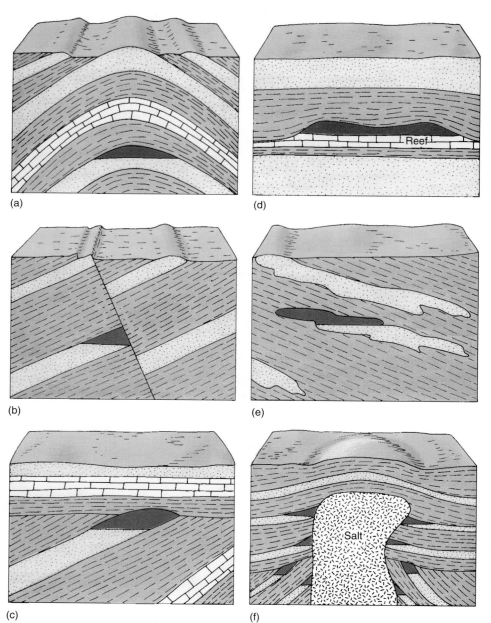

(a)

(b)

(c)

(d)

(e)

Reef

Salt

(f)

FIGURE 19.10

The common types of oil traps, drawn here in cross section, include (a) anticlines, (b) faults, (c) unconformities, (d) reefs, (e) sand lenses, and (f) salt domes. Oil accumulation is shown in dark green.

much is left. However, several estimates conclude that if we continue to use oil and gas at the current rate, we will run out of oil and natural gas in the United States early in the twenty-first century.

The picture on a worldwide scale is similar to that in America, although there is probably a greater percentage of undiscovered petroleum elsewhere than there is in the extensively explored United States. Nevertheless, when we consider how fast we are using oil and natural gas today we must conclude that the world supply will be consumed within about a hundred years—a very short time span in terms of human history. Advances in technology or desperation may eventually make it possible to extract petroleum more

efficiently from currently unprofitable reservoirs, but these too have finite lifetimes. What will replace them?

Oil Shales and Tar Sands During the Eocene epoch organic-rich sediments accumulated in large freshwater lakes that spread across portions of what are now Colorado, Utah, and Wyoming. Some of the shales that formed from them are source rocks for petroleum, rich enough to produce more than a 42-gallon barrel of oil per ton of rock. Geologists estimate that there is at least as much oil in those shales (called **oil shales**) as there is in all of the oil fields still left in the United States. However, because oil shale is a source rock rather than a reservoir rock, it is impossible to extract petroleum from it by conventional drilling and pumping methods. Instead, the rock must be mined from the Earth and then heated to "cook" the oil out of it. For many years pilot plants have been experimenting with ways to process the shale, but large-scale extraction operations have not begun. In part this is because the oil from processed shale is more expensive than traditional oil-field petroleum and in part it is because of environmental concerns. For instance, as oil shale is heated to extract oil, the shale expands, so that "used" shale takes up 20 percent more room than unprocessed shale. Immense volumes of rock must be used in any practical exploitation of the oil shale, so the problem of what to do with leftover shale is not a small one. Beyond this, the extraction process itself demands large amounts of water, already in short supply in the western states.

Another potential source of petroleum is the heavy, highly viscous hydrocarbon mixture that geologists call **tar** or **asphalt**, usually dispersed in loose sands known as **tar sands.** Large tar sand deposits have been discovered in Canada, Venezuela, and Russia. From these three localities plus several known small ones, tar sands appear to have the potential of yielding as much energy in the form of crude oil as all the oil and natural gas originally available in the traditional petroleum fields of the world. There may be between 250 billion and 500 billion barrels of recoverable oil in the Canadian deposits alone.

One technical problem in producing oil from tar sands is separating the sticky tar from the sands. Another problem is transportation; the hydrocarbons are too stiff to flow through pipelines. Furthermore, not many refineries are equipped to process this very heavy hydrocarbon.

If technical problems can be overcome, both oil shale and tar sands may one day prove to be profitable sources of petroleum. In the near future, however, these remain rich but untapped while energy companies exploit more conventional oil fields.

Other Energy Sources

Water Power Water, moving downhill under the force of gravity, has been used as a source of energy for at least 2,000 years and perhaps longer (Figure 19.11). Thanks to the hydrologic cycle, water has the advantage of being a renewable resource. Water power is clean and relatively cheap, but there are some drawbacks to its use. The dams generally necessary for the storage of water have limited lifetimes because sediments silt up the lakes formed behind them. They also flood a great deal of prized scenery and productive agricultural land and may change river-flow patterns in undesirable ways.

FIGURE 19.11 Hoover Dam, near Boulder City, Nevada. *(Frank M. Hanna, Visuals Unlimited)*

About 25 percent of the water-power capacity of the United States has already been developed, as has about 6 percent of the capacity worldwide. Africa and South America have the greatest water-power potential; thus far very little, about 1 percent, has been developed on these continents. It is estimated that if the world's water power were fully developed, it would meet only about 30 percent of the total demand for energy today.

Nuclear Power The nucleus of an atom, as we discussed in several earlier chapters, is made up of protons and neutrons. It is the **binding energy,** holding these particles together in the nucleus, that forms the basis for nuclear power. This energy can be released either by **fission,** breaking down the nucleus of an atom by striking it with a free neutron, or by **fusion,** joining the nuclei of two elements together. Fission has been used to create explosive energy releases in the atomic bomb, whereas fusion has been used in the hydrogen bomb. Scientists have been able to develop ways to control fission reactions to generate electric power, but they cannot yet control the fusion process.

For nuclear fission to take place spontaneously, there must be enough nuclei close to each other so that neutrons bombarding the area cannot fail to hit one and split it, producing more loose neutrons. The loose neutrons are important because they go on to smash into other atoms in a self-sustaining or **chain reaction.** This reaction, so explosive in the atomic bomb, is controllable and is used to heat water and drive electricity-generating turbines. By 1988 about 575 commercial nuclear reactors were operating worldwide (Figure 19.12).

To date, scientists have been able to harness fission reactions only in an isotope of uranium, ^{235}U, which, like the fossil fuels, is a nonrenewable resource. In the United States the richest uranium deposits occur in sedimentary rocks in New Mexico, Wyoming, and Texas. Rich deposits are also mined in South Africa, Tadzhikistan, and Australia. However, the reserves in these are limited.

There is a way to generate power from uranium without severely depleting the reserves. This is done by replacing ^{235}U with artificially produced atoms that can be split instead. Under controlled conditions, for example, the much more abundant ^{238}U may absorb a neutron in its nucleus and be transformed into ^{239}Pu (plutonium), which is fissionable and can sustain a chain reaction. In the same way, neutrons will convert ^{232}Th (an isotope of thorium), also not naturally fissionable, into fissionable ^{236}U. The idea, then, is to build a reactor in which loose neutrons from the decay of ^{235}U create radioactive ^{239}Pu or ^{236}U at least as fast as the ^{235}U is used up. This way of creating new radioactive fuels is called *breeding*, and the reactors in which they are used are called **breeder reactors.** They are not yet in common use.

FIGURE 19.12   Electrical power is generated by steam that is heated by controlled fission reactions in commercial nuclear power plants like this one, in Charleston, W. VA. *(E.R. Degginger, Earth Scenes)*

FIGURE 19.13   Steam from this geothermal field at Lardarello, Italy, is carried in pipes to drive turbines for electrical power generation.

Geothermal Power   Volcanoes obviously release immense quantities of geothermal energy, as do geysers and hot springs. This energy is currently being harnessed, although on a relatively small scale. For instance, most of the buildings in Reykjavik, the capital of Iceland, are heated by hot springs on the island. The Geysers, a hot-steam field north of San Francisco, generates electricity on a commercial scale; and the geothermal field of Larderello in central Italy has been producing power since before World War II (Figure 19.13).

Unfortunately, the heat flow at most places on the Earth is so small that general exploitation of geothermal energy is impractical. Only at those areas where heat flow is unusually high, such as plate boundaries and hot spots, can geothermal energy be tapped. In those areas where there are pools of geothermally produced hot water and steam, heated water may be tapped by drill holes and brought to the surface in a controlled fashion to produce electrical power. In some areas, such as the Jemez Mountains of New Mexico, however, hot igneous rocks are close enough to the surface that they make an ideal prospect for geothermal energy, yet no pools of heated water or steam exist. The problem is that overlying rocks are dry, mostly because they have a very low porosity. Pumping hot water or steam out of the ground is therefore impossible.

In New Mexico, however, scientists with the U.S. Department of Energy are trying an alternative approach. They have drilled a deep hole into the hot rock and then forced water into it under very high pressure, causing thousands of fractures to form. This process, called **hydrofracturing,** increased the porosity and the permeability of the rocks artificially. The next step was to drill a second hole into the fractured rock a short distance away. Now, when the scientists pump cool water into one hole, they can pump hot water out of the other, thus making use of the geothermal energy source.

One drawback to geothermal power is that it can only be used in those few places where there is a lot of heat just below the surface, and even there it is limited and nonrenewable. Also, the hot water dissolves mineral matter as it passes through rock. This dissolved material then reprecipitates in pipes in the generating installation, forming a scale that rapidly reduces plant efficiency and increases maintenance costs. Disposal of the mineral matter can also pose environmental hazards.

Sun, Wind, and Tides

If you have ever walked on asphalt pavement or along a sandy beach on a hot day, you are aware of the energy we receive from the sun. Architects and engineers make use of this energy for home heating in many of today's energy-efficient houses. Concentrating solar energy for more intensive purposes such as generating electricity, however, is difficult. The sun shines on each part of the Earth for only a portion of the day, and is easily blocked by clouds. Beyond these natural limitations, the greatest problem is a technological one. Most electricity is generated by burning fossil fuels or by using nuclear reactions to heat water for steam turbines. As warm as the sun's rays may feel on a summer day, their energy is too low to use for steam generation unless solar heat could be collected over an unrealistically large portion of the Earth's surface. Instead, we use solid-state electronic devices called **solar cells** that convert sunlight to electricity directly. These devices, distantly related to transistors and other semiconductor components, are commonly used to run wristwatches and pocket calculators that have low demands for power. Only occasionally are they used for more exotic purposes (Figure 19.14).

Wind, ultimately driven by uneven solar heating of the Earth's surface, is an important source of electrical power in some places, but usually only to meet the needs of individual homes or farms. A few **wind farms,** each with many windmills, have been estab-

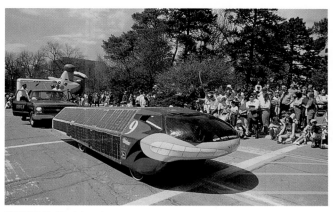

FIGURE 19.14   Solar energy, even with the best technology available, is still an unreliable source of power for running the family car. That does not prevent engineering students at Iowa State University from building experimental vehicles to see just how much power they can coax out of sunlight. *(Iowa State University Photo Service)*

lished, however. Sites for these are chosen carefully to take advantage of local topography that may channel the prevailing winds in valleys or mountain passes. The largest of these wind farms are in California and New Mexico (Figure 19.15).

The **tides** of the world, the last alternate source of energy we will consider, are produced by the gravitational pull of moon and sun, the moon exerting the greater influence. Tidal waters racing into and out of some estuaries can be used to generate electricity from water power. The Rance River estuary in northern France is producing tidally generated electricity, for example, as is a plant in northern Russia along the coast of the White Sea. The Passamaquoddy area in the Bay of Fundy on the Maine-Canadian border has long been cited as a possible place to install a tidal

FIGURE 19.15   Electrical energy is generated by the giant propellers at this wind farm in Altamont Pass, east of San Francisco. Projects such as these are only practical in places where the wind is steady and predictable, as in a mountain pass. *(J. Sohm, The Image Works)*

plant, although thus far it has not proved economically or politically feasible.

Mineral Deposits

Geologists use the words **mineral deposit** to describe any natural concentration of a valuable material in the Earth's crust, whether that material can be extracted profitably or not. The valuable material may contain a metallic element such as copper or gold, or it may be a material that is not valued for metals it may contain, but instead for its potential as an industrial commodity (for example, salt, building stone, or sand and gravel). Geologists thus classify mineral deposits as either **metallic** or **nonmetallic.**

Geologists use their knowledge of how the rock system works to help locate mineral deposits and then to work with mining engineers and skilled laborers to extract valuable materials from those deposits.

Mining

Mining is the removal of economically useful materials from the Earth, an enterprise that humanity has pursued for thousands of years. Materials may be mined either above or below ground. For instance, the surface process of **quarrying,** in which the objective is to remove rock in large, unfractured blocks, is used to extract building stone (Figure 19.16). **Open pit mining,** which leaves an excavation with characteristic 15-m-high steps on its sides (Figure 19.17), is also carried on from the surface, as is **strip mining,** a method in which large power shovels remove soil to expose coal or other materials in shallow sedimentary deposits. Other surface-mining techniques use running water to sort materials such as gold, diamonds, or tin-bearing minerals from nonlithified sands or gravels.

When most people think of mining, however, they usually think of excavation underground (Figure 19.18). **Underground mines** may extend a few dozen meters into a hillside or may extend to two or more kilometers below the surface. Many different ways exist to extract valuable minerals from underground, but most are based on the use of air-driven rotary percussion drills, explosives, and power shovels to excavate a system of shafts and tunnels. The excavations are planned to follow the richest material, but much valuable material must stay behind in unexcavated rock pillars that keep the tunnels from collapsing. Despite the danger of rockfall or of explosion due to accumulation of hydrocarbon gases, and the technical problems of providing light and air and of removing water, underground mining is the only option for exploiting all but the shallowest mineral deposits.

FIGURE 19.16 Some rock material, as in this Vermont granite quarry is used as it comes from the ground, needing only to be cut to the desired dimensions. Most of this granite will be cut into thin slabs and used for decorative facings on buildings, or for cemetery headstones. *(Jane Grigger)*

Geologic Processes That Form Mineral Deposits

Magmatic Processes By now you should be familiar with the fact that minerals crystallize from magma in a sequence. Some minerals, forming early in the cooling of the magma and being denser than the remaining melt, may settle toward the bottom of the magma chamber. This process of fractionation is responsible for the formation of several types of metallic mineral deposits. In the iron mining district of Kiruna in northern Sweden, for example, magnetite (Fe_3O_4) settled to the floor of a large intrusive igneous body as it crystallized. The result was a metal-rich layer that is over 60 percent iron on average. Similar segregations can occur in dikes, sills, layered igneous intrusions, or even in some lava flows. In addition to iron, metals commonly mined from these bodies include chromium, nickel, platinum, and cobalt, as well as many others.

FIGURE 19.17   The Bingham Canyon mine in Utah is an open-pit operation, mining a porphyry for its copper content. Each of the steps in the pit (called a "bench") is approximately 30 m high. *(Dr. John S. Shelton, from:* Earthquakes *by Bolt. Copyright © 1993 by W.H. Freeman and Company. Used with permission.)*

FIGURE 19.18   Anthracite coal is often mined underground from seams that are 1 to 3 meters thick. Unlike mining operations for metallic ores, which commonly require the use of hydraulic drills and high explosives to fracture the rock, coal mining can be done with machines such as this one that carve their way with rotary shovels. *(Science Vu-Nca, Visuals Unlimited)*

As a granitic magma cools, the fluids that remain in the final stages may produce the coarse-textured rock called *pegmatite,* as we described in Chapter 3. Most pegmatites simply contain large crystals of orthoclase and quartz with some mica. A few complex pegmatites, however, are enriched with scarce metals like beryllium, tantalum, niobium, and lithium. These may substitute for abundant metals in feldspars and micas, but often are concentrated enough in unusual pegmatitic minerals that, although rare, are the primary industrial sources for those metals. Pegmatites are also mined for precious and semiprecious gemstones such as beryl, topaz, and tourmaline (Figure 19.19).

Hydrothermal Processes   As we have seen, hydrothermal fluids consist of water that contains varying amounts of dissolved gas (largely CO_2), salts (predominantly chlorides of sodium, potassium, and calcium), and valuable metals. These fluids may originate in a magma or in surrounding sediments, or may be surface waters that have circulated through pore spaces and fractures. Hydrothermal solutions, heated by igneous or metamorphic activity, may carry ions from hot to cooler parts of the crust, where they can be precipitated to form mineral deposits that may be rich enough to mine.

Geologists do not fully agree on all the processes that cause metals to precipitate from hydrothermal solutions and create a mineral deposit. In some instances, it is clear that metals can be precipitated simply by cooling the fluid that carries them, just as sugar will gradually crystallize from a saturated solution once you remove it from the stove. The quicker the cooling, the more mineral matter is precipitated in a small area. Rapid changes in pressure, which can make a hot fluid start to boil, can also cause precipitation. In other cases, geologists have concluded that mineral matter precipitated as the hydrothermal fluid mixed and reacted chemically with a second fluid or with the rocks it invaded. Thus, a fluid containing dissolved metals might mix with another containing dissolved sulfur, generating a metal sulfide deposit.

Hydrothermal deposits can take many forms, depending on the type of rocks in them, their structural and tectonic setting, and the chemical compositions of both the rocks and the hydrothermal fluids involved. As a class, they include some of the most valuable and heavily studied mineral deposits. Many gold and silver mines, for example, are located in hydrothermal deposits, as are most of the major mines for copper, lead, zinc, and molybdenum. The largest hydrothermal deposits are **metal porphyry deposits,** named for their association with plutons of porphyritic granite or granodiorite and located in a belt from Alaska to Chile and in many other orogens around the world. Characteristically, the metal-bearing minerals in these deposits are very finely dispersed in the porphyry and surrounding sedimentary rocks, so that a great deal of rock must be processed to extract a relatively small amount of metal. (See Perspective 19.2.)

As you might expect from their association with magmatic and metamorphic events, hydrothermal deposits are found exclusively at plate boundaries or in orogens adjacent to them. The black smokers that we discussed in Chapter 6 are examples of modern-day deposition along the mid-oceanic ridge system, a plate boundary of a different kind. Geologists recognize many older deposits that were probably formed in this same way, which, because they are associated with volcanic activity at spreading centers and result in thick layers of metal sulfide minerals, are known as **volcanogenic massive sulfide deposits.**

FIGURE 19.19   Pegmatites are the source for many semi-precious gems like these from Oxford County, Maine. Crystals in a pegmatite grow to unusual sizes and may have superb clarity because they form very slowly. *(William E. Ferguson, State of Maine Tourist Bureau)*

FIGURE 19.20   The shiny, metallic layers in this sample of banded iron formation are almost pure hematite (Fe_3O_4). The dark red layers that alternate with the hematite bands are composed of chert, a finely crystalline form of quartz. The red color is due to small particles of hematite.

Sedimentary Processes

In some instances valuable minerals have been formed by the process of sedimentation. One widespread and heavily mined example of this type is called **banded iron formation,** also known among geologists by its abbreviation, **BIF.** The rock in a BIF consists of thin beds of iron minerals (typically hematite [Fe_2O_3], siderite [$FeCO_3$], or a hydrated iron silicate) alternating with beds of chert (Figure 19.20). Most BIFs were formed by chemical precipitation from the ocean during the late Archean and early Proterozoic eons, between 2 billion to 3 billion years ago, and are preserved only in ancient sedimentary basins in cratons. Typically, unaltered BIF contains about 15 to 30 percent iron. Unfortunately, the iron minerals are usually fine-grained and so intimately mixed with chert that they are difficult to separate. For that reason, iron is mined only from mineral deposits in which the average grain size has been increased by metamorphism—a variety of BIF known as **taconite**—or ones from which the chert has been removed by deep weathering.

Evaporites, which we discussed in Chapter 5, comprise another important class of sedimentary mineral deposits. Deposits of halite (NaCl) and gypsum ($CaSO_4 \cdot 2H_2O$), the most abundant evaporite minerals, have been found on every continent and in rocks of all ages. The most extensive evaporite deposits were formed during the Permian period, when shallow seas covered portions of Pangea. These are preserved today in widely separated basins from northern Europe to western Texas.

Weathering Processes

Weathering processes can help form very rich mineral deposits in either of two ways: They may leach valuable elements from one level in a soil or rock unit and concentrate them chemically in another, or they may leach the less valuable elements from a given area, thus leaving the valuable ones behind. Either process can produce a rich ore that is easily mined with bulldozers and power shovels.

The brilliant canary-yellow uranium ore, carnotite ($K_2(UO_2)_2(VO_4)_2 \cdot 3H_2O$), is an example of the first kind of weathering-produced deposit. Uranium, at low concentration, is common in many shales. Under some conditions, groundwater passing slowly through the shales can oxidize the uranium and leach it into more permeable sediments. As this groundwater comes in contact with organic matter, however, the dissolved uranium changes chemically to a less oxidized state, and uranium precipitates as carnotite or other rare minerals. The organic matter may be the remains of plants in modern soils, or it may be the hydrocarbons found in a coal- or petroleum-rich sediment. Even if there is too little uranium to form carnotite, this chemical process concentrates uranium well enough that a typical coal is slightly more radioactive than other sedimentary rocks.

Bauxite, the chief ore of aluminum, is an example of a material found in the second type of weathering-produced deposits. It is a rock composed of several

The Mines of Laurium

The city-state of ancient Athens attained its greatest glory during its so-called Golden Age in the fifth century B.C. when it was the artistic and intellectual center of the Western world. The age was Golden, but it was made possible in no small part by the silver extracted from state-owned mines at Laurium, 40 km southeast of Athens.

Rich veins of silver-bearing galena (PbS), deposited in fractures as hydrothermal solutions once percolated through the local limestone, had been worked for nearly 1,000 years previously. Under Athenian ownership the mines were leased by private citizens, who paid the state a royalty of about 4 percent of the value of the metal produced. The deposits were rich enough and so easily accessible that they required no sophisticated equipment to mine. Following the standard procedure of the day, mine owners used slave labor to break the rock with hand tools, and then crush it and separate the galena by hand. The silver could be extracted from the galena, where it was a minor substitute for lead atoms, by "roasting" the galena in a furnace. The metals would collect on the bottom of the furnace to form a puddle, on which the silver would float like a scum on the molten lead.

From the mines of Laurium came the wealth that built much of the combined Greek fleet that defeated the Persians at Salamis in 480 B.C. Athenian silver also did much to establish the monetary standard that dominated Mediterranean trade of the times. Intensive excavation, however, depleted the mines of Laurium quickly. Even by the early fourth century B.C., little silver was left. By the second century B.C., miners were reduced to searching through the rubble of discarded rock for the last remaining traces of silver-bearing galena.

hydrated aluminum oxide minerals, collectively described by the chemical formula $(Al_2O_3 \cdot nH_2O)$. The hydrolysis reaction that forms clays from feldspar (discussed in Chapter 4) is the first step in producing bauxite, but it continues, thereby leaching all of the remaining silica and converting the clays to hydrous aluminum oxides. Because weathering is most rapid and intense where the climate is hot and rainy, these kinds of deposits are typically found in tropical regions like the Caribbean, Malaysia, Indonesia, and equatorial Africa and South America. The tropical soils from which bauxite is mined may be as much as 30 m thick and nearly pure aluminum ore (Figure 19.21).

Hydraulic Processes: Placers Weathering sometimes frees minerals that are resistant enough to resist being abraded or decomposed themselves. These minerals can then be concentrated by running water in streams or by waves on beaches in deposits called **placers** (pronounced as if it were spelled *plassers*). Because placer minerals are heavier than the usual sand and gravel, these deposits may collect in ripples, cracks, or holes, or simply in places where the water is less energetic than usual, such as on sandbars. Gold and cassiterite (tin oxide) are two dense minerals that collect in this way. Diamonds, which resist both decomposition and scratching, survive as coarser gravels in some placers.

Placer deposits of gold are famous. The Gold Rush of 1849 that led to rapid immigration and nearly immediate statehood for California was set off by the discovery of gold in stream gravels at Sutter's Mill in the hills near Sacramento. In 1898 the discovery of gold in beach sands on the Seward Peninsula near the present site of Nome triggered the Alaskan Gold Rush. In both places, dispersed gold eroded from quartz veins had washed downstream and been concentrated in placers. Some miners, failing to understand how this natural process works, assumed that the gold must have come from a fantastically rich "Mother Lode" upstream. In fact, however, gold is much more concentrated in placers than in the source rocks from which it erodes (Figure 19.22).

Not all placers are in active streams or shorelines. The world's richest deposits of gold are found in the Witwatersrand of South Africa, where they occur in

FIGURE 19.21   Bauxite, the most common ore for aluminum, is a residual weathering product. It is mined in tropical regions, where the depth of intense weathering may be 30 to 40 meters. *(G. Prance, Visuals Unlimited)*

FIGURE 19.22   Panning for gold requires patience, but can yield great rewards. As the prospector swirls the pan, light grains of quartz and other minerals wash over the side, leaving denser particles of gold behind. This is an articifial way of doing what happens in a stream as a placer gold deposit forms. *(W. Kleck, Terraphotographics/BPS)*

Precambrian conglomerates of wide extent. The deposits give every indication of being ancient stream gravels and pebbles in which gold collected.

Metallic Resources

When the concentration of metals in a metallic mineral deposit is high enough to make mining profitable, geologists refer to the deposit as an **ore deposit,** and they call the rock from which metal is to be extracted an **ore.** The degree of concentration necessary to qualify as an ore varies from one metal to another. Geologists classify five metals (aluminum, iron, magnesium, titanium, and manganese) as **abundant metals,** meaning that the concentration of a metal in a typical ore is no more than four or five times as great as its concentration in the Earth's crust. For example, aluminum makes up 8 percent of the crust, and an aluminum ore is a rock that contains about 30 percent aluminum or more. The remaining metallic elements are the **scarce metals,** each so low in its crustal abundance that it usually substitutes randomly for atoms of abundant metals in common minerals rather than forming minerals of its own. A chemical analyst could find copper, for example, in almost any rock, but only dispersed as a substitute for individual atoms of iron, magnesium, or another of the abundant metals. It is not feasible to extract copper from a rock unless natural processes have concentrated enough copper to form minerals such as chalcopyrite ($CuFeS_2$) in which copper atoms are a fundamental part of the structure (Figure 19.23). Because copper atoms are so rare in the crust, they must be concentrated

much more than merely four or five times before chalcopyrite or other ore minerals form. A typical copper ore contains 0.5 to 0.8 percent copper, but that is 80 times more than the crust on average.

Other scarce elements must be concentrated even more in nature before they are profitable to mine. Zinc ore, for instance, must be 300 times richer than average rock, gold ore almost 400 times, and mercury ore 100,000 times (Figure 19.24). If a metal is extremely scarce, it is difficult to find mineral deposits that are enriched enough to be worth mining. Such metals are commonly extracted as **co-products** of mining for more abundant metals. For example, natural processes do not concentrate enough cadmium to form mineral deposits that can be mined profitably for that metal alone. Cadmium is always extracted instead from zinc ores, in which it is a minor constituent.

FIGURE 19.23   The copper ore mineral in this photograph is chalcopyrite, $CuFeS_2$, seen here in association with the carbonate mineral dolomite ($CaMg (CO_3)_2$). Copper is among the most common scarce metals, mined from a variety of deposits formed by hydrothermal and weathering processes.

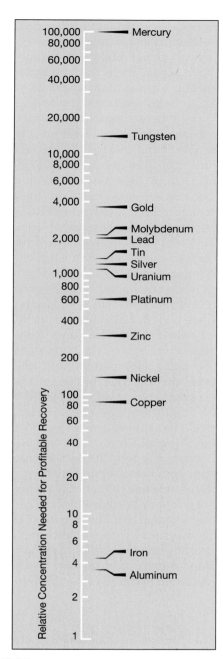

100,000
80,000
60,000
40,000
20,000
10,000
8,000
6,000
4,000
2,000
1,000
800
600
400
200
100
80
60
40
20
10
8
6
4
2
1

Mercury
Tungsten
Gold
Molybdenum
Lead
Tin
Silver
Uranium
Platinum
Zinc
Nickel
Copper
Iron
Aluminum

Relative Concentration Needed for Profitable Recovery

FIGURE 19.24   To be mineable, the metals in a mineral deposit must be concentrated to several times their average abundance in the crust. The scarcest metals, such as tungsten and mercury, must be concentrated by a factor of 10,000 or more to be worth mining. Deposits of abundant metals such as iron need only to be four or five times more concentrated than average rocks to make them profitable. Notice that the vertical scale on this diagram is logarithmic, so that each major division is ten times greater than the previous one. *(Modified from Brian J. Skinner,* Earth Resources. *Englewood Cliffs, N.J.: Prentice Hall, 1986)*

Nonmetallic Resources

Although the metals are more frequently associated with mining, the nonmetals are equally valuable and just as heavily mined. The nonmetals include mineral resources used in agriculture, the chemical industry, and building and construction. What follows is an overview meant to illustrate the wide range of non-metallic resources and the ways they are used.

Fertilizers   The increased productivity of agricultural land during the last 50 years has been partly due to the enormous increase in the use of chemical fertilizers, all of them synthesized from natural materials. The most important elements for the fertilizer industry are *nitrogen, potassium, phosphorous,* and *sulfur.*

Much of the nitrogen used in fertilizers is in the form of nitrates (compounds of nitrogen with oxygen and other elements). Nitrates are concentrated by weathering processes, but only in arid climates because they are highly soluble in water and readily washed away. Early in the twentieth century, most of the major nitrate mineral deposits were exhausted by mining, and very little is mined today. Instead, nitrogen from the atmosphere is combined industrially with hydrogen to form ammonia or related compounds, or with oxygen to form artificial nitrates.

Potassium is one of the most common elements in the Earth's crust, but it is unusable by plants and animals until turned into soluble form. Thus evaporite deposits, in which potassium has precipitated from seawater as a salt, are better sources from which to extract the potassium used in fertilizer than are deposits of more abundant potassium minerals such as orthoclase or the micas. The evaporite minerals supplying the bulk of our resources of potassium are sylvite (KCl) and carnalite ($KCl \cdot MgCl_2 \cdot 6H_2O$), two of the last minerals to precipitate in the marine evaporite series (see Chapter 5).

Phosphorus is extracted from the mineral apatite, ($Ca_5(PO_4)_3OH$). Most of the profitable deposits are in marine sedimentary rocks called *phosphorites*, but apatite has also been mined from igneous rocks.

Sulfur, in the form of sulfuric acid, is used heavily in the fertilizer industry to extract phosphorous from phosphorites and to put both nitrogen and potassium into usable forms. Sulfuric acid is also important in a wide variety of other industries. Until recently, much sulfur was mined as a native element or was extracted chemically from gypsum ($CaSO_4 \cdot 2H_2O$). As industries that used to release large amounts of sulfur oxide gases into the atmosphere have become more environmentally aware, however, an increasing amount of commercial sulfuric acid has come from toxic waste recovery systems.

Minerals for the Chemical Industry   The chemical industry, which synthesizes and purifies the fundamental compounds (such as acids, solvents, and salts) used in other manufacturing operations, depends on raw materials from the Earth. All of the minerals we have discussed in this chapter (as well as the many more that have gone unmentioned) provide elements for the myriad of chemicals eventually used to manufacture plastics, synthetic fibers, glass, and other commodities. We have already pointed to sulfuric acid as an example of a fundamental manufacturing compound, used well beyond the fertilizer industry. Halite (NaCl) is another example. It is most familiar as the salt for our table, but also serves as the raw material from which factories produce such diverse chemical products as chlorine gas, soda ash (Na_2CO_3), and sodium sulfate (Na_2SO_4), each of which is important in some other manufacturing operation.

Building Materials

Earth's crust provides us with a tremendous variety of materials for construction. Some material is used as it comes from the Earth, modified only in shape and size. Other rock goes through extensive change as it is converted into essentially new material.

Until the widespread use of concrete and steel in construction, *building stone* was used as the structural support in buildings (Figure 19.25). Today stone is used as an exterior facing and trim, as well as for interior paneling, steps, and decorative purposes (Figure 19.26). Most widely used are the sedimentary rocks limestone and sandstone, the igneous rock granite, and the metamorphic rocks marble and slate. These by no means exhaust the list of stones in buildings, however. Builders and architects have found uses for gneiss, basalt, hardened volcanic ash, conglomerate, peridotite, and many others. To be useful as building stone, a rock needs only to be attractive and durable.

Rock does not always have to be attractive to be useful in the building trades in other ways, of course. The dollar value of *crushed rock, sand,* and *gravel* produced each year in the United States is greater than the value of any mineral or energy resource except for petroleum extracted from the country's wells. Although they are "as common as dirt," sand and gravel are most useful if natural processes have sorted them into a narrow range of particle sizes and washed them free of clay. Instead of simply digging holes randomly, therefore, geologists look for places where running water has separated clastic sediments into fine-, medium-, and coarse-grained fractions. Sand and gravel are taken directly from stream beds, from the terraces of old streams (Figure 19.27), from the outwash of now-vanished glaciers, from beaches, and even from the continental shelves. Crushed rock is used primarily as an aggregate for concrete and as a base for streets and roads. Most crushed rock is limestone, but other common rocks such as granite, basalt, and well-cemented sandstone are also quarried for this purpose.

Sand, gravel, and crushed rock are necessary components in concrete, but we could not bind them together if we did not also have *cement,* a Roman invention of nearly 2,000 years ago. Roman cement was a mixture of quicklime (CaO), produced by heating limestone, and a particular volcanic ash called *pozzuolana,* named after the town of Pozzuoli near Naples. When water, sand, and gravel were added to this mix, the result was a hard and durable concrete.

The process for making cement was forgotten with the fall of the Roman Empire and was rediscovered barely two centuries ago. Today's cement is made from a mixture of limestone and clay or shale, or from an impure limestone with a high clay content. The raw materials are ground, heated almost until they melt, and then ground again to a fine powder. Cement manufactured by this simple process and used in concrete is an indispensable modern building material.

For indoor construction, a more practical choice than concrete in many buildings is *plaster,* most com-

FIGURE 19.25 Before the twentieth century, stone was the favored material for making large buildings like these in England. Because of its high compressive strength, rock can support immense weight without fracturing. *(Catherine K. Richardson)*

FIGURE 19.26 Sandstone and limestone (or marble) have long been favorite materials among decorative stonecutters. This carving decorates a building on the campus of Princeton University.

FIGURE 19.27   Sand and gravel operations are usually small, local businesses that take advantage of sources within a few kilometers of sites where their products will be used for making concrete. *(Cary Wolinsky, Stock Boston, Inc.)*

monly used now in prefabricated sheets of plaster board. Gypsum ($CaSO_4 \cdot 2H_2O$) is the raw material for plaster. In the manufacture of plaster, gypsum is heated, thus driving off some of the water and changing the rock's composition to $CaSO_4 \cdot H_2O$. The treated gypsum is then ground to a powder. When water is readded, the powder reverts back to gypsum, and a fine-grained, smooth-surfaced product results.

Another common building material is *clay*. Minerals of many types, ground to clay size, are used in making bricks and coarse pottery ware (Figure 19.28). Clay minerals such as kaolinite are generally sought for fine ceramics and for glazes on decorative tile and sculpture. Clay minerals are also used as fillers in paper, rubber, and plastics, and are used as a hydraulic medium during oil-drilling operations.

One final building material derived from the Earth is *glass*. You have learned that volcanic glass forms when a magma cools so rapidly that the atoms are unable to arrange in an ordered crystalline structure. Manufactured glass is essentially the same, although it is made from a "magma" of molten quartz sand plus soda ash and various oxides that control its hardness, strength, and color (Figure 19.29). Quartz is very plentiful, but it must be extremely pure if it is to be useful in making glass. It is mined both from loose sand and from sandstone.

FIGURE 19.28   Rooftops in towns and villages around the Mediterranean are commonly made of clay tile. These are in the town of San Gimignano, Italy. *(Pamela Hemphill)*

▶▶▶ EPILOGUE ◀◀◀

The use of mineral resources has increased dramatically during the present century. In fact, economists and politicians now recognize that some of these raw materials are being used up at such a rapid rate that they may not be available in the near future, or if they are, only at a very high cost. This raises a set of social, economic, and political problems that complicate the future not only of strategic minerals but also a great number of others. Crises in the petroleum market during the 1970s, 1980s, and 1990s are perhaps the most visible signs of the social effects of increased demand for and limited supply of nonrenewable resources. To the extent that geologists can help by exploring for fur-

FIGURE 19.29   Most formulas for manufacturing glass are closely guarded industrial secrets. Subtle changes in chemical composition of glass can produce new colors and change its hardness or other physical properties. Glassblowers like these in Murano, Italy, use these differences to produce decorative glassware, each piece of which is a unique work of art. *(Slvain Grandadam, Photo Reseachers, Inc.)*

ther mineral deposits or making better use of existing ones, they will play an important role in the century ahead.

During the past few decades other questions have been raised about the exploitation of natural resources and the disposal of waste products. Many of these are environmental questions that range from the detrimental effect of strip mining for coal to the health hazards related to the mining and use of asbestos. These are concerns that we will address in the final chapter of this book. The solutions to the problems raised in the environmental arena are not easily come by. Certainly, however, any adequate solutions will demand a recognition and understanding of the origin and the geologic and geographic distribution of natural resources, which this book has attempted to provide.

SUMMARY

1. The resources for a potentially valuable material include the total amount of the material in the crust, whether it is profitable to extract or not, even if it is only hypothetically known to exist. Reserves include only that portion of the resources that can be extracted today at a profit. Most energy and mineral resources are nonrenewable.

2. The energy we derive from the Earth ultimately comes from nuclear reactions or from the force of gravity. Nuclear reactions power the sun. When heat from the sun arrives at the Earth, it drives circulation in the atmosphere. Water evaporated by solar heat returns to the ocean under the force of gravity, thus providing a source of water power. Plants convert light from the sun into chemical energy by photosynthesis, and that energy can be stored in the Earth in fossil fuels. Nuclear reactions within the Earth generate geothermal energy, and controlled nuclear reactions in power plants generate electricity.

3. The fossil fuels (oil, gas, and coal) are derived from organic matter—the remains of plants and microscopic organisms—that were preserved in sediment. To avoid being destroyed by oxidation or being consumed by bacterial decay, the organic matter had to be buried rapidly in mud. Coal is a sedimentary rock composed of dead plants that accumulated in swamps, usually in a deltaic or lagoonal environment along an ancient coast. Petroleum (oil and natural gas) formed from the organic remains of simple marine plants and animals that accumulated in muddy sediments on the seafloor. Accumulation of oil and natural gas in the Earth requires a source bed, a permeable reservoir rock, an impermeable capping bed, and a trap to contain the hydrocarbons.

4. Potentially valuable mineral deposits form as the result of familiar geologic processes that can be grouped into five categories: magmatic processes, hydrothermal processes, sedimentation, weathering, and placer formation. Magmatic processes involve differentiation of a magma through crystal settling or other methods. Hydrothermal deposits form as heated water dissolves mineral matter from igneous or sedimentary parent rocks and then precipitates it in concentrated form in veins or as finely disseminated mineral matter in neighboring rocks. Chemical weathering concentrates mineral matter in residual minerals or transports leached mineral matter to other sites where it can be chemically precipitated. Sedimentary mineral deposits include evaporites, banded iron formation, and other resources formed as chemical sediments. Placer deposits are the product of mechanical segregation of dense or resistant minerals by a moving fluid, such as water flowing in a stream.

5. Geologists categorize valuable mineral deposits as either metallic (if the deposit is valuable primarily as the source for a metallic element) or nonmetallic (if the deposit is valuable primarily for some reason other than the metals it may contain). A valuable metallic mineral deposit is called an *ore deposit*. Nonmetallic deposits may be sources for industrial or agriculturally important minerals, or of a variety of building materials.

KEY WORDS AND CONCEPTS

abundant and scarce metals 492	bauxite 490	cap rock 482
asphalt 485	binding energy 485	chain reaction 486
banded iron formation (BIF) 490	breeder reactor 486	coal 477

QUESTIONS FOR REVIEW AND THOUGHT

19.1 What is the difference between the terms "resources" and "reserves"? What factors determine whether a particular deposit is part of the reserves for a useful mineral or energy source?

19.2 How do renewable and nonrenewable resources differ from each other?

19.3 In what type of depositional environment do coal deposits form? Why is this environment favored for the preservation of organic matter?

19.4 From where does the energy in coal or petroleum derive? How is it converted and stored in the fossil fuels? How is it released?

19.5 What structural, stratigraphic, or environmental features do geologists look for when they explore for coal?

19.6 In what kind of depositional environment does the organic matter that becomes petroleum accumulate? Why is this environment favored for the preservation of organic matter?

19.7 What are the advantages and disadvantages of water power? How much of its potential is used today?

19.8 How does a chain reaction in a nuclear fission power facility work? Where does the energy that is ultimately used to generate electrical power come from? What is the basic idea behind a breeder reactor?

19.9 What are the two alternative methods by which the Earth's geothermal energy can be tapped as a power source? What are the limitations of geothermal energy as an alternative to fossil fuels?

19.10 What methods are commonly used to remove valuable mineral resources from the Earth?

19.11 How can magmatic processes concentrate potentially valuable minerals in the Earth? Describe at least two different kinds of ore deposits that form directly from a magma.

19.12 How can metallic elements be carried and then deposited by hydrothermal fluids to form an ore deposit?

19.13 What is a banded iron formation (BIF) and in what kinds of rocks would a geologist expect to find one? How did it get there?

19.14 What are the two different general ways in which weathering processes may form an ore deposit?

19.15 What is a placer, and what kind of geological environment would a geologist explore to find one?

19.16 What criterion is used to distinguish the abundant metals from the scarce ones? What are the abundant metals?

19.17 What is a co-product? Which metals are commonly mined as co-products, and why?

Critical Thinking

19.18 A financial counselor has suggested that you invest your inheritance in Japanese Offshore Petroleum, a new company that plans to drill for oil in the ocean floor near the east coast of Japan. As a stockholder, you will also be entitled to a share of the profits from their subsidiary company, Pilbara Oil, which is going to be drilling for oil in Western Australia. Given what you have learned about plate tectonics and sedimentary environments, do you think that this sounds like a good investment? Why? If you do not know much about the places that the companies intend to drill, what questions would you like to have answered?

19.19 Look around the room. What different things containing material derived from rocks or minerals can you identify? Where did they probably come from?

20

Fire Island, on the south shore of Long Island, New York, after the northeast storm of March 7, 1962. (UPI/Bettmann)

People and Environmental Change

OBJECTIVES

▲▲▲

*As you read through this chapter, it may help if you focus
on the following questions:*

1. How have people become geologic agents?
2. How does agriculture affect the environment?
3. What landforms have been created by people?
4. How may mining adversely affect the environment?
5. How have people manipulated the hydrologic cycle?
6. What are some effects of the environment on people?

OVERVIEW

▲▲▲

As residents of this planet, people are directly related to the different geologic processes, systems, and cycles that we have discussed in this book. At various points we have drawn attention to some of these interactions between humans and the geologic environment. In this final chapter we examine more explicitly some of these interactions.

In the first half of this chapter we consider people as geologic agents. Often people can function harmoniously with their environment, not harming the land or natural processes and at the same time benefiting from Earth resources. At times, however, people's actions may have unplanned and often harmful results for themselves, or for the land, or both. This half of the chapter, then, examines the ways in which people affect the environment.

The latter half of the chapter explores the ways in which the geologic environment affects people. We have already seen, at various points, how major geologic hazards such as earthquakes and volcanic eruptions can have disastrous effects on people. We look briefly at some of these again, considering how people may be able to protect themselves from the worst effects of these hazards.

People as Geologic Agents

▲▲

Most people do not usually think of themselves as geologic agents, at least not on the same scale as floods, earthquakes, or volcanic eruptions. Yet ever since humans first appeared on the Earth, we have multiplied and spread out over the planet rapidly, and our technology has developed to a remarkable degree. In the process of building shelter, obtaining food, developing methods of transportation and communication, and providing amenities for ourselves we have left our imprint on the Earth, many times in unexpected ways. For instance, as agriculture has spread around the world and become increasingly intense, it has increased the rate of land erosion to ten times that of the rate in pre-agricultural time.

Recognition that people are themselves geologic agents and have an impact on the planet has been late in coming. Our role as participants in geologic processes began to be appreciated by a few only a century and a half ago. Since then we have learned in great detail what happens when humans move onto the land and how they influence the processes that operate there. Now, as the twentieth century nears an end, the human use and misuse of the environment has become an international concern that at times influences the highest levels of government policy. Even today, however, many refuse to acknowledge the effect they do have, refuse to take responsibility for their actions, or remain ignorant of the impact of those actions.

A few individuals have little effect on the environment. It is when "a few" are multiplied by millions—and today by billions—that humans become a major force in changing the environment. The immediate ancestors of modern humans, *Homo sapiens sapiens*, first appeared perhaps as long as 90,000 years ago. Our species expanded slowly at first, but then more and more rapidly. By the birth of Christ, 2,000 years

ago, the world population is estimated to have been about 200 million. By 1650 it had risen to 500 million. Two hundred years later, in 1850, it had reached a billion. By 1930 it had doubled and then doubled again to 4 billion by 1975. By 1992 the population had climbed to 5.5 billion and was predicted to reach 6.2 billion by 2000 (Figure 20.1).

After an origin we don't fully understand, humans spread rapidly, and by 15,000 years ago there was at least a sparse population on all the continents, including Australia. These populations all consisted of hunter-gatherers who affected the geologic environment very little, if at all. Then, beginning about 10,000 years ago, technological advances began to change our impact on the globe. Among these advances was the introduction of primitive agriculture, which began in the Middle East approximately 10,000 years ago. About 7,500 years ago, mining and metalworking were discovered. Sometime about 5,500 years ago the plow and irrigation were introduced into agriculture. With these developments people were poised to begin their role as agents of geologic change.

The domestication of plants and animals provided, for the first time, a dependable and continuing source of food and fiber. This made it possible for people to abandon the migrant life of hunter-gatherers and to establish permanent settlements. Introduction of metalworking yielded metal tools, which replaced less efficient stone tools and, along with the plow and irrigation, increased agricultural output. All of these changes meant a growing population that was more and more competent to modify the environment.

Impact of Some Human Activities on the Environment

▲▲

Humans have their greatest impact on the environment as participants in the Earth's surficial processes.

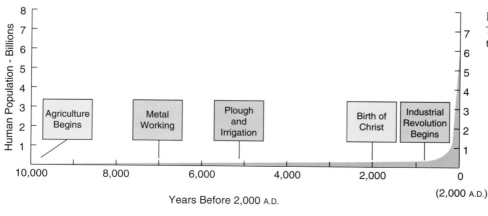

FIGURE 20.1
The growth of human population for the last 10,000 years.

We can modify, to a varying extent, the geologic processes such as mass movement, streams, ground water, coastal change, and the atmosphere. As we do so, we change the Earth's surface and introduce different landscapes that can differ widely from the natural terrain that we inherited. Additionally, humans have found that their use of the environment for one purpose almost invariably leads to other changes that may not have been foreseen and may be unwanted.

Agriculture

Agriculture has had enormous benefits for humans. It has freed them from the migrant life of hunters and gatherers and fed an ever-increasing world population. At the same time, our agricultural activity has had other results as well.

Agriculture involves a disruption of the soil and the replacement of a natural vegetative cover with one of human choosing. The type of disruption may vary from tearing up tough prairie cover, as happened in large parts of the American and Canadian plains, to clearing forest, as happened in eastern Canada and the United States. The crop that we plant, such as corn, wheat, hay, alfalfa, flax, or cotton, is almost always different from what would grow naturally in the same area. Crops, therefore, require a good deal of tending. The combined activities of preparing the ground for seed, cultivation, harvesting, and readying the field for the next planting keeps the soil exposed and without protective cover for several months of the year (Figure 20.2). In such a condition it is vulnerable to erosion by wind and water. It is not surprising, then, that soil erosion is 10 or more times greater on cropland than on undisturbed land.

Soil erosion almost invariably increases whenever land is opened to agriculture. An example of changes in erosion rate with changes in land use comes from

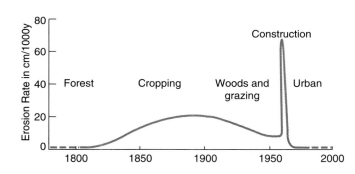

FIGURE 20.3   The results of a study near Washington, D.C., show the rate of soil erosion under five different uses of the land. Highest rates occurred during construction, but were short-lived. Land that had been farmed eroded at a maximum of 20 times that of forested and urbanized land. *(Modified from Wolman, M. G., "A Cycle of Sedimentation and Erosion in Urban Stream Channels," Geografiska Annaler, vol. 49-A, 1967, Fig. 1, p. 386.)*

a study near Washington, D.C. (Figure 20.3). When land was under forest the erosion rates were about 2 cm/1,000 years. As forested land gave way to farming the rate increased 20-fold. In the early part of this century farming gave way to grazing and to a return of woodland, which slowed erosion as the vegetative cover was partially replaced. As urbanization began, heavy construction equipment stripped the land of vegetation, reorganized it into new configurations, and thus produced a spectacular increase in erosion. Urbanization brought protection from erosion by buildings, roads, sidewalks, parking lots, and storm sewers, and the erosion rate dropped precipitously.

The increased volume of sediments produced by agriculture can clog ditches and small streams, making drainage less effective. Sediments carried to and deposited in lakes and reservoirs reduce the amount of water those lakes and reservoirs can hold, and thus shorten their effective lifetimes.

An increase in the rate of erosion is not the only effect of agriculture. The use of fertilizers, both natural and artificial, has increased agricultural production enormously. At the same time, however, fertilizers can contaminate both ground and surface waters. Waters high in soil nutrients, such as nitrogen and phosphorus, can seep into the ground and render ground water undrinkable. Surface water rich in these elements can feed lakes and cause their rapid aging, a process called **eutrophication,** from the Greek *eutrophos*, meaning "well-nourished." The rise in nutrients sets off a chain of events that include a rich growth of plant life, including algae, at and near the surface; reduction of bottom-dwelling plant life; reduction of game fish; and a decrease in the variety of life-forms. The waters may become cloudy and release disagreeable odors as bacteria feed on dead organisms. From the human perspective, a eutrophic lake is undesirable.

FIGURE 20.2   Agriculture exposes the soil to erosion for several months of the year. *(Roger V. Moseley)*

Another result of agricultural activity is a change in the habitats of both plants and animals. Preparing land for crops, whether by cutting down the forest or turning over the sod, alters the plant cover for the crop to be planted. This reduces the diversity of plant-forms in the area. Simultaneously, the animal population originally present changes as some species are displaced and others take their place.

Landforms Created by Humans

Before humans appeared, landscape was shaped exclusively by the surficial geologic processes, such as mass movement, running water, glaciers, and wind. The arrival of humans brought another process in landscape formation. We have been able to move large volumes of Earth material from one place to another. In the process we have created a variety of different landforms and modified others. Most of the landscape features we create are distinctive, being easily distinguishable from those formed by nonhuman processes. The result is a land surface different from that which preceded human activity. Some of these landforms may soon disappear, while others can persist for hundreds, even thousands, of years.

In the Middle East, particularly in Iran, some ancient settlements, many thousands of years old, are marked by artificial mounds called **tells,** from the Arabic *tall* for "small hill" (Figure 20.4). These hills consist of the debris of successive settlements in which the main construction material was mud brick. Rebuilding of a destroyed or abandoned settlement took place on top of the old rubble. When occupation of an area was continuous the process of repairing and renewing old structures also carried the settlement level higher.

Another hill made up of cultural debris is the modern **sanitary landfill.** The name is misleading because in a modern landfill, the debris does not fill an excavation but is built up into a ridge or hill. The hill is composed of individual units or cells of waste, which are enclosed by Earth material packed around them. In these landfills society's refuse accumulates rapidly and, in a generation or two a new and impressive feature of the landscape is born.

Agricultural practices often create new landforms, several of which are tied to specific activities. Terracing, for example, is a widely used technique in agriculture. The terraces turn a hillslope into narrow strips of level and arable land and slow the runoff of rainwater. The terraces can dominate a landscape while fields are in use (Figure 20.5) and can last for centuries when the land is abandoned. The clearing of stony land may result in long-lasting hillocks of stone in the resulting fields or along their margins, or in low ridges of tumbled stone walls that once enclosed now-abandoned fields.

Construction of transportation systems such as railroads, highways, and canals (Figure 20.6) account for easily recognized landforms. For instance, railroad embankments are distinctive ridges, as are the embankments that carry superhighways. The source of material to build these features is often close by, in rock quarries or in the linear excavations cut to create a manageable grade across steep slopes. These forms can persist for a long time. For example, Figure 20.7 shows a 2,500-year-old road-cut in central Italy. Certainly our constructions of the last 150 years, such as the road-cutting in Figure 20.8, can be expected to persist for thousands of years as well.

A striking change in the landscape of the United States is the increase in the number of lakes during the last 100 years. American engineers have placed dams on almost every river in the country, forming a myriad of lakes both large and small (Figure 20.9). Well-known examples include Lakes Powell and Mead on the Colorado River in Arizona, Franklin D. Roosevelt Lake on the Columbia River in Washington, and Lake Sakakawea in North Dakota on the Missouri River.

FIGURE 20.4   The top of the great tell of Chatal Hüyuk rises 25 meters above the Anatolian plain in extreme southeastern Turkey. Its base covers about 130,000 m². The mound grew during the period from about 6,000 BC to 2,000 BC. During this time, a succession of at least a dozen building levels occurred as mud brick structures crumbled or the population expanded. View from the west. Silhouettes of people along the top of the tell provides scale. Photo taken about 1935. *(Oriental Institute)*

FIGURE 20.5
Bhutan is a non-industrialized country. Despite the lack of mechanized equipment, however, farmers have built these agricultural terraces and substituted cultivated crops for the original plant cover. (Roger V. Moseley)

Figure 20.10, a map of Tennessee and parts of neighboring states, shows just how extensive large artificial lakes are in the region. Expect them to remain as permanent features until the dams fail or the lakes fill up with sediments. Even then, the dams and their lakes will leave recognizable marks of their former existence for a long time to come.

Mining

People have been mining minerals for thousands of years. With the introduction of large earth-moving equipment, however, the process has made a signifi-

FIGURE 20.6   The Delaware and Raritan Canal was begun in 1830, completed in 1834, and ceased operating in 1932. Now used for water supply and recreation, the canal is a 69-km-long ditch across the waist of New Jersey and connects the Delaware River with the Raritan River.

FIGURE 20.7   Road construction can leave its mark on the landscape for a long time. This road-cut near the town of Sovana, 60 km north of Rome, Italy, was excavated 2,500 years ago as part of an Etruscan road system. (Pamela Hemphill)

FIGURE 20.8   Interstate Highway 287 in New Jersey here runs through a large cutting excavated in resistant metamorphic rocks.

FIGURE 20.9   A dam impounding Lake Carnegie at Kingston, NJ., was built in 1906. Sedimentation in the lake made dredging necessary in the early 1970s.

cant impact on some parts of the landscape. Certainly the great open pit metal mines of the western United States are impressive landforms, and will not soon disappear. In addition, most of the material excavated from such holes is economically worthless and is dumped nearby. It then forms its own topography (Figure 20.11).

Strip mining, which peels off an overlying zone of Earth materials to get at the valuable material below, is most commonly used in mining shallow deposits of coal. The process involves removal of both soil and vegetation. Furthermore, the shape of the land is converted into a series of trenches and ridges. Concerted efforts are underway to reduce the impact of the process and to restore the land and its cover to some semblance of its former condition. Although placer mining is smaller in scale, it can produce its own mounds of debris (Figure 20.12) or disrupt stream

channels below the placer operation. Underground mining, too, can produce its characteristic surface results. The removal of rock from the subsurface leaves a void. If the overlying rock is weak it can collapse into the mine and produce a sinkhole-like depression at the surface. Underground mining also produces hills of useless rock.

Extraction of sand and gravel leaves behind depressions. Because there is no leftover material with which to back-fill the pits, the holes remain, filled sometimes with water, sometimes with trash, sometimes with both. The same is true of some quarries from which solid rock has been taken. Even if the quarrying operation leaves no pits behind, it can radically alter the hill or mountain where the quarrying takes place (Figure 20.13).

Another effect of our quest for minerals is water pollution. For instance, sulfur is often associated with

FIGURE 20.10   Artificial lakes are a dominant feature in Tennessee and neighboring states.

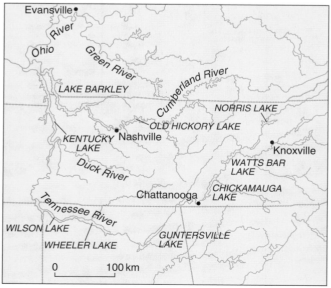

FIGURE 20.11   In many mining operations there is a large amount of useless rock. The hill against the skyline is rock left over from slate quarrying. The opening into one slate pit, 225 m deep, is in the immediate foreground. A second pit lies just in front of the long building.

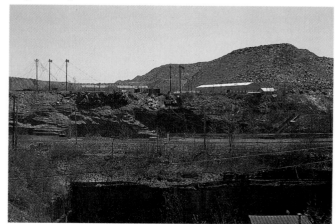

FIGURE 20.12   Placer mining involves the movement of a large volume of sand and gravel. The processed material in this Alaskan gold-dredging operation forms a distinctive pattern near Fairbanks, Alaska. *(Bradford Washburn)*

FIGURE 20.13   Rock in this northern New Jersey quarry comes from the cliff in the background. Rock is crushed and stored in piles in the foreground and middle distance. Note that the crushed rock assumes an angle of repose of about 40.°

the formation of coal. Piles of rock not rich enough to be considered usable are called **tailings** and may become the source of sulfuric acid as waters seep through the debris into the streamways. The resulting acidic waters can adversely affect both plant and animal life. Similarly, metal mining produces material from which both sulfur and toxic metals get into streams and ground water. Additionally, the large amounts of debris produced by mining can choke streams with sediments, change the character of the stream channels, and have adverse effects on water quality through increased turbidity.

Closely associated with mining of metals is **smelting,** the process of separating metal from ore. This operation can put both acidic and metallic fumes into the atmosphere and contaminate both ground and surface water. The effects have not been restricted to modern times. Evidence is increasing that smelting of silver ores by the ancient Greeks put large amounts of lead into the atmosphere as did similar activity by the Romans a few hundred years later.

Water Use

Engineering skills and equipment have allowed humans to manipulate parts of the hydrologic cycle in ways unplanned by nature. For instance, we build dams for our own immediate benefit, using them for purposes like flood control and hydroelectric power. Building a dam, however, can produce unintended effects.

The lakes behind dams serve as a local base level; thus, when a stream enters such a lake, it loses its velocity and begins to deposit its load. A delta then starts to grow into the lake. At the same time, deposi-

tion works its way upstream as the stream tries to maintain a gradient adjusted to carrying its water and sediments. Of course, some of the finer sediment is carried far into the lake and settles to the bottom. In some older lakes used as reservoirs deposition has been extensive enough to reduce significantly their capacity and hence their usefulness, as demonstrated by the Elephant Butte reservoir on the Rio Grande in New Mexico.

By the time water is released or flows from behind a dam it has lost most of its sediment. It is thus able to pick up and carry any loose sediment available and begins to erode the stream channel and banks below the dam. The Colorado River, as it leaves Lake Powell, illustrates this type of erosion. Before the construction of the dam the river ran muddy, and the sand and silt it carried built up the sandbars along the river (Figure 20.14). Since the construction of the dam, sediment has been caught in the lake, and the Colorado River now normally runs clear below the dam. The result has been the depletion of the sandbars along the canyon below the dam and the modification of both the river's aquatic life and the riverside habitats. Whether adjustments in the release of water from the lake will change the river back toward its condition prior to the dam is not yet clear.

The interception of sediment behind dams has other implications. Sediments that previously nourished deltas and fed the long shore currents that replenish beaches are drastically reduced. As a result, deltas do not receive the necessary sediments to keep them from sinking below a rising sea level or drowning because of the compaction of previously deposited sediments. In addition, beaches receive a reduced supply of sediments and therefore shrink. Some conservationists have taken the position that rivers and beaches

FIGURE 20.14   Sand eroded from this sandbar along the Colorado River in the Grand Canyon is not being replaced by the river. Sediments that would normally have replaced those eroded are now trapped behind the Glen Canyon Dam in Lake Powell. Human figures give scale. (*Pamela Hemphill*)

are part of a single system. They ask, therefore, that when dams along streams trap sediments and deprive nourishment to marine beaches, those responsible for the dams be held liable for starvation of the beaches and take steps to restore them.

Although people can create new lakes by building dams, we can also drain older lakes by manipulating their water supply. We have already seen this in the case of the Aral basin (see Perspective 18.1). The same thing is happening to the larger Caspian Sea to the west of the Aral. Here, upstream diversions from the Volga River, its main source of supply, are so great that the Caspian has been shrinking for years.

Human beings affect not only streams in the hydrologic cycle but also the ground water that they very frequently depend on. Ground water is like any other resource. It will be depleted if it is used more rapidly than it is replenished. In well-watered areas this depletion is, generally, not a problem. Where the recharge rate is low and the demand is high, however,

difficulties ensue. When use exceeds replenishment the ground water table will fall. Eventually the water supply will be so deep that it becomes uneconomical to pump. In such cases we are essentially mining ground water and exhausting our supply.

The Great Plains, east of the Rocky Mountains, provide a classic example of ground water depletion. An extensive aquifer, the Ogallala Formation, directly underlies parts of eight states and makes up what is known as the Great Plains Aquifer, one of the world's great ground water systems. It has provided water for up to 20 percent of the irrigated land in the United States. Recharge, however, is very low, on average about 1 cm/year. Pumping for irrigation began in earnest about 1940 and continued to increase through the 1970s. By 1980 the ground water level had fallen across most of the area, as shown in Figure 20.15.

FIGURE 20.15   Extensive pumping of groundwater for irrigation has seriously depleted the supply of the Great Plains Aquifer. This map shows the change in the level of the water table from about 1940 to 1980. (*U.S. Geological Survey, Geohydrology of the High Plains Aquifer in Parts of Colorado, Kansas, Nebraska, New Mexico, Oklahoma, Texas, South Dakota, and Wyoming, Professional Paper 1400-B, 1984.*)

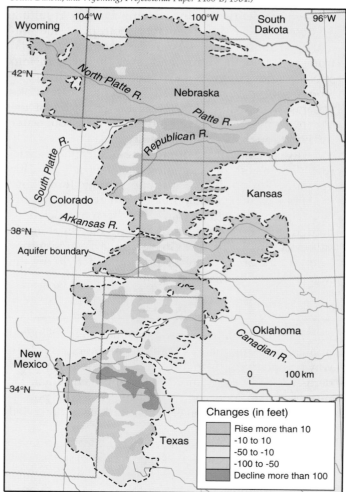

Falling ground water levels means higher costs for pumping, and in some places farmers, finding irrigation uneconomical, have reverted, particularly in the southern section, to **dry farming,** relying only on the semiarid region's uncertain rainfall. Widespread pumping continues in other places, although attempts have been made to control it. Usually the restrictions merely prohibit additional pumping but make no provision to reduce existing withdrawal. Because of the low recharge rate and the high demands still made on it, the future for the aquifer is bleak. It certainly will not be recharged in some places for hundreds of years. In the meantime, the Great Plains Aquifer demonstrates how effectively people can intervene in a natural system, in this case the hydrologic system.

The transfer of water from areas of surplus to areas of need is a common practice. In Canada, the world leader of these efforts, water transfers amount to about 4.5 percent of the country's streamflow. Most transfers there are made to create hydroelectric power. Even more ambitious plans exist, though. The North American Water and Power Alliance, for example, was proposed nearly 40 years ago and continues to be considered as the demand for water increases in the dryland areas. The plan would carry water from the western mountains of Canada to 7 Canadian provinces, 35 states in the United States, and 3 Mexican states. There are serious suggestions of exporting Alaskan water by marine pipeline to water-hungry California and to tow icebergs to Los Angeles. In Asia a plan has been suggested to transfer water from the northward draining Ob River basin 2,500 km south to the Aral Sea and its neighboring desert. The solution of the shrinking Caspian Sea lies, according to another plan, in exporting water from the rivers draining the northern slopes of Russia to the Volga River, and through it to the Caspian. Technical and economic feasibility, environmental effects, and political constraints may hinder all such plans.

One of the often overlooked effects that human activity can have on ground water results from the presence of underground storage tanks, whether domestic, commercial, or military. These tanks, which may contain anything from liquid fuels to toxic chemicals and radioactive waste, have a finite life. Eventually they corrode, and contents leak into the ground water. Removal of tanks and decontamination of the surrounding soil is expensive and often disruptive. Cleanup of the ground water is difficult to impossible. Often the only alternative is to let nature flush it out of the ground over the course of many years.

Changes in the Atmosphere

Industrial societies pollute the atmosphere by releasing gases and particulate matter into the air. Of the gases—sulfur dioxide from coal- and oil-burning power plants, and nitrous oxide from internal combustion engines used to drive everything from cars to power lawn mowers—are responsible for **acid rain,** better called **acid precipitation.** All precipitation is slightly acid because carbon dioxide in the atmosphere combines with water to form carbonic acid, a weak acid. Sulfur dioxide and nitrous oxides, however, combine with water to form sulfuric and nitric acids. These compounds increase the acidity of precipitation to dangerous levels in some places. Acid rain can damage trees and crops; it raises acidity of lake waters to a point that endangers aquatic life, and it hastens the deterioration of buildings. The problem is most extensive in areas of greatest concentration of industry and population such as in the eastern United States (Figure 20.16) and in Europe from the Atlantic coast westward into Russia.

The gas ozone (O_3) in the stratosphere plays a number of roles. Ozone absorbs incoming ultraviolet rays, thus limiting their penetration to the Earth's surface. The ozone layer of the stratosphere, at heights of 15 to 18 km in the polar latitudes and at around 25 km in the equatorial zone, has thinned in recent years. **Ozone holes** in the Antarctic and Arctic stratosphere have now developed. Without the ozone shield, the increased radiation received on the Earth's surface will heighten the incidence of skin cancer. Increased ultraviolet waves can affect phytoplankton in the upper zone of the ocean. These small plants are part of the base of the food chain for the rest of oceanic life, and their decline could have unfortunate consequences for marine organisms.

The major cause of ozone depletion has been traced to the release of **chlorofluorocarbons** (CFCs)

FIGURE 20.16 Increased levels of acid rain occur in the eastern United States and Canada. Normal rainfall is about 5 on the acidity scale. Values below 5 record increased acidity. (*U.S. Environmental Protection Agency*)

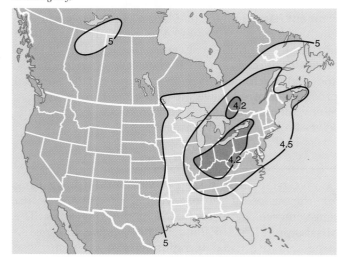

into the atmosphere. These have been used extensively in aerosol spray cans, refrigerants, and in the manufacture of foam plastic. The gases are diffused into the atmosphere, and solar radiation dissociates them, releasing chlorine atoms that react with ozone and destroy it. Governments and industry are making concerted efforts to reduce production and use of CFCs. In 1993, for the first time since the problem was identified, the rate of accumulation of CFCs in the atmosphere began to level off.

It is to human advantage to ensure that the ozone in the stratosphere is not depleted. In contrast, ozone near ground level is a component of some smog and is a health hazard. It forms in the lower atmosphere as the result of the reaction between sunlight and certain emissions from internal combustion engines, and includes nitrogen oxides and volatile organic compounds.

Impact of the Environment on Humans

Even though people can knowingly change their environment, we remain, like all living things, dependent on the environment for our existence. In fact, humans are the product of the changing environments of the Pleistocene. We have evolved through 20 or more alternations of glaciation and deglaciation, their accompanying retreats and advances of the sea, and rapidly changing climates, flora, and fauna. We prospered as we adapted to these repeated environmental changes. By the end of the last glaciation we had emerged as a major environmental force able to manipulate our environment and, thereby, that of other animals and of plants.

We have seen that human influence on the environment has been effective largely in the realm of the Earth's surface processes. We certainly have been unable to control, or even influence, the internal processes of the Earth, such as volcanism, earthquakes, and plate tectonics. Even in the area of the surficial processes, however, we are at the mercy of a number of environmental systems and processes. We find ourselves both slaves and masters.

Floods

We have already seen in Chapter 14 how, using the historical record, we can predict the recurrence interval between floods of differing sizes. The Mississippi-Missouri flood of 1993, however, was so large that the usual techniques of flood recurrence

broke down. We don't know whether it was so large that it could be expected to recur only once every 500 years, or whether it was larger or smaller. We do know, however, that it was the biggest flood that we have any record of in the area. By the time the floodwaters began to recede in early August 1993, preliminary damage estimates amounted to over $12 billion and were climbing, and flood-related deaths were said to be 52 (Figure 20.17a, b).

Certainly, the 1993 flood in the midwestern United States was not the worst recent flood in the world. In 1993, at the same time that the upper Midwest was being inundated, nearly half of Bangladesh, along the Ganges and Brahmaputra rivers at the head of the Bay of Bengal, was under water as a result of four weeks of heavy rains. Over 2,100 people lost their lives. An estimated six million others were stranded or homeless, and twice that number had been affected (Figure 20.18).

Another way to approach the prediction of flood hazard, in addition to using flood recurrence intervals, is to map the extent of the floodplain along a river. A river, over its lifetime, experiences a great many large floods. These floods leave their record behind both in the nature of sediments deposited and the shape of the land that they form. Mapping these sediments and landforms tells us where floods have been in the past, and that we should expect the river will rise there again in the future. We should thus be hesitant to use that land for purposes that a flood would damage.

Partial protection against floods takes many forms. A network of river-gauging stations and good meteorological information can predict the onset and progress of flooding and allow people to take emergency measures. In another approach some political units have outlawed construction on the flood plain as long as the construction is vulnerable to floods. Flood insurance is sometimes used as a measure of economic protection, but it is expensive and few people buy it. Levees have been erected to protect land adjacent to a rising river. Flood-control dams are constructed to hold excess water until the flood waters collected behind them can be let out slowly, thus reducing the flood crest. All these measures can give protection against some flood dangers, but the 1993 Mississippi flood showed that even the most extensive flood protection can fail.

Mass Movement

Landslides are the most hazardous of mass movements. Catastrophic landslides are, like other major geologic hazards, virtually impossible to predict. They move at high speeds and occur most often in areas of steep slopes in mountainous areas. Such slopes may be unstable and need only a small event to push them

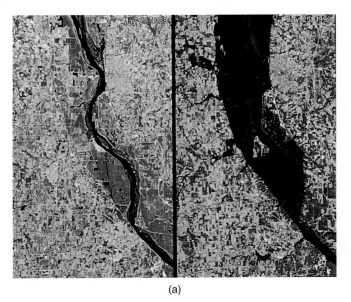

(a)

(b)

FIGURE 20.17   (a) The flood of the summer of 1993 in the upper Mississippi Valley was the great-est on record there. This pair of satellite pictures is along the Mississippi between Quincy, Illinois, and Hannibal, Missouri. On the left the Mississippi is seen during normal flow as a thin, dark, sinu-ous ribbon on May 27, 1989. On the right, in the image of July 25, 1993, the river has flooded to cover a broad band that is 8 km wide opposite Quincy, near the top of the picture. *(AP/Wide World Photos)* (b) This photo was taken from a NASA aircraft at 65,000 feet and shows the flooding at the conflu-ence of the Missouri and Mississippi rivers on August 6, 1993. The normal channels of the rivers are faintly and discontinuously outlined by dikes and levees that project above the flood. Alton, Illinois, is at the upper right. St. Louis, Missouri, and its suburbs occupy the most of the area in the center and lower portion of the image. This is an infrared photograph and vegetation appears in tones of red. Width of view about 32 km. *(AP/Wide World Photo)*

over a threshold into movement. They are often associ-ated with earthquakes, which serve to get them started. Large volcanic eruptions are not uncommonly accompanied by mudflows as they were during the Mt. Pinatubo eruption.

Hurricanes

Hurricanes are tropical storms of the western Atlan-tic in which winds have a minimum velocity of 117 km/hr and rotate in a counterclockwise circula-tion. In the western Pacific ocean they are called *typhoons* and in the Indian ocean *cyclones*. Whatever they are called, these storms are hazardous. Some are just more hazardous than others. A hurricane hazard scale runs from 1 to 5, from least to most severe. A number 1 hazard hurricane has winds between 117 and 153 km/hr. Winds in a number 5 hurricane are over 250 km/hr. A hurricane's course and rate of movement can be tracked by satellite and weather planes, so there is some warning about the approxi-mate time and place a hurricane will strike. This gives people time to protect themselves and their property. In fact, evacuation procedures along hurricane-prone

coasts have reduced deaths from hurricanes drasti-cally in the last two decades.

Hurricanes (Figure 20.19) are a major agent in fashioning beaches and shorelines. High winds are an obvious agent in the process. More important, how-

FIGURE 20.18   At the time of the 1993 Mississippi flood much of Bangladesh was also flooded. This low-level aerial view is near the capital city of Dhaka. *(Roger V. Moseley)*

FIGURE 20.19 Hurricane damage along the South Carolina coast. (Dean Abramson, Stock Boston, Inc.)

in 1908. It flattened the forest over an area half the size of New York City, started innumerable fires, and sent out a shock wave recorded around the world. Another reminder that we are not immune to encounters of meteorites, both large and small, was the unexpected close encounter of the Earth with an asteroid that passed through Earth's orbit on March 23, 1989. The object, designated 1989 FC, had an estimated diameter of about 1 km and passed within 765,000 km of the Earth. This asteroid orbits the Sun annually so we can expect future visits, sometimes closer than in 1989 and sometimes farther away. The prediction also is that 1989 FC will sometime impact the Earth. Given its size it will create disaster if it hits a densely populated area.

How serious is the possibility of a large meteorite crashing into the Earth? Military satellites have reported blasts the size of nuclear explosions at the outer edge of the Earth's atmosphere (Figure 20.20) caused by meteors too small to survive the trip through the atmosphere. A project called Spacewatch has for a number of years been searching the heavens for objects that pass through the Earth's orbit, and it has already discovered objects larger than those that satellites report exploding in the upper atmosphere. Furthermore, the celestial survey discovered the comet, Shoemaker-Levy, fragments of which bombarded Jupiter in July, 1994, producing fireballs larger than the Earth. The geologists and astronomers contributing to Spacewatch have speculated that objects that threaten to impact the Earth could be intercepted and destroyed with nuclear weapons before reaching us. This concept gives a new meaning for star wars.

Volcanic Eruptions

Some of the world's most productive soils have developed on deposits of volcanic ash. Their richness stems from the fact that nutrients for plants are easily available from the fine particles that make up the ash. Additionally, successive eruptions may periodically renew the soil base. Obviously, however, activity by any volcano can make a neighborhood a dangerous place to live. Thus the volcanic environment can have both favorable and unfavorable effects on people, as well as on other animals and on plants.

The general location of volcanic eruptions can be predicted. Experience has taught that most volcanic activity is concentrated along plate boundaries and at hot spots in the ocean. We know that the least destructive volcanoes are those with basaltic eruptions, and the most destructive are those that are fed by granitic magmas. The timing of a specific eruption may be predicted with continuous monitoring using sophisticated instrumentation, as was the case in the 1991 eruption of Mt. Pinatubo in the Philippines.

ever, is the **storm surge,** a ridge of water rising above normal high-tide level and sweeping over the shoreline along a front of 60 to 80 km. The storm surge can level sand dunes along a beach and break through barrier beaches. Additionally, rains accompanying hurricanes can be very heavy and trigger floods far inland.

Meteorites

With the arrival of space exploration, scientists have become more and more aware of the debris that is flying around in our Solar System. Much of it orbits the Sun in the belt of asteroids between Mars and Jupiter. There are asteroids and comets, however, that have paths that take them through the Earth's orbit. As the craters and meteorite fragments found in several places around the globe show, fragments of numerous meteors have found their way to the Earth. More recently, a stony asteroid weighing about 30 tons exploded over central Siberia

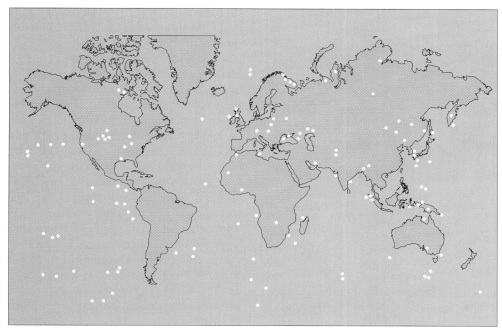

FIGURE 20.20   Satellite observations from 1975 through 1992 show 136 large blasts widely distributed in the upper atmosphere, as shown by the dots on a world map. They represented comets or asteroids too small to survive a passage through the atmosphere to the Earth. *(Beatty, J. Kelly, "Impacts Revealed," Sky & Telescope, 1944, vol. 87, no. 2, p. 26.)*

Although geologists may be able to predict some volcanic eruptions, there is nothing they can do to prevent eruptions. Furthermore, only in the case of some basaltic flows do we have any ability to influence the course of the eruptions. Lava will follow the low places in the topography; thus, it is occasionally possible to divert the flow by opening a new path to a still lower zone, thereby protecting some structures that might otherwise have been destroyed. Such tactics are limited at best and are completely inapplicable to andesitic and granitic eruptions. The only real protection from volcanic eruptions is to live in an area with no recent geologic or human history of volcanism.

Earthquakes

As we pointed out in Chapter 9 the only defense against earthquakes is preparedness. Buildings must be constructed to earthquake-proof standards (Figure 20.21), and the residents must be educated on what to do in the event of a large earthquake. But unexpected quakes can occur in tectonically quiet areas of the continents. If quakes of any serious magnitude were to strike an unprepared metropolitan district the loss of life and the property damage would be staggering.

Tsunamis

Prediction of tsunamis is even more uncertain than the prediction of earthquakes. As we found earlier, these destructive sea waves occur as a result of a sudden movement of the seafloor, usually as a result of an earthquake. They can also occur, however, because of a subsea volcanic eruption or a major submarine landslide. The Seismic Sea Wave Warning System, based in Hawaii, monitors earthquakes. Whenever there is a possibility that a quake has triggered a tsunami a warning goes to stations around the Pacific. For locations very close to the quake the warning will probably not be soon enough to allow evacua-

FIGURE 20.21   The best protection against earthquakes is preparedness, including the construction of buildings in accordance with earthquake building standards. This structure in Mexico City was not built to withstand earthquake shock, as the earthquake of 1986 demonstrated. *(Albert Copley, Visuals Unlimited)*

FIGURE 20.22   An earthquake off the Aleutian Islands on April 1, 1946 was felt around the shores of the Pacific Ocean. This view is of the tsunami breaking over Pier 1 in the harbor at Hilo, Hawaii. The man in the left foreground was one of 159 fatalities that resulted from the tsunami in the Hawaiian Islands. *(Notional Geophysical Data Center, Boulder, Colorado)*

tion of a coastal area. For those farther away such warnings can be real life-savers. Knowing the way a tsunami behaves can also help. First, a much larger than usual wave hits the shore. This is followed by a withdrawal of the sea for a distance much greater than the lowest of low tides. When that happens do not follow the retreating sea. Turn and run inland for the highest ground you can find before the sea surges back and floods well above the highest of high tides (Figure 20.22).

►►► EPILOGUE ◄◄◄

This chapter concludes our study of the changing Earth and its processes, cycles, and systems. We have discovered how elements go together to make up minerals, and how minerals combine to make up three great families of rocks. These rocks are constantly changing as their environments change. Internal forces of the Earth drive the global movements of the plates of the Earth's crust, and these movements in turn produce earthquakes and volcanic activity. All these operations expose the continents to the surficial processes of wind, water, and ice that continuously wear away its surface and carry the products of erosion to the oceans where they begin to form new rocks.

Geologists understand these Earth processes and systems well enough now to recognize that people themselves are geologic agents. A great deal is now known about what happens when humans move into an area, how we influence the processes that operate there, and how manipulation of those processes may in turn affect us. We have also learned that when we undertake to modify the environment in one way our actions can produce unforeseen environmental effects. In some instances these side effects have proved deleterious to us and harmful to other aspects of the environment. We are learning, therefore, that changes we make for our own benefit may have disadvantageous results. We are learning that we need to understand as much as we can about the results of proposed modifications of our environment. This has led to the realization that to minimize the undesirable environmental results as we bring about changes we need to work with natural systems and not against them. What we have learned about the Earth's processes through this book should help us toward that goal.

SUMMARY

1. People have become geologic agents because of their domestication of plants and animals, development of agriculture, discovery of ores and metalwork, and establishment of permanent settlements. These factors, combined with humans' rapid increase in numbers (today close to 6 billion), have made it possible for us to change habitats of animals and plants and for ourselves to move to every corner of the globe.

2. Agriculture demands the destruction of the original plant cover and the planting of a new, very different plant assemblage than the one replaced. Agriculture increases the rate of erosion at least 10 times over the pre-agriculture rates. Terracing changes the shape of the land, and nutrients from the fields can contaminate both surface and subsurface water.

3. Because we can move large amounts of Earth material we have created new landforms including hills of sanitary landfills, deep cuttings for railroads, canals, and highways. We have built artificial ridges for railroad and highway embankments. We have constructed dams that impound countless numbers of artificial lakes, some of immense size, and have fashioned artificial shores adjacent to most major maritime cities.

4. Mining, in addition to providing us with useful materials from the Earth, can leave behind the scars of open pit and strip mining, as can quarrying for sand, gravel, and rock. The surface may collapse into abandoned mines, leaving depressions at the surface. Acidic water from mining operations can pollute surface and ground water.

Toxic and acid fumes from smelting plants processing ores can pollute the atmosphere.

5. People have manipulated the hydrologic cycle largely through dams. Dams control water for irrigation, water supply, flood control, hydroelectric power, and recreation. These installations can have other results, not all of which are desirable. They can cause stream deposition and erosion where it did not occur before, deprive water from areas downstream, and starve streams, deltas, and beaches of needed sediments. People have also manipulated the hydrologic cycle by pumping ground water aquifers faster than infiltration can replenish them.

6. Modern humans developed in the rapidly changing environments of the Pleistocene. Today we are dependent on our environment for our existence. Although we can control and influence the environment to a limited extent, we are still mostly at the mercy of the environment. For instance, we cannot control the largest floods or effectively predict or stop earthquakes, volcanic eruptions, landslides, tsunamis, hurricanes, or meteorite impacts.

KEY WORDS AND CONCEPTS

acid precipitation 507	eutrophication 501	storm surge 510
acid rain 507	ozone hole 507	tailings 505
chlorofluorocarbons (CFCs) 507	sanitary landfill 502	tell 502
dry farming 507	smelting 505	

QUESTIONS FOR REVIEW AND THOUGHT

20.1 What has been the increase in human population from 10,000 years ago to the present?

20.2 In general numbers what is the relation of the amount of erosion of forested land to crop land? What causes the difference?

20.3 What are some of the landforms that people have created?

20.4 What are some of the effects of mining on the environment?

20.5 What changes might be brought about when a dam impounds a lake?

20.6 Why should the construction of a dam be blamed for erosion of marine beaches?

20.7 What is meant by "mining water"?

20.8 How do human activities lead to pollution of the atmosphere?

20.9 How do people protect against tsunamis?

20.10 What have people done to protect against hurricanes?

20.11 What are the least dangerous volcanic eruptions?

20.12 What are some ways that people can protect against floods?

Critical Thinking

20.13 Earthquakes and volcanoes, major threats to life and property, are the result of plate tectonics. Assume that it is possible to stop plate tectonics and hence volcanism and earthquakes. What would some other results be?

20.14 In discussions of glaciers and streams we found that water exists on Mars. If we were to establish a settlement on Mars, how could we exploit this water?

Appendix A
Periodic Table

Key:

26
Fe
55.847

— Atomic Number
— Atomic Symbol
— Atomic Mass

1 **H** 1.00797																	2 **He** 4.0026
3 **Li** 6.939	4 **Be** 9.0122											5 **B** 10.811	6 **C** 12.011	7 **N** 14.0067	8 **O** 15.9994	9 **F** 18.998	10 **Ne** 20.183
11 **Na** 22.990	12 **Mg** 24.312											13 **Al** 26.982	14 **Si** 28.086	15 **P** 30.974	16 **S** 32.064	17 **Cl** 35.453	18 **Ar** 39.948
19 **K** 39.102	20 **Ca** 40.08	21 **Sc** 44.956	22 **Ti** 47.90	23 **V** 50.942	24 **Cr** 51.996	25 **Mn** 54.938	26 **Fe** 55.847	27 **Co** 58.933	28 **Ni** 58.71	29 **Cu** 63.54	30 **Zn** 65.37	31 **Ga** 69.72	32 **Ge** 72.59	33 **As** 74.922	34 **Se** 78.96	35 **Br** 79.909	36 **Kr** 83.80
37 **Rb** 85.47	38 **Sr** 87.63	39 **Y** 88.905	40 **Zr** 91.22	41 **Nb** 92.906	42 **Mo** 95.94	43 **Tc** (99)	44 **Ru** 101.07	45 **Rh** 102.91	46 **Pd** 106.4	47 **Ag** 107.870	48 **Cd** 112.40	49 **In** 114.82	50 **Sn** 118.69	51 **Sb** 121.75	52 **Te** 200.59	53 **I** 200.59	54 **Xe** 200.59
55 **Cs** 132.91	56 **Ba** 137.34	*	72 **Hf** 178.49	73 **Ta** 180.95	74 **W** 183.85	75 **Re** 186.2	76 **Os** 190.2	77 **Ir** 192.2	78 **Pt** 195.09	79 **Au** 196.97	80 **Hg** 200.59	81 **Tl** 204.37	82 **Pb** 207.19	83 **Bi** 208.98	84 **Po** (210)	85 **At** (210)	86 **Rn** (222)
87 **Fr** (223)	88 **Ra** (226)	**															

* **Lanthanide Series** (*Rare Earths*)

57 **La** 138.91	58 **Ce** 140.12	59 **Pr** 140.91	60 **Nd** 144.24	61 **Pm** (147)	62 **Sm** 150.35	63 **Eu** 151.96	64 **Gd** 157.25	65 **Tb** 158.92	66 **Dy** 162.50	67 **Ho** 164.93	68 **Er** 167.26	69 **Tm** 168.93	70 **Yb** 173.04	71 **Lu** 174.97

** **Actinide Series**

89 **Ac** (227)	90 **Th** 232.04	91 **Pa** (231)	92 **U** 238.03	93 **Np** (237)	94 **Pu** (244)	95 **Am** (243)	96 **Cm** (247)	97 **Bk** (247)	98 **Cf** (251)	99 **Es** (254)	100 **Fm** (253)	101 **Md** (256)	102 **No** (254)	103 **Lw** (257)

Appendix B
Earth Data

TABLE B.1 Distribution of World's Estimated Supply of Water[a]

| | AREA, THOUSANDS OF | | VOLUME, THOUSANDS OF | | |
	KM²	MI²	KM³	MI³	TOTAL VOLUME, %
World (total area)	510,000	197,000	—	—	—
Land area	149,000	57,500	—	—	—
Water in land areas:					
Freshwater lakes	850	330	125	30	0.009
Saline lakes and inland seas	700	270	104	25	0.008
Rivers (average instantaneous volume)	—	—	1.25	0.3	0.0001
Soil moisture and vadose water	—	—	67	16	0.005
Groundwater to depth of 4,000 m	—	—	8,350	2,000	0.61
Icecaps and glaciers	19,400	7,500	29,200	7,000	2.14
Atmospheric moisture	—	—	13	3.1	0.001
World ocean	361,000	139,500	1,320,000	317,000	97.3
Total water volume (rounded)			1,360,000	326,000	100

[a]In part after R. L. Nace, *U.S. Geol. Surv. Circ.* 536, Table 1, 1967.

TABLE B.2 Composition of Seawater at 35 Parts per Thousand Salinity[a]

ELEMENT	μG/L	ELEMENT	μG/L	ELEMENT	μG/L
Hydrogen	1.10×10^8	Nickel	6.6	Praesodymium	0.00064
Helium	0.0072	Copper	23	Neodymium	0.0023
Lithium	170	Zinc	11	Samarium	0.00042
Beryllium	0.0006	Gallium	0.03	Europium	0.000114
Boron	4,450	Germanium	0.06	Gadolinium	0.0006
Carbon (inorganic)	28,000	Arsenic	2.6	Terbium	0.0009
(dissolved organic)	2,000	Selenium	0.090	Dysprosium	0.00073
Nitrogen (dissolved N_2)	15,500	Bromine	6.73×10^4	Holmium	0.00022
(as NO_3^-, NO_2^-, NH_4^-)	670	Krypton	0.21	Erbium	0.00061
Oxygen (dissolved O_2)	6,000	Rubidium	120	Thulium	0.00013
(as H_2O)	8.83×10^8	Strontium	8,100	Ytterbium	0.00052
Fluorine	1,300	Yttrium	0.003	Lutetium	0.00012
Neon	0.120	Zirconium	0.026	Hafnium	<0.008
Sodium	1.08×10^7	Niobium	0.015	Tantalum	<0.0025
Magnesium	1.29×10^6	Molybdenum	10	Tungsten	<0.001
Aluminum	1	Ruthenium	—	Rhenium	—
Silicon	2,900	Rhodium	—	Osmium	—
Phosphorus	88	Palladium	—	Iridium	—
Sulfur	9.04×10^5	Silver	0.28	Platinum	—
Chlorine	1.94×10^7	Cadmium	0.11	Gold	0.011
Argon	450	Indium	—	Mercury	0.15
Potassium	3.92×10^5	Tin	0.81	Thallium	—
Calcium	4.11×10^5	Antimony	0.33	Lead	0.03
Scandium	<0.004	Tellurium	—	Bismuth	0.02
Titanium	1	Iodine	64	Radium	1×10^{-13}
Vanadium	1.9	Xenon	0.047	Thorium	0.0015
Chromium	0.2	Cesium	0.30	Protactinium	2×10^{-10}
Manganese	1.9	Barium	21	Uranium	3.3
Iron	3.4	Lanthanum	0.0029		
Cobalt	0.39	Cerium	0.0013		

[a]Adapted from Karl K. Turekian, *Oceans*, p. 92, Prentice Hall, Englewood Cliffs, N.J., 1968.

TABLE B.3 Runoff from the Continents[a]

| | AREA, MILLIONS OF | | ANNUAL RUNOFF | | | |
| | | | TOTAL, $10^{15} \times$ | | DEPTH/UNIT AREA | |
	KM²	MI²	1	GAL	CM	IN
Asia	46.6	18.0	11.1	3.0	23.8	9.4
Africa	29.8	11.5	5.9	1.6	19.8	7.8
North America	21.2	8.2	4.5	1.2	21.1	8.3
South America	19.6	7.6	8.0	2.1	41.4	16.3
Europe	10.9	4.2	2.5	0.6	23.1	9.1
Australia	7.8	3.0	0.4	0.1	2.5	1.0
Total or (mean)	135.9	52.5	32.4	8.6	(24.9)	(9.8)

[a]Calculated from Data from D. A. Livingstone, *U.S. Geol. Surv. Prof. Paper* 440-G, 1963.

TABLE B.4 Composition of River Waters of the World[a]

SUBSTANCE	PPM
HCO_3	58.4
Ca	15
SiO_2	13.1
SO_4	11.2
Cl	7.8
Na	6.3
Mg	4.1
K	2.3
NO_3	1
Fe	0.67

[a]From Daniel A. Livingstone, "Data of Geochemistry," *U.S. Geol. Surv. Prof. Paper* 440-G, p. G-41, 1963.

TABLE B.5 Earth Volume, Density, and Mass

	AV. THICKNESS OR RADIUS, KM	VOLUME, MILLIONS OF KM³	MEAN DENSITY, G/CM³	MASS, $\times 10^{24}$ G
Total Earth	6,371	1,083,230	5.52	5,976
Oceans and seas	3.8	1,370	1.03	1.41
Glaciers	1.6	25	0.9	0.023
Continental crust	35	6,210	2.8	17.39
Oceanic crust	8	2,660	2.9	7.71
Mantle	2,883	899,000	4.5	4,068
Core	3,471	175,500	10.71	1,881

TABLE B.6 Average Composition of the Crust[a]

	AV. IGNEOUS ROCK, %	AV. SHALE, %	AV. SANDSTONE, %	AV. LIMESTONE, %	WEIGHTED-AV. CRUST,[B]%
SiO_2	59.12	58.11	78.31	5.19	59.07
TiO_2	1.05	0.65	0.25	0.06	1.03
Al_2O_3	15.34	15.40	4.76	0.81	15.22
Fe_2O_3	3.08	4.02	1.08⎫	0.54	⎰3.10
FeO	3.80	2.45	0.30⎭		⎱3.71
MgO	3.49	2.44	1.16	7.89	3.45
CaO	5.08	3.10	5.50	42.57	5.10
Na_2O	3.84	1.30	0.45	0.05	3.71
K_2O	3.13	3.24	1.32	0.33	3.11
H_2O	1.15	4.99	1.63	0.77	1.30
CO_2	0.10	2.63	5.04	41.54	0.35
ZrO_2	0.04	—	—	—	0.04
P_2O_5	0.30	0.17	0.08	0.04	0.30
Cl	0.05	—	Tr^c	0.02	0.05
F	0.03	—	—	—	0.03
SO_3	—	0.65	0.07	0.05	—
S	0.05	—	—	0.09	0.06
$(Ce, Y)_2O_3$	0.02	—	—	—	0.02
Cr_2O_3	0.06	—	—	—	0.05
V_2O_3	0.03	—	—	—	0.03
MnO	0.12	Tr^c	Tr^c	0.05	0.11
NiO	0.03	—	—	—	0.03
BaO	0.05	0.05	0.05	0.00	0.05
SrO	0.02	0.00	0.00	0.00	0.02
Li_2O	0.01	Tr^c	Tr^c	Tr^c	0.01
Cu	0.01	—	—	—	0.01
C	0.00	0.80	—	—	0.04
Total	100.00	100.00	100.00	100.00	100.00

[a]After F. W. Clarke and H. S. Washington, "The Composition of the Earth's Crust," *U.S. Geol. Surv. Prof. Paper* 127, p. 32, 1924.
[b]Weighted average: igneous rock, 95%; shale, 4%; sandstone, 0.75%; limestone, 0.25%.
[c]Trace.

TABLE B.7 Earth Size

	THOUSANDS OF KM
Equatorial radius	6.378
Polar radius	6.357
Mean radius[a]	6.371
Polar circumference	40.009
Equatorial circumference	40.077
Ellipticity [(equatorial radius—polar radius)/equatorial radius], 1/297	

[a]Term used by geophysicists to designate radius of a sphere of equal volume.

TABLE B.8 Earth Areas

	MILLIONS OF KM²
Total area	510
Land (29.22% of total)	149
Oceans and seas (70.78% of total)	361
Glacier ice	15.6
Continental shelves	28.4

Energy Units

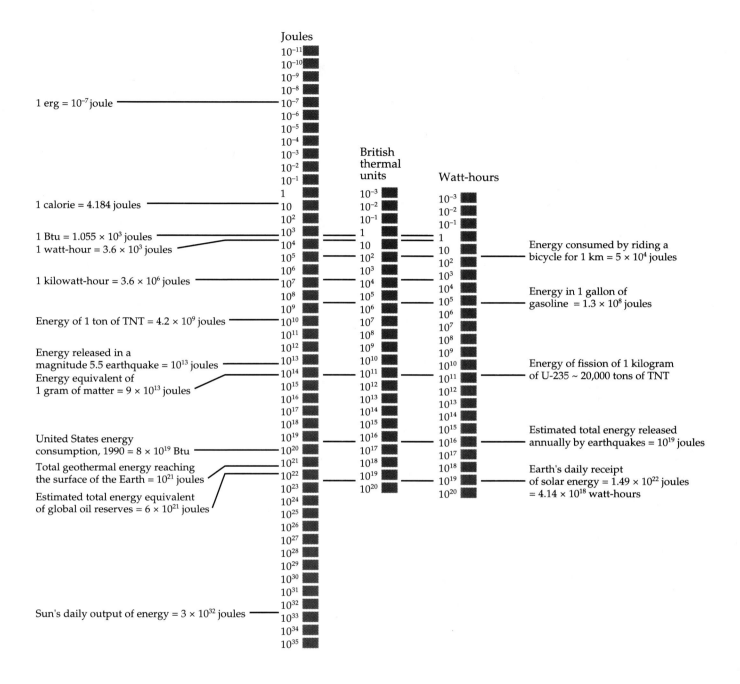

Joules

10^{-11}
10^{-10}
10^{-9}
10^{-8}
1 erg = 10^{-7} joule ——— 10^{-7}
10^{-6}
10^{-5}
10^{-4}
10^{-3}
10^{-2}
10^{-1}
1
1 calorie = 4.184 joules ——— 10
10^{2}
1 Btu = 1.055×10^3 joules ——— 10^{3}
1 watt-hour = 3.6×10^3 joules ——— 10^{4}
10^{5}
10^{6}
1 kilowatt-hour = 3.6×10^6 joules ——— 10^{7}
10^{8}
10^{9}
Energy of 1 ton of TNT = 4.2×10^9 joules ——— 10^{10}
10^{11}
10^{12}
Energy released in a
magnitude 5.5 earthquake = 10^{13} joules ——— 10^{13}
Energy equivalent of
1 gram of matter = 9×10^{13} joules ——— 10^{14}
10^{15}
10^{16}
10^{17}
10^{18}
United States energy
consumption, 1990 = 8×10^{19} Btu ——— 10^{19}
10^{20}
Total geothermal energy reaching
the surface of the Earth = 10^{21} joules ——— 10^{21}
10^{22}
Estimated total energy equivalent
of global oil reserves = 6×10^{21} joules ——— 10^{23}
10^{24}
10^{25}
10^{26}
10^{27}
10^{28}
10^{29}
10^{30}
10^{31}
10^{32}
Sun's daily output of energy = 3×10^{32} joules ——— 10^{33}
10^{34}
10^{35}

British
thermal
units

10^{-3}
10^{-2}
10^{-1}
1
10
10^{2}
10^{3}
10^{4}
10^{5}
10^{6}
10^{7}
10^{8}
10^{9}
10^{10}
10^{11}
10^{12}
10^{13}
10^{14}
10^{15}
10^{16}
10^{17}
10^{18}
10^{19}
10^{20}

Watt-hours

10^{-3}
10^{-2}
10^{-1}
1
10
10^{2}
10^{3}
10^{4}
10^{5}
10^{6}
10^{7}
10^{8}
10^{9}
10^{10}
10^{11}
10^{12}
10^{13}
10^{14}
10^{15}
10^{16}
10^{17}
10^{18}
10^{19}
10^{20}

Energy consumed by riding a
bicycle for 1 km = 5×10^4 joules

Energy in 1 gallon of
gasoline = 1.3×10^8 joules

Energy of fission of 1 kilogram
of U-235 ~ 20,000 tons of TNT

Estimated total energy released
annually by earthquakes = 10^{19} joules

Earth's daily receipt
of solar energy = 1.49×10^{22} joules
= 4.14×10^{18} watt-hours

Glossary

Where possible, definitions conform generally, and in some cases specifically, to definitions given in Robert L. Bates and Julia A. Jackson (eds.), *Glossary of Geology*, 3d ed., American Geological Institute, Alexandria, Virginia, 1987.

^{14}**C method** A method for determining the age in years of organic matter by calculating the amount of radioactive carbon still remaining, as compared to the stable isotope ^{12}C.

^{40}**K/^{40}Ar method** A method used for the dating of potassium-bearing rocks by using the ratio of radioactive ^{40}K to its daughter, ^{40}Ar.

aa A Hawaiian term for a lava flow that has a rough, jagged surface; *compare* **pahoehoe.**

ablation As applied to glacier ice, the process by which ice below the snow line is wasted by evaporation and melting.

absolute time Geologic time expressed in years before the present.

abundant metal Iron, aluminum, magnesium, manganese, and titanium. Ores of the abundant metals only need to be 3 to 5 times as metal-rich as average rock.

abyssal plain Large area of extremely flat ocean floor lying near a continent and generally over 4 km in depth.

acceleration The rate at which **velocity** changes, either by increasing or decreasing.

accretion The process by which the terrestrial planets grew, increasing their mass by gradually accumulating smaller bodies, called planetesimals.

acid mine drainage Water contamination by sulfuric acid produced by seepage through sulfur-bearing soil and tailings from coal and metal mining.

acid rain The acidity in rain due to gases from internal combustion engines and coal- and oil-burning power plants.

active layer The seasonally thawed zone above **permafrost.**

actualism *see* **uniformitarianism.**

aftershock An earthquake that follows and has its epicenter near a larger earthquake.

agate A siliceous rock with alternating bands of **chalcedony** and variously colored **chert.**

alluvial fan Land counterpart of a **delta.** An assemblage of sediments marking the place where a stream moves from a steep gradient to a flatter gradient and suddenly loses transporting power. Typical of arid and semiarid climates but not confined to them.

alpha decay The process of radioactive decay in which the nucleus of an atom emits an alpha particle. The new atom's **atomic number** is lower by two and its **atomic mass number** is reduced by four.

alpha particle A particle consisting of two protons and two neutrons, produced during alpha decay. Identical to the nucleus of a ^{4}He atom.

alpine glacier *see* **valley glacier.**

amygdaloidal A textural term describing volcanic rocks that contain numerous **amygdules.**

amygdule A gas cavity (**vesicle**) in volcanic rock that has been filled with mineral matter such as calcite, chalcedony, or quartz.

andesite A fine-grained volcanic rock of intermediate composition, consisting largely of plagioclase and one or more mafic minerals.

andesite line The geographic boundary between rocks of the Pacific Basin, which are basaltic, and those around the rim of the basin, which are in part andesitic.

angle of incidence The angle at which a ray of energy approaches a surface.

angle of reflection The angle at which a reflected ray of energy leaves a surface.

angle of refraction The angle at which a refracted ray of energy leaves a surface after passing through it.

angle of repose The maximum angle at which loose material will come to rest when added to a pile of similar material.

angular unconformity An **unconformity** in which the beds below the unconformity dip at a different angle from the beds above it.

anion An **ion** with a negative electrical charge. That is, an atom that has gained one or more electrons.

anticline A fold that is convex upward, or that had such an attitude at some stage of its development; *compare* **syncline.**

aphanitic A textural term meaning "fine-grained" that applies to igneous rocks.

aquifer A permeable region of rock or soil through which groundwater can move.

aquitard A material of low permeability that greatly slows the movement of groundwater.

arch Forms along a coast as wave erosion cuts through a headland.

Archean An eon of geologic time extending from about 3.9 billion years to 2.5 billion years ago.

arête A narrow, saw-toothed mountain ridge developed by glacier erosion in adjacent **cirques.**

arkose A sedimentary rock formed by the cementation of sand-sized grains of feldspar and quartz.

artesian well A well in which the water in the aquifer is under pressure, which raises the water above the point that the well first encounters it.

assemblage The collection of minerals that characterize a rock or a **facies.**

asthenosphere The weak or "soft" zone in the upper mantle just below the **lithosphere,** involved in plate movement and isostatic adjustments. It lies 70 to 100 km below the surface and may extend to a depth of 400 km. Corresponds to the seismic **low-velocity zone.**

astronomic theory of glaciation A theory based on the changing position of the Earth in its orbit around the Sun.

asymmetric rock knob or hill Bedrock forms with a gentle slope on one side created by glacial abrasion and a steep slope on the opposite side created by glacial plucking.

atoll A roughly circular reef with an occasional small, low coral sand island surrounding a shallow lagoon.

atom A building block of matter, the smallest particle that has the chemical characteristics of a particular **chemical element.** It contains a nucleus of protons and neutrons surrounded by a cloud of electrons.

atomic mass number The sum of the number of protons and the number of neutrons in an atom. Approximately equal to the mass of the atom.

atomic number The number of protons in an atom, a quantity that determines which element the atom represents. Example: All atoms of oxygen have 8 protons.

aureole A zone surrounding an igneous intrusion, in which **contact metamorphism** has taken place.

authigenesis The process by which new minerals form in a sediment or sedimentary rock during or after deposition.

axial plane A geometric plane that intersects the trough or crest of a fold in such a way that the limbs of the fold are more or less symmetrically arranged with reference to it.

axis The line formed by the intersection of the axial plane of a fold with a bedding plane, marking where the bed shows its maximum curvature.

back-arc basin The region between an **island arc** and the continental mainland, commonly with at least some oceanic crust on its floor.

back swamp A swamp that forms in the low-lying floodplain behind a **levee.**

backshore The area that lies between high-tide mark and the foot of the beach dune or the limit of effective wave action.

banded iron formation (BIF) A sedimentary mineral deposit dominated by iron oxides, carbonates, or silicates that were deposited chemically from seawater. Most BIFs were formed between 2.5 billion and 3.5 billion years ago. Their formation is related to the rise of oxygen in the atmosphere.

bankfull stage A stream discharge that just fills the stream channel.

bar 1. A mass of sand, gravel, or alluvium deposited on the bed of a stream, sea, or lake, or at the mouth of a stream. 2. A unit of pressure, approximately equal to atmospheric pressure at sea level.

barchan A crescent-shaped sand dune with horns pointing downwind.

barrier beaches or islands Long narrow beaches separated in many places from the mainland by lagoons.

barrier reef A coral reef separated from the mainland by a lagoon.

basalt A dark-colored extrusive igneous rock composed chiefly of calcium plagioclase and pyroxene. Extrusive equivalent of gabbro, underlies the ocean basins and comprises oceanic crust.

base flow Groundwater that enters a stream channel, maintaining stream flow at times when it is not raining.

base level The point below which a stream cannot cut. A temporary base level along a stream, such as a lake, may be removed by stream action. Ultimate base level is the ocean.

basin A synclinal structure, roughly circular in its outcrop pattern, in which beds dip gently toward the center from all directions.

batholith A large, discordant, and intrusive body of igneous rock.

bauxite A rock composed of aluminum hydroxides and impurities in the form of silica, clay, silt, and iron hydroxides. A residual weathering product, exploited as the primary ore for aluminum.

bay barrier A beach that cuts off a bay from the sea.

beach replenishment Rebuilding a beach by adding sand to it.

beach Temporary accumulation of sediments that collects between low- and high-tide marks.

bedding A collective term used to signify presence of beds, or layers, in sedimentary rocks and deposits.

bedding plane Surface separating layers of sedimentary rocks and deposits. Each bedding plane marks termination of one deposit and beginning of another of different character, such as a surface separating a sandstone bed from an overlying mudstone bed. Rock tends to break or separate readily along bedding planes.

bed load Material in motion along a stream bed.

bedrock Any solid rock exposed at the Earth's surface or overlain by unconsolidated material.

beheaded stream The headwaters of a stream that have been captured by another stream; *compare* **stream piracy.**

berm A small terrace in the **backshore** area of the coast with its terrace facing seaward.

beta decay The process of radioactive decay in which a neutron loses a beta particle, which is physically identical to an electron. This increases the **atomic number** of the atom by 1 by turning the neutron into a proton. The atom's **atomic mass number** stays the same because the total number of protons and neutrons remains the same. The most common form of radioactive decay.

BIF *see* **banded iron formation.**

binding energy The energy that holds the particles in the nucleus of an atom together. It is this energy, when released, that is used to generate nuclear power.

biogenic sediment Sediments produced directly by the life processes of plants or animals.

biogenic sedimentary rock A sedimentary rock composed primarily of **biogenic sediments.**

bioturbation The turning and mixing of sediments by organisms.

black smoker A vent on the seafloor from which hydrothermal fluids are emitted. Upon mixing with seawater and cooling, the fluids precipitate a cloud of fine-grained sulfide minerals resembling a cloud of black smoke.

blind valley A valley in **karst** that ends abruptly downstream at the point where its stream disappears underground as a **sinking stream.**

blowout An irregular depression excavated by wind, usually in previously deposited blown sand.

body wave Any seismic wave that travels through the body of the Earth, rather than along its surface; *compare* **surface wave.**

bond (ionic, covalent, Van der Waals, metallic) *see* **chemical bond.**

bottomset bed Layer of fine sediment deposited in a body of standing water beyond the edge of a growing **delta** and which is eventually built over by the advancing delta. Similarly, bottomset beds may accumulate in the wind shadow of a sand dune and be preserved beneath it as the dune advances.

boudinage A structure in which brittle beds bounded by more ductile ones have been divided into segments during metamorphism.

boulder train Clusters of **erratics** from same source, with some distinctive characteristic that makes their common source easily recognizable.

boundary The tectonic region in which two plates meet; *compare* **margin.**

Bowen's Reaction Series A series of minerals formed during crystallization of a magma, in which the formation of minerals alters the composition of the remaining magma. Mafic minerals comprise a discontinuous series, in which successive minerals form at the expense of earlier-formed ones. The plagioclase feldspars form in a continuous series, in which the composition of plagioclase becomes progressively sodium rich, but the crystal structure of the mineral does not change.

braided stream A stream with a complex tangle of converging and diverging channels separated by sandbars or islands.

branchwork cave Cave with passageways formed along bedding planes and with an areal pattern similar to that of surface streams.

breakwater A protective wall built offshore and usually parallel to the shore.

breccia A clastic rock in which the gravel-sized particles are angular in shape and make up an appreciable volume of the rock.

breeder reactor A nuclear reactor in which ^{238}U or ^{232}Th, which are not easily fissionable, absorb neutrons to become atoms of ^{239}Pu or ^{236}U, which can later be used as fuels in fission reactors. Breeder technology is not yet feasible.

brittle Structural behavior in which a material deforms permanently by fracturing.

brittle limit The stress limit beyond which a material fractures, rather than behaving in a ductile or elastic fashion.

burial metamorphism Takes place in an environment where pressure and temperature are barely more intense than during **diagenesis,** typically in a deepening sequence of sediments.

calcarenite A sandstone in which the sand-sized grains are calcite.

caldera A large, basin-shaped volcanic depression, more or less circular in form. Typically steep-sided, found at the summit of a **shield volcano.**

caliche Gravel, sand, or desert debris cemented by calcium carbonate, an accumulated product of chemical weathering in a dry climate; *compare* **claypan, fragipan, hardpan.**

calving The breaking away of ice from the front of the glacier when it ends in a lake or an ocean. Produces icebergs.

cap rock A comparatively impervious stratum immediately overlying an oil- or gas-bearing rock.

capacity The total amount of material a stream is able to carry under given conditions.

capillary water Water in the **zone of aeration** that holds to soil particles by surface tension of the water molecules for each other and for the soil particles.

carbonate conservation depth The water depth below which the calcium carbonate produced in the ocean is completely dissolved. There is no calcium carbonate deposition below this level.

carbonate rock A rock consisting primarily of a carbonate mineral such as calcite or dolomite, the chief minerals in limestone and dolostone, respectively.

cataclastic metamorphism Takes place in an environment where intense pressure due to shearing is common, as in a major fault zone.

cation An **ion** that has a positive electrical charge. That is, an atom that has lost one or more electrons.

cave A natural open space underground, large enough for a person to enter. Most commonly occur by the dissolution of soluble rocks, generally limestone.

cementation Process by which a binding, or cementing, agent is precipitated in spaces among individual particles of a deposit. Common cementing agents are calcite, quartz, and dolomite.

Cenozoic The current geologic era, which began 66.4 million years ago and continues to the present.

chain reaction A self-sustaining nuclear reaction, made possible when neutrons released by fission of some atoms in a nuclear reactor strike other atoms, causing them to fission as well.

chalcedony A cryptocrystalline form of quartz, microscopically fibrous with waxy luster. May be transparent or translucent, and with a uniform tint of white, gray, pale blue, and, less often, black.

chalk A variety of limestone made up in part of biochemically derived calcite, in the form of skeletons or skeletal fragments of microscopic oceanic plants and animals mixed with fine-grained calcite deposits of biochemical or inorganic-chemical origin.

chemical bond The interactions among the electrons of atoms that hold atoms together to form chemical compounds. If electrons cluster primarily around one atom of a pair, the bond is *ionic.* If they are shared more or less equally, it is *covalent.* If electrons move freely between atoms over an extended region, the bond is *metallic.* A weak electrostatic bond due to uneven distribution of electrons around atoms or groups of atoms is a *Van der Waals* bond.

chemical element A fundamental substance that cannot be further refined or subdivided by chemical means. All **atoms** of a chemical element have the same number of protons.

chemical remanent magnetism Acquired as magnetic minerals form and align themselves to the global magnetic field during diagenesis of a sedimentary deposit.

chemical sediment Sediment formed by chemical precipitation from water. Example: halite precipitated as the result of the evaporation of seawater.

chemical sedimentary rock A sedimentary rock made up of chemical sediments. Example: rock salt.

chemical weathering *see* **decomposition.**

chert A cryptocrystalline form of quartz, microscopically granular. Occurs as nodules and as thin, continuous layers. Duller, less waxy luster than chalcedony. Occurs in limestone, dolostone, and mudstones.

chlorofluorocarbons (CFC) Gases that can be dissociated by solar radiation, which releases chlorine, which in turn destroys ozone.

chute cutoff A narrow "short cut" across a meander bend, formed in flood as the main stream flow is diverted into a trough between point bars. Sometimes called simply a "chute."

cinder cone A conical volcano formed by the accumulation of **pyroclastic** debris around a vent.

cirque A steep-walled hollow in a mountain side, shaped like an amphitheater, or bowl, with one side partially cut away. Place of origin of a mountain glacier.

clastic Refers to rock or sediments made up primarily of broken fragments of preexisting rocks or minerals.

clay 1. The name for a family of finely crystalline sheet silicate minerals. 2. Fine-grained soil consisting of mineral particles, not necessarily clay minerals, that are less than 0.074 mm in their maximum dimension.

claypan A layer of stiff, compact, relatively impervious clay that is not cemented; *compare* **caliche, fragipan, hardpan.**

cleavage 1. of a mineral: The tendency of a mineral to split along planes determined by the crystal structure. 2. of a rock: *see* **slaty cleavage.**

coal Sedimentary rock composed of combustible matter derived from the partial decomposition of plant material.

coast A narrow strip of land along the margin of the ocean extending inland for a variable distance from low-water mark.

col Mountain pass formed by enlargement of two opposing **cirques** until their headwalls meet and are broken down.

column Pillar formed as a **stalactite** and **stalagmite** meet.

columnar jointing The type of jointing that breaks rock, typically basalt, into columnar prisms. Usually the joints form a more or less distinct hexagonal pattern.

compaction Reduction of pore space between individual particles as the result of overlying sediments or of tectonic movements.

competence The maximum size of particle that a stream can carry.

composite volcano *see* **stratovolcano.**

Comprehensive Soil Classification System (CSCS) The classification system in most common use by North American soil scientists. Categories are based on the chemical and physical characteristics of a soil; *compare* **USDA Soil Classification System.**

compression Squeezing a material from opposite directions.

concordant Lying parallel to, rather than cutting across, surrounding strata.

concretion A compact mass of mineral matter, usually spherical or disk-like in shape and embedded in a host rock of different composition. The mass forms by precipitation of mineral matter about a nucleus such as a leaf, or a piece of shell or bone.

conduction Heat transport by direct transfer of energy from one particle to another, without moving the particle to a new location; *compare* **convection, radiation.** Example: Heat is transported through a bar of steel by conduction.

cone of depression A downward distortion or dimple in the water table that forms as a well pumps water faster than it can flow through the aquifer.

conglomerate A clastic sedimentary rock composed of lithified beds of rounded gravel mixed with sand.

Constancy of Interfacial Angles The statement that the angles between congruent crystal faces on samples of a single mineral are always identical. A consequence of, and therefore evidence for the existence of, crystalline structure in minerals.

contact metamorphism Metamorphism genetically related to the intrusion (or extrusion) of magmas and taking place in rocks at or near their contact with a body of igneous rock.

continental arc A belt of volcanic mountains on the continental mainland that lie above a subduction zone; *compare* **island arc.**

continental crust The part of the crust that directly underlies the continents and continental shelves. Averages about 35 km in thickness, but may be over 70-km thick under largest mountain ranges.

continental deserts Located in continental interiors far from moisture-bearing winds.

continental divide A major drainage divide separating the drainage to one ocean from another.

continental drift The theory that explained the relative positions and shapes of continents, the formation of mountains, and other large-scale geologic phenomena as results of the lateral movement of continents. The crust of ocean basins was assumed to be relatively immobile; *compare* **plate tectonics, seafloor spreading.**

continental ice glacier An ice sheet that obscures all but the highest peaks of a large part of a continent.

continental rise The portion of the continental margin that lies between the **abyssal plain** and the **continental slope.** The continental rise is underlain by crustal rocks of the ocean basin.

continental shelf The portion of the continental margin that extends as a gently sloping surface from the shoreline seaward to a marked change in slope at the top of the **continental slope.** Seaward depth averages about 130 m.

continental slope That part of the continental margin that lies between the **continental shelf** and the **continental rise.** Slope relatively steep, 3° to 6°. The continental slope is underlain by crustal rocks of the continent.

convection Heat transport by moving particles, and the thermal energy that they carry, to a new location; *compare* **conduction, radiation.** Example: Heat rises through the atmosphere by convection.

convection cell A cyclical pattern of movement in a fluid body such as the ocean, the atmosphere, or the Earth's mantle, driven by density variations, which in turn are the result of differences in temperature from one part of the fluid to another.

convergent boundary A boundary between two plates of the Earth's crust that are pushing together.

co-product A mineral commodity that is recovered from a mining operation for some other mineral product. Example: Platinum is commonly a co-product of nickel mining.

coquina A coarse-grained, porous variety of clastic limestone made up chiefly of shells and shell fragments.

core Innermost zone of the Earth. Consists of two parts, an outer liquid section and an inner solid section, both chiefly of iron and nickel with about 10 percent lighter elements. It is surrounded by the mantle.

correlation Process of establishing contemporaneity of rocks or events in one area with rocks or events in another area.

crater 1. A steep-walled, usually conical depression at the summit or on the flanks of a volcano, resulting from the explosive ejection of material from a vent. 2. A bowl-shaped depression with a raised, overturned rim produced by the impact of a meteorite or other energetic projectile.

craton The stable portions of the continents that have escaped orogenic activity for the last 2 billion years. Made predominantly of granite and metamorphic rocks; *compare* **orogen.**

creep 1. The very slow, generally continuous downslope movement of soil and debris under the influence of gravity. 2. The movement of sand grains along the land surface.

crevasse 1. Breach in a natural **levee.** 2. Deep crevice or open fracture in glacier ice.

cross-cutting relationships Geologic discontinuities that suggest relative ages: A geologic feature is younger than the feature it cuts. Thus, a fault cutting across a rock is younger than the rock.

crust The upper part of the **lithosphere,** divided into **oceanic crust** and **continental crust.**

crystal The multisided form of a mineral, bounded by planar growth surfaces, that is the outward expression of the ordered arrangement of atoms within it.

crystal settling Gravitational sinking of crystals from the liquid in which they formed, by virtue of their greater density. A type of igneous **differentiation.**

crystal structure The regular and repeated three-dimensional arrangement of atoms or ions in a crystal.

crystalline 1. Having a crystal structure. 2. When referring to sedimentary rocks, *crystalline* designates a texture in which mineral crystals have formed in an interlocking pattern; *see* **nonclastic.** 3. As a generic term, geologists use the term "crystalline rocks" as a rough synonym for "igneous or metamorphic rocks."

cumulate An igneous rock that forms by **crystal settling.**

Curie point The temperature above which a mineral loses its magnetism.

current ripple mark An asymmetric ripple mark formed by wind or water moving generally in one direction. The steep face of ripple faces in direction of current.

cyclothem A series of beds, of interest because they include coal, which were associated with unstable shelf or interior basin conditions in which alternating marine transgressions and regressions occurred.

Darcy's law A formula describing the flow of water through an aquifer.

daughter An atom that results from the radioactive decay of a parent atom.

debris flow Fast-moving, turbulent mass movement with a high content of both water and rock debris. The more rapid debris flows rival the speed of rock slides.

decay rate The rate at which a population of radioactive atoms decays into stable daughter atoms. Rate often expressed in terms of **half-life** of the parent isotope.

decomposition (chemical weathering) Weathering processes that are the result of chemical reactions. Example: the transformation of orthoclase to kaolinite.

deflation A process of erosion in which wind carries off particles of dust and sand.

dehydration Any process by which water bound within a solid material is released. Example: Gypsum ($CaSO_4 \cdot 2H_2O$) becomes anhydrite ($CaSO_4$) by dehydration.

delta An assemblage of sediments accumulated where a stream flows into a body of standing water and its velocity and transporting power are suddenly reduced. A "delta plain" is the upper surface of a delta.

dendritic drainage A stream pattern that, when viewed on a map or from the air, resembles the branching pattern of a deciduous tree such as a maple or oak.

denudation The sum of the processes that result in the wearing away or the progressive lowering of the Earth's surface by weathering, erosion, mass wasting, and transportation.

depositional environment The nature of the environment in which sediments are laid down. They are immensely varied and may range from the deep ocean to the coral reef and the glacial lake of the high mountains. The nature of the depositional environment may be deduced from the nature of the sediments and rock deposited there.

depositional remanent magnetism Develops as magnetic minerals settle through water and align themselves in the Earth's magnetic field.

desert pavement A lag accumulation of pebbles or boulders that cuts off further **deflation.**

desertification A process of land degradation initiated by human activity, particularly in the zones along the margins of deserts.

detrital sedimentary rock A sedimentary rock made up of detrital sediments.

detrital sediments Sediments made of fragments or mineral grains weathered from preexisting rocks.

diagenesis All the physical, chemical, and biologic changes undergone by sediments from the time of their initial deposition, through their conversion to solid rock, and subsequently to the brink of metamorphism.

differential weathering Weathering that occurs at different rates, as the result of variations in composition and mechanical resistance of rocks, or differences in the intensity of weathering processes.

differentiation The process of developing more than one rock type, in situ, from a common magma.

dike A tabular igneous intrusion that cuts across the surrounding rock.

dilatancy An increase in the bulk volume of rock during deformation. Possibly related to the migration of water into microfractures or pores.

dip The angle that a structural surface such as a bedding plane or fault surface makes with the horizontal, measured perpendicular to the **strike** and in the vertical plane.

dip pole *see* **magnetic pole.**

dip-slip fault A fault on which the movement is parallel to the dip of the fault plane.

directed pressure Pressure applied predominantly in one direction, rather than uniformly.

discharge In a stream, the volume of water passing through a channel in a given time.

disconformity An **unconformity** in which the beds above the unconformity are parallel to the beds below the unconformity.

discordant Cutting across surrounding strata.

disintegration (mechanical weathering) The processes of weathering by which physical actions such as frost wedging break down a rock into fragments, involving no chemical change.

dissolution A chemical reaction in which a solid material is dispersed as ions in a liquid. Example: Halite (NaCl) undergoes dissolution when placed in water.

dissolved load Amount of material that water carries in solution.

distributary channels Stream channels that fan out from the upstream point of the **delta** and carry the sediments that build the delta.

divergent boundary Boundary between two crustal plates that are pulling apart.

dolostone A carbonate rock made up predominantly of the mineral dolomite, $CaMg(CO_3)_2$.

dome An uplift or anticlinal structure, roughly circular in its outcrop exposure, in which beds dip gently away from the center in all directions.

drag fold A minor fold produced within a weak bed or adjacent to a fault by the movement of surrounding rocks in opposite directions.

drainage basin The area from which a stream and its tributaries receive water.

drainage divide The line that separates one drainage basin from another.

drift Glacial deposits laid down directly by glaciers or laid down in lakes, the ocean, or streams as a result of glacial activity.

dripstone Calcium carbonate deposited from solution as water enters a cave through the zone of aeration. Forms **stalactites, stalagmites,** and other cave deposits.

drumlin Streamlined hill, largely of **till,** with blunt end pointing into direction from which ice moved. Occurs in clusters called drumlin fields.

dry farming Farming without irrigation in **drylands.**

drylands A general term for semiarid and desert lands.

ductile Structural behavior in which a material deforms permanently without fracturing.

dust bowl An area subject to dust storms, especially the south-central United States.

dust devil A small, dust-bearing whirlwind.

dust storm Large volume of dust-sized particles lifted high into the atmosphere.

Earth system System involving continuous interaction of the solid Earth, the atmosphere, the oceans, and living things.

earthflow A form of slow, but perceptible, mass movement, with high content of water and rock debris. Lateral boundaries are well-defined and the terminus is lobed. With increasing moisture content, grades into a mudflow.

eccentricity of the Earth's orbit A measure of the circularity of the Earth's orbit. It varies in cycles of about 100,000 and 400,000 years.

elastic Nonpermanent structural deformation during which the amount of deformation (strain) is proportional to the stress.

elastic rebound The statement that movement along a fault is the result of an abrupt release of a progressively increasing elastic strain between the rocks on either side of the fault.

electron A fundamental unit of matter, negatively charged and disposed in a cloud surrounding the nucleus of an atom.

electron capture Nuclear decay in which a proton in the nucleus acquires an electron from the outer cloud of the atom's electrons. This converts the proton to a neutron, reduces the number of protons in the nucleus by 1 and atomic number of the original element by 1. **Atomic mass number** remains constant because the total number of protons and neutrons is unchanged.

electron shell A characteristic energy level with which an electron is associated. Electrons occupy discrete shells within the cloud surrounding an atom's nucleus. These may be thought of, loosely, as if they represented orbits at distinct heights above the nucleus.

element see **chemical element.**

end moraine see **terminal moraine.**

eon The primary division of geologic time; from oldest to youngest, they are the Hadean, Archean, Proterozoic, and Phanerozoic eons.

epicenter The point on the Earth's surface that is directly above the **focus** of an earthquake.

epoch A division of geologic time next shorter than a **period.** Example: The Pleistocene epoch is in the Quaternary period.

equilibrium line On a glacier the line separating the **zone of accumulation** from the **zone of ablation.**

era A division of geologic time next smaller than the **eon** and larger than a **period.** Example: The Paleozoic era is in the Phanerozoic eon and includes, among others, the Devonian period.

erratic A stone or boulder, glacially transported from place of origin and left in an area of different bedrock composition.

esker A winding ridge of stratified drift. Forms in a glacial tunnel and, when ice melts, stands as ridge up to 15 m high and kilometers in length.

ETP curve see **Milankovitch curve.**

eustatic change in sea level A worldwide change in sea level, such as caused by melting glaciers.

eutrophication The process of aging of lakes by the addition of nutrients.

exfoliation The process by which concentric scales, plates, or shells of rock are stripped or "spall" from the bare surface of a large rock mass.

exfoliation dome A large dome-shaped form that develops in homogeneous crystalline rocks as the result of exfoliation.

exotic river A river that is able to maintain its flow through a desert because of water received from outside the desert.

extrusive Pertaining to igneous rocks or features formed from lava released on the Earth's surface.

facies see **metamorphic facies, sedimentary facies.**

failed rift A rift emanating from a **plate triple junction** along which minimal divergence has taken place.

fall When applied to mass movement of material, refers to free fall of material moving without contact with the surface.

fault The surface of rock rupture along which there has been differential movement of the rock on either side.

fault gouge Soft, uncemented, pulverized clay-like material found along some faults.

ferromagnesian Containing iron and magnesium, applied to the **mafic** minerals. Example: olivine.

fetch Distance over which wave-forming winds blow.

field capacity see **specific retention.**

fiery cloud see **nuée ardente.**

firn (névé) Granular ice formed by the recrystallization of snow. Intermediate between snow and glacier ice.

fjord Glaciated valleys now flooded by the sea.

fission The spontaneous or induced splitting, by particle collision, of a heavy atomic nucleus into a pair of fragments plus some neutrons. Controlled induced fission can be used as a source of nuclear power.

fission track dating Dating of minerals by fission tracks, lines of damage left in a mineral by spontaneous alpha emissions.

fissure eruption An eruption of lava that takes place from a fracture, usually without producing a cone.

flash flood A flood that rises and falls very rapidly.

flashy stream A stream with a high, short flood peak and short lag time.

flint A variety of **chert,** often black because of included organic matter.

flood Peak flow that tops the banks of a stream channel.

flood recurrence interval The number of years of record plus 1 divided by the rank of each maximum annual flood.

floodplain Area bordering a stream over which stream water spreads when the stream tops its channel banks.

flow folding A fold formed in relatively fluid rocks that have flowed toward a synclinal trough.

flowstone General term for deposits formed by dripping and flowing water on walls and floors of caves.

fluid inclusion A tiny cavity in a crystal, commonly 1 to 100 microns in diameter, containing liquid and/or gas. Formed by the entrapment of fluid during the growth or subsequent deformation of the crystal.

focus The point within the Earth which is the center of an earthquake, at which strain energy is first released and converted to elastic wave energy.

fold and thrust mountains Mountains, characterized by extensive folding and thrust faulting, that form at convergent plate boundaries on continents.

foot wall block The body of rock that lies above an inclined fault plane. *see* **hanging wall block.**

foreset bed Inclined layers of sediment deposited on the advancing margin of a growing delta or along the slip face of a sand dune.

foreshock A minor tremor that precedes an earthquake. An increase in seismicity may signal that a major release of strain energy is about to occur.

foreshore Lies between the low- and high-tide marks.

formation water The water, held in pore volume in sedimentary rocks, that has persisted with little change in composition ever since it was buried with the sediment.

fossil Evidence in rock of the presence of past life, such as a dinosaur bone, an ancient clam shell, or the footprint of a long-extinct animal.

fossil fuel A hydrocarbon (coal or petroleum) that can be extracted from the Earth for use as a fuel. Fossil fuels are nonrenewable energy sources.

fractional crystallization A sequence of crystallization from magma in which the earlier formed crystals are prevented from reacting with the remaining magma, resulting in a magma with an evolving chemical composition.

fragipan A dense layer of soil, containing silt and sand but no organic matter and little clay, whose extreme hardness and impermeability are due primarily to compaction; *compare* **caliche, claypan, hardpan.**

free oscillation A vibration of a body such as a bell or the Earth that continues without further influence after an initial event.

fringing reef A coral reef attached directly to the mainland.

frost wedging A type of **disintegration** in which jointed rock is forced apart by the expansion of water as it freezes inside fractures.

fusion The combination of two light nuclei to form a heavier nucleus, with the accompanying release of energy. This is the source of energy in a hydrogen bomb. If it could be controlled, it could serve as an alternative to **fission** in nuclear power generation.

gardening The constant and slow churning of the lunar **regolith** as the result of meteorite impacts.

geanticline An anticlinal structure presumed to form in the context of geosynclinal evolution. Not in current use since the development of plate tectonic theory.

geode Roughly spherical, hollow, or partially hollow accumulation of mineral matter. Can be a few centimeters to nearly 0.5 m in diameter. Outer layer of **chalcedony** lined with crystals that project toward the hollow center. Crystals, often perfectly formed, are usually quartz although calcite and dolomite and—more rarely—other minerals. Most commonly occur in limestone, and less often in shale.

geologic column The arrangement of rock units in the proper chronological order from youngest to oldest.

geologic time scale The chronological sequence of units of Earth time.

geology The science that deals with the study of the planet Earth—the materials of which it is made, the processes that act to change these materials from one form to another, and the history recorded by these materials; the forces acting to deform the outer layers of the Earth and create ocean basins and continents; the processes that modify the Earth's surface; the application of geologic knowledge to the search for useful materials and the understanding of the relationship of geologic processes to people.

geosyncline A downwarping of the Earth's crust, either elongate or basin-like, measured in scores of kilometers, in which sedimentary and volcanic rocks accumulate to thicknesses of thousands of meters. Not in current use since the development of plate tectonic theory.

geothermal energy Heat extracted from the Earth for use as a power source.

geothermal gradient The rate at which temperature increases with depth below the Earth's surface.

geyser A type of thermal spring that ejects water intermittently with considerable force.

glaciation The formation, advance, and retreat of glaciers and the results of these activities.

glacier A mass of ice, formed by the recrystallization of snow, that flows forward, or has flowed at some time in the past.

glacier ice Ice with interlocking crystals that makes up the bulk of a glacier.

glass An inorganic solid in which there is no **crystal structure.**

glassy A texture of extrusive igneous rocks that develops as the result of rapid cooling, so that crystallization is inhibited.

global warming The prediction that climate will warm as a result of the addition to the atmosphere of humanly produced greenhouse gases.

gneiss A coarse, foliated metamorphic rock in which bands of granular minerals (commonly quartz and feldspars) alternate with bands of flaky or elongate minerals (e.g., micas, pyroxenes). Generally less than 50% of the minerals are aligned in a parallel orientation.

gneissosity The style of foliation typical of gneiss.

Gondwana The southern portion of the late Paleozoic supercontinent known as **Pangea.** It means, literally "Land of the Gonds" (a people of the Indian subcontinent). The variant *Gondwanaland* found in some books, therefore, is a tautology.

gouge *see* **fault gouge.**

graben *see* **rift.**

graded bedding Type of sedimentary deposit in which individual beds become finer from bottom to top.

gradient Slope of a stream bed or hillside; the vertical distance of descent over horizontal distance of a slope.

granite Light-colored, coarse-grained, intrusive igneous rock characterized by the minerals orthoclase and quartz with lesser amounts of plagioclase feldspar and iron-magnesium minerals. Underlies large sections of the continents.

granitic belt A region of granitic rock, one of two characteristic regions within **cratons.**

granitization A metamorphic process by which solid rock is converted into granite by the addition or removal of material, without passing through a magmatic stage; *compare* **metasomatism.**

gravitational heating Planetary heating caused by the conversion of potential energy into heat. Associated with the **iron catastrophe.**

gravitational moisture Water in the **zone of aeration** that is moving down toward the **zone of saturation.**

graywacke (lithic sandstone) A variety of sandstone characterized by angular-shaped grains of quartz and feldspar, and small fragments of dark rock all set in a matrix of finer particles.

greenhouse gases Gases (primarily water and carbon dioxide, but also a variety of sulfur and nitrogen compounds and gaseous hydrocarbons) that trap the Sun's heat in the atmosphere.

greenstone An altered or metamorphosed mafic igneous rock that owes its dark color to the presence of chlorite, epidote, or amphiboles.

greenstone belt A region of greenstones, one of two characteristic regions within **cratons.**

groin A wall built out from the shore, usually perpendicular to it, to trap sand carried by **longshore currents.**

groove A broad, deep, generally straight furrow carved in bedrock by the abrasive action of debris embedded in a moving glacier. Larger and deeper than a glacial striation.

ground moraine Till deposited from main body of glacier during ablation.

ground-water table *see* **water table.**

ground water Water beneath the Earth's surface.

guyot *see* **seamount.**

habit A general term for the outward appearance of a mineral, defined by the relative sizes and arrangement of characteristic crystal faces.

Hadean The oldest eon in Earth history, extending from the origin of the Earth to about 3.9 billion years ago.

half-life The amount of time that it takes for one half of an original population of atoms of a radioactive isotope to decay.

hanging valley A valley whose mouth is high above the floor of the main valley to which it is tributary. Usually, but not always, the result of mountain glaciation.

hanging wall block The body of rock that lies below an inclined fault plane. *see* **foot wall block.**

hardness Resistance of a mineral to scratching, determined on a comparative basis by the **Mohs scale.**

hardpan A general term for a relatively hard layer of soil at or just below the ground surface, cemented by silica, iron oxide, calcium carbonate, or organic matter; *compare* **caliche, claypan, fragipan.**

head (hydraulic head) The level to which groundwater in the **zone of saturation** will rise.

heat flow The amount of thermal energy leaving the Earth per cm^2/sec.

heave In mass movement, the upward motion of material by expansion as, for example, the heaving caused by freezing water.

hiatus A gap or interruption in the continuity of the geologic record either because the record was never formed or because it was destroyed by erosion. It represents the time interval spanned by an **unconformity.**

high-level nuclear waste Radioactive waste from defense activities of the U.S. government and from spent fuel rods from nuclear reactors.

hinge fault A fault along which there is increasing offset or separation along the strike of the fault plane, from an initial point of no separation.

hoodoo A column or pillar of rock produced by **differential weathering** in a region of sporadic heavy rainfall, commonly facilitated by joints and by rock layers of varying hardness.

Hooke's law A statement of elastic deformation, namely that strain is directly proportional to stress.

horn The sharp spire of rock formed as glaciers in several **cirques** erode into a central mountain peak.

horst *compare* **rift.**

humus The generally dark, more or less stable part of the organic matter in a soil, so well decomposed that the original sources cannot be identified.

hydraulic conductivity Measure of **permeability** in Earth materials.

hydraulic gradient The slope of the **water table.** Measured by the difference in elevation between two points on the slope of the water table and the distance of flow between them.

hydraulic head *see* **head.**

hydrograph Graph of variation of stream flow over time.

hydrologic system (or hydrologic cycle) The pattern of water circulation from the ocean to the atmosphere to the land and back to the ocean.

hydrolysis A decomposition reaction involving water, in which hydrogen ions (H^+) or hydroxyl ions (OH^-) replace other ions. The result is a new residual mineral. Example: The addition of water to orthoclase produces kaolinite and releases K^+ and silica into solution.

ice sheet A broad, mound-like mass of glacier ice that usually spreads radially outward from a central zone.

ice shelf A floating ice sheet extending across water from a land-based glacier.

icecap A small ice sheet.

igneous rock A rock that has crystallized from a molten state.

inclined bedding Bedding laid down at an angle to the horizontal, as in many sand dunes.

inclined fold A fold whose axial plane is inclined from the vertical, but in which the steeper of the two limbs is not overturned; *compare* **overturned fold.**

inclusion (xenolith) A fragment of older rock caught up in an igneous rock.

index fossil A fossil that identifies and dates the strata in which it is typically found. To be most useful, an index fossil must have broad, even worldwide distribution and must be restricted to a narrow stratigraphic range.

index mineral A mineral formed under a particular set of temperature and pressure conditions, thus characterizing a particular degree of metamorphism.

inertia The tendency of a body to resist **acceleration.** A moving body tends to keep moving at a constant speed in the same direction, and a stationary body tends to remain in one place, unless acted upon by an outside force.

inner core The solid innermost part of the Earth's core with a diameter of a little over 1,200 km.

intensity A measure of the size of an earthquake in terms of the damage it causes.

interlobate moraine Ridge formed along junction of adjacent glacier lobes.

intrusive Pertaining to igneous rocks or features formed by the emplacement of magma in preexisting rock.

ion An atom that has an electrical charge, by virtue of having gained or lost electrons; *see* **cation, anion.**

ionic radius The effective distance from the center of an ion to the edge of its electron cloud.

ionic substitution The replacement of one or more ions in a crystal structure by others of similar size and electrical charge. Example: Fe^{2+} is interchangeable with Mg^{2+} in most ferromagnesian minerals.

iron catastrophe The period in the Hadean eon during which much of the iron in outer portions of the Earth migrated toward the center of the planet, producing the core and releasing large amounts of **gravitational heat.**

ironpan A **hardpan** in which iron oxides are the primary cementing agents.

island arc A curved belt of volcanic islands lying above a subduction zone; *compare* **continental arc.**

isochemical reaction A reaction in which chemical constituents of a rock are rearranged to form a new mineral assemblage, but no material is added to or lost from the rock as a whole. Applied generally to diagenetic or metamorphic environments.

isoclinal fold A fold in which the limbs are parallel.

isograd A line on a map joining points at which metamorphism took place under similar temperature and pressure conditions, as indicated by rocks belonging to the same **metamorphic facies.** Generally, the line separates two adjacent **metamorphic zones,** as indicated by specific **index minerals.**

isoseismal line A line on a map joining points of equal earthquake intensity.

isostasy The condition of equilibrium, comparable to floating, of units of the **lithosphere** above the **asthenosphere.**

isostatic change in sea level A sea level change due to change in load on Earth's crust.

jasper A red variety of **chert,** its color coming from minute particles of included hematite.

jet flow Flow in which fluid moves at high speed in jet-like surges as does water in free fall over a falls.

jetty Similar to a **groin** but built to keep sand out of a harbor entrance.

joint A surface of fracture in a rock, without displacement parallel to the fracture.

juvenile hydrothermal fluid A hot fluid, largely water, presumed to have been released from a magma.

kame Stratified drift deposited in depressions and cavities in stagnant ice and left as irregular, steep-sided hills when the ice melts.

kame terrace Stratified drift deposited between wasting glacier and adjacent valley wall. Stands as a terrace when glacier melts.

karst A landscape that develops from the action of groundwater in areas of easily soluble rocks. Characterized by caves, underground drainage, and sinkholes.

kettle Depression in ground surface formed by the melting of a block of glacier ice buried or partially buried by **drift.**

komatiite An ultramafic rock with a noncumulate texture, presumed to be extrusive.

laccolith A concordant igneous intrusion with a flat floor and a convex upper surface, usually less than 8 km across and from a few meters to a few hundred meters thick at its widest point.

lag time The delay in the response of stream flow between precipitation and flood peak.

lahar A **mudflow** composed chiefly of **pyroclastic** material on the flanks of a volcano.

laminar flow Fluid flow in which flow lines are distinct, and parallel and do not mix; *compare* **turbulent flow.**

lateral continuity The extent of a rock unit over a considerable but definite area.

lateral moraine Moraine formed by valley glaciers along valley sides.

laterite A highly weathered red soil rich in iron and aluminum oxides. Typically formed in a tropical to temperate climate where intense chemical weathering is common.

Laurasia The northern portion of the late Paleozoic supercontinent called **Pangea.**

lava Molten rock that flows at the Earth's surface.

lava dome A steep-sided rounded extrusion of highly viscous lava squeezed out from a volcano and forming a dome-shaped or bulbous mass above and around the volcanic vent. The structure generally develops inside a volcanic crater.

lava flood (plateau basalt) A term applied to large areas of basaltic lava presumably extruded from fissures.

lava lake A lake of lava, usually basaltic, in a volcanic caldera.

layered complex An intrusive igneous body in which there are layers of varying mineral content.

levees Banks of sand and silt along a stream bank built by deposition in small increments during successive floods.

limb The portions of a fold that are away from the hinge; the "sides" of the fold.

limestone A sedimentary rock composed mostly of the mineral calcite, $CaCO_3$.

linear dune Long, straight dune with slip faces on each side.

lineation A general term applying to any linear feature in a metamorphic rock.

liquefaction The transformation of a soil from a solid to a liquid state as the result of increased pore pressure.

lithic sandstone *see* **graywacke.**

lithification The process by which an unconsolidated deposit of sediments is converted into solid rock. Compaction, cementation, and recrystallization are involved.

lithophile Said of an element that has a greater chemical affinity for silicate rocks than for sulfides or for a metallic state. Example: aluminum.

lithosphere The rigid outer shell of the Earth. It includes the crust and uppermost mantle and is on the order of 100 km in thickness.

lithostatic stress The confining (nondirected) pressure imposed by the weight of overlying rock.

littoral current *see* **longshore current.**

load Of a stream, the amount that it carries at any one time.

loess Deposits of wind-borne dust.

longshore current (littoral current) A current that flows parallel to the shore just inside the surf zone. Also called the *littoral current.*

Love wave A seismic **surface wave** that has a horizontal (side-to-side) component but no vertical component.

low-level nuclear waste (TRU) Comes largely from national defense utilities and includes contaminated lab coats, gloves, and laboratory equipment.

low-velocity zone The seismic region within the upper mantle that corresponds to the **asthenosphere.**

luster The manner in which light reflects from the surface of a mineral, described by its quality and intensity.

mafic Referring to an igneous rock with significant amounts of one or more **ferromagnesian** minerals, or to a magma with significant amounts of iron and magnesium.

magma Molten rock, containing dissolved gases and suspended solid particles. At the Earth's surface, magma is known as **lava.**

magma ocean A global-scale ocean of magma, according to some calculations several hundred kilometers deep, thought to have existed during the final stages of **accretion** as the Earth was forming.

magnetic anomaly The amount by which a measurement of the local magnetic field intensity exceeds or falls below the intensity of the global magnetic field.

magnetic chron Time during which magnetic polarity is dominantly normal or dominantly reversed.

magnetic declination Angle of divergence between true north and magnetic north. Measured in degrees east or west of true, or geographic, north.

magnetic equator Lies halfway between the north and south magnetic poles.

magnetic inclination The angle of dip of the compass needle as it varies from horizontal at the magnetic equator to vertical at the magnetic poles.

magnetic polarity The direction, north (normal) or south (reversed), that a magnetic needle points.

magnetic polarity time scale A chronology based on the shifting polarity of the Earth's magnetic field.

magnetic pole The point on the Earth's surface where a magnetic needle points vertically downward (north magnetic pole) or vertically upward (south magnetic pole).

magnetic stratigraphy A stratigraphic sequence based on the magnetic polarity of the rocks.

magnetic subchron A period during a magnetic chron when the magnetic polarity is the opposite from that of the magnetic chron.

magnitude A measure of the strength of an earthquake based on the amount of movement recorded by a seismograph; *compare* **Richter scale.**

mantle That portion of the Earth below the crust and reaching to about 2,780 km, where a transition zone of about 100-km thickness separates it from the core.

marble A metamorphic rock composed largely of calcite. The metamorphic equivalent of limestone.

margin The tectonic region that lies at the edge of a continent, whether it coincides with a plate **boundary** or not.

mass movement The downslope movement of material under the influence of gravity.

maze cave Caves in which passageways have interconnecting loops that form a maze-like pattern.

meander A sharp bend, loop, or turn in a stream's course. When abandoned, called a meander scar or an **oxbow.**

medial moraine Formed by the merging of lateral moraines as two valley glaciers join.

mélange (clastic wedge) A mappable body of rock characterized by blocks and fragments of all sizes, embedded in a sheared matrix. A tectonic mélange commonly forms in the upper portions of a subduction zone, where crustal rock is crushed and sheared.

mesosphere A zone in the Earth between 400 and 670 km below the surface separating the upper mantle from the lower mantle.

Mesozoic An era of time during the Phanerozoic eon lasting from 245 million to 66.4 million years ago.

metal porphyry deposit A mineral deposit genetically related to a pluton of porphyritic rock, commonly granodiorite. Scarce metals are typically enriched by the passage of hydrothermal fluids through rocks surrounding the intrusion, with the result that a metal-rich halo forms there.

metamorphic facies A set of metamorphic mineral assemblages, repeatedly associated in space and time, such that there is a constant and therefore predictable relationship between mineral composition and chemical composition. That relationship is a consequence of conditions of temperature and pressure under which the assemblages are stable.

metamorphic rock A rock changed from its original form and/or composition by heat, pressure, or chemically active fluids, or some combination of them.

metamorphic zone A mappable region in which rocks have been metamorphosed to the same degree, as evidenced by the similarity of mineral assemblages in them.

metamorphism The processes of recrystallization, textural, and mineralogical change that take place in the solid state under conditions beyond those normally encountered during diagenesis.

metasomatism The metamorphic processes that occur as a result of the passage of chemically active fluids through a rock, adding to or removing constituents during metamorphism.

microplate *see* **terrane.**

migmatite A composite rock composed of igneous and metamorphic materials, the result of partial melting at the upper limit of metamorphism.

Milankovitch curve (ETP curve) A graph representing the amount of solar radiation received at the Earth's surface at a particular latitude and time and based on the variations in the Earth's orbital motions.

mineral A naturally occurring inorganic solid that has a well-defined chemical composition and in which atoms are arranged in an ordered fashion.

mineral deposit Any natural concentration of a valuable material in the Earth's crust, whether that material can be extracted profitably or not.

Modified Mercalli scale A commonly used scale of earthquake **intensity.**

Mohorovičić discontinuity (Moho) The sharp seismic velocity discontinuity that separates the crust and the mantle.

Mohs scale The ten-point scale of mineral **hardness,** keyed arbitrarily to the minerals talc, gypsum, calcite, fluorite, apatite, orthoclase, quartz, topaz, corundum, and diamond.

molecule The smallest unit of matter that has the chemical and physical properties of a particular chemical compound.

momentum transfer In a rock slide the forward transfer of energy by the collision of one block of rock with the next block forward. The process makes possible progressively more rapid movement of material in downslope positions.

monocline A simple fold, described as a local steepening in strata with an otherwise uniform dip.

moraine Landform made largely of **till.**

mountain glacier *see* **valley glacier.**

mud cracks Cracks, generally polygonal, caused by the shrinking of a deposit of clay or silt under surface conditions.

mudflow Form of mass movement similar to a debris flow but containing less rock material.

mudstone A fine-grained detrital sedimentary rock made up of clay- and silt-sized particles.

mylonite A chert-like rock without cleavage but with a banded or streaky structure produced by extreme shearing of rocks that have been pulverized and rolled during intense dynamic metamorphism.

nappe A sheet of rock that has moved over a large horizontal distance by thrust faulting, recumbent folding, or both, so that it lies on rocks of markedly different age or lithologic character.

neck cutoff Occurs as a river cuts through the narrow neck of a meander. Sometimes called simply a "cutoff."

neutron A particle in the nucleus of an atom, which is without electrical charge and with approximately the same mass as a proton.

névé *see* **firn.**

nivation Erosion beneath and around edges of a snow bank. Results can foreshadow a cirque.

nodule A small, irregular, knobby-surfaced rock body that differs in composition from the rock that encloses it. Formed by the replacement of the original mineral matter. Quartz in the form of flint or chert is the most common component. Most common in limestone and dolostone.

nonclastic A term applied to sedimentary rocks that are not composed of fragments of preexisting rocks or minerals. The term "crystalline" is more commonly used.

nonconformity An **unconformity** that separates profoundly different rock types, such as sedimentary rocks from metamorphic rocks.

normal fault A **dip-slip fault** on which the **hanging wall block** is offset downward relative to the **foot wall block;** *compare* **reverse fault.**

normal polarity Time when the compass needle points to the magnetic north pole.

north magnetic pole The point on the Earth where the north-seeking end of a magnetic needle, free to swing in space, points directly down.

nuclear power Power generated by controlled **fission** or (potentially) **fusion** reactions, the heat from which is used to produce steam and drive turbines.

nucleus (atomic) The center of an atom, containing both protons and (except for ^1H) neutrons.

nuée ardente (fiery cloud) A dense, hot (sometimes incandescent) cloud of volcanic ash and gas produced in a **Peléan eruption.**

oblique-slip fault A fault with both **dip-slip** and **strike-slip** components of movement.

obliquity of the Earth's ecliptic Tilt of the Earth's rotational axis in relation to the plane in which the Earth circles the Sun. Cycles range from about 21.5° to 24.5° and back to 21.5° every 41,000 years.

oceanic crust That part of the Earth's crust underlying the ocean basins. Composed of basalt and having a thickness of about 5 km.

oil shale A mudrock that will yield liquid or gaseous hydrocarbons upon distillation.

oolite Spheroidal grains of sand size, usually composed of calcite and thought to have formed by inorganic precipitation.

open pit mining Surficial mining, in which the valuable rock is exposed by removal of overlying rock or soil.

ophiolite An assemblage of mafic and ultramafic rocks and associated marine cherts and their metamorphic equivalents.

ore The naturally occurring material from which a mineral or minerals of economic value can be extracted at a profit.

ore deposit A continuous well-defined mass of material of sufficient **ore** content to make extraction economically feasible; *compare* **mineral deposit.**

original horizontality Refers to the condition of beds or strata as being horizontal or nearly horizontal when first formed.

orogen Linear to arcuate in plan, intensely deformed crustal belt associated with mountain building; *compare* **craton.**

orogeny The process of mountain building.

outer core The outermost part of the core. It is liquid, about 1,700 km thick, and separated from the inner, solid core by a transition zone about 565 km thick.

outwash Beds of sand and gravel laid down by glacial meltwater.

outwash plain A plain underlain by outwash.

overbank deposits Sediments deposited from floodwater on the floodplain.

overturned fold An inclined fold in which one limb has been tilted beyond the vertical, so that the stratigraphic sequence within it is reversed; *compare* **inclined fold.**

oxbow An abandoned **meander.**

oxbow lake A lake in an abandoned meander.

oxidation The **decomposition** process by which iron or other metallic elements in a rock combine with oxygen to form residual oxide minerals.

ozone hole Decrease of ozone in the stratosphere.

P-wave (primary wave, compressional wave) A seismic **body wave** that involves particle motion, alternating compression and expansion, in the direction of wave propagation. It is the fastest seismic wave; *compare* **S-wave.**

pahoehoe A Hawaiian term for a basaltic lava flow with a smooth, or ropy surface; *compare* **aa.**

paleomagnetism Study of the Earth's past magnetism as it is recorded in the rocks.

paleosol A buried soil horizon of the geologic past.

Paleozoic An era of geologic time lasting from 570 million to 245 million years ago.

Pangea A supercontinent that existed from about 300 million to 200 million years ago, and included most of the Earth's continental crust.

parabolic dune A sand dune that is parabolic in plan with slip-face convex downwind.

parent A radioactive element whose decay produces stable daughter elements.

partial melting The igneous process in which a rock begins to melt at the lower end of its melting interval, yielding a magma with a chemical composition different from the bulk composition of the parent rock.

pascal A unit of pressure, equal to 1/100,000 of a **bar.**

pedalfer A generic term to describe soils typically formed in a humid region. Characteristically having an accumulation of iron and aluminum oxides and hydroxides.

pedocal A generic term to describe soils typically found in an arid or semiarid region. Characteristically having an accumulation of carbonates, particularly calcite.

pegmatite An extremely coarse-grained igneous rock with interlocking crystals, usually with a bulk chemical composition similar to granite but commonly containing rare minerals enriched in lithium, boron, fluorine, niobium, and other scarce metals. Pegmatites are also the source for many gem-quality precious and semiprecious stones.

pegmatitic Having the texture of a pegmatite.

pelagic ooze A deep ocean sediment consisting of at least 30% skeletal remains of calcareous or siliceous microorganisms, the rest being clay minerals.

Peléan eruption A type of volcanic eruption characterized by **nuées ardentes** and the development of **lava domes.**

peneplain Low, gently rolling landscapes produced by long-continued erosion.

perched water table A water table that develops at a higher elevation than the main water table.

peridotite An ultramafic igneous rock, the major constituent of the mantle.

periglacial Refers to conditions in a near glacial climate.

period In the geologic time scale a unit of time less than an **era** and greater than an **epoch.** Example: The Tertiary period was the earliest period in the Cenozoic era and included, among others, the Eocene epoch.

permafrost Soil conditions prevailing in area whose mean annual temperature is 0°C.

permafrost table The depth in a permafrost region at which the maximum temperature reaches 0°C.

permeability The capacity of material to transmit water or other fluids.

petroleum A general term including both oil and natural gas.

phaneritic A textural term meaning "coarse-grained" that applies to igneous rocks.

Phanerozoic The most recent eon of geologic time beginning 570 million years ago and continuing to the present.

phenocryst Any relatively large, conspicuous crystal in a **porphyritic** igneous rock; *compare* **porphyroblast.**

phyllite A metamorphosed mudstone with a silky sheen, more coarse-grained than a slate and less coarse-grained than a schist.

piedmont glacier A glacier that spreads out at the foot of mountains, formed by the coalescence of two or more valley glaciers.

pillow A structure observed in certain igneous rocks extruded into water, characterized by discontinuous, close-fitting, pillow-shaped masses, commonly 30 to 60 cm across.

pipe A vertical conduit through the Earth's crust below a volcano, through which magma has passed.

pirate stream A stream that captures the headwaters of another stream.

placer A surficial mineral deposit formed by mechanical concentration of valuable minerals from weathered debris, usually through the action of stream currents or of waves.

plate A rigid segment of the Earth's lithosphere that moves horizontally and adjoins other plates along zones of seismic activity. Plates may include portions of both continents and ocean basins.

plate boundaries Zones of seismic activity along which plates are in contact. These may coincide with continental **margins,** but usually do not. Movement between plates is predominantly horizontal, and may be divergent, convergent, or side-by-side.

plate tectonics A theory of global tectonics according to which the lithosphere is divided into mobile **plates.** The entire lithosphere is in motion, not simply those segments composed of continental material; *compare* **continental drift.**

plate triple junction A point from which three rifts emanate at roughly 120° angles. Example: the Afar triangle in East Africa.

playa A broad, flat desert basin, often containing an ephemeral playa lake.

plucking (quarrying) A process of erosion in which the glacier pulls loose pieces of bedrock.

plume The movement of water along flow lines from a point source of groundwater pollution toward its eventual emergence at the surface.

plunging fold A fold in which the axis is inclined at an angle from the horizontal.

pluton An igneous intrusion.

pluvial lake A lake formed during a pluvial period.

pluvial period Time when a dryland area had greater effective moisture than at present.

pocket beach Small, narrow beach, usually crescentic, at head of a bay or small inlet.

point bar Accumulations of sand and gravel deposited in slack water on inside of a winding or meandering river.

polar deserts Deserts in which most moisture is locked up in ground ice and unavailable as liquid water.

polar glacier A glacier whose temperature throughout is always below freezing.

polish A smooth, polished surface imparted to some rock types by glacier abrasion.

polymetamorphism A series of events in which two or more metamorphic episodes have left their imprint on the same rocks.

polymorphism The circumstance in which two minerals with different crystalline structures have identical chemical compositions. Example: diamond and graphite.

porosity The percentage of material occupied by pore space.

porphyritic A texture of an igneous rock in which large crystals (**phenocrysts**) are set in a matrix of relatively finer-grained crystals or of glass.

porphyroblast A large crystal of a mineral such as garnet or staurolite set in a matrix of much finer-grained minerals in a metamorphic rock; *compare* **phenocryst.**

potentiometric surface The level to which water will rise in an **artesian** system when its confining **aquitard** is pierced.

pothole A hole or basin cut into bedrock of a stream by the abrasive action of pebbles and sand swirled by turbulent stream flow.

Precambrian An informal term to include all geologic time from the beginning of the Earth to the beginning of the Cambrian period 570 million years ago.

precession of the equinox The wobble of the Earth as it spins changes the direction in which its axis of rotation points. One wobble takes about 23,000 years.

pressure melting The phenomenon causing increased melting of ice by increase of pressure.

Principle of faunal and floral succession Groups of animals and plants have succeeded one another in a definite and discernible order.

prograde A succession of metamorphic conditions, each of which is at a higher temperature and/or pressure than the preceding one.

Proterozoic The geologic eon lying between the Archean and Phanerozoic eons, beginning about 2.5 billion years ago and ending about 0.57 billion years ago.

proton A fundamental particle of matter. Provides a positive charge in the nucleus of an atom.

pyroclastic Pertaining to clastic material formed by volcanic explosion or aerial expulsion from a volcanic vent.

quarrying 1. The process by which building stone, usually in blocks or sheets, is extracted from the Earth. 2. *see* **plucking.**

quartz arenite A sandstone in which the sand grains are predominantly quartz.

quartzite A metamorphic rock consisting largely of interlocking quartz grains; the metamorphic equivalent of a sandstone or chert.

radial drainage A pattern in which streams radiate outward from a high central zone.

radiation Heat transport without the intervention of matter, as in the transport of heat from the Sun to the Earth; *compare* **conduction, convection.**

radioactivity The spontaneous decay of the nucleus of an element. It involves the change in the number of protons in the nucleus and therefore creates an atom of a new element.

radiocarbon ^{14}C derived from ^{14}N as cosmic ray bombardment adds a neutron to its nucleus and the nucleus emits a proton. Radiocarbon decays back to ^{14}N by **beta decay.** Half-life is $5,730 \pm 30$ years.

rain shadow deserts Deserts formed by blocking moisture-bearing winds with mountain barriers.

ramp The planar surface sloping seaward from the foot of the **shore face.**

rapids Turbulent streamwater flow down a steep gradient, but not as steep as in a waterfall.

Rayleigh wave A type of seismic **surface wave** in which particles follow a backward elliptical orbit in a vertical plane.

reaction rim A peripheral zone around a mineral grain, composed of another mineral.

recessional moraine Ridges of glacial **till** marking halt and slight readvance of glacier during its general retreat.

rectangular drainage A pattern in which a stream and its tributaries follow courses marked by nearly right-angle bends.

recumbent fold A fold in which the axial plane is horizontal.

refraction 1. Bending of waves or rays of energy, e.g., seismic waves. 2. As applies to the near shore environment, the bending of wave crests as they approach the shore.

regional metamorphism Metamorphism affecting an extensive region, associated with **orogeny.**

regolith A layer of unconsolidated fragmental rock material.

relative time Dating of rocks and geologic events by their positions in chronological order without reference to number of years before the present.

remanent magnetism Magnetism acquired by a rock at some time in the past.

reserves That portion of the **resources** for a valuable mineral commodity that can be extracted from the Earth at a profit today.

reservoir rock Any porous and permeable rock that yields oil or natural gas.

residual (resistant) mineral A mineral that persists in soil after weathering, either because it was resistant to weathering or because it was formed during the weathering process.

residual soil A soil presumed to have developed in place as the product of decomposition and disintegration of **bedrock.**

resources The **reserves** of a valuable mineral commodity plus all other mineral deposits that may eventually become available, even those that are presumed to exist but have not yet been discovered and those that are not economically or technologically exploitable at the moment. The total mineral endowment ultimately available for extraction.

retrograde A succession of metamorphic conditions, each one of which is at a lower temperature and/or pressure than the preceding one.

reverse fault A dip-slip fault on which the **hanging wall block** is offset upward relative to the **foot wall block;** compare **normal fault.**

reversed polarity Time when a magnetic needle points to the south pole.

rhyolite A fine-grained, silica-rich igneous rock, the extrusive equivalent of granite.

Richter scale A commonly used measure of earthquake **magnitude** based on a logarithmic scale. Each integral step on the scale represents a tenfold increase in the extent of ground shaking, as recorded on a seismograph.

rift (graben) A valley caused by extension of the Earth's crust. Its floor forms as a portion of the crust moves downward along **normal faults.**

rip current Carries excess water in the **longshore current** out through the surf zone where it dissipates.

ripple marks Small waves produced by wind or water moving across deposits of sand or silt.

ripple marks of oscillation Symmetrical ripple marks formed by oscillating movement of water such as may be found along the coast just outside the surf zone.

rock An aggregate of one or more minerals in varying proportions.

rock avalanche see **rockslide.**

rock cleavage see **cleavage.**

rock cycle The concept of a sequence of events involving the formation, alteration, destruction, and re-formation of rocks as a result of geologic processes and which is recurrent, returning to a starting point. It represents a closed system; compare **rock system.**

rock flour Finely divided rock material ground by glacial action and fed by streams fed by melting glaciers.

rock glacier A mass of ice-cemented rock rubble found on slopes of some high mountains. Movement is slow, averaging 30 to 40 cm/yr.

rock record The history recorded in rocks.

rockslide (rock avalanche) A slide involving a downward and usually sudden movement of newly detached segments of bedrock sliding or slipping over an inclined surface of weakness such as a bedding plane, fault plane, or joint surface.

rock system The concept of a sequence of events involving the formation, alteration, destruction, and re-formation of rocks as a result of geologic processes. Unlike the rock cycle, it is an open system and does not return to a starting point; compare **rock cycle.**

rock varnish A thin, shiny veneer of clay minerals and iron and manganese oxides deposited on some rocks in a desert environment.

rock waste Angular fragments of rock. Forms a **talus** if abundant enough.

rockfall The sudden fall of one or more large pieces of a rock from a cliff.

roundness The degree to which a sedimentary particle's corners and edges are rounded.

runoff The precipitation that runs directly off the surface to stream or body of standing water.

S-wave (secondary wave, shear wave) A seismic **body wave** that involves particle motion from side-to-side, perpendicular to the direction of wave propagation. S-waves are slower than P-waves and cannot travel through a liquid; compare **P-wave.**

salinization A process by which salts accumulate in soil.

saltation A process of sediment transport in which a particle jumps from one point to another.

salt-water invasion Displacement of fresh surface or groundwater by the advance of salt water.

sand dune An accumulation of wind-driven sand into a distinctive shape.

sand sea A large area completely, or nearly completely, covered with sand dunes.

sandstone A clastic sedimentary rock in which the particles are predominantly of sand size, from 0.062 mm to 2 mm in diameter.

sandstorm A blanket of wind-driven sand with an upper surface about a meter above ground level.

sanitary landfill An artificial hill formed by the refuse of present-day civilization.

schist A strongly foliated, coarsely crystalline metamorphic rock, produced during regional metamorphism, that can readily be split into slabs or flakes because more than 50% of its mineral grains are parallel to each other.

schistosity The foliation in a schist, due largely to the parallel orientation of micas.

seafloor spreading Process by which ocean floors spread laterally from crests of main ocean ridges. As material moves laterally from the ridge, new material replaces it along the ridge crest by welling upward from the mantle.

seamount (guyot) A volcanic mountain on the seafloor. If flat-topped, it is a guyot.

seawall A wall at the shore and parallel to it for protection against wave erosion.

sedimentary facies An accumulation of deposits exhibiting specific characteristics and grades laterally into other sedimentary accumulations that were formed at the same time but exhibit different characteristics.

sedimentary rock Rock formed from the accumulation of sediment, which may consist of fragments and mineral grains of varying sizes from preexisting rocks, remains or products of animals and plants, the products of chemical action, or combinations of these.

seismic gap A segment of an active fault zone that has not experienced a major earthquake during a time period when most other segments of the zone have. Generally regarded as having a higher potential for future earthquakes.

seismic sea wave (tsunami) A sea wave produced by any large-scale, short-duration disturbance on the seafloor, commonly a shallow submarine earthquake but possibly also a submarine slide or volcanic eruption.

seismic tomography A technique for three-dimensional imaging of the Earth's interior by using a computer to compare the seismic records from a large number of stations. Similar in concept to a CAT scan used for medical purposes.

seismograph An instrument that detects, magnifies, and records vibrations of the Earth, especially earthquakes.

seismology The study of earthquakes, and of the structure of the Earth by both natural and artificially generated seismic waves.

seismoscope An instrument that merely indicates the occurrence of an earthquake.

self-exciting dynamo In reference to the Earth, the suggestion that movements in the fluid core may help initiate the Earth's magnetic field.

shadow zone A region 100° to 140° from the epicenter of an earthquake in which, owing to refraction from below the core-mantle boundary, no direct seismic waves can be detected.

shale A mudstone that splits or fractures readily.

shatter cone A distinctively striated conical structure in rock, ranging from a few centimeters to a few meters in length, believed to have been formed by the passage of a shock wave following meteorite impact.

shear Rock deformation involving movement past each other of adjacent parts of the rock and parallel to the plane separating them.

shear strength The resistance of a body to shear stress.

shear stress The stress on an object operating parallel to the slope on which it lies.

sheeting A type of jointing produced by pressure release (**unloading**) or **exfoliation.**

shield volcano A volcano in the shape of a flattened cone, broad and low, built by very fluid flows of basaltic lava.

shock lamellae Closely spaced microscopic planes, distinct from cleavage planes, that occur in shock-metamorphosed minerals and that are regarded as important indicators of shock metamorphism.

shock metamorphism Metamorphism induced in rock by the passage of a high-pressure shock wave acting over a period of time from a few microseconds to a fraction of a minute. The only known natural cause of shock metamorphism is the hypervelocity impact of a meteorite.

shore Seaward edge of coast between low tide and effective wave action.

shore face The concave section of the beach from high-tide mark down to the ramp between 5 and 20 m offshore.

shoreline The line separating land and water. Fluctuates as water rises and falls.

shore platform A surface of erosion that slopes gently seaward from a cliff base to the low-tide mark.

sial The upper layer of the continental crust, so called because it is rich in silica and aluminum oxide; *compare* **sima.**

sialic Enriched in sial.

silica Silicon dioxide (SiO_2) as a pure crystalline substance makes up quartz and related forms such as **flint** and **chalcedony.** More generally, silica is the basic chemical constituent common to all silicate minerals and magmas.

silica tetrahedron The basic structural unit of which all silicates are composed, consisting of a silicon atom surrounded symmetrically by four oxygen atoms. The structure, therefore, has the form of a tetrahedron with an oxygen atom at each corner.

sill A tabular igneous intrusion that parallels the planar structure of the surrounding rock.

sima The oceanic crust, also the lower layer of the continental crust, so called because it is enriched in silica and magnesium oxide; *compare* **sial.**

sinkhole Depression in ground surface caused by collapse into a cave below.

sinking stream A stream that empties into the underground into a cave, usually through a **sinkhole.**

slate A compact, fine-grained metamorphic rock that has **slaty cleavage.**

slaty cleavage *see* **cleavage.**

slickenside A polished and smoothly striated surface that results from friction along a fault plane.

slide A mass movement in which material maintains continuous contact with the surface on which it moves.

slip face Steep face on lee side of sand dune.

slump Downward and outward rotational movement of Earth materials traveling as a unit or series of units.

smelting The process of removing metal from **ore.**

snow line The elevation at which snow persists throughout the year.

snowfields Expanses of snow that lie above the snow line.

soil All unconsolidated materials above **bedrock.** Natural earthy materials on the Earth's surface, in places modified or even made by human activity, containing living matter, and supporting or capable of supporting plants out-of-doors.

soil horizon A layer of soil that is distinguishable from adjacent layers by characteristic physical properties such as texture, structure, or color, or by chemical composition.

soil moisture ground water in the **zone of aeration.**

soil structure The combination of soil particles into aggregates or clusters, which are separated from adjacent aggregates by surfaces of weakness.

soil texture The physical nature of the soil, according to its relative proportions of sand, clay, and silt.

sole mark Develops as an irregularity on the bottom of a stratum. It is a cast of a depression on the top surface of the immediately underlying bed.

solifluction Turbulent movement of saturated soil or surficial debris.

sorting The range of particle sizes in a sedimentary deposit. A deposit with a narrow range of particle sizes is termed "well-sorted."

south magnetic pole The point on the Earth where a north-seeking magnetic needle free to swing in space points directly up.

specific gravity The ratio of the density of a material to the density of water.

specific retention (field capacity) The amount of **capillary water** retained in a soil after the drainage of **gravitational moisture.**

sphericity A descriptive term to describe how close a particle's shape is to a sphere.

spit A sandy bar built out from the land into a body of water.

spoil Overburden or non-ore removed in mining or quarrying.

spreading axis (spreading center) A region of divergence on the Earth's surface, as at a rift.

spreading pole A rotational pole around which a plate appears to rotate on the Earth's surface.

spring Occurs at the intersection of the water table with the ground surface.

stack An isolated, steep-sided, rocky mass or island just offshore from a rocky headland, usually on a shore platform.

stalactite An icicle-shaped accumulation of **dripstone** hanging from cave roof.

stalagmite A post of **dripstone** growing up from a cave floor.

star dune A sand dune built by winds alternating through several directions. Builds vertically rather than migrating and growing laterally.

stick-slip A jerky, sliding motion associated with fault movement.

stock A small **batholith.**

stoping A process of magmatic intrusion that involves detaching and engulfing pieces of the surrounding rock so that the magma moves slowly upward.

storm surge A ridge of high water associated with a hurricane and which floods over the shore.

strain Change in the shape or volume of a body as a result of stress.

strain rate The rate at which a body changes shape or volume as a result of stress.

strain seismograph A **seismograph** that is designed to detect deformation of the ground by measuring relative displacement of two points.

stratification The accumulation of material in layers or beds.

stratified drift Debris washed from a glacier and laid down in well-defined layers.

stratigraphy The succession and age relation of layered rocks.

stratovolcano (composite volcano) A volcano that is composed of alternating layers of lava and pyroclastic material, along with abundant dikes and sills. Viscous, intermediate lava may flow from a central vent. Example: Mt. Fuji in Japan.

streak The color of a mineral in its powdered form, usually obtained by rubbing the mineral against an unglazed porcelain tile to see the mark it makes. A mineral harder than the tile must be pulverized by crushing.

stream capture *see* **stream piracy.**

stream order A classification of the relative hierarchy of stream segments in a drainage network.

stream piracy (stream capture) The natural diversion of the headwaters of one stream into the channel of another stream that has greater erosional activity and flows at a lower level.

stream terrace A relatively flat surface along a valley, with a steep bank separating it either from the floodplain or from a lower terrace.

strength The ability to withstand a stress without permanent deformation.

stress The force per unit area acting on any surface within a solid; also, by extension, the external pressure that generates the internal force.

striations Scratches, or small channels, gouged by glacier action. These occur on boulders, pebbles, and bedrock. Striations along bedrock indicate direction of ice movement.

strike The compass direction of the intersection between a structural surface (e.g., a bedding plane or a fault surface) and the horizontal.

strike-slip fault (transcurrent fault) A fault on which the movement is parallel to the fault's **strike.**

strip mining **Open pit** mining, typically for coal.

subduction zone A narrow, elongate region in which one lithospheric plate descends relative to another.

sublimation The process by which matter in the solid state passes directly to the gaseous state without first becoming liquid.

subtropical deserts Deserts in zones of descending air between 25° and 30° north and south latitude.

superimposed stream A stream that was established on a new surface and then, as it cut downward, maintained its course despite encountering different lithologies in the process.

superposition A statement of relative age in layered rocks: In a series of sedimentary rocks that has not been overturned, the topmost layer is always the youngest and the bottommost layer is always the oldest.

surf Produced as a wave steepens and falls forward as the wave nears the shore.

surface of discontinuity In sand dune formation the surface between quiet air of the wind shadow and the rapidly moving air above.

surface wave *compare* **body wave.**

surging glacier A glacier that moves rapidly (tens of meters per day) as it breaks away from the ground surface on which it rests.

suspended load The amount of material a stream carries in suspension.

suspension A method of sediment transport in which the turbulence of a fluid is able to keep particles supported in the fluid.

suture The line of juncture where continental rocks on two converging plates meet. Example: The region in the Himalayas where the Eurasian and Indian-Australian plates meet.

swash and backwash Uprush of a wave onto the beach followed by the return flow of the water down the beach slope in the intervals between waves.

swells Persistence of wind-formed waves after wind ceases.

syncline A fold that is convex downward, or that had such an attitude at some stage in its development; *compare* **anticline.**

taconite A bedded ferruginous chert containing at least 25% iron. A potential iron **ore.**

tailings Washed or milled ore that is too poor to be further treated.

talus A slope built up by the accumulation of rock waste at the foot of a cliff or ridge.

tar A thick brown to black viscous organic liquid, too thick to migrate easily through most porous sediment.

tar sand A sand containing tar or asphalt, from which the hydrocarbons may potentially be extracted by distillation.

tarn A lake in the bedrock basin of a **cirque.**

tell An artificial hill formed by the debris of successive human settlements.

temperate glacier A glacier whose temperature throughout is at, or close to, the pressure point of ice, except in winter when it is frozen for a few meters below the surface.

tensile fracture A fracture caused by tensional stress in a rock.

tension A stress that tends to pull a body apart.

tephra A general term for all **pyroclastic** material.

terminal moraine (end moraine) Ridge of till marking farthest extent of glacier.

terrane (microplate) A fragment of the lithosphere, smaller than a plate, that forms a portion of an accreted terrane **margin.**

texture The general appearance of a rock as shown by the size, shape, and arrangement of the materials composing it.

"The present is the key to the past" A shorthand reference to the principle of **uniformitarianism.**

thermal conductivity A measure of the ability of a material to conduct heat.

thermal gradient *see* **geothermal gradient.**

thermal spring A spring whose temperature is 6.5°C or more above mean annual air temperature.

thermoremanent magnetism The magnetism of a mineral that is acquired as it cools below its **Curie point.**

threshold of movement The point at which a slope or slope material crosses from a condition of stability to one of instability and movement begins.

thrust fault A **reverse fault** on which the dip angle of the fault plane is 15 degrees or less.

thrust sheet A body of rock above a large-scale thrust fault.

tidal delta A **delta** formed at both sides of a tidal inlet.

tidal inlet Waterway from open ocean into a lagoon.

tidal power Power generated by harnessing the energy of tidal motion in the ocean.

till (unstratified drift) Glacial **drift** composed of rock fragments that range from clay to boulder size and randomly arranged without bedding.

topset bed Layer of sediments deposited over surface of a delta, nearly horizontal and covering the tops of the inclined **foreset beds.**

transcurrent fault *see* **strike-slip fault.**

transform boundary A plate boundary in which plates on opposite sides of the boundary move past each other in opposite directions.

transform fault A plate boundary that ideally shows pure strike-slip movement. Associated with the offset segments of mid-ocean ridges.

transported soil A soil that has been moved from the site of its parent rock.

transverse dune A long, straight dune, perpendicular to direction of wind.

trap 1. Any barrier to the upward migration of petroleum, allowing it to accumulate. 2. Any dark-colored extrusive igneous rock. A reference to the tendency of basalt and similar rocks to form **columnar joints.**

travel-time diagram A plot of seismic wave travel time against distance on the Earth's surface from the epicenter of an earthquake.

travertine (tufa) Variety of limestone that forms **stalactites** and **stalagmites** and other deposits in limestone caves (**dripstone**) and the mouths of hot and cold calcareous springs.

trellis drainage A drainage pattern in which a stream and its tributaries resemble the pattern of a vine on a trellis.

trench A long, narrow, steep-walled, often arcuate depression in the ocean floor, much deeper than the adjacent ocean and associated with a **subduction zone.**

triangulation The method of locating an **epicenter** by determining how far it lies from three widely separated seismographs.

troughs and bars Linear features in unconsolidated sediments at the foot of the **shore face,** the result of breaking waves.

TRU *see* **low-level nuclear waste.** TRU stands for transuranic waste, in reference to radioactive elements heavier than uranium.

truncated spur The beveled end of a ridge separating two valleys where they join a larger glaciated valley. Glacier of main valley has eroded back the end of the ridge.

tsunami *see* **seismic sea wave.**

tufa *see* **travertine.**

tuff A general term for all consolidated **pyroclastic** rock. Not to be confused with **tufa.**

turbidite Sedimentary deposit settled out of turbid water carrying particles of widely varying grade size. Characteristically displays graded bedding.

turbulent flow Fluid flow in which the flow lines are confused and mixed. Fluid moves in eddies and swirls; *compare* **laminar flow.**

U-shaped valley A valley carved by glacier erosion and whose cross-valley profile has steep sides and a nearly flat floor, suggestive of a large letter "U".

unconformity A buried erosion surface separating two rock masses.

uniformitarianism The principle that applies to geology our assumption that the laws of nature are constant. As originally used it meant that the processes operating to change the Earth in the present also operated in the past and at the same rate and intensity and produced changes similar to those we see today. The meaning has evolved, and today the principle of uniformitarianism acknowledges that past processes, even if the same as today, may have operated at different rates and with different intensities than those of the present. The term "actualism" is sometimes used to designate this later meaning.

unloading The release of confining pressure associated with the removal of overlying material. May result in expansion of rock, accompanied by the development of joints or **sheeting.**

unstratified drift *see* **till.**

USDA Soil Classification System A classification of soils on the basis of the processes and conditions by which they form; *compare* **Comprehensive Soil Classification System.**

valley glacier (alpine glacier, mountain glacier) Streams of ice that flow down valleys in mountainous areas.

valley train Outwash plain contained within valley walls.

varve A pair of sedimentary units, one coarse-grained, the other fine-grained, interpreted as representing one year of sedimentation.

velocity Distance of travel in unit of time.

velocity profile A plot of seismic velocity against depth in the Earth.

ventifact A pebble, cobble, or boulder faceted by wind-driven sand.

vesicle A cavity in a lava, formed by the entrapment of a gas bubble during solidification of the lava.

vesicular A textural term applied to an igneous rock containing abundant vesicles, formed by the expansion of gases initially dissolved in the lava.

viscosity The internal resistance to flow in a liquid.

volcanic ash The dust-sized, sharp-edged, glassy particles resulting from an explosive volcanic eruption.

volcanic cinder A **pyroclastic** fragment, 0.5 to 2.5 cm in diameter, formed as magma spatters into the air during a volcanic eruption and cools as it falls to Earth.

volcano A vent in the surface of the Earth from which lava, ash, and gases erupt, forming a structure that is roughly conical.

volcanogenic massive sulfide deposit A mineral deposit of metallic sulfides formed directly through processes associated with volcanism, commonly in a submarine setting.

vulnerable mineral A mineral that does not easily resist **decomposition.**

Wadati-Benioff zone An inclined plane, roughly coincident with a **subduction zone,** along which the foci of earthquakes cluster.

Waste Isolation Pilot Plant A pilot plant near Carlsbad, New Mexico, for the storage of **low-level nuclear waste.**

water gap A gap in a ridge or mountain through which a stream flows.

water power Power generated through the agency of moving water.

water table The surface between the **zone of saturation** and the **zone of aeration.**

waterfall The perpendicular or very steep descent of a stream.

wave base A depth equal to one half the wavelength of waves in deep water, below which stirring due to wind is negligible.

wave crest The top of a wave.

wave height The vertical distance between the crest and adjacent trough of a wave.

wavelength The distance between two successive wave crests or troughs.

wave trough The low spot between two successive waves.

weathering The process by which Earth materials change when exposed to conditions at or near the Earth's surface and different from the ones under which they formed; *compare* **decomposition, disintegration.**

welded tuff A **pyroclastic** rock in which glassy clasts have been fused by the combination of the heat retained by the clasts, the weight of overlying material, and hot gases.

well An artificial intersection of the surface and the water table.

Wilson Cycle The opening and closing of ocean basins through plate tectonics.

wilting point The stage at which all water available to plants has been used.

wind farm An area in which a large number of windmills have been erected to generate electrical power.

wind gap An abandoned **water gap.**

wind power Power generated by using the force of the wind.

wind shadow An area of quiet air in lee of an obstacle. Zone of sand accumulation in lee of sand dune.

xenolith *see* **inclusion.**

X-ray diffraction The diffraction of a beam of X-rays by the three-dimensional periodic array of atoms in a **crystal structure.** The identity and arrangement of atoms in the structure can be determined by interpreting the angles at which X-rays are scattered by the structure and the intensities of scattered beams.

yardang Sharp, irregularly crested ridges carved by wind and oriented parallel to wind.

yazoo-type river A tributary stream unable to enter a main stream because of natural **levees** along the main stream. It flows in a **back swamp** area, parallel to the main stream until it finds an entry to the main stream.

yield point The stress limit at which permanent deformation takes place in a non-brittle material.

Yucca Mountain Site Site in Nevada proposed for the storage of high-level nuclear waste.

zone of ablation The area of wastage in a glacier.

zone of accumulation 1. The B horizon in a residual soil. 2. The area in which ice accumulates in a glacier.

zone of aeration Zone immediately below the ground surface within which pore spaces are partially filled with water and partially filled with air.

zone of flow The zone in a glacier that flows by deforming along planes of weakness in the ice crystals.

zone of fracture The near surface zone in a glacier that behaves like a brittle substance.

zone of leaching The upper horizons in a soil, through which gravitational moisture travels, removing soluble decomposition products.

zone of saturation The zone below the **zone of aeration** in which all pore spaces are filled with water.

Index

tsunamis and, 511–12
  World Wide Seismic Network (WWSN), 271
Earth system, 6–7
Earthy luster, 41
East African rift, 288, 367, 368
East China Sea, 300
East Pacific Rise, 147, 148, 290, 304
Eccentricity of Earth's orbit, 419, 422–23
Ecliptic, obliquity of, 419, 423
Eclogite, 141, 226
Eclogite facies, 143
Elastic deformation, 235
Elasticity, 206
Elastic rebound, 204–5, 206, 241
Electron capture, 163–64, 166
Electronic amplifier in seismograph, 211
Electrons, 4, 31
Elements
  chemical, 4, 30
  isotopes of, 30–31
  noble, 31
Elephant Butte reservoir, New Mexico, 505
Eleuthera, Grand Bahama Banks, 116
End moraine, 411–12
Energy
  binding, 485
  elastic rebound and release of, 204–5
  ray of, 207, 208
  sources of, 15
  weathering and, 78
Energy resources, 472, 475, 476–88
  consumption in U.S. of, 477
  fossil fuels, 476–85
  geothermal power, 477, 486–87
  nuclear power, 477, 485–86
  sun, wind, and tides, 476, 487–88
  water power, 476, 485
Eniwetok atoll, 440
Entisols, 98, 101
Environment(s), 4
  depositional, 106, 125–27
  human activity and, 472–73, 500–508
  impact on humans, 508–12
  metamorphic, 141–48
  plate tectonics and, 316–17
  stress, 241
  urban, 472–73
Eocene Epoch, 164
Eons, 160, 164
Epicenter, earthquake, 212, 214, 271, 272
Epidote, 132, 144
Epidote-amphibolite, 141
Epoch, 161, 164
Equator, magnetic, 172
Equilibrium line, 403
Equinoxes, precession of, 419, 423
Eras, 161, 164
Erosion, 15, 471
  agriculture and soil, 501
  on cratons, 316
  denudation, 88–89
  differential, 325
  by glaciers, 406–10, 411
  human activity and, 323, 327
  isostasy theory and, 263
  of laterites, 97
  mass movement and, 322, 324, 325
  by meandering stream, 362, 363
  missing records and, 157–60
  mountain-building and, 316
  peneplains, 316
  plate tectonics and, 316–17
  by rivers, 16
  by streams, 350–51, 352

valley enlargement and, 356, 357
water, 465, 505
water gaps and, 355–56
wave, 325, 429–30
by wind, 18, 448, 453–55, 462, 465
See also Weathering
Erratic, 413
Esker, 413, 414
Eskola, Pennti, 140, 141
Estuaries, glaciation and, 415–16
Ethyl alcohol, 32, 33
ETP curve, 422, 423
Eurasian Plate, 13, 14, 279, 308, 309
Eustatic change of sea level, 432, 433
Eutrophication, 501
Evaporation of seawater, precipitation of salts from continuing, 117, 118
Evaporites, 117, 118, 490, 493
Everest, Mt. (Chomolungma or Sagarmatha), 282, 301, 303
Ewing, Maurice, 271
Exfoliation, 80
Exfoliation dome, 80
Exotic rivers, 463
Exotic streams, 466–67
Extinct volcano, 63
Extrusive igneous rock bodies, 59, 63–68

F

Facies
  metamorphic, 138–41, 143
  sedimentary, 127
Faeroes, 67
Failed rift, 288, 290, 291
Faiyum depression, Egypt, 18
Fall (free-fall of material), 324, 328
Fall River, California, 380
Farallon Plate, 304, 305, 306
Fault gouge, 205, 206
Faults, 13, 244–48, 290
  cross-cutting rule and, 155
  dip-slip, 244–46
  earthquakes on, 204, 205
  evidence of movement on, 247–48
  hinge, 244, 247, 249, 253, 255
  normal, 244, 245, 247
  oblique-slip, 244, 247, 249
  oil trap in, 484
  reverse, 244, 245, 248
  strike-slip (transcurrent), 244, 245
  thrust, 244, 245, 248
  transform, 15, 244, 246, 273, 303, 436
Fault scarp, 412
Faunal and floral succession, principle of, 157
Fayalite, 35
Feldspars, 46
  decomposition of, 84–85, 86, 94
  orthoclase, 43, 46, 84, 86, 87
  plagioclase, 35, 46, 48, 54–55, 60, 84, 87, 243
  potassium, 60, 81, 84
  weathering of, 84–85
Fernandina Volcano, Galapagos Islands, 64
Fernow Experimental Forest, West Virginia, 348
Ferromagnesian minerals, 43, 56, 85, 86, 94
Fertilizers, 493, 501
Fetch, 428
Field capacity, 375
Fiery clouds, 66
Finger Lakes, New York, 408–9

Fire, earthquakes and, 221
Fire Island, New York, 498
Firn, 400, 401
First Law of Thermodynamics, 183
Fission, 485–86
Fission track dating, 169
Fissure eruptions, 66–68, 69
Fissures, 64
Fissure springs, 381
Fjords, 408
Flash floods, 463
Flashy streams, 348, 350
Flint, 46, 117
Flood basalts, 67
Flood-control dams, 347, 350
Floodplains, 16, 347, 363
  loess deposits and, 456–57
Flood recurrence interval, 348–49, 508
Floods, 346–48, 473
  flash, 463
  impact of, 508
  lava, 66–68, 69
  protection against, 442–43, 508
  stream, 346–48
  in Venice, Italy, 394
Flow
  stream, 342–48
  surface heat, 188–89
  zone of, 405–6
Flow folds, 250, 254
Flowing, 324, 328
Flowstone, 390, 392
Fluid inclusions, 136, 139
Fluorite, 34, 38, 40, 46, 47
Focus, earthquake, 204, 209, 272–74
Fold-and-thrust mountains, 295, 296, 297, 306, 309
Folds, 13, 248–54
Foliation, 133, 134, 135, 143–44
Food chain, path of radiocarbon in, 168
Foot wall block, 245, 246, 247
Foraminifera in deep-sea sediments, 421
Foreset beds, 364, 365
Foreshocks, 217
Foreshore, 431, 432
Formation water, 135, 138
Forsterite, 35
Fossil fuels, 476–85
  coal, 477–80, 505
  oil and natural gas, 480–84
  oil shale and tar sands, 484–85
Fossils, 175, 176
  continental drift theory and, 267–68
  correlation of rock units by, 156–57
  seafloor, zonation of, 420–21
  in sedimentary rocks, 123, 124
  trace, 124
  of Triassic Period, 290, 294
Fractional crystallization, 55–56
Fractionation, 488
Fracture(s), 38, 204, 242–48
  brittle, 237
  depth of weathering and, 86
  faults. *See* Faults
  joints, 242–44
  porosity and, 377, 378
  tensile, 247–48, 249
  zone of, 405, 406
Fragipan, 96
Frank, Alberta, rockslide, 327–29, 330, 331
Franklin D. Roosevelt Lake, 502
Franz Josef glacier, New Zealand, 406
Free oscillations, 209
Freshwater, 315

Terrane, 306
Terrebonne Parish, Louisiana, 436
Tertiary mountain ranges, 270
Tertiary Period, 164
Tests, 114
Tethys Sea, 301–3, 306
Teton Range, Wyoming, xxviii
Texture
　of igneous rocks, 57–60
　of metamorphic rocks, 133, 134
　of sedimentary rocks, 108–9
　soil, 94
Tharp, Marie, 271, 273
Theory of the Earth with Illustrations and Proofs (Hutton), 2, 3
Thermal conductivity of rock, 186
Thermal gradient, 173, 188, 190
Thermal springs, 11, 117, 382–83, 486
　geysers, 11, 383–84, 486
Thermoremanent magnetism, 174
Thin sections, 58–59, 112–13
Thorium, 165
Thorium-232, 165, 167, 186, 194, 486
Threshold of movement, 320, 321
Thrust fault, 244, 245, 248
Thrust sheets, 295
Tiber River, 88, 89n
Tibetan (Qinghai-Xizhang) Plateau, 303, 449
Tidal delta, 434
Tidal inlet, 434
Tidal power, 477
Tides, energy from, 487–88
Till (unstratified drift), 410, 411–13
Timber Cove, California, 426
Time
　metamorphism and, 136–37
　soil formation and, 93
　strain rate and, 241
　See also Geologic time
Tokyo, 1923 earthquake in, 218–19
Tonga Trench, 272, 274, 290
Topset beds, 364, 365
Trace fossils, 124
Trailing (passive) continental margins, 287–90, 292
Traleika glacier, 402
Transcurrent fault, 244, 245
Transform fault, 15, 244, 246, 273, 303, 436
Transform margins (boundaries), 13, 15, 18, 287, 289, 303–6
Translocation, 94
Transportation of materials
　by streams, 92, 351–53
　by wind-driven waves, 429–30
Transportation systems, construction of, 502, 503
Transuranic waste (TRU), 394
Transverse dune, 458, 459, 460
Trap, oil, 483, 484
Trap rock, 61
Travel-time diagram, 212, 213
Travertine, 116, 117
Trellis stream pattern, 355, 356
Trench, oceanic, 14, 126, 272
Triangulation, 212
Triassic Period, 164, 290, 294
Tributaries, 345, 353, 364
Trilobites, 124
Troughs, 431, 432
Truncated spur, 407, 412, 416
Tsunamis, 221–23, 511–12
Tufa, 116
Tupai, 440
Turbidites, 119–20
Turbidity currents, 120

Turbulence of streams, 343–46, 352, 353
Turbulent flow, 343
Turtle Mountain, Alberta, 328, 330, 331
Typhoons, 509

U

Ultimate base level, 349, 351
Ultisols, 98, 101
Ultramafic magma, 52, 61
Uncomphagre River, 342
Unconformities, 158
　angular, 158–59, 160
　in Grand Canyon, 159–60, 161
　oil trap in, 484
Underground mines and mining, 488, 504
Underground water. *See* Ground water
Uniformitarianism, principle of, 2, 270
　applying, 125, 356, 399, 416
United States
　coal fields of, 478–79, 480
　oil and natural gas, 482–83, 484
　Pleistocene glaciation in, 418, 420
　stream channels in, 342–46, 347, 355
　uranium deposits in, 486
U.S. Army Corps of Engineers, Denver earthquakes caused by, 205–6, 207
U.S. Department of Agriculture (USDA)
　soil classification system, 96–100
United States Forest Service, 348
United States Geological Survey, 217, 218
Universal Time, 211
Unloading, 80, 81, 243
Unstratified drift, deposits of, 410, 411–13
Unzen, Mt., 198
Uraninite, 102
Uranium, 11, 165, 490
Uranium-235, 165, 167, 486
Uranium-238, 165, 167, 169, 194
Urban environment, 472–73
Urbanization
　erosion and, 501
　stream flow and, 347–48, 350
U-shaped glaciated valley, 407, 410, 412, 415, 416

V

Vaiont dam in Italian Alps, 338
Valle Marineris, Mars, 337, 338
Valley glaciers, 402
　moraines formed by, 412–13
Valleys
　blind, 386, 388
　glaciated, 407–9, 410, 412, 415, 416
　hanging, 408, 416
　U-shaped, 407, 410, 412, 415, 416
　V-shaped, 407, 416
　See also Stream valleys
Valley trains, 413, 414
Vanderford fjord, Antarctica, 408
Van der Waals bonds, 32–33
Varves, 169–70, 413–14
Vatna glacier, Iceland, 402, 403
Vegetation
　decomposition rate and, 82
　soil formation and, 93–94
　stream flow and changes in, 348
　See also Plants and animals

Velocity
　of ground water, 377
　of streams, 343–46, 347, 350–53
　wave, 206–7
　of wind, 449, 452
Velocity profile of seismic waves, 223, 224
Venice, Italy, 394–95
Ventifacts, 454
Venus, 12, 18, 459–62
Vertisols, 98, 101
Vesey's Paradise, 380
Vesicles, 58
Vesicular texture, 58
Vesuvius, Mt., 65, 66
Victoria, Lake, 368
Vine, Fred, 277, 278, 280
Virgin River, Utah, 352
Viscosity, 187
　of magma, 52–53
Vishnu Schist, 130, 159
Vitreous luster, 41
Volcanic activity, 5
Volcanic dust, 62
Volcanic eruptions, 473
　basalt plateaus and, 66–68, 69
　impact on humans of, 66–67, 510–11
　location of, 510
　mudflows and, 330
　plate tectonics and, 14–15
　submarine, 66, 285–86
Volcanic (extrusive) igneous rocks, 59, 63–68
Volcanic glass, 495
Volcanic islands
　coral reef coasts around, 439–40
　subsidence of, 440
Volcanoes, 11, 63–68
　active, 63, 66–67
　classification, 64–66
　defined, 63
　dormant, 63
　earthquakes associated with belts of, 212
　eruption of. *See* Volcanic eruptions
　extinct, 63
　formation of, 295–97
　geothermal power from, 486
　lava. *See* Lava
　on Mars, 286
　"Ring of Fire," 197, 198, 272
　stratovolcanoes, 65–66, 198, 286, 295, 296, 297, 300
　in subduction zones, 297–98, 300
Volcanogenic massive sulfide deposits, 489
Volga River, 506, 507
Volume change, jointing caused by, 242–43
Von Laué, Max, 35, 36
Von Rebeur-Paschwitz, Ernst, 210
V-shaped stream valleys, 407, 416
Vulnerable minerals, 84–85

W

Wadati, K., 272–73
Wadati-Benioff Zone, 272–75, 276, 290, 297
Warm springs, 382–83
Warm Springs, Georgia, 382, 383
Wasatch Mountains, 311
Waste Isolation Pilot Plant (WIPP), 393–94
Water, 4
　artesian, 381–82, 383, 394
　capillary, 375
　deformation and, 240
　different states of, 34

erosion and, 465, 505
formation, 135, 138
gravitational, 375
hard, 381
human use of, impact on environment of, 505–7
hydrologic system, 6, 7, 9–10, 16, 342, 400, 505–7
liquefaction and, 220–21
on Mars, 369
mass movement and, 320–22, 326, 329, 330, 332, 333
melting and, effect of, 196–99
physical difference between air and, 449–50
role in shaping Earth's surface, 16–18
running, 16, 463–64
    See also Rivers; Streams
seawater, 117, 118, 136, 138
soft, 381
triggering events in earthquake and, 205–6
See also Ground water
Waterfalls, 357–61, 362
Water gaps, 355–56, 357, 358, 437
Water pollution, 390–94, 501
Water power, 476, 485
Water table, 17, 506
    ground water and, 374, 375, 376, 381
    perched, 381, 386
Wave base, 428
Wave crest, 428, 430
Wave erosion, 325
Wave height, 428
Wavelength, 428
Waves
    infrared, 186
    seismic, 206–9, 210
    seismic sea, 221–23, 511–12
    wind-formed, 18, 428–31
Wave trough, 428
Wave velocity, 206–7
Weathering, 77–92, 314, 471
    chemical. *See* Decomposition
    of common rocks and minerals, 84–86
    defined, 78, 90

depth of, 86–87
in deserts, 462–63
differential, 90–92
ease of mineral, 87–90
energy and, 78
hydrologic systems and, 9–10
mechanical. *See* Disintegration
of metamorphic rock, 7, 90, 91
mineral deposits formed due to, 490–91
paleosols and ancient, 100–102
processes of, 78–84
progress of, 86–92
rock cycle and, 8
of sedimentary rocks, 5–6
spheroidal, 82–84, 243
valley enlargement and, 356–57
Wegener, Alfred, 259, 264–71, 274–75, 278
Welded tuff, 63
Wells, 381
    artesian, 381, 382
Werner, Abraham, 72
Wet partial melting, 197–99
Whin Sill, 69–70
White Nile, 368
Wilson, J. Tuzo, 244, 273, 280, 309
Wilson cycle, 309–11
Wilting point, 375
Wind, 314–15
    air movement, 449–50
    coastal processes and, 431
    convection and, 187
    deposition by, 18, 448, 455–59
    duststorms and sandstorms caused by, 450–53, 465
    erosion by, 18, 448, 453–55, 462, 465
    hurricanes and, 509–10
    on Mars and Venus, 459–62
    shorelines and, 18
    velocity of, 449, 452
Wind Cave, South Dakota, 385
Wind-driven waves, 18
Wind farms, 487
Wind-formed waves, 428–31
Wind gaps, 358
Wind power, 476
Wind River Canyon, Wyoming, 324, 357

Wind shadow, 457, 458
Wine, aging in caves, 390
Wisconsin glaciation, 420
Wollastonite, 142
Woodworth Glacier Tosauma Valley, Alaska, 415
World Wide Seismic Network (WWSN), 271
Wrangell Mountains, Alaska, 335
Wright, Frank Lloyd, 218

X

Xenoliths, 71
X-ray diffraction, 35, 36
X-rays, 185

Y

Yardangs, 454
Yazoo River, 364
Yazoo-type rivers, 364
Yellow River, China, 456
Yellow Sea, China, 456
Yellowstone National Park, 198, 199
Yield point, 237
Yosemite Falls, 360–61
Yosemite Park, California, domes of, 80
Yuba River, California, 342
Yucca Mountain Site, Nevada, 393

Z

Zaire-Lualaba River, 369
Zeolite, 141
Zinc ore, 492
Zircon, 162, 286–87
Zoisite, 43